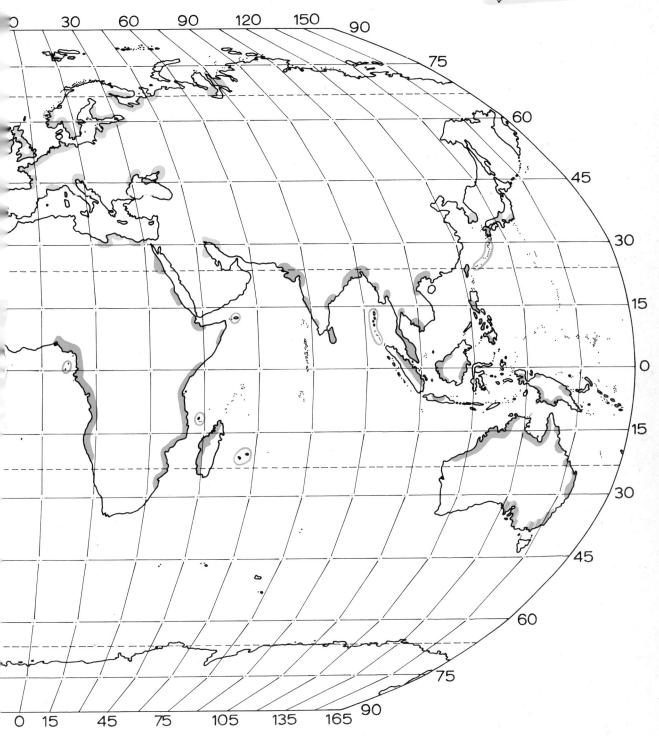

ECOSYSTEMS OF THE WORLD 1

WET COASTAL ECOSYSTEMS

ECOSYSTEMS OF THE WORLD

Editor in Chief:
D.W. Goodall

CSIRO Division of Land Resources Management, Wembley, W.A. (Australia)

I. TERRESTRIAL ECOSYSTEMS

    A. Natural Terrestrial Ecosystems
1. Wet Coastal Ecosystems
2. Dry Coastal Ecosystems
3. Polar and Alpine Tundra
4. Swamp, Bog, Fen and Moor
5. Shrub Steppe and Cold Desert
6. Coniferous Forest
7. Temperate Deciduous Forest
8. Natural Grassland
9. Heath and Related Shrubland
10. Temperate Broad–Leaved Evergreen Forest
11. Maquis and Chaparral
12. Hot Desert and Arid Shrubland
13. Savannah and Savannah Woodland
14. Seasonal Tropical Forest
15. Equatorial Forest
16. Ecosystems of Disturbed Ground

    B. Managed Terrestrial Ecosystems
17. Managed Grassland
18. Field Crop Ecosystems
19. Tree Crop Ecosystems
20. Greenhouse Ecosystems
21. Bioindustrial Ecosystems

II. AQUATIC ECOSYSTEMS

    A. Inland Aquatic Ecosystems
22. Rivers and Stream Ecosystems
23. Lake and Reservoir Ecosystems

    B. Marine Ecosystems
24. Intertidal and Littoral Ecosystems
25. Ecosystems of Estuaries and Enclosed Seas
26. Coral Reefs
27. Ecosystems of the Continental Shelves
28. Ecosystems of the Deep Ocean

    C. Managed Aquatic Ecosystems
29. Managed Aquatic Ecosystems

ECOSYSTEMS OF THE WORLD 1

WET COASTAL ECOSYSTEMS

Edited by

V.J. Chapman

Department of Botany
University of Auckland
Auckland (New Zealand)

ELSEVIER SCIENTIFIC PUBLISHING COMPANY

Amsterdam — Oxford — New York 1977

ELSEVIER SCIENTIFIC PUBLISHING COMPANY
335 Jan van Galenstraat
P.O. Box 211, Amsterdam, The Netherlands

Distributors for the United States and Canada:

ELSEVIER NORTH-HOLLAND INC.
52 Vanderbilt Avenue
New York, N.Y. 10017, U.S.A.

Library of Congress Cataloging in Publication Data
Main entry under title:

Wet coastal ecosystems.

   (Ecosystems of the world ; 1)
   Includes bibliographical references and indexes.
   1. Tidemarsh ecology. 2. Coastal ecology.
I. Chapman, Valentine Jackson. II. Series.
QH541.5.S24W47   574.5'26   77-342
ISBN 0-444-41560-2

Copyright © 1977 by Elsevier Scientific Publishing Company, Amsterdam

All rights reserved. No part of this publication may be reproduced, stored in a retrieval system or transmitted in any form or by any means, electronic, mechanical, photocopying, recording or otherwise, without the prior written permission of the publisher, *Elsevier Scientific Publishing Company, P.O. Box 330, Amsterdam, The Netherlands.*

Printed in Great Britain

PREFACE

This is the first volume in the general series on *Ecosystems of the World*. In this volume a review is made of coastal wetlands, which comprise marine and brackish salt marshes of temperate regions and mangrove swamps in tropical and subtropical regions of the world. There is a preliminary consideration of the general nature of such wetlands followed by an account of the physiographic conditions under which they may and do develop. This is followed by a general account of the climatic conditions and soils associated with these wetlands, and the fauna to be found in them. The latter, of course, are integral to any ecosystem and there are further accounts in the regional chapters. The volume closes with an account of the uses to which these wetlands have been and are currently put to and the changes such uses involve.

As editor I want to extend my thanks to the writers of the various chapters not only for their willingness to contribute, but also in their acceptance of suggestions I have made. I also want to thank Dr. D. W. Goodall, Editor-in-Chief of the series, for his help and courtesy, and Mrs. Raylee Johnstone for her work in preparing the index.

V.J. CHAPMAN
Professor of Botany
University of Auckland

LIST OF CONTRIBUTORS TO THIS VOLUME

W.G. BEEFTINK
Delta Institute for Hydrobiological Research
Vierstraat 28
Yerseke (The Netherlands)

F. BLASCO
Institut de la Carte Internationale du Tapis Végétal
39 Allées Jules Guesde
31400 Toulouse (France)

V.J. CHAPMAN
Department of Botany
University of Auckland
Private Bag
Auckland (New Zealand)

FRANKLIN C. DAIBER
College of Marine Studies and Department of Biological Sciences
University of Delaware
Newark, Del. 19711 (U.S.A.)

T. HOSOKAWA
Professor Emeritus of Kyushu University
8–9 4 Choume
Chuouku
810 Fukuoka (Japan)

KEITH B. MACDONALD
Department of Geological Sciences
University of California
Santa Barbara, Calif. 93106 (U.S.A.)

FRED B. PHLEGER
Scripps Institution of Oceanography
P.O. Box 1529
La Jolla, Calif. 92037 (U.S.A.)

WILLIAM H. QUEEN
Chesapeake Research Consortium
University of Maryland
College Park, Md. 20742 (U.S.A.)

ROBERT J. REIMOLD
Marine Resource Extension Program
University of Georgia
P.O. Box 517
Brunswick, Ga. 31520 (U.S.A.)

PETER SAENGER
Botany Department
University of Queensland
St Lucia, Qld. 4067 (Australia)

MARION M. SPECHT
Botany Department
University of Queensland
St Lucia, Qld. 4067 (Australia)

RAYMOND L. SPECHT
Botany Department
University of Queensland
St Lucia, Qld. 4067 (Australia)

J.A. STEERS
Flat 47
Gretton Court
Girton
Cambridge, CB3 OQN (U.K.)

H. TAGAWA
Biological Institute
Kagoshima University
Kagoshima (Japan)

GERALD E. WALSH
U.S. Environmental Protection Agency
Gulf Breeze Environmental Research Laboratory[1]
Gulf Breeze, Fla. 32561 (U.S.A.)

HEINRICH WALTER
Botanisches Institut der Universität
7000 Stuttgart–Hohenheim 70 (Germany)

ROBERT C. WEST
Louisiana State University
Baton Rouge, La. 70803 (U.S.A.)

M.A. ZAHRAN
Department of Botany
Faculty of Science
Mansoura University
Mansoura (Egypt)

[1] Associate laboratory of the National Environmental Research Center, Corvallis, Ore.

CONTENTS

PREFACE V

LIST OF CONTRIBUTORS VII

Chapter 1. INTRODUCTION
    by V.J. Chapman 1

Definitions 1
Geographical distribution 2
Algal communities 17
Fauna 18
Past origins 19
Principal features 22
Biogeographical relations 23
Succession 25
Productivity 25
External relations 26
Conclusion 27
References 27

Chapter 2. PHYSIOGRAPHY
    by J.A. Steers 31

Introduction 31
Norfolk marshes 31
Marshes and change of sea level 37
San Francisco marshes 40
Areas with both salt marsh and mangal . . 46
Mangals 46
Isostasy and eustasy 54
Preservation of salt marshes 56
Coastal changes and salt marshes 57
Postscript 59
Acknowledgement 59
References 59

Chapter 3. CLIMATE
    by H. Walter 61

Climate types 61
Mangal climates 64
Salt-marsh climates 66
References 67

Chapter 4. SOILS OF MARINE
    MARSHES
    by F.B. Phleger 69

Introduction 69
Origin and composition of the sediment . . 69
Soil structures 72
Populations of foraminiferids 73
Populations of Ostracoda 74
Environmental parameters 76
References 76

Chapter 5. SALT-MARSH ANIMALS:
    DISTRIBUTIONS RELATED
    TO TIDAL FLOODING,
    SALINITY AND VEGETA-
    TION
    by F.C. Daiber 79

Introduction 79
Protozoa 81
Nematoda 83
Mollusca 84
Arthropoda 86
Chordata 91
Mangrove animals 105
Discussion and conclusions 105
References 106

Chapter 6. THE COASTAL SALT MARSHES OF WESTERN AND NORTHERN EUROPE: AN ECOLOGICAL AND PHYTOSOCIOLOGICAL APPROACH
by W.G. Beeftink 109

Introduction 109
Environmental characteristics 112
Phytogenic interrelationships 118
European salt-marsh communities: floristic ecology, composition and distribution . . 120
Zonation and dynamic trends in vegetation . 139
Algal ecology of the salt marsh 141
Ecology of salt-marsh animals 144
Ecosystem approach to the salt marsh . . . 147
Acknowledgements 149
References 149

Chapter 7. MANGALS AND SALT MARSHES OF EASTERN UNITED STATES
by R.J. Reimold 157

Introduction 157
Northern marshes 160
New England marshes 161
Coastal Plain marshes 161
References 164

Chapter 8. PLANT AND ANIMAL COMMUNITIES OF PACIFIC NORTH AMERICAN SALT MARSHES
by K.B. Macdonald 167

Introduction 167
Geological setting 168
Environmental variables 169
Regional review 170
Phytogeography 177
Numerical analysis 179
Introduced plant species 180
Productivity 181
Animal communities 181
Synthesis 185
Acknowledgements 187
References 187

Chapter 9. TIDAL SALT–MARSH AND MANGAL FORMATIONS OF MIDDLE AND SOUTH AMERICA
by R.C. West 193

Introduction 193
Salt-marsh formations 193
Mangal formations 199
Conclusion 211
References 211

Chapter 10. AFRICA A. WET FORMATIONS OF THE AFRICAN RED SEA COAST
by M.A. Zahran 215

Introduction 215
Vegetation types 217
Soil 224
Classification 225
Biogeography 227
Zonation 228
References 230

Chapter 11. AFRICA B. THE REMAINDER OF AFRICA
by V.J. Chapman 233

North Africa 233
West Africa 233
South Africa 236
East Africa 237
References 239

Chapter 12. OUTLINES OF ECOLOGY, BOTANY AND FORESTRY OF THE MANGALS OF THE INDIAN SUBCONTINENT
by F. Blasco 241

Introduction 241
General features 241
Local diversity 247
Conclusions 258
References 258

CONTENTS

Chapter 13. WET COASTAL FORMATIONS OF INDO–MALESIA AND PAPUA–NEW GUINEA
by V.J. Chapman 261

Introduction 261
Indo–Malesia 261
Papua–New Guinea 267
Vietnam 268
Land animals 269
References 270

Chapter 14. MANGALS OF MICRONESIA, TAIWAN, JAPAN, THE PHILIPPINES AND OCEANIA
by T. Hosokawa, H. Tagawa and V.J. Chapman 271

Micronesia (T. Hosokawa) 271
Taiwan (T. Hosokawa) 278
Mangals in southwest Japan (H. Tagawa) . 281
The Philippines (V.J. Chapman) 287
New Caledonia and Oceania (V.J. Chapman) 288
References 290

Chapter 15. MANGAL AND COASTAL SALT–MARSH COMMUNITIES IN AUSTRALASIA
by P. Saenger, M.M. Specht, R.L. Specht and V.J. Chapman 293

Australia (P. Saenger, M.M. Specht and R.L. Specht) 293

New Zealand (V.J. Chapman) 310
Appendix: species lists 311
References 339

Chapter 16. EXPLOITATION OF MANGAL
by G.E. Walsh 347

Introduction 347
Early uses 347
Forest management and products 348
Mariculture 353
Tannin production 353
Agricultural use of mangal soil 355
Discussion 356
References 358

Chapter 17. HUMAN USES OF SALT MARSHES
by W.H. Queen 363

Introduction 363
Marsh uses 363
Marsh value 366
Marsh protection 367
Conclusion 367
Acknowledgement 368
References 368

AUTHOR INDEX 369

SYSTEMATIC INDEX 379

GENERAL INDEX 401

Chapter 1

INTRODUCTION

V.J.CHAPMAN

DEFINITIONS

Wet coastal formations include salt marshes up to the limit of extreme high watermark, mangrove vegetation up to the same limits, and brackish representatives of both vegetation types up estuaries to the limit of tidal influence. Below the salt marshes and mangrove forests there frequently occur beds of eelgrass or sea meadows of *Posidonia*, *Halophila*, *Thalassia*, etc., but these belong to the sublittoral and will be considered in a separate volume of this series.

Maritime salt marsh, as distinct from inland salt marsh, is essentially confined to the temperate regions of the world. Salt marsh or salt flat can occur in the tropics, especially in arid or monsoonal regions, and in such cases it is to be found generally landward of the mangal belt. In both hemispheres there is obviously a zone in which both mangal and salt marsh can be found intermingled. Such areas include the Gulf Coast of the U.S.A., central Florida, southern Australia, New Zealand and southern Japan.

Mangrove formations are perhaps best described as "Mangal" (MacNae, 1968; Chapman, 1975), leaving the term mangrove for the individual genera and species. Mangal reaches its optimum development in the tropics but it does extend into the subtropics reaching its geographic limits in some regions that are appropriately described as warm temperate, e.g., New Zealand and southern Japan.

In recent years concern for preservation of the natural environment has caused a reappraisal of the value of both mangrove and salt marshes. In the past they have been regarded as suitable repositories for city waste and for a wide range of reclamation projects. The total world acreage of mangrove and salt marsh has been greatly reduced during the present century. The value of both ecosystems as the base of estuarine and offshore fisheries is now regarded highly though our information is not so good as it should be. Little or no information is available about the effect of man's manipulations of the hydrology of salt marshes and mangals upon their survival and their ecosystems. The question is continually being asked whether we can afford to alienate any more acres and increasingly the answer seems to be "no". Important studies are under way in various countries to follow nutrient cycling, especially of nitrogen and phosphorus (Chapman, 1974b, 1976). Oil spillage has become a serious marine pollutant but recent work here has shown that unless spillage is really serious (Chapman, 1974a) most species survive and the marine muds do, in fact, denature the oil. Cooling water from power stations poses a more serious threat and can result in non–productive marshland (Nelson–Smith, 1972). Increasing the ambient water temperature by 5°C increases the density of red mangrove prop roots (Kolehmainen, 1973), whilst the density of organisms living on that species' prop roots is inversely related to water temperatures above 34°C (Kolehmainen and Morgan, 1972).

The use of herbicides to defoliate and destroy mangrove vegetation has assumed considerable prominence in recent years. It is now evident that after the destruction of mangal in Vietnam the areas remained uncolonised for at least six years and probably more (Walsh, 1974). The combination of 2–4D plus Picloram effectively prevented recolonisation, though admittedly a large proportion of the defoliated mud flats were at a high level and the soil may have hardened to a point where seedlings would no longer be trapped. Because of possible future importance of the use of herbicides, Walsh et al. (1973) have made a study of the effect of herbicides on red mangrove seedlings. They found that red mangrove is very sensitive to herbicides and that seedling development is inhibited. This can make restoration difficult if low residues from sprays remain in the soil.

The general distribution of mangal and salt marsh species, and hence communities, is broadly related to seven basic features of the maritime environment. These are as follows:

(1) *Air temperature*. This determines the northern and southern limits of mangrove species in the Northern and Southern hemispheres, respectively. Generally mangal flourishes where the temperature in the coldest month does not fall below 20°C and where the range is around 10°C. *Avicennia germinans* is capable of tolerating a minimum of 12.7°C in Florida and 10°C in Brazil. *Avicennia marina* tolerates a temperature of 15.5°C in the northern Red Sea and var. *resinifera* a winter temperature of 10°C in New Zealand (see Fig. 1.1). The southern and northern limits of salt-marsh species are related to either winter or summer temperatures in the Northern and Southern hemispheres, respectively. *Spartina townsendii*, for example, is probably limited, as it has extended south in the Atlantic, by an insufficiently low winter temperature; and the same is probably true in the northern plantings in New Zealand. The northward spread of *Iva*, *Batis* and *Sesuvium* is determined by low winter temperatures, whilst the southern limits of *Salicornia stricta* and *S. ramosissima* are probably set by high summer temperatures.

(2) *Protected coastline*. Neither salt marsh nor mangal can develop on an exposed coast where wave action prevents establishment of the seedlings. Bays, lagoons, estuaries and shores behind spits and offshore islands are all favoured localities.

(3) *Shallow shores*. The shallower and more extensive the shallows the greater is the extent of the mangal and salt-marsh communities. On steeply shelving shores only fringe communities develop.

(4) *Currents*. Since seeds, seedlings or plant portions are carried by currents these can be responsible for distribution of species along coasts. It is fairly evident that the spread of mangrove species has taken place in this manner, and the appearance of *Spartina townsendii* spontaneously in new localities must be attributed to this factor.

(5) *Salt water*. Whilst there is increasing evidence that most mangrove and salt–marsh species are not obligate halophytes there is evidence that a number of them have their optimal growth in the presence of some additional sodium chloride. Species of *Salicornia* and *Rhizophora* are probably obligate halophytes, growth being poor or reduced under glycophytic conditions.

(6) *Tidal range*. The greater the tidal range the greater is the vertical range available for communities and one tends to find a wider range of communities on coasts with large tidal ranges. Such communities are also likely to be more extensive in area.

(7) *Substrate*. This varies widely in composition and it can play a part in determining community composition. A sandy salt–marsh soil favours the development of *Puccinellia* in Europe but inhibits *Spartina townsendii*. A sandy mangal soil in the West Indies is usually colonised by *Laguncularia* rather than *Avicennia* (Chapman, 1944). Heavy loams affect water movement and some species cannot tolerate water-logging. In Indo-Malesian mangal *Bruguiera parviflora* usually indicates such a site.

GEOGRAPHICAL DISTRIBUTION

Figure 1.1 illustrates the general geographical distribution of the principal salt–marsh and mangal areas throughout the world. The more detailed accounts for the different regions will be found in later chapters. At this point it may be noted that the main areas represent coasts where there is protection from severe wave action or where major rivers debouch into the sea (cf. p. 263). Establishment and maintenance of either type of vegetation is only possible where there is sufficient lack of wave action to enable seedlings or seeds to put out roots before being removed by the tide. Whilst mangrove species will grow on coral (see p. 209), no extensive formations are to be found in such a habitat. The most extensive salt marshes and mangal are associated with regions where there is abundant silt brought down by the rivers or where the geological nature of the land makes it subject to considerable erosion that results in much silt suspension. Pure sand is not generally a favourable

INTRODUCTION

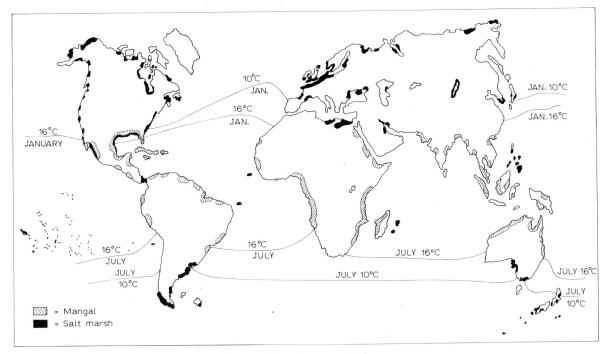

Fig. 1.1. World distribution of salt marshes and mangal (after Chapman, 1975).

substrate for either type of vegetation because it is usually sufficiently mobile to inhibit seedling establishment. Stable sand areas, however, can be colonised by mangrove (Fig. 1.2) and salt marsh is common where sand is admixed with the silt. It may be noted that sandy salt marshes are generally associated with regions (e.g., southeast England, Wales, Denmark) where the land is rising slowly in relation to sea level. In such areas there is an increasing exposure of sand flats, and drying winds at low tide blow sand onto the marshes.

A. Salt–marsh formations

Both salt marshes and mangal exhibit the same general features, i.e., a bare seaward mud or sand flat (which may carry *Zostera*, *Halophila*, etc., or algae), then an area of various types of low–growing vegetation (salt marshes) or an area of various types of tree vegetation (mangal) leading either to upland vegetation or, in the case of estuaries, via a transitional zone to fresh–water swamp or swamp forest (mangal areas). The transition to fresh–water swamp is commonly characterised by species of *Typha*, *Scirpus*, *Cyperus* and *Phragmites*. The transition to swamp forest is usually characterised

Fig. 1.2. Young *Avicennia germinans* on a sand bar in Jamaica (after Chapman, 1944).

by the presence of the palm *Nypa fruticans* and the fern *Acrostichum aureum*. Other species may mark the transition zone but they differ in various parts of the tropics (see p. 235).

Drainage creeks penetrate both salt marsh and mangal (cf. p. 42) as well as major and minor rivers. Some salt marshes (but not mangal) have bare patches known as pans that arise in different ways (see p. 32).

One feature of wet coastal formations is the widespread occurrence of the dominant genera throughout the world. On the salt marshes, *Salicornia*, *Spartina*, *Juncus*, *Arthrocnemum*, *Plantago* are all widespread in both hemispheres. Among the mangroves *Rhizophora*, *Avicennia*, *Acrostichum* and *Bruguiera* are all widespread genera, especially the first three. The worldwide spread of these dominant genera exerts some influence upon the type of classification that can be employed for wet coastal formations. The present writer is of the opinion that the Montpellier classification represents the best available at the present time despite some of the criticisms that have been levelled at it (Poore, 1955; Chapman, 1974a) (see also p. 120).

Table 1.1 (from Chapman, 1974a) sets out the principal orders and subunits that can be recognised on maritime salt marshes.

The genus *Salicornia* is cosmopolitan and occurs

TABLE 1.1

Distribution of maritime salt-marsh communities

	Atlantic Europe	Great Britain	Baltic	Scandinavia	Arctics	N. Canada	W. Mediterranean	E. Mediterranean	Caspian	Atl. Canada/U.S.A.	Pacific U.S.A.	Brazil	Japan	Australia	New Zealand	S. Africa
A. Coeno–Salicornietalia																
1. Thero–Salicornion																
Salicornietum strictae	+	+	?	+									+			
Salicornietum patulae	+	+	+	+									+			
Salicornietum brachystachyae	+	+											+			
Astereto–Salicornietum	+	+														
Suaedeto–Salicornietum	+	+			+											
Salicornieto–Spartinetum strictae	+															
Salicornietum rubrae										+						
2. Salicornion fruticosae																
Salicornietum radicantis							+	+								
Salicornietum fruticosae							+	+								
Salicornietum ambiguae											+?					
Salicornietum australi														+	+	
Salicornieto–Distichlidetum											+					
Salicornieto–Suaedetum californicae											+					
Bato–Salicornietum										+	+					
Salicornio–Sesurietum										+	+					
3. Arthrocnemion																
Arthrocnemetum glauci		+[1]														
Arthrocnemetum africani																+
Arthrocnemetum halocnemoidis														+?		
B. Suaedetalia																
1. Thero–Suaedion																
Suaedetum maritimae	+	+	+	+	+					+					+	
Salsoletum sodae									+							

INTRODUCTION

TABLE 1.1 *(continued)*

	Atlantic Europe	Great Britain	Baltic	Scandinavia	Arctics	N. Canada	W. Mediterranean	E. Mediterranean	Caspian	Atl. Canada/U.S.A.	Pacific U.S.A.	Brazil	Japan	Australia	New Zealand	S. Africa
Suaedeto–Atriplicetum													+			
Suaedeto–Asteretum	+	+														
Suaedeto–Spergularietum	+	+														
2. Fruti–Suaedion																
Suaedetum fruiticosae	+	+					+	+								
Suaedetum depressae										+	+					
Suaedetum erectae										+						
Suaedetum californicae											+					

C. Coeno–Puccinellietalia

1. Puccinellion phryganodis																
Puccinellietum phryganodis					+	+							+			
Puccinellietum coarcticae																
Puccinellietum kurilensae													+			
Puccinellio–Spergularietum salinae				+												
2. Puccinellion maritimae																
Puccinellietum maritimae	+	+	+	+												
Puccinellietum americanae										+						
Puccinellietum distantis	+		+	+												
Puccinellietum giganti											+					
Puccinellio–Salicornietum	+	+		+												
Puccinellio–Suaedetum	+	+		+												
Puccinellio–Spartinetum	+	+								+						
Puccinellio–Asteretum	+	+												+		
Puccinellio–Limonietum	+	+								+						
Puccinellio–Festucetum	+	+		+												
Puccinellio–Halimionetum	+	+														

D. Artemisietea vulgaris

1. Halo–Artemision																
Artemisietum maritimae	+	+														
Artemisieto–Limonietum virgati							+									

E. Halostachyetea

1. Halostachyon																
Halocnemetum strobilacei							+	+						+?		
Halocnemetum caspiae									+							
Kalidietum caspici									+							

F. Coeno–Juncetalia

1. Juncion																
Juncetum maritimi	+	+	+	+						+					+	+
Juncetum baltici			+	+						+						
Juncetum gerardii	+	+	+	+						+						

TABLE 1.1 (continued)

	Atlantic Europe	Great Britain	Baltic	Scandinavia	Arctics	N. Canada	W. Mediterranean	E. Mediterranean	Caspian	Atl. Canada/U.S.A.	Pacific U.S.A.	Brazil	Japan	Australia	New Zealand	S. Africa
Juncetum acuti							+	+								
Juncetum roemeriani										+						
Juncetum bufoni			+	+						+						
Junceto–Festucetum rubrae	+	+	+	+												
Junceto–Caricetum extensae	+	+														
Junceto–Stipetum textifoliae														+	+	
Junceto–Fimbristyletum										+						
Junceto–Sporoboletum virginici										+						
Junceto–Puccinellietum	+	+	+	+						+						
Eleochareto–Juncetum				+	+											
2. Leptocarpion																
Leptocarpeto–Juncetum														+	+	
Leptocarpeto–Baumetum juncei														+	+	
Leptocarpeto–Plagianthetum														+	+	
3. Baccharion																
Baccharetum halimifoliae										+						
Bacchareto–Ivetum orariae										+						
Plucheo–Baccharetum										+						
G. Atriplexitalia																
1. Atriplicion																
Atriplicetum littoralis	+	+		+						+				+	+	
H. Halo–Phragmitetalia																
1. Phragmition																
Halo–Phragmitetum	+	+		+						+						
Scirpeto–Phragmitetum	+	+		+						+						
Halo–Typhetum	+	+		+						+				+	+	
I. Bolboschoenetea maritimi																
1. Bolboschoenion maritimi																
Scirpetum maritimi	+	+		+						+					+	
Scirpetum robusti										+						
Scirpetum americani										+						
Blysmetum rufi			+													
J. Halo–Caricetalia																
1. Magnocaricion paleaceae																
Caricetum paleaceae					+	+								+		
Caricetum mackenziei					+	+								+		
Caricetum ramenskii					+	+								+		
Caricetum subspathaceae					+	+										

INTRODUCTION

TABLE 1.1 *(continued)*

	Atlantic Europe	Great Britain	Baltic	Scandinavia	Arctics	N. Canada	W. Mediterranean	E. Mediterranean	Caspian	Atl. Canada/U.S.A.	Pacific U.S.A.	Brazil	Japan	Australia	New Zealand	S. Africa
2. Halocladion																
Cladieto–Plagianthetum														+	+	
Cladieto–Cyperetum ustulati														+	+	
Baumetum juncei														+	+	
Eleocharetum	+			+												
K. Spartinetalia maritimae[2]																
1. Spartinion europeae																
Spartinetum maritimae	+	+														
Limonio–Spartinetum maritimae							+	+								
Limonio–Spartinetum townsendii	+	+	+													
Astereto–Spartinetum maritimae	+	+														
2. Spartinion americanae																
Spartinetum alterniflorae	+	+	+							+						
Spartinetum patentis										+						
Spartinetum cynosuroidis										+						
Spartinetum foliosae											+					
Spartinetum brasiliensi												+				
Spartineto–Distichlidetum										+						
3. Distichlion																
Distichlidetum spicatae										+						
Distichlideto–Salicornietum										+						
Distichlideto–Frankenietum											+					
Distichlideto–Limonietum											+					
L. Limonietalia																
1. Limonion galloprovincialis																
Limonietum vulgari	+	+														
Limonietum ferulacei							+									
Limonietum latifolii							+									
Limonietum bellidifoliae	+	+														
Suaedeto–Limonietum vulgari	+	+														
Astereto–Triglochinetum	+	+														
Triglochinetum maritimi	+	+									+					
2. Limonion gmelinii																
Limonieto–Artemisetum									+							
3. Limonion occidentale																
Limonietum nashii											+					
Limonietum carolinae										+						
Asteretum subulati										+						
Plantagineto–Limonietum nashii										+						
4. Limonion septemtrionale																
Limonietum pacifici													+			
Limonietum brasiliensi														+		

TABLE 1.1 (continued)

	Atlantic Europe	Great Britain	Baltic	Scandinavia	Arctics	N. Canada	W. Mediterranean	E. Mediterranean	Caspian	Atl. Canada/U.S.A.	Pacific U.S.A.	Brazil	Japan	Australia	New Zealand	S. Africa
Limonietum japonici													+			
Triglochineto–Potentilletum pacificae											+					
5. Armerion																
Armerieto–Festucetum	+	+														
Armerieto–Limonietum	+	+														
Plantaginetum maritimae	+	+														
Plantagineto–Puccinellietum	+	+														
M. Coeno–Festucetalia																
1. Festucion																
Festucetum rubrae	+	+	+	+												
Festucetum arundinaceae							+									
Festuceto–Agrostidetum stoloniferae	+	+		+												
Agrostidetum stoloniferae	+	+		+												
Festuceto–Glaucetum	+	+														
Festuceto–Poetum							+									
Festuceto–Caricetum glareosae								+								
Plantagineto–Agrostidetum stoloniferae	+	+														
N. Halostachyetea																
1. Halimionion																
Halimionetum portulacoidis	+	+					+	+								
Halimioneto–Crypsidetum schoenodis							+	+?								
Halimioneto–Artemisietum monogynae							+	+								

[1] Atlantic Africa.
[2] Based on Beeftink and Géhu (1973).

in one of two phases — communities dominated by the annual species, such as *S. stricta* and *S. rubra*, and communities dominated by perennial shrubby species, such as *S. radicans* and *S. australis*. The former group are essentially Northern Hemisphere species and generally occur on low salt marsh as primary colonists or else as colonists of pans or bare depressed areas at higher levels where water may stand (Fig. 1.3). The latter group occur in both hemispheres on temperate and tropical coasts. They may be primary colonists, e.g., *S. australis* in New Zealand, or occur at higher levels, e.g., *S. ambigua*. The allied shrubby perennial genus *Arthrocnemum* is a primary colonist on salt marshes from the Mediterranean to Australia, and in the Mediterranean *Halocnemum strobilaceum* can also replace *Salicornia* as a primary colonist.

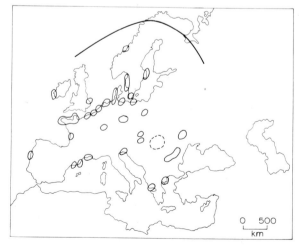

Fig. 1.3. Distribution in Europe of Thero–Salicornion and Thero–Suaedion associations (after Beeftink, 1965).

INTRODUCTION

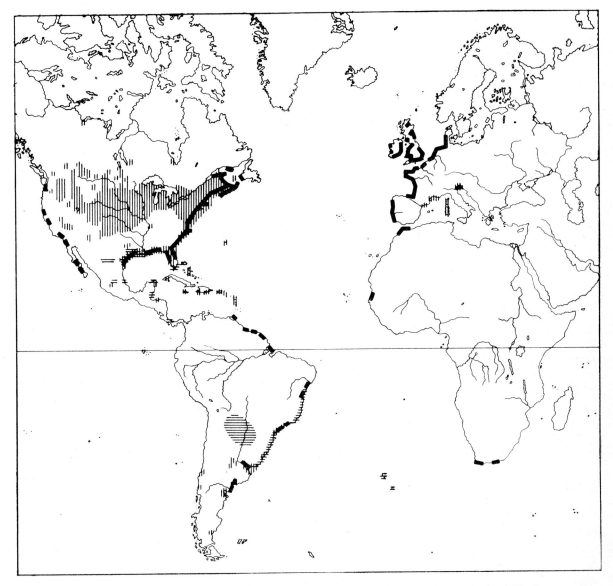

Fig. 1.4. World distribution of *Spartina*. Black: complex of species placed in Spartinetea maritimae; horizontal hachures: complex of interior species; vertical hachures: complex of second group of interior species.

The genus *Suaeda* is equally widespread and also occurs in two phases — communities dominated by the annual species such as *S. maritima* and *S. novae-zelandiae* and communities dominated by perennial shrubby species such as *S. fruticosa* and *S. californica*. The communities of annual *Suaeda* species are usually to be found on mud patches within the salt marsh or else as primary colonists with or without *Salicornia* on salt pans (see p. 127) (Fig. 1.3). The perennial species are generally not primary colonists and they tend to occur at the higher levels on the marshes.

Both the Coeno–Puccinellietalia and Spartinetalia maritimae (see Beeftink and Géhu, 1973) are widespread in the Northern Hemisphere (Fig. 1.4). The former does not appear to have spread into the Southern Hemisphere but the latter has invaded South America. The action of man in introducing *Spartina townsendii* and *S. alterniflora* into Southern Hemisphere countries (cf. Bascand, 1970) may

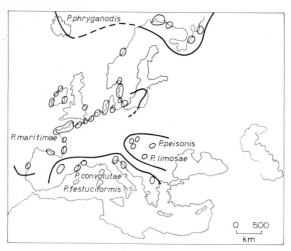

Fig. 1.5. Distribution in Europe of the Puccinellion alliance (after Beeftink, 1965).

result in the spread of representative associations in the future. In the Puccinellietalia, *Puccinellia phryganodes* is widespread around the Arctic (Fig. 1.5), *P. maritima* is predominant in the European Atlantic region, *P. americana* on the western Atlantic, and *P. kurilensis* in the North Pacific.

In Europe the original *Spartina* community was dominated by *S. maritima*. With the introduction of *S. alterniflora* from the western Atlantic and the subsequent development of the hybrid complex *S. townsendii—S. anglica* the entire pattern has changed. The original two parents have been virtually eliminated and the two hybrids now form the predominant and widespread association. In the New World the *Spartinetum glabrae*[1], *Spartinetum pilosae*[1], *Spartinetum patentis* and *Spartinetum cynosuroidis* are widespread along the Atlantic seaboard from the St. Lawrence to Florida. On the Pacific coast there is a single widespread association, the *Spartinetum foliosae*. In South America on the Brazilian sea coast *Spartina brasiliensis* forms a distinct and widespread association. Within the same order the *Distichlidetum spicatae* is equally widespread all down the Atlantic seaboard of Canada and the U.S.A.

The Limonietalia is represented on the North Atlantic and North Sea coasts by the *Limonietum vulgari* and *Limonietum bellidifoliae* and in the Mediterranean region by the Limonion gmelinii

---

[1] *Spartina glabra* and *S. pilosa* are forms of *S. alterniflora*.

alliance. Beeftink (1972) restricts the Limonietalia to the Mediterranean but this cannot be sustained in view of the *Limonietum vulgari* community in East Anglia. On the western Atlantic coast the corresponding alliance is the *Limonietum occidentale* with the *Limonietum nashii* in the northern and the *Limonietum carolinae* in the southern U.S.A. In South America they are replaced by the *Limonietum brasiliensi*. In the North Pacific the corresponding association is the *Limonietum pacifici* whilst on the West African coast there is the *Limonietum linifolii*.

In the Coeno–Festucetalia the alliance Festucion maritimae is widespread in Europe as also is the Halimionion alliance in the Halostachyetalia. In the Halo–Artemision the association *Artemisietum maritimae* is common throughout European salt marshes. In the Caspian Sea the four associations of the alliance Halostachyon are to be found and of these the *Halocnemetum strobilacei* extends into the Mediterranean.

Within the order Coeno–Juncetalia associations of the alliance Juncion are widespread on salt marshes in temperate regions. The association *Juncetum maritimi* occurs on both sides of the North Atlantic and also in Australasia. Because of its occurrence outside of Europe it is difficult to accept Beeftink's (1972) concept of the Juncetalia maritimi as a Mediterranean order. The *Juncetum gerardii* and *Juncetum baltici* also occur on both sides of the North Atlantic but are not present in the Southern Hemisphere. The *Juncetum bufoni* is primarily confined to the Baltic, the *Juncetum acuti* to the Mediterranean and the *Juncetum roemeriani* to the southern Atlantic states of the U.S.A. In Australasia the Leptocarpion alliance is a characteristic and widespread feature, as also the Halo-Typhetum. The latter is also found in the Northern Hemisphere.

The alliance Baccharion is essentially typical of the eastern Atlantic marshes of the U.S.A. The eco–Phragmition also occurs here as well as in Europe. Similarly, the associations of the Bolboschoenion maritimi are to be found on both sides of the Atlantic. In the Halo–Caricetalia the first three associations are widespread on circum-Arctic marshes and marshes at high latitudes in the Northern Hemisphere. In the Southern Hemisphere the alliance Halocladion is widespread on temperate Australasian marshes.

INTRODUCTION

Arising from the relatively small number of species and communities found on salt marshes one can note broad geographical areas in which there is a substantial uniformity in the vegetation. In some cases, subdivision can be based on temperatures or upon soil type. With this as background it seems that the major groups of maritime salt marshes are as follows (cf. Chapman, 1974b):

(1) *Arctic Group*

The Arctic Group is represented by the salt marshes of the Canadian and American Arctic, Greenland, Iceland, northernmost Scandinavia, and Arctic Russia. The environment is extreme, especially in winter, and the marshes appear to be somewhat fragmentary as a result. Succession is very simple. The vegetation is dominated by the grass *Puccinellia phryganodes*, though species of *Carex*, especially *C. subspathacea* and *C. maritima*, play an important part at higher levels. There are probably subgroups here. Macdonald (Chapter 8) recognises Arctic Alaska and west Alaska subgroups.

(2) *Northern European Group*

This group includes marshes from the Iberian Peninsula northward around the English Channel, North Sea coasts, and the Baltic Sea, as well as those on the Atlantic coasts of Eire and Great Britain. Whilst there is a fundamental similarity in physiognomy and floristic composition, nevertheless different communities can be recognised in specific areas. These subareas are related either to soil differences or, in the case of the Baltic, to salinity differences. Throughout the region there is a dominance of annual *Salicornia* species, *Puccinellia maritima*, *Juncus gerardi*, and the general salt-marsh community (see Chapter 6).

Scandinavian Subgroup. Scandinavian marshes are characterised by a high proportion of sand in the soil and have developed on a rising coastline where newly exposed shore sand can be blown inland. They are found in Scandinavia, Schleswig-Holstein, the west coast of Britain from the Severn to northern Scotland, on the east coast of Eire, and from Kincardine to Inverness on the east coast of Scotland. These marshes are dominated by grasses, especially *Puccinellia maritima*, *Festuca rubra* and *Agrostis stolonifera*, and for this reason are grazed by domestic animals. Grazing probably results in the elimination, or control, of other low herbs such as *Aster*, *Limonium*, *Triglochin*, etc. Since the publication of *Salt Marshes and Salt Deserts of the World* (Chapman, 1960), a number of contributions have been made to the literature of the Scandinavian Subgroup. Foremost among these have been the papers by Dalby (1970), west Wales marshes; Gillner (1960), Swedish marshes; Gimingham (1964), east Scottish marshes; Packham and Liddle (1970), Dee marshes; and Taylor and Burrows (1968), Dee marshes.

North Sea Subgroup. North Sea marshes have much more clay and silt in the soil and there tends to be a wider range of communities. Grasses are not so dominant and the general salt-marsh community plays a greater role in the successions. These marshes are generally associated with a subsiding coastline, but their character is currently changing through the introduction or natural spread of *Spartina townsendii* and *S. anglica*. North Sea marshes are represented by the marshes of eastern England, southeast Scotland, northern Germany and the Low Countries. Recent significant publications on the marshes of the North Sea Subgroup have been those by Koster (1960) and Beeftink (1965, 1966).

Baltic Subgroup. The Baltic Subgroup differs from the Scandinavian and North Sea Subgroups by the presence of some species, e.g., *Carex paleacea*, *Juncus bufonius*, *Desmoschoenus bottnica*, that occupy dominant places in the succession. The brackish nature of the Baltic also permits *Scirpus* to be the primary colonist.

English Channel Subgroup. Marshes in this subgroup were probably comparable with the North Sea Subgroup, but, since the appearance of *Spartina townsendii* and *S. anglica* towards the end of the last century and their subsequent spread, these marshes physiognomically now look more like those of the U.S.A. In Poole Harbour, the original home of *S. townsendii*, aerial photography has been employed to see whether there have been any recent changes in the vegetation (Bird and Ranwell, 1964; Ranwell, 1964). The total area of *Spartina* has not altered materially though there has been some seaward spread of *Scirpus* (0.6 m per annum), *Glaux, Phragmites* (1.2 m per annum), *Puccinellia*, and *Aster tripolium*.

Southwest Ireland Subgroup. This subgroup is still regarded as a separate entity because of the peaty nature of the soil and the simplicity of the succession. It is a group that requires further study.

(3) *Mediterranean Group*

Whilst Mediterranean marshes show some affinity with those of northern Europe, there are species (e.g., of *Arthrocnemum* and *Limonium*) which are characteristic. The eco–climax is dominated by *Juncus acutus* and the Mediterranean is the only region where this species attains such prominence. It is perhaps convenient to recognise western and eastern subgroups though obviously there is a gradual transition from one to the other. The marshes of the Caspian would appear to belong to a separate subgroup.

Western Mediterranean Subgroup. The marshes of southern France, particularly of the Rhone Delta (Knoerr, 1960; Corré, 1962; Molinier et al., 1964) are typical of this subgroup. The primary colonists are *Salicornia herbacea* agg. or *Salsola soda* plus *Suaeda* or a *Kochio–Suaedetum*. *Salicornia fruticosa* is also prominent. Other characteristic communities are the *Limonieto–Artemisietum, Limonieto–Limoniastretum, Crithmo–Staticetum,* and *Halimionetum portulacoidis*.

Eastern Mediterranean Subgroup. Whilst many of the dominants are the same as in the western subregion, there are eastern species present, such as *Halocnemum strobilaceum, Petrosimonia crassiflora, Bupleurum gracile,* and *Suaeda altissima*.

Caspian salt marshes. Annual *Salicornia* or *Halocnemon* (Tagunova, 1960) are the primary colonists of the Caspian marshes with thickets of *Salsola soda, Suaeda,* or *Puccinellia gigantea* behind (Agadzhanov, 1962). *Kalidium caspicum* and *Anabasis aphylla* are other typical species.

(4) *Western Atlantic Group*

Marshes in this group extend from the St. Lawrence down to Florida where there is a transition to mangrove. They are subdivided essentially on differences in mode of formation, though there are floristic differences as well in respect to the southern subregion. The salt–marsh flora of the Atlantic coast of North America is discussed in detail by Reimold (Chapter 7).

Bay of Fundy Subgroup. This subgroup is characterised by the significant role of *Puccinellia americana*, and *Juncus balticus* var. *littoralis* at the highest levels, and by a transition to bog rather than fresh–water swamp. These marshes are formed in front of a weak rock upland.

New England Subgroup. New England marshes are formed in front of a hard rock upland, and the soil is a peat rather than a muddy clay, as in the other two subgroups. *Puccinellia americana* is not a dominant, and the transition is to reed swamp.

Coastal Plain Subgroup. Recent descriptions of coastal plain marshes have been given by Adams (1963) and Davis and Gray (1966), whilst Blum (1963, 1968) has described the algal communities. The marshes have developed in front of a soft rock upland.

(5) *Pacific American Group*

The marsh communities of the American Pacific coast are entirely different from those elsewhere and the successions appear to be very simple. The marshes extend from California to southern Alaska. They are discussed by Macdonald in Chapter 8. He recognises Subarctic, Temperate and Dry Californian groups and it is a matter of opinion whether these should be groups or subgroups.

(6) *Sino–Japanese Group*

Recent accounts (Ito, 1961, 1963; Ito and Leu, 1962; Umezu, 1964; Miyawaki and Ohba, 1965) have added much valuable information about this region. Primary communities are dominated by *Triglochin maritima, Salicornia brachystachya,* or *Limonium japonicum*, though in the far north *Suaeda japonica* with *Atriplex gmelini* occupy this position with *Limonium tetragonum* behind. Southwards the equivalent zone is occupied by *Zoysia macrostachya* or *Puccinellia kurilensis*. In the northern part a relationship to the Arctic region is evidenced by the presence of *Carex ramenskii*. It is probable that this region should be subdivided into northern and southern subgroups (see Chapter 14).

(7) Australasian Group

The Australasian Group is characterised by some specifically Southern Hemisphere species such as *Salicornia australis*, *Suaeda nova–zelandiae*, *Triglochin striata*, *Samolus repens*, and *Arthrocnemum* species. Some recent descriptions are those by Sauer (1965) and Clarke and Hannon (1967). Floristic similarities justify regarding them as one region, but there are sufficient floral differences to render it desirable to subdivide the group (see Chapter 15).

Australian Subgroup. This subgroup is characterised by *Arthrocnemum arbuscula*, *A. halocnemoides*, *Hemichroa pentandra*, and *Suaeda australis*. At higher levels *Atriplex paludosa*, *Limonium australis*, and *Frankenia pauciflora* are important.

New Zealand Subgroup. Nearly all species of the New Zealand Subgroup are found on Australian salt marshes with the exception of those listed above.

(8) South American Group

This group is characterised by species of *Spartina* (*S. brasiliensis*, *S. montevidensis*), *Distichlis*, *Heterostachys*, and *Allenrolfea* which are not found elsewhere. These marshes are in need of further study (see Chapter 9).

(9) Tropical Group

Tropical marshes generally occur at high levels behind mangrove swamps and are flooded only by extreme tides. Typical species are *Sesuvium portulacastrum* and *Batis maritima*. The arid marshes of Baja California form a subgroup (see p. 176).

B. Mangal formations

Since the present author also believes that the Montpellier classification is the most appropriate for dealing with mangal vegetation, Table 1.2 sets out the proposed units and also their distribution throughout the world.

TABLE 1.2

Distribution of principal mangal associations

	Lower Calif.	C. America	W. Indies/Florida	Brazil/Uruguay	W. Africa	E. Africa	India	Burma/Indo–China	N. Australia	Pac. Islands	Philippines
RHIZOPHORETEA											
A. Rhizophoretalia											
1. Rhizophorion occidentale											
Rhizophoretum manglae	+	+	+	+	+					+	
Rhizophoretum racemosae		+		+	+						
Rhizophoretum harrisonii		+			+						
2. Rhizophorion orientale											
Rhizophoretum mucronatae						+	+	+	+	+	+
Rhizophoretum apiculatae							+	+		+	+
Rhizophoretum stylosae								+	+	+	+
3. Bruguierion											
Bruguieretum gymnorrhizae						+	+	+	+	+	+
Bruguieretum parviflorae								+	+		+
Bruguieretum cylindricae								+	+		+
Bruguieretum sexangulae						+		+			
Bruguiereto–Xylocarpetum								+	+		

TABLE 1.2 *(continued)*

	Lower Calif.	C. America	W. Indies/Florida	Brazil/Uruguay	W. Africa	E. Africa	India	Burma/Indo-China	N. Australia	Pac. Islands	Philippines
4. Ceriopion											
Ceriopetum tagalae						+	+	+	+	+	+
Ceriopeto–Aegiceretum corniculatae							+	+			
5. Kandelion											
Kandelietum candeli							+	+		+[1]	+
6. Avicennion occidentalis											
Avicennietum germinansae	+	+	+	+							
Avicennietum africanae					+						
Avicennietum germinans/A. schauerianae			+	+							
7. Avicennion orientalis							+	+			
Avicennietum albae							+	+	+	+	+
Avicennietum officinalis							+	+			+
Avicennietum marinae						+	+	+	+[2]	+	+
Avicennietum albae–marinae								+	+		
Avicennieto–Excoecarietum										+	+
B. Sonneratietalia											
1. Sonneration											
Sonneratietum albae						+	+	+	+	+	+
Sonneratietum caseolariae							+				
Sonneratietum apetalae							+		+	+	+
Sonneratio–Camptostemonetum										+	+

COMBRETETEA

A. Combretalia

1. Conocarpion
| *Conocarpetum erectae* | | + | + | + | + | | | | | | |

2. Laguncularion
| *Lagunculariaetum racemosae* | | | + | + | + | + | | | | | |

3. Lumnitzion
Lumnitzeretum racemosae							+	+	+		+
Lumnitzeretum littorale								+	+	+	+
Lumnitzereto–Xylocarpetum obovatae										+	+

B. Xylocarpetalia

1. Xylocarpion
Xylocarpetum granatae						+	+	+	+	+	+
Xylocarpetum moluccensis						+[3]	+	+	+	+	+
Xylocarpetum australasicae									+		
Xylocarpetum benadirensae						+					

INTRODUCTION

TABLE 1.2 (continued)

	Lower Calif.	C. America	W. Indies/Florida	Brazil/Uruguay	W. Africa	E. Africa	India	Burma/Indo-China	N. Australia	Pac. Islands	Philippines
C. Pellicieretalia											
1. Pellicierion											
Pellicieretum rhizophorae		+									
D. Acrostichetea											
1. Halo–Acrostichion											
Acrostichetum aureae		+			+		+				+
Acrostichetum speciosae									+?		
E. Excoecarietalia											
1. Excoecarion											
Excoecarietum agallochae							+	+	+	+	
Excoecarieto–Xylocarpetum australasicae									+		
2. Acanthion											
Acanthetum ilicifoliae							+	+			+
F. Heritietalia											
1. Herition											
Heritieretum minori							+	+			
Heritieretum littorali										+	+
G. Aegiceretalia											
1. Aegicerion											
Aegiceretum corniculatae									+	+	+

[1] Japan. [2] S. Australia also. [3] Madagascar.

The associations, with the sole exception of two (*Rhizophoretum manglae* and *Acrostichetum aureae*) are restricted to either the New World (plus West Africa) or to the Old World. This is especially evident in the two alliances Rhizophorion and Avicennion. In the former, three associations are restricted to the New World and three to the Old World. Other species of *Rhizophora* are known but they are not recorded as forming distinct associations. In the Avicennion three associations are restricted to the New World and five associations to the Old World. Similarly there are other species of *Avicennia* but they are not reported as forming distinct associations. Two associations in the Rhizophorion occidentale, the *Rhizophoretum racemosae*, and *Rhizophoretum harrisonii*, have a distribution on the two sides of the Atlantic, a feature also exhibited by the *Conocarpetum erectae* and *Laguncularietum racemosae*, both within the order Combretalia. The one association reported here that is restricted to the West African mangal is the *Avicennietum africanae*.

Within the alliance Bruguierion the *Bruguieretum gymnorrhizae* is the most widespread, the other three associations being essentially confined to Indo–Malesia, northern Australia and the Philippines, with one *(Bruguieretum sexangulae)* extending to the Indian subcontinent. The *Ceriopetum tagalae* has the same distribution as the *Bruguieretum gymnorrhizae* and the *Ceriopeto-Aegiceretum* a similar distribution to that of the *B. sexangulae*. The *Kandelietum candeli* is comparable but extends north into southern Japan.

Only two of the *Sonneratia*–dominated associations possess a wide distribution *(Sonneratietum albae, Sonneratietum caseolariae)*, the other three being restricted to single geographical regions. The associations dominated by species of *Xylocarpus* are similar, with two associations widely distributed and two restricted to small geographical areas. Other highly area–specific associations are the *Pellicieretum rhizophorosae* and *Acrostichetum speciosae*. Finally, in the associations listed in Table 1.2 the segregated distribution of the *Heritieretum minori* and *Heritieretum littorali* should be noted.

Just as there were broad geographical areas of salt marshes with substantially uniformity of vegetation, so there are comparable areas of mangal. The mangal areas differ, however, in that segregation does not seem possible on either temperature or soil type. It seems more closely related to past geographical spread (see p. 21). With this background it seems that the major groups of mangal are as follows:

(1) *New World Group*

This includes all the mangal to be found in North, Central and South America. It is characterised through the major part played in seres by *Rhizophora mangle, Avicennia germinans, Conocarpus erecta* and *Laguncularia racemosa*. It is not completely uniform and at least the following subgroups can be recognised (see p. 199 in Chapter 9).

West Central American Subgroup. This mangal is characterised by the additional presence of *Pelliciera rhizophorae, Rhizophora racemosa, R. harrisonii,* and *Avicennia bicolor*.

Gulf of California Subgroup. Sparse mangal with only *Rhizophora mangle, Conocarpus erectus* and *Avicennia germinans*.

East Central American Subgroup. Very similar to West Central America but lacking *Pelliciera*.

Caribbean Subgroup. The mangal on the West Indies and in Florida generally lacks *Rhizophora racemosa* and *R. harrisonii. Avicennia schaueriana* also makes its appearance here.

Atlantic South American Subgroup. Rhizophora racemosa and *Avicennia schaueriana* are additional species in the seres, whilst *Conocarpus erecta* is commonly absent.

(2) *West African Group*

The mangal on this coast is typified by *Avicennia africana, Rhizophora mangle, Conocarpus erecta,* and *Laguncularia racemosa,* with species of *Osbornia* marking the transition to fresh water. There does not seem any justification for subdivision (see Chapter 11).

(3) *East African Group*

The mangal here is characterised by the predominance of *Rhizophora mucronata, Avicennia marina, Bruguiera gymnorrhiza, Sonneratia alba* and *Ceriops tagal.* It is not so rich in species as the next group, but it possesses one species, *Xylocarpus benadirensis,* not found elsewhere. It is conveniently divided into two subgroups:

Red Sea—Persian Gulf Subgroup. Characterised by a very sparse mangal with essentially only *Avicennia marina* and *Rhizophora mucronata*.

Madagascan Subgroup. Same features as for the group.

(4) *Indo–Malesian Group* (see Chapter 12 and 13)

This comprises all the mangal from Pakistan, India, Burma, Malaysia, Indonesia, northern Australia, Celebes and Papua–New Guinea. A large number of species are to be found throughout the region and the seres are correspondingly the most complex. The group can probably be subdivided as follows:

Indian Subgroup. This subgroup is characterised by the presence of *Sonneratia apetala* in abundance and *Heritiera minor.* Some of the species found in Indo–Malesia do not extend into the area.

Burma—Indo–Malesian Subgroup. Characterised by possessing the largest number of species in any mangal.

INTRODUCTION

North Australian—Papuan Subgroup. This subgroup lacks the predominance of some species, e.g., *Rhizophora apiculata, Bruguiera sexangula, Sonneratia ovata,* and *Acanthus ilicifolius,* whilst specific to the area are *Xylocarpus australasicus* and *Acrostichum speciosum.*

Philippines Subgroup. This subgroup is characterised by the additional presence of *Camptostemon philippinensis* and *Heritiera littoralis.*

(5) *Australasian Group* (see Chapter 15)

This group involves the mangal to be found in southwest, south and southeast Australia together with New Zealand. Mangrove species are few, as most are excluded because of low winter temperatures, and salt marsh is almost equally significant. *Avicennia marina* var. *resinifera* is predominant together with *Aegiceras corniculatum* in Australia but not in New Zealand.

(6) *Oceanian Group* (see Chapter 14)

This group generally contains fewer associations as many species have not spread westwards. *Heritiera littoralis* is a characteristic species. This group should perhaps contain the southern islands of Japan where *Kandelia candel* reaches its northern limit.

ALGAL COMMUNITIES

It has been shown that the phanerogamic communities are limited in number and species composition because the specifics of the habitat only allow certain species to survive. The particular nature of the salt–marsh and mangal environment has imposed similar limitations upon the algal communities which are also widespread.

Since less information is available it is not profitable to illustrate the world distribution of the algal communities. The following communities have been recognised by Chapman (1974a) based on personal studies and those of other workers.

General Chlorophyceae Community

This is widespread and occurs in a variety of forms. Ropes of *Enteromorpha, Lola,* and *Rhizoclonium* occur on bare sand flats and on the ground, especially if sandy, between the phanerogams at all levels. The community can sometimes conveniently be subdivided into a lower and a higher variant based on differences in species composition. Other variants have also been recorded dominated by individual species, e.g., *Enteromorpha prolifera* var. *pilifera, E. clathrata* var. *prostrata*. The community has been recorded from Europe, Iceland, eastern and western North America, India and New Zealand.

General Myxophyceae Community

This is another widespread community found on bare mud flats and on muddy soils between phanerogams and extending up to the highest levels. It is also common on the mud banks of creeks. Species of *Oscillatoria, Phormidium, Lyngbya, Microcoleus,* are all involved and there is sometimes an accompanying diatom flora. The community, in one form or another, has been recorded from Europe, Iceland, eastern North America, India and New Zealand. The autumn Myxophyceae recorded in earlier works (references in Chapman, 1960) is probably best regarded as a variant of this.

Vernal Ulothrix Community

This is essentially a spring community found on the soil between plants on lower marshes. It has been recorded from Europe and eastern North America.

Bostrychia—Catenella Community

This is very widespread and is found at the bases of plants on high marsh, especially in Juncetea, but it also is associated with *Halimione* in Europe. As well as the two named genera, *Caloglossa leprieurii* and *Lola capillaris* may be associated. It has been recorded from Europe, eastern North America and Australasia. The community is also widespread throughout the tropics and subtropics on the pneumatophores and trunk bases of mangrove. In the mangal the composition is more commonly *Bostrychia—Caloglossa.*

Enteromorpha nana Community

This community is widespread, occurring generally on the bases of plants on low marsh, e.g., *Spartina* species, *Salicornia*, but it also occurs on the lower leaves of *Avicennia marina* var. *resinifera* in New Zealand mangrove swamps.

Limnicolous Fucaceae Community

Variants of this community have been recorded from salt marshes in Europe and eastern North America where the species involved have been ecads of *Fucus* or *Pelvetia* spp. The plants are generally free–living or mud–embedded, reduced in size from the original parents and generally reproduce only by vegetative means. The community is commonly found associated with low marshes. Baker (1950) has shown that the plants retain the juvenile three–sided form of the apical cell. In the New Zealand mangrove a variant of this community with ecad *libera* of *Hormosira banksii* has been recorded (Chapman and Ronaldson, 1958).

Vaucherietum Community

This occurs on mud flats and mud banks of creeks, where the dominant species is *Vaucheria thuretii,* or else on the mud of higher marshes in Europe where the dominant is *V. sphaerospora.* The low mud community has been recorded from both sides of the Atlantic and also from New Zealand.

Gelatinous Myxophyceae Community

This is characterised by species of *Nostoc* and *Rivularia* and is commonly found on high marshes on both sides of the Atlantic. The *Rivularia/Phaeococcus* community described in earlier works (Chapman, 1960, and references) is probably best regarded as a variant.

The general Chlorophyceae and Myxophyceae communities are both widespread in tropical and subtropical mangal as well as the *Bostrychia-Catenella* community. One additional community recorded from New Zealand mangal and not found in salt marsh is the *Gracilaria secundata* community. This also occurs on bare mud flats and may occupy considerable areas. The plants are originally attached to shells but are capable of a free–living existence. It would appear that algal communities of mangal and of salt marshes in Australia, South America and the North Pacific generally are in need of much further study.

FAUNA

The fauna of mangrove swamps has been studied in some detail so far as the Indo–Pacific region is concerned (MacNae, 1968) but considerably less information is available for other areas (Chapman, 1975). Many of the animals are clearly visitors from adjacent upland (e.g., birds, mammals, insects) and the basic fauna is generally associated with the marine muds. Here also many of the organisms are immigrants from the seaward bare mud flats. This is particularly true of the salt–marsh fauna.

In Tuléar (Madagascar) the faunas of the compact and more muddy mangal soils are regarded as distinct (Chapman, 1975), balanoids, mytilids and oysters (*Crassostrea* spp.) being predominant at low levels and at higher levels the molluscs *Cerithidea* spp. and *Melampus* sp. *Cerithidea* spp. with *Batallaria* also occur in Brazilian mangal with the last–named genus also in the mangal of Palao Island in the Pacific.

In a study of the mangal in Careel Bay (New South Wales in Australia), Hutchings and Recher (1974) recorded sixteen species of Mollusca with the oyster *Saccostrea cucullata* and the littorinid *Littorina scabia* as very abundant. The maximum molluscan density they recorded was 1690 per square meter. Eleven species of crab were noted, some being quite abundant. The six polychaete species were primarily confined to the creeks.

The late Dr. MacNae (1968) considered that there are five distinct mangal habitats occupied by specific faunal components to which the present writer (Chapman, 1975) added a sixth. These are as follows:

(1) *Tree canopy*. The animals here are essentially visitors and comprise birds (at least some 45—50 species) and mammals of which there may be fourteen or more species. A large number of insects also occur here, the great majority being species of mosquitoes (*Aedes* spp.) and midges (*Culicoides* spp.).

(2) *Rot holes in branches*. Water accumulates in

such places and they are ideal for mosquito and midge larvae.

(3) *Soil surface.* Many of the animals here also live in the soil or spend part of their time up the mangrove trees. Mud-skippers from the channels flip their way across the soil and there are also the hermit crabs. Snails, such as *Amphibola crenata* (New Zealand) leave their tracks on the mud but do go up onto the trunks of the trees.

(4) *In soil.* Nereids, snails and crabs predominate in this habitat. Species of the crab genera *Sesarma*, *Helice* and *Ilea* are abundant and the mounds of the mud lobster, *Thalassina*, are a feature of some tropical mangal. A wide range of snail genera are common, such as *Nerita* and *Pythia* in Indo–Malesia.

(5) *Permanent and semi-permanent pools.* These provide the homes for the larvae of mosquitoes and midges. In Southeast Asia tadpoles of the frog *Rana cancrivora* are also to be found as well as the adults. Small crabs of the genus *Ilyoplax* are common in habitats of such pools.

(6) *Channels.* These are the home for crocodiles and alligators as well as for a fairly wide range of fish, including the mud-skippers. In East African mangal two water monitors can be found in the channels whilst in Indo–Malesia one or two snakes have been recorded.

Although there is considerable uniformity, especially in respect of genera to be found in mangal, MacNae (1968) believed that at least six zoogeographical regions could be identified in the African, Indo–Malesian and Pacific mangal (Fig. 1.6). This figure also serves to show that mangal in the Atlantic has not received the same zoological attention. Golley et al. (1962) listed two snails, four crabs, numerous insects and some nine birds to be found in Puerto Rican red mangrove forest. Undoubtedly the list for New World mangals would be much larger, but this is an area where work is required. Details of the fauna, where known, will be found in the various regional chapters.

On salt marshes the fauna is very much less important except perhaps for that found in the creeks which are really an extension of the adjacent estuary or lagoon. None of them is of major significance and only a few of minor influence (Chapman, 1974b). In some cases large numbers of a few species may be important (Hutchings and Recher, 1974). Molluscs occur at low levels, e.g., *Hydrobia ulvae*, and also in salt pans, in both habitats green algae providing their principal food. Larger molluscs can occur on low *Salicornia* marshes, e.g., *Amphibola crenata* in New Zealand, where the burrowing crab *Helice crassa* also occurs. In Careel Bay (Australia) eight molluscs, three crustaceans and three nereids were recorded (Hutchings and Recher, 1974). The pans, with their supply of water, form a breeding ground for mosquitoes and midges and the same is true of drainage ditches in the Atlantic coast marshes of the U.S.A. (see p. 159). Rabbits will invade salt marshes for the grasses and some phanerogams, and domestic animals, if not kept off the marshes, are also likely to take advantage of the fodder, especially if there is a high proportion of grasses. Five to seven sheep per hectare can be accomodated on Lancashire marshes (Chapman, 1974b).

Regional information is so scanty that it is impossible to define zoogeographical regions for salt marshes though Bilio (1965) has shown that the marshes of the Baltic and North Sea coasts are zoologically distinct. When a more detailed study comes to be made of the salt-marsh fauna it is likely the potentially different habitats will be fewer than those currently used for mangal.

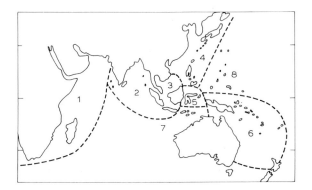

Fig. 1.6. Old World zoogeographical regions (after MacNae, 1968). Legend: *1* = western division; *2* = west–central; *3* = east–central; *4* = northeast; *5* = Borneo, Celebes, Moluccas; *6* = New Guinea; *7* = Western Australia; *8* = western Pacific Islands.

PAST ORIGINS

The present distribution of mangal associations is obviously related to the origin and spread of the

TABLE 1.3

Distribution of dominant mangrove species forming communities

Dominant community	Gulf of Mexico/Caribbean	Florida	W. Panama, Mexico Colombia, Ecuador	N. Brazil	E. Brazil	W. Africa	Red Sea	E. Africa	W. India	E. India	Ceylon	Bay of Bengal	Irrawaddy	Indo-Malesia	Papua/New Guinea	Philippines	Ryukyu Islands	Australia
Avicennia germinans	—	—	—	—	—													
Rhizophora mangle	—	—	—	—	—													
Conocarpus erectus	—	—	—	—	—													
Laguncularia racemosa	—	—	—	—	—													
Rhizophora racemosa			—	—	—													
R. harrisonii			—		—													
Avicennia germinans/A. schaueriana				—														
Pelliciera rhizophorae				—														
Avicennia africana						—												
Laguncularia racemosa/Conocarpus erecta						—												
Avicennia marina							—	—	—	—	—	—	—	—	—	—	—	—[1]
Avicennia marina/Rhizophora mucronata							—											
R. mucronata								—	—	—								
R. mucronata/Bruguiera spp.								—										
Sonneratia alba								—					—	—				
Ceriops tagal								—	—	—	—	—						
Sonneratia apetala									—	—								
Xylocarpus obovatus/Heritiera littoralis									—									
Bruguiera gymnorrhiza									—					—	—			
Mixed mangrove									—	—	—	—	—	—	—			—
Avicennia alba										—	—	—						
Rhizophora mucronata/R. apiculata												—	—	—				
Avicennia spp./*Excoecaria agallocha*												—	—	—				
Lumnitzera racemosa												—	—	—				
Aegiceras corniculata/Ceriops tagal												—						
Avicennia officinalis												—	—					
Excoecaria agallocha												—	—	—	—			
Nypa fruticosa/Acrostichum aureum												—	—	—	—	—		
Bruguiera cylindrica													—					
Bruguiera parviflora													—					
Sonneratia caseolaris/Nypa fruticans													—	—				
Rhizophora stylosa/R. apiculata																		—
R. stylosa																		—
Excoecaria agallocha/Xylocarpus obovatus													—					
Excoecaria agallocha/Heritiera littoralis													—					
Heritiera fomes												—	—					
Lumnitzera racemosa/Ceriops tagal															—	—		
Aegiceras/Aegilitis spp.																—		
Aegiceras corniculatum/Camptostemon schultzii																—		
Aegiceras corniculatum																—		
Bruguiera gymnorrhiza/Heritiera littoralis																—		
Kandelia candel																	—[2]	
Rhizophora apiculata														·········				·····

[1] *A. marina* var. *resinifera* in New Zealand. [2] Also in southern part of Kyushu Island (Japan).
Stippled lines represent areas where species are also present.

INTRODUCTION

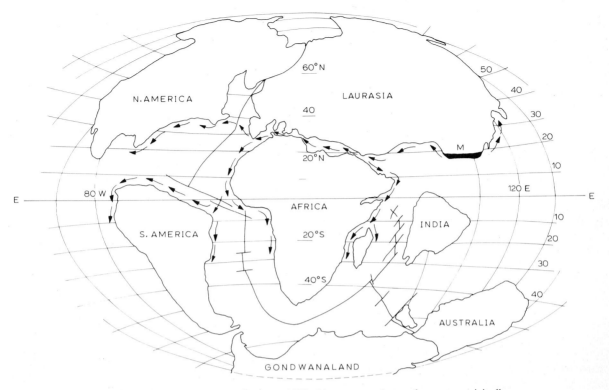

Fig. 1.7. World at the end of the Cretaceous (after Chapman, 1976). M = suggested area of mangrove originally.

major species (Table 1.3). Attempts (Ding Hou, 1960; Van Steenis, 1962) to explain the distribution in terms of the present land masses are not satisfactory and the present writer believes that only by accepting continental drift can a satisfactory explanation be provided. At the end of the Cretaceous angiosperm evolution was in progress, and if the Tethys Sea remained in communication with the Mediterranean a migration route along the shores of Laurasia (Fig. 1.7) would be available. Since *Avicennia* and *Rhizophora* are the only two genera common to both the Old and New Worlds it can be argued that these were the first two mangrove genera to evolve (Chapman, 1976). From the Mediterranean the ancestral forms could have passed into the young and relatively narrow Atlantic Ocean, down to the coast of Africa and across from Spain to North America, from whence they could have spread south to Florida and South America. If the other Old World genera evolved later this migration route would not be available; the currents around South Africa do not favour the migration of floating seedlings from east to west, and at this time the climate was also probably too cold (Chapman, 1976). The presence of *Nypa* pollen in the London clay flora is a further argument for the migration route via the Mediterranean and for a climate that would be warm enough to support both *Rhizophora* and *Avicennia*.

Past continental drift may also be helpful in accounting for some of the salt–marsh communities dominated by widespread genera. Thus, the genus *Salicornia* may well have arisen early in the course of evolution and before North America had effectively split off from Eurasia. This would account for its widespread distribution in both New and Old Worlds. The relatively late (geologically speaking) approach of Australia to Laurasia (Fig. 1.8) may well explain why the genus is not widespread in Australasia and there will also have been much less time for speciation to have occurred. Other genera which appear to have arisen in the Old World and then spread to America, probably before it was too separate from the Eurasian land mass, are *Limonium*, *Suaeda*, *Puccinellia* and *Juncus*. The genus *Spartina*, on the other hand, would seem to have had its origin in the New World where there are a number of species, and

Fig. 1.8. World 35 million years ago (after Chapman, 1976).

the Atlantic appears to have been wide enough by then to have been an effective barrier until man arrived with his ships. Prior to that time only *S. maritima* was to be found in Europe and it may have migrated via Iceland and the Faeroes at a time when the climate was warmer than at present.

PRINCIPAL FEATURES

Salt–marsh coastal formations are essentially characterised by low–growing flowering herbs, especially grasses. There are some annuals (e.g., annual species of *Salicornia, Suaeda, Atriplex*) but the great majority are perennials. Large areas are occupied by grasses, species of *Spartina, Puccinellia, Distichlis,* and rushes of the genus *Juncus*. Many of the communities consist of but one species, or essentially of a single species. At low levels where primary colonisation is taking place there may be extensive swards of species of *Salicornia* (Europe, Atlantic U.S.A., Australasia). Other communities are generally of lesser extent (e.g., those with *Armeria, Plantago, Limonium, Cotula, Selliera*) and frequently there is a mixture of species. Such a general salt–marsh community has been recognised in Europe, U.S.A. and Australasia (Chapman, 1938, 1940, 1960; Chapman and Ronaldson, 1958).

The perennial species of *Salicornia* are commonly so low–growing that their recognition as shrubs can be overlooked. This, however, is not the case with *Baccharis, Iva* and *Pluchea* which form shrubby communities at the transition zone between marsh and upland on the Atlantic coast of the U.S.A. or else along creek banks. The shrubby genera *Halocnemon, Arthrocnemum* and *Allenrolfea* are primarily found on salt marshes in the subtropics and tropics, whilst the low shrubby *Sesuvium portulacastrum* and *Batis maritima* form salt marsh frequently associated with mangrove swamps.

Mangal formations are wholly arborescent, though at the northern and southern geographical limits the trees become shrubby (e.g., New Zealand and Louisiana). As in the case of the salt marsh many of the communities are unispecific. A characteristic feature of the mangal are the intricate tangles of prop roots where *Rhizophora* species predominate, of short aerial pneumatophores with *Avicennia* and *Sonneratia* and the knee roots associated with *Bruguiera* species. At the appropriate season mangal is also noted for the pendulous seedlings, especially the very large ones of *Rhizophora* and *Bruguiera* with smaller ones of *Ceriops, Kandelia* and *Aegiceras*. In addition there are the bean–like seedlings of the species of *Avicennia*. Undergrowth in mangal is generally restricted to seedlings of the component species, except where there is a transition to brackish conditions, where the fern *Acrostichum* and the palm *Nypa* can form a rather impenetrable growth. Very few lianes are recorded from mangal (e.g., *Derris* spp. in the Old World and *Cydista equinoctialis* in the New World) and epiphytes are generally few, though species of the ant–harbouring genera *Hydnophytum* and *Myrmecodia* are regular features of some mangal (Philippines, Indonesia).

The fauna of wet coastal formations is generally non–specific and appears to be similar to that of the open seaward mud and sand flats. It is composed primarily of arthropods and molluscs with a variety of fish in the creeks and rivers[1]. The soil microfauna and flora has not, as yet, been ade-

---

[1] MacNae (1968) has given a good general account of the fauna of African, Indo–Malesian and Australian mangal, details of which will be found in later chapters. A comparable study is wanted for salt marshes.

quately studied and until this has been done the significance of both components to the formations cannot be properly assessed.

Studies of life–form spectra (see Chapman, 1960, 1974a for references) has shown that on salt marshes it is essentially Raunkiaer's hemicryptophyte group that is dominant, with therophytes as the next most important group. An exception to the latter is found in the marsh formations of New England and New Zealand (Chapman, 1960), where the percentage of therophytes is very low. In Europe therophytes are predominant in the pioneering associations and they become replaced by hemicryptophytes and geophytes in later stages (Table 1.4).

Table 1.5 sets out comparable data for specific

TABLE 1.4

Life–form spectra of salt–marsh formations (Chapman, 1960)

Area	Nano–phanerophytes	Chamaephytes	Hemicryptophytes	Geophytes	Hydrohelophytes	Therophytes	Parasites
S. California	—	14	31	9	—	43	3
New England	3.5	3.5	60.5	11	7	14.5	—
Europe	—	2	37.5	20	1	39.5	—
New Zealand	8	4	56	8	8	16	—

TABLE 1.5

Life–form spectra of salt–marsh communities (Chapman, 1960)

Region	Community	Chamaephytes	Hemicryptophytes	Geophytes	Therophytes
California	*Spartinetum gracilis*	40	20	—	40
Maryland	*Spartinetum alterniflorae*	—	60	—	40
New Jersey	*Spartinetum alterniflorae*				
Denmark	*Salicornietum strictae*	—	—	—	100
Sweden	*Salicornietum strictae*	—	57.5	11.5	31
Denmark	*Puccinellio–Plantaginetum*	—	69	—	31
Sweden	*Plantaginetum*	6	57.5	12.5	24
Denmark	*Juncetum gerardii*	—	74.5	19.5	6
Sweden	*Juncetum gerardii*	—	28	8	30
Sweden	general salt marsh	5	52	10	32
Sweden	*Suaedetum maritimae*	—	50	17	43
Denmark	*Puccinellio–Asteretum*	—	76	—	24

communities of the salt–marsh formations. With the exception of the Danish *Juncetum gerardii* the general pattern is the same, hemicryptophytes being predominant with therophytes as the second important group.

No comparable life–form spectra have been published for mangal formations, but it is evident that phanerophytes would always account for the greatest percentage and that in many cases the figure would be 100% (cf. p. 277, Table 14.4, MM, M and N).

BIOGEOGRAPHICAL RELATIONS

Table 1.6 sets out the percentage of the different elements in the maritime salt–marsh floras of certain regions. This table demonstrates a common affinity between the two New World formations with the European and New Zealand salt marshes as two separate distinct entities.

Table 1.7 sets out the percentage of the different elements in mangal formations of certain regions.

The relationship of comparable ecological com-

TABLE 1.6

Percentage of different geographical elements in salt-marsh formations (from Chapman, 1974a)

Region	Local	N. Hemisphere	S. Hemisphere	Cosmopolitan	Tropics	America	Mediterranean	S.E. Europe
Europe	34	33	—	25	—	—	—	1
E. U.S.A.	22	35	—	10	—	33	—	—
W. U.S.A.	39.5	10.5	2	13	8.5	27	—	—
New Zealand	20	18	38	18	3	3	—	—

TABLE 1.7

Percentage of different geographical elements in mangrove formations

	Local	Cosmopolitan	Pantropic	Tropic America	Old World tropics	Tropic Atlantic	Boreal America	S. America	Temp. Old World
Ecuador	52	0	17.5	8.7	0	17.5	4.3	0	0
Atlantic New World	26	3.7	7.4	22.2	0	11.1	18.5	11.1	0
West Africa	36.3	0	9	9	9	27.3	0	0	9
East India	21.9	2.4	7.3	0	68.3	0	0	0	0
Malaysia	24.4	0	4.2	0	71.4	0	0	0	0
Philippines	26	0	7.4	0	66.6	0	0	0	0

Based on information from the following sources: *Index Kewensis;* Chapman, 1975; *Flora Malesiana* (Van Steenis); *Useful Plants of the Philippines* (Brown); *Florida Mangroves* (Davis, Carn. Inst. Wash. Publ., 1940); Diels, *Bibliotheca Botanica,* 116, 1937; *Mangrove Forests of the Malay Peninsula* (Watson, *Mal. For. Rec.*, 1928).

munities to each other can be expressed mathematically as a degree of affinity. Various ways of producing this figure have been proposed. In Raabe's (1952) procedure the basis of comparison is the mean species numbers, i.e. those species with a constancy of more than 50%. On this basis the absolute affinity is arrived at as follows:

$$\text{Absolute affinity} = 100 \left[1 - \frac{1}{n} \sum_{i=1}^{n} \frac{|a_i - b_i|}{\max(a_i, b_i)} \right]$$

where a_i, b_i are the constancy values for the ith species in the two associations, respectively.

Figs. 1.9 and 1.10 set out the absolute affinities for maritime examples of the *Puccinellietum maritimae* and *Juncetum gerardii*. In the former case it would seem that the affinity (using this basis) is of a low order (e.g., Schleswig–Holstein and southwestern Sweden). In general the *Puccinellietum* affinities tend to decrease as one passes from south to north and from west to east. The *Juncetum*

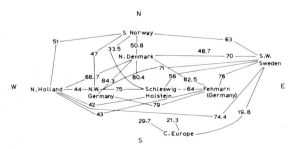

Fig. 1.9. Absolute affinity diagram for the European *Juncetum gerardii*.

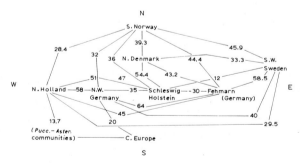

Fig. 1.10. Absolute affinity diagram for the European *Puccinellietum maritimae*.

INTRODUCTION

gerardii appears to be much more homogeneous and in most cases the affinity is high (communities with values over 50% are regarded as identical). Further work of this nature for both salt–marsh and mangal associations is much to be desired.

SUCCESSION

Wet coastal formations are essentially dynamic and there is a gradual progression from bare sand or mud flat through a range of different plant communities up to the point where there is a transition, through brackish–water vegetation, to fresh–water swamp or fresh–water swamp forest in the tropics. Alternatively there may be a sharp boundary between the highest coastal community and terrestrial upland community in places where coastal formations abut onto upland. In such cases one can recognise the highest maritime community as the sere climax (Chapman, 1960). In yet other places the maritime community may abut onto a shingle, shell or boulder beach in which case again a sere climax can be recognised with a sharp transition to the beach communities. Some authors (Egler, 1952; Thom, 1967; Lugo and Snedaker, 1974) consider that the zonation does not necessarily represent a successional sequence. In such cases salinity (see below) is regarded as a competition eliminator. At present insufficient information is available to show whether zonation is always successional. The range of communities which can be recognised in passing from the primary colonists to the sere climax or to the fresh–water transition communities are related to the changing environmental factors — in particular changes in soil type, frequency of tidal inundation, salinity changes and variations in drainage (see pp. 32, 48). Collectively the changes in the communities represent a succession or sere. In most cases there is usually a major sequence which represents the principal succession, but generally there are a number of variations related to local environmental factors or to man's activities in burning (Gulf of Mexico marshes) or felling (mangal). Other variations in the succession may be related to differences in soil type or drainage pattern. Numerous examples of such successions will be found in the regional chapters.

The transition via brackish–water swamp to fresh water is usually characterised by reed communities. In Europe *Phragmites communis*, *Scirpus* spp. and *Typha* are all taxa that can tolerate brackish water and mark the transition. In the case of mangal, *Typha* and *Nypa* are genera that again characterise the transition to fresh–water swamp or forest. Certain forest trees in the Gulf of Mexico (see p. 210) and in Indo–Malesia (see p. 266) can also mark the transition to fresh–water swamp forest.

Lugo and Snedaker (1974) have proposed another division for mangal and in certain circumstances their classification could be useful. The types are determined by local tide patterns and terrestrial surface drainage:

(1) Fringe forest. Generally best on steep shorelines.

(2) Riverine forest. This can be fronted by fringe forest occupying the steep creek banks.

(3) Overwash forest. Small low islands and spits bear *Rhizophora mangle* which obstructs tidal flow so that the trees are overwashed at high tide.

(4) Basin forest occurs in drainage depressions channelling terrestrial runoff toward the coast.

(5) Dwarf forest. Occurs on the landward edge where salinities can be high and also at biogeographical limits.

PRODUCTIVITY

Productivity of salt marshes has come to be regarded as high relative to other natural and artificial ecosystems. Odum (1974) has pointed out that tidal energy subsidises the solar energy available to plants growing on salt marshes. Odum notes that in natural and fuel–subsidised solar–powered ecosystems the average productivity is of the order of 20 000 kcal. m^{-2} yr^{-1}. Data obtained from Georgia salt marshes show a range of 16 000 kcal. m^{-2} yr^{-1} for low marsh with frequent tidal inundation to 3000 kcal. m^{-2} yr^{-1} for high marsh with only occasional tidal flooding. It is only low to mid marsh, therefore, where one can expect high productivity. Apart from the studies on U.S.A. Atlantic coast marshes, this is an area that needs much further investigation.

Productivity of mangrove swamps is also really unexplored, and this is serious in view of the value attached to both ecosystems relative to reclamation proposals for other uses. Golley et al. (1962) made a study of the productivity of a *Rhizophora mangle* stand in Puerto Rico and concluded that with a productivity of 8 g organic matter m^{-2} day^{-1} the forest is more fertile than most marine and terrestrial communities. Whether this conclusion has a universal application must await further studies.

Lugo and Snedaker (1974) have listed the factors that regulate mangrove productivity, but the same factors also operate on salt marshes. These factors are as follows:

(1) *Tidal:* (a) Transport of oxygen to roots. (b) Physical exchange of soil water with flooding tides, removal of toxic sulphides and reduction of soil–water salinity. The efficiency of this factor depends on soil composition and soil spaces. (c) Tidal flushing in relation to sediment deposition or soil erosion. (d) Drainage after flooding tides can carry down nutrients from detritus.

(2) *Water chemistry:* (e) Control of osmotic pressure gradient by total salt content and its effect on plant transpiration rate. (f) Provision of high macronutrient content of soil water. (g) Allochthonous macronutrients in wet season runoff may dominate the macronutrient budget. This would be important in monsoonal areas (see p. 248).

A study of the Florida mangroves by Carter et al. (in Lugo and Snedaker, 1974) showed that salinity affected the gross productivity of the red, black and white mangroves.

Gosselink et al. (1974) have made a first attempt to evaluate eastern U.S.A. salt marshes. Chapman (1974a) has pointed out that, whilst not everyone would agree with either their premise or their values, nevertheless this does represent an important first step. No comparable estimates have been made for mangrove forest though where these are commercially exploited there should be no difficulty in obtaining an answer. The results for the eastern U.S.A. salt marshes are shown in Table 1.8.

TABLE 1.8

Evaluation of eastern U.S.A. salt marshes

	Annual return per ha (U.S. $)	Capital value per ha (U.S. $)
1. Fish and fish food	250	5 000
2. Oyster culture (max. return)	2 200	44 000
3. Sewage effluent treatment	6 200	124 000
4. Life support value (based on gross–primary production)	10 250	205 000

Items 1 and 3 or 2 and 3 are compatible.

EXTERNAL RELATIONS

Many of the halophytic species that occur in maritime coastal formations are capable of growing in fresh–water areas providing there is not continuous flooding. Some salt–marsh species, such as *Plantago maritima,* grow to much larger plants under glycophytic conditions. Experimental work tends to the belief that very few of the genera of wet coastal formations require excess sodium chloride for satisfactory growth, i.e., are obligate halophytes. Among the salt–marsh plants, species of *Salicornia* would seem to be obligate halophytes, and also *Spartina alterniflora.* Among mangrove species, *Rhizophora* species are certainly obligate halophytes. It would seem that one reason why many of these so–called halophytes will not survive under glycophytic conditions is that they are very slow growing and hence cannot compete with the more rapidly growing glycophytes. There is also some evidence that many of them are sun plants and hence do not survive under shade. Since they grow more slowly they are eliminated by the shade cast by the taller glycophytes. This general observation seems to apply to species of salt–marsh and mangal formations. Examples of mangrove species growing up to elevations of 150 m in the absence of competition have been reported by Van Steenis (1963). The present writer has also seen mangroves growing in Calcutta Botanic Gardens and in fresh

INTRODUCTION

waters in Jamaica where trees reached maturity before competition appeared.

Many of the coastal salt–marsh species (though not the mangrove species) occur in inland salt deserts, and even where they do not occur other species of the maritime genera are to be found — e.g., *Salicornia utahensis* around Great Salt Lake, *Spartina montevidensis* in the Salinas Grandes of Argentina, and *Puccinellia limosa* in European inland salt–desert areas. There is therefore a clear relationship with inland salt deserts, but even if some individual species may be the same there are sufficient species restricted to the inland areas, especially *Suaeda, Salsola, Haloxylon, Sarcobatus, Anabasis,* to render the two types of habitat floristically distinct. This has also been established by using absolute affinity values and comparisons of floristic elements (Chapman, 1960).

CONCLUSION

Salt marshes and mangals are both examples of an open ecosystem with respect to both energy and matter. Both these ecosystems are sensitive to external influences, and where they impinge on mud or sand flats they can affect the surrounding ecosystems significantly in the same way that they can be affected by adjacent systems such as fresh–water marsh or riverine forest. Lugo and Snedaker (1974) have constructed a simple model illustrating the structural and functional attributes of the mangrove ecosystem. With very slight amendments the same structure and attributes apply to the salt–marsh ecosystem (Fig. 1.11). These are two major compartments (above–ground structures and muds); four external energy sources (sun, tides, upland runoff and fresh–water swamp and riverine forest); five processes (primary productivity, plant respiration, mud respiration, recycling of mineral nutrients and export of organic matter to other contiguous ecosystems); seven potential stress factors (channelisation, drainage and siltation, hurricanes, herbicides and oil, harvesting). Hurricanes primarily affect mangal, and oil so far only affects salt marsh.

REFERENCES

Adams, D.A., 1963. Factors influencing vascular plant zonation in North Carolina saltmarshes. *Ecology*, 44: 445—456.

Agadzhanov, S.D., 1962. Solonchakovaya rastitel' nost' primorskikh peskov Azerbaidzhana. (The saliniferous vegetation of the coastal sand plain of Azerbaidzhan.) *Uch. Zap. Azerb. Gos. Univ., Ser. Biol. Nauk*, 1962 (2): 3—12.

Baker, S.D., 1950. *An Investigation Into the Morphology of Saltmarsh Fucoids*. Thesis, Auckland University, Auckland, 115 pp.

Bascand, L.D., 1970. The roles of *Spartina* species in New Zealand. *Proc. N.Z. Ecol. Soc.*, 17: 33—40.

Beeftink, W.G., 1965. De zout vegetatie van ZW–Nederland beschouwd in Europees verband. (Saltmarsh communities of the SW Netherlands in relation to the European halophytic vegetation.) *Meded. Landbouwhogesch. Wageningen*, 65 (1): 1—167.

Beeftink, W.G., 1966. Vegetation and habitat of the salt marshes and beach plains in the south–western part of the Netherlands. *Wentia*, 15: 83—108.

Beeftink, W.G., 1972. Übersicht über die Anzahl der Aufnahmen europäischer und nordafrikanischer Salzpflanzengesellschaften für das Projekt der Arbeitsgruppe für Datenverarbeitung. In: R. Tüxen (Editor), *Grundfragen und Methoden in der Pflanzensoziologie*. Junk, The Hague, pp. 371—396.

Beeftink, W.G. and Géhu, J.M., 1973. Spartinetea maritimae. In: R. Tüxen (Editor), *Prodrome des groupements végétaux d'Europe*. J. Cramer, Lehre, 43 pp.

Bilio, M., 1965. Die Verteilung der aquatischen Bodenfauna und die Gliederung der Vegetation im Strandbereich der deutschen Nord– und Ostseeküste. *Bot. Gothob.*, 3: 25–42.

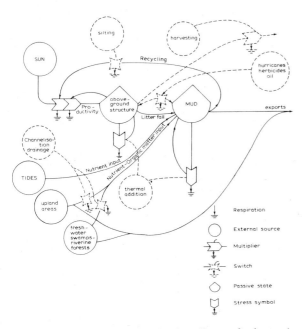

Fig. 1.11. Structural and functional attributes of salt–marsh ecosystem (modified from Lugo and Snedaker, 1974).

Bird, E.C.F. and Ranwell, D.S., 1964. *Spartina* saltmarshes in Southern England. IV. *J. Ecol.*, 52: 255—266.

Blum, J.L., 1963. Interactions between certain salt marsh algae and angiosperms. *Bull. Ecol. Soc. Am.*, 44: 92.

Blum, J.L., 1968. Salt marsh *Spartinas* and associated algae. *Ecol. Monogr.*, 38: 199—221.

Chapman, V.J., 1938. Studies in salt marsh ecology. I—III. *J. Ecol.* 26(1): 144—179.

Chapman, V.J., 1940. Studies in salt marsh ecology. VI—VII. *J. Ecol.*, 28(1): 118—152.

Chapman, V.J., 1944. 1939 Cambridge University Expedition to Jamaica. *J. Linn. Soc. Bot. (Lond.)*, 52: 407—535.

Chapman, V.J., 1960. *Salt Marshes and Salt Deserts of the World*. Hill, London, 392 pp.

Chapman, V.J., 1974a. *Salt Marshes and Salt Deserts of the World*. Cramer, Lehre, 2nd ed., 392 pp. (complemented with 102 pp.).

Chapman, V.J., 1974b. Salt marshes and salt deserts of the World. In: R.J. Reimold and W.H. Queen (Editors), *Ecology of Halophytes*. Academic Press, New York, N.Y., pp. 3—22.

Chapman, V.J., 1975. *Mangrove Vegetation*. Cramer, Lehre, 425 pp.

Chapman, V.J., 1976. Mangrove biogeography. *Proc. Symp. Mangrove, Honolulu, 1974, Vol. 1*. Univ. Florida, pp. 3—22.

Chapman, V.J. and Ronaldson, J.W., 1958. The mangrove and salt marsh flats of the Auckland Isthmus. *N.Z. Dep. Sci. Ind. Res., Bull.*, No. 125: 79 pp.

Clarke, L.D. and Hannon, N.J., 1967. The mangrove swamp and salt marsh communities of the Sydney District. I. Vegetation, soils and climate. *J. Ecol.*, 55: 753—781.

Corré, J.J., 1962. Une zone de terrains salés en bordure de l'étang de Mangio. *Bull. Serv. Carte Phytogéogr. Sér. B.*, 6(2): 1—105; 7(1): 9—48.

Dalby, D.H., 1970. The salt marshes of Milford Haven, Pembrokeshire. *Field Stud.*, 3(2): 297—330.

Davis, L.V. and Gray, I.E., 1966. Zonal and seasonal distribution of insects in North Carolina salt marshes. *Ecol. Monogr.*, 36(3): 275—295.

Ding Hou, 1960. A review of the genus *Rhizophora*. *Blumea*, 10(2): 625—634.

Egler, F.E., 1952. Southeast saline Everglades vegetation, Florida, and its management. *Vegetatio*, 3: 213—265.

Gillner, V., 1960. Vegetations- und Standortsuntersuchungen in den Strandwiesen der Schwedischen Westküste. *Acta Phytogeogr. Suec.*, 43: 1—198.

Gimingham, C.H., 1964. Maritime and submarine communities. In: J.H. Burnett (Editor), *The Vegetation of Scotland*. Oliver and Boyd, Edinburgh, pp. 67—143.

Golley, F., Odum, H.T. and Wilson, R.F., 1962. The structure and reproduction of a Puerto Rican red mangrove forest in May. *Ecology*, 43(1): 9—19.

Gosselink, J.G., Odum, E.P. and Pope, R.M., 1974. *The Value of the Tidal Marsh*. Publ. No. LSU–SG–74–03, Center for Wetland Resources, Louisiana State University, Baton Rouge, La., 30 pp.

Hutchings, P.A. and Recher, H.F., 1974. The fauna of Careel Bay with comments on the ecology of mangrove and sea–grass communities. *Aust. J. Zool.*, 18(2): 99—128.

Ito, K., 1961. On the saltmarsh communities of Notsuke–Zaki (Notsuke sand beach), Prov. Nemuro, Hokkaido, Japan — Ecological studies on the saltmarsh vegetation in Hokkaido, Japan, 4. *Jap. J. Ecol.*, 11(4): 154—159.

Ito, K., 1963. Study on the vegetation of the salt marshes in Eastern Hokkaido, Japan. *Sapporo Bull. Bot. Gard. Hokkaido Univ.*, 1: 1—102.

Ito, K. and Leu, T., 1962. Ecological studies on the salt marsh vegetation in Hokkaido, Japan, 5. *Jap. J. Ecol.*, 12(1): 17—20.

Knoerr, A., 1960. Le milieu, la flore, la végétation, la biologie des halophytes dans l'archipel de Rion et sur la côte sud de Marseille. *Bull. Mus. Hist. Nat. Mars.*, 20: 89—175.

Kolehmainen, S., 1973. *Ecology of Sessile and Free–Living Organisms on Mangrove Roots in Jobos Bay*. Rep. Puerto Rico Nuclear Center, Mayaguez, Puerto Rico.

Kolehmainen, S. and Morgan, T., 1972. Mangrove root communities in a thermally altered bay in Puerto Rico. *Proc. Ann. Meet. Am. Soc. Limnol. Oceanogr.*, 34th.

Koster, J.T.L., 1960. Caribbean brackish and fresh water Cyanophyceae. *Blumea*, 10: 325—366.

Lugo, A.E. and Snedaker, S.C., 1974. The ecology of mangroves. *Ann. Rev. Ecol. Syst.*, 5: 39—64.

MacNae, W., 1968. A general account of the fauna and flora of mangrove swamps and forests in the Indo–West Pacific region. *Adv. Mar. Biol.*, 6: 73—270.

Miyawaki, A. and Ohba, T., 1965. Studien über Strand–Salzwiesengesellschaften auf Ost–Hokkaido (Japan). *Sci. Rep. Yokohama Natl. Univ., Sec. 11*, 12: 1—25.

Molinier, R., Viano, Le Forestier, and Devaux, J.P., 1964. Etudes phytosociologique et écologiques en Camargue et sur le plan du Bourg. *Ann. Fac. Sci. Mars.*, 36: 3—100.

Nelson–Smith, A., 1972. Effects of the oil industry on shore life in estuaries. *Proc. R. Soc. Lond., Ser. B.*, 180: 487—496.

Odum, E.P., 1974. Halophytes, energetics and ecosystems. In: R.J. Reimold and W.A. Queen (Editors), *Ecology of Halophytes*. Academic Press, New York, N.Y., pp. 599—602.

Packham, J.R. and Liddle, M.J., 1970. The Cefni salt marsh, Anglesey, and its recent development. *Field Stud.*, 3(2): 331—356.

Poore, M.E.D., 1955. The use of phytosociological methods in ecological investigations. I—III. *J. Ecol.*, 43: 226—269; 606—651.

Raabe, E.W., 1952. Über den "Affinitätswert" in der Pflanzensoziologie. *Vegetatio*, 4: 53—68.

Ranwell, D.S., 1964. *Spartina* salt marshes in Southern England. II, III. *J. Ecol.*, 52(1): 79—94; 95—106.

Sauer, J., 1965. Geographic reconnaissance of Western Australian seashore vegetation. *Aust. J. Bot.*, 13: 39—70.

Tagunova, L.N., 1960. The relation between soil and plant cover of the northeastern shore of the Caspian Sea and salinity and moisture conditions. *Byull. Mosk. O.–Va. Ispyt. Prir., Itd. Biol.*, 65: 61—76 (in Russian) (*Ref. Zh. Biol.*, 1961, No. 7V242, transl.).

Taylor, M.C. and Burrows, E.M., 1968. Studies on the biology of *Spartina* in the Dee estuary, Cheshire. *J. Ecol.*, 56(3): 795—810.

Thom, B.G., 1967. Mangrove ecology and deltaic geomorphology, Tabasco, Mexico. *J. Ecol.*, 55: 301—343.

Umezu, Y., 1964. Über die Salzwasserpflanzengesellschaften in der Nähe von Yukuhasi, Nordkyushu, Japan. *Jap. J. Ecol.*, 14(4): 153—160.

Van Steenis, C.G.G.J., 1962. The distribution of mangrove plant genera and its significance for palaeo–geography. *Verh. K. Ned. Akad. Wet., Ser. C.*, 65(2): 164—169.

Van Steenis, C.G.G.J., 1963. Miscellaneous notes on New Guinea plants. VII. *Nova Guinea*, 12(1).

Walsh, G.E., 1974. Mangroves: A review. In: R.J. Reimold and W.H. Queen (Editors), *Ecology of Halophytes*. Academic Press, New York, N.Y., pp. 51—174.

Walsh, G.E., Barrett, R., Cook, G.H. and Hollister, T.A., 1973. Effect of Herbicides on seedlings of the Red Mangrove, *Rhizophora mangle* L. *Bioscience*, 23(6): 361—364.

Chapter 2

PHYSIOGRAPHY

J.A. STEERS

INTRODUCTION

Salt marshes are found in those parts of the world where the temperature is suitable for plant growth and where the physical conditions of the coast afford areas of shallow and sheltered water. These conditions also apply to mangals, the main difference being that mangals are associated with the warmer parts of the earth. In subtropical and warm temperate areas of the globe salt marshes and mangal intermingle. In this sense it is correct to regard the mangal as the tropical equivalent of the salt marsh, but there are significant differences. Salt–marsh vegetation never attains any height (some of the grasses within American marshes may reach 1 m) whereas a mangrove forest can reach 30 m or more (see p. 51). Both types of vegetation have been regarded as promoting land accretion but it is by no means certain that mangals have the same ability as have salt marshes to cause accretion. It is a common sight to see mangroves, especially *Rhizophora*, apparently "walking" out to sea. Closer inspection all too often reveals deepish water immediately in front of them, and it is then seen that the seaward edge may be an eroded edge. Within a mangrove swamp accumulation of plant debris and, to some extent, of fine sediments takes place, but there is not always, indeed seldom, a close parallel between the precise means of upward growth of the surface in the two habitats.

Although this chapter is primarily designed as a general account of salt–marsh physiography, it may be helpful to begin with a discussion of salt marshes in England and Wales and then to try and show how marshes in some other countries differ from them. This is not to insist in any way that British marshes are standard, but because it is useful to work from a marsh area with which I am familiar in order to have a standard with which to compare the various ways in which marshes can arise.

NORFOLK MARSHES

Development

The finest development of marshes in Britain is on the coast of Norfolk between Hunstanton and Blakeney (Steers, 1960, 1969). Nearly all this stretch is fronted by shallow flats on which the waves have built barrier beaches and spits. In general, these features are built of shingle, often with a considerable admixture of sand, and have extended to the west. In doing so they have thrown out recurved ends. The outer beaches are slowly migrating landwards under wave action and in consequence the older recurves have been partly over–ridden and now join the main ridge more or less at a right angle. Behind the main ridges and in between the laterals are ideal places for marshes to originate. Since the barriers enclose a low flat area which is filled at high water and almost emptied at low springs, the marshes often develop equally well on the landward side of the flat.

Each inflowing tide brings with it abundant fine sediment, fragments of vegetation and other materials in suspension. At the turn of the tide there is a short period of minimum, or even no, motion and some of this load is deposited, especially near the margins of the flat and also on the slightly higher parts of it. The flat itself is part of the very gently sloping sand floor of the sea and is not, of course, entirely smooth. In early stages, except in sheltered places, as much material may be carried out or redistributed on the ebb as is brought in by the flood. In the long run, however, deposition increases and the level of the flat, particularly near its margins and in other favourable places, gradually rises. During this period seeds of halophytes are carried in by the tide and it is only a matter of time before they begin to take root. Eventually plants, at first scattered, begin to appear and once this starts and the plants spread they begin to increase the deposition of sediment, and so upward growth is promoted. The early stages in this process may be much helped by algae which often spread on the original sand flat in the form of mats and afford admirable traps for sediment and seeds. In those places between two laterals — which may often be close together — sedimentation may be rapid since the enclosed space is sheltered and encourages deposition. This is the main reason why small enclosed marshes (= closed marsh) of this sort are often noticeably higher than the adjoining open flats and marshes (= open marsh).

Two important factors must be introduced at this stage. First, the nature of the tides; and second, the nature of the substratum on which the plants have to grow. In eastern England, and around the Atlantic generally the tides are semi–diurnal and often the range (the vertical distance between the high and low waters of comparable tides — neaps, springs or storm tides) may be considerable. If, let us say, the range is 3 m and, assuming that there is no isostatic or other movement affecting the coast, the marshes could eventually reach a thickness of about 3 m but, as will be explained below, this need not apply to the whole marsh. There are many places where the range is greater than 3 m; on parts of the Norfolk coast extreme spring tide range is about 6 m, and in the upper parts of the Bristol Channel about twice this value. In the Bay of Fundy the range is 18 m or more. Thus, the thickness of marsh sediment on a relatively stable coast can vary greatly with the tidal range.

The vegetation on a marsh depends to a considerable extent on the substructure. Plants such as *Puccinellia* and *Salicornia* may begin to grow on more or less pure sand, and may locally form close mats of some size. As they grow, they trap more and more of the fine sediment brought in by the tides, and the substratum becomes muddy or silt–covered. At this stage, in Norfolk, the sea aster (*Aster tripolium*) comes in together with *Suaeda maritima*. Whilst this is happening another major feature of marsh scenery is developing. It will be appreciated that in the early stages the tidal water, especially on a calm day, flows and ebbs as a sheet, drops some of its load as already described, and in time builds up the margins of the flat and also the slightly higher parts of it. As time goes on these areas expand laterally and sooner or later a time comes when the tidal flow and ebb becomes somewhat restricted into rather vaguely delimited but nevertheless distinct channels. With increasing growth of the marsh these become more and more defined and eventually become part of the creek system which is a characteristic feature of a salt marsh. We shall return to creek development later on; meanwhile it should be noted that as the height of the marsh surface rises it will not be covered by every tide. With this change there is also a change in the vegetation; *Aster* and *Salicornia* may continue to flourish but other plants which cannot tolerate submergence at every tide make their appearance (cf. 113).

Creeks and salt pans

As plant growth expands over a surface it ultimately forms an edge to the draining creeks. The appearance of creeks varies with the nature of the marsh. In Norfolk where the marshes are mainly formed of mud, the creeks are usually steep–sided, although slips of one sort or another modify parts of them (Fig. 2.1). In Cardigan Bay at, for instance, Talsarnau the mud is very thin, a layer resting on sand, and the cross–section of many creeks shows an inner and narrow channel and above a wider channel with fallen "blocks" of sandy mud held together by vegetation (Fig. 2.2). In some bigger creeks the banks may be sloping and partially covered by "blocks" eroded by the tidal streams in the creeks. Moreover, the much more luxuriant and varied vegetation in the Norfolk marshes stands in marked contrast to the largely grass marshes in the west. Other cross–sections may be found in *Spartina* marshes. Different patterns will be found in U.S.A. and elsewhere.

On the marshes the vegetation sooner or later

PHYSIOGRAPHY

Fig. 2.1. Creek bordered by *Halimione*, Scolt Head Island (Photo J.A. Steers).

encloses small ponds which are known as salt pans (Fig. 2.3). They are perhaps best seen in middle–high marshes, but can be found at higher and lower levels. Sometimes an elongated pan is found and, if it is examined, it may well prove to be part of an old creek which for some reason or other became blocked or dammed so that a part of it was left without outlet. Pans usually have a well–defined margin, possibly produced, or rather maintained, by the miniature waves which occur in them when the wind is strong, and partly by the swirl of the tide entering the pan during the rise of a big tide. The sharp margin is likely to be best seen in pans where a layer of sand underlies the mud or silt on which the vegetation is growing. One other feature that may be noticed in Norfolk pans is that they often contain a mass of the free-living seaweed, *Pelvetia canaliculata* var. *libera*. The point, however, that must be made is that this alga and also *Bostrychia scorpioides* are found almost all over the marshes if one looks at the surface on which the vegetation is growing. It will then be realised that they often form a close mat and this, taken together with the higher–growing plants, gives ideal conditions for the trapping of sediment and upward growth of the marshes.

We have outlined so far the conventional views about the formation of salt pans. Steers (1960), Pestrong (1965), Packham and Liddle (1970), and Chapman (1974) have all agreed in general with the views first explained by Yapp et al. (1917). Pethick (1969), however, finds good reason to doubt if it is a complete explanation for all primary pans. He studied many of the Norfolk marshes and had no difficulty in separating primary and channel pans. He approached the problem statistically and came to the conclusion that the density of pans increases with the height of the marsh — which implies that their density increases as the marshes develop. Pethick argues that the basic

Fig. 2.2. Creek showing effect of undercutting and slumping of edges, Talsarnau (Photo J.A. Steers).

point of divergence between the original theory of Yapp et al. and the results of his (Pethick's) analysis is that the latter indicates that salt pans are formed on mature marshes as well as on the initial marsh surface. Yapp himself called attention to a statement by Warming (1904) that "marsh turf could be broken through by the presence of putrefying masses of algae or *Zostera* or by the treading of cattle". Ranwell (1972) states that tidal litter thrown on to higher marshes will remain at least for the current growing season, and that this litter will give rise to bare patches, some of which may later develop into incipient salt pans. In Norfolk, Pethick's analysis indicates that pans are most numerous near the seaward edge of the marsh where much litter is dropped. Chapman (1938a), when he was working on the Atlantic marshes of the United States, found that snow patches lying for long periods on the marshes led to the formation of rotten spots. Pethick is in no way trying to disprove the old and classic view of pan formation, but rather to show that it is inadequate to explain the complete distribution of salt pans in Norfolk. What is now wanted is a full photographic case history of a Pethick-type pan.

The Norfolk marshes are fairly characteristic of the east coast in general. They are formed of compact mud or silt and, except in the early stages of development, are firm and easy to walk over. In some places, e.g., Hamford Water in Essex, the composition is much more liquid and it would be, in places, only too easy to sink in them. This is not

PHYSIOGRAPHY

Fig. 2.3. Scolt Head Island, salt pans on Plover Marsh (Photo J.A. Steers).

unlike some of the marshes on the south coast, some of which are built of soft watery mud. Those on the west coast, however, and some of the Scottish marshes, are largely, sometimes entirely, sandy.

Upward growth of marshes

Measurements of the upward rate of growth of marshes have been made in various places. In the earliest stage marsh growth is irregular. Sediment may be laid down by one tide and swept away by the next, so that any measurements in such places are almost impossible. When, however, vegetation has taken root conditions become more stable. Since the lower marshes, or the lower parts of a large marsh, are covered most frequently by the tides it follows that the rate of sedimentation is usually highest in those parts, always assuming a fairly close plant cover. In general, the rate decreases with the increasing height of the marsh, so that on the highest marshes covered only by high spring or storm tides the rate is much slower. It is in these places in eastern England that plants such as *Juncus maritimus, J. gerardi, Artemisia maritima* and *Plantago maritima* abound. The incoming tide flows up the creeks and, when it has reached the level of the marsh surface, spreads rapidly over it. The role of *Halimione portulacoides* is interesting in this connection. It is a low-growing plant which follows the creek edges and because of its bushy nature traps some of the silt overflowing the creek banks (Fig. 2.1). For this reason the banks

TABLE 2.1

Rate of sedimentation on the marshes of Scolt Head Island
All stations put down in September 1935

Station	Missel Marsh			Lower Hut Marsh			Upper Hut Marsh
	1937	1947	1957	1937	1947	1957	1947
1	0.20 cm			0.70 cm	6.5 cm	11.0 cm	2.4 cm
2	1.25	8.5 cm	20.0 cm	1.18	7.5		
3	1.25	7.5	17.0	1.42	8.0		2.0
4	1.75	12.5	22.0	1.05	7.0		1.4
5	1.78	10.8		1.40	8.7		1.2
6	1.75	9.0		1.50			1.5
7	1.48	9.4		1.15	7.0	12.0	2.5
8	1.88	10.6		1.37	7.6	13.0	1.0
9	1.70	9.4	14.0	1.68	8.5		1.7
10	2.50	14.0	23.0	1.72	8.4	14.0	3.5
11				1.45	8.0	12.0	4.0
12	1.67	10.0		1.80	6.4	10.0	6.0
13	1.00		14.0	1.50	6.5		
14	1.72	10.0	20.0	1.32	9.0	15.0	
15	1.48	10.0		1.50	7.4	14.0	
16	1.55	11.3	15.0	1.40	7.3	12.0	6.8
17	1.45	11.5		1.30	6.5	11.0	
18	1.60	10.0		1.52	8.8	14.0	3.8
19	1.68	9.0		1.65	9.0	17.0	
20	1.72	11.0	19.0	1.12	5.8		1.5
21	1.80	10.5	23.0	1.35	6.0		
22	1.63	10.5	20.0	1.52	8.5		
23	1.60	11.5	17.0	1.12			
24	2.05	11.5					
25	2.47	11.0					
26	1.85	9.8	18.0				
27	1.68	10.0					
28	1.70	12.0					
29	1.70	11.0					
30	1.95	12.0	19.0				
31	1.65	11.0					
32	1.60	10.5	17.0				
33	1.50	11.5					
34	1.60	12.5	20.0				
35	0.75	10.5	22.0				
36							
37							

Measurements were made in June 1937, June 1947, August 1957.
Horizontal lines in tables mean a creek is crossed: sedimentation is often higher at stations close to a creek. Between stations 22 and 37 there were numerous tiny creeks, pans and runnels.
Station 1 on Lower Hut Marsh line is 24 cm lower than Station 1 on the Upper Hut Marsh line.
Station 1 on Missel Marsh line is 16 cm lower than Station 1 on Lower Hut Marsh line.
Station 1 on Missel Marsh line is 40 cm lower than Station 1 on Upper Hut Marsh line.
No. 1 Station on each line is indicated on map (Fig. 2.4).
A blank means no record; presumably marking material had been swept away.
In 1957 (22 years interval) several stations were either lost or spoiled, so no reliable record was made.

PHYSIOGRAPHY

may be just a little higher than the adjacent marsh.

It is comparatively easy to measure upward growth. If a patch, about a metre in diameter, of some distinct sediment is put on the marsh surface where it is well vegetated, and left until a tide has washed it off the plants, a thin layer remains on the surface. Any minor irregularities can easily be smoothed out by hand. In course of time the sediment will be covered by mud or silt brought in by successive tides and, after some years, the thickness of the accretion can easily be measured. To obtain a reasonable idea of upward growth on different marshes, and different parts of a large marsh, all that is necessary is to set out lines of patches, mark them, and after a given time measure the upward growth. The increase will vary with tidal range, nature of vegetation and various local factors such as the open or closed nature of the marsh, and also the type of the sediment carried on to it; on some marshes partly enclosed by dunes, blown sand may play a significant role. Tables 2.1 and 2.2 give some results for the Dovey Estuary (Richards, 1934)

TABLE 2.2

Rate of sedimentation on the Dovey (Dyfi) marshes (Richards, 1934)

Line 1; about 120 m long

Association	Accretion (cm in 100 lunar months)	Land height (m)	No. of observations
Puccinellietum	6.61	0.50	8
Transition *Puccinellietum—Armerietum*	5.53	0.47	7
Armerietum	3.53	0.65	30
Transition *Amerietum—Festucetum*	2.00	0.87	1
Festucetum	1.75	0.99	4
Juncetum	2.03	0.82	3

Line 2; about 75 m long

Association	Accretion (cm in 54 lunar months)	Land height (m)	No. of observations
Puccinellietum	3.67	0.35	6
Transition *Puccinellietum—Armerietum*	6.93	0.50	2
Armerietum	4.09	0.51	15
Festucetum	3.38	0.70	6
Transition *Festucetum—Juncetum*	4.60	0.70	1
Juncetum	2.25	0.91	1

These marshes are at the mouth of the Dovey (Dyfi), Wales. A sand and shingle spit encloses them. They are much more sandy than the marshes at Scolt Head Island.

in Wales and Scolt Head Island in Norfolk, a map of which is reproduced as Fig. 2.4 (Steers, 1960). They could be matched by similar findings made in Denmark and elsewhere (Nielsen, 1935).

In some places sedimentation is far more rapid. In the Wash, especially alongside the outfalls of the rivers it may reach 7 cm a year. In the Bristol Channel, the Bay of Fundy and on marshes on the east coast of the United States high figures are reached. *Spartina* marshes are usually sloppy but the rate is generally high (e.g., Poole Harbour).

MARSHES AND CHANGE OF SEA LEVEL

It is essential to have a number of measurements of the thickness of marsh deposits if any reliable estimates are to be made about possible changes of

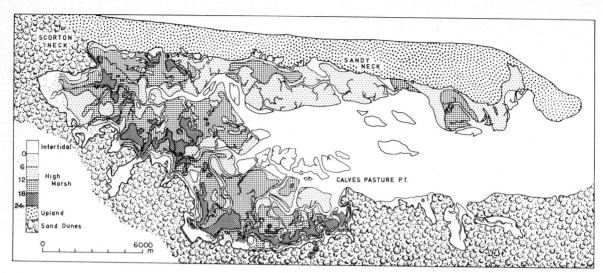

Fig. 2.5. The Barnstable Estuary, showing distribution of depth of peat in the high marsh (after Redfield). Contour intervals 6 ft ≡ 1.83 m.

sea level relative to the land. Borings made on the Scolt Head marshes showed that there were pockets where the mud was a good deal thicker than elsewhere. It was clear that these occurred in shallow hollows of the surface on which the mud was deposited, and their greater depth could not be taken as a true indication of change of level. One of the most interesting discussions on this topic is that by Redfield (1967) on the salt marsh in the Barnstable Estuary, Massachusetts (Fig. 2.5). The estuary is some 10 km long and contained within a sand pit, Sandy Neck, on the north and the upland of the Sandwich moraine on the south. It has been carefully sounded and the deposits dated by ^{14}C methods, and its evolution over the last 4000 years can be traced with some precision. "...mean high-water level has risen, relative to the land, at an average rate of 3.3×10^{-3} feet per year during the last 2100 years. Prior to that time the average rate of rise was 10×10^{-3} per year extending back for at least 3700 years". The earliest certain salt-water peat occurs at 7 m below present mean high water and is about 3660 years old. Deeper basins, 7–9 m, may have contained peat about 4000 years old. There is no doubt that this marsh has developed during a rise of sea level relative to the land, and we may assume that others in this part of the coast have evolved similarly.

Chapman (1938b) made some interesting calculations on the rate of marsh development in Norfolk, based on the rates of sedimentation during 21 months. Later measurements covering 141 months suggest that Chapman's conclusions, as he himself recognised, are open to criticism, and he has recently (Chapman, 1974) recalculated the rate. Table 2.3 shows how the data can be analysed relative to the communities. In Norfolk the primary succession is generally *Salicornietum* → *Asteretum* → late *Asteretum* → *Limonietum* → *Armerietum* → *Plantaginetum* → *Juncetum*[1].

Chapman thought that in Norfolk about 200 years were required for a marsh to pass from the beginning of the *Salicornietum* to the *Juncetum* stage if the coast were stable. There are, however, so many variable factors that any rigid conclusions are impossible at this stage. Chapman himself (1959) considered some of these, including possible change of level. This presents difficulties in East Anglia. In the Thames Estuary a downward movement of the order of 20—30 cm per century since Roman times is fairly well authenticated. The amount may decrease a little farther north, but archaeological evidence at Great Yarmouth strongly suggests that in the thirteenth century sea level was lower than it is today at that place. Green and

---

[1] Community terms are useful, but sometimes they can too easily over-emphasise the changes in vegetation; in no sense are the stages in the succession abrupt.

TABLE 2.3

Ratio of marsh development in Norfolk (Chapman, 1974)

Vegetation	No. of sites	Range of height referred to Island Zero Level			Av. rate of accretion 21 months	Av. rate of accretion 1 year
Salicornietum	8	−27 cm to	12 cm	= 39 cm	1.19 cm	0.68 cm
Asteretum	22	12 to	40	= 28	1.71	0.98
Late Asteretum	4	40 to	58	= 18	1.57	0.90
Limonietum	5	58 to	70	= 12	1.41	0.80
Armerietum	5	70 to	88	= 18	0.63	0.36
Plantaginetum	3	88 to	100	= 12	0.73	0.42

The rate of development of any one marsh thus passes through a number of phases, with accretion gradually becoming less as height increases and tidal submergences become less frequent. Moreover, bearing in mind the figures of the same table, a very rough estimate is possible of the time taken under average conditions for a marsh to develop from the *Salicornietum* to the *Juncetum* stage:

Vegetation	Av. max. depth of silt at conclusion of phase (cm)	Av. time required for the accumulation of depth of silt found at conclusion of phase, if coast stable
Salicornietum	39	ca. 44 years
Asteretum	67	ca. 28
Late Asteretum	85	ca. 20
Limonietum	98	ca. 15
Armerietum	116	ca. 47
Plantaginetum	128	ca. 31

Hutchinson (1960) concluded from the evidence obtained in the excavations on Yarmouth dunes that extreme low water mark of the twelfth—early thirteenth century now lies rather more than 5.3 m below Ordnance Datum. They refer to this episode as the Saxo–Norman Marine Regression. It is difficult to think that such a movement could be limited only to the Yarmouth area; local explanations such as slumping were found inadequate. The marshland of north Norfolk is 50—80 km away; can we assume that it was unaffected? Unfortunately as yet no further evidence of this regression is known, but it is clear that figures based on a downward movement of the order of 30 cm a century on the north Norfolk coast must be regarded as extremely improbable.

It is known that the general downward movement in the Thames Estuary is, if one may say so, balanced by an upward movement in the Solway Firth and in parts of Scotland. The Solway marshes present interesting problems, and certain relatively minor steps or small cliffs occur. These are interpreted by Marshall (1962) as primarily erosion features, whereas the Geological Survey sees in them indication of changes of level. On the other hand, between northern Wales and Morecambe Bay meticulous work by Tooley (1974) shows that there is an imperceptible gradient in raised beach level between these two places. From Morecambe Bay the gradient reaches 4.5 m in 100 km toward the Solway. These observations make one chary of accepting any figures of marsh evolution based on relative changes of level of land and sea unless they are, as in Barnstable Marsh, Massachusetts, based on detailed and reliable dating. This caution does not invalidate Chapman's views on the time taken for the various vegetation stages of a marsh to evolve in Norfolk, but it does imply that we must know far more about eustasy and isostasy before venturing to draw general conclusions. This caution applies to northern lands in general. In southern Sweden, on the west coast, in Öresund and in the Baltic, tides are very small and occasionally absent (Gillner, 1965), but changes in the water level of up to 1 m can be produced by the wind in winter on the west coast, and somewhat similar conditions are found in Öresund. In the Baltic tides are insignificant, but all Sweden is an area where isostatic movement plays an important role. Marshes develop, and their vegetation is generally similar to that on the other side of the North Sea, but since tidal conditions are of such minor importance marsh thickness, especially on a rising (isostatically) land area, is minimal, In the southwestern part of The Netherlands, however, conditions are different (Beeftink, 1966). Apart from the changes made by man the area is, and has been for long periods of geological time, subsiding. Transgressions have taken place and have broken through the old coastal barriers which were discontinuous because of the river mouths. In general, ingression of the sea meant the shifting inland of the border between salt and fresh water, and the deposition of mainly marine sediments. At other times peat formation took place. After the Roman period sand and clay were deposited in deep erosion channels and clay was laid down on the peat areas. Since the tidal range in this part of Holland varies from about 3.8 m near the Belgian frontier to about 1.5 m near the The Hook, and reaches 5.0 m near Antwerp it will be realised that the level of the marshes and peat is irregular and rises inland. It is, therefore, difficult if not impossible to make satisfactory estimates of sea–level fluctuations based simply on marsh deposits.

The marshes on the west coast of Denmark, especially in and near the Skalling Peninsula have been made classic by Niels Nielsen (1935) and his colleagues (Schou, 1967) and successors (Fig. 2.6). They are all built on sand flats, and they form the northern part of the sands and marshes that extend with breaks to The Netherlands. The tidal range decreases towards Esbjerg where it is about 1.5 m. Nielsen proved that the outer marshes of the Skalling area are very recent and that silting began as late as about 1900. Since that time spectacular changes have taken place in the vegetation which resulted in a succession like that in Norfolk and unusual in Scandinavia. *Spartina townsendii* has spread to places some distance from Ho Bugt where it was planted. These few examples show how much variation in development can take place around the North Sea. They demonstrate also the care that must be taken in generalising. Each area needs to be studied in detail and although the types of sedimentation, the spread of the plants, the variations in exposure and tidal range, and the vertical changes of sea level relative to the land are all similar in essence, they nevertheless express themselves in different modes or at different speeds, so that conclusions drawn from any one area are only comparable in the most general way with any other area[1].

SAN FRANCISCO MARSHES

In recent years interest has been taken in the evolution of marshes from other points of view. Pestrong (1965) in a paper on the development of drainage patterns on tidal marshes introduced new ideas and based his work on some marshes at the southern end of San Francisco Bay. This is a sheltered locality and the marshes are built on and grade into alluvial fans which have been produced by streams from the Santa Cruz Mountains. Of the rock formations which underlie the bay, only one, the top one, is under the marshes—and, in fact only the soft layer of the "Younger Bay Mud" which, nevertheless, may reach a thickness of anything between 6 and 20 m. In this part of the world the tides are mixed, two high and two low waters occurring each day. The diurnal inequality is most marked when the declination of the moon is at a maximum, and it disappears when the moon is over the Equator. The sun, of course, also exerts a modifying influence. In this bay a rise in sea level, relative to the land, of about 8 cm has been noted during the last half century. A considerable amount of fresh water also enters the bay from the San Joaquin—Sacramento river system.

Careful analysis was made of the hydrological characteristics within tidal channels and the rates of movement of water up and down the channels were obtained at different depths. *Salicornia* is the most widely established plant and is found especially in dense growth along the channels; *Spartina* is common on the bottom and sides of shallow gullies (cf. p. 161). The more significant difference is that the root system of *Salicornia* makes the soil in which it is growing resistant, and often is the reason why channel banks in this zone are undercut. The *Salicornia* areas are much firmer than those

---

[1] The Danish west coast marshes are discussed in many papers in the annual volumes of *Medd. Skalling Lab*.

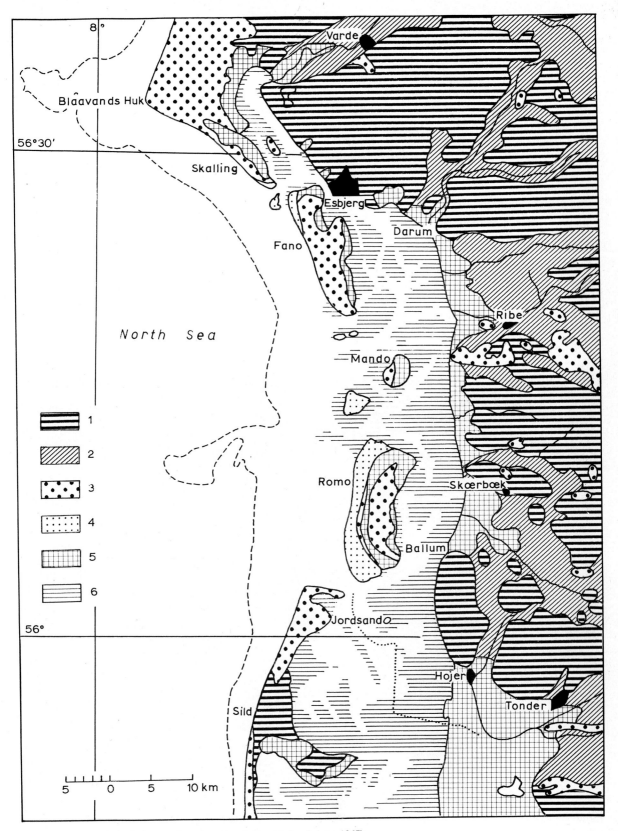

Fig. 2.6. Geomorphological map of southwestern Jutland (after Schou, 1967).
Legend: *1* = moraine landscapes from the Riss glacial period; *2* = outwash plains; *3* = dune landscapes; *4* = foreshore plains and "high sands"; *5* = salt marshes; *6* = tidal flats (wadden).

under *Spartina*. Since this is such a sheltered environment, the sediments are all fine-grained. Pestrong makes some comparisons with the Wadden Sea in northern Holland, and emphasises that in San Francisco Bay "the channels continue out on to the high flats from the marshes since the channel banks are composed of sediment cohesive enough to maintain their slope without the aid of vegetation".

Small channels begin to develop in the high mud flats as a result of the ebb flows. These, following from sheet erosion a little way below the higher salt marsh, are parallel with the slope; later new rills form at angles to the original ones. These lengthen upstream as a result of headward erosion (Fig. 2.7). They form the earliest primary channels, and on a mud flat undergoing dissection they will be somewhat widely spaced and empty into some hollow or major channel. As plants colonise the upper flats, the creeks deepen as a result of accretion on the marsh; it is argued that this will cause the heads of the channels to cut backward, and the tributary creeks will do likewise as a result of their steeper gradient to the main creek. With the increased growth of *Spartina*, water draining off the flats will take more sinuous courses, and so small tortuous channels may originate. Lateral erosion and, therefore, some small migration of the meanders will take place as a result of the faster ebb currents. In these minor creeks the current at flood is minimal or non-existent since the rising tide ponds back any water in the creeks. Pestrong follows earlier writers in his views on the origin of pans.

D.R. Stoddart (personal communication, 1974) reported that he and Pethick made measurements of the water velocities in the creeks at Wells (Norfolk). Their results do not always confirm

Fig. 2.7. Head cutting of marsh creek (Photo R. Brinsdon).

those of Pestrong. They found on one occasion the highest velocities on the flood, and movement at the ebb was almost imperceptible — and this on a bank–full tide. This means that we need careful observations on different marshes under varying winds and other conditions, and a full appreciation of local factors.

Pestrong has followed Horton's lead and adopted quantitative methods in classifying creeks on the marshes. The small upper ends of tributary creeks are designated first–order; two of these unite to make a second–order segment; a third–order segment may be formed by the junction of two second orders, and so on. The master channel is always of the highest order. In general, first–order channels are shortest, and the segment length increases as the order increases. Since a marsh surface is almost flat it is difficult to recognise any boundaries between areas drained by major creeks. The sinuosity of a creek "is the ratio between the sinuous length of the channel segment and the straight line distance between the ends of the segment". The pattern of tributary and main creeks will change with various local features; the tributaries may have a dendritic, trellis, or rectangular pattern. In general, on the San Francisco Bay marshes the junction angle is usually nearer to a right angle when the difference between the tributary and the main creek is considerable.

To follow Pestrong in detail in his chapter on the hydraulic geometry of drainage systems on tidal marshes is perhaps unnecessary; suffice to say that it is probably the most important attempt to tackle this subject. Comparison is made between tidal–creek systems and ordinary terrestrial streams. The migration of channels is thought to be caused mainly by lateral undercutting of the more resistant bank stiffened by *Salicornia* and its roots. Undercutting leads to slumping, and in these marshes creek banks are usually steeper where there is *Salicornia* than where *Spartina* is growing. The erosion in all channels is more severe during the ebb flows, but the difference is less marked in the larger channels.

Near the area investigated, Pestrong found unconsolidated bay muds reaching 8.5 m below the marsh surface, and in places the tidal flat sediments are more than 18 m in thickness. Since the drainage channels begin on the upper tidal flats it seems as if their gradients are distinctly steeper than, for example, those found along creeks in the Norfolk marshes; moreover, marshes in an enclosed bay are somewhat different in their nature and evolution from those on an open coast, especially where barrier beaches have been built by waves and where marshes have developed behind them.

Pethick (1969) has analysed certain Norfolk marshes in the light of Pestrong's work. It must be remembered that all these marshes have a sand substructure and that only locally does the mud/silt deposited on it exceed 1.8 m in thickness. Moreover, sand is often blown over the marshes so that the marsh sediment may be distinctly inhomogeneous. Pethick takes the open western marsh, the Cockle Bight at Scolt Head Island, as an example of early development. The upper part now shows fairly definite creeks, and considerable changes have taken place in it since 1945. Minor creeks are embryonic. A little farther east is another marsh which before the war had advanced to a stage somewhat beyond that of the Cockle Bight today. Missel Marsh has a well–developed creek system, and headward cutting is excellently shown because the middle and lower parts developed into a marsh well before the upper part which, until quite recent years, was a bare gently sloping sand flat (Fig. 2.8). The drainage to the mud–marsh creeks from this area after a flooding tide pours into the heads of the minor creeks and back–cutting has been rapid. Similar features can be found at, for instance, Warham and Blakeney, two other localities on the Norfolk coast. Some of the creeks are by no means sinuous, especially on the marsh between the Hood and the Yankee ridge at Blakeney. At Warham the creeks are slightly sinuous; at Missel Marsh (above) they are distinctly so. On the biggest (Hut Marsh) and some of the older marshes (e.g., Plover Marsh) at Scolt Head Island the main creeks are relatively simple, the minor ones very sinuous (Fig. 2.9).

The vegetation on these marshes is usually thick (see p. 33) and some of the marshes may be as much as 500 years old. This is little more than a guess, but it is worth pointing out that in Plover Marsh, one of the oldest, the main creek may be 5 m deep, whereas on Hut Marsh the creek depths are less, even where they join the main channel separating the island from the mainland. The amount of sand incorporated in the marshes has undoubtedly a noticeable influence on permeability

Fig. 2.8. Creeks and shingle ridges, Scolt marshes (Photo Dr. St. Joseph, by permission Air Photo Unit, Cambridge University).

in the early stages of growth. As the mud layer increases drainage becomes more and more of a surface phenomenon, allowing for the occasional thin outcrops of sand in creek banks. Just as in San Francisco Bay, the larger and deeper creeks show a two-way flow, and lateral erosion can take place in them at flood or ebb. In small creeks and "headwaters" only the ebb is effective. Pethick sums up his general view: "in early stages undefined central creek systems extend over the whole marsh surface as surface discharge increases. This rapid development leads to the formation of the meandering low order streams, indicative of one-way flow regimes. These, in turn, merge with the characteristic tidal two-way flow of the larger channels. As the marsh increases in height so discharge is slowly

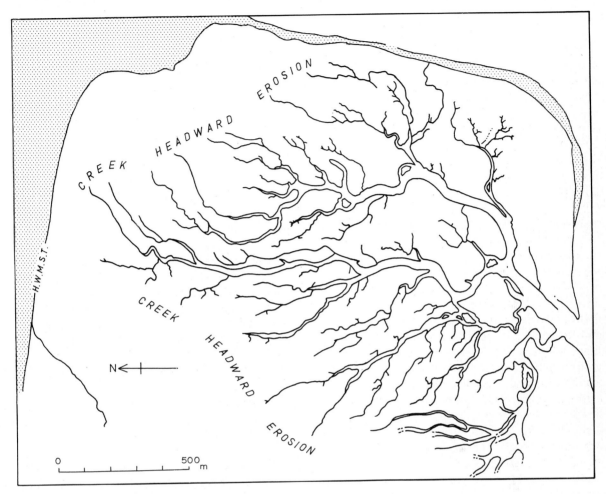

Fig. 2.9. Creek system on a salt marsh (after Pethick, 1969).

reduced; the two-way flow system encroaches on the whole drainage pattern and many channels are abandoned, forming extensive areas of elongated salt pans."

The point that needs emphasis is that salt-marsh areas are not all alike. Norfolk and San Francisco Bay show this, and although the views of Pestrong and others can be adapted to other areas, it is essential that full weight should be given to the details of each marsh — the tides, the type, and means of sedimentation, the floor on which the sediments rest, the slope of this floor and the irregularities on it, the spread of vegetation and other factors. One final example may be added. In the Wash near the Witham outlet, Evans (1965) recognised six morphological subzones which were arranged parallel to the shoreline: (1) the salt marsh, plant-covered, and passing into (2) the higher mud flats, a broad zone projecting seawards along creeks; (3) inner and *Arenicola* sand flats — the inner flats have a smooth or rippled muddy surface, the *Arenicola* flats well rippled, and abundant worm casts; (4) lower mud flat coinciding with the more steeply sloping part of the intertidal zone; (5) lower sand flats, usually very flat; (6) creeks and bordering areas — the creeks frequently show levees. Profiles levelled across the flats indicate some 7 m of sediment, taking the base at the surface of the eroded Holocene clay. Sea level has been rising relative to the land, and all the sediments come from seaward.

AREAS WITH BOTH SALT MARSH AND MANGAL

Salt marshes can be found along all continental and island shores as long as physical conditions favour their formation. However, as we pass from colder through temperate to tropical regions there are some changes in vegetation. Since this chapter is primarily concerned with physiographical matters the purely vegetational changes will not be discussed. However, sooner or later the change from an ordinary marsh to a mangrove swamp takes place, and it will be helpful to enlarge on this point. There are many places where such a development can be examined but we shall consider three — Florida, New Zealand and the southeastern part of Australia. The northern, or southern, limit of mangroves is determined by the increased number of frosts which will kill the trees. In Florida a number of salt–marsh areas extend as narrow zones between the mangals and the fresh–water marshes, and transitions from true salt–marsh, through brackish–marsh to fresh–water conditions are common (p. 162). In parts of the southwest coast "the depth of the peat, shell and marl deposits indicate a general submergence of these shores, or else a rise in the sea level, over a long period of time that amounts to about 7 to 10 feet. These deep peat deposits, some over 12 feet, are now interpreted as indicating a rise in sea level" (Davis, 1940). In fact there are few, if any, places of such extent that are so favourable to swamp development of one type or another as is Florida.

Auckland, New Zealand, is close to the southern limit of mangrove growth (Chapman and Ronaldson, 1958) (cf. also p. 310). New Zealand is a noticeably seismic area, and variations in the relative levels of land and sea have often taken place for that reason, so that actual measurement of marsh thicknesses do not have any easy interpretations, the more so because in the Auckland district there are several ancient bowl–like craters, some of which are now drowned. The geology at Pollen Island, Mangere and Traherne Island is generally similar — surface mud overlying mottled clay (or locally blue mud), fresh–water peat (at Pollen Island) and soft blue marine clay. The peat has been dated at more than 20 000 years and at Pollen Island it appears to pass beneath a 12–m terrace which makes the Rosebank Peninsula. It may, therefore, be of the order of 120 000 years old.

Table 2.4, taken from their paper, sums up Chapman and Ronaldson's point of view.

In Australia there are several marginal salt marsh—mangal areas near Sydney that will serve also as an example (cf. p. 297). The marshes occur in two main ways, extensive mud flats often with striking zonation, and in narrow strips in arms of creeks. Unfortunately many have been reclaimed or altered by man. There is — or was — no great development of halophytic vegetation. The parent material of the sediment is permeable sandstone so that the amount of sediment is small. Deep water occurs close inshore; there are no big rivers, the tidal range does not exceed 1.8 m, and it is a high–energy coast.

Ecologically and in other ways these three areas are very different and it is not worthwhile to try to make correlation between them about relative movements of sea level. What is clear is that in each case one mangal is left to "compete" with the salt marsh in which several genera are common to all areas (cf. also p. 130).

MANGALS

Florida

Mangrove swamps and forests present several interesting physiographical problems which have by no means received conclusive answers. One of the most significant statements was made by Davis (1940): "Only the assumption of a slowly rising sea level can explain how mature mangroves could grow to the very edge of deep peat deposits facing bays and estuaries 6 to 10 feet in depth, for seedlings do not begin growth in water 6 or more feet deep." He goes on to say that a sinking of the coast could produce the same effect. How then are we to explain the seaward, or riverward, *apparent* growth of mangroves.

It may be helpful to discuss the matter in relation to different environments. Davis was writing of Florida. Apart from the marginal areas referred to on this page, the main mass of mangrove growth in Florida is probably the most extensive in America (Fig. 2.10). The area is flat and much broken up and intersected by channels, and spits of sand are common on the outer coast. The waters are generally shallow and they are floored with

PHYSIOGRAPHY

TABLE 2.4

Past changes in sea level at Rosebank (after Chapman and Ronaldson, 1958)

Age	Changes in sea level	Sea level (m)	Physiography
400 years ago	fall	present	present marsh formation
700—800	rise	+ 1.5	erosion of marsh
800—2300	steady	present	first marsh formed
11 000	rise (Flandrian)	− 24.4	
30 000	max. (drop)	− 82.3	
Würm (L.g.3)	fall	− 30	
Last interstadial 2	rise	− 12	erosion of 12 m terrace
Würm (L.g.2)	fall	− 70	back to present lines
Last interstadial 1	rise	+ 1.2	
Würm (L.g.1)	fall	− 100	
Late Monastirian	fall	+ 7.7	
150 000 yr Main Monastirian	max. rise	+ 18	12 m terrace built
Prior to 150 000 yr Pre-Monastirian	rising	lower than present	Pollen Island peat laid down

Assuming the relative accuracy of the figures it is interesting to note their estimation of the present rates of accretion at Pollen Island:

	Average annual increase (mm)	Increase per 100 years if same rates prevail (cm)
Avicennia	1.66	16.6
Avicennia + *Salicornia*	1.4	14.0
Salicornia	1.0	10.0
Juncus + *Leptocarpus*	0.3	3.0

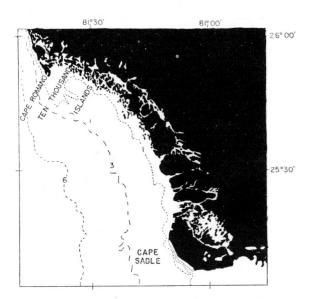

Fig. 2.10. Map of southwest Florida showing Ten Thousand Islands area and southwest coast (from Chapman, 1975). Contour lines in fathoms (1 fathom ≡ 1.8 m).

sands, marls, and shell fragments on which the young mangroves take hold. Silting occurs owing to the joint action of various agencies, and there is no doubt that the aquatic vegetation, including the mangroves, helps in this. But there is also no doubt that, if, as is not uncommon, mangrove swamps face open water affected by considerable wave and current action, they began to grow when the waters were shallower and less rough. In such an area the pioneer, at least in Florida, would be *Rhizophora mangle* and with it there could be low-growing vegetation including, for instance, *Thalassia testudinum* and *Syringodium filiforme*. Then, in course of time, the vegetation spreads and may become zoned. If, however, the coast remains stable for long periods of time the zonation is likely to be narrow. Whether the coast is stable or otherwise, in the course of time it is almost certain that the mangroves themselves will have been building soils; this, however, is not universally accepted. But it is an accepted fact that land increase takes place in suitable environments in existing mangrove swamps; riparian rights are certainly regarded as significant by all landowners in such places. In some localities there are deep peat soils which imply a gradual upward growth of the surface, and Davis (1940) called attention to one swamp that in thirty years had developed from an open shoal to a thick forest. He was, nevertheless, careful to point out that marl soils do *not* imply long–term deposition within mangrove swamps, since they (the marls) are mainly of marine origin. The general vegetational succession in Florida from *Rhizophora* to *Conocarpus* (cf. p. 162) is regarded as the result of the upward growth of the surface as the swamps increase in age. Stratification is not always conspicuous except where marls occur — that on the surface on which the trees originated and the occasional layer or lenses swept in by the waves. It is reasonable to argue that, where the peat layer is much thicker than the tidal range, there has been a movement of sea level relative to the land. The fibrous peat itself is nearly all organic in origin, and largely consists of the remains of mangroves and associated plants. Sedimentary peat, marls, and sand also contain some mangrove material. Mangrove swamps, like salt marshes, are cut up by creeks. The creek banks are not steep like those in marshes, and nearly always are formed of roots and not sediments. It may well be that primary colonisation by mangroves was in patches or clumps and that as they grew upwards glades — perhaps a better name than creeks — were formed. Hydraulic creek geometry would have little or no meaning in mangals. A great deal more work is required on the subject; why, for example, do mangal creeks meander? Another interesting point was made by Stoddart — salt marshes are relatively rare in the tropics and when they occur are seldom if ever found on the open coast, but behind some sort of protection, which may well be a belt of mangroves.

Deltaic areas

In deltaic areas mangroves often play a dominant role. Thom (1967) has given an interesting account of the Tabasco Delta in Mexico. The Caribbean in general is a sea in which the tidal range is very small and in the Tabasco Delta stream discharge is of much greater importance than the tides, which there have a range of 40—50 cm. The lengthening of channels and distributaries and the resulting shift of sedimentation effectively govern what Thom calls "the dynamic ecology of (the) mangroves...". If a marine current runs across the delta, spits or barriers are likely to grow with the current. Six stages of morphology and vegetational change were recognised[1]:

(1) The development of a new distributary; gradual emergence of levees and the beginning of grass and shrub growth on them.

(2) Widening of channels and levees; effects of wave action; probable growth of mud flats. Mangroves, especially *Avicennia,* and herbs begin to colonise the flats, and vegetation may help in sedimentation (Fig. 2.11).

(3) Continued growth of channels and levees. Discharge increases; more rapid spread of vegetation on levees and flats. *Rhizophora* and *Laguncularia* encroach on *Avicennia*. There may also be some slow subsidence of the delta.

(4) Diversions may take place upstream and so cause abandonment of some distributaries. If subsidence continues together with compaction of the levees *Avicennia* is likely to spread at expense of levee thicket, and eventually *Rhizophora* comes in.

(5) The levee thicket may disappear, and peat spreads on to back slopes of levees, outer parts of *Rhizophora* belt may be wave–eroded.

(6) Mangal peat spreads over subsided alluvial landforms, parts of which may show up by difference of vegetation (Fig. 2.12).

Probably somewhat similar changes occur in other deltas. In the Lower Mississippi deposition could not keep pace with the rise of sea level, or rather with the isostatic sinking of the land relative to sea level. This means that the sea transgressed older fluvial deposits, and earlier deltas disappeared under the Gulf of Mexico waters. This delta is marginal to mangal areas (see p. 193), but the many shifts of the river in the last 3000—4000 years are suggestive of what may happen elsewhere.

The Irrawaddy Delta (Burma) offers certain other variations. A distinction must be drawn

---

[1] This implies that mangals are a consequence of rather than a cause of physiographical patterns — but Thom is discussing a deltaic environment where initial patterns might well precede vegetation.

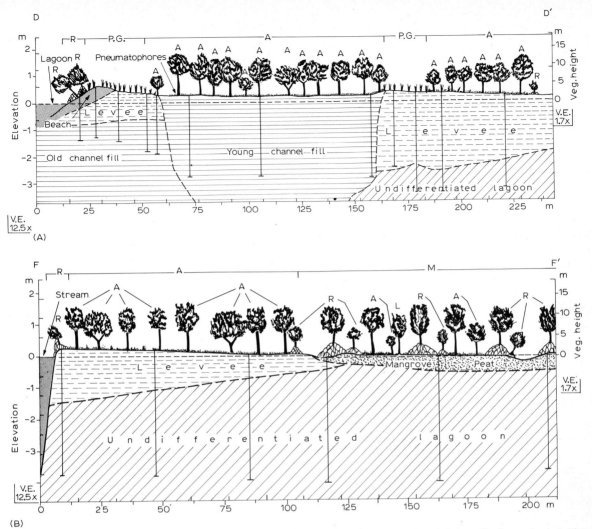

Fig. 2.11. A. Inorganic channel fill and levee habitats, east of Laguna de Cocal, Tabasco (Mexico). B. Interdistributary basin habitats, lower Rio Cuxcuchapa (after Thom, 1967). R = *Rhizophora*; A = *Avicennia*; L = *Laguncularia*; $P.G.$ = "pajon" grass; M = mixed mangrove forest.

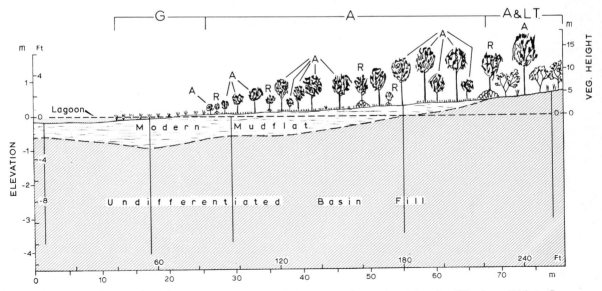

Fig. 2.12. Prograding mud–flat habitat, Laguna del Carmen. R = *Rhizophora*; A = *Avicennia*; LT = levee thicket; G = grass. Vertical lines: bore holes (after Thom, 1967).

between the tidal forests and the mangrove swamps which are on the fringes (Stamp, 1925). This delta was surveyed from the air, the photographs being taken at 3000 m. The rivers and creeks in the western part of the delta are mainly salt; those in the centre and east are fresh to brackish. The Irrawaddy carries much silt and mud, and deposition takes place both out to sea and within the delta. It seems to be at a maximum where there is a "rough equality" between the volume of fresh water and salt water. Thus, in the rainy season it takes place lower down the channels, and especially in those which are mainly fresh water. Once the new mud is exposed at low water, vegetation may get a hold — first grasses and other plants and sooner or later kambala *(Sonneratia apetala)*. The root system of this tree also helps to collect sediment and if the level rises sufficiently the kambala is killed and replaced by the kanazo *(Heritiera fomes)*. The whole delta is remarkably flat except for the Myaungmya rock ridge and outer sand ridges. It is only along the sea face and along the banks of creeks near the sea where mangroves flourish.

There are extensive mangrove forests around the Malay Peninsula including an almost continuous belt from Kedah to Singapore, but exploitation and other changes have taken place since Watson wrote (1928). He was insistent that the mangroves follow silting rather than cause it. There is much sheltered water between Sumatra and the mainland and in places, under natural conditions, the deep soft mud is almost impassable. It is colonised by *Avicennia alba* with *A. intermedia* on somewhat firmer ground.

West coast of Colombia

There is no need to consider all major mangal areas in other continents. There are numerous examples in Africa, more particularly on the west; in the east, except between Tanga and Moçambique, they are often poorly developed. Attention must, however, be given to an important study by West (1956) on the Pacific coast of Colombia, not only because of the luxuriance of the development, but even more because West is one of the few writers who have treated the subject from a physiographical point of view. Mangals fringe the coast for about 650 km from Cabo Corrientes (Fig. 2.13).

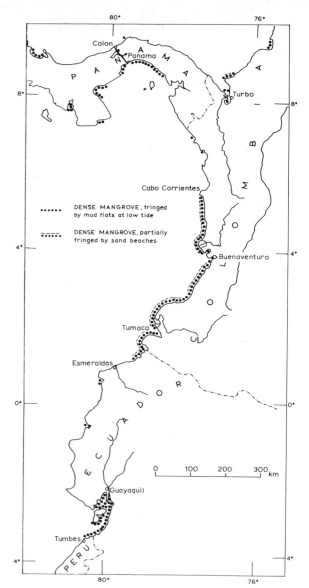

Fig. 2.13. Extent of mangrove swamps along coast of northwestern South America (after West, 1956).

They grow on part of a coastal plain which is, in effect, a composite delta formed by streams from the Andes. For this reason it is very irregular in detail, and in three places cliffs reach the shore and there may be a mangrove fringe only 1.5—3 m wide; elsewhere it may reach 25 km. West recognises four main belts: (1) shoal water and mud flats immediately off the coast; (2) discontinuous sand beaches; (3) mangrove forest, 1—5 km wide; (4) fresh–water tidal swamp. This sequence is very like that in Malaya, West Africa and Guiana. The tidal range is considerable, reaching 3—4 m at springs, and penetrates inland some distance in the low–lying coastal plain and also implies a wide alluvial, and mangrove, belt. The prevalent south-westerly winds and corresponding wave action are largely responsible for a number of coastal land forms. The area has been compared with that of the Ten Thousand Isles in Florida, and they are certainly similar in the size of the trees which may reach 30 m in height and 90 cm in diameter.

One zonal feature that is conspicuous is the almost continuous inland waterway behind the beaches. It is formed by lagoons and channels which run roughly parallel to the coast, and separated from it by the beaches built through wave action. This channel joins together the lower parts of the river estuaries. There is no doubt that a good deal of fine soft mud, both deposited and redistributed by tidal currents gathers on channel margins between the intricate root systems of *Rhizophora* and *Avicennia*. This mud is blue–black in colour, mainly of grains less than 0.02 mm and rich in organic matter. Occasionally, sand patches are found within swamps. They may be inhabited and cultivated. They are almost certainly parts of former beach ridges, and are called *firmes*. Some of them contain evidence of pre–Colombian Indian sites.

Despite this and other apparent indications of the progradation of the coast, it is not to be assumed that mangroves cause the advance. It may indeed be agreed that they entrap fine sediment, but West is almost certainly correct in thinking that the precipitation of particles by electrolytic action at the contact of river and salt water is a more potent cause. He asserts that on the coast "it is observed" that land must have fully emerged before it is overgrown by mangrove and, what is more, *Avicennia* precedes *Rhizophora* because the the latter only thrives in quiet brackish and salt water in sheltered places or where beaches or barriers of one sort or another protect it from wave action. *Avicennia*, on the other hand, seems more tolerant of highly salt water and can grow on soil containing much coarse quartz sand. Both genera prefer fine sediment and so are found where sedimentation occurs irrespective of the vegetation. In other words, vegetation is *not* the major factor in coastline advance. West argues that "like most low, alluvial coasts, the mangrove littoral of Colombia is unstable.... Despite its instability, in general, the coast appears to be advancing, as evidenced by remnants of beaches (firmes) in the present mangrove belt. Owing to the lack of historical data, determination of the rate of advance is not possible, but it has probably been extremely slow; certainly it is not comparable to the rapid advance of the deltaic marginal shorelines of south-eastern Asia, where wave and current action is much less and stream load much greater." This conclusion, that mangals are not significant in this area in coastal progradation, is in accord with the findings of investigators in other parts of the world.

Coral reefs and islands

One other habitat must be considered — the role of mangal on certain coral islands. Although mangals are not limited to the islands within the Great Barrier Reef of Queensland, I will refer to them first because I am more familiar with them. The finest growth in Queensland is in the Cairns–Innisfail district (cf. p. 301), and may be associated with the fact that this is also the area of highest rainfall (MacNae, 1966).

There is an interesting contrast between the Innisfail and Townsville areas. In the first the mangals and coastal trees or forest are contiguous, in the latter a semi–arid saline sand and mud flat lies behind the mangals. This pattern seems characteristic of some tropical areas and is presumably the result of differences of climate (cf. Chapter 3).

The low–wooded islands are peculiar to the Great Barrier Reef. They are low coral platforms, situated within the main barrier, and the considerable wave action in the "lagoon" has built on them a sand cay usually near the leeward margin of the reef, and shingle ridges (ramparts) on the windward margin. Between these is the flat surface of the reef. Depending mainly upon the shape and size of the individual reef there is considerable variation in the detailed physiography. The cay and ramparts may be some distance apart; there may be more than one cay and there certainly may be more than one rampart, and even more than one group of them; sometimes the ramparts and cay are contiguous and little or nothing is left of the reef flat. In one case, Hope Isles, the cay is on one reef, the rampart on another. But in all the same features and the same general arrangement are easily traced.

The best known of these low–wooded islands is Low Isles, the headquarters of the 1928 expedition to the Great Barrier Reef (Stephenson and Stephenson, 1931) and visited and examined on several subsequent occasions, including that of the Royal Society and Queensland Universities expedition in 1973 (Fig. 2.15). The mangroves grow at a somewhat lower level than they do on the mainland. Coral debris, whether scattered on the flat or in the form of ramparts or ridges, is cemented at least in part. In front of the ramparts there is a hard–rock platform with, in places, basset edges. These stand up rather like the sods on the upper side of a plough furrow and represent all that remains of former ridges. It is in this area that dwarf *Aegialitis* and *Avicennia marina* bushes are common, and also occasional *Rhizophora stylosa*. The crests of the ramparts stand high and are only covered in storms and by spray; they carry some *Osbornia octodonta* but much more *Aegialitis*, abundant *Pemphis acidula* and some *Thespesia populnea*. D.R. Stoddart (personal communication, 1971) suggests *Osbornia* is now on the ridges only because it has grown up through them as waves and storms have

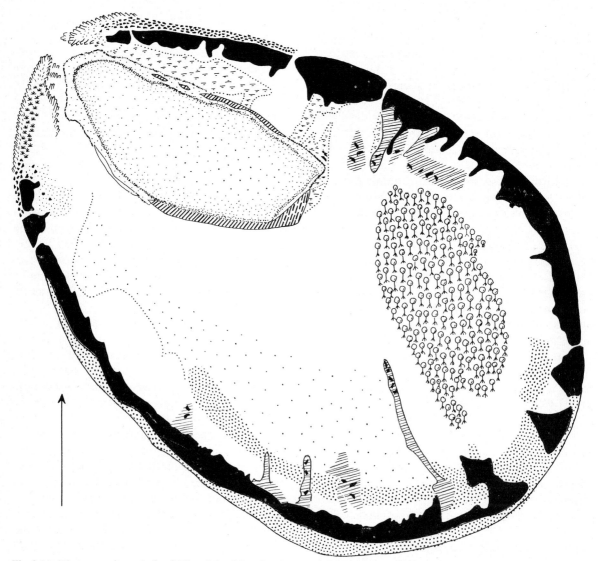

Fig. 2.14. Diagrammatic analysis of Three Isles (after Stephenson and Stephenson, 1931). (Scale: max. E—W = 1560 m, max. N—S = 1070 m.)

PHYSIOGRAPHY

pushed them inwards. The belt of *Osbornia* now (1973) in front of the rampart at Low Isles was not mapped in 1928—29, nor by Fairbridge (1950). It was then inside the ramparts. On the leeside is the reef flat — it is *not* a lagoon. At about high water mark trees of *Avicennia marina, Excoecaria agallocha, Bruguiera gymnorrhiza,* and some *Rhizophora stylosa* flourish. At a slightly lower level is the main mass of mangroves, particularly *R. stylosa*, which reaches 10 m. Its root system remains submerged at most tides. The mangroves are spreading on the reef flat and present a markedly park-like appearance. In the mangal area there is plenty of mud, but it is a surface layer, sometimes about 0.6 m deep or a little more. MacNae (1966) argues that this lack of deep mud explains the absence of most other species. However, since the mangroves take root in cracks and crevices of the coral debris, even where mud is completely absent, they nevertheless help to accumulate mud once their root system becomes of any size. Since some mud of terrigenous origin, as distinct from that of local and organic origin, was found by Fairbridge and Teichert, they suggested that it is swept to the island when floods occur in the Daintree River. In all the other low-wooded islands I have seen (Steers, 1938) the general picture is the same. Low Isles is the farthest south of this type (Fig. 2.15), but they continue nearly all the way to Cape York. It is, I think, clear that the mangals, in all cases, have developed to their greatest extent behind shingle ramparts, apart from *Avicennia* and *Aegialitis* which, as noted above, may also occur on the windward side. In other words, they have colonised a

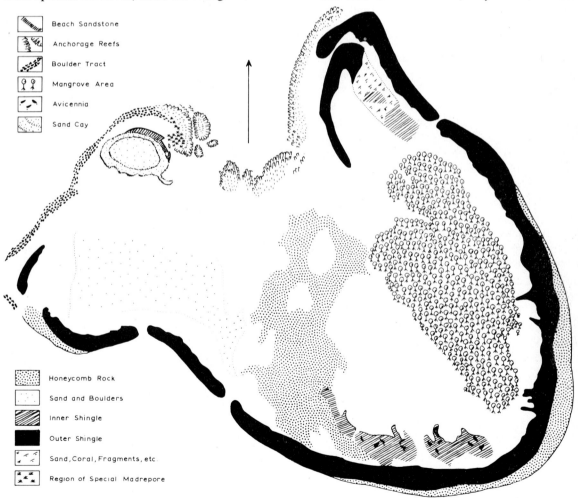

Fig. 2.15. Diagrammatic analysis of Low Isles reef for comparison with Fig. 2.14 (after Stephenson and Stephenson, 1931). (Scale: max. E—W = 1765 m, max. N—S = 1450 m.)

relatively sheltered area; they have spread by seeds from the mainland, and the mud is a later addition, although it is certainly possible that thin spreads of organic and possibly some of terrestrial origin had gathered on the reef flat at any early stage. To some extent the mangals stabilise the ramparts, but it is only in storms that waves have the power to drive them back into the mangals. Except when the ridges are, for some reason, lightly cemented it is improbable that they will advance noticeably and it is the mangals that suffer. Once again, the mangals do not play the part that is all too often assumed for them. Mangals characterise many other reefs and islands. In 1939 I mapped several islands around Jamaica. Two, Pigeon and Salt islands, are worth mention. Both are rims of shingle and sand with an almost complete surround of mangals which grow mainly within the ridge and fall to the lagoon inside. However, far more extensive mangrove growth has been described by Stoddart (1962) on Turneffe Reefs off the coast of British Honduras (Fig. 2.16). The reef is 50 km long north—south, and has a maximum width of about 16 km. At first sight, from the sea, it appears to be almost surrounded by mangals. The greater part of the east side is fringed by a discontinuous sand ridge up to 1.5 m high and 180 m broad. There are coconut plantations on it. Westwards (i.e., to the lagoon) they give place to dense *Rhizophora* and some *Avicennia*. On the east side the mangals are on dry land; the west part is wet — i.e., the western side of the lagoon. Some *Rhizophora* trees have reached the sand ridge and even the reef flat to the east. The west side of the reef is rather similar, but the mangal belt is narrower. The ridges of sand and shingle on the east form a series of long narrow islands. On the west islands are absent. This general pattern can be matched elsewhere, but once again the mangals are behind beaches or ramparts and do not directly face the sea as salt marshes so often do.

Stoddart (1963) was able to re–visit British Honduras after Hurricane Hattie (1961). It is relevant to note that in exposed places *Rhizophora* was completely defoliated. In sheltered areas the effect was less. Individual *Rhizophora* trees showed great stability "typified by the survival of a dead *Rhizophora* ring at Big Calabash Cay II and Blackbird Cay (both on the east side of Turneffe Reef), where the enclosed island disappeared under wave attack".

Jennings and Coventry (1973) in the Fitzroy Estuary, Western Australia, show that barrier ridges have passed through mangrove swamps as a result of wave action. This implies that mangals are not always protective (cf. also Chapman and Ronaldson, 1958; Chapman, 1974). J. Sauer, quoted in the same paper, records that in Mauritius littoral vegetation was washed out from behind a belt of mangals that remained undisturbed.

Many other papers have been written on salt marshes and mangrove swamps, and each one illustrates certain local phenomena. It is, however, unlikely that, even if it were possible in one chapter to consider the greater part of the extensive literature, new points of major significance would arise. Local conditions must play an important part, but principles are unlikely to be affected. There is no doubt that ordinary salt marshes demonstrate accumulation, often active and rapid; depending upon their environment they usually show also considerable increase in the height of the ground so that it is a comparatively simple process to build embankments and make extensive reclamations. To some extent this is also true in some mangal areas, but the raising of the ground level and, possibly, reclamation are, as has been shown, rather secondary effects and have not proceeded *pari passu* with the growth of salt marshes. The extension of mangal areas has perhaps been the result of wave or current action in lengthening or prograding a shoal of mud or sand, and the mangroves have followed. This is particularly noticeable in deltas. But there is no doubt that once a mangrove forest is developing it will greatly affect the process of sedimentation within it.

ISOSTASY AND EUSTASY

Comment has been made on the general relationship of marsh thickness to tidal range. It is convenient to explore this point somewhat more fully in relation to isostasy and eustasy. It is probably impossible to claim that any shoreline is completely stable. All coastlines in areas that were affected by the Quaternary glaciation are still "alive". Usually this means that areas that were under the ice are still steadily moving upwards — Scandinavia, Scotland, and much of the northern part of America illustrate this. But the areas south of the

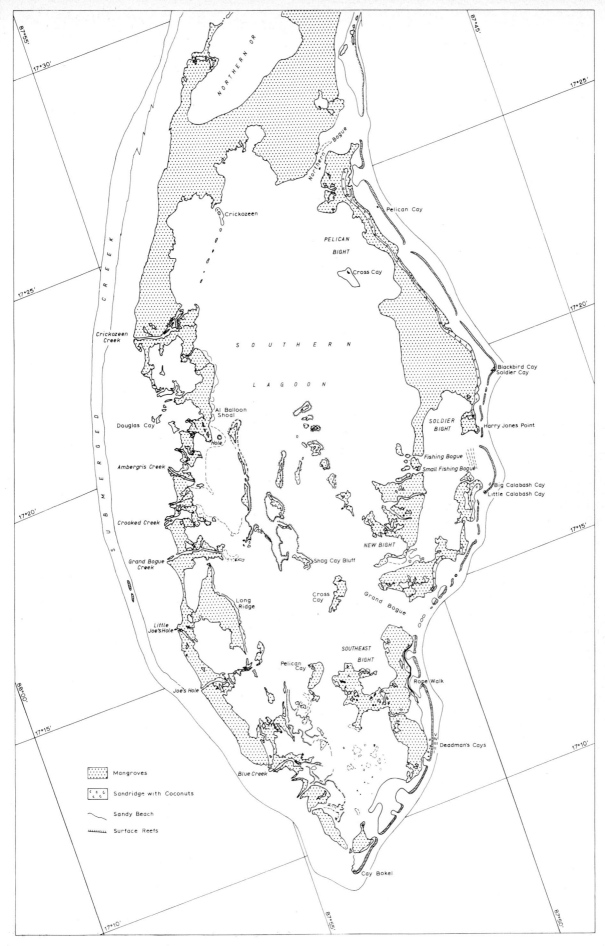

Fig. 2.16. Southern end of Turneffe Island (after Stoddart, 1962).

ice were also affected. To oversimplify the matter: it will be appreciated that, if a great mass of ice presses down on those parts of the world on which it rested, it will also cause a bulging out of the parts beyond its margins. It is like pressing hard on a large rubber ball; the downward pressure under your hand causes some upward and outward pressure elsewhere. This is a greatly oversimplified statement, but it will help to emphasise that when the ice was at a maximum the parts of the globe around its margins rose up to some extent, and now that the ice has melted they are sinking and the part that was under the ice is rising. In some such way as this we can account for the downward movement in southeastern England, and the upward movement in the Solway area and Scotland, and, *mutatis mutandis*, in many other areas.

We do not know whether we are in an inter-glacial or post–glacial period. There is some reason to think that apart from any movement of the land of an isostatic nature, there is a slight rise in sea level. If we could consider this alone it would mean a very slight rise, probably to be measured in millimetres in a century, of the level of the oceans and interconnected seas. Since the isostatic movements are much more significant we need, for the present purpose, only assume that the eustatic movement slightly modifies the isostatic one in any area — it may add to or subtract from it according to local circumstances. Some fifty years ago there was animated discussion of Wegener's theory of continental displacement. Several writers discussed other possibilities of moving continents, but until after the 1939—45 war there was a little belief in, and indeed great scepticism about, the possibilities of continental movement. In recent years, owing to the great discoveries in oceanic structure, the whole attitude to this problem has changed. Nowadays, the views held by those best able to express opinions on plate tectonics make it clear that any shoreline, irrespective of simple glacial isostasy and eustasy, is probably far from stable. Additionally many regions are seismic and movements of a more local nature may be considerable. It is true that, apart from certain seismic disturbances, all these movements are slow, but they are there and also long continued.

With these points in mind it becomes increasingly difficult to rely too much on such phenomena as marsh thickness for evidence of coastal movement.

Douglas Johnson, many years ago, in his *New England—Acadian Shoreline,* showed how easy it was for the sea to flood low–lying areas of forest or vegetated land without any vertical movements taking place. If man had not intervened in The Netherlands and Britain after the severe flooding of 1953 it would be interesting to speculate on what would have happened to the flooded land. There seems little doubt that in some places like Barnstable Marsh (p. 38) we can derive reliable information from the depth and thickness of marsh deposits; but if a coast is slowly subsiding, relative to sea level, as seems to be the case in Norfolk, what are we to expect? The tidal range is about 6 m at spring tides, the outer shingle and sand barriers are in several places advancing over the marsh and presumably compressing it. The surface of the flats on which the marshes rest was not entirely level in the first instance, so that measurements of thickness vary for that reason. We may certainly conclude that slow subsidence will allow a thick marsh to accumulate, and we may prove it to be so if we can make a considerable number of deep sections and analyse the cores by ^{14}C and other methods.

PRESERVATION OF SALT MARSHES

This brings us to another matter of great importance. A salt marsh is in many ways similar to a submerged forest. It is true that "forest" implies vegetation very unlike that of a salt marsh, but some peaty deposits at just below sea level are very similar to salt marshes. To take but one example: at or immediately adjacent to Scolt Head Island in Norfolk four peat deposits were analysed by H. and M.E. Godwin (1960). The best example is that at Judy Hard, a locality inside the island. This peat certainly represents a change of level of some 6 m. It is Boreal to Early Atlantic in age. The peats on the foreshore just west of the island are later in age, probably Hallstatt. They are at a higher level, but all contain tree pollen and indicate that, despite a close topographical relationship with existing marshes, they are not themselves remains of earlier marshes. They indicate comparatively recent relative movement of land and sea, but they cannot be regarded as direct ancestors of the present marshes. Can we perhaps assume a very long future for the existing marshes during which time they will pass through a *Phragmites* stage, such as is locally present at Thornham, a few miles away, to a much later stage of land vegetation? If that were to happen, would an investigator at some future time be able to trace the basal marsh deposits or would they not, following the present evolution of the barrier beaches, have been obliterated by the inward cutting of the sea?

If, on the other hand, a coast is rising and salt–marsh deposits are slowly being raised, they are likely to be thin and sooner or later dispersed as a result of ordinary terrestrial erosion. For a time they may afford interesting ground for controversy concerning actual uplift or wave erosion to explain certain very minor cliff–like features, as in those of the Solway marshes. If an uplifted marsh is to be preserved for any length of time it is more than likely it will owe its preservation to some sudden, possibly seismic, movement. It is more than probable that any marshes formed on an isostatically rising coast are likely to be thin, and equally likely to be destroyed in the process of rising.

If we turn back for a moment to submerged forests, we may well consider how and why they are so well preserved. They are known to depths of 60 m or more in dock excavations; they are usually thin beds underlain and overlain by marine deposits. They themselves are by no means resistant to erosion. How then were they preserved? Their sedimentary covers apparently imply some wave or current action and it seems odd that, unless the vertical movement bringing the forest beds to their present positions were sudden, they were not eroded away. An ordinary salt marsh with no true tree vegetation could presumably disappear even more easily. If the coast on which a raised beach exists is very flat, it can be easily inundated by severe storms or tsunamis. It is, for example, possible to imagine that shingle and sand driven landwards in Lincolnshire in 1953 could have buried a salt marsh or a high–lying submerged forest and so preserved it, especially if it had been a natural and uninhabited coast. In other parts of the world it is conceivable that this could happen. But even if many marshes are in estuaries or to some extent protected by barrier beaches and dunes, it still remains extremely difficult to understand how submerged forests have been preserved and why salt marshes have not been eroded away if the relative levels of land and sea change as they clearly have done. Probably, we may assume the parts that do remain are remnants of more extensive beds. The 1953 storm inundated the Norfolk marshes just as it inundated the Lincolnshire coast, but the marshes hardly suffered, and it would require a succession of such storms to cover them even with a thin layer of sand washed over them from the beach. What has been said about mangrove swamps seems to indicate that where they face moderately deep water they are being eroded. It seems very improbable that salt marshes or mangrove swamps are likely to throw much, if any, light on vertical movements of the coast except in favourable local conditions. The emphasis in the discussion has been on change of level. Reference to p. 54 makes it clear that hurricane damage to mangals may cause defoliation but trunks often remain. When I visited Night Island (Queensland) in 1936 there were numerous bare trunks and nearly all the island had been devastated. In other places mangals have been replaced by what is, in fact, a salt marsh — *Batis, Sesuvium* and *Salicornia*.

COASTAL CHANGES AND SALT MARSHES

The growth of salt marshes and mangrove swamps may cause profound changes on a coast. Parts of estuaries and embankments, and the lagoons behind barrier beaches can, in a relatively short time, be converted into dry land, or land that can be enclosed for reclamation. The long history of the conversion of the Fenland of Britain from open water swamps in pre–Roman and Romano–British times to its present condition is a major instance. The Wash is all that remains of the former large inlet which reached almost to Cambridge. There are many other examples in Britain and other countries, and in an old country with a long recorded history the relations between physiographic change and settlement can often be traced. As an example we may take the estuary of the Chester Dee (Marker, 1967). This estuary was at one time more important commercially than the Mersey; now it is almost filled up with deposits (Fig. 2.17). It is 8 km wide at the mouth, and extends in all 30 km to Chester. It appears to lie in an over–deepened fault trough. The estuary is an inlet in Liverpool Bay, and glacial deposition in this bay

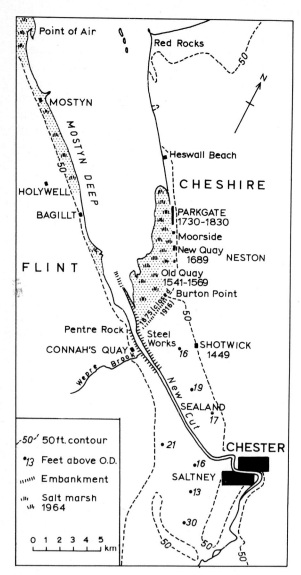

Fig. 2.17. Changes in the Dee Estuary, its progressive silting and salt–marsh development (after Marker, 1967). Contours and altitudes are in feet (1 ft = 0.3048 m).

has left an abundance of material which has been carried into the estuary by tidal and longshore action. Since, unlike the Mersey, the estuary widens regularly seaward, scour is relatively unimportant. Chester was a Roman port at the head of the estuary; the town grew throughout the Norman period, but silting was affecting it in the early Middle Ages, and continued to do so. In 1541 a new haven was built at Neston, but was abandoned in 1569. Collins' chart of 1689 is the first reliable map; Old Quay (1541) and New Quay (1689) are shown (see Fig. 2.17). Marsh was developing near Burton Point. In 1730 Parkgate became the port for Chester until it was finally replaced by Liverpool in 1830. The New Cut was made in 1732, and its embankment was extended in 1819, and the deep–water channel was made to follow the west bank. In 1834 the Mostyn–Holywell Bank was fronted by about 400 m of marsh. The upper 8 km of the estuary were fully reclaimed by 1840 — hence the name Sealand. This reduced scour and so marsh developed between Connah's Quay and Shotwick, and this was inned in 1875. The present embankments were completed in 1916.

Silting, therefore, has continued since the eleventh century; a good sandy beach existed at Parkgate up to 1938, but marsh had covered it by 1947, and throughout it has developed rapidly, and has extended northwards to Heswall where the beach is being covered. All this represents a seaward growth of 5.5 km in 26 years. Marker (1967) states: "Between 1938 and 1947, 523 mm of silt and mud have been laid down on top of the beach sand and 40 feet out from the seawall at Parkgate. A further 237 mm were deposited between 1947 and 1963.... It appears that between 612 and 627 mm of silt and mud have collected on this marsh in the past 26 years" — roughly 25 mm a year. If man does not disturb it, the process will continue and a sward will replace the estuary, but such an entirely natural evolution is most improbable. This instance has been taken because it can be related to history in a direct manner; it is, however, only one instance. Morecambe Bay, Poole Harbour and Southampton Water are others (cf. Chapman, 1974); and it would not be difficult to find parallels in other countries. Given favourable circumstances silting followed by growth of salt marshes can bring about great changes in a coastline, and in the Dee and some other examples any isostatic movement that may now have, or may have had, some effect on the locality in question is negligible in comparison with the filling.

It would be valuable to have details of a corresponding growth in a mangrove area. Unfortunately most mangrove areas are in places where there is probably not a long *record* of change. Moreover, since mangroves follow, rather than initiate, silting, the changes that seem most probable are likely to be in places where, because perhaps of delta growth or progradation produced by wave action, mangroves have spread and so filled up part or all of a shallow inlet or other feature. If climatic conditions had favoured mangroves they could have spread over the Dee flats; it is very unlikely that they could have played the part *Spartina* has played in Southampton Water.

POSTSCRIPT

This chapter was written before Professor Chapman's book *Mangrove Vegetation* was available. I was able to see galley proofs of the first chapter. Chapman emphasises that silting precedes the growth of mangals and that they help accretion once they have established. When mangals abut on terrestrial vegetation there is often an open salina (see p. 51) or belt of herbs or shrubs, including *Sesuvium*. This belt is probably present because the trees cannot tolerate "too high a percentage of salt unless the watertable is very close to the surface". Later, it is argued that the absence of mangals from arid areas is not so much because of lack of rainfall, but because the absence of rivers means an absence of silt which can build up the sea floor to the levels at which mangroves can grow. Also along much of the coast of western South America the Humboldt Current is undoubtedly a deterrent. This does not contradict the view expressed on p. 51, and on some arid coasts occasional sheet flooding can sweep much material in to the sea.

ACKNOWLEDGEMENT

My thanks are due to Dr. David Stoddart for much constructive help and criticism.

REFERENCES

Beeftink, W.G., 1966. Vegetation and habitat of the saltmarshes and beach plains in the south—western part of the Netherlands. *Wentia*, 15: 85—108.
Chapman, V.J., 1938a. Coastal movement and the development of some New England saltmarshes. *Proc. Geol. Assoc. Engl.*, 49: 373—384.
Chapman, V.J., 1938b. Marsh development in Norfolk. *Trans. Norfolk Norwich Nat. Soc.*, 14(4): 394—397.
Chapman, V.J., 1959. Studies in saltmarsh ecology. IX. *J. Ecol.*, 47: 619—639.
Chapman, V.J., 1974. *Salt Marshes and Salt Deserts of the World*. Cramer, Lehre, 2nd ed., 392 pp. (complemented with 102 pp.).
Chapman, V.J., 1975. *Mangrove Vegetation*. Cramer, Lehre, 425 pp.
Chapman, V.J. and Ronaldson, J.W., 1958. The mangrove and saltmarsh flats of the Auckland Isthmus. *N.Z. Dep. Sci. Ind. Res., Bull.*, No. 125: 79 pp.
Davis, J.H., 1940. The ecology and geologic role of mangroves in Florida. *Carnegie Inst. Wash. Publ.*, No. 517: 303—412.
Evans, G., 1965. Intertidal flat sediments and their environments of deposition in the Wash. *Q. J. Geol. Soc. Lond.*, 121: 209—245.
Fairbridge, R.W., 1950. Recent and Pleistocene coral reefs of Australia. *J. Geol.*, 58: 330—401.
Fairbridge, R.W. and Teichert, C., 1947. The rampart system of the Low Isles. *Rep. Great Barrier Reef Comm.*, 6: 1—16.
Gillner, V., 1965. Saltmarsh vegetation in southern Sweden. *Acta Phytogeogr. Suec.*, 50: 97—104.
Godwin, H. and Godwin, M.E., 1960. Pollen analysis of peats at Scolt Head Island, Norfolk. In: J.A. Steers (Editor), *Scolt Head Island*. Heffer, Cambridge, pp. 73—84.
Green, C. and Hutchinson, J.N., 1960. Archeological Evidence. In: J.M. Lambert et al. (Editors), *The Making of the Broads*. Royal Geographical Society, London, Research Series No. 3, Part III, pp. 113—146.
Jennings, J.N. and Coventry, R.J., 1973. Structure and texture of a gravelly barrier island in the Fitzroy estuary, Western Australia, and the role of mangroves in the shore dynamics. *Mar. Geol.*, 15(3): 145—167.
MacNae, W., 1966. Mangroves in eastern and southern Australia. *Aust. J. Bot.*, 14: 67—104.
Marker, M.E., 1967. The Dee Estuary: Its progressive silting and saltmarsh development. *Trans. Inst. Br. Geogr.*, 41: 65—71.
Marshall, D.R., 1962. The morphology of the Upper Solway saltmarshes. *Scott. Geogr. Mag.*, 78: 81—99.
Nielsen, N., 1935. Eine Methode zur exakten Sedimentations—Messung. *K. Dan. Vidensk. Selsk. Biol. Medd.*, 12(4): 1—97.
Packham, J.R. and Liddle, M.J., 1970. The Cefni saltmarsh, Anglesey, and its recent development. *Field Stud.*, 3(2): 331—356.
Pestrong, R., 1965. The development of drainage patterns on tidal marshes. *Stanford Univ. Publ., Geol. Sci.*, 10: 1—87.
Pethick, J., 1969. Drainage in tidal marshes. In: J.A. Steers, *The Coastline of England and Wales*. Cambridge University Press, London, pp. 725—730.
Pethick, J., 1974. The distribution of salt pans on tidal saltmarshes. *J. Biogeogr.*, 1: 57—62.
Pidgeon, I., 1940. The ecology of the central coast area of New South Wales. III. Types of primary succession. *Proc. Linn. Soc. N.S.W.*, 65: 221—249.
Ranwell, D.S., 1972. *Ecology of Saltmarshes and Sand Dunes*. Chapman and Hall, London, 258 pp.
Redfield, A.C., 1967. The ontogeny of a saltmarsh estuary. In: G.H. Lauff (Editor), *Estuaries. Am. Assoc. Adv. Sci., Publ.*, 83: 108—114.
Redfield, A.C. and Rubin, M., 1962. The age of saltmarsh peat and its relation to recent changes in sea level at Barnstaple, Mass. *Proc. Natl. Acad. Sci. U.S.*, 48(10): 1728—1735.

Richards, F.J., 1934. The saltmarshes of the Dovey Estuary. IV. The rates of vertical accretion, horizontal extension and scarp erosion. *Ann. Bot.*, 48: 225—259.

Schou, A., 1967. Estuarine research in the Danish moraine archipelago. In: G.H. Lauff (Editor), *Estuaries. Am. Assoc. Adv. Sci., Publ.*, 83: 129—148.

Stamp, L.D., 1925. The aerial survey of two Irrawaddy delta forests. *J. Ecol.*, 13(2): 262—276.

Steers, J.A., 1938. Detailed notes on the islands surveyed and examined by the geographical expedition to the Great Barrier Reef in 1936. *Rep. Great Barrier Reef Comm.*, 4(3): 51—96.

Steers, J.A., 1939. Sand and shingle formation in Cardigan Bay. *Geogr. J.*, 94(3): 209—227.

Steers, J.A., 1960. *Scolt Head Island*. Heffer, Cambridge, 269 pp.

Steers, J.A., 1969. *The Coastline of England and Wales*. Cambridge University Press, London, 750 pp.

Stephenson, T.A. and Stephenson, A., 1931. *The British Museum (Nat. Hist.) Great Barrier Reef Expedition 1920—29*. Br. Mus. (Nat. Hist.), London, Vol. 3, No. 2.

Stoddart, D.R., 1962. Three caribbean atolls: Turneffe Islands, Lighthouse Reef, and Glover's Reef, British Honduras. *Atoll Res. Bull., Wash.*, No. 87.

Stoddart, D.R., 1963. Effects of Hurricane Hattie on the British Honduras reefs and cays, October 30—31, 1961. *Atoll Res. Bull., Wash.*, No. 95.

Thom, B.G., 1967. Mangrove ecology and deltaic geomorphology, Tabasco, Mexico. *J. Ecol.*, 55: 301—343.

Tooley, H.J., 1974. Sea level changes during the last 9000 years in North—west England. *Geogr. J.*, 140(1): 18—42.

Warming, E., 1904. Bidrag til vadernes, Sandenes og Marskens Naturhistorie. *Mem. Acad. R. Sci. Lett. Dan., Copenh.*, 7ᵉ Sér., Sec. Sci., 2.

Watson, J.G., 1928. Mangrove forests of the Malay Peninsula. *Malay. Forest Rec.*, No. 6: 245 pp.

West, R.C., 1956. Mangrove swamps of the Pacific coast of Colombia. *Ann. Assoc. Am. Geogr.*, 46: 98—121.

Yapp, R.H., Johns, D. and Jones, O.T., 1917. The saltmarshes of the Dovey Estuary. II. The saltmarshes. *J. Ecol.*, 5: 65—103.

Chapter 3

CLIMATE

H. WALTER

CLIMATE TYPES

Salt marshes and mangals are restricted to moist saline soils of the sea coasts, particularly in the tidal zone. They are therefore an azonal vegetation for which the soil constitution is primarily decisive, but the climate is not wholly without significance. It plays, rather, a differentiating role which will be discussed more closely here.

We shall use, as a basis for the climatic division of the earth, the over eight thousand climate diagrams collected in the *Klimadiagramm–Weltatlas* (Walter and Lieth, 1960—1967). Using these diagrams one can distinguish nine main climate types, which correspond to climate zones I—IX and are characterised as follows (Walter et al., 1975):

I. *The equatorial zone* between approximately 10°N and 5—10°S. This zone has a daily season, i.e., the mean daily temperature variations are greater than the difference between temperature means of the warmest and coolest months of the year. In this zone, annual precipitation is high in the typical situation. Monthly rainfall exceeds 100 mm, and the rainfall curve shows two equinoctial maxima (zenithal rains). (See Fig. 3.1A.)

II. *The tropical summer–rainfall zone* north and south of zone I to approximately 25—30°N and S with already a well–identifiable annual march of temperature, and heavy rains in the summer and an extreme drought during the cooler season of the year. With increasing distance from the equator, this drought season becomes longer, and the amount of rainfall decreases concomitantly.

III. *The subtropical dry zone* of the deserts, still further poleward, with descending air masses. The air masses are warmed during their descent so that it rarely rains. Direct solar radiation is very strong; similarly strong is the outgoing radiation during the night. This causes extreme variations between day and night temperatures and can result in rare nocturnal frosts. Annual precipitation is less than 200 mm and in extreme deserts less than 50 mm. (See Fig. 3.3A.)

IV. *The transition zone with winter rainfall* and a long summer–drought season at approximately 35—40°N and S. This is the typical Mediterranean climate. (See Fig. 3.3B.)

V—VIII. These are the temperate zones, predominantly of the Northern Hemisphere, with cyclonic rainfall throughout the year. Rainfall decreases with distance from the oceans. Therefore, we have moist oceanic and dry continental climates. The following zones can be distinguished:

V. *A warm temperate climate* without a really cold winter and typically with a high humidity, particularly in the summer (Figs. 3.1B, 3.2B).

VI. *The typical temperate climate* as prevailing in Central Europe, with relatively short cold winters and warm, but not hot, summers and with moderate humidity; or in the oceanic region with winter cold nearly absent and with cool summers and rather high humidity. (See Figs. 3.4A, 3.5A.)

VII. *The arid temperate climate* of continental regions with strong temperature contrasts between summer and winter and with low rainfall. Within this climate one can recognise three degrees of aridity: (1) VII — the semi–arid steppe climate with a summer–dry season and occasional drought; (2) VIIa — the arid semi–desert climate with a very pronounced drought season and a short humid

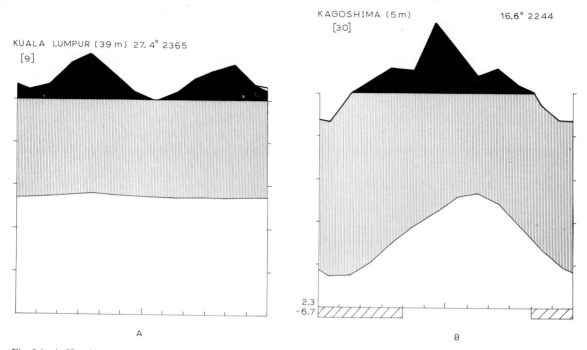

Fig. 3.1. A. Humid equatorial climate (type I) in Malaya. Abscissa: months (January to December); ordinate: each graduation mark corresponds with 10°C, or 20 mm precipitation, respectively. Where precipitation in the month exceeds 100 mm, the excess is represented in black, at a reduced scale of 1:10; thus, precipitation in April is over 300 mm. B. Warm temperate very humid climate (type V) in south Japan without cold annual period, but with occasional frosts.

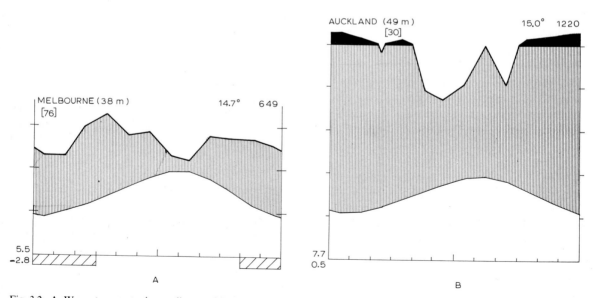

Fig. 3.2. A. Warm temperate, but a climate of light rainfall (type V), in southeast Australia (months in the Southern Hemisphere recorded from July to June). B. Warm temperate climate in the north of New Zealand (type V) (months as with A).

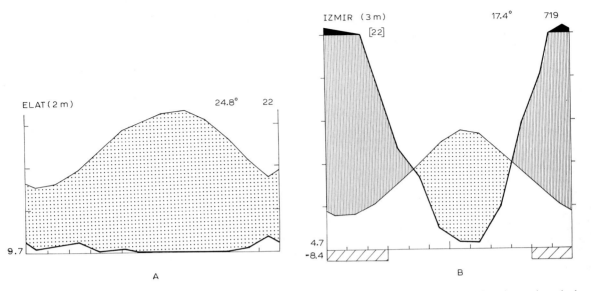

Fig. 3.3. A. Extreme desert climate (type III) at the Gulf of Aqaba. B. Climate type IV in west Anatolia with winter rain and a long summer period of aridity.

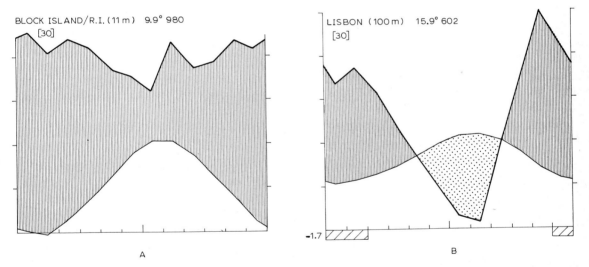

Fig. 3.4. A. Humid moderate climate (type VI) on the east coast of U.S.A. B. Climate type IV in south Portugal. Winter milder and aridity in summer somewhat less pronounced than in Fig. 3.3B.

season; (3) VII(rIII) — an extremely arid climate with low rainfall like in III and drought conditions lasting through the entire year, but with cold winters as in type VII.

VIII. *The cold temperate or boreal climate* with cool, moist summers and very cold winters, which last for more than six months. This climatic zone is practically absent in the Southern Hemisphere, while it occupies a very large area in the Northern Hemisphere. (See Fig. 3.5B.)

IX. *The Arctic zone* which, in terms of land available for plant growth, is likewise almost entirely restricted to the Northern Hemisphere, with low precipitation that is uniformly low

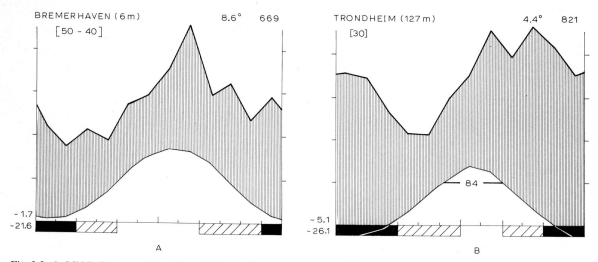

Fig. 3.5. A. Middle European climate (type VI) with winter of three months. B. Oceanic boreal climate (type VIII) on the west coast of Norway. Period of growth with daily averages over 10°C only 84 days.

through the year. There are short, cool summers lacking nights, but long cold winter nights. (See Fig. 3.6A, B.)

These main climate types either blend into one another without sharp divisions or they exhibit more or less strong modifications through special, more local conditions (e.g., monsoon winds, mountainous relief, etc.), all of which is indicated more precisely in the work cited above. An explanation of the basis of presentation of the climate diagrams will be found in various references (Walter, 1971, p. 53; Walter, 1973a, p. 20; Walter et al., 1975).

MANGAL CLIMATES

If one examines the distribution of salt marshes and mangals on the world map (see Fig. 1.1), then it is immediately obvious that mangals occur only in the tropical region, approximately between 32°N and 38°S, i.e., in the climatic region of types I, II, partly also type III, and, only on the eastern

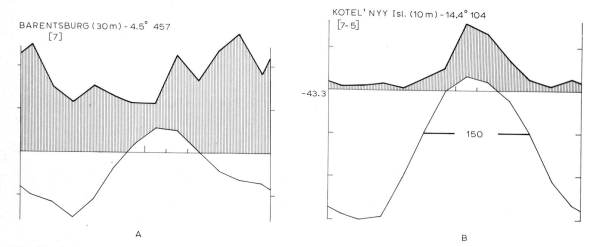

Fig. 3.6. A. Arctic climate in the group of islands at Spitsbergen (type IX). July mean under 10°C. B. Extreme Arctic continental climate (type IX) on an island in the Arctic Sea north of East Siberia. Only 150 days with daily means over −10°C.

border of the continents, partially up to type V. The restriction to these regions is conditioned by the sensitivity of mangroves to frost. Only *Avicennia*, which is distinguished by a particularly high concentration of cell sap, does not perish so easily from cold and occurs even near Auckland in New Zealand as *Avicennia marina* var. *resinifera*. Here occasional frosts of −2°C are tolerated each winter.

On sea coasts from the entire tropical region right up into the Arctic and Antarctic one finds salt marshes exclusively composed of herbaceous halophytes. Only in the transition area of climate type IV (without cold winters, but with isolated cold snaps) do some partly lignified shrubs occur in these salt marshes (e.g., *Salicornia fruticosa, Arthrocnemum glaucum*), whilst on the Black Sea *Halocnemon strobilaceum* among others is also to be found.

Temperature conditions thus limit the spread of mangroves, whereas rainfall conditions are more decisive for the sequence of zones in the tidal region. On flat coasts protected from wave action and in estuaries, the duration of inundation with seawater during flood tides decreases from the sea landward. This leads in humid and arid climates to opposite results:

(1) In the humid climate the rain leaches the salty soil more effectively between two successive flood tides, so that the transition from the saline soils on the outer edge of the mangals or salt marshes to the non–saline soils on the landward side, not flooded by salt water, is effected quite gradually. The most salt–resistant mangrove species are therefore found on the outer edge, while on the landward side a brackish–water mangal with *Acrostichum aureum* or the *Nypa* palm passes easily into fresh–water marshes with *Pandanus* and other species.

(2) In the arid climate region, or in climate zone II with a pronounced period of aridity, evaporation of the water from the surface of the soil is perceptible during the time between two successive inundations, and this leads to an even stronger increase in the salt concentration of the soil water on the landward side. This particularly relates to coastal mangal which is not influenced by river water. The concentration of the soil solution rises, for example, at Tanga in East Africa, in a climate with a long period of aridity, from 25.5 atm on the seaward edge up to 41 atm on the landward edge (Walter, 1971, fig. 91; Walter, 1973a, fig. 35). For this reason one finds in this case the most salt–resistant mangrove species on the landward edge. These are species of *Avicennia*, which form under extreme conditions only small bushes (cf. p. 238). The adjacent landward zone is a sand surface without vegetation, which is flooded only twice yearly by the equinoctial tides, but otherwise lies dry, so that the salt solution in the soil attains to saturation point, except when heavier rains occur and it is completely leached. Such extreme deviations in concentration cannot be endured by the roots of any plant. Behind this inner border of the tidal zone without vegetation, there then starts, only slightly higher and sharply demarcated, the non–halophytic vegetation. Very often there are coconut palm plantations.

One can also observe such areas without vegetation on the arid sea coasts of North Venezuela. But at Chichirivichi, where there is a somewhat longer period of rain, one finds growing on them groups of *Bromelia humilis, Opuntia wentiana* or *Acanthocactus*, genuine succulents which can endure no salt. Closer examination showed that these plants grow on small sand drifts. This sand is leached out during each period of rain so that the succulents can take up water to fill their water–storage tissue. During the period of aridity, if the sand dries out or the salt concentration in it rises, the succulents have no absorbing roots and use the stored water; thus, the salt soil corresponds to a certain extent to a flat rock soil, on which one often finds these succulents. They occupy the otherwise vegetationless surfaces on the inner edge of the mangroves, because they are protected on these from competition by faster–growing plants (Walter, 1973b, pp. 410—412, figs. 276—277). (See Fig. 3.7.)

These differences in salt conditions of the soil within the tidal zone in humid and arid climatic regions are also noticeable on sea coasts outside the tropics where the zonation within the salt marshes is also correspondingly altered. However, up to the present no attention has been paid to this problem, so that it is not possible to quote any concrete examples.

The temperature requirements of the various mangrove species are not the same; even in a continuously humid climate the number of man-

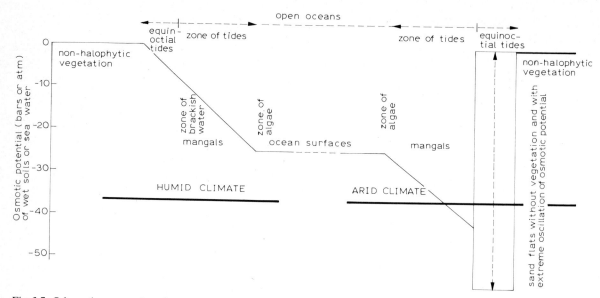

Fig. 3.7. Schematic presentation of mangrove zonation in humid and arid climatic regions in relation to the osmotic potential of the soils.

grove species decreases with increased distance from the Equator. The greatest abundance of species (see Table 1.2) is found on the coasts of Malaya, Indonesia and New Guinea, in the regional climate type I (Fig. 3.1A). Further to the north even more species are lacking with the falling temperatures, until in the warm temperate climate of south Japan (Fig. 3.1B) there is only one species, *Kandelia candel*. South of the Equator a single species of mangrove, the *Avicennia marina* var. *resinifera* already mentioned, remains as the least frost–tender in southern Australia (Fig. 3.2A) and northern New Zealand (Fig. 3.2B). *Rhizophora mucronata* and *Avicennia marina* penetrate into the arid climate zone III on the Red Sea coast, and the latter species even into the Gulf of Aqaba (Fig. 3.3A). The lack of mangroves on certain sections of the west coast of North America as well as Africa is related to the low water temperatures of the cold ocean currents which wash their shores (see p. 202).

SALT–MARSH CLIMATES

Salt marshes in the climate zone outside the tropics can locally show such a high concentration of salt in the soil that areas without vegetation can occur, as, for example, in the region of the Mediterranean in the Bay of Izmir with a long summer period of aridity. Even on the American east coast with a humid climate (Fig. 3.4A) Steiner (1934) found small pans locally dried out in summer on which only the very salt–tolerant *Salicornia mucronata* could grow.

The cell sap concentrations of *Salicornia* can be more than 50 atm, but even this species discolours red toward the end of summer through β–cyanin formation, which is a sign of incipient damage, and then perishes.

The salt marshes also show impoverishment of species with increasing latitude and decreasing temperature. One can trace them quite clearly, for example, on the European coasts of Portugal through middle Europe and Norway up to Spitsbergen. The shortening of the growth period and the increasing coldness of the winter is very clearly apparent from the climate diagrams (Figs. 3.4B—3.6A).

On the west coast of Sweden near Göteborg the salt marshes show, according to Gillner (1960; see also Walter, 1968, pp. 902—906), quite a complicated zonation: (1) *Ruppietum maritimae*, which only rarely appears out of water at low tide; (2) *Salicornietum*, which occupies the lowest part of the salt marsh; (3) *Puccinellietum maritimae* with *Triglochin maritimum*, *Aster tripolium*, *Scirpus maritimus* as well as *Spergularia marginata*, which leads to (4) *Juncetum gerardii* with *Glaux latifolia* and *Plantago maritima*, after which *Agrostis stolonifera*, *Atriplex latifolia* and *Armeria maritima* form the landward, non-saline marsh with many other common meadow plants.

Because of the humid climate the transition from the halophytic grouping to the non-halophytic is effected quite gradually.

On the old coasts of the Polar Sea on Kolguyev Island a depauperate salt marsh occurs according to Pohle (1907; see also Walter, 1974, fig. 22). *Aster tripolium, Pisum maritimum, Ligusticum scoticum* are not present, as they are on the coasts of the White Sea. Among the salt-marsh grasses are to be found *Dupontia fischeri, Arctophila (Colpodium) fulva* as well as the salt-tolerant *Carex glareosa* and *Calamagrostis deschampsioides*. *Matricaria ambigua*, which is related to *M. inodora* but is a perennial, is particularly abundantly represented; in addition there is *Stellaria humifusa* and a particular strain of *Potentilla anserina*.

In the extreme Arctic climate of Spitsbergen (Fig. 3.6A), Hofmann (1968; see also Walter, 1974, pp. 24—26) refers to only one association — poor *Puccinellietum phryganodis* with *Stellaria humifusa* and in one place *Carex ursina*, likewise with numerous small *Cochlearia officinalis* plants.

This description agrees with that of Gorodkov (1956) for Kotel'nyy Island (see also Walter, 1974, pp. 27—28), one of the New Siberian Islands at 74—75°N with an extremely cold climate (Fig. 3.6B). The first pioneer on the saline muddy soils of the sea coast is the species *Puccinellia phryganodes*, which forms sterile offshoots; on somewhat drier soil with lower salt content *Cochlearia groenlandica* and the grass *Dupontia fischeri* occur. If inundation by sea water occurred only during storms in winter, then other species like *Stellaria humifusa, Pleuropogon sabinii, Ranunculus hyperboreus* and *Arctophila fulva* became associated with an even more predominant *Puccinellia*. Here the real salt marsh ends. At somewhat higher levels in the absence of silt deposits there is a narrow intermediate zone with *Phippsia algida, Alopecurus alpinus, Deschampsia brevifolia, Juncus biglumis* and *Caltha arctica* leading to a grass sward of *Dupontia fischeri* with 40—70% coverage and in which mosses *Distichum* and *Bryum* and the alga *Nostoc* form a light cover on the soil. These meadows become free of snow only at the beginning of July.

Our short survey shows that the climate itself is of significance both for the azonal vegetation of the salt marshes and mangals and also for species distribution on a large scale as well as patterns on a small scale.

REFERENCES

Chapman, V.J., 1975. *Mangrove Vegetation.* Cramer, Lehre, 425 pp.

Chapman, V.J. and Ronaldson, J.W., 1958. The mangrove and saltmarsh flats of the Auckland Isthmus. *N.Z. Dep. Sci. Ind. Res., Bull.*, No. 125: 79 pp.

Gillner, V., 1960. Vegetations- und Standortsuntersuchungen in den Strandwiesen der schwedischen Westküste. *Acta Phytogeogr. Suec.*, 43: 1—198.

Gorodkov, B.N., 1956. Rastitel'nost' i pochvy ostrova Kotel'nyi (Novosibirskie O-va). In: B.A. Tikhomirov (Editor), *Rastitel'nost' krainego severa SSSR i eë osvoenie*, 2. Akad. Nauk S.S.S.R., Moscow, pp. 7—132.

Hofmann, W., 1968. *Geobotanische Untersuchungen in Südost-Spitzbergen, 1960.* Steiner-Verlag, Wiesbaden, 83 pp.

Pohle, R., 1907. Vegetationsbilder aus Nordrußland. *Vegetationsbilder*, 5(3—5): Tab. 16—33.

Steiner, M., 1934. Zur Ökologie der Salzmarschen der nordostlichen Vereinigten Staaten von Nordamerika. *Jahrb. Wiss. Bot.*, 81: 94—202.

Walter, H., 1968. *Die Vegetation der Erde, II.* Fischer, Stuttgart, 1001 pp.

Walter, H., 1971. *Ecology of Tropical and Subtropical Vegetation.* Oliver and Boyd, Edinburgh, 539 pp.

Walter, H., 1973a. *Vegetation of the Earth in Relation to Climate and the Eco-Physiological Conditions.* Springer, Berlin, 237 pp.

Walter, H., 1973b. *Die Vegetation der Erde, I.* Fischer, Stuttgart, 3rd ed., 743 pp.

Walter, H., 1974. *Die Vegetation Osteuropas, Nord- und Zentralasiens.* Fischer, Stuttgart, 452 pp.

Walter, H., Harnickell, E. and Mueller-Dombois, D., 1975. *Climate-Diagram Maps of the Individual Continents and the Ecological Climate Regions of the Earth.* (Supplement to *Vegetation Monographs*) Springer, Berlin, 36 pp.

Walter, H. and Lieth, H., 1960—1967. *Klimadiagramm-Weltatlas.* Fischer, Stuttgart.

Chapter 4

SOILS OF MARINE MARSHES

FRED B. PHLEGER

INTRODUCTION

Marine marshes develop on low–lying marginal marine areas which are being aggraded. A marsh will develop where relatively fine–grained soil accumulates in the intertidal zone at an elevation relative to tides where marsh plants can colonize and propagate. In places where lunar tides have been adequately analyzed the marsh is above the tide level of mean lower high water and the bare intertidal flats occur below this elevation. This level may be modified by wind–induced tides and by water levels related to river flow in either lagoons or estuaries (cf. p. 32).

Coastal lagoons, estuaries and deltas are the most common places in which marshes develop (Fig. 4.1). This is because such environments usually contain abundant sediment in suspension which tends to be deposited around the margins where current velocities are low. Such areas are thus sediment traps which tend to become filled with marshes except where water velocities are sufficient to maintain channels for water circulation.

The sedimentary substrate of a marine marsh is important as the soil in which the characteristic marsh plants become established and proliferate. The soil furnishes some nutrients to the plants, and this is especially important in estuaries and coastal lagoons where significant increments of new sediment are added during flood conditions on the rivers. River water contains abundant nutrients, as evidenced by high organic productivity in coastal lagoons or the open ocean near river effluents (Phleger and Lankford, 1957; Lankford, 1959). Such nutrients are derived from the soils of their drainage basins, and fertilize the soils of marshes which occur in estuaries and hyposaline lagoons.

Where daily flooding by lunar tides is important, nutrients from sea water are added to the marsh on each flood tide. Thus, the hypersaline lagoons which have been studied and their associated marshes are highly productive because of the net transport of sea water into the areas (Phleger, 1960b; Ayala–Castañares and Segura, 1968).

ORIGIN AND COMPOSITION OF THE SEDIMENT

The immediate source of sediment to provide soils of marine marshes may be the land or the sea, or both. In estuarine and lagoonal areas with river flow into them most of the sediment being supplied to marshes is transported and deposited by river flow. It has been shown by Meade (1969), however, that estuaries on the U.S. Atlantic Coastal Plain are being filled by a landward movement of sediment caused by flow of sea water along the bottom. Some of this material was apparently derived from offshore, based on its mineral components, and was transported into the estuaries by tidal flow of sea water along the bottom. Favejee (1951) and others have demonstrated by mineralogical analyses that the sediments in the Dutch Wadden Sea have come largely from the North Sea. Wind–borne silt and sand may be a significant source of sediment supplied to marshes in arid climates.

The mineral composition of the sediment varies and will depend upon the composition of the source materials available in the drainage basin or in the adjacent open ocean. Organisms contribute significant amounts of material to the marsh soil in the form of plant debris which eventually becomes peat and calcareous shell material. Some marsh soils

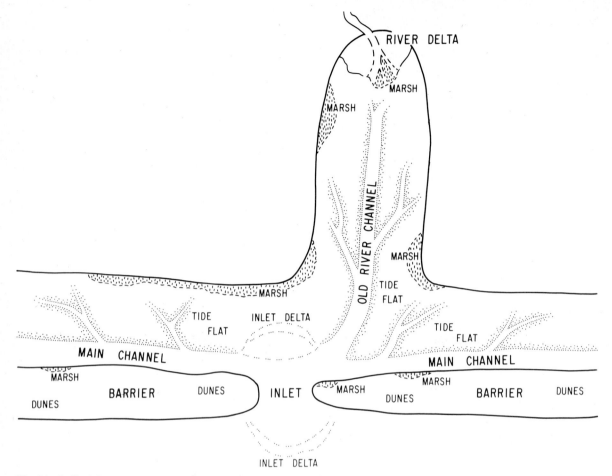

Fig. 4.1. Stylized diagram of a coastal lagoon showing types of places where marshes develop (after Phleger, 1969b).

are composed of a large percentage of calcareous material derived from organisms (Scholl, 1963).

The grain size will depend largely upon what sizes are available. Marshes, however, do accumulate the finest sediment available in the system due to the low current velocities in the very shallow water. Deposition of fine-grained sediment also is aided by abundant plants in marshes which act as a baffle to further reduce water velocities. The finest sediments are deposited at the end of the tidal excursions where there is slack water or at the shallow edge of the estuary where dead water also occurs. This has been clearly shown by studies in the Dutch Wadden Sea by Postma (1954, 1961) and Van Straaten and Kuenen (1957), who demonstrate the mechanism for deposition of the finest silts and clays on the marshes by tidal action. Where lunar tides are small and river flow is erratic, such as in a semi-arid climate, wind tides may be important in flooding marshes and providing sediment. This has been observed especially along the coast of south Texas in the Gulf of Mexico where onshore winds may raise water level in coastal lagoons such as San Antonio Bay and Laguna Madre 1 m or more.

The nature of the soil varies from place to place in a regular pattern in all the marshes which I have seen. Levees border the channels within a marsh and the coarsest sediment accumulates here. Fine sand and silt make up the levee soil in the marshes of southern California and northern Mexico which I have examined, and the levees provide a relatively firm footing. Levees or barriers of relatively coarse sediment also commonly occur at the outer edge of a marsh in areas where flooding is frequent due to tides or river flooding. Such barriers may be augmented in height by wave action, as in upper Laguna Ojo de Liebre in Mexico (Phleger, 1969a) and on the delta of the Guadalupe River in San Antonio Bay, Texas. Away from the levees the sediment in most marshes is very fine-grained mud (silt and clay). Ephemeral salt pans may occur in the highest marsh where flooding is only occasional, and these usually contribute gypsum to the local sediment as well of fragments of algal pads.

In the Mugu Lagoon marsh, California, a mid-latitude grass marsh, Warme (1971) has described the low marsh sediments as averaging 50% sand and 50% mud and containing shells, foraminifera and vegetable fibers; the middle and upper marsh is 100% mud, mostly clay; and the salt pans contain silt and clay with traces of sand and gypsum crystals.

The marsh sediments in the southwest shore of San Francisco Bay have been described by Pestrong (1965). The sediments are predominantly clays with 15—25% silt, and the grain size decreases from the bay to the land edge of the marsh. The coarsest sediments are in the channels and creeks, but the size here also is a function of the distance from the bay. Typical channel sediments on the bayward edge contain 5% sand, 15% coarse shell fragments and organic debris, 15% silt and 65% clay. On the land edge of the channel most of the sediments are 15% silt and 85% clay. The levees also have relatively coarse sediment and they become somewhat desiccated due to exposure. Likewise, the upper *Salicornia* marsh compared to the lower *Spartina* has a somewhat lower moisture content, greater permeability, higher shear strength, higher penetrative resistance and higher organic content. These differences undoubtedly are a result of more exposure of the high marsh than the low marsh. It was observed that the *Spartina* plants in the low marsh act as sediment traps.

In the Mission Bay marsh, California, the sediment is uniformly mud in both the high and low marsh, except on the levees bordering the single channel where there is 10% or more of very fine sand.

Sediments of extensive mangrove marshes in Tabasco, Mexico, have been described by Thom (1967). The marshes fringe all of the coastal lagoons and estuaries which are essentially continuous in the coastal lowlands. Some of the estuarine channels in this area are filled with inorganic sediment, generally very fine-grained mud. Others are filled with organic muck which may consist of floating debris which is trapped by mangrove roots, especially those of *Rhizophora mangle*. Some of the organic debris also is produced and deposited in place, and peat deposits 1—3 m thick have filled some of the abandoned channels and interdistributary areas. The fine-grained muds have an appreciable amount of organic materials and are reducing in most places. Levees are composed of silty sand to clayey silt, are compact relative to the channel deposits, and may be oxidized and contain iron and calcareous nodules.

Marsh soils are not everywhere very fine muds as characterize most of the Tabasco marshes and also other areas. Shell produced in place provides a coarse component in many areas, and coarse to fine sand may be introduced into the marsh in some coastal areas. Sediments of two areas of mangrove swamp in southeastern Florida have been studied by Scholl (1963). In the Ten Thousand Islands area surface soils of these swamps are mostly calcareous quartz—fine sands to very fine sands and silts, usually containing abundant organic debris. The calcareous component is biogenic being derived from indigenous shells. Detrital quartz constitutes 60–70% of most of these sediments. The amount of organic material in some places may be as much as 50–70%, particularly on the landward side of the area. In Whitewater Bay on the other hand, the sediments are mostly molluscan shells and shell fragments with abundant organic detritus from mangroves and also contain 5–15% quartz. The grain size is dependent upon the abundance of coarse shell fragments, and the organic content is high.

In Ojo de Liebre Lagoon, in the arid part of Baja California, Mexico, I have seen small areas of *Spartina* marsh within the dune fields of the lagoon barrier. These occur in interdune flats which are flooded with each higher high tide and are wet even at low tide. The substrate for the marsh plants is made up of a small amount of silt and some plant debris and contains a large amount of dune sand. Extensive *Spartina* and *Salicornia* marshes are in the same area between Ojo de Liebre Lagoon and the adjacent Guerrero Negro Lagoon. Silt and some clay have been deposited in this location at the end of the excursion of tidal currents in both lagoons. A considerable amount of sand also has been deposited in the marsh by wind transport from the dune field on the lagoon barrier. The amount of sand being blown into Guerrero Negro Lagoon has been estimated at about 400 000 m^3 yr^{-1} (Phleger, 1965). A considerable volume of this sand is deposited on the extensive marshes in this lagoon. Shaler (1886) also recognized the importance of wind-blown sand in a marsh on Plum Island, Massachusetts, in his pioneer studies of marshes on the east coast of the United States.

SOIL STRUCTURES

Marsh soils may have alternating laminae of coarse and fine sediment in certain locations, but most sediments are considerably reworked and otherwise modified by the roots of marsh plants and certain animals. Crab burrows are abundant especially between low and high water in tidal areas and can be observed especially in banks of marsh channels. Biogenic reworking can concentrate sand and produce fecal pellets from organisms which ingest sediment to extract food. Mounds of small balls of sediment are common around crab burrows on the surface of many marshes. It has been shown that different marsh environments, based on salinity, substrate and vegetation, may be burrowed by different crabs (Allen and Curran, 1974) near Beaufort, North Carolina, and near Sapelo Island, Georgia (Frey and Howard, 1969). The latter authors stress the importance of roots of marsh plants in binding the sediment and making it firm.

In the Wash, England, the outer portion of the marsh sediments are alternating laminae of sand and mud (Evans, 1965). Farther inland these laminae decrease and finally disappear, and most structures are due to decay of plant roots and filling of them with sand. Allen (1965) illustrates structures from the mangrove swamps of the Niger Delta which range from laminated sediments in the floors of the tidal channels to mottled sediments on the mud banks and interchannel flats which are due to mangrove roots and possibly also to burrowing by animals. Cores from the marshes on the central Louisiana coast have an occasional silt lamination but otherwise show mixed sediment, caused by burrowing by organisms and abundant plant remains (Coleman, 1966). Cores from marshland in southwestern Louisiana show similar structures (Byrne et al., 1959).

The abundant burrows in much of a marsh soil act to give good drainage to some of the area. This drainage can be observed in a marsh channel during an ebb tide. Tributaries of the meandering channel system often originate in pools within the marsh. These pools characteristically have standing water even at low tide. Often, there are no marsh plants in these pools, suggesting that the stagnant water in the soils has produced an adverse environment.

The pH of the soil surface water varies, depending upon the time of day, and also upon the stage of the tide in lagoonal marshes where there is marked tidal influence (Phleger and Bradshaw, 1966). Variations measured in marsh water in Mission Bay, California, range from 6.8 to 8.3 in marsh area which receives only marine water, and the monthly mean pH ranges from 7.3 to 7.6 (Figs. 4.2, 4.3; Bradshaw, 1968). The pH of the soil water was shown to decrease with depth at four stations in this marsh and the Eh went from strongly positive at the surface to strongly negative at a depth of a few centimeters (Table 4.1). Where open burrows are present the oxygen values, pH and Eh must be similar to the surface sediment, at least locally within a burrow.

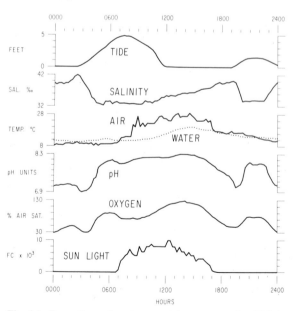

Fig. 4.2. Recordings of environmental parameters for 24 h in the Mission Bay marsh, California (after Phleger and Bradshaw, 1966).

SOILS OF MARINE MARSHES

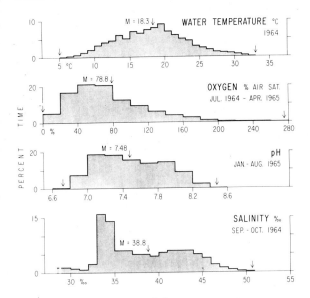

Fig. 4.3. Long-period values of water temperature, oxygen, pH and salinity in the Mission Bay marsh, California (after Phleger and Bradshaw, 1966).

POPULATIONS OF FORAMINIFERIDS

There are two groups of shelled microorganisms which are common to abundant in marshes having marine influence. Ostracoda are calcareous bivalved crustaceans which occur also in fresh-water marshes, lakes and rivers. The foraminiferids are protozoans which occur in all marine and brackish-water environments, are very abundant and have a great variety of living and fossil species (several thousand). Foraminiferids have a test, which in most species is composed of calcium carbonate; but in many forms the test is made up of agglutinated sand or silt-size particles. The empty tests thus become a part of the sediment and in some cases compose most of the sediment. They have been intensively studied because they are useful tools in studying geological and oceanographic problems, and the ecology and distribution of modern assemblages are very well known (Phleger, 1960a; Murray, 1973).

The following marsh species are listed by Murray (1971);

abundant and dominant cosmopolitan species:

Ammotium salsum
Arenoparrella mexicana
Miliammina fusca
Trochammina inflata
Jadammina macrescens
Jadammina polystoma

confined to hypersaline marshes:

Pseudoeponides andersoni
Discorinopsis aguayoi
Glabratella sp.
Glomospira sp.
Textularia earlandi

TABLE 4.1

Vertical distribution of pH and Eh at four stations in Mission Bay marsh, 21 January 1966 (after Bradshaw, 1968)

Depth (mm)	Station A Instrument locality in tidal stream; ebb tide; greenish mud; 15.2°C; 15.05 h		Station B Shallow pool; clayey silt; abundant *Batis*, *Salicornia* and *Enteromorpha*; 15.5°C; 16.35 h		Station C *Spartina* beneath algal mass; 15.0°C; 16.35 h		Station D Mud flat, shallow pool; 15.0°C; 15.45 h	
	pH	Eh (mV)	pH	Eh (mV)	pH	Eh (mV)	pH	Eh (mV)
Overlying water	7.9	+380	8.8	+390	8.3	+150	7.5	+370
					7.0	−50		
Sediment surface	8.3	+5	8.5	+205	—	—	7.9	0
			9.3	+200				
5 mm	7.8	−12	8.2	−20	—	—	7.6	−23
12	7.4	−55	7.6	−54	—	−62	7.4	−43
20	7.2	—	7.0	—	—	—	7.2	−55
25	7.1	—	6.7	—	6.2	−132	7.0	—
60	7.0	—	6.5	—	—	—	—	—
120	6.8	−105	6.1	−145	6.1	−153	6.7	−122

confined to hyposaline marshes:

Protelphidium tisburyense
Protelphidium anglicum
Tiphotrocha comprimata
Ammobaculites sp.
Pseudoclavulina sp.
Haplophragmoides spp.
Elphidium articulatum

The following species which also commonly occur in marshes should be added to the above lists:

Protoschista findens
Ammoastuta inepta
Ammonia beccarii
Palmerinella palmerae

The majority of these species have agglutinated tests and a lesser number have calcareous tests. Pure assemblages of agglutinated species appear to be limited mostly to hyposaline marshes. Some hypersaline marshes have an essentially pure assemblage of calcareous species. The agglutinated species are the ones which appear to be most characteristic of foraminifera in most marshes. Some of the species or variants of them appear to be essentially world-wide in distribution and others appear to be geographically restricted in distribution (cf. p. 81).

It may be assumed that the size of the standing stock (number of individuals per unit area) of any group of organisms reflects in a general way the organic productivity in any area or environment. Data are available for the standing stock of benthic foraminifera in marine marshes from several areas (Murray, 1973), principally in North America and Europe. Data are also available on standing stocks of benthic foraminifera in other environments from numerous areas. Where rates of fixation of organic carbon also have been measured large standing stocks of foraminifera do in fact reflect high organic productivity. Examples are in the southeastern Mississippi Delta (Lankford, 1959; Thomas and Simmons, 1960) and in Ojo de Liebre Lagoon (Phleger and Ewing, 1962). In other areas large standing stocks of foraminiferids and large standing stocks of fish and/or shrimp occur together (Phleger and Lankford, 1957; Phleger, 1960b; Phleger and Ayala-Castañares, 1972).

Many samples from marine marshes contain very large standing stocks of foraminiferids in comparison to most other environments, and a few samples have rather low standing stocks. Very high organic production is indicated for most marine marshes relative to most other marine environments. High organic productivity is obviously also indicated by the dense plant growth in marshes (cf. p. 25). Blue-green algae on the marsh may also be important in fixation of nitrogen.

Relative rates of deposition of sediment are indicated by the percent of the total foraminiferal population which is alive at the time of collection. Large percents of living specimens suggest fast deposition and low percentages suggest slow rates of deposition. Marsh faunas have generally high percentages of living forams in the total population (living and dead specimens) in comparison to most other environments, indicating rapid deposition of the sediment composing the soil substrate. Rapid states of deposition also are indicated by direct observations and measurements.

A zonation of foraminifera into high-marsh and low-marsh assemblages can be recognized in many areas, and these may be distinguishable from marsh channel and adjacent tide-flat faunas (Fig. 4.4). In Mission Bay, California, low and high marsh areas are related to specific tide levels: (1) the boundary between the tide flat and the low marsh is at mean lower high water; (2) the boundary between the low marsh and the high marsh is mean higher high water; and (3) the upper limit of the high marsh is highest high water (Phleger, 1970; see Fig. 4.5). It is believed that there are similar relationships in other marshes where the principal flooding is by ocean tides, as in San Francisco Bay (Hinde, 1954).

POPULATIONS OF OSTRACODA

Ostracods are not so abundant in most marine marsh samples as foraminifera; moreover, they have been described from few areas and it is not possible to generalize on the faunas. Curtis (1960) reports that samples from a marsh in the Mississippi Delta contained no benthic ostracods and did have unidentifiable fragments of pelagic species. Benda and Puri (1962) list the following species in the Cape Romara area, Florida, as characteristic of their marsh river assemblage:

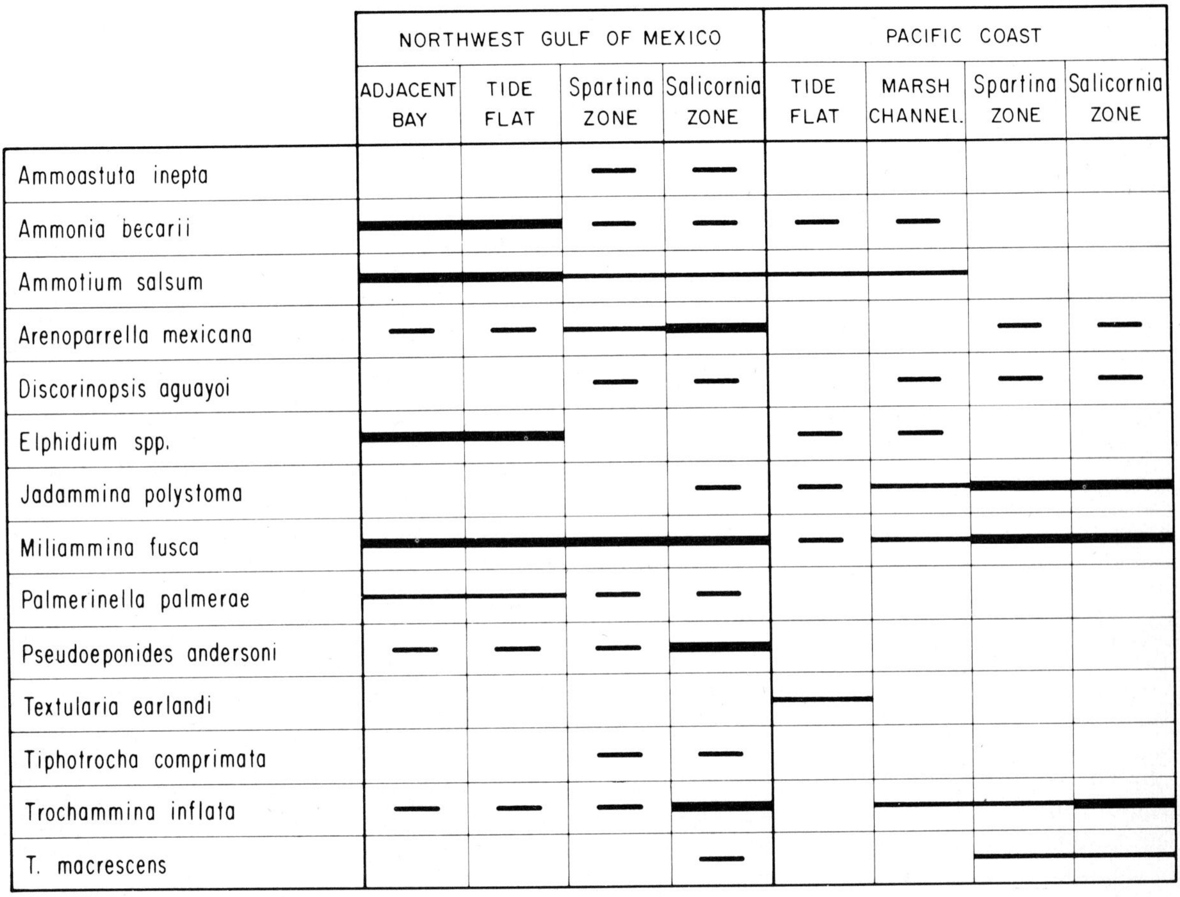

SOILS OF MARINE MARSHES

	NORTHWEST GULF OF MEXICO				PACIFIC COAST			
	ADJACENT BAY	TIDE FLAT	Spartina ZONE	Salicornia ZONE	TIDE FLAT	MARSH CHANNEL	Spartina ZONE	Salicornia ZONE
Ammoastuta inepta			—	—				
Ammonia becarii	▬	▬	—	—	—	—		
Ammotium salsum	▬	▬	▬	▬				
Arenoparrella mexicana	—	—	—	▬			—	—
Discorinopsis aguayoi			—	—		—	—	—
Elphidium spp.	▬	▬			—	—		
Jadammina polystoma				—	—		▬	▬
Miliammina fusca	▬	▬	▬	▬		—	▬	▬
Palmerinella palmerae	—	—	—	—				
Pseudoeponides andersoni	—	—	—	▬				
Textularia earlandi					▬	▬		
Tiphotrocha comprimata			—	—				
Trochammina inflata	—	—	—	▬		▬	▬	▬
T. macrescens				—		▬	▬	▬

Fig. 4.4. Generalized zonation of foraminifera within marshes in the northwestern Gulf of Mexico and Pacific coast of the United States (after Phleger and Bradshaw, 1966).

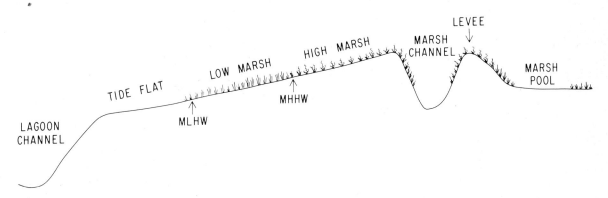

Fig. 4.5. Diagrammatic cross-section of tide levels and marsh zones in Mission Bay, California (after Phleger, 1970). MLHW: mean lower high water; MHHW: mean higher high water.

Candona sp. A
Cyprideis sp.
Cytherura sp. F
Physocypria pustulosa

Five additional species also occur, but are not considered characteristic of this environment:

Anomocytheridea cf. *A. floridana*
A. sp. A
Cyprideis sp. A
Cytherura forulata
Perissocytheridea brachyforma

The mangal island assemblage is characterized by the following:

Actinocythereis subquadrata
Anomocytheridea cf. *A. floridana*
Cyprideis floridana
Cytherella cf. *C. lata*
Cytherura johnsoni
Loxoconcha cf. *L. levis*
Neocaudites cf. *N. nevianii*
Paracypris sp. A
Puriana sp. A

Swain (1955) gives *Cypridopsis,* a fresh–water genus, as the typical genus in his fluvial facies in San Antonio Bay, Texas. Benson (1959) records the following from a marsh and estuarine environment bordering Todos Santos Bay, Mexico:

Cyprideis stewarti
C. miguelensis
C. sp.
Cypridopsis vidua

ENVIRONMENTAL PARAMETERS

In the Mission Bay marsh, California, several environmental parameters were recorded for long periods of time in an attempt to determine some of the important factors affecting the sediment in and on which the organisms live (Phleger and Bradshaw, 1966). Most of the measurements were taken of the water in the main channel of this small marsh during flood, ebb and slack tide. There is no runoff in this area, and the water is sea water which is modified within the lagoon and the marsh.

The significant variable factors in this area are temperature, salinity, pH and oxygen, and important variables contributing to these parameters are tidal regime, sunlight and the season of the year. The daily and longer term ranges (Figs. 4.2, 4.3) of these factors are significant and surprisingly large. Salinity variations are correlated with the tidal phase, being highest at the end of the ebb tide. It is apparent, therefore, that salinity is enriched on the marsh. Evaporation can be one cause for this increase, primarily during daylight hours. High salinities also occur where the ebb is at night (Fig. 4.2) when evaporation is relatively insignificant indicating that there is another mechanism causing such an effect. It has been suggested that much or most of the salinity increase is a result of the metabolism of the halophytic marsh plants, which secrete excess salt. A typical daily range of salinity was from about 33‰ to 42‰ (parts per thousand), and the long–term range was from less than 30‰ (during heavy rain) to more than 50‰.

Oxygen values ranged from about 30% to 130% of air saturation during one 24–h period and showed a long–term range from less than 20% to 275%. Variations in oxygen content of the water result from variations in rate of photosynthesis of the marsh flora, being highest late in the day at the end of an ebb tide. The pH follows oxygen closely; the yearly range was from less than 6.8 to about 8.3, and the daily range was similar.

REFERENCES

Allen, J.R.L., 1965. Late Quaternary Niger Delta and adjacent areas: sedimentary environments and lithofacies. *Bull. Am. Assoc. Pet. Geol.*, 49: 547—600.

Allen, E.A. and Curran, H.A., 1974. Biogenic sedimentary structures produced in lagoon margin and salt marsh environments near Beaufort, North Carolina. *J. Sediment. Petrol.*, 44: 538—548.

Ayala–Castañares, A. and Segura, L.R., 1968. Ecologia y distribucion de los foraminiferos Recientes de la Laguna Madre, Tamaulipas, Mexico. *Univ. Nac. Auton. Mex., Inst. Geol., Bol.*, No. 87: 1—89.

Benda, W.K. and Puri, H.S., 1962. The distribution of Foraminifera and Ostracoda off the Gulf Coast of the Cape Romano area, Florida. *Trans. Gulf Coast Assoc. Geol. Soc.*, 12: 303—341.

Benson, R.H., 1959. Ecology of Recent ostracodes of the Todos Santos Bay region, Baja California, Mexico. *Kansas Univ. Paleontol. Contrib.*, Arthropoda, art. 1: 80 pp.

Bradshaw, J.S., 1968. Environmental parameters and marsh foraminifera. *Limnol. Oceanogr.*, 13: 26—38.

Byrne, J.V., LeRoy, D.O. and Riley, C.M., 1959. The chenier plain and its stratigraphy, southwestern Lousiana. *Trans. Gulf Coast Assoc. Geol.*, 9: 1—23.

Coleman, J.M., 1966. Recent coastal sedimentation, central Louisiana coast. *Coastal Stud. Inst., La. State Univ., Tech. Rep.*, 29: 1—73.

Curtis, D.M., 1960. Relation of environment energy levels and ostracod biofacies in east Mississippi Delta area. *Bull. Am. Assoc. Pet. Geol.*, 44: 471—494.

Evans, G., 1965. Intertidal flat sediments and their environments of deposition in the Wash. *Q. J. Geol. Soc., Lond.*, 121: 209—245.

Favejee, J.Ch.L., 1951. The origin of the Wadden mud. *Meded. Landbouwhogesch. Wageningen*, 51(5): 113—141.

Frey, R.W. and Howard, J.D., 1969. A profile of biogenic sedimentary structures in a Holocene barrier island — salt marsh complex, Georgia. *Trans. Gulf Coast Assoc. Geol. Soc.*, 19: 427—444.

Hinde, H.P., 1954. The vertical distribution of salt marsh phanerogams in relation to tide levels. *Ecol. Monogr.*, 24: 209—225.

Lankford, R.R., 1959. Distribution and ecology of foraminifera from east Mississippi Delta margin. *Bull. Am. Assoc. Pet. Geol.*, 43: 2068—2099.

Meade, R.H., 1969. Landward transport of bottom sediments in estuaries of the Atlantic coastal plain. *J. Sediment. Petrol.*, 39: 222—234.

Murray, J.W., 1971. Living foraminiferids of tidal marshes: a review. *J. Foraminiferal Res.*, 1: 153—161.

Murray, J.W., 1973. *Distribution and Ecology of Living Benthic Foraminiferids*. Crane, Russak and Company, New York, N.Y., 274 pp.

Pestrong, R., 1965. *Tidal Flat Sedimentation at Cooley Landing, S.W. San Francisco Bay*. Tech. Rep. under the Office of Naval Research Contract NONR–4430(00) with Stanford University.

Phleger, F.B., 1960a. *Ecology and Distribution of Recent Foraminifera*. Johns Hopkins University Press, Baltimore, Md., 297 pp.

Phleger, F.B., 1960b. Foraminiferal populations in Laguna Madre, Texas. *Sci. Rep. Tohoku Univ., 2nd Ser. (Geol.)*, 33(4): 83—91.

Phleger, F.B., 1965. Sedimentology of Guerrero Negro Lagoon, Baja California, Mexico. In: W.F. Whittard and R. Bradshaw (Editors), *Submarine Geology and Geophysics. Colston Papers*, No. 17. Butterworths, London, pp. 205—237.

Phleger, F.B., 1967. Marsh foraminiferal patterns, Pacific coast of North America. *Ann. Inst. Biol., Univ. Nac. Auton. Mex., Ser. Cienc. Mar Limnol.*, 1: 11—38.

Phleger, F.B., 1969a. A modern evaporite deposit in Mexico. *Bull. Am. Assoc. Pet. Geol.*, 53: 824—829.

Phleger, F.B., 1969b. Some general features of coastal lagoons. In: A. Ayala–Castañares and F.B. Phleger (Editors), *Coastal Lagoons, A Symposium*. Universidad Nacional Autonoma de Mexico, Mexico, pp. 5—26.

Phleger, F.B., 1970. Foraminiferal populations and marine marsh processes. *Limnol. Oceanogr.*, 15: 522—534.

Phleger, F.B. and Ayala–Castañares, A., 1972. Ecology and development of two coastal lagoons in northwest Mexico. *Ann. Inst. Biol. Univ. Nac. Auton. Mex., Ser. Cienc. Mar Limnol.*, 1: 1—20.

Phleger, F.B. and Bradshaw, J.S., 1966. Sedimentary environments in a marine marsh. *Science*, 154: 1551—1553.

Phleger, F.B. and Ewing, G.C., 1962. Sedimentology and oceanography of coastal lagoons in Baja California, Mexico. *Bull. Geol. Soc. Am.*, 73: 145—182.

Phleger, F.B. and Lankford, R.R., 1957. Seasonal occurrences of living benthonic foraminifera in some Texas bays. *Contrib. Cushman Found. Foraminiferal Res.*, 8: 93—105.

Postma, H., 1954. *Hydrography of the Dutch Wadden Sea*. Thesis, University of Groningen, Groningen, 106 pp.

Postma, H., 1961. Transport and accumulation of suspended matter in the Dutch Wadden Sea, Netherlands. *Neth. J. Sea Res.*, 1: 148—190.

Scholl, D.W., 1963. Sedimentation in modern coastal swamps, southeastern Florida. *Bull. Am. Assoc. Pet. Geol.*, 47: 1581—1603.

Shaler, N.S., 1886. Sea–coast swamps of the eastern United States. *U.S. Geol. Surv., Sixth Ann. Rep. of the Director*, pp. 359—398.

Swain, F.M., 1955, Ostracoda of San Antonio Bay, Texas. *J. Paleontol.*, 29: 61—643.

Thom, B.G., 1967. Mangrove ecology and deltaic geomorphology, Tabasco, Mexico. *J. Ecol.*, 55: 301—343.

Thomas, W.H. and Simmons, E.G., 1960. Phytoplankton production in the Mississippi Delta. In: F.P. Shepard, F.B. Phleger and Tj. H. Van Andel (Editors), *Recent Sediments, Northwest Gulf of Mexico*. Am. Assoc. Pet. Geol., Tulsa, Okla., pp. 103—116.

Van Straaten, L.M.J.U. and Kuenen, Ph.H., 1957. Accumulation of fine grained sediments in the Dutch Wadden Sea. *Geol. Mijnbouw*, N.S., 19: 329—354.

Warme, J.E., 1971. Paleoecological aspects of a modern coastal lagoon. *Univ. Calif. Publ. Geol. Sci.*, 87: 1—131.

Chapter 5

SALT-MARSH ANIMALS: DISTRIBUTIONS RELATED TO TIDAL FLOODING, SALINITY AND VEGETATION

FRANKLIN C. DAIBER

INTRODUCTION

Numerous authors have discussed the influence of tidal flooding and salinity on the distribution and zonation of marsh vegetation. Chapman (1960) and Ranwell (1972, 1974) have very ably summarized much of the existing work on the subject. Although there is an extensive literature pertaining to salt-marsh animal distributions, it has hitherto only been partially reviewed and brought together (Daiber, 1974).

There are distinctive plant-animal relationships existing in tidal marshes. Shanholtzer (1974) comments on this with particular reference to the vertebrates, identifying direct and indirect associations. The direct interactions involve spatial and physical utilization of vegetation, providing a habitat volume and structural foundation for feeding, reproductive and roosting activities of various marsh land vertebrates. Thermal effects and refuge from predation are also cited. The indirect effects involving nutrient cycling and seed dispersal become evident some time after the actual contact between plant and animal.

Various workers have listed numerous marine invertebrate species associated with tidal marshes; but when Kraeuter and Wolf (1974) eliminated those found in the creeks, pools and similar areas, only a few species remained to dominate the marsh surface. The direct effect of these few species on the vegetation appears minimal; but the opposite interaction can include protection from predation, a stabilizing platform or a food source. Kraeuter and Wolf consider indirect effects as those that ameliorate various environmental fluctuations such as wave action, current velocities, humidity, temperature and light.

Man has a considerable influence on the vegetation and animals of tidal marshes, particularly through water-level regulation. Marshes have been drained for insect control or agricultural usage or flooded for wildlife management purposes (Daiber, 1974). The impact of drainage and lower water tables on marsh invertebrate faunas has been summarized in work carried out on the Mispillion marshes of Delaware (Bourn and Cottam, 1950). Dramatic reductions of all forms of macroinvertebrates were noted. The *Spartina alterniflora* zone showed reductions ranging from 39.3 to 82.2%. Populations in the *Spartina patens* region were decreased by 41.2 to 97.3% and those occurring in the *Scirpus robustus* area by 49.6 to 97%. Bourn and Cottam asserted that such reductions had direct effects on waterfowl by limiting available food supplies.

Nicol (1936) has divided the fauna of the marsh surface in English salt marshes into two groups: (1) animals visiting the marsh to feed; (2) animals living permanently on the marsh. Nichol finds this latter group of particular interest in terms of respiration rather than the impact of salinity, most species being air-breathing Insecta. There is a characteristic fauna associated with the marsh surface, and for those English marshes it consisted of such species as *Orchestia gammarella* (Crustacea), *Paragnathia maxillaris* (Isopoda), *Poliera marina* (Insecta) and *Dichirotrichus pubescens* (Insecta). Nichol comments on the distinctive faunas associated with tidal marsh pools. The fauna of those ponds with salinities lower than 5‰ was dominated by various insects, *Aedes detritus, Helophorus viridicollis,* and *Agabus bipustulatus,* and the crustacean *Gammarus duebeni*. Pools with higher salinities (15—20‰) were dominated by

Gobius microps (Pisces), *Neomysis vulgaris* (Crustacea), *Corophium volutator* (Crustacea) and *Nereis diversicolor* (Polychaeta). Other species were found to be common in particular pools. Nichol speaks to what has become a well-established dictum that salt–marsh animals cannot escape changing salinities. They must tolerate the osmotic effect on body fluids brought on by changing salinities.

Half a world away, Paviour–Smith (1956) adds to the examination of marsh surface communities by reporting work done at Hoopers Inlet on the Otago Peninsula. In this New Zealand salt meadow, animal zonations, food pyramids and trophic categories were defined. There was a broad area near the water edge dominated by *Salicornia australis*, merging into a narrow band of salt meadow which in turn gave way to grass and herbs. Paviour–Smith identified the boundary between salt and grass meadows as an ecotone characterized by tussocks of *Poa caespitosa* and *Scirpus nodosus*. The salt meadow possessed four plant species and, on the basis of total dry weight, consisted of: *Cotula dioica*, 1.9%; *Samolus repens*, 8.6%; *Selliera radicans*, 31.0%; *Scirpus cernuus*, 35.2%; and unidentified roots, 23.3%. *Samolus* was more abundant in the lower reaches of the *Salicornia* zone, *Selliera* and *Scirpus* in the intermediate region, while *Cotula* did not appear until the salt meadow proper (see p. 311).

The wide overlap of marine and terrestrial animals was the most striking feature of the transect across the *Salicornia* zone, salt meadow and into the grass land. In the lower *Salicornia* area, Collembola, Coleoptera, cyclorraphous larvae, Hemiptera, lepidopterous larvae, mites, spiders, and oligochaetes occurred along with the amphipods, polychaete worms and crabs. In the salt–meadow ecotone, the amphipod *Orchestia chiliensis* was still present with a more completely terrestrial fauna. Paviour–Smith reiterates the fact that these animals must adapt to or tolerate varying conditions of water and salinity. The arthropods had their maximum abundance in the drier grass meadow while the soil microfauna and arthropods were both active in the transitional salt–meadow zone.

The mesofauna of the salt meadow were mostly full–time annual residents whose seasonal variation was due to breeding cycles. Temperature may be the most important physical factor influencing breeding among the arthropods. Temperature and population curves were similar with increases from October to April with a peak from January to March. While there were many irregularities in the microfauna, Paviour–Smith showed an overall tendency toward a larger soil population during the summer and early fall months (January and February) than in the winter and early spring (July to September). The animal numbers were approximately 7.6 million m^{-2} with a zoomass of 32.4 g m^{-2} when the microfauna was at its maximum. At the same time the approximate total biomass was 1712 g m^{-2}, of which bacteria comprised 9.9 mg, and plants 1680 g. During the cooler summers when the mesofauna was at its maximum the zoomass was approximately 25.3 g m^{-2} with about two million animals. Yearly averages were estimated to be 3.3 million animals with a zoomass of 24.9 g m^{-2}.

Generalities directed to salt–marsh vegetation and animal zonations have been put forth by Teal (1962). He identified five regions consisting of: (1) creek bank, a muddy or sandy region between low water and the lower limits of *Spartina*; (2) streamside marsh, 1–3 m wide comprising tall *Spartina* just above the bare bank; (3) levee marsh with intermediate *Spartina*; (4) short *Spartina*, a region consisting of short widely spaced plants; and (5) *Salicornia* marsh, sandy areas near land where *Salicornia* is conspicuous. Although Teal confined his observations to the Georgia marshes, these categories can be applied to most tidal marshes. These distinctions reflect the interaction of tidal inundation, drainage, salinity, substrate and marsh height on vegetation zonation. Because these parameters come together in differing combinations, one should expect some variation in vegetational patterns from marsh to marsh. The animals associated with these five vegetative zones were placed into three general categories by Teal: (1a) terrestrial animals living in the marsh, (b) terrestrial or freshwater species living on the landward edge; (2) aquatic species with their center of abundance in the estuary, (a) estuarine species limited to the marsh low water level, (b) species in the streamside marsh, (c) estuarine species found well into the marsh; (3) marsh species with aquatic ancestors, (a) species with planktonic larvae, and (b) species living entirely within the marsh. Teal recorded

57% of the aquatic species (2a and 2b) occupying the lowest portions of the marsh where exposure at low tide is the shortest. Those on the marsh are living at one edge of their species distributions, their numbers maintained by migrations from the water and by adaptations and tolerance levels. Those living above the mud are subject to great environmental stress, and those penetrating farthest into the marsh have adopted burrowing habits. The remainder are either tolerant enough to inhabit the entire marsh or are most common on the marsh itself. The terrestrial species comprise almost half the marsh fauna, yet have made few adaptations for marsh living and are much less important in community energetics than the aquatic species (Teal, 1962).

MacDonald (1969) and Phleger (1970) have examined marsh invertebrate faunas on a much larger scale than previous workers. MacDonald has identified distinctive molluscan faunas in the tidal creeks and *Spartina—Salicornia* salt marshes of the North American Pacific coast. There are one or two widely distributed and very abundant species, with the remaining populations represented by small numbers of patchily distributed species. The creek faunas which are largely infaunal in nature contain more species and have a more variable composition than do the marsh faunas which are all epifaunal and independent of sediment type. The marshes are very uniform; five species make up 96.9% of the samples, with the remaining 3.1% more frequently found in the tidal creeks.

The following sections will summarize the interaction of tidal inundation and storm flooding, drainage, salinity, marsh height and vegetational zonation with animal distributions in tidal marshes.

PROTOZOA

Specific studies concerned with the Protozoa appear to be confined to the Ciliata and Foraminifera. Borror (1965) describes the external morphology, internal anatomy, behavior and taxonomy of ciliates collected from New Hampshire tide marshes. The algal mats associated with the borders of marsh tidal pools often had a rich, diverse ciliate fauna. Preliminary observations indicated ciliate abundance equal to or superior to some psammolittoral habitats.

Phleger and Walton (1950) call attention to the role of environmental factors affecting Foraminifera distributions including tidal action, nature and movement of bottom materials, presence of marsh grass and relative organic production. The role of environmental parameters is reiterated by Phleger (1965), Phleger and Bradshaw (1966), Phleger (1970), and Matera and Lee (1972) when salinity, oxygen levels, temperature, pH and light duration are added to the list (see also pp. 73, 74).

Phleger and Walton (1950) identified two major foraminifera facies, one found in Barnstable Harbor characterized by *Trochammina inflata*, and secondly an adjacent nearshore Cape Cod Bay facies dominated by *Proteonina atlantica* and *Eggerella advena*. Three subfacies comprised the Barnstable Harbor fauna; on the high marsh, intertidal flats and in the channels. The greatest total populations of harbor Foraminifera were measured in the high marsh (*Spartina patens* and *S. glabra* = *S. alterniflora*) consisting principally of *Trochammina inflata* and *T. macrescens*. The greatest frequency of *Miliammina fusca* occurred in the high marsh while *Armorella sphaerica*, *Webbinella* (?) sp. and *Valvulineria* sp. were confined to this zone suggesting an indigenous fauna. They noted the highest organic content of the sediments was found in the high marsh and offered two explanations for large marsh populations: (1) the environment was able to support large numbers; or (2) Foraminifera were washed in. Phleger and Walton found little evidence to support the latter thesis, since there were few broken shells and the high marsh was not subjected to frequent flooding and strong currents. The former seemed more credible as the sediments contained large accumulations of grass, thus providing shelter and food matter.

Phleger (1965) identified the general marsh foraminiferal assemblage of Galveston Bay, Texas, to be dominated by *Ammotium salsum* and *Miliammina fusca* with common *Ammonia beccarii, Arenoparrella mexicana,* and *Trochammina inflata*. There were somewhat smaller frequencies of *Ammoastuta inepta, Elphidium* spp., *Tiphotrocha comprimata* and *Trochammina macrescens*. Matera and Lee (1972) described the Foraminifera in the epiphytic communities of a Long Island marsh as being very patchy, 2.6% of the samples accounted for 56.4% of the total Foraminifera encountered. The standing crop and species composition changed with the summer. One peak occurred in early summer with *Protelphidium tisburyense* as dominant on the epiphytes. *Elphidium incertum* was prevalent during a second peak in late July—early August. Over the entire summer, *Elphidium incertum* and *Protelphidium tisburyense* comprised 46.6% and 25.6%, respectively, of the epiphytic Foraminifera community. There were fewer species but more individual Foraminifera in the benthos than the epiphytic community for this time. *Ammotium salsum* (10.9%), *Elphidium incertum* (31.8%) and *Trochammina inflata* (49.6%) dominated the benthos and bloomed successively in the psammolittoral community. The latter two were correlated with vertical and horizontal changes in grain size; *Elphidium* distribution clustered around a grain size of 0.1 mm whereas *Trochammina* clustered around a median size of 0.46 mm.

For many Foraminifera species, Matera and Lee (1972) found no continuity between the epiphytic and the psammolittoral communities. *Protelphidium tisburyense* was an early summer dominant epiphyte, but rare in the benthic community. *Trochammina inflata* was dominant in coarser deeper sediments, but rare in the epiphytic community. Both *Elphidium incertum* and *Ammotium salsum* were abundant in both communities, and Matera and Lee considered them as general–list species.

While some Foraminifera species are being reported from a wide geographic area, Phleger (1965, 1970) reminds us that each marsh environment has its distinctive foraminiferal assemblages. These groupings reflect the environmental differences delineated by plant zonations within that marsh. Phleger further suggests that microenvironments within each zone may affect distributions within the zone. The great variability so characteristic of tidal marshes, induced primarily by tidal flushing, temperature variations, sunlight and plant metabolism (Phleger and Bradshaw, 1966), may thus account for the marked fluctuation in population numbers. A limited number of organisms can tolerate such variability, resulting in few species indigenous to marshes as well as few species in a marsh compared to other marine environments (Phleger, 1970; Murray, 1973). Phleger's examination of the south Texas coast shows most samples contain 4—5 species with a range of 1—16 per sample. In contrast, the bay or ocean samples average about 10 species with a range of 2—20 species per sample. The least diverse populations are found in the high marshes in most areas and the numbers of species increase moving across the low marsh toward the bay. Murray's (1973) report of low species diversity for marsh Foraminifera included in the estuarine environment support this evidence. The flooding marine water with its diverse calcareous population probably has an influence on the low marsh fauna (cf. p. 19).

Salinity has different effects on various Foraminifera. Populations of *Arenoparrella mexicana*, *Haplophragmoides hancocki*, *Tiphotrocha comprimata* and *Trochammina macrescens* decrease with increasing salinity while *Jadammina polystoma* and *Trochammina inflata* expand their numbers along an increasing salinity gradient. Populations belonging to *Ammobaculites dilatatus*, *Ammotium salsum*, *Miliammina fusca* and *Protelphidium tisburyense* respond to other parameters than changing salinity (Parker and Athern, 1959). Salinity has been considered by Murray (1973) to be a major environmental factor when trying to establish criteria for recognizing estuarine environments using Foraminifera. On tidal marshes, salinity is highly variable and the species assemblage might include *Elphidium articulatum* and *Miliammina fusca* but typically would consist of *Jadammina macrescens* and *Trochammina inflata*. Species diversity on the marsh surface was found to be $\alpha < 2$, being considerably lower than diversities for other estuarine environments examined by Murray.

The Gulf of Mexico assemblages differ between *Spartina* low and *Salicornia* high marshes where the standing stocks are generally large but vary between 10 and 1500 per 10 ml of surface sediment 1 cm thick. Samples from tidal flats have twice the average standing stock of the low marsh and four times the populations of the high marsh. Rates of deposition are inferred to be high but are lower in the high marsh than in the low marsh or tidal flats (Phleger, 1970).

On the Pacific coast of North America, Phleger distinguishes between assemblages of Foraminifera from the tidal flats, marsh creek channels and the marshes. In some cases low– and high–marsh assemblages could be differentiated. In contrast to the Gulf coast, standing stocks are higher on high marshes than on the tidal flats but vary greatly. High standing stocks imply high rates of production, possibly due to abundant supply of nutrients and trace materials from land runoff, tidal flooding and continual replenishment of fine-grain soil (cf. p. 71). The reference to the generally low production in New Zealand marshes may be due to coarser-grained sediments than in most marshes.

Phleger goes on to say marshes often contain mixtures of lagoon or bay assemblages as well as endemic marsh species (see Fig. 4.4). The dominance of one group depends on the hydrology of the marsh and lagoon or bay, and the position of fauna in the marsh. Phleger considers 11 species with world-wide similarities as endemic to marshes although not all are found in any one marsh or in all geographic regions: *Ammoastuta inepta, Arenoparrella mexicana, Discorinopsis aguayoi, Haplophragmoides* spp., *Jadammina polystoma, Miliammina fusca, Protoschista findens, Pseudoponides andersoni, Tiphotrocha comprimata, Trochammina inflata, Trochammina macrescens. Trochammina inflata* has been present in every marsh study. *Miliammina fusca* and *Jadammina polystoma* and/or *Trochammina macrescens* occur in most marshes. In addition to those universally distributed forms, tidal marshes contain certain species that are more or less restricted to particular geographic areas.

Because of the hydrographic continuum that exists between tidal marshes and bays or lagoons, it is not surprising for Phleger (1970) to report the frequent displacement of marsh species into other habitats. This is a reflection of the abundance and persistence of runoff. *Miliammina fusca* which is common in the marshland bordering the Mississippi Sound is also found in the adjacent bays, lagoons and continental shelf area, all of which have a high rainfall and high runoff. The same species is confined to the marsh bordering the Guerrero Negro Lagoon in Mexico where runoff is zero and there is a net water flow into the lagoon and marsh.

There is an interesting relationship between zonation and the magnitude of runoff as influencing the abundance of agglutinated tests (Phleger, 1970). The low marsh in many places contains a fauna dominated by calcareous tests with a few agglutinated ones. On the Pacific coast of North America, the marsh Foraminifera have primarily agglutinated tests. This is an area of high runoff where undiluted sea water containing calcareous species does not flood the marshes. In contrast, areas such as Mission Bay, California, and Guerrero Negro Lagoon, Mexico, that receive undiluted sea water and have no runoff possess a marsh fauna with calcareous tests.

Fewer foraminiferal species have been recorded from areas of high runoff (Phleger, 1970) and this is thought to be due to the exclusion of marine water and not primarily temperature. Most places with high runoff have 10—11 species in contrast to much more diverse fauna in the absence of runoff; 25 species in a Baja California hypersaline marsh and more than 50 species in a temperate hypersaline lagoon and marsh in the Coorong, South Australia. Phleger reports the low diversity in two subarctic marshes may be due to low temperature or high runoff or a combination of the two.

NEMATODA

The nematodes of the salt marsh have received some attention. Teal and Wieser (1966) sampled an area between low tide level and land in a Sapelo, Georgia, marsh, reporting numerical densities as great as 16 million m^{-2} and weighing 7.6 g m^{-2}. Densities were reported to be ten times greater than those at Woods Hole, Massachusetts, where Wieser and Kanwisher (1961) found patchy distributions with some densities ranging from 1.4 to 2.1 million m^{-2}. The biomass of the Sapelo Island samples was small, reflecting the small size of the dominant species. This biomass more closely resembled the sublittoral zone of the Woods Hole area than that of the salt marsh. Teal and Wieser (1966) found Sapelo Island nematodes generally most abundant and with the greatest vertical distribution at sites closest to the water. Food supply was suggested as an important factor limiting distribution and abundance, the low tide area receiving the greatest amount of detritus. Wieser and Kanwisher (1961) inferred vertical distribution was related to the reduction of available space caused by the mat of *Spartina* roots. Most of the nematode population was concentrated in the upper 4 cm with a maximum penetration of 8 cm.

MOLLUSCA

Pelecypoda

The ribbed mussel, *Modiolus demissus,* is a much more obvious member of the marsh fauna than the previously described forms. The mussel has been examined in the Sapelo Island marshes of Georgia where random samples gave an estimated population density of 7.8 m^{-2} for the inhabited portions of the marsh. They were most dense near the heads of small creeks, an average of 52 animals m^{-2} being common in certain marsh types (Table 5.1). The tall *Spartina* (1—3 m) Creek Head marsh, which occupies only 6% of the total marsh area, contained 46% of the population by weight. The medium *Spartina* (0.5—1.0 m) levee had a reported density one–seventh of the tall *Spartina* Creek Head marsh, but the levee occupied 3.5 times as much area and contributed the second largest percentage of total weight (Table 5.1) (Kuenzler, 1961).

Clumped distributions within uniform marsh areas were a striking feature of the Sapelo Island mussel populations in that sample variances were larger than the means, indicating non–random distributions. Clumps were larger and more numerous in densely populated marsh types such as the creek heads where they often exceeded 1 m in diameter or paralleled the creeks for many meters.

Kuenzler reported the vertical distribution to be approximately 200—240 cm above mean low water. The center of biomass was situated at 220 cm with the maximum density probably at this same level. It was calculated that a mussel at the 220–cm elevation would be covered by 77% of the flood tides, or about 18% of the time. The greater density by weight at low and medium elevations was thought to be made possible by longer feeding times. When comparing growth rates, they were always greater in the low–marsh types (short *Spartina,* low elevation; tall *Spartina,* creek heads) than in higher marsh types (medium *Spartina,* levee; *Salicornia—Distichlis;* medium *Spartina,* medium elevation).

Lent (1967a,b, 1968, 1969) elaborates on the significance of intertidal distribution and air–gaping in the ribbed mussel. He compares those mussels typically found in tidal marsh muds with those located on exposed surfaces such as bridge pilings. There was a significant difference in the shell length : height ratios, bridge mussels being longer than marsh mussels. There was no difference between marsh mussels from the different geographic locations of Delaware and Georgia. Lent (1967b) attributes the difference between habitats to the crowding effect of clumping in the marsh.

Lent (1967b) found excellent allometric relationships between the shell height and all weights for both bridge and marsh animals. He did not find

TABLE 5.1

Population distribution and density of *Modiolus demissus* from the Sapelo Island marshes. (From Kuenzler, 1961, by permission of the editor of *Limnol. Oceanogr.*)

No.	Marsh type description	% of total marsh	Number in 0.2 m^2		Density m^{-2}		% Total	
			mean	variance	number	weight	number	weight
1	muddy creek banks	4	0		0	0	0	0
2	tall *Spartina*, edge	9	0		0	0	0	0
3	medium *Spartina*, levee	21	1.14	12.32	5.7	4.35	17	22
4	short *Spartina*, low elevation	16	1.24	9.29	6.2	3.0	15	12
5	short *Spartina*, high elevation	20	0.08	0.11	0.4	0.15	1	1
6	*Salicornia* spp., *Distichlis spicata*	9	0.23	0.37	1.1	0.40	2	1
7	*Juncus roemerianus*	4	2.64	26.4	13	6.15	9	7
8	tall *Spartina*, creek heads	6	10.4	146	52	31.6	47	46
9	medium *Spartina*, medium elevation	11	1.11	5.72	5.6	4.55	9	12
Total		100					100	100
Mean			1.33		6.66[1]	4.11		

[1] The difference in density of 6.66 m^{-2} and 7.82 m^{-2} reported in the text was due to the mat not being sieved. Mussels less than 25 mg were not recovered but constituted 18% of the total number.

any relationship between the shell weight:meat weight ratios and intertidal height. Since the mussel is exposed during part of the tidal cycle and air-gapes during exposure, respiration could be carried on, and such exposure would thus not reduce meaty tissue weight as rapidly. Air-gaping, by increasing the efficiency of aerial respiration, would permit a greater landward penetration in the intertidal zone.

In his 1968 paper, Lent points out that median survival times in gaseous environments are proportional to the volume of oxygen present. Lent reported oxygen tension during exposure at 15 mm Hg and 110 mm Hg while inundated (fig. 6, 1968). Lower rates of respiration in air were explained by these lower oxygen tensions of the mantle cavity fluid in air-gaping mussels. Although the rate was lower in air than when submerged, more oxygen was consumed because of extended exposure (82% aerial exposure; Kuenzler, 1961). Mussels survive for long periods exposed as long as they are protected from desiccation.

Desiccation and temperature changes are environmental hazards with which the exposed mussels are confronted. Lent (1969) considers the mussel to possess a high tolerance to dehydration and a very high enzyme thermostability. Temperature has a significant effect on the rate of desiccation at low but not at high humidity (Lent, 1968). Desiccation results from water loss by air-gaping and is a physical phenomenon in which a surface to volume relationship causes small mussels to reach a median lethal weight loss of 36—38% more rapidly than large mussels.

Mussels can survive over a temperature range of at least 56°C with a recorded minimum of −22°C (Lent, 1969). An upper LD50 for a 10–h heat stress fell between 36.4 and 37.8°C, with the large mussels more labile to thermal stress than small animals (Lent, 1968). Such temperature tolerance would explain the geographic distribution ranging from Prince Edward Island in the north to South America.

Salinity limits are wide but not firmly established. The lower limit may be as low as 3‰. Mussels can survive water losses of 36—38% by desiccation and have at least 71% of their tissue water frozen (Lent, 1969). Therefore the upper limit must be high.

Mussels living in moist marsh mud occupy a

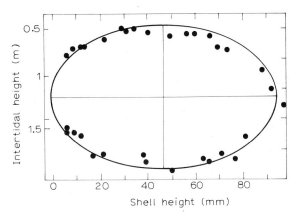

Fig. 5.1. Intertidal height as a function of shell height for 33 mussels on the periphery of the distribution of Canary Creek bridge (from Lent, 1968, by permission of the editor of *Biol. Bull.*).

higher intertidal height than those living on exposed bridge pilings. When the intertidal position of bridge mussels is examined as a function of shell size the population falls on an ellipse (Fig. 5.1). The intertidal range is greatest for medium–size mussels and is reduced for both larger and smaller groups. Lent (1968) considers this ellipse to be a geometric form within which the natural population can live. The upper surface is generated by the physical factors of desiccation and thermal stress whereas predation and competition are biotic factors determining the lower surface.

Lent (1969) proposes air-gaping in the mussel to be a significant behavioral adaptation which permits aerial respiration and penetration of the high intertidal zone. In addition, it is physiologically and biochemically adapted in that it is both eurythermal and euryhaline. These two adaptations provide a tolerance toward desiccation, salinity variation, thermal stress, and possibly anaerobic conditions. However, there is no obvious morphological adaptation; thus, the pelecypod body plan precludes further landward penetration.

Gastropoda

The coffee bean snail, *Melampus bidentatus*, is an abundant member of the tidal marsh fauna and has received the attention of numerous workers over the years. Hauseman (1932) found few snails where tides flood the tall marsh grass and where fiddler crabs are common, with snails more numerous above the high-tide mark. According to Holle (1957), *Melampus* can be found only in salt marshes flooded by normal tides and avoids tidal submersion by climbing grass stems or piles of debris. In contrast, Hackney (1944) reported *Melampus* as common on the mud flats of Beaufort, North Carolina.

Dexter, in a series of papers on the marine mollusks of the Cape Ann, Massachusetts, region (1942, 1944, 1945), related *Melampus* distributions to vegetation. In 1942, he often found individuals in small groups under solid objects, occasionally finding the snail in the upper margins of thatch grass or *Spartina glabra* (= *S. alterniflora*) marsh. Small clusters of *Melampus* were found in the *Spartina patens* marshes of Little River (1944). Dexter was of the opinion in his 1945 paper that the high marshes of *Spartina patens*, located for the most part above mean high water, were the only important habitat for *Melampus bidentatus*. In his study of the Poropotank River area of Virginia, Kerwin (1972) found the coffee bean snail associated with the brackish-water marshes dominated by *Spartina alterniflora—Scirpus robustus* stage and *Spartina cynosuroides*; and with the salt marshes dominated by *Spartina alterniflora* (short form), *S. patens* and *Distichlis spicata*. The highest density was 144 m^{-2} for the *Distichlis* zone. The percent occurrence was: *Distichlis*, 87.5%; *Spartina patens*, 75.0%; and *S. alterniflora* (short form), 56.3%.

Russell-Hunter et al. (1972) report the snail primarily in the *Spartina patens—Juncus—Distichlis* zone, in the *Distichlis* zone and in the upper levels of the area dominated by the dwarf form of *Spartina alterniflora*. They go on to say *Melampus* is largely found in the upper two-thirds of the zone lying above mean high water of neap tides (MHWN) and below the mean high water of spring tides (MHWS), but is also found in the zone lying between MHWS and the extreme upper limit worked by any tides (Fig. 5.2). They describe the snail as occupying the upper 12% of the intertidal zone which is bathed for only 8 of the 354.4 h (or 2.3%) of each semilunar cycle. Apley et al. (1967), Apley (1970), and Russell-Hunter et al. (1972) describe the reproductive cycle of the snail as being synchronized to summertime spring tides.

ARTHROPODA

Crustacea

There are a number of crustacean species associated with tidal marshes. Fiddler crabs, being unique, have generated considerable interest. The less obvious amphipods and isopods have received little attention. In an early report, Smallwood (1905) found the salt-marsh amphipod *Orchestia palustris* to be distributed over the entire marsh except in the *Spartina polystachya* (= *Spartina cynosuroides*) zone and the typically fresh-water regions. It was abundant in the drift of salt-water plants, under some kind of cover and in the more saline portions of the marsh. The amphipod *Orchestia gammarella* and the isopod *Paragnathia maxillaris* have been described by Nicol (1936) as characteristic faunal members of the surface of English marshes.

In his study of the thermo-saline tolerances of the isopod *Sphaeroma rugicauda*, Marsden (1973) expects the species to be able to colonize marsh areas infrequently covered by tides. He goes on to say that suitable vegetation prevents the penetration of the isopod into a more marine environment.

In light of M.S. May's (1974) work on amphipods as possible agents for detritus formation and the apparent paucity of information, it seems that amphipod and isopod biology and populations in

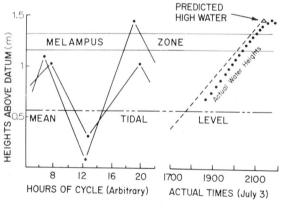

Fig. 5.2. The relation of the vertical zone (1.19 to 1.34 m above datum) occupied by natural populations of *Melampus* at Little Sippewisset, typical ranges of spring (−0.01 to 1.45 m) and neap (0.27 to 1.00) tides, and the observed time course of actual tidal heights during one high water of springs. Note that the mean level of all tides (MTL) does not necessarily correspond to mean sea level (MSL). (From figure 8 in Russell-Hunter et al., 1972, by permission of the editor of *Biol. Bull.*).

salt marshes could be fruitful areas of investigation.

Fiddler crabs are an obvious faunal component of tidal marshes. Pearse (1914) identified three characteristics strongly associated with these crabs, having watched thirteen species in North and South America and in the Philippines: (1) fiddler crabs are diurnal, whereas other crabs are nocturnal as well; (2) a substrate specificity is displayed — this gives rise to zonation in habitat selection; (3) fiddler crabs feed on tidal flats during low tide with greatest activity on the ebbing flow. Frantic activity by the crabs was reported by Pearse as the tide began to flood. The crabs hastily gather mud pellets to plug the burrow entrance as the tide floods over the marsh surface. Crabs remain in their burrows during high tides and prolonged flooding. They open their burrows but will remain in them during extended periods of low tides which tend to dry out the tidal flats. Drying tends to inhibit the feeding process for the crabs.

Marsh crabs can be divided into three behavioral groups associated with tidal cycles (Teal, 1959).

(1) *Eurytium limosum, Panopeus herbsti,* and *Sesarma reticulatum* are active when the tide is high, or when the sky is cloudy and the sun is not shining. When the marsh is uncovered at low tide they are in their burrows, near the top, either exposed to air or in the water. Crichton (1960) describes *Sesarma reticulatum* clearing mud from its burrow as the tide recedes, and depositing it at either side of the entrance where a complete canopy is frequently formed.

(2) *Uca pugnax* and *U. minax* are active principally when the tide is out and do feed under water.

(3) *Uca pugilator* and *Sesarma cinereum* are active only in the air. *U. pugilator* is in its burrow at high tide while *S. cinereum* climbs above the water level.

The first group is found in the low marsh at the low tide level and the second group occupies an intermediate zone. The third group, with one exception, is located at the high tide level. Teal found colonies of *Uca pugilator* just above the low tide level on the creek banks of the Georgia marshes. Crabs in the first group build burrows with numerous openings which allow the burrows to fill with water. *Uca pugnax* and *U. minax* dig burrows with single openings but seldom plug them. *Uca pugilator* almost always plugs its burrow entrance at high tide and thus remains in a pocket of air. However, Teal considers the air pocket incidental to the primary need for a hiding place.

Salinity plays some part in fiddler crab distributions. Kerwin (1971) found *Uca minax* widely distributed in brackish and salt marshes of the Poropotank River, Virginia, where the salinity was greater than $2‰$ and less than $16‰$. The description of the habitat for the *U. minax* in Solomons Island, Maryland (Gray, 1942), which included *Spartina cynosuroides*, would suggest this to be a brackish–water species. *Uca pugnax* was reported from high salinity ($21—29‰$) whereas *U. minax* was present in lower salinities and even in fresh waters of three Delaware tidal streams (Miller and Maurer, 1973). Their distributions overlapped along the salinity gradients of these rivers. They were equally abundant at $12‰$ on the Broadkill River, at $11‰$ on the Mispillion River and $8‰$ on the Murderkill River. Miller and Maurer consider these midpoints may be areas where neither species is favored by competitive advantage or where both are equally adapted to salinity regardless of competition. Something else other than salinity must be playing a role in such distributions. Students of the present author reported numerous *U. pugnax* and few *U. minax* from one area of the Blackbird Creek marsh in Delaware where the stream salinity was $4‰$ and $5‰$. This is supported by Kerwin's comment that low salinities and/or a lack of suitable substrate appear to inhibit ecesis of *U. pugnax* and *U. minax* in the Virginia marshes he examined.

Tidal flooding plays a profound role in the lives of these crabs beyond the daily ebb and flow. *Uca minax* digs a burrow downward to the point where the chamber is at low–tide water level (Gray, 1942). This species excavates new burrows during prolonged low–water periods or when the burrows are filled with water from extended flooding. In either case, the burrow is dug to a depth to assure flooding of the chamber. Teal (1958) suggests predation may be influential in restricting the distribution of *Uca pugnax* at the lower intertidal levels. This species frequently remains outside its burrows as the tide is flooding, while *Uca pugilator* almost never does. The former is therefore more subject to predation by aquatic organisms.

Whereas Pearse (1914), Gray (1942) and Teal

(1958) point out the importance of substrate and tidal flooding in the lives of fiddler crabs, specific associations between crabs and vegetation are also evident. Gray describes *Uca minax* habitat at Solomons, Maryland, as a high–water sand and mud substrate sparsely covered by *Spartina cynosuroides* and *Paspalum* sp. *Uca minax* is found in association with fifteen species of marsh plants, occurring with five species in more than 20% of the collecting periods in the Poropotank River marshes of Viginia: *Spartina alterniflora*, 87.8%; *Scirpus robustus*, 32.9%; *Distichlis spicata*, 31.7%; *Spartina patens*, 24.4%; and *S. cynosuroides*, 20.7%. High population densities are reported in the tall and short *S. alterniflora* and at the edge of the *D. spicata—S. patens* community with densities ranging from 0 to 76 burrows m^{-2} (Kerwin, 1971).

Schwartz and Safir (1915) describe the burrowing of *Uca pugnax* where there is dense vegetational cover, whereas *U. pugilator* burrows in sparse vegetation or where it is absent. Teal (1958) found this species in most vegetation types of Georgia marshes. It was most numerous in the short marshes where the *Spartina* is 10—50 cm tall on firm substrate containing varying amounts (0—70%) of sand and the medium marshes with *Spartina* 1 m tall on soil that is soft with small amounts of sand (0—10%). Gray (1942) and Teal (1958) found *Uca minax* in the short *Spartina* high marshes. *Uca pugilator* was found primarily in the *Salicornia—Distichlis* marsh and on the tidal creek banks.

The large burrows of the marsh crab *Sesarma reticulatum* have been reported within 5 m of the tidal water of Canary Creek marsh of Delaware and associated with *Spartina alterniflora* (Daiber and Crichton, 1967). Crichton (1960) considers this marsh crab responsible for patches of bare mud along the creeks and ditches as it crops the cord grass and undermines the roots. Frequently these bare areas are swathes separating several meters of lush tall *Spartina alterniflora* beside the creek from the medium–size *S. alterniflora* further back on the marsh surface. At other times, they border directly on the bare mud of the stream bank. It is in these swathes that the burrows are the largest and most numerous. The number of burrow openings range from one to ten per square meter with five to six being an average. Crichton considers this burrowing activity to enhance the erosive action of the water and keep the banks low enough so the tides can flood over the marsh surface.

Insecta

Insects are a common component of the tidal marsh fauna, occupying a variety of habitats. Those noxious to man are of considerable economic importance. Much work has been devoted to the control of salt–marsh mosquitos, beginning with Smith (1902) who recognized the relationship between mosquito distributions and zonation of salt–marsh vegetation. He advocated filling those depressions on the marsh surface adjoining the uplands where mosquitos breed, or ditching so tidal action could drain such areas.

Tidal inundation has been identified as a factor limiting the distribution of *Aedes* species of mosquitos on Delaware salt marshes (Connell, 1940). Young larvae fail to appear in portions of the marshes flooded by tides as frequently as 25 days per lunar month. Abundant breeding can be expected only in portions of *Spartina alterniflora* marsh where the frequency of tidal inundation is less than 8 days per lunar month. Table 5.2 depicts

TABLE 5.2

Average number of *Aedes* larvae and pupae per dip from the salt marshes of Egg Island, New Jersey (from Ferrigno, 1958)

Vegetation	Mean number of larvae and pupae per dip		
	A. cantator	*A. sollicitans*	total
Spartina patens	0.31	6.95	7.26
S. alterniflora	0.01	0.37	0.38
Mixed	0.29	2.42	2.71

SALT–MARSH ANIMALS

TABLE 5.3

Mosquito breeding at stations affected by feeding snow geese in New Jersey marshes (from Ferrigno, 1958)

Description of eat-outs		Water depth (cm)	Number of dips	Number of larvae—pupae	
percent denuded within 21 m radius of station marker	vegetation type			per dip	total
A 10—30	*Spartina patens*	0—5	520	12.71	6607
B 10—40	*S. alterniflora*	0—15	1355	1.56	2119
C 50—90	*S. patens—S. alterniflora*	2—20	460	0.05	25
D 50—90	*S. alterniflora*	2—15	490	0.03	
E 100 (Ponds)	*S. patens—S. alterniflora*	12—60	1401	0.002	3

the relationships between vegetation and insect numbers for *Aedes sollicitans* and *Aedes cantator*; most of the mosquitos were taken in the *Spartina patens* zone. Ferrigno (1958) demonstrates the relationship between flooding and water depth on the success of mosquito breeding caused by snow geese feeding activity (Table 5.3). In addition, the restriction of regular tidal flushing in vegetational zones for agricultural purposes has an impact on the numbers of mosquitos produced (Table 5.4). Diking enhances the growth of salt hay, *S. patens*, by preventing flooding. Later reflooding produces great broods of mosquitos, mostly *Culex salinarius* and *Aedes sollicitans*.

There appears to be some uncertainty about the relationship between tidal flooding and vegetation zonation on one hand and the distributions of larval greenhead flies (Tabanidae). Hansens (1952) found them in the wetter portions of the salt marsh, while Gerry (1950) described development as taking place primarily in the upper marsh reached only by the higher tides. The latter went on to say evidence indicates that larvae originate in the creeks from which they migrate to thatch piles at

TABLE 5.4

The numbers of mosquito larvae and pupae sampled from natural and diked salt marshes of the Caldwalder Tract, New Jersey (a portion of table 2, Ferrigno, 1959)

Vegetation type	Average number larvae—pupae per dip		Seasonal totals	
	Culex salinarius	*Aedes sollicitans*	dips	larvae—pupae
Undiked				
Spartina alterniflora	0	0.0001	8280	1
S. patens	0.11	2.74	1080	3293
Panicum virgatum	0.003	0	600	2
Woodland swamp	0.02	0.01	840	620
Diked				
S. alterniflora	0.26	4.22	360	1701
S. patens	0.72	3.54	2760	13376
S. cynosuroides	2.94	4.66	240	1988
P. virgatum	0.21	0.75	1320	2219
Distichlis spicata	0	3.52	600	2761
Juncus gerardi	0	2.86	240	780
Typha	2.01	0.21	120	707

the head of the marsh. Jamnback and Wall (1959) found the larvae associated with several species of salt-marsh vegetation but most abundant in *Spartina alterniflora* and *S. patens*. Contrary to Gerry (1950), they believed that the larvae could survive in water for a long time. Olkowski (1966) found most larvae among the *S. alterniflora* of Delaware marshes with fewer individuals present as ground elevation increased toward the *Spartina patens* zone. The mature larvae seemed to be dispersed to higher ground by tidal action.

Insects in general display an association between the extent of tidal flooding and vegetational zonation. Tide-elevation influences, primarily the length of the hydroperiod, determine the distribution of insect dominants (Davis and Gray, 1966). The Homoptera dominated and the Diptera ranked second at most stations set up within the North Carolina marshes examined by Davis and Gray. Hemiptera, Orthoptera, Coleoptera and Hymenoptera were present at all stations but in lower numbers. Homoptera numbers decreased as other orders increased with the transition from low marsh to high marsh elevations. It should be pointed out, however, that Davis and Gray found considerable variation in average number of insects from station to station, especially within the low-marsh zones of *Spartina alterniflora* and *Juncus roemerianus*. Of the high-marsh stations, *Distichlis spicata* had a greater insect fauna in numbers and species than *Spartina patens*, and the average number did not vary greatly from station to station as it did for the low-marsh locations.

The length of the hydroperiod was not believed to limit the total size of the insect aggregations (Davis and Gray, 1966). There was no evidence that any insect in the study allowed itself to be inundated by the tide. They escaped by swimming, hopping, walking on the surface film, or by flying. Davis and Gray (1966) concluded that the size of animal assemblages is determined by food and shelter.

There were many more insects in the *Distichlis* zone than for *Juncus*, although both have short hydroperiods. The low-marsh *S. alterniflora* and the *Spartina—Salicornia—Limonium* zones which were frequently flooded had more insects than did *Spartina patens* which was flooded only on storm tides. The more frequent branching of *Distichlis* and larger leaf crevices of *Spartina alterniflora* provide more hiding places. Not many insects can handle the coarse fibers of *Juncus* for food. Presumably, the Orthoptera which were more evident are better able to cope with these tougher *Juncus* tissues.

In the description of the *Spartina*-feeding insects of the Poole Harbour, Dorset, salt marshes, Payne (1972) portrayed a gradual decline in numbers of the grasshopper, *Chorthippus albomarginatus*, as the marsh sloped toward the water. *C. albomarginatus* was reported to be a good swimmer but appeared to deposit its eggs at the base of *Spartina* stems where developing nymphs were apt to drown. The grasshopper *Conocephalus dorsalis* was associated with the higher *Spartina* plants as it was quite susceptible to drowning. The hemipteran *Euscelus obsoletus*, fed upon by the omnivorous *C. dorsalis*, was found as far as the water's edge, presumably not affected by inundation.

Davis and Gray (1966) found the Hymenoptera (ants) more evident on the high ground of the *Spartina patens* zone of North Carolina marshes. Woodell (1974) describes the impact on vegetation found on ant hills located on the border between the landward dune and the marsh at Scolt Head Island, Norfolk. These ant hills have been present for at least thirty years, flooded by high spring tides and completely inundated on occasion. The ants survive due to the well-drained shingle beneath the shallow marsh soil, along with improved drainage on the ant hills themselves. The grasses *Puccinellia maritima* and *Festuca rubra* are present on these ant hills due to their tolerance to rabbit grazing. The dwarf shrub, *Frankenia laevis*, a Mediterranean species at the northern limit of its range, flourishes on the ant habitations due to good drainage. This shrub is found only on the south slopes of the ant hills, indicating ant activity produces profound microclimatic effects.

Arachnida

The arachnids have received intensive attention with Barnes' (1953) study of spider distributions in the non-forest maritime communities in Beaufort, North Carolina, and Luxton's (1964, 1967a, b) examination of English salt-marsh Acarina. Barnes describes closely related populations of spiders for the three estuarine communities: (1) *Spartina alterniflora*, (2) *Spartina—Distichlis—Salicornia*, and (3) *Juncus roemerianus*, with differences being largely variations in densities. The greatest densities and species numbers were found in the first zone and dropped off sharply at the proximity of the high–tide mark with the *Juncus* zone having a very sparse population. The low spider densities in the higher two zones may be explained by the structure of the vegetation. The *Spartina—Distichlis—Salicornia* zone was reported by Barnes as often drying out in the summer time, being an area less frequently inundated by the tides. The plants do not grow higher than 45 cm, thus there is less space for web building. *Juncus* grows to 150 cm, but the lack of branching reduces the spaces where spiders can build webs.

Acarina zonation is very marked in the horizontal plane (Luxton, 1964). However, the majority of the truly intertidal Acarina do not possess any morphological adaptations that would fully explain their ability to withstand the unstable conditions of salt–marsh turf, especially in the lower marsh. There are no special respiratory modifications, yet salt–marsh soil oribatids do not seem to suffer unduly from prolonged immersion (up to 12 h in the laboratory). Salinity gradients do not display the same abrupt changes seen in the sharp demarcations of acarine populations. The only abrupt changes come at the limitations of the various tide levels. Thus, it would seem that tidal effects are of fundamental importance in delimiting distributional patterns. In order to establish populations in a marsh, especially the lower reaches, the Acarina need to possess all or some of the following special features: (1) ability to withstand high osmotic pressures in the soils; (2) ability of the eggs and juveniles in particular to withstand the sudden high salinities of tidal inundations; (3) a means whereby the immediate offspring can be protected from the flushing action of tides and rains. Those possessing viviparity have a distinct advantage over egg–layers.

Luxton (1967a, b) identified three main mite communities, one for each vegetation zone. The acarine community in the *Juncus maritimus* zone is essentially terrestrial in nature and origin. The salt meadow fauna seems to consist of two major components: (1) *Festuca rubra* turf (above HWST mark), which can be considered a transitional community with species able to exploit the salt-meadow niche but not able to deal with the rigorous conditions in the lower marsh; and (2) the *Puccinellia maritima* zone (below HWST mark), which is composed of truly haline species that have developed means to withstand the difficult conditions of a tidal marsh.

Luxton (1967b) suggested the egg–laying microarthropod species are controlled primarily by the mechanical effects of tides. He noted that pH is an important regulating factor (either directly or indirectly) on viviparous oribatids as well as the Mesostigmata and Prostigmata, but offered no explanation. There was no significant correlation between tide level, pH, salinity and soil water content on the one hand, and the acarine population densities in the *Festuca rubra* marsh on the other. It is suggested that the landward region is less rigorous since tides rarely reach it. On the other hand, in the *Puccinellia maritima* zone, the acarine populations showed strong correlations with pH and salinity, but most markedly with tide level. The denser oribatid populations were associated with the highest tide levels, while pH had the greatest effect on the Megostigmata and Prostigmata. Luxton reported that correlations between densities of both oribatid and prostigmatid populations on the one hand and both salinity and tide level on the other are independent of the water content of the soil. *Punctoribates quadrivertex*, an egg–laying mite influenced by the tides, was reported most abundant in the *Puccinellia maritima* zone. This suggests that the eggs are laid out of harm's way, possibly in the axils of salt–marsh plants.

CHORDATA

Pisces

Harrington and Harrington (1961) describe the fish feeding activity on a subtropical Florida salt marsh, where an abrupt tidal–pluvial flooding gave fish sudden access to the marsh surface. Prior to the flooding there had been a prolonged drought, driving fish from the marsh and permitting extensive mosquito egg deposition. A mosquito hatch was synchronized with the flooding. Resident marsh fish, the cyprinodontiforms, consumed mosquito larvae and pupae, turning to other things only after the mosquitos were gone. Among the transients (immigrant young of larger species) moving onto the flooded marsh surface, only the mullet *Mugil cephalus* fed to some extent upon mosquitos. This work clearly indicates that flooding tidal waters, by permitting access to the marsh surface for marsh fish species, are important in salt–marsh mosquito control.

The fishes associated with tidal marshes have a very wide tolerance to salinities. Harrington and Harrington (1961) call attention to this when they cite references reporting maximum salinity tolerances for the cyprinodontiform fishes at the levels of 80—90‰. Schmelz (1964), in a study of the mummichog, *Fundulus heteroclitus,* reported salinities up to 41‰ from surface pools on Canary Creek marsh in southern Delaware. Later (1970) he described the eggs of the striped killifish, *Fundulus majalis,* as hatching at salinities of 72—73‰. However, stress was evident about 35‰ in that newly hatched larvae were smaller than those hatched from lower salinities. Warlen (1964) reported the sheepshead minnow, *Cyprinodon variegatus,* from marsh surfaces and drainage ditches of southern Delaware with salinities ranging between 14 and 31‰. Zilberberg (1966) found no correlation between species abundance and salinity for fish in a northwest Florida coastal marsh. All of this demonstrates tidal marsh fishes to be euryhaline.

Ichthyoplankton of tidal streams has been examined on a seasonal basis, along a salinity gradient and over several tidal cycles, for Delaware tidal creeks (Daiber, 1962, 1963a, b). A biweekly night–time mid–flood sampling program was carried out over fifteen months from June 1961 into September 1962. Collections were made for periods of 0.5 h by suspending 30 cm No. 0 mesh nets in the flooding current, surface and bottom. The movement of the tidal front was such that three streams could be sampled successively during a single flooding tide. Another sampling program during the summer of 1962 involved 0.5-h collections each hour over periods of 6, 12, 24, and 48 h using the same nets.

Table 5.5 depicts the numbers of eggs and larvae taken in the three creeks at various dates. The majority of eggs and larvae were collected during the summer months and in the higher salinities of Canary Creek and Little River. This, in light of the fact that the Appoquinimink is a much larger stream with a greater velocity of flow, suggests that more spawning occurs in the higher salinities of the lower half of the bay. Table 5.6 indicates both eggs and larvae were more abundant at the surface during these collections. However, hourly sampling over full tidal cycles showed no consistent pattern. From one hour to the next, or between collection dates, the distribution of eggs and larvae between top and bottom net varied, with a possible tendency for greater numbers in the bottom net. Because of the general turbulence of tidal creeks, any differences between top and bottom may not be as great as it would be in quieter waters.

The several collections of 6, 12, 24, and 48 h duration show the Appoquinimink again had a minor role: the greatest numbers of eggs and larvae were taken from the more saline waters of Canary Creek and Little River (Figs. 5.3—5.5; note the difference in scale used for the three figures). During each collecting period, the numbers fluctuated from one creek to the other. This was also evident in the biweekly collections (Table 5.6). This

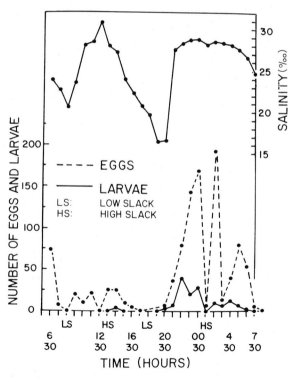

Fig. 5.3. Numbers of fish eggs and larvae (average of top and bottom nets) from Canary Creek, Delaware, during a 24–h period of June 21—22, 1962.

TABLE 5.5

Biweekly plankton collections from three tidal creeks in Delaware from June 1961 into September 1962

Date	Canary Creek			Little River			Appoquinimink Creek		
	salinity (‰)	eggs	larvae	salinity (‰)	eggs	larvae	salinity (‰)	eggs	larvae
25.vi.61		341	0		0	38		1	1
30.vi		160	24		30	46		0	2
17.vii	15.88	33	15	5.21	40	29	3.66	1	20
31.vii	26.22	0	95	12.81	1	72	4.60	0	5
14.viii	27.63	8	13	12.00	2	7	4.36	0	7
28.viii	18.40	3	0	10.72	0	5	4.25	2	7
13.ix		0	1		2	18		0	11
27.ix		0	0		0	1		0	10
11.x		1	0		0	2		0	4
25.x		0	0		0	1		0	1
9.xi		0	0		0	0		0	0
24.xi		0	0		0	0		0	0
9.xii		0	0		0	0		0	0
20.i.62		0	0		0	0		0	0
5.ii		0	5		0	1		0	0
24.ii		0	0		0	2		0	3
23.iii		0	7		0	19		0	7
8.v		7	1		0	53		1	1
25.v		8	0		5	1081		0	20
8.vi	28.40	168	193	11.98	0	179	4.26	0	6
21.vi	27.53	74	14	17.04	1550	65	4.43	0	4
6.vii	28.52	36	5	13.60	0	12	4.98	2	2
19.vii	29.37	7	37	17.76	0	0	6.10	0	8
6.viii		2	2		0	15		0	18
20.viii		0	6		0	30		1	33
5.ix		4	0		0	14		0	3
Totals		852	418		1630	1690		8	173

TABLE 5.6

The vertical distribution of fish eggs and larvae from three Delaware tidal creeks during the mid-flood stage of tide

Date	Canary Creek				Little River				Appoquinimink Creek			
	top net		bottom net		top net		bottom net		top net		bottom net	
	eggs	larvae	eggs	larvae	eggs	larvae	eggs	larvae	eggs	larvae	eggs	larvae
17.vii.61	33	10	0	5	1	22	39	7	1	20	0	0
31.vii	0	62	0	33	0	64	1	8	0	5	0	0
14.viii	3	13	5	0	0	7	2	0	0	7	0	0
28.viii	0	0	3	0	0	4	0	1	2	6	0	1
13.ix	0	1	0	0	0	6	2	12	0	11	0	0
23.iii.62	0	0	6	1	—	—	0	19	0	7	0	0
8.v	2	1	5	0	0	23	0	30	0	5	1	1
25.v	8	0	0	0	5	581	0	500	0	13	0	7
8.vi	79	152	89	41	0	58	0	121	0	5	0	1
21.vi	67	7	7	7	957	36	593	29	0	4	—	—
6.vii	17	0	19	5	0	12	0	0	1	2	1	0
19.vii	2	27	5	10	0	0	0	0	0	8	0	0
6.viii	0	2	2	0	0	9	0	6	0	18	0	0
20.viii	0	6	—	—	0	15	0	15	1	32	0	1
5.ix	4	0	0	0	0	3	0	11	0	3	0	0
Totals	215	279	141	102	963	840	637	759	5	146	2	11

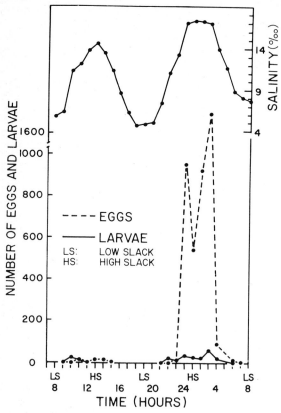

Fig. 5.4. Numbers of fish eggs and larvae (average of top and bottom nets) from Little River, Delaware, during a 24-h period of June 21—22, 1962.

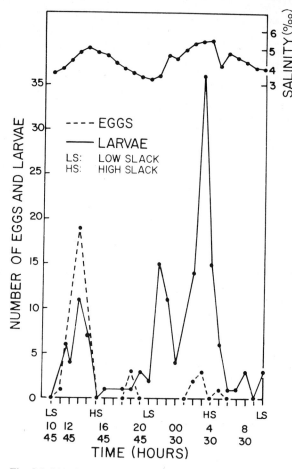

Fig. 5.5. Numbers of fish eggs and larvae (average of top and bottom nets) from Appoquinimink Creek, Delaware, during a 24-h period of June 21—22, 1962.

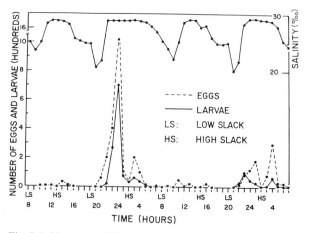

Fig. 5.6. Numbers of fish eggs and larvae (average of top and bottom nets) from Canary Creek, Delaware, during a 48-h period of June 6—8, 1962.

difference in abundance between the two streams suggests that eggs and larvae are "bunched" and enter a creek from the bay only if they are in the immediate vicinity. Even though dispersal undoubtedly occurs, the apparent concentration could result if the eggs and larvae were transported largely within the water mass in which they were spawned or concentrated by eddies and currents.

The eggs and larvae were few in number or even non-existent in the collections made during daylight hours, late stages of ebb, and early stages of the flooding tide (Figs. 5.3—5.6). Their generally reduced numbers during daylight suggests light as a factor in their distribution. Photoperiodism may explain the fluctuation in the numbers of larvae with locomotor capabilities but does not clarify the movement of eggs. One explanation suggests the stage of development may affect buoyancy or a change in numbers. Eggs of the sciaenid *Bairdella icistia* fertilized in low–salinity water (15‰) were larger and more buoyant than eggs fertilized in high salinities (33‰) (R.C. May, 1974). The salinity of the medium during the first 5—7 min had a lasting effect on egg buoyancy. Subsequent transfer to a different salinity influenced buoyancy but the capacity for adjustment was limited. Predation may be an important factor as demonstrated by Moore (1968) in his studies of *Menidia menidia,* the Atlantic silverside, from lower Delaware Bay. He reported this fish to be a daylight feeder with fish larvae making up 4.4% and 18.4% (by weight) of the stomach contents for June—July and August 1967 collections, respectively. Fish eggs composed less than 1% of the total recorded for both time periods. In addition, many eggs were observed in various ctenophores collected at these times.

Salinity generally reached its highest peak near mid–flood on the night tide and then leveled off (Figs. 5.3–5.6). At the beginning of this salinity plateau, there was a very marked increase in the numbers of eggs and larvae for both the top and the bottom nets. At mid–ebb, as the salinity first began to decrease, there was a sharp decline in the numbers of eggs and larvae taken.

The decrease in numbers at slack water before ebb tide was undoubtedly an artificial condition. Water velocity decreases with the approach of slack conditions and the nets hang vertically in the water and do not "fish" well. Thus, there may be no actual decrease in eggs and larvae at this time, as compared to the period prior to and immediately following slack water (Figs. 5.3—5.6).

A series of 6-h flooding–tide collections spanning a month of time were made during the early evening hours or after dark. These data suggest that there was not a uniform number of eggs and larvae present. The disparity in numbers is also demonstrated in Figs. 5.3—5.5 and again in Fig. 5.6. This suggests that spawning had certain peaks of intensity.

Since the movement of planktonic stages of fish is primarily dependent on water flow, it should be possible to ascertain the extent of landward dispersion by determining the penetration of high–salinity water along the axes of tidal streams. During daylight spring tides, flooding waters penetrate 0.5 to 0.75 of the length of small tidal streams such as Canary Creek (4.4 km long). In contrast, high salinities frequently exist the full length of the creek during the night flood tide. On the following ebb, the fresher water found at the head of the creek at high slack has moved downstream beyond the mouth of the creek out into the bay, substantially flushing the high–salinity water from the length of the creek (Daiber, 1963b).

Uniformly distributed high–salinity water penetrates larger streams like Broadkill River 4—6.4 km, not much farther than the total length of Canary Creek. There is a sharp decline in salinity beyond this point, the distance and time of penetration being dependent upon wind velocity and direction and the amount of fresh–water runoff. The flushing time for the lower portions of these tidal streams has been calculated to be half a tidal cycle (DeWitt and Daiber, 1973).

These various observations of Delaware tidal creeks have demonstrated several things. The great majority of the planktonic fish eggs and larvae were present during June and July. Most of them were found in tidal creeks in the lower half of Delaware Bay where salinities are higher. This pelagic phase was not uniformly distributed in space or time; large numbers of eggs and larvae may enter one creek but not another. It is suggested that little spawning occurred in the creeks. The greatest numbers of eggs and larvae entered the creeks on the night flood tide, as the highest salinity was reached for that particular tidal cycle. These eggs and larvae remained with this water mass being carried inland to the extent of high–salinity intrusion and were on their way out of the creeks by the time the salinity began to decline. Invariably there were fewer eggs and larvae taken on the ebb tide than on the flood.

Aves

There is a definite relationship between bird distributions in a salt marsh and the interplay between vegetational zonation, tidal flooding and salinity as it affects feeding and reproductive activities. Urner (1935) identified the avifauna specialized to nest and feed in salt marshes in relation to the marsh wetness, creek and pond depth, salinity and accessibility of the tides. Both he and Ferrigno (1961) called attention to changes in bird populations following any alterations in these various parameters.

The resident clapper rail, *Rallus longirostris waynei,* of the Georgia salt marshes displays distinct distributional preferences within the dominant grass, *Spartina alterniflora* (Oney, 1954). This grass can be divided into three categories. A Tall Grass zone is located on the edge of natural drainage ditches, creeks and rivers with a soft mud substrate and an average plant density of 130 m^{-2}, 120—300 cm tall. The Medium Grass zone is found on the gentle levee slope away from the creek bank, has a firmer soil with an average plant density of 320 m^{-2}, 60—120 cm tall. The Short Grass zone is found in the lowest parts of the Georgia marshes, has a high sand content with an average plant density of 430 m^{-2}, 12—60 cm tall.

The square back marsh crabs, *Sesarma,* the primary food of the rail, were most abundant in the Tall Grass zone. The fiddler crabs, *Uca,* and the periwinkle, *Littorina irrorata,* of secondary and tertiary food value, were found in the Medium and Short Grass zones, respectively. However, during his three-year study, Oney found a definite nesting preference for the Medium Grass zone: of 118 nests measured, 30 were in the Tall Grass zone, 87 (or 74%) were in the Medium Grass zone and only one nest was recorded for the Short Grass zone. Furthermore, his prior studies in 1949 and 1950 indicated the bird's preference for the medium-type grass bordering the Tall Grass zone along a small ditch or creek. The average distance from a nest to the creek at low tide was 140 m with a range from 1 to 360 m. The average nest was 6 m from a change in vegetation zone with a range of 1—24 m. Earlier, Stewart (1951) had called attention to the importance of the edge between vegetation zones in placement of clapper rail nests. Working in the eastern shore marshes of Maryland, he found a high correlation between nest densities and the amount of edge between the tall (and dense) and the short (and sparse) growth form of *Spartina alterniflora.* Lower correlations were derived from pure stands of short and tall *Spartina.* Stewart recorded nest density in the best edge at 6.2 ± 0.7 ha^{-1} with nests within 4.5 m of the creeks.

Several references call attention to the height of nest placement and the impact of inundation on hatching success of the rails. In his early work on the life histories of North American marsh birds, Bent (1963) located the second nestings of the clapper rail, *Rallus longirostris crepitans,* on higher drier ground covered with only a few centimeters of water at high tide. He went on to say that most nests were built in small clumps of grass along creek banks in soft wet mud: nest heights varied 20—30 cm above the mud, probably high enough to escape ordinary high tides but not spring tides. King rail *(Rallus elegans)* nests were usually found in the shallow-water portion of the marsh (10—60 cm of water). The nest height above water was dependent on water depth; the shallower the water, the lower the nest (Meanley, 1969). Most nests of the resident clapper rail of Georgia were 20—23 cm off the ground, the average distance from the ground to egg level being 37 cm with extremes of 22 and 75 cm. These nests could be covered by as much as 30—48 cm of water and still support a hatch (Oney, 1954). Earlier, Stewart (1952) found 1951 hatching success of the first sets of eggs in the Chincoteague, Virginia, marshes to be less than 45% compared to 94% in 1950. Due to high storm tides during the early part of the nesting season, production of young was greatly staggered. Ferrigno (1957) gives us a better picture of storm-tide flooding impact on hatching success (Table 5.7). On May 10, 1955, there had been a high tide with strong winds but little nest damage since the birds

TABLE 5.7

Clapper rail nest census during 1955 and 1956 in Cape May County, New Jersey (from Table 1, Ferrigno, 1957)

Area	Number of observed nests		Number destroyed (by tides)		Number hatched successfully	
	1955	1956	1955	1956	1955	1956
Coney's	29	20	9	4	20	16
Keye's	7	10	3	1	4	9
Total	36	30	12	5	24	25

had just begun to nest. High tides of June 8 to 11 resulted in virtually complete nest destruction. There was subsequent re-nesting as evidenced by a later hatching peak. There were high lunar tides in 1956 but with no wind. Although 14% of the nests were destroyed by a flood tide on June 8, 83% of the first nests hatched successfully in that year.

Changes in vegetational cover can affect clapper rail nesting. Ferrigno (1957) reported a decrease in observed nests on the Coney's area but an increase on the Keye's area, both in Cape May County, New Jersey. He attributed the decrease at Coney's to the development of large barren areas which had been formerly covered with *Spartina alterniflora*. Possible causes were either adverse weather conditions, whatever that means, or over-population of fiddler crabs killing the vegetation by burrowing activities.

The clapper rail has been described primarily as a resident of the more saline low marshes (Stewart, 1951; Oney, 1954). In contrast, the king rail, *Rallus elegans*, is essentially an inhabitant of the fresh and brackish marshes, numbers varying with vegetation (Meanley, 1969). They are common in the coastal marshes of Louisiana, abundant in the South Carolina low-country fresh and brackish marshes, especially where the giant cut grass, *Zizaniopsis miliacea*, and the fiddler crab, *Uca minax*, are present. The grass provides good nesting cover with a nest density as high as 2.5 ha^{-1}. In the Chesapeake Bay area, the king rail is most abundant where the big cord grass, *Spartina cynosuroides*, is dominant, providing good year-round shelter. Those plants found in fresher areas such as *Scirpus olneyi*, *Typha* and wild rice, *Zizania aquatica*, provide cover, although the wild rice breaks down during the winter.

Both rail species occur in transition areas, especially in the lower reaches of brackish marshes where interbreeding sometimes produces hybrids. In Delaware both the king and clapper rails were taken in the Broadway meadows located between Flemings Landing and Woodland Beach. In that section at Taylors Gut where mixed populations occurred, the dominant vegetation was *Spartina alterniflora* and *Scirpus robustus* with the salinity range reported at 5.7—7.2‰. Inland at Flemings Landing the dominant vegetation was *Spartina patens* and *Distichlis spicata* with a salinity range of 3.7—4.4‰ and only the king rail was observed.

Only clapper rails were found at Woodland Beach with *S. alterniflora* and *S. robustus* as the dominant vegetation and a salinity range of 7.5—7.6‰.

While both rails can be identified with the low marshes, Bent (1929) described the eastern willet, *Catoptrophorus semipalmatus*, as a decidedly coastal bird, seldom seen far from coastal marshes, beaches and islands. He described the nesting area as being on sand islands overgrown with tall grass or on dry uplands close to marshes. Vogt (1938) was a bit more specific in his observations in the Fortesque, New Jersey, marshes. He recorded most nests in the dense *Spartina patens* of the high marsh, near the ecotone with *Spartina alterniflora* or *Typha angustifolia*. In addition, Vogt found the willet most abundant where *S. patens* was regularly mowed or burned, presumably making it easier for the birds to feed. This clamorous bird was also found to be abundant where wintering brant or greater snow geese *(Branta bernicla, Chen hyperborea)* had grazed *S. patens* in the absence of eelgrass *(Zostera marina)*. Stewart and Robbins (1958) reported breeding densities of 0.26 ha^{-1} in brackish hay marsh during a 1956 survey of Dorchester County, Maryland. In contrast to earlier observations these workers described the habitat as a strip along the tidal creek 200 m wide.

The casual visitor to a salt marsh seldom sees the secretive rails as they slip through the vegetation but during the first half of the summer a visitor can be announced by the clattering presence of the willet. However, as the summer progresses these clamorous calls decline to silence although the bird can still be seen flying over the expanses of grass or walking along a creek bank. Black ducks *(Anas rubripes)* can be put to flight during the early summer nesting season but waterfowl typically are most obvious during the fall migration flights and over the wintering areas.

Thirteen major types of waterfowl habitat have been categorized by Stewart (1962) in the upper Chesapeake Bay area, six in open tide-water areas, five marsh types and two in the coastal plain interior which were designated as river bottoms and impoundments. The five marsh habitats were distinguished by salinity distributions which in turn had an influence on vegetational composition. Stewart considers the brackish estuarine bays to be the most important habitat for waterfowl populations as a whole. Fresh estuarine bays, brackish

TABLE 5.8

Principal (P) and secondary (S) waterfowl species associated with various types of upper Chesapeake Bay marshes during 1958—59 (derived from Stewart, 1962)

Species		Coastal embayed marsh		Salt estuarine-bay marsh		Brackish estuarine-bay marsh		Fresh estuarine bay marsh		Estuarine-river marsh	
		P	S	P	S	P	S	P	S	P	S
Branta canadensis	(Canada goose)	×			×	×		×			×
Anas rubripes	(black duck)	×		×		×		×		×	
Anser caerulescens	(snow goose)		×								
Anas platyrhynchos	(mallard)		×		×		×		×	×	
Anas carolinensis	(green-wing teal)		×		×	×			×	×	
Anas discors	(blue-wing teal)		×		×	×			×	×	
Spatula clypeata	(shoveler)		×		×		×		×		
Mareca americana	(American widgeon)		×		×		×		×		×
Lophodytes cucullatus	(hooded merganser)		×				×		×		×
Mergus merganser	(common merganser)								×		×
Aythya collaris	(ring neck duck)								×		×
Fulica americana	(American coot)		×						×		×
Anas acuta	(pintail)		×		×		×		×	×	
Anas strepera	(gadwall)				×		×		×		
Olor colombianus	(whistling swan)								×		
Aix sponsa	(wood duck)									×	
		2	9	1	8	4	6	2	11	6	6
			11		9		10		13		12

estuarine marshes and estuarine river marshes also attract large numbers of waterfowl while other habitats attract fewer species. Table 5.8 records those ducks and geese associated with the five tidal marsh types. While Stewart considers the brackish estuarine bays to be the most valuable waterfowl habitat, the marshes bordering such bays had fewer recorded species (10) than the less saline fresh estuarine–bay marshes (13) and estuarine–river marshes (12). The three marshes with one or two designated principal species had higher numbers of secondary species. It is also evident that nine species were restricted to certain marsh habitats whereas seven were recorded from all five types. The black duck was the most ubiquitous, being classed as a principal species for all five marsh habitats. Making an arbitrary selection of those seven species found throughout the marshes and using population estimate size data available from Stewart, Table 5.9 depicts the population distributions for four of the seven species. The widely distributed black duck seems to be more associated with the open bays or agricultural lands during migration and wintering. Large breeding populations were found along the shallow brackish estuarine bays and associated marshes. In contrast, while the recorded population was low (Table 5.9) the great majority (99%) of green–wing teal *(Anas carolinensis)* were found in the tidal marshes. Stewart indicates the same pattern for the blue–wing teal *(Anas discors)*. It should be pointed out that waterfowl numbers within a particular type of habitat vary from location to location.

Vegetation dominance and distribution as well as salinity levels have an influence on waterfowl distribution by providing food and nesting sites. *Spartina alterniflora* is the predominant plant in the coastal embayed salt marshes located back of the barrier beaches. The black duck is the only species that is common and widely distributed while the Canada goose is numerous in local areas. The black duck, mallards and green–wing teal are most numerous along tidal creeks and guts. Canada and snow geese are most numerous on extensive cord grass areas or mud flats. Pintails, shovelers, blue–wing teal and American widgeon

TABLE 5.9

Population distribution of waterfowl associated with various types of upper Chesapeake Bay marshes during 1958—1959 (derived from Stewart, 1962, tables 44, 54, 76, 84)

Species		Total population (× 1000)	Percent of total population[1]				
			coastal embayed marsh	salt estuarine-bay marsh	brackish estuarine-bay marsh	fresh estuarine-bay marsh	estuarine-river marsh
		Area (ha):	8500	45 700	19 000	12 000	27 000
Anas rubripes	(black duck)	317	12	10	3	3	5
Branta canadensis	(Canada goose)	—					
Anas platyrhynchos	(mallard)	148	6	1	3	5	7
Anas carolinensis	(green-wing teal)	10	32	23	13	6	25
Anas discors	(blue-wing teal)	—					
Mareca americana	(American widgeon)	196	8	5	4		—
Anas acuta	(pintail)	—					

[1] The percentage figures are averaged values derived from five observation periods from October 2, 1958 through March 16, 1959. The difference between the total percentage figure for each waterfowl species and 100% indicates the birds were observed on the open waters of the adjoining bays and areas other than in the marshes.

(*Mareca americana*) prefer open ponds with poor drainage or stable water levels in artificially created ponds. Scattered pairs of black ducks along with a few mallards and blue–wing teal nest in these marshes. They prefer marsh islands rather than shore–zone marshes, and all nests found by Stewart (1962) were in the drier more elevated areas.

The salt estuarine–bay marshes are characterized by a high salinity and narrow tidal fluctuation. Widgeon grass, *Ruppia maritima*, salt grass, *Distichlis spicata*, salt–marsh cord grass, *Spartina alterniflora*, salt–meadow cord grass, *S. patens*, the bulrush *Scirpus robustus*, and *Juncus roemerianus* are the common plant species. The black duck is the only common and widely distributed waterfowl. All others are scarce.

The brackish estuarine–bay marshes comprise a complex mosaic of ponds, creeks and marshes. The principal plant species are *Ruppia*, *Distichlis*, *Spartina cynosuroides*, *S. alterniflora*, *S. patens*, *Scirpus olneyi* and *Juncus roemerianus* depending on the locale. There is a diverse fauna with raccoons and crows as important waterfowl predators. As in the coastal marshes, the black ducks and green-wing teal are generally distributed, showing a definite preference for tidal creeks and ponds in drainage systems with marginal mud flats exposed at low tide. The American widgeon and gadwalls (*Anas strepera*) concentrate on stable ponds with beds of widgeon grass or musk grass (*Chara* sp.). The hooded mergansers are restricted to the larger tidal creeks, whereas Canada geese prefer the larger ponds. Large numbers breed in these marshes, with the blue–wing teal restricted almost entirely to the marsh meadow (presumably *Spartina patens*).

The fresh estuarine–bay marshes have next to the smallest total recorded area but a much greater diversity of vegetation. Along with this vegetation variety, Stewart recorded the greatest numbers of waterfowl species, though eleven of the thirteen were classed as secondary (Table 5.8) and most of the population were located on the bays adjoining these fresh estuarine–bay marshes (Table 5.9) rather than in the marshes themselves. As with some of the other marsh types, the Canada geese are found on the larger ponds while the black ducks and teals are found in well–drained areas of creeks and ponds with exposed mud flats.

The vegetation varies with salinity in the estuarine–river marshes of the upper Chesapeake. This is an area with a great variety of emergent vegetation and possesses a greater tidal fluctuation than some other marsh types. The greatest concentrations of waterfowl, including many dabbling ducks, are located here between the fresh– and brackish–water habitats where excellent cover and food is available.

An examination of the numerous tables in Stewart (1962) depicting waterfowl food habits in the upper portion of Chesapeake Bay discloses a relationship between the horizontal distribution of salinity and vegetation types and food habits. A greater variety of food species were consumed as salinities decreased from one marsh type to another. Presumably this is a reflection of increased diversity and availability associated with less saline areas. However, there may be more than availability playing a part in the food habits of waterfowl associated with tidal marshes. By using two criteria, timing of usage and relative quantity of food per unit area remaining after usage, Owen (1971) ascertained the winter feeding habits of white-fronted geese *(Anser albifrons)* in England to be highly selective in the choice of vegetation zone. By both criteria, the *Agrostis* zone, which is the lowest grassy area and dominated by *Agrostis stolonifera*, was preferred to all others. *Lolium perenne, Festuca rubra, Hordeum secalinum* and *Juncus gerardi* are found at higher elevations, and were consecutively grazed later in the season. Selection of a zone was not absolute, however; when large numbers of geese were present all zones were occupied. Owen presents evidence that plants may be selected on basis of protein, fiber and carbohydrate content. If so, *Puccinellia maritima* was considered to be preferred by white-fronted geese over *Agrostis* and *Festuca*.

In addition to those birds already described, tidal flooding and marsh elevation influence the distributions and activities of numerous other bird species. The laughing gull, *Larus altricilla*, placed most of its nests where the marsh was elevated from 0 to 0.2 m above mean high water with *Spartina alterniflora* taller than 0.6 m. Some nests were found in shorter vegetation and at lower elevations. The gulls would not nest where the grass had been cut though they had nested at the same site in previous years. They would nest on piles of dead grass brought together by tidal action or by raking (Bongiorno, 1970).

Adequate feeding grounds and suitable nesting cover within easy flight of the feeding grounds provide two components of a suitable habitat for Macgillivray's seaside sparrow, *Ammospiza maritima macgillivraii* (Tomkins, 1941). This is the breeding seaside sparrow of South Carolina and Georgia, and in this locale these two requirements are often separated by a short distance. Food requirements are far stronger than nesting needs in determining habitat limitations, and population shifts are caused by erosion of feeding sites. Feeding grounds are the wet banks of salt creeks where *Spartina alterniflora* grows rankest, the ponds that are at the heads of salt creeks and the patches of *S. alterniflora* located on the outer beaches that are flooded by tides. Tidal flooding of the feeding grounds makes them unsuitable for nesting sites. Nests may be built in a variety of places; a few centimeters above the mud in *Sporobolus—Paspalum*, 1 m in *Spartina* or *Juncus*, and up to 1.5 m in *Baccharis*. Nesting preferences appear to follow this same order with the taller shrubbery the least desirable site.

Much attention has been focused on the song sparrows *(Melospiza melodia)* of the salt marshes of the San Francisco Bay region (Marshall, 1948; Johnston, 1956a, b). These investigators discussed habitat, abundance, annual cycles, population structure and maintenance, and concluded that tidal flooding and vegetational zonation were the determining influences. Song sparrow distributions are more circumscribed than vegetational distributions in that the birds are less tolerant of drying (Marshall, 1948). *Scirpus acutus* and *Typha latifolia* are dominant plants in the brackish marshes comprising a large portion of the region. *Scirpus californicus* and *Typha latifolia* grow upstream beyond the tidal flow while *Scirpus acutus*, being more tolerant of salt, grades into the *Spartina* (largely *S. foliosa*) marshes. There was no interruption of linear sequence of breeding song sparrows through these transitions, and birds of any intermediate zone did not sort out into different plant associations in the brackish marshes (Marshall, 1948).

Both Marshall (1948) and Johnston (1956a, b) noted that the song sparrow occupies territories strung out along the tidal sloughs, exercising definite preferences for vegetation types and not dispersing over the marsh surface. Marshall calls attention to differences in territorial spacing. Sparrow pairs were about 44 m apart along the bayshore and about 48—62 m between pairs along the brackish marsh sloughs, depending on the width of fringe plants, 13 and 4.5 m respectively. The birds used the tallest *Scirpus acutus* in the centers of patches of this species for song and calling perches. Plants on levees were visited, but territory headquarters were always at slough margins. Flooded areas behind these levees, or any stagnant areas where tidal flow had been cut off, were avoided, as were low, dense-growing stands of *Scirpus campestris*. The birds would visit drained *Salicornia* patches but did not establish territories in such locations.

Marshall (1948) goes on to describe the relationship between song sparrow distributions and the salt-marsh vegetation dominated by *Spartina foliosa, Salicornia ambigua* and *Grindelia cuneifolia. Spartina* is found at the lowest elevations, *Salicornia* is covered only by the highest tides while *Grindelia* grows on the levees and elevated banks of the sloughs in the *Salicornia* zone. Song sparrow pairs were spaced at 60 m, single file, along each bank in the *Spartina* marsh (Marshall, 1948) with an average territory width of 9 m (Johnston, 1956b). In the *Salicornia—Grindelia* marsh, pairs were 25—90 m apart, being closer together where the *Grindelia* was widest (Marshall, 1948). Johnston (1956b) seems to disagree somewhat since he found larger territories at the heads of the sloughs where the height and amount of vegetation was reduced, especially *Grindelia,* than in the mud marsh with its lusher vegetation. He describes the *Spartina* zone territories as being larger than those in the *Salicornia* and, during periods of high numbers, actual densities were 20—25 pairs per ha.

The song sparrows utilize the tallest *Spartina* plants for singing perches. They are vertically limited by vegetation height, being absent from *Spartina* less than 45 cm high. They utilize the tops of the *Grindelia* bushes in the *Salicornia* marsh, finding concealment under the vegetation or the characteristic overhanging banks of the sloughs. They forage on the ground in such locations having dropped down through a *Grindelia* plant from a singing perch (Marshall, 1948).

The breeding season is influenced by tidal heights (Johnston, 1956a) with the salt-marsh song sparrow *(Melospiza melodia maxillaris)* breeding about two weeks earlier than the upland form. This seems to be an adaptation to marsh life, the birds nesting mainly during the lower tidal conditions existing in March. Late nests suffered high mortality from the extended high run of tides in April to June. Nests were placed off the ground, most commonly in *Salicornia, Grindelia, Distichlis* or *Spartina.* Any nest less than 12 cm above the ground would be flooded out. Average nest height above ground was 24 cm for the whole marsh and 30 cm in the low marsh. The birds usually tried to put the nests in the highest vegetation but not always. Johnston is of the impression that predators may exert some selective pressure on the upper limit of nest height. No nests were used more than once, nests of the season being scattered about in the territory. Generally an increase in vegetation height during the growing season enabled the birds to nest at successively higher levels. Such action paralleled the increased height of tides during the breeding season.

Marshall (1948) described the song sparrow to be the only ground foraging bird of the San Francisco Bay salt marshes. In fact, the species is limited to areas covered by tides where flows are unimpeded. Both Marshall and Johnston stressed that song sparrow distributions are limited by vegetation, tidal water, accessibility of the ground for foraging, habitat selection and the sedentary nature of the populations. Such characteristics favor the maintenance of integrity and distinctness of song sparrow populations in the salt marshes of the San Francisco Bay region.

Sibley (1955) described the responses of salt-marsh birds to extreme tides in the San Francisco Bay area at times other than nesting periods. White-crowned, savannah and song sparrows *(Zonotrichia leucophrys, Passerculus sandwichensis* and *Melospiza melodia)* and long-billed marsh wrens *(Telmatodytes palustris)* were concentrated along the levees. When flushed they did not fly out over the flooded marsh. Willets *(Catoptrophorus* spp.*)* and the least and western sandpipers *(Erolia minutilla* and *Ereunetes mauri)* settled on floating debris. The clapper, sora and Virginia rails *(Porzana carolina* and *Rallus limicola),* normally not seen at low or average tidal conditions, were quite evident clinging to *Spartina* stems or huddled in clumps of emergent vegetation. Predators avail themselves of these concentrations as California and ring-billed gulls *(Larus californicus* and *L. delawarensis),* marsh hawks *(Circus cyaneus),* and short-eared owls *(Asio flammeus)* were reported hunting along the levees and over clumps of emergent vegetation.

There appears to be some relationship between avian distributions in tidal marshes and the ability to utilize saline water. Very few terrestrial species of birds can drink sea or estuarine water (Bartholomew and Cade, 1963, p. 518). Bartholomew and associates (Cade and Bartholomew, 1959; Poulson and Bartholomew, 1962; Bartholomew and Cade, 1963) demonstrated subspecific differences in this ability among the savannah sparrows, *Passerculus sandwichensis*. Two west coast subspecies, *P. s. beldingi* and *P. s. rostratus*, are restricted to salt marshes. Three subspecies that have typical avian salt–water responses are migratory and nest in or near fresh-water habitats. The two salt–marsh subspecies handle the salt response in different ways; *P. s. rostratus* reduces drinking or may go without water for long periods, whereas *P. s. beldingi* drinks large quantities of salt water. *P. s. rostratus* has the most clearly defined salt response and is the most sharply distinguishable subspecies. *P. s. beldingi* shows an overlap of response with the migratory forms. It is capable of consuming large quantities of distilled water in captivity. It can also maintain a fairly constant salt–water consumption up to a $0.6M$ NaCl salinity (100% sea water) representing 84–112% of body weight. At higher concentrations ($0.7M$ NaCl) consumption declined to 75% of body weight. The salt–marsh races of the song sparrow, *Melospiza melodia maxillaris* and *M. m. samuelis*, display a similar pattern of consumption but they cannot maintain their body weight on full–strength sea water. These various authors presume the salt–marsh subspecies get most of their water needs from water condensing on the marsh vegetation from the frequent coastal fogs of the San Francisco area.

The long–billed marsh wren, *Telmatodytes palustris griseus*, is a permanent resident of the marshes from South Carolina to northern Florida (Kale, 1967). During the breeding season, it is found only in the tall *Spartina alterniflora* where fresh water rarely occurs except during a heavy rain. Dew does wet the plants but it is very salty. When birds were supplied with estuarine water, fresh water or none there was no loss of weight. Instead of adjusting to saline water, the wrens changed food habits by turning to more succulent foods when estuarine water or no water was offered.

Mammalia

A number of small mammals have been associated with the fringes of salt marshes (Paradiso and Handley, 1965). Shure (1970) found a definite relationship between small–mammal distributions and the topographically controlled pattern of barrier beach vegetation of a New Jersey marsh habitat. The meadow mouse *Microtus pennsylvanicus* was the most abundant, particularly associated with the dense herbaceous cover of the marshes of the bay shore dominated by *Spartina patens*. *Zapus hudsonius*, the meadow jumping mouse and *Peromyscus leucopus*, the white–footed mouse, were fairly abundant. The masked shrew, *Sorex cinereus*, and the house mouse, *Mus musculus*, were taken occasionally [1].

All of these are inconspicuous marsh fringe inhabitants. The muskrat, *Ondatra zibethica*, is a much more obvious animal, conspicuous by its houses dotting the marsh surface and its aquatic runs among the vegetation. Much has been written about its activities both in fresh– and brackish–water marshes. Ditching for mosquito control which has lowered the water levels is reported to have an adverse effect on vegetation needed for muskrat food and house construction (Stearns, et al., 1939, 1940). Important muskrat foods such as *Scirpus olneyi* and *Spartina cynosuroides* have been replaced by species such as *Hibiscus oculiroseus* (marsh mallow), *Kosteletzkya virginica* (salt–marsh mallow), *Solidago sempervirens* (seaside golden rod), *Bidens trichosperma* (tick seed sunflower) and *Aster novi–belgii*, presenting a brilliant picture but of no value to the muskrat (Fig. 5.7).

Muskrat food and populations decrease as salinity increases (Harris, 1937; Dozier, 1947; Dozier et al., 1948). The muskrats prefer the less saline types of vegetation such as the three–square sedges, *Scirpus olneyi, S. robustus, S. americanus*, and the cat–tails, *Typha*. Less favored foods are *Spartina cynosuroides, S. alterniflora, S. patens, Distichlis spicata* and *Juncus roemerianus*. Dozier et al. (1948) reported that the heaviest muskrats, with average weights of 1.02 and 1.03 kg, consumed plants of Group I (*Scirpus olneyi, S. americanus, Typha*; salinity 0—5‰) and Group II (*S. olneyi*,

---

[1] In Norfolk rabbits from the dunes used regularly to feed upon the marsh grasses (Editor).

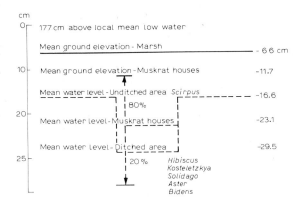

Fig. 5.7. Water levels in ditched and unditched tidal marshes as related to muskrat houses and vegetation (from Stearns et al., 1939).

Typha; 5—10‰). Those feeding in higher-salinity areas on plants of Group IV (*S. patens*, *S. alterniflora*, *S. olneyi* and *Typha*; 15—25‰) and Group V (*S. patens*, *S. alterniflora* and *Juncus roemerianus*; 25—43‰) had average weights of 0.98 and 1.00 kg, respectively. On the surface, this weight differential does not appear significant. However, even causal observation reveals the divergence in plant distributions and numbers of muskrat houses along a salinity gradient. Dozier (1947) made a special point about the impact of salinity on tide-water muskrat production. The animals were found in the upper reaches of tidal streams and were abundant where tidal influence was reduced. Diked areas or marshes that were flooded by high-salinity water during storms or through evaporation tended to have reduced muskrat populations through loss of food plants and drinking water.

Palmisano (1972) portrayed the interaction between vegetation types, salinity, extent of flooding and drought on the distribution and abundance of the muskrat in Louisiana coastal marshes. Palmisano recognized four major plant communities: (1) saline marshes adjacent to the Gulf of Mexico dominated by relatively few salt-tolerant species; (2) brackish marshes forming a broad zone of moderate salinity where plant growth was vigorous; (3) intermediate, slightly brackish marshes; and (4) fresh-water marshes. Muskrats occurred in all the coastal marshes examined; however, population densities varied greatly. Brackish marshes composed of a mixed community of *Scirpus olneyi* and *Spartina patens* were a preferred habitat (Table 5.10). Although Palmisano recorded approximately equal percentage values for southeastern and southwestern Louisiana, population densities were much higher in the southeast. The saline marshes in the southeast contained populations equal to overall average density. The saline marshes of the southwest were poor muskrat habitat, possibly because of their well-drained nature. The intermediate marshes had below-average population densities but high or average levels in restricted areas adjacent to the brackish marshes. The fresh-water marshes exhibited the lowest densities of any type; 31.4% of the total area examined supported only 4.1% of the total houses counted. Populations were recorded at their highest during periods of high precipitation and low salinity. The greatest house counts were

TABLE 5.10

Area of marsh vegetative types in coastal Louisiana, percent of area surveyed and percent of muskrat (*Ondatra zibethica*) houses counted (modified from tables 2, 3, 4, Palmisano, 1972)

Marsh type	Southwestern			Southeastern			Total		
	area (ha)	percent surveyed		area (ha)	percent surveyed		area (ha)	percent surveyed	
		area	houses		area	houses		area	houses
Saline	18 417	2.6	1.0	304 792	22.2	20.8	323 208	12.4	14.2
Brackish	187 756	37.5	70.3	292 275	36.1	70.0	480 031	36.8	72.6
Intermediate	149 214	31.5	25.0	114 561	7.2	3.7	263 775	19.4	9.1
Fresh	158 139	28.4	3.7	336 212	34.5	5.5	494 352	31.4	4.1
Total	513 526	100.0	100.0	1 047 840	100.0	100.0	1 561 366	100.0	100.0

recorded in February when water levels were generally high, temperatures low and the spring breeding season about to begin.

Palmisano reported wide population fluctuations in localized areas, with rapid increases for three to four years followed by sharp declines to almost zero in a few months. Such a decline was observed in the brackish marshes in December 1971 following two dry summers. The other marsh types were not as severely affected by the drought, with populations actually increasing in a few locations. These marshes served as important reservoirs for muskrats during periods of stress in the brackish marshes. In December 1971, when populations were below normal, non–brackish marshes accounted for over 50% of the houses recorded. During normal years less than 20% were recorded outside the brackish–marsh zone.

Mammals respond to storm tides on a salt marsh much as do the birds; Norway rats *(Rattus norvegicus)* and meadow mice are often found in the same clumps of grass with birds (Sibley, 1955). Extremely high tides that drive small mammals from cover occur during the winter months in the San Francisco Bay area (Johnston, 1957; Fisler, 1965). At such times only 10—12 cm of vegetation remain exposed, mostly *Grindelia cuneifolia*, a woody perennial growing along the elevated banks of tidal sloughs. During flooding conditions, these mammals use the emergent vegetation, floating debris, or move to higher ground. Since high ground is some distance from centers of mammal populations, most animals use emergent vegetation or floating debris. Some are lost and a number of young are drowned in the spring (Johnston, 1957). Swimming ability plays a part. The shrew *Sorex vagrans* swims and dives well and most nests are on high ground more than 2 m above mean sea level. The meadow mouse *Microtus californicus* swims and dives well (Johnston, 1957) but prefers to hide in what emergent vegetation exists (Fisler, 1965). Since the mouse does not normally move out of its home range during and after a storm tide, only a rare combination of factors would reduce or displace a population. Fisler found few mice in the *Spartina foliosa*, more in *Salicornia ambigua* and the largest concentrations in the *Grindelia* growing on the levees. This preference for high ground by *M. californicus* agrees with the association between *Microtus pennsylvanicus* and high–marsh *Spartina patens* in New Jersey (Shure, 1970). However, Johnston found *Microtus* nests at all levels of the marsh surface, the breeding season occurring only during extreme spring tides. The Norway rat was described by Johnston (1957) to be an excellent swimmer and diver, but suffering considerable losses of young because its nests are located on the ground. In contrast, the harvest mouse, *Reithrodontomys raviventris,* is a poor swimmer; its fur wets easily and it does not take to the water readily. It normally nests above ground in old sparrow nests and Johnston did not observe any nests being flooded out.

As with some salt–marsh song birds, there appears to be a relationship between small–mammal distributions and their ability to utilize salt water in a tidal marsh. There are three possible sources of drinking water for a harvest mouse: (1) dew, (2) juices of succulent plants such as *Salicornia* with a water content of 88%, and (3) sea water. The harvest mouse is restricted to the brackish and salt marshes of San Francisco Bay. One subspecies, *R. r. halicoetis,* is better able to survive higher salt concentrations and greater dehydration than related forms *R. r. raviventris* and *R. megalotis* but none can survive indefinitely. Food and water consumption are greatly curtailed by drinking sea water (Fisler, 1963). In contrast to Fisler's observations, MacMillen (1964) found *Reithrodontomys megalotis* can readily utilize undiluted sea water for drinking, by virtue of an extreme urine concentrating capacity. This mechanism is probably an adaptation for aridity, since this species is reported by MacMillen to be primarily distributed inland and thus preadapted for drinking sea water. Haines (1964) found the harvest mouse drank 19% of its body weight or 2.4 g day^{-1}. When water was restricted, the animal could get along on 0.8 g day^{-1} and Haines was of the opinion that atmospheric dew could provide the 0.8 g day^{-1} to meet metabolic needs. The vole *Microtus pennsylvanicus* relies on snow, rain or dew for water (Getz, 1966) since it is unable to tolerate a salinity above 0.30M NaCl or approximately half that of sea water (0.56M NaCl = 35‰). Getz investigated the water content and salinity of some marsh plants as possible sources of drinking water. He concluded that the voles could get useable water from green *Spartina patens* if other sources were unavailable. *Spartina alterniflora* had a salinity of approximately $0.35M$ NaCl, which would make its tissue water unsuitable.

MANGROVE ANIMALS

Accounts of the various animals associated with mangroves will be found in relevant chapters of this book (Chapters 1, 8—12) as well as in a major survey by MacNae (1968).

DISCUSSION AND CONCLUSIONS

Salt—marsh animals find themselves in a particularly harsh intertidal zone. They must possess structural, physiological or behavioral capabilities that enable them to adjust to or avoid wide-ranging levels of salinity, temperature, humidity, desiccation and inundation. They must also adapt to the physical and chemical nature of the substrate. These parameters have both direct and subtle effects on the animals' ability to acquire food and shelter as well as on reproduction and care of the young.

Few species have tolerance limits broad enough to accommodate such variable conditions. Many of the papers cited (Phleger, 1970; Murray, 1973; Kraeuter and Wolf, 1974) call attention to this. Where conditions are less harsh, or a greater variety of habitats exist, a larger number of species can be identified as associated with tidal wetlands (Barnes, 1953; Davis and Gray, 1966; Stewart, 1962). As a general rule, more species can be found in those shallow water areas seldom exposed to the air and those terrestrial regions seldom inundated even by storm tides than in the salt marshes proper.

The animals of a salt marsh can be divided into several categories (Nicol, 1936; Teal, 1962). There are those found in the creeks and lower intertidal areas whose centers of distribution are located elsewhere in the estuarine environment. At the other extreme, there are essentially terrestrial animals that live part or all of their lives near the upland margins. In between can be found the few marsh animals associated with the marsh surface itself. Some, like the fiddler crabs and ribbed mussel, still retain direct links to the aquatic environment through planktonic larvae. As Teal (1962) points out, few of the terrestrial animals contribute much to the general marsh economy. Rather, it is the aquatic species which are largely responsible for energy transfer via the detritus food chain.

Some animal species of the salt marshes have close allies in various other ecological habitats, such as the Acarina (Luxton, 1967a, b), the Foraminifera (Phleger, 1970), the Insecta (Davis and Gray, 1966) or the song sparrows (Marshall, 1948). However others, including fiddler crabs, *Melampus* and the clapper rail, are unique to the tidal marshes. While certain animals are endemic to salt marshes, they can have very wide geographic distributions (Foraminifera: Phleger, 1970; *Uca*: Pearse, 1914). Conversely, some species or subspecies, although still endemic to salt marshes, can have very restricted geographic distributions. The song sparrows (Johnston, 1956a, b) and harvest mice (Fisler, 1963; MacMillen, 1964) of the San Francisco Bay marshes, or the Georgia subspecies of clapper rail (Oney, 1954), illustrate this point.

Zonation within a marsh is largely determined by salinity gradients, degree of inundation and character of the substrate. The interaction of these influences is often complex but each one may be the limiting factor for a specific organism. Salinity affects the distribution of a wide range of species including Foraminifera, fiddler crabs, fish eggs and larvae, marsh birds and muskrat. It acts directly on the osmotic regulatory device of some animals through immersion, and on others through the drinking mechanism. The extent of inundation contributes another limitation which acts on vegetative zonation — thus, in turn, determining animal distributions through the location and kinds of food, and the availability of nest and home sites. Studies of the ribbed mussel, coffee bean snail, clapper rail, savannah sparrow and muskrat illustrate this phenomenon. Sediment type and particle size of the substrate influence distribution of burrowing animals. This is particularly important for the fiddler crabs which dig extensive burrows in the mud.

Marsh animals have developed a variety of ways to survive in the intertidal area of a salt marsh. The behavior of fiddler crabs has been correlated with tidal stage (Pearse, 1914; Teal, 1959). All species dig burrows to escape desiccation and some even plug the openings against the flooding tide. The ribbed mussel's ability to air-gape has enabled it to invade the intertidal area; but physical stresses and body plan will prevent further penetration (Lent, 1969). Some of the mites have penetrated outward into the intertidal zone by developing a viviparous reproduction. Egg–laying forms insure some success by depositing their eggs in crevices and crannies of the vegetation where tidal waters will not dislodge them (Luxton, 1967b). Many insects and *Melampus,* the coffee bean snail, escape inundation by crawling up grass stems, while some ants live inside the grass stems (Davis and Gray, 1966; Russell–Hunter et al., 1972). The timing of the nesting season enables the salt marsh song sparrow to escape the high waters of the seasonal spring tides. This species also places its nest off the ground to avoid the ordinary high tide as do the clapper and king rails and the marsh wren (Johnston, 1956a; Meanley, 1969; Oney, 1954). Excessive spring or storm tides still destroy many nests and young, but the high reproductive potential and re–nesting habits enable these species to recoup such losses. This is equally true for the Norway rat which nests directly on the ground (Johnston, 1957).

Several species have adaptations to overcome the lack of fresh drinking water. The marsh wren turns to more succulent foods for its needs (Kale, 1967). The harvest mouse and some subspecies of song sparrows have developed osmotic capabilities for handling salt water (MacMillen, 1964; Cade and Bartholomew, 1959). A few birds abstain.

REFERENCES

Apley, M.L., 1970. Field studies on life history, gonadal cycle and reproductive periodicity in *Melampus bidentatus* (Pulmonata: Ellabiidae). *Malacologia*, 10(2): 381—397.

Apley, M.L., Russell–Hunter, W.D. and Avolizi, R.J., 1967. Annual reproductive turnover in salt marsh pulmonate snail, *Melampus bidentatus. Biol. Bull.*, 133(2): 455—456 (abstract).

Barnes, R.D., 1953. The ecological distribution of spiders in non–forest maritime communities at Beaufort, North Carolina. *Ecol. Monogr.*, 23(4): 315—337.

Bartholomew, G.A. and Cade, T.J., 1963. The water economy of land birds. *Auk*, 80: 504—539.

Bent, A.C., 1929. Life histories of North American shore birds. *U.S. Natl. Mus. Bull.*, 146(I—IX): 1—412.

Bent, A.C., 1963. Life histories of North American marsh birds. *U.S. Natl. Mus. Bull.*, 135: 490 pp.

Bongiorno, S.F., 1970. Nest–site selection by adult Laughing Gulls *(Larus atricilla). Anim. Behav.*, 18: 434—444.

Borror, A.C., 1965. New and little–known tidal marsh ciliates. *Trans. Am. Microsc. Soc.*, 84(4): 550—565.

Bourn, W.S. and Cottam, C., 1950. Some biological effects of ditching tidewater marshes. *U.S. Fish Wildl. Serv., Res. Rep.*, No. 19: 17 pp.

Cade, T.J. and Bartholomew, G.A., 1959. Sea water and salt utilization by Savannah sparrows. *Physiol. Zool.*, 32(4): 230—238.

Chapman, V.J., 1960. *Salt Marshes and Salt Deserts of the World.* Hill, London, 392 pp.

Chapman, V.J., 1974. *Salt Marshes and Salt Deserts of the World.* Cramer, Lehre, 2nd ed., 392 pp. (complemented with 102 pp.).

Connell, W.A., 1940. Tidal inundation as a factor limiting distribution of *Aedes* spp. on a Delaware salt marsh. *Proc. N.J. Mosq. Exterm. Assoc.*, 27: 166—177.

Crichton, O.W., 1960. Marsh crab: Intertidal tunnel–maker and grass eater. *Estuarine Bull.*, 5(4): 3—10.

Daiber, F.C., 1962. Role of the tide marsh in the lives of salt water fishes. *Del. Board Game Fish Comm., Ann. D.–J. Rep.*, F–13–R–3: 25 pp. (mimeograph).

Daiber, F.C., 1963a. The role of the tide marsh in the lives of salt water fishes. *Del. Board Game Fish Comm., Ann. D.–J. Rep.*, F–13–R–4: 22 pp. (mimeograph).

Daiber, F.C., 1963b. Tidal creeks and fish eggs. *Estuarine Bull.* 7(2,3): 6—13.

Daiber, F.C., 1974. Salt marsh plants and future coastal salt marshes in relation to animals. In: R.J. Reimold and W.H. Queen (Editors), *Ecology of Halophytes.* Academic Press, New York, N.Y., pp. 475—510.

Daiber, F.C. and Crichton, O., 1967. Caloric studies of *Spartina* and the marsh crab *Sesarma reticulatum* (Say). *Ann. Pittman–Robertson Rep., Del. Board Game Fish Comm., Proj. W–22–R–2*, Job No. 4, 20 pp.

Davis, L.V. and Gray, I.E., 1966. Zonal and seasonal distribution of insects in North Carolina salt marshes. *Ecol. Monogr.*, 36(3): 275—295.

DeWitt, P. and Daiber, F.C., 1973. The hydrography of the Broadkill River estuary, Delaware. *Chesapeake Sci.*, 14(1): 28—40.

Dexter, R.W., 1942. Notes on the marine mollusks of Cape Ann, Massachusetts. *Nautilus*, 56(2): 57—61.

Dexter, R.W., 1944. Annual fluctuations of abundance of some marine mollusks. *Nautilus*, 58(1): 20.

Dexter, R.W., 1945. Zonation of the intertidal marine mollusks at Cape Ann, Massachusetts. *Nautilus*, 58(2): 56—64.

Dozier, H.L., 1947. Salinity as a factor in Atlantic Coast tide water muskrat production. *Trans. N. Am. Wildl. Conf.*, 12: 398—420.

Dozier, H.L., Markley, M.H. and Llewellyn, L.M., 1948. Muskrat investigations on the Blackwater National Wildlife Refuge, Maryland, 1941—1945. *J. Wildl. Manage.*, 12(2): 177—190.

Ferrigno, F., 1957. Clapper rail study. In: J.W. Aldrich et al. (Editors), *Investigations of Woodcock, Snipe, and Rails in 1956. U.S. Fish Wildl. Serv., Spec. Sci. Rep., Wildl.*, 34: 81—85.

Ferrigno, F., 1958. A two–year study of mosquito breeding in the natural and untouched salt marshes of Egg Island. *Proc. N.J. Mosq. Exterm. Assoc.*, 45: 132—139.

Ferrigno, F., 1959. Further study on mosquito production on the newly acquired Caldwalder Tract. *Proc. N.J. Mosq. Exterm. Assoc.*, 46: 95—102.

Ferrigno, F., 1961. Variations in mosquito—wildlife associations on coastal marshes. *Proc. N.J. Mosq. Exterm. Assoc.*, 48: 193—203.

Fisler, G.F., 1963. Effects of salt water on food and water consumption and weight of harvest mice. *Ecology*, 44(3): 604—608.

Fisler, G.F., 1965. Behavior of salt marsh *Microtus* during winter high tides. *J. Mammal.*, 42: 37—43.

Gerry, B.I., 1950. Salt marsh fly control as an adjunct to mosquito control in Massachusetts. *Proc. N.J. Mosq. Exterm. Assoc.*, 37: 189—193.

Getz, L.L., 1966. Salt tolerance of salt marsh meadow voles. *J. Mammal.*, 47(2): 201—207.

Gray, E.H., 1942. Ecological and life history aspects of the red-jointed fiddler crab, *Uca minax* (Le Conte), region of Solomons Island, Maryland. *Md. Board Nat. Resour., Dep. Res. Educ.*, 51: 3—20.

Hackney, A.G., 1944. List of mollusca from around Beaufort, North Carolina, with notes on Tethys. *Nautilus*, 58(2): 56—64.

Haines, H., 1964. Salt tolerance and water requirements in the salt-marsh harvest mouse. *Physiol. Zool.*, 37(3): 266—272.

Hansens, E.J., 1952. Some observations on the abundance of salt marsh greenheads. *Proc. N.J. Mosq. Exterm. Assoc.*, 39: 93—98.

Harrington Jr., R.W. and Harrington, E.S., 1961. Food selection among fishes invading a high subtropical salt marsh: from onset of flooding through the progress of a mosquito brood. *Ecology*, 42(4): 646—666.

Harris, E.S., 1937. Muskrat culture and its economic significance in New Jersey. *Proc. N.J. Mosq. Exterm. Assoc.*, 24: 20—25.

Hauseman, S.A., 1932. A contribution to the ecology of the salt-marsh snail, *Melampus bidentatus* Say. *Am. Nat.*, 66: 541—545.

Holle, P.A., 1957. Life history of the salt marsh snail *Melampus bidentatus* Say. *Nautilus*, 70: 90—95.

Jamnback, H. and Wall, W.J., 1959. The common salt marsh Tabanidae of Long Island, New York. *Bull. N.Y. State Mus.*, No. 375: 77 pp.

Johnston, R.F., 1956a. Population structure in salt marsh song sparrows. I. Environment and annual cycle. *Condor*, 58: 24—44.

Johnston, R.F., 1956b. Population structure in salt marsh song sparrows. II. Density, age structure, and maintenance. *Condor*, 58: 254—272.

Johnston, R.F., 1957. Adaptation of salt marsh mammals to high tides. *J. Mammal.*, 38(4): 529—531.

Kale II, H.W., 1967. Water sources of the long-billed marsh wren in Georgia salt marshes. *Auk*, 84(4): 589—591.

Kerwin, J.A., 1971. Distribution of the fiddler crab (*Uca minax*) in relation to marsh plants within a Virginia estuary. *Chesapeake Sci.*, 12(3): 180—183.

Kerwin, J.A., 1972. Distribution of the salt marsh snail (*Melampus bidentatus* Say) in relation to marsh plants in the Poropotank River area, Virginia. *Chesapeake Sci.*, 13(2): 150—153.

Kraeuter, J.N. and Wolf, P.L., 1974. The relationship of marine macroinvertebrates to salt marsh plants. In: R.J. Reimold and W.H. Queen (Editors), *Ecology of Halophytes*. Academic Press, New York, N.Y., pp. 449—462.

Kuenzler, E.J., 1961. Structure and energy flow of a mussel population in a Georgia salt marsh. *Limnol. Oceanogr.*, 6(2): 191—204.

Lent, C.M., 1967a. *Effects and Adaptive Significance of Air-Gaping by the Ribbed Mussel, Modiolus (Arcuatula) demissus (Dillwyn)*. Dissertation, University of Delaware, Newark, Del., 77 pp.

Lent, C.M., 1967b. Effect of habitat on growth indices in the ribbed mussel, *Modiolus (Arcuatula) demissus. Chesapeake Sci.*, 8(4): 221—227.

Lent, C.M., 1968. Air-gaping by the ribbed mussel, *Modiolus demissus* (Dillwyn): effects and adaptive significance. *Biol. Bull.*, 134: 60—73.

Lent, C.M., 1969. Adaptations of the ribbed mussel, *Modiolus demissus* (Dillwyn) to the intertidal habitat. *Am. Zool.*, 9: 283—292.

Luxton, M., 1964. Some aspects of the biology of saltmarsh Acarina. *Acarologia C.R. 1er Congr. Int. Acarol., 1963*, pp. 172—182.

Luxton, M., 1967a. The ecology of salt marsh Acarina. *J. Anim. Ecol.*, 36(2): 257—277.

Luxton, M., 1967b. The zonation of salt marsh Acarina. *Pedobiology*, 7: 55—66.

MacDonald, K.B., 1969. Quantitative studies of salt marsh faunas from the North American Pacific coast. *Ecol. Monogr.*, 39(1): 33—60.

MacMillen, R.E., 1964. Water economy and salt balance in the western harvest mouse, *Reithrodontomys megalotis. Physiol. Zool.*, 37(1): 45—56.

MacNae, W., 1968. A general account of the fauna and flora of mangrove swamps and forests in the Indo-West Pacific region. *Adv. Mar. Biol.*, 6: 73—270.

Marsden, I.D., 1973. The influence of salinity and temperature on the survival and behavior of the Isopod *Sphaeroma rugicauda* from a salt marsh habitat. *Mar. Biol.*, 21(2): 75—85.

Marshall Jr., J.T., 1948. Ecologic races of song sparrows in the San Francisco Bay region. I. Habitat and abundance. *Condor*, 50: 193—215.

Matera, N.J. and Lee, J.J., 1972. Environmental factors affecting the standing crop of foraminifera in sublittoral and psammolittoral communities of a Long Island salt marsh. *Mar. Biol.*, 14(2): 89—103.

May, M.S., 1974. Probable agents for the formation of detritus from the halophyte, *Spartina alterniflora*. In: R.J. Reimold and W.H. Queen (Editors), *Ecology of Halophytes*. Academic Press, New York, N.Y., pp. 429—440.

May, R.C., 1974. Factors affecting buoyancy in the eggs of *Bairdiella icistia* (Pisces: Sciaenidae). *Mar. Biol.*, 28(1): 55—59.

Meanley, B., 1969. *Natural History of the King Rail*. N. Am. Fauna, No. 67. U.S. Fish and Wildl. Serv., Bur. Sports Fish. Wildl., 108 pp.

Miller, K.G. and Maurer, D., 1973. Distribution of the fiddler crabs, *Uca pugnax* and *Uca minax*, in relation to salinity in Delaware rivers. *Chesapeake Sci.*, 14(3): 219—221.

Moore, C.J., 1968. *The Feeding and Food Habits of the Silversides Menidia menidia (Linnaeus)*. Thesis, University of Delaware, Newark, Del., 65 pp.

Murray, J.W., 1973. Recognition of estuarine environments using foraminiferids. *J. Geol. Soc. Lond.*, 129(4): 456.

Nicol, E.A., 1936. The ecology of a salt marsh. *J. Mar. Biol. Assoc. U.K.*, 20: 203—261.

Olkowski, W., 1966. *Biological Studies of Salt Marsh Tabanids in Delaware*. Thesis, University of Delaware, Newark, Del., 116 pp.

Oney, J., 1954. *Final Report: Clapper Rail Survey and Investigation Study*. Ga. Game Fish Comm., 50 pp.

Owen, M., 1971. The selection of feeding site by white–fronted geese in winter. *J. Appl. Ecol.*, 8(3): 905—917.

Palmisano, A.W., 1972. The distribution and abundance of muskrat *(Ondatra zibethica)* in relation to vegetative types in Louisiana coastal marshes. *Proc. Ann. Conf. Southeast. Assoc. Game Fish Comm.*, 26: 1—31.

Paradiso, J.L. and Handley Jr., C.O., 1965. Check list of mammals of Assateague Island. *Chesapeake Sci.*, 6(3): 167—171.

Parker, F.L. and Athearn, W.D., 1959. Ecology of marsh foraminifera in Poponesset Bay, Massachusetts. *J. Paleontol.*, 33: 333—343.

Paviour–Smith, K., 1956. The biotic community of a salt meadow in New Zealand. *Trans. R. Soc. N.Z.*, 83: 525—554.

Payne, K.T., 1972. A survey of the *Spartina* feeding insects in Poole Harbour, Dorset. *Entomol. Mon. Mag.*, 108(1295—1297): 66—79.

Pearse, A.S., 1914. Habits of fiddler crabs. *Ann. Rep. Smithson. Inst.*, 1913: 415—428.

Phleger, F.B., 1965. Patterns of marsh foraminifera, Galveston Bay, Texas. *Limnol. Oceanogr.*, 10: R169—184.

Phleger, F.B., 1970. Foraminifera populations and marine marsh processes. *Limnol. Oceanogr.*, 15: 522—534.

Phleger, F.B. and Bradshaw, J.S., 1966. Sedimentary environments in a marine marsh. *Science*, 154: 1551—1553.

Phleger, F.B. and Walton, W.R., 1950. Ecology of marsh and bay foraminifera, Barnstable, Massachusetts. *Am. J. Sci.*, 248(4): 274—295.

Poulson, T.L. and Bartholomew, G.A., 1962. Salt balance in the savannah sparrow. *Physiol. Zool.*, 35: 109—119.

Ranwell, D.S., 1972. *Ecology of Salt Marshes and Sand Dunes*. Chapman and Hall, London, 258 pp.

Ranwell, D.S., 1974. The salt marsh to tidal woodland transition. *Hydrobiol. Bull., Netherlands Hydrobiol. Soc.*, 8(1/2): 139—151.

Russell–Hunter, W.D., Apley, M.L. and Hunter, P.D., 1972. Early life history of *Melampus* and the significance of semilunar synchrony. *Biol. Bull.*, 143(3): 623—656.

Schmelz, G.W., 1964. *A Natural History of the Mummichog, Fundulus heteroclitus (Linnaeus) in Canary Creek Marsh*. Thesis, University of Delaware, Newark, Del., 65 pp.

Schmelz, G.W., 1970. *Some Effects of Temperature and Salinity on the Life Processes of the Striped Killifish, Fundulus majalis (Walbaum)*. Dissertation, University of Delaware, Newark, Del., 104 pp.

Schwartz, B. and Safir, S.R., 1915. The natural history and behavior of the fiddler crab. *Cold Spring Harbor Monogr.*, 8: 1—24.

Shanholtzer, G.F., 1974. Relationship of vertebrates to salt marsh plants. In: R.J. Reimold and W.H. Queen (Editors), *Ecology of Halophytes*. Academic Press, New York, N.Y., pp. 463—474.

Shure, D.J., 1970. Ecological relationships of small mammals in a New Jersey barrier marsh habitat. *J. Mammal.*, 51: 267—278.

Sibley, C.G., 1955. The responses of salt–marsh birds to extremely high tides. *Condor*, 57: 241—242.

Smallwood, M.E., 1905. The salt marsh amphipod: *Orchestia palustris*. *Cold Spring Harbor Monogr.*, 3: 3—21.

Smith, J.B., 1902. The salt marsh mosquito, *Culex sollicitans*, Wlk. *Spec. Bull. N.J. Agric. Exp. Station*, 10 pp.

Stearns, L.A., MacCreary, D. and Daigh, F.C., 1939. Water and plant requirements of the muskrat on a Delaware tide water marsh. *Proc. N.J. Mosq. Exterm. Assoc.*, 26: 212—221.

Stearns, L.A., MacCreary, D. and Daigh, F.C., 1940. Effects of ditching on the muskrat population of a Delaware tidewater marsh. *Univ. Del. Agric. Exp. Station Bull.*, 225: 55 pp.

Stewart, R.E., 1951. Clapper rail populations of the Middle Atlantic States. *Trans. N. Am. Wildl. Conf.*, 16: 421—430.

Stewart, R.E., 1952. Clapper rail studies. In: J.W. Aldrich et al. (Editors), *Investigations of Woodcock, Snipe and Rails in 1951*. U.S. Fish Wildl. Serv., Spec. Sci. Rep. Wildl., 14: 56—58.

Stewart, R.E., 1962. Waterfowl populations in the upper Chesapeake region. *U.S. Fish Wildl. Serv., Spec. Rep. Wildl.*, 65: 208 pp.

Stewart, R.E. and Robbins, C.S., 1958. Birds of Maryland and the District of Columbia. N. Amer. Fauna No. 62. *U.S. Fish Wildl. Serv., Bur. Sports Fish. Wildl.*, 401 pp.

Teal, J.M., 1958. Distribution of fiddler crabs in Georgia salt marshes. *Ecology*, 39: 185—193.

Teal, J.M., 1959. Respiration of crabs in Georgia salt marshes and its relation to their ecology. *Physiol. Zool.*, 32(1): 1—14.

Teal, J.M., 1962. Energy flow in the salt marsh ecosystem of Georgia. *Ecology*, 43(4): 614—624.

Teal, J.M. and Wieser, W., 1966. The distribution and ecology of nematodes in a Georgia salt marsh. *Limnol. Oceanogr.*, 11: 217—222.

Tomkins, I.R., 1941. Notes on Macgillivray's Seaside sparrow. *Auk*, 58: 38—51.

Urner, C.A., 1935. Relation of mosquito control in New Jersey to bird life of the salt marshes. *Proc. N.J. Mosq. Exterm. Assoc.* 22: 130—136.

Vogt, W., 1938. Preliminary notes on the behavior and ecology of the eastern willet. *Proc. Linn. Soc. N.Y.*, 49: 8—42.

Warlen, S.M., 1964. *Some Aspects of the Life History of Cyprinodon variegatus Lacepede 1803, in Southern Delaware*. Thesis, University of Delaware, Newark, Del., 40 pp.

Wieser, W. and Kanwisher, J., 1961. Ecological and physiological studies on marine nematodes from a small salt marsh near Woods Hole, Massachusetts. *Limnol. Oceanogr.*, 6: 262—270.

Woodell, S.R.J., 1974. Anthill vegetation in a Norfolk salt marsh. *Oecologia*, 16(3): 221—225.

Zilberberg, M.H., 1966. Seasonal occurrence of fishes in a coastal marsh of northwest Florida. *Publ. Inst. Mar. Sci. Univ. Texas*, 11: 126—134.

Chapter 6

THE COASTAL SALT MARSHES OF WESTERN AND NORTHERN EUROPE: AN ECOLOGICAL AND PHYTOSOCIOLOGICAL APPROACH

W.G. BEEFTINK

INTRODUCTION

The purpose of this chapter is to give an overall view of the biotic components of the West- and North-European salt marshes including the Atlantic coasts of the Iberian Peninsula. It endeavours to focus attention on the ecological features in pattern and process characteristic of salt marshes, and on their variability within this European coastal region. It gives some remarks on functional relationships within the European salt-marsh ecosystem.

The salt marshes are first considered as a border environment and their dominant diagnostics are briefly sketched. Among the organisms, phanerogams and their communities are first discussed, followed by remarks on algal and animal ecology.

Emanating from the Zürich—Montpellier tradition of phytosociology the author makes use of the concepts, terms and methods of this school for the description and classification of vegetation (Braun-Blanquet, 1964; Shimwell, 1971; Westhoff and Van der Maarel, 1973). For the nomenclature of the community types (syntaxa) reference is made to Beeftink (1968). Plant nomenclature is according to the Flora of the British Isles (Clapham et al., 1962) and the check-list of British marine algae (Parke and Dixon, 1968), insofar as authorities are omitted.

Definition and range

There are three main obvious attributes of coastal salt marshes: (1) they consist of alluvial sediments deposited on the shore by the sea; (2) they are subjected to tidal or weather-effected inundation by more or less diluted sea water, the moisture conditions of the soil varying from continuously waterlogged to at least temporarily moist; (3) their soil has a chloride content varying in space and time and, consequently, they are occupied by plant communities completely or mainly consisting of halophytes (see pp. 3—13).

In the vertical range the salt marsh emerges from sand and mud flats mostly inhabited by many benthic organisms and locally covered by seaweeds and eelgrasses *(Zostera)*. From about the level of mean high water neap tide (MHWN), but depending on the degree of exposure, the mud flats are colonised by terrestrial halophytic vegetations, first by *Salicornia* and *Spartina* species. Between this level and that of mean high water (MHW) other species enter and the vegetation cover is usually closed. The marsh finds its upper limit where the tidal or non-tidal saline influences are so much diminished that halophytes are in the minority or totally lacking. This upper limit is found between the levels of mean high water spring (MHWS) and extreme high water spring (EHWS), usually the level where in winter plant litter is thrown up in strandlines.

For hard substrates the coastal belt is subdivided into biologically defined zones by various authors (e.g., Du Rietz, 1940, 1947; Den Hartog, 1959a). The boundaries between these zones are then related to tidal water levels as much as possible. Although it is more difficult to do this for soft substrates this sytem may also be applied as a general characterization of the vertical range of salt-marsh biota (Fig. 6.1).

Distribution and types

Compared with the rocky coasts and dune formations the European salt marshes occupy a

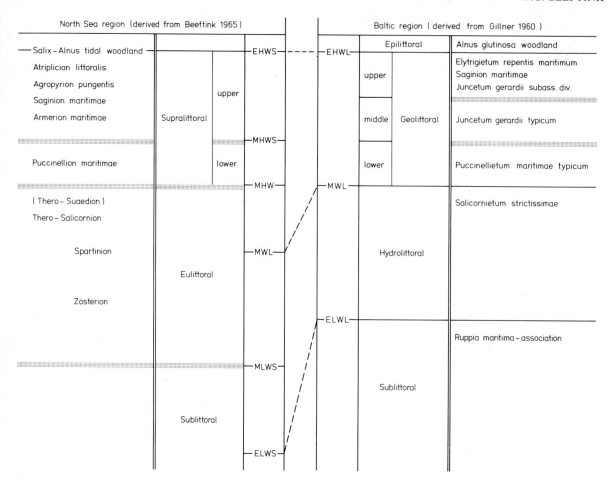

Fig. 6.1. Zonation of phanerogamic syntaxa in the alluvial littoral belt of the North Sea and Baltic regions in relation to critical water levels.

minor place in the total coast length. Generally the marshes show a disjunct distribution of limited areas, but on certain coasts they occupy vast parts of the shore, e.g., in the Wadden area. Salt marshes are primarily found in and around river mouths (estuaries), but also in bays, Wadden areas, lagoons and on beach plains protected by sand or shingle spits (see p. 31). Dependent on the conditions of salinity and tidal range we can distinguish between six types of salt–marsh formations: (1) estuarine, (2) Wadden (see below), (3) lagoonal, (4) beach plain, (5) bog, and (6) polderland type.

Salt marshes of the estuarine type are represented in every West- and North-European river mouth. Except in the Baltic where rivers discharge into a tideless brackish water body, this type is characterized by largely fluctuating salinities (temporarily or permanently brackish soil conditions) and mostly by strong tidal currents of the estuarine water body (preponderance of creek and gully formations in the geomorphology). In deltas the banks of the river channels consist generally of natural levees backed by low-lying, swampy or flooded depressions often occupied by swamp vegetation, which may build up peat deposits. In dry climatic regions, such as in the Mediterranean, or in dry years when repeated evaporation of water from enclosed back-swamp depressions leads to increasing soil salinity, some of these depressions may become converted to salt marsh or even to unvegetated saline flats (Bird, 1968).

The Wadden type is found behind a chain of barrier islands in front of the coastline. Where the tides flowing in from entrances meet, deposition of silt is common behind the barrier islands and against the mainland. For marsh genesis wave action is here of equal importance with tidal currents (see p. 32). Consequently, terrace formations are common as a consequence of local alternating erosion and sedimentation processes (Fig. 6.2). In Europe this type is found in the Dutch—German—Danish Wadden area and on the south side of Scolt Head Island (Great Britain).

The lagoonal type comes into existence where spit- or *Lido*-like sand or shingle barriers have been built up parallel to the coastline, leaving only a narrow entrance to the tides. This type is well illustrated by the salt marshes of the Fleet at Chesil Beach on the Dorset coast, by those on the inner side of the *Nehrungen* on the southern coast of the Baltic, and by the Venice and Trieste lagoons in the northern Adriatic.

The next type includes beach plain and salt marshes protected by sand or shingle laterals, sometimes surmounted by dunes, and admitting the tides unimpeded. This type stands out by its sandy soil and usually shows transitions to adjacent dune formations. Accretion is here due jointly to the tides and the wind. Examples are found in the Wadden Islands — for instance the Boschplaat, Terschelling; on the north coast of Norfolk — Blakeney Point and Scolt Head Island; and in the Baltic — parts of the Graswarder near Heiligenhafen (Schmeisky, 1974).

The bog type, rather common on the New England coast, occurs also in Europe, where it is known from southern England and southwestern Ireland (Chapman, 1941) and from the Baltic (Steinführer, 1955; Piotrowska, 1974; Schmeisky, 1974). The formation of salt–marsh peat is mostly ascribed to a gradual subsidence of the coastline compensated for by the accretion of former fresh-water or brackish bogs, brought about by a heavy deposition of autochthonous plant debris (e.g., *Spartina, Phragmites*) under a limited supply of silt (Chapman, 1960; Schmeisky, 1974). Other peat formations are known from still less saline conditions: from depressions with stagnant water

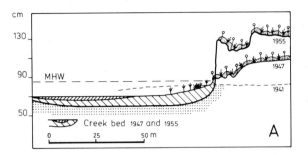

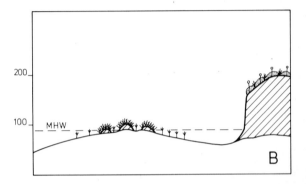

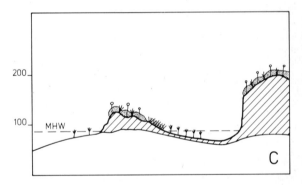

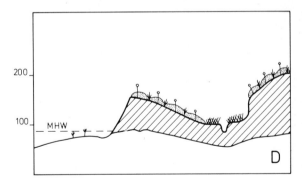

Fig. 6.2. Formation of a terraced salt marsh. A, the development of an offshore channel at the windward side of a new salt marsh in the Danish Wadden area. B, initial salt-marsh formation on a high-lying tidal flat off the channel. C and D, continuing development of a new terrace, the channel filling up with mud (A: after Jakobsen, 1964; B—D: after Jakobsen, 1954).

⊥ Salicornia europaea

〰 Puccinellia maritima

🌿 Mixed vegetation

exposed to the peripheral influence of oligohaline[1] to fresh–water conditions of estuaries (Lambert et al., 1960; Zonneveld, 1960), and from permanently wet to moist dune valleys almost completely cut off from the sea by sand or shingle barriers. Although peat formation is thus generally favoured by a diminishing sea–water influence, the importance of the heavy Atlantic rainfall for the formation of salt–marsh peat on the Irish coast needs further investigation.

Finally, the polderland type is represented along former creeks, in bottom land and other depressions remaining saline in the reclaimed marshland. The habitats are common in estuarine polder areas, such as in the southwestern Netherlands, and also occur in other embanked marshlands now under cultivation (e.g., in the Danish, northwest German and Dutch polderland reclaimed from the Wadden area, the Fens of East Anglia and the Romney Marshes of Kent).

ENVIRONMENTAL CHARACTERISTICS

Soil profile and soil components

In general the foundation on which a salt marsh originates is a sand flat, a part of the sea or Wadden floor before accretion sets in. On the Scottish and Scandinavian coasts, however, the substrate is often composed of moraine washed down from the mainland during postglacial periods (Gillner, 1960; Gimingham, 1964). The salt–marsh profile rising up above this bedrock differs mainly in two ways: (1) levees silted up to maximum levels, such as creek banks and the outer edge of mature marshes, have mostly an unbroken soil profile starting at the base with sand, and passing via more or less silty and clayey stages to silty or sandy conditions in the surface soil; (2) depressions, enclosed by creek–bank levees or formed as back–marshes in terraces, show mostly a discontinuity in their soil profile consisting of a sudden transition from silty sand to more or less heavy clay. This discontinuity is due to a period of stagnation in deposition, during which the sedimentation conditions have been changed, transforming the tidal marsh from a convex to a concave one (see p. 35). These properties are important for understanding many other soil conditions such as ground–water fluctuations, salinity, etc.

Besides mineral particles of different size the marsh soil consists of carbonates and organic matter. Along the European coasts there are many gradient situations in carbonate content. The most important one is their decrease from 35% and more on the coasts of Brittany and Normandy to a figure between 0.5 and 5% on the west coast of Jutland (Jacquet, 1949; Verhoeven, 1963). But also in estuaries gradients may occur; in the Scheldt estuary, for instance, there is a decrease of a third of the total amount from the mouth upwards to Rupelmonde in the oligohaline zone (Beeftink, 1965).

There are no indications that under normal salt–marsh conditions these differences in carbonate content have an influence on the structure and composition of vegetation. In local studies, however, the carbonate content can be an indicator of the degree of waterlogging, as under such conditions decalcification takes place by formation of bicarbonates or by solution as a result of acid formation under anaerobic conditions (Brümmer et al., 1971). Waterlogging occurs especially in depressions with heavy clay content, and decalcification may therefore be related to this clay content. These results are in strong contrast with the clay—carbonate relations in freshly deposited silt (Fig. 6.3).

Internal reworking of these partly decalcified deposits may, however, ultimately lead to the formation of salt–marsh soils totally free from carbonates. In the southwestern Netherlands such formations are found in some localities. Although this too has normally no important ecological consequences, if periodical flooding with tidal water has been impeded — either naturally or by man — a considerable fall in pH may occur, as the soil is unable to neutralize sulphuric acid liberated from oxidation of sulphides. On such combined acid and saline soils the halophytic vegetation is poor in species, but it is not yet sure whether this is the result of these acid conditions or of other environmental factors, such as grazing pressure.

---

[1] The salinity ranges used are euhaline (22—16.5‰ chlorinity, mean values at limits), polyhaline (16.5—10‰), mesohaline (10—3‰), oligohaline (3—0.3‰), and fresh water (< 0.3‰). See for terminology *Symposium on the Classification of Brackish Waters, Venice* (Anonymous, 1959).

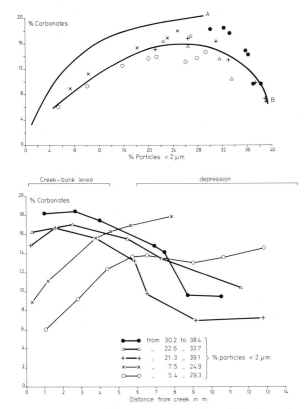

Fig. 6.3. Carbonate contents of salt-marsh soils in the southwestern Netherlands in relation to clay content and geomorphology. Above: *A*, freshly deposited sediments from the Grevelingen (after Verhoeven, 1963); *B*, salt-marsh deposits from the Kaloot near Vlissingen (surface soil, 0—4 cm). Below: as *B*, but carbonate content related to distance from the creek. The five types of symbols represent different transects.

Organic debris — either autochthonous or washed ashore — is almost exclusively found in the surface layers of the soil. There, it plays an important role in covering the soil surface and supplying nutrients while it is decaying and before it is washed away. Beeftink (1965, 1966) distinguished between two types of plant litter rafted onto the marsh: (1) debris of higher plants and detached large Phaeophyceae floating on the surface of the flood water and washed ashore in concentrated belts at the spring and storm-flood lines; and (2) other algae floating deeper in the flood-water layers and washed ashore in a broad zone scattered over the lower salt marsh. On both types specific halophytes and other salt-tolerant nitrophilous plants establish, dependent on the salinity level, on the quantities and regularity with which the organic matter is supplied, and on the magnitude of the deposition of water-borne silt or wind-borne sand.

Apart from mineral deposits with or without deposition of organic debris, peaty formations occur (see also p. 111). Bilio (1964—67, 1965, 1966), comparing the salt marshes of both coastal regions of Germany, distinguished the salt-marsh areas of the southwestern Baltic as "Verlandungsgebiete" (land formation mainly due to the deposition of autochthonous plant debris) from those of the southeastern North Sea called "Anlandungsgebiete" (land formation mainly due to sedimentation of allochthonous mineral materials). This distinction characterizes at the same time two extreme types of soil development in salt marshes, the corresponding differences in soil texture, water capacity, etc., being of particular importance in determining different environmental conditions for the soil fauna (see p. 115).

Water régime

The border character of the salt marsh involves both above-ground and underground water relations. Tides and rainfall affect the conditions above ground, to an extent dependent on the proximity and drainage capacity of the watercourses. The underground water relations are influenced by those above ground, and by hydraulic drainage and seepage conditions.

The tides influence the zonation pattern in vegetation by local or regional variations in their period, amplitude and phase. Although quantitative data are scanty and other environmental factors (climate, salinity) interfere, some indications are evident. Tutin (1942) pointed out that the time at which low water of spring tides occurs, as well as the occurrence of double tides, influence the ultimate height to which *Zostera marina* can grow. Inversions of the zonation of the *Puccinellietum maritimae* and the *Halimionetum* are probably linked up with the magnitude of the tidal range, increasing from 1.5 m in the western Wadden area to ca. 12 m on the Breton coasts. A tidal range of about 8 m is the inversion point; wider ranges generally cause settling of the *Halimionetum* on lower levels than the *Puccinellietum*. The reverse is the case at narrower tidal ranges (Beeftink, 1965). Telescoping of the zonation may perhaps also be ascribed to the magnitude of the tidal range, and to the vertical difference between the high–water levels of neap and spring tides. There are, for instance, indications that in the southwestern Netherlands *Spergularia media* enlarges its vertical range to lower salt–marsh levels when the difference between these high-tide levels is wide (Beeftink, 1965).

Rainfall is an important factor insofar as it inundates enclosed depressions and influences the salinity of the surface soil. Stagnant rainwater can be found in dune slacks and beach plains. As they are also partly cut off from the sea flood water can stay behind if high–tide exceeds a threshold level. Under such conditions especially *Juncus gerardi, Juncus maritimus, Glaux maritima* and *Agrostis stolonifera* are dominant, but also *Salicornia europaea* agg. and *Spergularia marina* can develop depending on salinity. The leaching effect of rainfall starts as soon as the tidal influence is retreating owing to accretion, but persists to a large extent when the surface soil becomes temporarily aerated. Germination of many halophytes seems to be activated by temporary desalination of the surface soil (Adriani, 1958; Waisel, 1972). The sudden increase in number of species always encountered when the salt marsh has been built up above MHW level may be closely connected with this combined temporary aeration and desalination processes.

Salinity

In the mud flats and lower salt marshes the soil salinity is closely related to that of the flooding water bodies. In the upper marshes rising above MHW level, the influence of climatic conditions grows considerably with accretion progress (Fig. 6.4). At those levels, the lower frequency of flooding increases the influence of the mutually opposing effects of evapotranspiration and precipitation on salinity, subjecting it to great variability. In most places along the West-European coasts annual precipitation exceeds annual evapotranspiration, and thus flooding of the marsh mostly results in an input of salts. In many places, however, the relatively high evapotranspiration during the summer months causes the salinity to rise to extremely high values in the surface soil, sometimes as far as the saturation point with the result that salt crusts may form. Such localities are found in depressions drying out in summer, and in sandy places in the

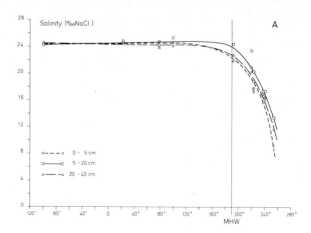

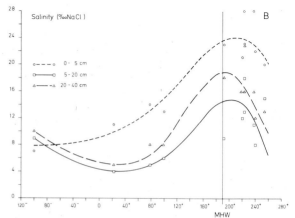

Fig. 6.4. Soil salinity in a tidal marsh of the Eastern Scheldt in relation to height with respect to the tides. Median values (A) and maximal amplitudes (B) of 24 monthly soil samples (May 1965—April 1967).

supralittoral zone, especially when flooding with sea water is succeeded by a period of strong desiccation. In the first case *Salicornia europaea* and *Spergularia marina* are mostly dominant; *Suaeda maritima* may sometimes develop along soil cracks. The other conditions, especially found in Armerion communities on sandy soils, cause an open vegetation, as many species die back owing to high soil salinity, and others are introduced: *Sagina maritima, Parapholis strigosa, Plantago coronopus, Pottia heimii,* etc.

In the higher salt marshes the opposing effects of precipitation and evapotranspiration are maximal if the soil salinity, and that of the flooding water bodies, is relatively high. Therefore, there is a zone below which flood frequency is too high to allow evapotranspiration to cause high soil–salinity values, and above which this frequency is too low to neutralize the leaching effect of rainfall. In the southwestern Netherlands, maximum fluctuations are normally found at 25—35 cm above MHW, but exceptionally somewhat lower, at 15 cm (Beeftink, 1965, 1966). Under estuarine conditions, in the upstream direction this zone shifts to lower tidal levels (Table 6.1). Similar shifts occur in the zonation pattern of the vegetation (Beeftink, 1965, 1966). Species thus enduring increasing submergence in the upstream direction are *Spergularia marina, Puccinellia distans, Juncus gerardi, Glaux maritima, Agrostis stolonifera, Spartina townsendii* agg., and *Atriplex hastata*. On the other hand, many riparian species, originating from the fluvial area, penetrate into the brackish environment, but are limited to progressively higher levels as one proceeds downstream: *Phragmites communis, Scirpus maritimus, Cirsium arvense, Rumex obtusifolius, Galium aparine, Solanum dulcamara,* etc.

Remane (1955), dealing with the vertical distribution of benthic organisms in different parts of the Baltic and in the North Sea, used the term *Brackwassersubmergenz* (brackish–water submergence) for a downward shift in limits of distribution with decreasing salinity. Bilio (1964–67, 1965) extended this concept to the aquatic fauna of supralittoral biotopes, and added examples of corresponding changes in the distribution of salt–marsh plants on the coasts of the North Sea and the Baltic, which recall those mentioned above for the Dutch estuaries and the earlier observations of Gillham (1957) in the Exe Estuary.

Vertical shifts in vegetation pattern, however, should not be referred to differences in salinity and precipitation—evaporation relations without a detailed knowledge of the influence of geographical and local differences in other environmental conditions, such as tidal range (Beeftink, 1965), soil characteristics (Bilio, 1965) and perhaps grazing intensity. For instance, whether the occurrence of Armerion species, such as *Armeria maritima* and *Glaux maritima,* on lower levels in Ireland (Moore et al., 1970) than on continental coasts can be referred to the Eu–Atlantic climatic conditions of that country, cannot be decided without further investigation.

Nutrients

The nutritional state of salt–marsh soils depends on a series of external and internal processes, of which the most important are: (1) input of nutrients adsorbed on the clay and silt particles deposited, stored in organic debris washed ashore, and dissolved in the flooding water bodies; (2) spatial distribution of these nutrients brought in by tidal actions; (3) mineralisation and humification in organic materials; (4) fixation of nutritional elements into clay minerals and living materials; and (5) output of organic debris into offshore water bodies.

Little is still known of most of these effects in European salt marshes, contrary to their American counterparts, and only some general remarks may be made. Most nutritional elements (potassium, phosphorus, sulphur, iron, trace elements) are supplied by the flood tides. Nutrient transportation from the mainland to the salt marsh, indeed, cannot be excluded, but this depends very much on the local situation and is supposed to be generally slight.

TABLE 6.1

Maximal fluctuations in chloride concentrations (‰ NaCl) in the soil solution of the main salt-marsh associations occurring in the Scheldt estuary (after Beeftink, 1965)

Distance from Vlissingen (km):	8.1	20.2	32.7	45.1	53.6	57.5
Estuarine chlorinity at high tide, mean values (‰)	13.9	12.3	11.1	9.6	7.5	6.5
Atriplici – Elytrigietum pungentis	13.6	11.5	—	9.3	—	11.5
Artemisietum maritimae	10.9	17.0	16.4	9.2	10.3	10.8
Halimionetum portulacoidis or *Agrostis stolonifera* sociation	14.5	11.4	21.4	—	—	9.8
Puccinellietum maritimae	16.1	21.4	25.7	20.2	—	17.0

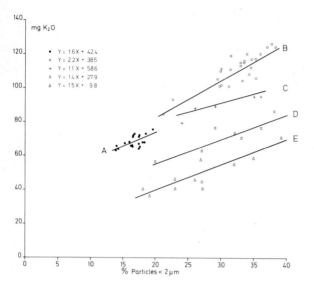

Fig. 6.5. Potassium content of salt marshes in relation to clay content and salinity in the Scheldt estuary. *A*, Kaloot near Vlissingen (polyhaline); *B*, Ossendrecht near the Dutch—Belgian frontier (mesohaline); *C*, Galgeschoor near Lillo (mesohaline); *D*, near former fort Pipe de Tabac (oligohaline); *E*, south of Antwerp and near Rupelmonde (oligohaline to fresh).

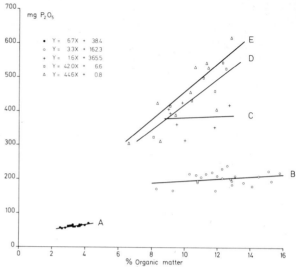

Fig. 6.6. Phosphorus content of salt marshes in relation to organic matter in the Scheldt estuary. *A—E* as for Fig. 6.5.

Potassium content is closely related to the percentage of clay particles ($< 2 \mu m$) in the soil. Input, therefore, will take place together with the silt deposited and by percolation of floodwater into the soil profile. In the Scheldt estuary it decreases with decreasing salinity of flooding tides (Fig. 6.5). Probably this relationship is connected with the fact that sea water generally contains considerably more potassium than river water. Moreover, as Zonneveld (1960) pointed out, in the fresh–water tidal zone and further upstream much potassium is fixed into the space lattices of clay minerals. This shift of potassium from the adsorption complex into the space lattices may already be initiated in the brackish environment.

Phosphorus content appears to be more closely related to humus or organic–matter content than to percentage of clay particles. In the brackish water bodies of non–polluted estuaries the phosphorus content may be slightly higher than in the marine and fresh–water zones, owing to mass mortality of marine as well as limnetic plankton organisms. The much higher values found in the brackish and fresh–water zones of the Scheldt estuary, however, are determined by large amounts of urban sewage and industrial waste discharged into the river and its tributaries (Bakker and De Pauw, 1974, 1975). The phosphorus content found in the salt–marsh soils of the Scheldt estuary are closely related to these figures (Fig. 6.6).

The nitrogen status appears to be a much more complex environmental factor. Firstly, it is represented both as cations (NH_4^+) and as anions (mainly NO_3^- and NO_2^-). Secondly, it originates from different sources: (1) supplied by the flooding water; (2) fixed from the air by blue–green algae and bacteria growing in the marsh; and (3) liberated from plant debris either produced by the standing vegetation or washed ashore by the flood tides.

The magnitude of the input of nitrogen by flooding depends on the origin of the water body with respect to estuarine conditions, and on its degree of pollution. Whereas the N/P ratio of normal sea water is about 15, N/P ratios in the Scheldt estuary, for instance, rise to more than ten times this value (Bakker and De Pauw, 1974). Just as in the case of phosphorus content, this abnormal rise must be ascribed to the high pollution rate.

The products of algal and bacterial nitrogen fixation seem to be important as nutrient sources for higher salt–marsh plants. Small, but consistent, levels of nitrogen fixation by bacteria are found in many parts of the marsh, but much more is stored by blue–green algae with a maximum in late summer and early winter (Jones, 1974).

Since, in the salt marsh, mineralization predominates greatly over humification processes, plant debris produced on the spot or rafted onto the marsh will be another source of nitrogen. Its distribution over the marsh, however, is generally very unequal, and mostly concentrated on strand lines. There, it smothers the original vegetation, resulting in even higher concentrations of mineralization products. Table 6.2 gives an impression of the values of some nitrogen fractions, together with other soil properties, in relation to elevation and vegetation.

With respect to plant growth, the availability of nitrogen and phosphorus has gained some attention recently. Experiments made by Pigott (1969) on cultured plants of *Salicornia europaea* and *Suaeda maritima* showed that plants of both species grown in soil cores from the upper marsh showed an increased growth response to the addition of nitrogen and phosphorus combined, whereas an addition of phosphorus alone had no significant effect compared with the controls. As these species are annuals not germinating until April, it is clear that rapid growth can only be maintained if there is a plentiful supply of these mineral nutrients. Assuming from studies made by D.W. Jeffrey (see Pigott, 1969) that there is a general similarity in the concentrations of soluble nitrogen compounds and phosphate at all tidal levels, Pigott (1969) suggests that the significant difference between the bare mud and the upper part of the marsh is the extent to which the sediment is exploited by the roots of the perennial species already present. The rapid shoot growth of the perennial halophytes during early summer may therefore exhaust the nitrogen available in the rooting zone, at the same time affecting the growth of the annuals.

Nitrogen deficiency in the upper marsh, already suggested by Feekes (1936), was also found in *Suaeda* seedlings by Stewart et al. (1972), who at the same time found very low amounts of nitrate reductase compared with those present at the seaward edge of the marsh. Tyler (1967) studied the effects of the addition of nitrogen and phosphorus

TABLE 6.2

Clay, humus, salinity and nitrogen contents in the surface soil (0—5 cm) of a salt marsh near Krabbendijke, Eastern Scheldt, The Netherlands (soil samples 1964—65; sea-water salinity 14—17.5‰; MHW 1.90 m above ordnance level; clay, humus and total N in % dry soil; salinity in ‰ NaCl per l soil solution; total mineral N and ammonium N extractable with N KCl in mg per kg dry soil)

Topographic feature	Dominance of	Metres above ordnance level	Clay (< 2 μm)	Humus	Salinity	Total N	Total mineral N	Ammonium N
Mud flat	*Spartina maritima*	0.92	9.6	2.1	20.3—32.2	0.06—0.10	3.7—5.3	1.9—3.0
Low salt marsh	*Puccinellia maritima*	2.09	23.9	14.6	20.0—36.4	0.59—0.68	24.8—72.3	12.3—29.3
	Triglochin maritima	2.16	14.3	7.5	16.0—38.2	0.35—0.39	14.1—22.3	7.2—18.6
Medium salt marsh	*Plantago maritima* / *Armeria maritima*	2.23	14.9	12.8	11.4—37.1	0.54—0.63	18.2—28.9	13.3—21.5
	Halimione portulacoides	2.19	26.1	9.9	10.0—36.0	0.41—0.47	14.1—21.2	7.1—13.6
Creek-bank levees	*Artemisia maritima* / *Festuca rubra*	2.35	15.3	8.2	2.1—27.1	0.40—0.48	18.1—43.2	7.5—20.2
Strand lines with decomposed plant litter	*Atriplex littoralis*	2.46	22.6	32.4	4.4—35.9	2.71—3.04	106.6—184.0	64.9—145.3
Strand lines with fresh plant litter	*Atriplex littoralis*	2.62	17.3	38.8	3.1—25.2	0.31—2.75	31.5—1179.5	21.1—763.8

on Baltic shore–meadow vegetation. Addition of ammonium chloride resulted in an increase of 30% in the productivity of the field layer, whereas a nitrate fertilizer had little effect. As possible reasons for these effects he suggests that clayey soils may have a lower capacity for NO_3^- than for NH_4^+ ions, or that the vegetation might be better adapted to NH_4^+ than to NO_3^- ions.

However, the question remains how far biomass or degree of luxuriance, and thereby nitrogen and phosphorus conditions, are critical indicators for the criteria of stability and longevity of this ecosystem. For more physiological and biochemical approaches to the halophyte problem the reader is referred to Epstein (1969), Jefferies (1972), Waisel (1972) and Queen (1974).

PHYTOGENIC INTERRELATIONSHIPS

In salt marshes the botanical interrelationships appear to be obvious in the fields of symbiosis, epiphytism and parasitism. Without entering far into symbiosis questions, it can be considered evident that bacterial and fungal activities play an important part in the functioning of the metabolism of the entire ecosystem (cf. literature in Chapman, 1974). Apart from the mineralizing function of microbial organisms at all trophic levels for food (energy) transfer, representatives of the fungi have also been found to live as mycorrhizae, mostly endotrophic types, associated with halophytes such as *Armeria maritima, Aster tripolium, Glaux maritima,* and *Plantago maritima* (Mason, 1928; Fries, 1944). Other species, such as *Halimione portulacoides,* show a specialised mycoflora of which the nature of interdependence is still unclear (Dickinson, 1965).

Pugh (1974) covered the results of his studies on Gibraltar Point (Great Britain) and those of other authors, on leaf– and root–surface fungi of salt–marsh plants and on soil fungi in the rhizospheres of halophytes. On leaf surfaces, fungi were rarely recorded on halophytes liable to regular submergence by the tides, but they are common on sand–dune plants living above the tidal belt. Evidence is accumulating that such fungi are washed off the leaves when covered by the tides. On the roots of salt–marsh plants there is a relative abundance of sterile hyphae associated with high levels of water in the mud as a result of repeated inundation by the sea.

In the rhizospheres of halophytes Pugh (1974) found a decrease in numbers of fungi isolated, and in numbers of species, as the plants are sampled on lower levels with respect to the tides. He supposed this decrease to be related to the increased waterlogging and anaerobic state of the soil with increased duration of tidal cover, and to a smaller amount of organic matter. This corresponds with his finding that the majority of the fungi found are terrestrial. However, *Dendryphiella salina,* which only occurs in areas subjected to sea–water cover, increases in abundance up the shore in developing marshes, and then decreases in abundance in mature saltings where it is replaced by *Gliocladium roseum.*

Various species of bacteria play a very important role in the phosphorus and nitrogen cycles. Daiber (1972) points out that sulphur bacteria have a profound influence on the quantities of phosphate ion available for use by other organisms. According to Jones (1974), nitrogen fixation by bacteria and blue–green algae occurs in all zones of the salt marsh and contributes substantial amounts of available combined nitrogen to the soil. The process is stimulated in the rhizosphere of *Puccinellia maritima.* Apart from this *Puccinellia* effect the greatest levels of nitrogen fixation are found in the lower zones of the marsh, despite its lower combined nitrogen level in comparison with that of the upper zones. The higher levels of nitrogen in the upper marsh are probably a result of the addition of algae and plant litter washed ashore and rafted into the marsh, together with a nutrient supply from the faeces of cattle, sheep or rabbits. The role of microbes in mineralizing plant litter from *Spartina* has been emphasised by Burkholder (1956) and Burkholder and Bornside (1957). Microbial conversion to *Spartina*–marsh litter suggests also that such activities are a key to continual production of Vitamin B12 essential in the life cycles of plants and animals (Burkholder and Burkholder, 1956).

Among the plant–to–plant relationships, a number of instances of fungal parasitism are known (Table 6.3). Some of these parasites, especially *Uromyces* and *Erysiphe* species, can show epidemic outbreaks especially after extreme environmental circumstances, such as prolonged drought periods

TABLE 6.3

Some parasite—host relationships in European halophytes found in the southwestern Netherlands* and derived from literature

Host	Parasite	Taxonomic group	References
Aster tripolium	*Phomopsis achilleae* var. *asteris* Grove	Coelomycetes	Ellis (1960)
	**Puccinia asteris* Duby	Uredinales	Ellis (1960)
	**Ramularia asteris* (Phill. et Plowr.) Lind.	Moniliales	
Armeria maritima	*Uromyces armeriae* Lév.	Uredinales	Ellis (1960)
Artemisia maritima	*Puccinia absinthii* DC	Uredinales	Ellis (1960)
Elytrigia pungens	**Claviceps purpurea* (Fr.) Tul.	Pyrenomycetes	
	Leptosphaeria discors (Saccardo et Ellis) Saccardo et Ellis	Loculoascomycetes	Wagner (1969)
	**Urocystis agropyri* (Reuss.) Schröt.	Ustilaginales	
	Ustilago hypodytes (Schlecht.) Fr.	Ustilaginales	Ellis (1960)
Festuca rubra	*Claviceps purpurea* (Fr.) Tul.	Pyrenomycetes	Ellis (1960)
Glaux maritima	*Uromyces lineolatus* (Desm.) Schroet.	Uredinales	Ellis (1960)
Halimione portulacoides	*Ascochytula obiones* (Jaap) Died.	Coelomycetes	Ellis (1960)
Juncus maritimus	*Leptosphaeria discors* (Saccardo et Ellis) Saccardo et Ellis	Loculoascomycetes	Wagner (1969)
	Phoma neglecta Desm.	Coelomycetes	Ellis (1960)
	Septoria junci Desm.	Coelomycetes	Ellis (1960)
Limonium vulgare	*Erysiphe polygoni* DC	Pyrenomycetes	Ellis (1960)
	Phoma exigua Desm.	Coelomycetes	Ellis (1960)
	Phoma statices Tassi.	Coelomycetes	Ellis (1960)
	**Uromyces limonii* (DC) Lév.	Uredinales	Ellis (1960)
Plantago maritima	**Erysiphe lamprocarpa* (Wallr.) Duby	Ascomycetes	
Ruppia maritima	*Melanotaenium ruppiae* G. Feldm.	Plasmodiophorales	Feldmann (1959)
Salicornia europeae	*Pleospora herbarum* (Pers.) Rabenh.	Pyrenomycetes	
Salicornia stricta	*Stagonosporopsis salicorniae* (Magn.) Died.	Coelomycetes	Ellis (1960)
	Uromyces salicorniae Lév. ex Cooke	Uredinales	Ellis (1960)
Scirpus maritimus	*Uromyces lineolatus* (Desm.) Schroet.	Uredinales	Ellis (1960)
Spartina spp.	*Leptosphaeria discors* (Saccardo et Ellis) Saccardo et Ellis	Loculoascomycetes	Wagner (1969)
Spartina townsendii	**Claviceps purpurea* (Fr.) Tul.	Pyrenomycetes	
	Lulworthia medusa (Ell. et Ev.) Cribb et Cribb	Pyrenomycetes	Lloyd and Wilson (1962)
Spergularia media	*Cystopus lepigoni* de Bary	Phycomycetes	Ellis (1960)
	**Uromyces caryophyllacearum* (Wallr.) Cif. et Biga	Uredinales	
	Uromyces sparsus (Schum. et Kunze) Lév.	Uredinales	
Suaeda maritima	**Uromyces chenopodii* (Duby) Schroet.	Uredinales	Ellis (1960)
	**Uromyces giganteus* Speg.	Uredinales	
Triglochin maritima	*Asteroma juncaginacearum* Rabh.	Coelomycetes	Ellis (1960)
	**Leptosphaeria juncaginacearum* (Schröt.) Munk	Pyrenomycetes	
	Plasmodiophora maritima G. Feldm.	Plasmodiophorales	Feldmann (1958)
Zostera marina	*Labyrinthula* sp. div.	Hydromyxomycetes	Den Hartog (1970)
	Lulworthia halima (Diehl et Mounce) Cribb et Cribb	Pyrenomycetes	Den Hartog (1970)
	Ophiobolus maritimus (Saccardo) Saccardo	Pyrenomycetes	Den Hartog (1970)
Zostera noltii	*Plasmodiophora bicaudata* J. Feldmann	Plasmodiophorales	Feldmann (1956)

* The author is greatly indebted to the Department of Mycology of the Plant Protection Service (PD), Wageningen (the Netherlands) for the determinations.

occurring in summertime or as a consequence of embankment. Only one case is known of parasitism among phanerogams: *Cistanche lutea*, a member of the Orobanchaceae, feeding on *Halimione portulacoides* and *Salicornia fruticosa*, is found in communities of those species on the creek–bank levees of Portuguese salt marshes (Table 6.4).

Finally, epiphytism is common among green and blue–green algae growing on the stem of phanerogams (see p. 143).

TABLE 6.4

Sequence of phytocoenoses developing on an estuarine salt marsh near Lagos (south coast, Portugal) with *Cistanche lutea* Hoffgg. et Link (Orobanchaceae) growing on creek-bank levees) May 13, 1972)

Number of relevé	1	2	3	4	5	6	7	8
Surface in m^2	4	6	6	45	20	20	20	15
Cover in %	65	100	70	98	95	90	90	90
Spartina maritima	4	2	—	—	—	—	—	—
Salicornia perennis	—	5	4	—	—	—	—	—
Salicornia fruticosa	r	+	—	2	1	3	+	—
Halimione portulacoides	r	r	+	5	4	2	1	2
Suaeda splendens	—	—	1	1	+	+	—	1
Cistanche lutea	—	—	—	2	2	2	+	(+)
Suaeda fruticosa	—	—	—	r	4	3	3	2
Inula crithmoides	—	—	—	—	+	—	2	3
Puccinellia maritima	—	—	—	—	—	+	—	—
Limoniastrum monopetalum	—	—	—	—	—	—	3	3
Atriplex halimus	—	—	—	—	—	—	—	(+)

Estimation of cover abundance according to Braun–Blanquet (1964): r = rare, only one or some individuals and very small cover; + = few individuals and small cover; 1 = rather many individuals, cover less than 5%; 2 = individuals numerous and cover less than 5%, or 5—25% cover irrespective of the number of individuals; 3 = 25—50% cover; 4 = 50—75% cover; and 5 = 75—100% cover.

EUROPEAN SALT-MARSH COMMUNITIES: FLORISTIC ECOLOGY, COMPOSITION AND DISTRIBUTION

In Europe vegetation analysis has been practised using different systems of vegetation description. On the continent the technique most often used is that of the Zürich—Montpellier school of phytosociology. Only in Scandinavia have other floristic systems been followed: the Raunkiaerian school, mainly adhered to in Denmark, the Uppsala school of Professor Du Rietz, and some intermediate methods developed by Nordhagen and Dahl using elements of both the Zürich—Montpellier and Uppsala schools. In the British Isles a tradition arose based more upon structural–functional features than upon floristic composition, culminating in the work of Tansley (1939). For comparison and criticism of these schools the reader is referred to Shimwell (1971) and Westhoff and Van der Maarel (1973).

An overall coverage of the coastal salt marsh communities of Europe therefore gives the best results if use is made of the data obtained from studies carried out with the methods of the Zürich—Montpellier school of Braun–Blanquet and related approaches. It is on this basis that the communities are dealt with in the following pages.

In Tables 6.5 and 6.6 a general view is given of the syntaxonomical relationships of communities occurring in southwestern Netherlands and on the Swedish west coast. The geographical distribution of the higher syntaxa in Europe is outlined in Fig. 6.7 (see also pp. 4—8).

A. *Zostera communities* (alliance Zosterion)

Communities of marine aquatic species of the genus *Zostera*.

Associations. Differently interpreted: mostly two, viz. *Zosteretum marinae* and *Zosteretum nanae* or *Zosteretum nano–stenophyllae*, but sometimes three, viz. *Zosteretum marinae stenophyllae* and *Zosteretum nanae* besides *Zosteretum marinae* (Harmsen, 1936; Den Hartog, 1958; Beeftink, 1965, 1968).

Floristic composition and structure. These are homogeneous communities composed of one (or two) *Zostera* species. In the *Zostera marina* beds of the brackish fjords of Denmark, in estuaries, and in the Baltic Sea some other phanerogams occur, viz. *Potamogeton pectinatus*, *Ruppia* species, and *Zannichellia palustris,* as well as some Characeae and algae, such as *Furcellaria fastigiata* (Ostenfeld,

WESTERN AND NORTHERN EUROPE

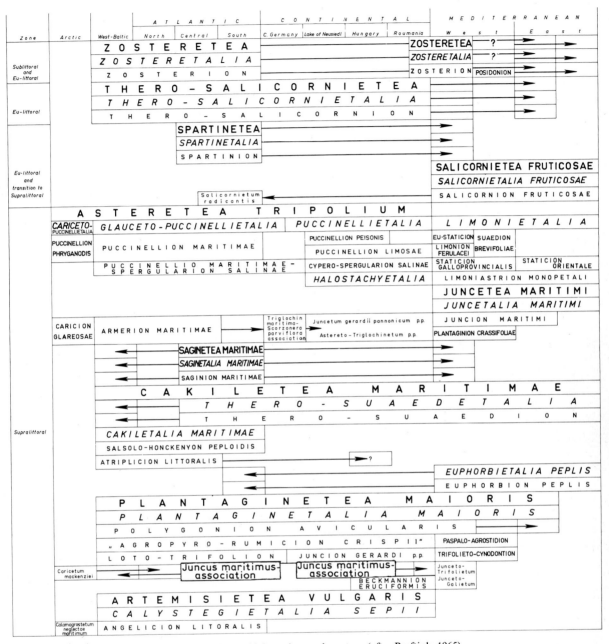

Fig. 6.7. Geographical distribution of the European higher salt-marsh syntaxa (after Beeftink, 1965).

1908; Philip, 1936; Gillner, 1960; Kornaś et al., 1960). *Zostera marina* is often associated with a great number of epiphytic algae and animals (Ostenfeld, 1908; Van Goor, 1919, 1921; Kornaś et al., 1960). Under brackish conditions the *Zostera noltii* (= *Z. nana*) communities are often found together with *Ruppia spiralis* and sometimes with *Zannichellia palustris*. Epiphytic organisms are rare. Detached and free-living algae dragged on by the tidal currents, such as *Ulva lactuca, Porphyra umbilicalis, Chaetomorpha linum,* and *Enteromorpha* spp., are often found caught in the seagrass communities.

Distribution. *Zostera marina* is widely distributed, and extends far into the Arctic zone, reaching the

TABLE 6.5

Classification of salt–marsh communities in the southwestern Netherlands according to the Braun–Blanquet method (from Westhoff and der Maarel, 1973)

Classes	Thero–Salicornietea	Spartinetea		Asteretea tripolii	
Orders	Thero–Salicornietalia	Spartinetalia		Glauco–Puccinellietalia	
Alliances	Thero–Salicornion	Spartinion		Puccinellion maritimae	
Associations	Salicornietum strictae	Spartinetum maritimae	Spartinetum townsendii	Puccinellietum maritimae	Halimion portulaco
Column	1	2	3	4	5
Number of relevés	14	24	30	124	40
Character–taxa of the associations					
Salicornia europaea coll.[1]	100(1—3)	33(+ —2)	14(+ —2)	78(+ —2)	40(+ —
Spartina maritima	14(+)	100(2—4)	—	6(+)	—
Fucus vesiculosus f. volubilis	—	79(+ —5)	7(2—3)	4(+ —2)	—
Spartina townsendii agg.	86(+ —1)	79(+ —2)	100(3—5)	50(+ —2)	35(+ —2
Puccinellia maritima[2]	25(+ —1)	33(+ —1)	30(+ —1)	100(3—5)	92(+ —2
Halimione portulacoides	8(+)	—	33(r—1)	83(+ —2)	100(3—5)
Artemisia maritima	—	—	—	5(+ —1)	—
Armeria maritima	—	—	—	6(+ —1)	—
Carex extensa	—	—	—	—	—
Puccinellia distans	—	—	—	—	—
Puccinellia fasciculata	—	—	—	—	—
Puccinellia retroflexa	—	—	—	—	—
Scirpus maritimus var. compactus[3]	—	—	27(+ —2)	—	—
Faithful taxa of Puccinellion maritimae					
Bostrychia scorpioides	—	17(+ —4)	14(+ —2)	21(+ —4)	37(+ —4
Character–taxa of Armerion maritimae					
Juncus gerardi	—	—	—	2(+)	—
Festuca rubra f. litoralis	—	—	—	14(+ —2)	62(+ —1
Glaux maritima	—	—	10(r—1)	50(+ —2)	15(+ —1
Parapholis strigosa	—	—	—	—	—
Agrostis stolonifera var. compacta subvar. salina	—	—	—	—	—
Character–taxon of Puccinellio–Spergularion salinae					
Spergularia salina	—	—	—	—	—
Character–taxa of Glauco–Puccinellietalia					
Spergularia media	—	—	7(r— +)	73(+ —2)	42(+ —2)
Limonium vulgare ssp. vulgare	8(+)	8(+)	30(r— +)	73(+ —2)	60(+ —2)
Character–taxa of Asteretea tripolii					
Aster tripolium	50(+ —2)	33(+ —2)	77(r—2)	98(+ —2)	97(+ —2)
Triglochin maritima	8(+)	—	27(r—2)	86(+ —4)	62(+ —2)
Plantago maritima	—	—	20(r—1)	65(+ —4)	65(+ —2)
Other taxa					
Suaeda maritima	50(+ —1)	12(+ —2)	37(+ —1)	64(+ —2)	60(+ —2)
Atriplex hastata	—	—	77(+ —2)	25(+ —2)	5(+)
Elytrigia pungens	—	—	7(r— +)	1(+)	22(+)
Lolium perenne	—	—	—	—	—
Plantago coronopus	—	—	—	—	—
Phragmites communis	—	—	—	—	—

Addenda

Column 1: Zostera noltii 29(+ —2); Column 7: Centaurium pulchellum 17(+ —2), Carex distans 5(+), Sagina maritima 2(+), Solanum dul mara 2(+); Column 8: Centaurium pulchellum 57(r—2), Carex distans 29(r—2), Juncus maritimus 29(r—1), Lotus tenuis 29(2), Hippophae rha noides 29(r— +), Trifolium fragiferum 14(r), Sonchus arvensis 14(r), Trifolium repens 14(r), Centaurium littorale 14(+)⁰; Column 9: Polygon. aviculare 41(+ —2), Elytrigia repens 22(+ —1), Potentilla anserina 8(+ —1), Plantago major 24(+ —2), Leontodon autumnalis 5(+), Trifoli. repens 11(+)⁰, Coronopus squamatus 8(+ —2), Matricaria inodora 11(+)⁰, Bromus mollis 8(+), Ranunculus sceleratus 16(+ —1)⁰, Poa ann

[1] In the alliances Thero–Salicornion and Spartinion represented by S. stricta Dum.
[2] Preferential character–taxon of the association; also selective character taxon of the alliance Puccinellion maritimae and exclusive charact. taxon of the order Glauco–Puccinellietalia.
[3] Also character–taxon of the alliance Halo–Scirpion.
N.B. The superscript ⁰ is a convention meaning that the taxon is represented by stunted individuals.

...erion maritimae			Puccinellio–Spergularion salinae			Halo–Scirpion
...misietum ...itimae	Juncetum gerardii 7 64	Junco–Caricetum extensae 8 7	Puccinellietum distantis 9 37	Puccinellietum fasciculatae 10 10	Puccinellietum retroflexae 11 12	Halo–Scirpetum maritimi 12 19
(+−1)°	42(+−1)°	14(+)°	35(+−2)°	70(r—2)°	100(1—2)°	—
	—	—	—	—	—	—
(+)	20(+−2)	—	8(+)°	—	17(r)	37(+−2)
(+−2)	31(+−2)	14(+)	62(+−1)	90(r—3)	25(r—1)	26(1—3)
(+−3)	62(+−1)	—	5(+)	—	—	—
(+−3)	59(+−1)	—	8(+−1)	—	—	—
(+)	90(+−3)	14(r)	—	—	—	—
	—	100(1—4)	100(1—5)	20(r—2)°	17(r−+)	—
	—	—	—	100(2—4)	17(r)	—
	—	—	—	—	100(1—4)	—
	—	—	32(+−2)°	10(+)°	8(r)°	100(3—5)
(+−2)	—	—	—	—	—	—
	86(+−5)	100(2—4)	14(+−2)	30(+)	—	5(+)
(3—5)	97(+−5)	86(+−3)	35(+−3)	10(r)	—	11(+)
(+−1)	98(+−3)	100(2—3)	35(+−3)	30(r—2)	—	—
(+−1)	64(+−3)	43(+−2)	19(+−2)	10(+)	8(+)	—
	11(+−1)	100(+−3)	65(+−2)	30(+−2)	—	42(+−4)
—	—	—	97(+−3)	80(+−2)	100(r—2)	—
2(+−2)	47(+−1)	14(+)	19(+−1)	20(1—2)	8(+)	—
5(+−2)	91(+−3)	57(r—2)	3(+)°	—	—	—
0(+−2)	72(+−2)	71(r—1)	65(+−3)	100(r—3)	100(+−4)	74(+−2)
9(+−1)	77(+−2)	29(+)	5(+−1)	50(r—2)	8(r)	16(+−1)
2(+−3)	98(+−3)	100(+−2)	11(+)	—	8(+)	—
1(+−1)	20(+−1)°	—	38(r—1)°	30(r—1)°	8(1)°	—
1(+)	5(+)°	—	62(+−2)°	40(r)°	—	63(1—3)
1(+−2)	28(+−2)	—	30(+−2)	—	—	32(+−2)
—	2(+)	—	30(+−3)	10(r)	—	—
—	2(+)	29(+)	11(+−1)	10(1)	25(+)°	—
—	2(+)	43(r−+)°	30(+−3)	10(r)°	67(r—2)°	11(+−2)

(+−2), *Cochlearia officinalis* 11(+), *Festuca arundinacea* 8(+), *Cirsium arvense* 5(+−1), *Poa trivialis* 3(3), *Hordeum secalinum* 3(1), *Taraxa-*
...m sp. 5(+), *Sonchus arvensis* 3(+), *Poa pratensis* 3(+), *Solanum nigrum* 3(+), *Senecio vulgaris* 3(+), *Anagallis arvensis* 3(+), *Leontodon*
...dicaulis 3(+); Column 10: *Centaurium pulchellum* 10(+), *Plantago major* 20(r)°, *Matricaria inodora* 10(r)°, *Bromus mollis* 10(+), *Sagina mari-*
...na 40(+−2), *Juncus bufonius* 40(r—2), *Hordeum marinum* 10(r), *Samolus valerandi* 10(r); Column 11: *Bromus mollis* 8(r); Column 12: *Ra-*
nculus sceleratus 5(+), *Cochlearia officinalis* 5(+), *Atriplex littoralis* 5(+).

TABLE 6.6

Classification of salt–marsh communities on the Swedish west coast (derived from Gillner, 1960)

Classes	Ruppietea		Thero–Salicornietea		Asteretea tripolii				
Orders	Ruppietalia		Thero–Salicornietalia		Glauco–Puccinellietalia				
Alliances	Ruppion maritimae		Thero–Salicornion		Puccinellion maritimae		Puccinellio-Spergularion salinae	Halo-Scirpion	
Associations	*Ruppia maritima* ass.	*Eleocharetum parvulae*	*Salicornietum strictissimae*	*Salicornietum europaeae*	*Puccinellietum maritimae*		*Puccinellia distans* ass.	*Scirpetum maritimi*	*Caricetum paleaceae*
Subassociations					subass. with *Salicornia europaea*	typical subass.			
Column	1	2	3	4	5	6	7	8	9
Method[1]	B–B	H.S	H.S	H.S	H.S	H.S	B–B	B–B	B–B
Number of relevés	7	16	35	23	30	35	11	21	12
Character–taxa of the associations									
Ruppia maritima	100(2—4)	60(+—2)	23(+—1)	—	—	—	—	10(+)	—
Eleocharis parvula	14(+)	100(2—5)	—	—	—	—	—	5(+)	—
Salicornia strictissima	14(+)	6(+)	100(+—5)	—	—	80(+—1)	—	—	—
Salicornia europaea	—	—	—	100(1—5)	100(+—4)	73(+—3)	23(+—1)	27(+)	—
Suaeda maritima	—	—	—	78(+—5)	73(+—3)	23(+—1)	9(+)	—	—
Puccinellia maritima	—	13(+—1)	3(+)	30(+—1)	100(4—5)	100(5)	55(+—1)	10(+)	—
Limonium humile	—	—	—	—	7(1)	6(1)	—	—	—
Halimione pedunculata	—	—	—	4(+)	50(+—4)	3(+)	—	—	—
Aster tripolium	—	—	3(1)	—	—	71(+—3)	—	19(+)	17(+)
Spergularia media	—	—	3(+)	—	20(+—1)	49(+—1)	—	—	—
Spergularia marina	—	—	—	4(1)	67(+—4)	9(+—1)	100(1—3)	—	—
Puccinellia distans	—	—	—	—	—	—	91(1—3)	—	—
Juncus bufonius	—	—	—	—	—	—	65(+—2)	—	—
Polygonum aviculare	—	—	—	—	—	—	73(+—1)	—	—
Scirpus maritimus	—	38(+—2)	—	—	—	—	—	100(1—5)	67(+)
Carex paleacea	—	—	—	—	—	—	—	10(+)	100(2—)
Eleocharis uniglumis	—	—	—	—	—	—	—	5(+)	58(+)
Carex mackenziei	—	—	—	—	—	—	—	—	—
Artemisia maritima	—	—	—	—	—	—	—	—	—
Limonium vulgare	—	—	—	—	—	—	—	—	—
Armeria maritima	—	—	—	—	10(+—1)	—	—	—	—
Leontodon autumnalis	—	—	—	—	—	—	—	—	—
Trifolium fragiferum	—	—	—	—	—	—	—	—	—
Sagina nodosa	—	—	—	—	—	—	9(+)	—	—
Centaurium pulchellum	—	—	—	—	—	—	—	—	—
Centaurium vulgare	—	—	—	—	—	—	—	—	—
Carex recta	—	—	—	—	—	—	—	—	—
Scirpus rufus	—	—	—	—	—	—	—	—	—
Sagina maritima	—	—	—	—	—	—	—	—	—
Cochlearia danica	—	—	—	—	—	—	—	—	—
Plantago coronopus	—	—	—	—	—	—	—	—	—
Sedum acre	—	—	—	—	—	—	—	—	—
Sagina procumbens	—	—	—	—	—	—	—	—	—
Elytrigia repens	—	—	—	—	—	—	—	—	—
Character–taxa of Armerion maritimae									
Plantago maritima	—	—	—	—	50(+—2)	49(+—1)	55(+)	—	—
Glaux maritima	—	—	—	—	10(+—2)	14(+—2)	36(+)	—	8(+)
Juncus gerardi	—	—	—	—	—	—	73(+—1)	—	17(+)
Festuca rubra	—	—	—	—	—	—	—	—	8(1)
Potentilla anserina	—	—	—	—	—	—	—	—	—
Poa pratensis	—	—	—	—	—	—	—	—	—
Trifolium repens	—	—	—	—	—	—	9(+)	—	—
Character–taxa of Asteretea tripolii									
Triglochin maritima	—	38(+—1)	6(+—2)	—	17(+—1)	40(+—3)	27(+—1)	38(+—1)	83(+)
Agrostis stolonifera	—	—	—	—	—	—	100(+—1)	38(+—3)	100(+)
Other taxa									
Atriplex latifolia	—	—	—	—	10(+—1)	3(+)	82(+—2)	33(+)	42(+)
Tripleurospermum maritimum	—	—	—	—	—	—	18(+)	—	—
Cochlearia officinalis	—	—	—	—	—	—	—	—	—
Eleocharis pauciflora	—	—	—	—	—	—	—	—	—
Triglochin palustris	—	—	—	—	—	—	—	—	—
Phragmites communis	—	6(+)	—	—	—	—	46(+—1)	—	—
Scirpus tabernaemontani	—	—	—	—	—	—	—	43(+—5)	17(+)
								48(+—3)	50(+)

Addenda

Column 1: *Ruppia spiralis* 14(+), *Zannichellia palustris* ssp. *pedicellata* 14(1); Column 2: *Zannichellia palustris* ssp. *pedicellata* 6(2), *Potamogeton pusillus* 19(1—2); Column 7: *Plantago major* 27(+), *Ranunculus sceleratus* 9(+), *Chenopodium glaucum* 9(+), *Alopecurus geniculatus* 54(+—1); Column 11: *Myosotis laxa* ssp. *caespitosa* 9(+); Column 12: *Vicia cracca* 11(+), *Atriplex littoralis* 33(+—1), *Galium aparine* 11(+), *Stellaria media* 11(+), *Rumex crispus* 33(+), *Sonchus arvensis* 33(+), *Angelica archangelica* var. *litoralis* 11(+); Column 13: *Odontites littoralis* 11(+), *Rhinanthus serotinus* 11(+), *Lotus corniculatus* 11(+), *Lotus tenuis* 22(1—2), *Atriplex littoralis* 11(+), Column 14: *Odontites littoralis* 33(1—4), *Rhinanthus serotinus* 25(+—2), *Carex distans* 3(1), *Cerastium holosteoides* 3(1), *Festuca pratensis* 6(+), *Lotus corniculatus* 3(1), *Lotus tenuis* 8(+—1), *Plantago major* 11(+—1), *Taraxacum* 22(+), *Vicia cracca* 3(1); Column 15: *Cerastium holosteoides* 25(+—1), *Plantago major* 8(+), *Puccinellia retroflexa* 8(1), *Silene maritima* 8(+), *Sagina subulata* 8(+); Column

[1] Estimation of cover-abundance according to Braun-Blanquet (B–B) or according to Hult and Sernander (H.S).
Numbers indicate presence (%) and maximal fluctuations in cover-abundance of the species.

				Juncetum gerardii					Saginetea maritimae	
									Saginetalia maritimae	
									Saginion maritimae	
		Caricetum mackenziei	Artemisietum maritimae	typical subass.	subass. with Odontites littoralis and with Leontodon autumnalis	subass. with Centaurium	Caricetum rectae	Scirpetum rufii	Sagino maritimae—Cochlearietum danicae	Elytrigietum repentis maritimum
		11	12	13	14	15	16	17	18	19
		H.S	B–B	B–B	H.S	B–B	B–B	B–B	B–B	H.S
		11	9	9	36	4	12	11	12	42
		—	—	—	—	—	—	—	—	—
		—	—	—	—	—	—	—	—	—
		—	22(+—1)	—	—	—	—	—	—	—
(+)		—	39(+—1)	22(+)	—	25(+)	—	—	8(1)	—
		—	22(+—1)	—	—	—	—	—	—	—
		—	—	—	—	—	—	—	—	—
		—	—	—	—	—	—	—	—	—
(+—1)		—	—	—	—	—	—	—	—	5(+—1)
(+—1)		36(1—2)	—	—	—	—	—	—	—	—
(2—3)		82(+—2)	—	—	—	—	83(+—2)	36(+—1)	—	—
		100(4—5)	—	—	—	—	—	—	—	—
		—	67(3—4)	—	—	—	—	—	—	—
		—	44(1—3)	—	—	—	—	27(+)	100(+—2)	5(+)
		—	44(+—1)	67(+—1)	14(1—4)	100(+—2)	42(+—1)	73(+—1)	42(+—1)	21(+—1)
		—	—	11(+)	89(+—2)	50(+—1)	—	—	—	2(+)
		—	—	—	53(+—5)	75(1)	—	—	—	—
		—	—	11(1)	6(+)	100(+—1)	—	18(+)	8(+)	—
		—	—	—	3(+)	75(+—1)	—	18(+)	25(+—1)	—
		—	—	—	—	75(+—1)	—	9(+)	8(+)	2(+)
		46(1—2)	—	—	8(+—2)	—	100(2—4)	9(+)	—	—
(+)		18(1—2)	—	—	3(1)	—	—	100(2—3)	67(+—2)	—
		—	—	—	—	—	—	—	58(+—2)	—
		—	—	—	—	—	—	—	50(+—3)	—
		—	—	—	—	—	—	—	58(+—2)	—
		—	—	—	—	—	—	—	67(+—1)	—
		—	33(+—2)	11(+)	17(+—2)	25(+)	—	27(+—1)	17(+)	83(+—5)
		—	—	—	8(+—1)	—	—	—	—	—
(+)		—	89(+—2)	100(+—2)	72(+—3)	100(+—2)	50(+—2)	100(+—2)	92(+—1)	21(+—1)
(+—1)		—	78(+—2)	89(+—2)	67(+—2)	100(+—2)	67(+—2)	100(+—2)	42(+)	12(+—1)
(+—2)		—	78(+—2)	100(+—4)	100(1—5)	100(2)	100(+—2)	91(1—2)	75(+—1)	19(+—1)
		—	100(2—3)	100(1—4)	97(+—2)	100(1—2)	92(+—2)	100(+—2)	100(+—2)	88(+—3)
		—	44(+—2)	33(1)	75(+—4)	50(+—1)	42(+—1)	65(+—2)	33(+—1)	79(+—5)
		—	33(+—1)	—	56(+—2)	—	67(+—1)	65(+—1)	17(+)	45(+—1)
		—	—	—	56(+—5)	—	92(+—2)	36(+—2)	—	10(+—1)
(+—1)		9(1)	67(+—1)	89(+—1)	44(+—2)	25(+)	83(+—2)	82(+—2)	—	2(+)
(+—3)		65(+—3)	89(+—2)	100(1—3)	100(+—2)	100(1—2)	100(+—1)	100(+—1)	92(+—1)	26(+—1)
		—	56(+—1)	22(+)	—	—	—	—	33(+)	36(+—3)
		—	33(+)	—	—	—	—	—	50(+)	7(1—2)
		—	22(+)	11(+)	—	—	—	—	50(+—2)	—
		18(+)	—	—	—	—	—	45(+—1)	—	—
(+—1)		36(+—2)	—	—	22(+—2)	—	50(+—1)	46(+—1)	—	—
(+—1)		18(1—2)	11(1)	22(+—1)	22(+—1)	—	33(+)	—	—	5(+—1)
(+—1)		46(+—2)	—	—	—	—	—	—	—	—

dontites littoralis 25(+), *Rhinanthus serotinus* 25(+—1), *Carex nigra* 9(+), *Eriophorum angustifolium* 17(+), *Galium palustre* 9(+), *Montia fontana* ssp. *lamprosperma* 9(+); Column 7: *Carex oederi* 45(+), *Carex nigra* 36(+), *Juncus articulatus* 36(+—1); Column 18: *Carex oederi* 25(+), *Lotus corniculatus* 25(1), *Linum catharcticum* 25(+); Column 19: *Carex stans* 2(+), *Cerastium holosteoides* 2(+), *Lotus corniculatus* 12(1—2), *Taraxacum* 14(+—1), *Vicia cracca* 28(+—3), *Atriplex littoralis* 14(+—1), *Stellaria media* 2(+), *Rumex crispus* 10(+1), *Sonchus arvensis* 39(+—3), *Angelica archangelica* var. *litoralis* 5(+—1), *Angelica sylvestris* 2(+), *Anthriscus sylvestris* 5(+—2), *Artemisia vulgaris* 2(+), *Cirsium arvense* (+), *Equisetum arvense* 2(+), *Festuca pratensis* 7(+—1), *Galeopsis bifida* 17(+—2), *Linaria vulgaris* 5(1), *Carex disticha* 2(+).

White Sea and Cheshskaya Guba in northern Russia. Its southern limit is reached in Spain near Gibraltar. In the Mediterranean it is mainly restricted to some northern localities (south France, north Adriatic, Aegean seas), but it is common in the Black Sea. In the Baltic it extends up to the Gulf of Bothnia. *Zostera noltii* is distributed from southern Norway and the British Isles southward to Mauritania. It occurs locally in the Mediterranean and the Black Sea, and in the Baltic it reaches Kiel Bay, Heiligenhafen and the island of Bornholm (Den Hartog, 1970).

Ecology. The *Zosteretum marinae* is essentially a sublittoral association penetrating into the intertidal belt at the most to about mean sea level, dependent on the water–holding capacity of the substratum. It develops in more or less sheltered localities on substrates ranging from soft mud to firm sand. The upper limit of the area occupied by *Zostera marina* seems to depend greatly on the degree of desiccation during low tide of the part of its stems just emerging from the substratum (Beeftink, 1965). From the French Channel coasts northwards the rhizomes are killed by frost in the eulittoral zone. Consequently, the association is summer–annual there *(Zosteretum marinae stenophyllae)*. In estuaries *Zostera marina* does not penetrate upstream beyond the point where the average salinity at high tide is about 14‰ Cl$^-$. In bays and seas, where salinity fluctuations are much less pronounced, the species is found up to isohalines of 6—7‰ (former Zuyderzee: Van Goor, 1919, 1921) or even of 3‰ Cl$^-$ (Baltic Sea: Luther, 1951).

The *Zosteretum nanae* develops in intertidal flats of mud or fine sand rich in detritus, between the MLWN and MHWN tide levels. The characteristic species is less susceptible to frost damage and is less euryhaline than *Zostera marina,* reaching average annual isohalines of 9—10‰ Cl$^-$ at high tide both in estuaries and in the Baltic. From the fully marine West–European coasts to well into the Baltic its habitat shifts downward with decreasing salinity from the intertidal to the permanently submerged zone. *Zostera noltii* thus exhibits an excellent example of what Remane (1955) has called "total brackish–water submergence". For more ecological data see Den Hartog (1970).

Dynamics. Both associations usually are in a state of continual change, being built up in one place and broken down in another. They are, therefore, both not an initial stage in a succession series, but a succession series in themselves (Den Hartog, 1970), each of them perhaps in a cyclic–dynamic state or in a long–term dynamic equilibrium. This agrees with the results of transplantings with *Zostera* species carried out by Ranwell et al. (1974), suggesting that growth of *Zostera noltii* is favoured in areas where a close balance between the forces of erosion and accretion occurs.

Table 6.7 gives examples of vegetation dynamics within the *Zosteretum nanae.* The "wasting disease" in *Zostera marina,* which destroyed most of the Atlantic *Zostera* meadows between 1930 and 1935, had precursors of different intensity back to 1854 (Cottam, 1934, 1935). From many reports, recently summarized by Rasmussen (1973), it appears that the last epidemic of the thirties started in North

TABLE 6.7

Vegetation dynamics in a *Zostera noltii* community growing on a mud flat in the Eastern Scheldt, southwestern Netherlands (salinity of the flood water poly- to euhaline; mean tidal range 360 cm; soil surface 105 cm below MHW level; quadrat size 30 m^2

Year:	1965	1967	1968	1969	1970	1971	1972	1973	1974
Cover in %:	20	50	70	70	70	30	40	20	3
Zostera noltii	02	05	07	07	07	03	04	02	m!
Zostera marina	—	—	—	—	r	p	a	—	—
Spartina townsendii	—	(r)	r	—	(r)	r	—	—	—
Hydrobia ulvae	m	m!	m!!	m!	m!	m!!	m!	p	m
Littorina littorea	r	r	—	—	—	—	—	—	—

Estimation of cover–abundance according to Doing Kraft (1954): Letters: cover less than 5%, r = 1—3 individuals, p = 4—15 individuals, a = 16—40 individuals, and m = >40 individuals, m! and m!! = individuals (very) numerous. Numbers: 01 = cover 5—15%, 02 = cover 15—25%, etc., 09 = 85—95%, and 10 = 95—100%.

America, crossed the Atlantic and spread from northwestern France and south England to the north (The Netherlands: summer 1932; Denmark: autumn 1932; Norway: 1933—34; White Sea: 1960—61) and to the south (southwestern France to the Iberian Peninsula: 1932—33). According to Den Hartog (1970) it seems most likely that a *Labyrinthula* species (Protozoa) caused the damage, but certain fungi have also been stigmatized. Recently, however, McRoy (1966) and Rasmussen (1973) claims that this catastrophic elimination was the result of extremely high water temperatures in those coastal waters.

B. *Annual Salicornia communities* (alliance Thero-Salicornion)

Primary or secondary pioneer communities of annual *Salicornia* species growing on tidal mud flats and in depressions occurring in the salt marsh and in inland saltings (Table 6.5, col. 1; Table 6.6, col. 3, 4).

Associations. The characteristic and dominant taxon, *Salicornia europaea* agg., is very variable but not yet fully understood from the morphological and genetic points of view. A number of subspecific taxa or more distinct species can be distinguished growing in different habitats and consequently excluding each other. Therefore, associations could at present only be described locally: *Salicornietum strictae* (e.g., Christiansen, 1955; Beeftink, 1965), *Salicornietum strictissimae* (Gillner, 1960), *Salicornietum patulae* (Christiansen, 1955).

Floristic composition and structure. These are communities poor in species, and optimally consisting of summer annuals only. Algae washed ashore by the tides and caught by the phanerogams may be abundant, but under very sheltered conditions marsh fucoids (e.g., *Fucus vesiculosus* ecad *volubilis*) and other algae (*Vaucheria* sp. div., *Bostrychia scorpioides*) may be associated (Géhu, 1960).

Distribution. These communities occur from the Lofoten (Gillner, 1955) and the Shetland Islands southwards to the west coasts of France, and locally on the Portuguese coasts. In the Baltic eastwards to the Polish coasts (Piotrowska, 1974), and locally to well into the Gulf of Bothnia (Leiviskä, 1908). In the Mediterranean locally along the northern coasts (Frei, 1937; Oberdorfer, 1952; Pignatti, 1966; Wolff, 1968).

Ecology. These are pioneer communities of the halosere on tidal mud flats and in low salt marshes of euhaline and polyhaline seas and estuaries, as well as on banks of more or less stagnant saline inland seas and lakes (Baltic Sea, Black Sea); they are also found in inland saltings. They occur within the tidal reach on clayey to sandy mud flats normally subject to daily submergence (eulittoral). In the salt marsh proper they are confined to saline depressions discharging excess of water. The soil is only superficially aerated (0.5—1 cm), or totally reduced, and often black-coloured underneath owing to concentrations of sulphides. Soil salinity is mostly close to that of the flooding water, but may increase considerably in dry summer periods. The *Salicornietum strictae* is distributed exclusively within the daily reach of the tides, and seems to avoid high soil salinities as it is not found on the warm temperate coasts. Beyond the reach of the tides (in the higher parts of the salt marsh, on the Baltic coasts and in inland saltings) other *Salicornia* taxa occur, and, consequently, other associations develop, such as the *Salicornietum patulae* (Christiansen, 1955; Piotrowska, 1974).

Dynamics. These communities are ephemeral owing to dominance of annuals. Overlapping zones with the lower-developing *Zostera* communities may occur, though rarely, as well as transition zones to higher-marsh associations, for instance the *Puccinellietum maritimae*. Other species from the latter associations are found in the *Salicornietum*, but mostly sparsely. *Suaeda maritima* penetrates *Salicornia* communities where algae gathered by the tides have been washed ashore or caught by the phanerogams, or — in tideless areas — where many bird droppings provide high nutrient levels. In the succession, development towards *Puccinellietum maritimae* communities is obvious (Table 6.8), but also other lines of succession may be conceivable, for instance towards the *Spartinetum maritimae* and *Halimionetum portulacoidis*. *Spartina townsendii* agg., which in recent times has been invading a large part of the European coasts, has

TABLE 6.8

Transformation of a *Salicornia stricta* community into a *Puccinellia maritima* community in a salt marsh bordering the Eastern Scheldt, southwestern Netherlands (salinity of the flood water polyhaline; mean tidal range 298 cm; soil surface 10 cm below MHW level; quadrat size 15 m^2)

	Year: 1965 Cover in %: 70	1966 45	1967 70	1968 98	1969 98
Salicornia stricta	07	03	03	02	p
Puccinellia maritima	a	01	03	08	10
Spartina townsendii	p	a	m	m	m
Aster tripolium	p	a	a	01	01
Spergularia media	—	—	—	r	r
Hydrobia ulvae	m!!	m!!	m!!	m!!	m!!
Littorina saxatilis	p	a	p	r	a
Littorina littorea	r	—	—	r	—

For the meaning of cover–abundance symbols see Table 6.7.

occupied many of the habitats of the association on the tidal mud flats.

Allied vegetation types. In the higher parts of the *Salicornia* belt, where the soil is eutrophicated by algae deposited on the mud, *Salicornia* is usually intermingled with, or even replaced by, *Suaeda maritima* populations. Under brackish conditions, especially on heavy clay in the mesohaline parts of estuaries, a belt of the mostly biennial *Aster tripolium* develops in this zone. *Aster* may show here a very luxuriant growth owing to estuarine eutrophication. This community may also develop in depressions and on low creek banks of the brackish marsh, but only when it is not grazed (compare the "Creek *Asteretum*" of Chapman, 1934). Under more exposed conditions, especially at the slightly eroded edge of the marsh, the perennial *Salicornia perennis* (= *S. radicans*) may develop as communities. This *Salicornietum radicantis*, which is also known from the north Mediterranean, is distributed from Norfolk and Wales southward to the south coast of Portugal. It is ecologically closely related to the *Halimionetum portulacoidis*, as these two associations are often found adjoining one another, or even forming a mosaic (Beeftink, 1965).

Also broadly in the *Salicornia* zone but generally somewhat lower and under brackish conditions, *Eleocharis parvula* communities occur (Gillner, 1960; Table 6.6, col. 2). This species has a wide but very disjunct area of distribution, extending from southern Finland and southern Scandinavia to northern Spain and central Germany, but is probably very rare at the moment as it is often associated with Littorellion species indicating a kind of oligotrophy (Beeftink, 1965).

C. *Spartina communities* (alliance Spartinion)

These are communities of *Spartina* species growing on tidal mud flats and on salt bottom land in polder areas (Table 6.5, col. 2, 3).

Associations. On the European coasts three associations occur: (1) *Spartinetum maritimae* (incl. *Limonio–Spartinetum maritimae* in the Venice lagoons); (2) *Spartinetum alterniflorae;* and (3) *Spartinetum townsendii,* parallel to the three dominant *Spartina* taxa which usually exclude one another totally. For an elaborate description of the European *Spartina* communities see Beeftink and Géhu (1973).

Floristic composition and structure. These communities are very poor in species, usually forming dense monospecific stands of one of the hemicryptophytic *Spartina* species. Algae are generally rare except in the least dense *Spartinetum maritimae,* in which marsh fucoids and the rhodophyte *Bostrychia scorpioides* may be common locally.

Distribution. The *Spartinetum maritimae* extends from southern England and the southwestern Netherlands southwards to the Moroccan west coast and, isolated, in the lagoons of the Gulf of

Venice. *Spartina alterniflora* was introduced from North America in Southampton Water, Rade de Brest, the Adour and Bidassoa estuaries, and in some neighbouring localities in northern Spain. *Spartina townsendii* agg. is a neophyte and originated as a male–sterile primary hybrid *(S. x townsendii)*, a product of the crossing of *S. alterniflora* and *S. maritima* in Southampton Water about 1860—70. By doubling of the chromosomes this primary hybrid developed into a fertile amphidiploid *(S. anglica)*. The *Spartinetum townsendii* is distributed from Ireland, the west and east coasts of Scotland, and Skallingen to southwestern France and the Rio de Pontevedra (northwestern Spain). The hybrid has been introduced in many parts of the world for use in tidal–land reclamation (Ranwell, 1967). Its area of distribution can therefore still expand, both spontaneously and on the initiative of man.

Ecology. The associations develop optimally in waterlogged mud under conditions of accretion. The *Spartina maritima* communities form a rather narrow belt closely below MHW level, bordering euhaline and polyhaline tidal water bodies of bays, lagoons and estuaries. They withstand some erosion after establishment, in which case algae such as *Fucus spiralis* and *Enteromorpha* attached to the emerging roots, shells, etc., may be common. Contrary to the American *Spartina alterniflora* communities, the European ones evidently prefer mesohaline and oligohaline tidal water bodies. Their preference for lower and more fluctuating salinities is also shown by the presence of *Limonium humile* and the almost complete absence of the less euryhaline phanerogams. The *Spartina townsendii* communities develop in a broad and dense belt mainly below MHW level bordering euhaline to mesohaline tidal water bodies. They occur also in the lower parts of the salt marsh proper, and even on sandy flats and beaches building low primary dunes. In polder areas, *Spartina townsendii* is established in shallow euhaline and polyhaline pools and ditches communicating with the tidal water by means of seepage. The association is more resistant against erosion than any other pioneer community of the tidal flats. Its susceptibility to frost, however, can promote erosive processes from The Netherlands northwards along the continental coasts.

Dynamics. In Europe the *Spartinetum maritimae* is the only *Spartina* association which has from older times occupied a functional place in the salt–marsh ecosystem. It is also the least effective in promoting sedimentation processes. Owing to its pioneer character, transitions from the *Salicornietum strictae* are rare. On the West–European coasts the *Spartinetum maritimae* is followed in succession by *Puccinellia maritima* communities. Southwards, at the southwestern French and Iberian coasts, it is also followed by *Salicornia perennis* or by *Halimione portulacoides* stands, but this latter succession can also be superseded by a change into *Salicornia perennis* stands, owing to a greater exposure to wave and tidal action. Succession of *Spartina alterniflora* communities towards the indigenous salt–marsh communities seems to be hardly possible, owing to its exotic origin and to the way this vegetation changes the original environmental conditions once it is firmly established. Likewise, the *Spartinetum townsendii* seems to have still not consolidated its position in the succession series from eulittoral to supralittoral salt–marsh communities. Ousting *Spartina maritima*, *Salicornia europaea* and *Zostera nana* stands as well as the lower parts of the *Puccinellia maritima* belt (Table 6.9), its vigorous and dense colonization of the tidal flats initiate rapid accretion processes and thus a different soil genesis, blocking the lines of succession towards salt–marsh ecosystems of higher levels. The "die–back" phenomena observed in southern England in the oldest *Spartina townsendii* meadows — whatever their cause may be — seem to be a spectacular indication of their dysfunction in coastal ecosystems (Beeftink and Géhu, 1973; Beeftink, 1976).

D. *Scirpus maritimus communities* (alliance Halo-Scirpion)

These are pioneer communities of *Scirpus maritimus* growing on brackish tidal marshes and in saline bottom land in polder areas (Table 6.5, col. 12; Table 6.6, col. 8).

Associations. There is a single association: Halo–Scirpetum maritimae.

Floristic composition and structure. These communities have *Scirpus maritimus* var. *compactus*

TABLE 6.9

Puccinellietum maritimae gradually ousted by *Spartina townsendii* agg., Groene Strand, Oostvoorne, southwestern Netherlands (quadrat size 28 m^2)

	Year: 1961	1962	1963	1964	1965	1966
Cover in %:	95	70[1]	95	95	95	100
Puccinellia maritima	09	07[1]	09	07	05	03
Spartina townsendii	a	a	02	05	06	07
Aster tripolium	a	a	m	m	01	m
Triglochin maritima	r	r	p	p	r	r
Salicornia europaea	m	r	–	–	–	–
Suaeda maritima	a	r	–	–	–	–
Plantago maritima	a	p	r	–	–	–
Spergularia media	p	–	(r)	(r)	–	–
Atriplex hastata	–	(r)	p	a	p	p
Glaux maritima	–	–	(r)	r	–	–
Scirpus maritimus	–	–	–	p	p	p

[1] Cover reduced because of surface water stagnating for a long time.
For the meaning of cover-abundance symbols see Table 6.7.

as dominant, associated with halophytes and salt-tolerant species, such as *Aster tripolium, Triglochin maritima, Puccinellia maritima, Atriplex hastata,* and *Agrostis stolonifera.* Other *Scirpus maritimus* communities associated with reed-swamp species, and distributed along fresh-water and oligohaline water bodies, are classified into the order Phragmitetalia and fall outside the scope of this chapter.

Distribution. These communities occur locally from west Norway (Nordhagen, 1923) and Great Britain to Brittany and north Portugal (Beeftink, unpublished), and in the transition area between North Sea and Baltic Sea (Dahl and Hadač, 1941; Gillner, 1960), and in brackish pools and ditches of coastal polder areas (Beeftink, 1965).

Ecology. These are pioneer communities in the eulittoral zone of the mesohaline parts of estuaries and other contact zones between bodies of salt and fresh water, for instance in localities where fresh water flows down into the salt marsh, or where it wells up from adjacent sedimentary or rocky outcrops. In the estuaries the association substitutes for the *Spartina* communities growing seaward, and *Scirpus lacustris, S. triqueter* and *Phragmites* communities developing in the oligohaline and fresh-water tidal zones.

Dynamics. From the seaward side the first estuarine *Scirpus maritimus* stands develop as isolated clones in the upper *Spartina townsendii* meadows. Further upstream the *Scirpus* belt extends at the expense of *Spartina*, while its upper limit is usually equilibrated with a *Phragmites* belt. However, grazing, very common in this marsh type, causes *Phragmites* and *Scirpus* stands to be superseded sooner than *Spartina townsendii* communities, and favours *Puccinellia maritima* and *Agrostis stolonifera*.

E. *Communities of the lower salt marsh* (alliance Puccinellion maritimae)

These communities develop on the salt-marsh belt between a little below MHW level and the level of MHWS, and are characterized by the presence (or dominance) of *Puccinellia maritima* and *Halimione portulacoides* (Table 6.5, col. 4, 5; Table 6.6, col. 5, 6).

Associations. Three associations can be recognized, viz.: (1) *Puccinellietum maritimae;* (2) *Plantagini–Limonietum;* and (3) *Halimionetum portulacoidis.*

Floristic composition and structure. Association 1 is very variable and can be subdivided into a typical subassociation 1a characterized by *Halimione portulacoides, Limonium vulgare* and (locally) the rhodophyte *Bostrychia scorpioides;* a subassociation *pholiurietosum* (1b) defined by the presence of

Parapholis strigosa, Halimione pedunculata, Agrostis stolonifera f. *subarenaria,* and, locally, *Elytrigia juncea;* a subassociation *agrostidetosum* (1c) characterized by *Agrostis stolonifera* subvar. *salina, Glaux maritima* and *Juncus gerardi.* Association 2 is characterized by combined co–dominance of *Limonium vulgare* and *Plantago maritima,* and association 3 by dominance of *Halimione portulacoides.* Subassociation 1a in particular is found in different variants and developmental stages (Beeftink, 1965). In the Eu–Atlantic salt marshes of the Irish and Scottish coasts, species which elsewhere grow in the upper–marsh belt (alliance Armerion maritimae) descend into the *Puccinellietum maritimae.* Examples are *Glaux maritima, Armeria maritima* and *Artemisia maritima.*

Distribution. The *Puccinellietum maritimae* is distributed from the Lofoten (Gillner, 1955), south Iceland (Jónsson, 1914), and the Faeröer to the Portuguese coasts (Beeftink, unpublished), but in the southern part of its range it is limited to brackish estuaries. In the Baltic it extends eastward to the Gotland coast. The *Plantagini–Limonietum* has a much more disjunct distribution, extending from the North Sea coasts southward to Normandy and northern Brittany. The area of distribution of the *Halimionetum portulacoidis* reaches its northern limit on Skallingen (Denmark), where *Halimione* was recently established (Iversen, 1952–53; Beeftink, 1959), and probably in Northumberland and Ayr (Great Britain) and the east and south coasts of Ireland. Southward the association is found on the west and south coasts of Portugal (Fontes, 1945; Beeftink, unpublished).

Ecology. The *Puccinellietum maritimae* develops in the lower parts, especially in depressions, of salt marshes, particularly if grazed, but is also found in wet saline parts of polder areas. From this association the subassociation 1a is distributed in the eu– and polyhaline zones within the reach of the tides; subassociation 1b occurs in dune slacks partly cut off from the tides; and subassociation 1c is exclusively found under mesohaline conditions. The *Plantagini–Limonietum* is an ungrazed form of the former association, especially developing where silt is deposited on high sand flats and against the faces of dunes and beach banks. The *Halimionetum portulacoidis,* finally, develops on the clay soils of creek–bank levees at a relatively high level, where a rather high salinity is combined with a relatively intense aeration. The two latter associations occur exclusively within the reach of the tides.

Dynamics. Grazing intensity, and the resulting cutting and damage of shoots and stems, and trampling of the soil surface, seem to be very important factors in the dynamic interrelationships of these associations. In succession three series have been found (Figs. 6.8, 6.9): Under more or less waterlogged conditions, succession starts from *Salicornia* or *Spartina maritima* communities and changes via association 1 and possibly 2 into the *Halimionetum portulacoidis* or Armerion communities. The second series is found on higher and thus more aerated ground, such as creek–bank levees, mostly starting as a *Vaucheria* or a *Salicornia–Suaeda* community, and passing via the *Halimionetum* to Armerion or *Elytrigia pungens* communities. On beach plains and in dune slacks a third series is found. There, the *Puccinellietum maritimae* evolves usually from *Glaux maritima, Spergularia* spp. or *Salicornia europaea* communities, and changes over into Armerion communities. In Fig. 6.10 the susceptibility of *Halimione* to severe winter frost in 1962–63, and its subsequent recovery, is illustrated in the vegetation dynamics of some permanent quadrats in the southwestern Netherlands.

Allied vegetation types. In the utmost northern and southern parts of the European coast eco-vicarious communities are found belonging to the adjacent phytogeographic regions. In the arctic and subarctic regions the alliance Puccinellion maritimae is replaced by the Puccinellion phryganodis from which the arctic and presumably circumpolar associations *Puccinellietum phryganodis* and *Caricetum subspathaceae* are recognized, as well as the low arctic and Amphi–Atlantic *Puccinellietum coarctatae* and the subarctic (North-European) associations (or subassociations?) *Triglochino–Puccinellietum phryganodis, Triglochino–Puccinellietum coarctatae* and *Triglochino–Caricetum subspathaceae* (De Molenaar, 1974). Moreover, there is a boreal transition zone between this arctic vegetation of the low salt marsh and its West-European counterpart, especially in brackish environments (Beeftink, 1972): the boreal—subarctic

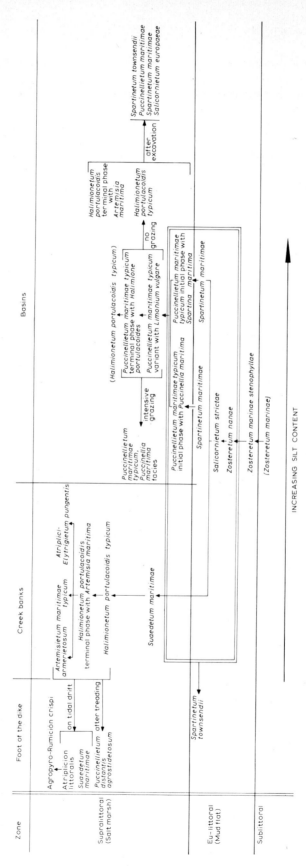

Fig. 6.8. Survey of the principal succession series on mud flats and salt marshes in the southwestern Netherlands (after Beeftink, 1965).

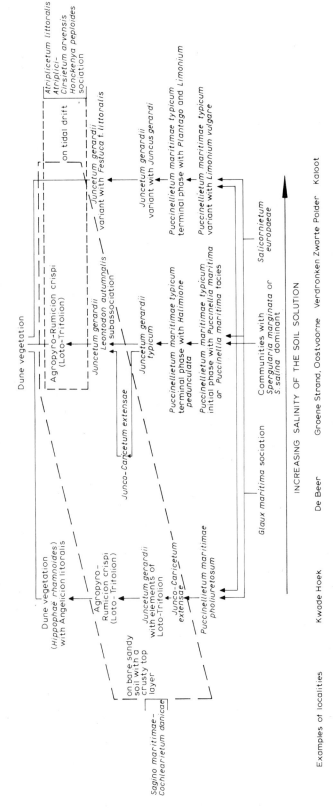

Fig. 6.9. Survey of the principal succession series on beach plains and dune slacks in the southwestern Netherlands (after Beeftink, 1965).

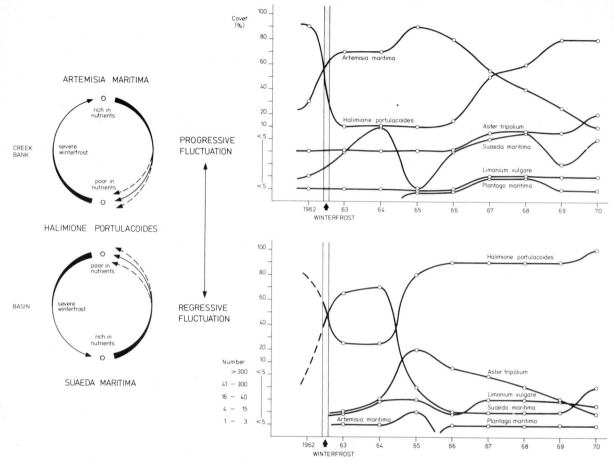

Fig. 6.10. Progressive and regressive fluctuations in *Halimione portulacoides* communities induced by severe winter frost in 1962—63 on a salt marsh in the southwestern Netherlands (from Beeftink, 1973, by permission of Pudoc, Wageningen). The upper part of the vertical scale is in per cent, the lower part in numbers per sample area (32 m^2 and 15 m^2, respectively).

Caricetum salinae incl. *Caricetum rectae* (Nordhagen, 1954; Gillner, 1960; Table 6.6, col. 16), and the more boreal *Caricetum paleaceae* (Gillner, 1960; Table 6.6, col. 9). At the southern end of the European coasts, communities belonging to Mediterranean vegetation types wedge out into the Atlantic salt–marsh vegetation, such as those in which *Salicornia fruticosa* and, in the upper parts, *Limoniastrum monopetalum* dominate. There, other Mediterranean species, for instance *Salicornia perennis*, *Inula crithmoides*, *Limonium ferulaceum* and *Suaeda fruticosa* are of minor importance only in otherwise Atlantic communities.

For this belt, but also for that of the upper salt marsh, communities with *Cotula coronopifolia*, a composite probably introduced from South Africa, must be mentioned. This species found its habitat in the beds of rivulets discharging in the salt marsh, subject to violent fluctuations in salinity and water content.

F. *Communities of the upper salt marsh* (alliance Armerion maritimae)

Communities growing on the salt–marsh belt between the level of MHWS and the storm–flood zone consist of a closed vegetation of mostly grasses or rushes, and not eutrophicated with organic debris to any extent. Characteristic species are *Armeria maritima*, *Festuca rubra* f. *litoralis*, *Glaux maritima*, *Juncus gerardi*, *Hordeum marinum*, *Carex distans*, and *Agrostis stolonifera* var. *compacta* subvar. *salina* (Table 6.5, col. 6, 7, 8; Table 6.6, col. 12—17).

Associations. The four associations included are: (1) *Artemisietum maritimae;* (2) *Juncetum gerardii;* (3) *Junco–Caricetum extensae;* and (4) *Scirpetum rufi.*

Floristic composition and structure. Association 1 is rather variable and can be subdivided into a typical subassociation 1a, a subassociation *armerietosum* (1b) characterized by *Armeria maritima* and *Juncus gerardi,* and a subassociation *agrostidetosum* (1c) defined by the presence of *Agrostis stolonifera* subvar. *salina, Atriplex hastata, Lolium perenne,* etc. Association 2 is very variable and characterized by the combined presence of *Armeria maritima, Juncus gerardi, Limonium vulgare, Spergularia media,* and other halophytes. Dominant taxa may be *Juncus gerardi, Festuca rubra* f. *litoralis, Plantago maritima, Glaux maritima,* and *Agrostis stolonifera* subvar. *salina.* The association can be subdivided into a typical subassociation 2a consisting nearly exclusively of halophytes, and a subassociation *Leontodontetosum autumnalis* (2b) with *Leontodon autumnalis, Agrostis stolonifera* subvar. *salina, Carex distans, Trifolium fragiferum, Lotus tenuis,* etc. These species are considered elements of the very complex alliance Agropyro-Rumicion crispi (suballiance Loto–Trifolion) characterizing border situations between the extremes of a number of soil factors, such as salt—fresh, wet—dry, basic—acid, rich—poor in nutrients, etc. (e.g., Westhoff and Den Held, 1969). Association 3 is characterized by *Carex extensa, C. punctata* and *Odontites littoralis,* and association 4 by *Scirpus rufus.* The latter two syntaxa especially may form transitional communities (*Junco–Caricetum extensae* subassociation with *Scirpus rufus*). In the Baltic area other subdivisions of this alliance have been proposed (e.g., Fröde, 1958; Gillner, 1960; Piotrowska, 1974).

Distribution. The *Artemisietum maritimae* extends from Skallingen (Denmark) and Aberdeen and Dumbarton (Great Britain) to northern Brittany, although *Artemisia maritima* itself extends southward to the Portuguese coasts. It occurs in the Baltic eastward to Gotland (Englund, 1942), Hiddensee (Fröde, 1958) and Rügen (Preuss, 1911—12). The *Juncetum gerardii* is widely distributed: in the north it extends to the Arctic region (Gillner, 1955), but allied communities are also described from the European coasts of the Arctic Ocean (Korchagin, 1935; Kalela, 1939; Nordhagen, 1954); in the south it extends to the French (Allorge, 1941; Corillion, 1953) and north Portuguese coasts (estuary of the Rio Lima: Beeftink, unpublished). It occurs in nearly the whole Baltic area, occupying a central position in the salt–marsh vegetation. The *Junco–Caricetum extensae* is rather rare, and extends from south Norway (Høeg and Lid, 1949) southwards to the Iberian Peninsula (Allorge, 1941), but allied communities are found on the north Portuguese coasts (Beeftink, unpublished). It also occurs in the Baltic to Fehmarn, but allied communities occur along the Swedish coasts eastward to Uppland (Almquist, 1929) and along the German coasts to the island of Hiddensee (Fröde, 1958). The *Scirpetum rufi,* finally, is known from west Norway (Nordhagen, 1923), the Wadden Isles (Westhoff and Den Held, 1969) and in the Baltic from the Swedish coasts (Almquist, 1929; Sterner, 1933; Gillner, 1960), the Gulf of Bothnia (Leiviskä, 1908) and the German Baltic coasts (Libbert, 1940; Preuss, 1911—12).

Ecology. Among these associations the *Artemisietum maritimae* is most under the influence of the tides: it develops on sandy to silty creek–bank levees and other elevations in the marsh easily reached by the tides, and therefore on soils in which a rather high salinity level is combined with a proportionally high aeration rate. Moreover, it is confined to flooding from euhaline to polyhaline water bodies, developing only fragmentarily in the mesohaline zone.

The *Juncetum gerardii* evolves optimally in the same salinity zones as the *Artemisietum* does, but mainly on the sandy soils of dune slacks, or of marshes enclosed by beach banks, etc., but subjected to flooding. The salinity level generally varies considerably (but never rises to extremely high values), and may temporarily fall to almost zero. The soil moisture, too, may vary considerably, as the association tolerates stagnant flood water and precipitation water as well as periods of desiccation. Under generally wet conditions *Juncus gerardi* predominates, on dryer places *Festuca rubra* is usually dominant, whereas *Agrostis stolonifera* comes to the fore when the salinity level is lower. Grazing has a considerable impact on structure and floral composition of this association, owing to consumption preference and trampling pattern of cattle and sheep. Under the influence of brackish flood water (e.g. in the Baltic) or of fresh–water seepage from dunes or rocky outcrops many temporarily salt-tolerant species become prominent while most less euryhaline halophytes have disappeared. Those communities are here classified into the subassociation *leontodontetosum autumnalis,* but, dependent on soil factors such as salinity, humidity, and texture, and on the grazing pressure, this vegetation unit may vary considerably (Fröde, 1958; Gillner, 1960; Piotrowska, 1974). The *Junco–Caricetum extensae* and, still more, the *Scirpetum rufi* are found in dune slacks, etc., where sea water only has access when tides are extremely high, and where fresh–water seepage from the dunes through the subsoil can easily take place. The *Scirpetum rufi* prefers a more clayey soil than the *Junco–Caricetum extensae,* especially when it occurs in depressions without open drainage courses. Gillner (1960) supposes that *Scirpus rufus* is nitrophilous, but Westhoff (1947) points out that the species is disseminated epizoochorously on the legs of cattle, and therefore grows optimally in the neighbourhood of cattle tracks.

Dynamics. The *Artemisia maritima* communities evolve from the *Halimionetum portulacoidis* when creek–bank levees and other elevations in the marsh accumulate sediment and enlarge. Usually *Festuca rubra* comes first, but in cases of deeper drainage, for instance through withdrawal of ground water owing to the control of water in neighbouring polder areas, *Artemisia maritima* colonizes the *Halimione* communities. However, creek–bank levees and other elevations are usually bound to collect organic debris washed over the marsh during flood. In that case *Elytrigia pungens* will soon establish, often accompanied by *Atriplex hastata* and *A. littoralis* (association *Atriplici–Elytrigietum pungentis*).

The *Juncetum gerardii,* and sometimes the *Junco–Caricetum extensae* usually follows *Puccinellia maritima* communities growing on beach plains and in dune slacks. They may also develop from *Glaux* or *Juncus gerardi* populations colonizing bare sand. During gradual accretion, or simply after gradual retreat of the tidal influence through isolation processes, species characteristic of the alliance Agropyro–Rumicion crispi penetrate, forming the subassociation *leontodontetosum autumnalis.*

Allied communities. Just as in the lower salt-marsh belt, the communities of the upper marsh also have eco-vicarious counterparts in the utmost northern and southern parts of the European coasts. In the Arctic and Subarctic regions they are classified into the alliance Caricion glareosae, from which the associations *Caricetum glareosae, Agrosto–Caricetum glareosae* and *Potentillo–Caricetum rariflorae* have been described (De Molenaar, 1974). On the boreal coasts a *Juncus gerardi— Gentiana detonsa* association has been distinguished (Gillner, 1955), and, in more brackish environments, the *Caricetum mackenziei* (Nordhagen, 1954; Table 6.6, col. 11). On the Portuguese coasts Armerion communities are rare and mostly limited to brackish environments of estuaries and lagoons. Conspicuous species in these communities are *Juncus maritimus, Carex extensa, Samolus valerandi, Cotula coronopifolia,* and many other species, most of which are characteristic of Agropyro–Rumicion crispi communities.

G. *Communities of cattle tracks, enclosed areas, etc.* (alliance Puccinellio–Spergularion salinae)

These are ephemeral communities on saline, clayey to sandy soils characterized by a certain form of instability through treading, seepage, temporarily stagnant water, sudden drainage, embankment, and similar factors. Characteristic species: *Spergularia marina* (= *S. salina*) and various *Puccinellia* species, such as *P. distans, P. fasciculata, P. capillaris* (= *P. retroflexa*), *P. pseudodistans* (Table 6.5, col. 9, 10, 11; Table 6.6, col. 7).

Associations. There are three associations, viz.: (1) *Puccinellietum distantis,* (2) *Puccinellietum fasciculatae,* and (3) *Puccinellietum retroflexae* (Table 6.5).

Floristic composition and structure. These are open communities, mostly soon giving place to Puccinellion or Armerion communities after stabilization of the environment, or to glycophytic vegetation of grasslands or tall herbage. Therefore, they often contain elements of these communities.

The *Puccinellietum distantis* is floristically the most variable one: a subassociation *atriplicetosum* (1a) with *Atriplex hastata* and species such as *Lolium perenne*, *Plantago major*, *Polygonum aviculare* and *Potentilla anserina*; a subassociation *pholiurietosum* (1b) with *Agrostis stolonifera salina*, *Parapholis strigosa* and *Salicornia ramosissima*; and a subassociation *juncetosum* (1c) with *Juncus bufonius* agg., *Triglochin maritima*, *T. palustris*, and *Scirpus maritimus*. Communities without the *Puccinellia* species mentioned but dominated by *Spergularia marina* frequently occur.

Distribution. The *Puccinellietum distantis* extends probably from the southeastern coast of Norway southward to the west coast of France, but is apparently rare on the west coasts of Great Britain and in Ireland (Perring and Walters, 1962). In the Baltic it is probably distributed on all coasts (Beeftink, 1965). For the *Puccinellietum fasciculatae* only the northern limit of its distribution area is known: the SW Netherlands, southern England and southern Ireland. Furthermore, it may be found in Brittany (Corillion, 1953). The distribution area of the *Puccinellietum retroflexae* is also only partly known: it is probably scattered throughout the whole Baltic area, and from the Lofoten down to the southwestern Netherlands, but not in Great Britain or Ireland.

Ecology. Instability of the environmental factors is caused by highly fluctuating water and salt content of the soil, and a high mineralization level of nutrients. Mesohaline conditions are optimal, but it also occurs under poly- and euhaline conditions. Subassociation 1a develops within the reach of the tides on cattle tracks, and after removal of the turf layer for the purpose of lawn–laying or reinforcement of sea walls. It is also found in tideless habitats characterized by seepage of saline water. Subassociation 1b occurs in dune slacks in which some silt has been deposited by the tides. Subassociation 1c develops optimally, but only for a short time, on mud and sand flats isolated from the sea — e.g., by a barrage. The *Puccinellietum fasciculatae* is very rare within reach of the tides, and is only found where salinity of the flooding water fluctuates greatly. It succeeds subassociation 1c at the permanently moist and poikilohaline edges of embanked mud and sand flats, if the salinity of the stagnant open water exceeds a mean annual value of about 9‰ Cl$^-$, but it can also occur on cattle tracks in saline grasslands. The *Puccinellietum retroflexae* seems only to develop under less unstable conditions of water and salt content.

Dynamics. The associations establish themselves on heavily disturbed ground, either permanently or for a short period only. In the latter case they are soon overgrown by Puccinellion or Armerion communities when salinity remains at a constant level, or by glycophytic vegetation of grasslands or tall herbage when salinity drops. Thus, the *Puccinellietum retroflexae* is the most ephemeral of the three associations, and the *Puccinellietum fasciculatae* is most permanent.

Allied communities. De Molenaar (1974) suggests that the arctic, probably circumpolar *Sagino–Phippsietum algidae* would be vicarious to the communities of this alliance.

H. *Therophytic communities in the storm–flood zone* (Saginion maritimae)

Open communities, mainly consisting of therophytes and hemicryptophytes, grow in open spots (glades) among a vegetation mainly of hemicryptophytes in the zone in which storm floods contact fresh–water land formations (Table 6.6, col. 18). Characteristic species are *Sagina maritima*, *Cochlearia danica*, *Plantago coronopus*, *Catapodium marinum*, *Bupleurum tenuissimum*, *Parapholis strigosa*, *Pottia heimii*, *Amblystegium serpens* var. *salinum*, *Rhynchostegiella compacta* var. *salina*, probably also *Hordeum marinum* (Tüxen and Westhoff, 1963; Beeftink, 1965; Westhoff and Den Held, 1969).

Associations. In the West-European salt marshes there is only one association of this type: *Sagino maritimae—Cochlearietum danicae*; vicarious associations are the *Sagina maritima—Catapodium marinum* association (rocky coasts) and the *Sagina maritima—Tortella flavovirens* associations (west Mediterranean).

Floristic composition and structure. These are ephemeral communities on open spots in the closed original vegetation, or on shallow soils between rocky formations, and therefore often developed in fragmentary form. In full development, however, they exhibit catenas from outer to the inner marshes and from the lower to the upper parts in the storm–flood zone, in which *Parapholis* starts at the seaward edge, and *Catapodium* and *Bupleurum* finish on the landward side (Westhoff et al., 1961).

Distribution. These communities occur from southwestern Sweden, Denmark and the German north coasts to western France.

Ecology. The communities occupy the unstable border situations — mostly very narrow — between the halo–(salt, wet) and xerosere (fresh, dry), such as on slopes of beach banks and dune ridges, also on sandy soils of dune slacks, deltaic deposits and embanked tidal flats. They occur locally in the aerohaline zone (salt spray) of the West–European rocky coasts where a shallow sandy to silty soil has been deposited on the rocks. A characteristic of their habitat are the fluctuating conditions ranging from wet and fresh combined (in case of a rainy spell) to extremely dry and high salinity (after flooding or spraying with sea water and subsequent abrupt desiccation). Under such circumstances surface often turns into a thin crust of amalgamated soil particles, perhaps promoted by blue–green algal activities.

Dynamics. The presence of Saginion communities depends greatly on the combined impact of storm floods in spring, and continental (warm and dry) climatic periods in summer. Under these conditions the original Armerion and Koelerion (dune) vegetation is burned either by excessive soil salinity, or by too low soil humidity, or — mostly — by both combined. In the open places that have arisen in this manner the Saginion communities develop between some remaining elements, under which stunted forms of *Armeria maritima*, *Plantago maritima*, *Elytrigia pungens*, *Festuca rubra*, or *Agrostis stolonifera* usually persist. In cool and humid summers, however, the original vegetation seems to be able to recover soon, covering the open spots. Therefore, the Saginion communities on the salt–marsh or lower dune formations may show concertina–like dynamic relations with the original vegetation, or may shift considerably in time with respect to tidal level, depending on the height of storm floods.

Allied communities. From southern England and western France southward to the Portuguese coasts, species such as *Hordeum marinum*, *Frankenia laevis*, *Limonium bellidifolium*, *L. occidentale*, *Hutchinsia procumbens* and *Tortella flavovirens* are found in the habitat of the association, while *Cochlearia danica* seems to become rarer (Beeftink, 1965). Some of these communities have been classified into the *Frankenio–Staticetum lychnidifoliae* (Lemée, 1952). It may also be noted that such communities may show relations with the *Atriplici–Elytrigietum pungentis* to be discussed in the following section (Tüxen and Westhoff, 1963).

I. *Communities on plant debris washed ashore* (Thero–Suaedion, Atriplicion littoralis and Agropyrion pungentis)

Halo–nitrophilous communities of annuals and/or perennials are formed on alluvial substrates enriched with tidal drift of plant debris, rafted onto the marsh but not covered — or at most mixed — with sand. Characteristic species are *Suaeda maritima*, *Bassia hirsuta*, *Atriplex littoralis*, *A. hastata*, *Matricaria maritima*, *Elytrigia pungens*, *Suaeda fruticosa* (= *S. vera*) and *Lepidium latifolium* (Beeftink, 1965, 1968; Géhu, 1968).

Associations. The classification is not yet fully worked out: most community types occur in the storm–flood zone of sand and shingle beaches, and of rocky coasts (Géhu, 1968), and have been omitted here. When the plant debris has been fully or nearly fully leached by heavy precipitation, other associations develop (Nordhagen, 1940). They have to be classified into quite other vegetation units, viz. the class Artemisietea vulgaris, comprising nitrophilous tall herb and weed communities of border zones.

The associations described from European salt–marsh and adjacent fixed dune formations are *Suaedetum maritimae*, *Suaedo–Kochietum hirsutae*, *Salsoletum sodae* and *Suaedetum splendentis* (Thero–Suaedion), *Atriplicetum littoralis*, *Atriplex littoralis—Erysimum hieraciifolium* association (Atriplicion littoralis), and *Atriplici–Elytrigietum pungentis, Suaedetum fruticosae* and *Lepidietum latifolii* (Agropyrion pungentis).

Floristic composition and structure. In these respects, the communities are widely divergent owing to differences in life–form spectra: they constitute either an open *(Suaeda maritima, Matricaria maritima)* or a more closed *(Atriplex* spp.) vegetation, predominantly of therophytes, or a closed vegetation mostly of tall, sometimes tussocky, herbage of hemicryptophytes *(Elytrigia pungens, Lepidium latifolium, Beta maritima, Rumex crispus,* etc.).

Distribution. The *Suaedetum maritimae* extends from southwestern Sweden to France and northwestern Spain, and is also found locally in the western Mediterranean including the northern Adriatic. The *Suaedo–Kochietum hirsutae* has a very disjunct area of distribution, mainly within that of the former association (Beeftink, 1965). The *Atriplicetum littoralis* is found from northern Norway to northern France including Great Britain and parts of the Baltic coasts, where it runs out into the *Atriplex littoralis—Erysimum hieraciifolium* association (Tüxen, 1950). Finally, the *Atriplici–Elytrigietum pungentis* occurs from Skallingen, Denmark (Beeftink, 1959) and Great Britain to the French and, locally, the Portuguese coasts.

In Great Britain and Ireland *Elytrigia pungens* decreases towards the west, giving place to *Juncus maritimus* (Chapman, 1941; O'Reilly and Pantin, 1957).

Ecology. These communities may be very unstable or only slightly so, depending on the quantity and frequency in which plant debris is supplied. There are two kinds of tidal drift: (1) that consisting of filamentous and sheet–like algae floating in the flood water and washed ashore, scattered in a broad zone over the lower salt marsh; and (2) that consisting of debris of higher plants and detached fucoid brown algae floating on the flood–water surface, and washed ashore in concentrated belts on the spring and storm–flood strand lines of creek–bank levees, sand and shingle bars, dune ridges, dike faces, etc. On soils covered by the first type especially, *Suaeda maritima* develops. On the second type the other species mentioned grow, *Atriplex* species and *Elytrigia pungens* usually predominating (Beeftink, 1966). It is suggested that the very large share of *Spartina* foliage and stems locally in the total composition of debris during recent decennia has stimulated the occurrence of both *Atriplex* species, especially *A. hastata* (Beeftink, 1965).

The annuals are adapted to horizontal migration over short distances parallel to the lines along which the debris is deposited. The perennials, however, are able to endure several years without input of organic material to any extent. Obviously, this irregular frequency of nutrient supply can be absorbed by storage in the ecosystem itself, though, on the other hand, therophytes seem largely to take up the extra supply of nutrients during the first growing season after accumulation of the organic material. Apparently, debris composition and age decide the composition and structure of the vegetation.

Dynamics. Depending on the magnitude and distribution of the accumulated organic material, and on the frequency with which it is washed ashore at the same spot, the communities discussed in this section have dynamic liaisons with the "basic" salt–marsh vegetation. These communities may be considered "veiling" or "carpeting" ones (*Schleier* and *Teppich* communities, respectively) as they are apt to veil (in a vertical direction) or cover (in a horizontal direction) the original vegetation (Tüxen, 1950).

ZONATION AND DYNAMIC TRENDS IN VEGETATION

It often seems difficult to distinguish between structure in space and time in vegetation studies. Although zonation can be the spatial expression of succession, studies on subsequent stages of vegetation development over a long period are needed in order to gain insight into succession. However, long–term studies on vegetation succession, including that of salt marshes, are still rare; consequently, succession diagrams presented by various authors seem to have many speculative elements. The schemes outlined in Figs. 6.8 and 6.9 should also be interpreted with this criticism in mind. Even floristically related series of communities, logically exhibiting a sequence of stages corresponding, for instance, to desalinizing or accretion processes will not necessarily represent a natural succession series, as species formerly rare or even absent may have been introduced meanwhile. The possible objection that the introduction of species may also form an aspect of succession is not denied; but this should be viewed as a warning that present–day zonation is not necessarily an exact copy of former successional stages (see p. 25).

Succession and the development of the salt marsh depend on: (1) the relationship of land to sea level, viz. the relative emergence, subsidence or stability of the coastline (Chapman, 1960, 1974); (2) the rate at which the marsh is being built up by deposits of sand and silt (Richards, 1934; Ranwell, 1964); (3) the influence and frequency of climatic extremes, for instance gales (Packham and Liddle, 1970) and extremely cold winters (Ranwell, 1968; Beeftink, 1975); and (4) the availability of seed stocks and the introduction of exotics (Ranwell, 1968). Some of these factors transform the marsh vegetation from one type to another gradually and according to highly predictable trends. Others of a more "catastrophic" nature change the vegetation in an unpredictable direction (Ranwell, 1968); such factors include gales and cold winters, but also the advent of *Spartina townsendii* in the nineteenth century, sudden movements of the limits of distribution areas of plants and animals (e.g., the northward extension of *Halimione portulacoides* and *Elytrigia pungens*), and epidemic diseases (e.g., the wholesale destruction of *Zostera marina* in 1932—34).

Some authors emphasize that succession may pass through threshold conditions at which transformation can occur with relatively great rapidity (Ranwell, 1968; Packham and Liddle, 1970). In respect of stability, thresholds seem to be most obvious. After crossing a critical stability, sand and mud flats can become colonized very rapidly providing seed stocks of the appropriate colonists are available. Wiehe (1935) found that *Salicornia* requires a threshold time of two or three days undisturbed by the tides to prevent dragging of the seedlings from their primordial root systems. Examples are known where sand flats and beaches became sheltered by the formation of an offshore bar or by adjacent reclamation activities and were transformed from bare sandy silt via pioneer

Puccinellia and *Juncus gerardi* marsh into *Juncus maritimus* and *Festuca rubra* marsh in about one decade (Ranwell, 1968) (Fig. 6.11). According to Brereton (1971) the initial stages of succession at Foryd Bay, Wales, start with random colonization of *Salicornia stricta* and *Puccinellia maritima*. This initial random distribution is followed by an aggregated distribution. Both the scale and intensity of aggregation increase up to the point where clump coalescence occur. Following clump coales-

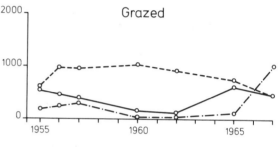

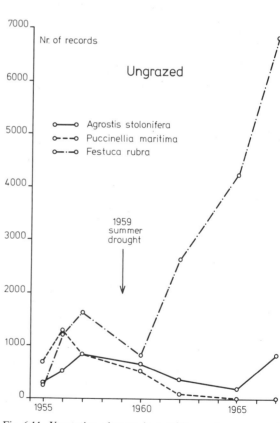

Fig. 6.11. Vegetation changes in a salting pasture measured within and outside a grazing enclosure over a ten–year period on Fenning Island marsh, Bridgewater Bay, Somerset (from Ranwell, 1968).

cence there was some evidence that both the scale and intensity of aggregation decrease. It seems that this sequence of changes found by Brereton (1971) is also common in other succession stages, such as in the establishment of *Halimione portulacoides, Festuca rubra, Elytrigia pungens,* and other species coming in as dominants. In practice other species, such as *Aster tripolium, Salicornia* spp., *Suaeda maritima, Spergularia* spp., can generally fluctuate in numbers of individuals depending more on occasional environmental conditions than on threshold conditions, or, alternatively, as a consequence of the death of other species (compare Fig. 6.13). The distinction between these two types of species may not be absolute, but rather a matter of frequency of the threshold conditions in question. In the second type of species these conditions may simply be much rarer.

The time required for succession varies much, as the scarce evidence already shows: The development of a *Juncus maritimus* marsh from a bare sand flat can take about ten years (Ranwell, 1968), while within five years a *Spartina* marsh is found completely replaced at its upper limits by *Puccinellia maritima* where the marsh is grazed, and by *Phragmites communis* where it is ungrazed (Ranwell, 1961) (see also pp. 10 and 57).

Succession in salt–marsh vegetation is generally considered with the vertical direction in mind, i.e., starting from species and communities colonizing the lower levels of the beach (with respect to MW or MHW) and proceeding to others establishing on higher levels. However, as Bilio (1965) pointed out, "horizontal" succession can also occur, extending the salt–marsh vegetation of different levels simultaneously to still uncolonized parts of the beach. As an example observations on a sand bar protecting a small inlet near Kiel (western Baltic) are referred to, where the accretion of the distal parts of the bar increased the stability of the more proximal sections, thus providing favourable conditions for the simultaneous horizontal progression of salt–marsh plants on all levels of the beach on the inlet side of the bar.

ALGAL ECOLOGY OF THE SALT MARSH

Compared with studies on the phanerogamic communities, investigations on algal communities growing on the soft substrates of salt marshes and beach plains are few. This is remarkable as, besides the phanerogams, the benthic algae occupy an important place in the salt–marsh ecosystem (see pp. 17—18). Perhaps algal ecologists failed to cross some methodological thresholds inherent in the study of most cryptogamic communities, viz.: (1) taxonomic difficulties and the impossibility of carrying out identifications in the field on many algal taxa; (2) the problems of homogeneity and minimal area in such communities; (3) their large seasonal and other environment–induced fluctuations in species composition and structure; and thus (4) the problem of how to classify these algal communities into approved abstract community types (syntaxa) such as sociations (consocies, associes) and associations (Carter, 1933; Chapman, 1960; Nienhuis, 1975). It is also noteworthy that, up to now, benthic algal studies have been mainly restricted to salt marshes around the North Sea: First on the coasts of Great Britain (Chapman, 1940) and later in The Netherlands.

Earlier ecological studies on salt–marsh algae may briefly be mentioned here: Baker (1912) and Baker and Bohling (1915—17) studied marsh fucoids, mainly in eastern England; Cotton (1912) distinguished four algal community types in the salt marshes of Clare Island (Ireland); Nienburg (1927) described five zones of algae and phanerogams in the Wadden Island of Sylt (Germany). Carter (1932, 1933) was the first thoroughly to study algal communities with a view to characterizing structural conditions. She found twelve community types on salt marshes in the Dovey and Thames estuaries.

After the initial period indicated above, Chapman (1938—41, 1964) greatly advanced the study of salt–marsh ecology by his work in the Norfolk marshes. Considering conditions of submergence and exposure, he pointed out that algae occurring in the upper marsh are peculiarly suited to withstand desiccation. He suggested that the tidal factor has a greater influence upon phanerogams than upon the algae, especially since the occurrence of the latter may be determined by other factors such as the nature of the substrate, light and space relations (see also Carter, 1933). The results of his investigations on algal distribution in space and time are illustrated in Fig. 6.12. Comparing his

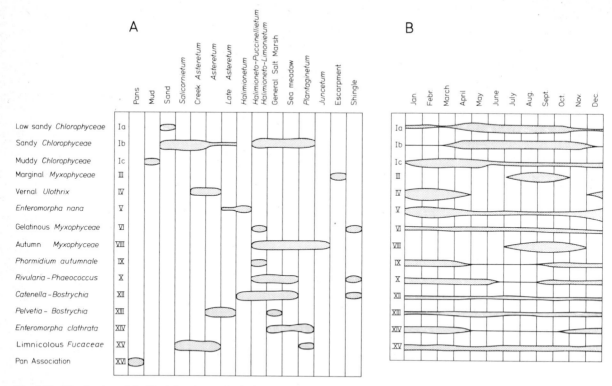

Fig. 6.12. Distribution of the Norfolk salt–marsh algal communities in space and in time (after Chapman, 1939).

studies with those of Carter (1932, 1933), Cotton (1912) and Rees (1935), Chapman (1939, 1959) proposed a unifying nomenclature as a future basis to be used in further studies on algal communities of the salt marsh, although he was pessimistic about finding a common scheme of algal distribution as this is determined very largely by special local conditions (see also pp. 17—18).

With the exception of a single study by Häyrén (1956) dealing with algal communities in Swedish and Finnish coastal marshes, floristic studies on the classification of algal communities in this habitat have hardly started in the Baltic. Probably, the approach of Du Rietz (1930) emphasizing dominance in the study of vegetation, instead of floristic composition, led here to an under–estimation of algae as compared with phanerogams.

Chapman's approach, however, encouraged Dutch algologists to draw the logical conclusion that ecological studies on algae should be carried out in accurate plot and gradient analyses repeated seasonally for one or more years. An adaptation of the method of the Zürich—Montpellier tradition of phytosociology, developed for the study of phanerogamic communities, proved to suit this accurate algal ecological research very well (Nienhuis, 1970; Polderman and Prud'homme van Reine, 1973). The centre of interest in these Dutch studies is that the minimal area of benthic algal communities is about 100 times smaller than that of the phanerogamic communities in which they are living (see Nienhuis, 1972). Consequently, there may be large floristic and structural differences in the algal vegetation within a homogeneous phanerogamic community. Furthermore, two or even more algal community types can be distinguished at the same spot, owing to the occurrence of stratified but mutually independent synusiae, though some of these may be closely confined to specific seasons (Nienhuis, 1970, 1972). It appears, therefore, that, for quantitative studies in algal ecology, permanent sample plots are essential as a means of investigating seasonal periodicity, irregular changes in cover and abundance owing to variations in weather conditions, and succession (Nienhuis, 1972).

In this manner the distribution, periodicity and morphology of Dutch *Vaucheria* species have been surveyed (Nienhuis and Simons, 1971; Simons and Vroman, 1973; Simons, 1974). Polderman and Prud'homme van Reine (1973) described communities with *Chrysomeris ramosa* occurring in unstable mesohaline salt–marsh habitats in The Netherlands and Denmark. Polderman (1974) compared the algal vegetation in the *Juncetum gerardii* and the *Puccinellietum fasciculatae* growing in a seepage area on the landward side of the dike near Vlissingen and studied its periodicity. Nienhuis (1970), investigating the algae in the Grevelingen, an estuarine arm in the southwestern Netherlands, arrived at eight community types usually exhibiting very vague relations with phanerogamic communities (Fig. 6.13). Nienhuis (1972) pointed to a negative correlation between the growth of *Entero-morpha* species and that of *Triglochin maritima*, light being probably the limiting factor for the algae, except in winter when both *Enteromorpha* and *Triglochin* show minimal development. Cooperative phenomena are indicated in the more stratified communities, where *Halimione portulacoides* and *Limonium vulgare* provide shade and a supporting substrate for the rhodophytes *Bostrychia scorpioides* and *Catenella repens*.

With respect to the environmental relations of benthic algae, Nienhuis (1973) found also that water conditions (frequency of tidal floodings, fluctuations in the soil–moisture content) are most important in determining their distribution pattern, and that extreme salinities, going hand in hand with intense desiccation in the surface soil, can modify this pattern considerably. Stability of the sediment is a prerequisite for development and

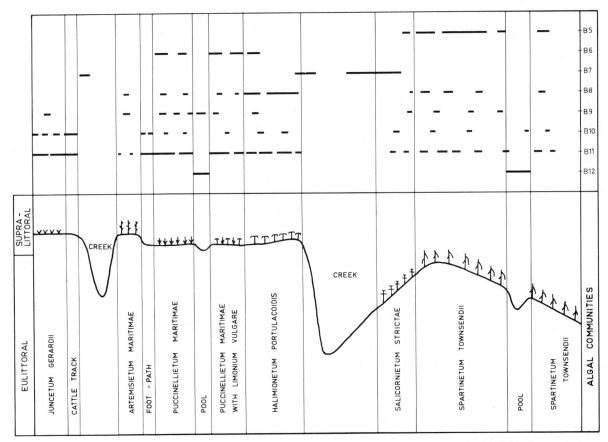

Fig. 6.13. Distribution of salt–marsh algal communities on the Springersgors, Grevelingen (The Netherlands). B5 = *Fucus vesiculosus* f. *volubilis* sociation; B6 = *Bostrychia scorpioides* sociation; B7 = *Vaucheria thuretii—V. sphaerospora* sociation; B8 = *Blidingia minima* sociation; B9 = vernal *Ulothrix* sociation; B10 = communities dominated by Cyanophyceae; B11 = general Chlorophyceae community; B12 = vegetation in salt–marsh pools. (After Nienhuis, 1970.)

survival of algal populations, and may be a limiting factor in the zone around the MHW level — that is, in the lower marsh. However, most of the algal species involved are listed as semi–terrestrial and show great ecological potential for adaptation (Nienhuis, 1973).

ECOLOGY OF SALT–MARSH ANIMALS

The animals living in the salt marsh can be considered in five categories, viz. mammals, birds, arthropods, molluscs, and a number of meiofaunal groups such as nematodes, flatworms and protozoans.

Mammals usually live in the salt marsh only during part of their life cycles. They visit the marsh for food, and only small mammals such as the rabbit *(Oryctolagus cuniculus)*, brown hare *(Lepus capensis)*, root vole *(Microtus ratticeps)* (Van Wijngaarden, 1961, 1969), and brown rat *(Rattus norvegicus)* (Drummond, 1960) find habitats for other activities in the uppermost marshes where the tides will hardly ever reach them. Only the muskrat *(Ondatra zibethicus)*, introduced into Europe in 1905, is adapted to lower levels (Martin et al., 1951; Koenders, 1968).

As human interference replaced many of the original macroherbivores by cattle and sheep, other mammals are limited to the smaller ones mentioned above. Grazing by either domestic or wild animals seems to be limited by the tides, as the upper marshes are more intensively grazed than the lower ones and grazing intensity on the latter increases considerably when tides have been excluded by barrages. The influence of grazing is a very complex one, depending on (1) the selectivity of grazing; (2) the treading pattern of the animals in relation to geomorphology and vegetation types; and (3) the pattern of manuring by faeces. The effects of consumption and the overall influence of grazing on salt–marsh species are summarized in Table 6.10 (see also Fig. 6.11). For dwarf shrubs, such as *Halimione portulacoides,* the influence of grazing by cattle manifests itself especially in breakage of the branches.

There seems to be a conspicuous relationship between grazing intensity of salt marshes and the climatic factor: In the Eu–Atlantic and Subatlantic cool–temperate parts of the European coasts (Ireland, Great Britain, southern Scandinavia and the Low Countries) grazing is, or was at least until recently, much practised. Southwards, however, in the warm–temperate region, grazing becomes an ever rarer phenomenon, and in the Portuguese salt marshes it seems to be practised only in the very brackish ones. Although the proportion of grasses in the vegetation cover tends to decrease in a southward direction, and that of woody Chenopodiaceae increases, it is supposed that an increasing salinity level in the salt–marsh ecosystem, owing to shifting climatic influences, could be a limiting factor for cattle grazing, rather than a shift in the composition of the vegetation.

Birds may be found in salt marshes in great quantities, either breeding (sea–gulls, terns, oyster–catcher [*Haematopus ostralegus*], redshank [*Tringa totanus*], sky lark [*Alauda arvensis*]) or migrating and overwintering (geese, ducks). Their influence on salt–marsh vegetation is scanty but selective: brent geese *(Branta bernicla)* feed on *Zostera, Puccinellia* and green algae (Campbell, 1946; Ranwell and Downing, 1959; Burton, 1961; Lebret, 1965; Wolff et al., 1967), grey–lag *(Anser anser)*, and occasionally bean geese *(Anser fabalis)* feed on the root tubers of *Scirpus maritimus* (Lebret, 1965). At high tides the numbers of shore birds roosting on the marsh may be very high. High concentrations of bird droppings on the edge of the marsh may promote growth and development of *Suaeda maritima, Cochlearia officinalis, Plantago coronopus* and other species (Gillham, 1956).

Arthropods are represented in great numbers, both of species and of individuals. The main groups are insects, arachnids and Crustacea. These groups have in the main been investigated only incidentally in salt marshes, but a few authors have reported thorough studies (Heydemann, 1960—64; Amanieu, 1969). The insects and arachnids are largely confined to the supralittoral zone. Only the Coleoptera *Bledius spectabilis* and *B. arenarius* seem to be adapted to eulittoral conditions. The Crustacea, on the contrary, have their greatest species richness in the eulittoral zone.

In the Bay of Arcachon (France), Amanieu (1969) found that in the salt marsh, contrary to the mud flats, endogeous macrofauna are entirely absent. The vagrant epigeous species (mainly insects, amphipods, isopods, arachnids, and mol-

TABLE 6.10
Effects of grazing by cattle and sheep on salt-marsh halophytes in the southwestern Netherlands (Beeftink and Daane, unpublished)

Degree of consumption			Overall influence of grazing		
low	moderate	high	favouring	indifferent	harmful
Moderate grazing					
Salicornia europaea	Puccinellia distans	Aster tripolium	Salicornia europaea	Suaeda maritima	Scirpus maritimus
Suaeda maritima	Plantago maritima	Puccinellia maritima	Puccinellia maritima	Aster tripolium	Atriplex spp.
Limonium vulgare	Scirpus maritimus	Triglochin maritima	Puccinellia distans	Triglochin maritima	
Spergularia spp.	Phragmites communis	Festuca rubra	Spergularia marina	Limonium vulgare	
Artemisia maritima	Elytrigia pungens	Agrostis stolonifera	Juncus gerardi	Plantago maritima	
Armeria maritima		Juncus gerardi		Spergularia media	
Spartina townsendii		Atriplex spp.		Artemisia maritima	
Halimione portulacoides				Halimione portulacoides	
Glaux maritima					
			Spartina townsendii		
			Phragmites communis		
			Elytrigia pungens		
			Agrostis stolonifera		
			Festuca rubra		
			Glaux maritima		
			Armeria maritima		
Intensive grazing					
Salicornia europaea	Puccinellia distans	Aster tripolium	Salicornia europaea	Suaeda maritima	Scirpus maritimus
Suaeda maritima	Plantago maritima	Puccinellia maritima	Puccinellia maritima	Triglochin maritima	Atriplex spp.
Limonium vulgare	Spartina townsendii	Triglochin maritima	Spergularia marina	Plantago maritima	Aster tripolium
Spergularia spp.	Halimione portulacoides	Festuca rubra	Armeria maritima	Spergularia media	Spartina townsendii
Artemisia maritima	Glaux maritima	Agrostis stolonifera		Puccinellia distans	Phragmites communis
Armeria maritima		Juncus gerardi		Juncus gerardi	Limonium vulgare
		Atriplex spp.			Halimione portulacoides
		Scirpus maritimus			Artemisia maritima
		Phragmites communis			
		Elytrigia repens			
			Elytrigia pungens		
			Agrostis stolonifera		
			Festuca rubra		
			Glaux maritima		

luscs) exhibit considerable degrees of mobility, migrating according to tidal rhythms. In summer, many species are attracted to the trophically rich *Puccinellietum maritimae*. In winter frequent floodings cause many animals to die, and the populations retreat towards the storm–flood zone. According to Amanieu (1969), the salt marsh is one of the most characteristic zones where littoral and terrestrial faunas meet and mix. Other invertebrate data are enumerated by Serventy (1960) for Scolt Head Island (Great Britain).

In northwestern Germany the insect and arachnid faunas were investigated thoroughly in the salt marsh (Heydemann, 1960, 1964b; Abraham, 1970) and on sea–walls (Heydemann 1961, 1963), as well as in cases where the marsh had been embanked (Heydemann, 1962, 1964a). These studies have also been reviewed in the European context (Heydemann, 1967). Strenzke (1955) described Collembola faunas from the German North Sea and Baltic coasts. English data on insect fauna are given by Ellis (1960) and Stebbings (1971).

Luxton (1964, 1967) investigated salt–marsh Acarina and found that mite numbers reached their peak in August. Fluctuations in these numbers were attributed to changes in soil moisture, originating both from precipitation and the tides. Summer minima were induced by drought, and winter minima by cold or excessive moisture. Some mite species were restricted to specific intertidal zones, others showed no preference. Luxton (1964) demonstrated that, although these animals can withstand immersion in sea water for up to twelve weeks without apparent harm, they showed distinct preferences for specific salt–marsh fungi as food; thus salinity could affect acarine zonation indirectly through the food. Duffey (1970) has emphasized habitat selection by spiders on a salt marsh in Gower (Great Britain). He demonstrated the ability of those species more characteristic of the salt marsh than other habitats to establish themselves in normally suboptimal environments, when conditions improved for short periods. Recently, Woodell (1974) has drawn attention to air–trapping problems for survival of *Lasius flavus* in anthills on the salt marsh, and the environmental consequences of these hills for the vegetation.

Among insects living on salt–marsh plants the beetle *Mecinus collaris*, which is found in the flower stems of *Plantago maritima*, must be mentioned. Also Cercopidae (froghoppers) and Aphidae may be numerous. Among the latter *Staticobium limonii* can shade the inflorescences of *Limonium vulgare* to the point that flowering is prevented. In one of the oldest *Spartina* marshes in Poole Harbour, Payne (1972) drew attention to the development of a new, balanced animal community including a herbivorous bug (*Euscelis obsoletus*, Hemiptera, Cicadellidae), a carnivore (*Dolichonabis lineatus*, Hemiptera, Nabidae) feeding on the bug, and an omnivorous grasshopper (*Conocephalus dorsalis*, Orthoptera, Tettigonidae) feeding on the bug and on *Spartina*.

Talitrid and gammarid amphipods were investigated ecologically by Den Hartog (1963, 1964) in the southwestern Netherlands. Heath (1975) described factors affecting the temperature and salinity conditions of an estuarine salt marsh of the river Tyne, in relation to the ecology of the isopod *Sphaeroma rugicauda*, a common inhabitant of such marshes in Western Europe.

European salt–marsh molluscs have interested many authors but ecological data are few. *Littorina littorea* and *L. saxatilis* live on rocky coasts, walls, stones, etc., but also on mud flats and in the lower marshes. *Hydrobia ulvae* is also an intertidal snail, but is usually concentrated in great masses in the lower marsh around the MHW line (Newell, 1962, 1964, 1965). *Assiminea grayana* lives optimally in the brackish lower marshes from Denmark and southeastern England southward to France and perhaps Spain, and *Ovatella myosotis* occurs on the creek–bank levees of the marshes bordering euhaline and polyhaline water bodies (McMillan, 1949; Meyer, 1955; Mörzer Bruijns et al., 1959). Finally, the sacoglossans *Alderia modesta* and *Limapontia depressa* can be found mostly living on *Vaucheria* spp. (Nicol, 1935; Engel et al., 1940; Den Hartog, 1959b).

Among the aquatic meiofauna inhabiting the salt–marsh soil, nematodes, turbellarians, oligochaetes, harpacticoid copepods and halacarids are the most important taxonomical groups (Bilio, 1963a, b, 1964—67, 1965). Studies restricted to single groups of this fauna were published by Den Hartog (1963, 1964, 1964—66, 1968) and Bilio (1966, 1967) for the turbellarians, and by Lorentzen (1969) for the nematodes. Phleger (1970) investigated foraminiferal populations of lower and upper salt marshes in Norfolk and in the southwestern

Netherlands and found large standing stocks reflecting abundant food and high organic production. Referring to the aquatic fauna of the upper 3 cm of the soil, Bilio (1965) found more important changes in community composition between lower and upper *Puccinellietum* than between upper *Puccinellietum* and *Juncetum gerardii*. This difference in zonation pattern between fauna and vegetation is explained by the superficial position of the soil stratum mainly inhabited by the fauna studied, compared with the vertical extension of the roots of the plants used to characterize the syntaxonomic units of the vegetation in question. The influence of the adjacent sea on the surface layer of the soil is certainly much more enduring in the lower *Puccinellietum* than in all higher zones; thus, corresponding differences can be expected in the life conditions of aquatic organisms inhabiting this stratum.

ECOSYSTEM APPROACH TO THE SALT MARSH

Compared with North America very little attention has been paid to the study of trophic structure, material budgets, recycling and energy-flow in the European salt marsh. A survey is given by Ranwell (1972) in his excellent book on the ecology of salt marshes and sand dunes. There seems to be no reason to suppose that European interrelationships in salt marshes are in principle different from those in North America. The most important vegetational difference between the North American and European salt marshes is the great share of *Spartina* species in the former, while *Puccinellia* species and dwarf shrubs (especially Chenopodiaceae and *Artemisia*) are in the minority. *Spartina* species produce a good deal of detritus and, moreover, the main American one *(S. alterniflora)* can extend on the mud down to below mid-tide level (Johnson and York, 1915), contrary to its European counterparts, especially *S. townsendii* agg., which in the southwestern Netherlands ultimately reach down to 1 m below MHW or 80—100 cm above mid-tide level (Beeftink, 1965; see also Morley, 1973). Therefore, the amounts of detritus derived from the ecosystem and introduced into the adjacent aquatic ecosystem may be much larger under American than under European conditions. In contradistinction to the views of American ecologists (Odum, 1971) that the salt-marsh ecosystem is a nutrient supplier to the coastal water body, a very recent and first attempt to measure the detritus balance of a salt marsh in the southwestern Netherlands gave the provisional impression that the annual output of detritus and dissolved nutrients to the aquatic ecosystem is much less than their input in the reverse directions. Transport of detritus during storm floods, however, had to be left out of consideration because under these circumstances no measurements could be made. The annual output of detritus from the marsh washing ashore on the dike face under storm-flood conditions, might be about 10 % of the total above-ground biomass of the marsh (unpublished data).

Following Paviour-Smith (1956) and Teal (1962), who produced food-web and energy-flow diagrams for New Zealand and North American salt marshes respectively, a scheme is given in Fig. 6.14 which may characterize the trophic levels of the European salt-marsh ecosystem. In view of this scheme and following the literature available only some general trends in the ecosystem approach will be outlined here:

(1) *The pathways in the nitrogen and phosphorus cycles* between the tidal-marsh soil and adjacent aquatic ecosystem on the one side and the living world on the other is highly controlled by close proximity of oxidized and reduced conditions. According to North American experience, microbial processes are essential for nitrogen and sulphur cycles, while the mobility state of phosphorus varies on account of changes brought about in iron compounds by soil reduction and oxidation processes and also owing to the content of Ca and Mg ions (Patrick and Khalid, 1974). The nitrogen pathway is little known (see also p. 116 f.). Nitrogen is not always available in the salt marsh in optimal concentrations, while its chemical state (NH_4—NO_2—NO_3), important for uptake by the plants, may depend highly on oxido-reduction conditions in the soil. Nevertheless, the amount of nitrogen fixed by bacteria seems to be significant to the vegetation and probably contributes to the general high productivity of the salt marsh. Experiments with dosages of sludge and urea fertilizers carried out in North America by Van Raalte et al. (1974) indicate that the micro-organisms responsible for nitrogen fixation prefer to obtain their nitrogen directly from the added fertilizer. Apparently, the

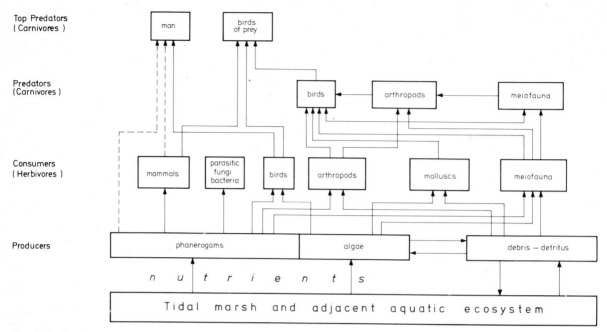

Fig. 6.14. Hypothetical diagram of trophic levels and food–chain pathways in the salt–marsh ecosystem (after Beeftink, 1976). The complicated role of microbial organisms as decomposers and as food for many animals ingesting detritus or feeding directly on bacteria is not considered since for the salt marshes there is almost no information on this subject.

bacteria and algae would avoid the energy–requiring fixation processes, and still achieve growth rates as great as or greater than those obtained with fixation as the main nitrogen source. This ability of salt–marsh micro–organisms to use varied sources of combined nitrogen may be involved in making the salt marsh an effective remover of nitrogen from coastal waters (Valiela et al., 1973). Sewage contamination may therefore change the structure of the nitrogen cycle in a salt–marsh environment, replacing atmospheric nitrogen as a nitrogen input (Van Raalte et al., 1974).

Phosphorus seems to be readily exchanged between the flood water and the sediments, especially in relation to their organic compounds, and hence there is a large reservoir of phosphorus in the muds (see p. 116). American ecologists found that *Spartina alterniflora* translocates significant quantities of phosphorus from the salt–marsh soil through the roots which penetrate deeply into the anaerobic sediments, towards the leaves (Pomeroy et al., 1969, 1972). Then, with tidal inundation, measurable quantities of phosphorus are released in the flood water (Reimold, 1972) as is also found with *Zostera* (McRoy and Barsdate, 1970). Thus, according to American experience, there is a documented pathway whereby phosphorus moves from the sediment through *Spartina* (and perhaps other salt–marsh species) into the flooding waters. Another branch in this phosphorus cycle is through debris of *Spartina*, etc., and detritus and filter feeders into the estuarine and coastal waters, as is illustrated by Pomeroy et al. (1969, 1972).

(2) *Food–chain and energy–flow studies* point to two principal pathways, viz.: (a) the grazing food chain, in which living plants are the primary energy source for the consumers; and (b) the detritus food chain, in which dead and decaying organic materials decomposed by bacteria, together with benthic blue–green algae, are the primary energy source (Odum, 1971). According to W.E. Odum (1970) most food chains in shallow estuaries in America are based upon macro–plant detritus and benthic and epiphytic micro–algae, rather than upon phytoplankton. In such areas phytoplankton production is about zero owing to turbidity in the water body (Ragotzkie, 1959), and zooplankton thus is relatively unimportant in energy transfer. Thus, the estuarine animals must be living on the exported marsh production (Teal, 1962; W.E. Odum, 1970). Hence events taking place in salt marshes may subsequently influence the growth of consumers in

adjacent water bodies, either by decreasing or increasing the detritus export of the marshes (Jefferies, 1972), or by influencing the phosphorus flux (Reimold, 1972). The number of studies on the share of the two food chains in the salt–marsh ecosystem is very limited and restricted to North American salt marshes.

Among arthropods the salt marsh harbours certain insects as dominant grazers, as well as some other insects associated with the detritus complex (Odum and Smalley, 1959; Marples, 1966). Spiders are the important carnivores in both the grazing and detritus food chain (Marples, 1966). Unfortunately, there appear to be no studies on the share of birds and mammals directly feeding on salt–marsh vegetation. According to Teal (1962) the American *(Spartina)* salt marshes achieve community stability through involving a few species with broad diets rather than many species with restricted diets.

Little is known about the food relations between the animals of the aquatic and of the aerial phase of the soil environment. Bilio (1967), examining the food relations of salt–marsh Turbellaria of Northern and Western Europe, found little evidence for such connections. Turbellarians were shown to feed mainly on nematodes, oligochaetes and diatoms. Turbellarians themselves are fed upon by aquatic predators only where frequent inundations provide favourable conditions for the penetration of aquatic macrofauna elements, such as the polychaete *Nereis diversicolor,* from adjacent mud flats. Bilio states that with increasing elevation above mean–water level the length of the purely aquatic food chains decreases and finally, i.e., in the intermediate and upper salt marshes, becomes restricted to meiofauna predators.

(3) *The turnover values,* either of the particular nutritive elements or as the ratio between the annual production and the maximum standing crop, may vary considerably. Firstly, algal turnover is much more rapid than phanerogamic (e.g., *Spartina*) turnover (Teal, 1962). Secondly, as many salt–marsh phanerogams are hemicryptophytes or therophytes, the turnover rate of above–ground biomass seems to be higher than that of the root systems. The latter has been estimated as 2—5 years in herb vegetation (Ketner, 1972). Thirdly, the turnover patterns of the elements differ considerably, from an almost complete annual turnover of potassium, for instance, to the very incomplete turnover of iron, lead, and nickel (Tyler, 1971). According to Tyler (1971) these patterns, however, are probably of general occurrence in meadow ecosystems and depend mainly on the chemical properties of the particular element.

(4) *The influence of man* on the salt–marsh ecosystem may vary widely, from no impact at all to a total destruction by excavation or industrial and urban occupation. In between there are several forms of exploitation all of which consider the marsh more or less as a natural resource: (a) by gathering edible products such as periwinkles *(Littorina littorea),* samphire *(Salicornia europaea)* and sea–aster *(Aster tripolium);* (b) by tending cattle, sheep or domestic geese, or by cropping for hay–making, both mostly combined with the construction of a drainage–furrow system adapted to the local geomorphology; and (c) by open–cast mining consisting of cutting and removal of turf and clay–winning. For a more detailed description of these forms of human impact, and their consequences on the salt–marsh ecosystem the reader is referred to Beeftink (1976) and to Chapter 17 by Queen.

ACKNOWLEDGEMENTS

The author wishes to express his gratitude to Dr. K.F. Vaas, Director of the Delta Institute for Hydrobiological Research, Yerseke (The Netherlands), for his help in formulating the English text, and to his colleagues Dr. M. Bilio, Dr. P.H. Nienhuis and Dr. W.J. Wolff for discussing parts of the text pertaining to their respective fields of interest.

REFERENCES

Abraham, R., 1970. Ökologische Untersuchungen an Pteromaliden (Hym., Chalcidoidea) im Grenzraum Land—Meer an der Nordseeküste Schleswig-Holsteins. *Oecologia,* 6: 15—47.

Adriani, M.J., 1958. Halophyten. In: W. Ruhland (Editor), *Handbuch der Pflanzenphysiologie.* Springer, Berlin, pp. 709—736.

Allorge, P., 1941. Essai de synthèse phytogéographique du Pays Basque. *Bull. Soc. Bot. Fr.,* 88: 291—356.

Almquist, E., 1929. Uplands vegetation och flora. *Acta Phytogeogr. Suec.,* 1: XII + 622 pp.

Amanieu, M., 1969. Recherches écologiques sur les faunes des plages abritées de la région d'Arcachon. *Helgol. Wiss. Meeresunters.*, 19: 455—557.

Anonymous, 1959. The Venice system for the classification of marine waters according to salinity. In: *Symposium on the Classification of Brackish Waters, 1958, Venice. Arch. Oceanogr. Limnol.*, Suppl. (supplementary sheet).

Baker, S.M., 1912. On the brown seaweeds of the salt marsh. I. *J. Linn. Soc. Bot.*, 40: 275—291.

Baker, S.M. and Bohling, M.H., 1915—17. On the brown seaweeds of the salt marsh. II. *J. Linn. Soc. Bot.*, 43: 325—380.

Bakker, C. and De Pauw, N., 1974. Comparison of brackish plankton assemblages of identical salinity ranges in an estuarine tidal (Westerschelde) and stagnant (lake Veere) environment (SW–Netherlands). I. Phytoplankton. *Hydrobiol. Bull.*, 8: 179—189.

Bakker, C. and De Pauw, N., 1975. Comparison of plankton assemblages of identical salinity ranges in estuarine tidal and stagnant environments. II. Zooplankton. *Neth. J. Sea Res.*, 9: 145—165.

Beeftink, W.G., 1959. Some notes on Skallingens salt marsh vegetation and its habitat. *Acta Bot. Neerl.*, 8: 449—472.

Beeftink, W.G., 1965. De zoutvegetatie van ZW–Nederland beschouwd in Europees verband. *Meded. Landbouwhogesch. Wageningen*, 65: 1—167.

Beeftink, W.G., 1966. Vegetation and habitat of the salt marshes and beach plains in the south–western part of the Netherlands. *Wentia*, 15: 83—108.

Beeftink, W.G., 1968. Die Systematik der europäischen Salzpflanzengesellschaften. In: R. Tüxen (Editor), *Pflanzensoziologische Systematik*. Junk, The Hague, pp. 239—272.

Beeftink, W.G., 1972. Übersicht über die Anzahl der Aufnahmen europäischer und nordafrikanischer Salzpflanzengesellschaften für das Projekt der Arbeitsgruppe für Datenverarbeitung. In: R. Tüxen (Editor), *Grundfragen und Methoden in der Pflanzensoziologie*. Junk, The Hague, pp. 371—396.

Beeftink, W.G., 1973. Ecologie en vegetatie met betrekking tot het Deltaplan. In: *De Gouden Delta*. Pudoc, Wageningen, pp. 81—109.

Beeftink, W.G., 1976. Salt marshes. In: R.S.K. Barnes (Editor), *Handbook on the Applied Ecology and Management of the Coastline*. Wiley, Chichester (in press).

Beeftink, W.G. and Géhu, J.M., 1973. Spartinetea maritimae. In: R. Tüxen (Editor), *Prodrome des groupements végétaux d'Europe, I*. Cramer, Lehre, 43 pp.

Bilio, M., 1963a. Die biozönotische Stellung der Salzwiesen unter den Strandbiotopen. *Verh. Dtsch. Zool. Ges. München*, 1963: 417—425.

Bilio, M., 1963b. Die Zonierung der aquatischen Bodenfauna in den Küstensalzwiesen Schleswig–Holsteins. *Zool. Anz.*, 171: 328—337.

Bilio, M., 1964—67. Die aquatische Bodenfauna von Salzwiesen der Nord– und Ostsee. I, II, III. *Int. Rev. Ges. Hydrobiol.*, 49: 509—562; 51: 147—195; 52: 487—533.

Bilio, M., 1965. Die Verteilung der aquatischen Bodenfauna und die Gliederung der Vegetation im Strandbereich der deutschen Nord– und Ostseeküste. *Bot. Gotob.*, 3: 25—42.

Bilio, M., 1966. Charakteristische Unterschiede in der Besiedlung finnischer, deutscher und holländischer Küstensalzwiesen durch Turbellarien. *Veröff. Inst. Meeresforsch. Bremerhaven*, Sonderband II: 305—318.

Bilio, M., 1967. Nahrungsbeziehungen der Turbellarien in Küstensalzwiesen. *Helgol. Wiss. Meeresunters.*, 15: 602—621.

Bird, E.C.F., 1968. *Coasts*. The M.I.T. Press, Cambridge, Mass., 246 pp.

Braun–Blanquet, J., 1964. *Pflanzensoziologie. Grundzüge der Vegetationskunde*. Springer, Vienna, 3rd ed., 865 pp.

Brereton, A.J., 1971. The structure of the species populations in the initial stages of salt–marsh successions. *J. Ecol.*, 59: 321—338.

Brümmer, G., Grünwaldt, H.S. and Schroeder, D., 1971. Beiträge zur Genese und Klassifizierung der Marschen. II. Zur Schwefelmetabolik in Schlicken und Salzmarschen. *Z. Pflanzenernähr. Düng. Bodenkd.*, 128: 208—220.

Burkholder, P.R., 1956. Studies on the nutritive value of *Spartina* grass growing in the marsh areas of coastal Georgia. *Bull. Torrey Bot. Club*, 83: 327—334.

Burkholder, P.R. and Bornside, G.H., 1957. Decomposition of marsh grass by aerobic marine bacteria. *Bull. Torrey Bot. Club*, 84: 366—383.

Burkholder, P.R. and Burkholder, L.M., 1956. Vitamin B_{12} in suspended solids and marsh muds collected along the coast of Georgia. *Limnol. Oceanogr.*, 1: 202—208.

Burton, P.J.K., 1961. The Brent Goose and its food supply in Essex. *Ann. Rep. Wildfowl Trust*, 12: 104—112.

Campbell, J.W., 1946. The food of Wigeon and Brentgoose. *Br. Birds*, 39: 194—200; 226—232.

Carter, N., 1932. A comparative study of the alga flora of two salt marshes. I. *J. Ecol.*, 20: 341—370.

Carter, N., 1933. A comparative study of the alga flora of two salt marshes. III. *J. Ecol.*, 21: 385—403.

Chapman, V.J., 1934, 1960. The plant ecology of Scolt Head Island. In: J.A. Steers (Editor), *Scolt Head Island*. Heffer, Cambridge, pp. 85—163 (1st and 2nd ed.).

Chapman, V.J., 1938—41. Studies in salt–marsh ecology. Sections I—VIII. *J. Ecol.*, 26: 144—179; 27: 160—201; 28: 118—152; 29: 69—82.

Chapman, V.J., 1959. Les "sociétés" des algues des marais salés et des mangroves. In: *Ecologie des algues marines*. Centre Natl. Rech. Sci., Paris, pp. 153—165.

Chapman, V.J., 1960. *Salt Marshes and Salt Deserts of the World*. Hill, London, 392 pp.

Chapman, V.J., 1964. *The Algae*. MacMillan, London, 472 pp.

Chapman, V.J., 1974. *Salt Marshes and Salt Deserts of the World*. Cramer, Lehre, 2nd ed., 392 pp. (complemented with 102 pp.).

Christiansen, W., 1955. Salicornietum. *Mitt. Flor.–Soziol. Arbeitsgem.*, N.F., 5: 64—65.

Clapham, A.R., Tutin, T.G. and Warburg, E.F., 1962. *Flora of the British Isles*. Cambridge University Press, London, 1591 pp.

Corillion, R., 1953. Les Halipèdes du Nord de la Bretagne. *Rev. Gén. Bot.*, 60: 609—658; 707—775.

Cottam, C., 1934. The eelgrass shortage in relation to waterfowl. *Trans. Am. Game Conf.*, 20: 272—279.

Cottam, C., 1935. Wasting disease of *Zostera marina*. *Nature*, 135: 306.

Cotton, A.D., 1912. Clare Island survey. Part 15. Marine algae. *Proc. R. Ir. Acad.*, 31: 1—178.

Dahl, E. and Hadač, E., 1941. Strandgesellschaften der Insel Ostøy im Oslofjord. Eine Pflanzensoziologische Studie. *Nytt Mag. Naturv.*, 82: 251—312.

Daiber, F.C., 1972. Ecology of coastal marshes. *Md. State Med. J.*, May: 70—72.

Dalby, D.H., 1970. The salt marshes of Milford Haven, Pembrokeshire. *Field Stud.*, 3: 297—330.

Dalby, R., 1957. Problems of land reclamation. 5. Salt marsh in the Wash. *Agric. Rev.*, 2: 31—37.

De Molenaar, J.G., 1974. Vegetation of the Angmagssalik district, Southeast Greenland, I. Littoral vegetation. *Medd. Grønl.*, 198(1): 1—79.

Den Hartog, C., 1958. De vegetatie van het Balgzand en de oeverterreinen van het Balgkanaal. *Wetensch. Meded. K.N.N.V.*, 27: 1—28.

Den Hartog, C., 1959a. The epilithic algal communities occurring along the coast of the Netherlands. *Wentia*, 1: 1—241.

Den Hartog, C., 1959b. Distribution and ecology of the slugs *Alderia modesta* and *Limapontia depressa* in the Netherlands. *Beaufortia*, 7: 15—36.

Den Hartog, C., 1963, 1964. The amphipods of the Deltaic region of the rivers Rhine, Meuse and Scheldt in relation to the hydrography of the area. I, II, III. *Neth. J. Sea Res.*, 2: 29—67; 407—457.

Den Hartog, C., 1963. The distribution of the marine triclad *Uteriporus vulgaris* in the Netherlands. *Proc. K. Ned. Akad. Wet., Ser. C*, 66: 196—204.

Den Hartog, C., 1964. Proseriate flatworms from the Deltaic area of the rivers Rhine, Meuse and Scheldt. I—II. *Proc. K. Ned. Akad. Wet., Ser. C*, 67: 10—34.

Den Hartog, C., 1964—66. A preliminary revision of the *Proxenetes* group (Trigonostomidae, Turbellaria). I—VI and supp. *Proc. K. Ned. Akad. Wet., Ser. C*, 67: 371—407; 68: 98—120; 69: 97—163 and 557—570.

Den Hartog, C., 1968. An analysis of the Gnathorhynchidae (Neorhabdocoela, Turbellaria) and the position of *Psittacorhynchus verweyi* nov. gen. nov. sp. in this family. *Proc. K. Ned. Akad. Wet.*, 71: 335—345.

Den Hartog, C., 1970. The sea-grasses of the world. *Verh. K. Ned. Akad. Wet. Afd. Nat.*, 59(1): 1—275.

Dickinson, C.H., 1965. The mycoflora associated with *Halimione portulacoides*. III. Fungi in green and moribund leaves. *Trans. Br. Mycol. Soc.*, 48: 603—610.

Doing Kraft, H., 1954. L'analyse des carrés permanents. *Acta Bot. Neerl.*, 3: 421—424.

Drummond, D., 1960. The food of *Rattus norvegicus* Berk. in an area of seawall, saltmarsh and mudflat. *J. Anim. Ecol.*, 29: 341—347.

Duffey, E., 1970. Habitat selection by spiders on a salt marsh in Gower. *Nat. Wales*, 12: 15—23.

Du Rietz, G.E., 1930. Vegetationsforschung auf soziations-analytischer Grundlage. In: E. Abderhalden (Editor), *Handbuch der biologischen Arbeitsmethoden*, Abt. XI, 5(2). Urban and Schwarzenberg, Berlin, pp. 293—480.

Du Rietz, G.E., 1940. Das limnologisch–thalassologische Vegetationsstufensystem. *Verh. Int. Ver. Theor. Angew. Limnol.*, 9: 102—110.

Du Rietz, G.E., 1947. Wellengrenzen als ökologische Äquivalente der Wasserstandslinien. *Zool. Bidr. Uppsala*, 25: 534—550.

Elliott, J.S.B., 1930. The soil fungi of the Dovey salt marshes. *Ann. Appl. Biol.*, 17: 284—305.

Ellis, E.A., 1960. An annotated list of the fungi. In: J.A. Steers (Editor), *Scolt Head Island*. Heffer, Cambridge, pp. 179—182.

Engel, H., Geerts, S.J. and Van Regteren Altena, C.O., 1940. *Alderia modesta* (Lovén) and *Limapontia depressa* Alder et Hancock in the brackish waters of the Dutch coast. *Basteria*, 5: 6—34.

Englund, B., 1942. Die Pflanzenverteilung auf den Meeresufern von Gotland. *Acta Bot. Fenn.*, 32: 1—282.

Epstein, E., 1969. Mineral metabolism of halophytes. In: I.H. Rorison (Editor), *Ecological Aspects of the Mineral Nutrition of Plants*. Blackwell, Oxford, pp. 345—355.

Feekes, W., 1936. De ontwikkeling van de natuurlijke vegetatie in de Wieringermeerpolder, de eerste groote droogmakerij van de Zuiderzee. *Ned. Kruidkd. Arch.*, 46: 1—295.

Feldmann, G., 1956. Développement d'une plasmodiophorale marine: *Plasmodiophora bicaudata* J. Feldm., parasite du *Zostera nana* Roth. *Rev. Gén. Bot.*, 63: 390—421.

Feldmann, G., 1958. Une nouvelle espèce de plasmodiophorale parasite de *Triglochin maritimum* L.: *Plasmodiophora maritima* nov. sp. *Rev. Gén. Bot.*, 65: 634—651.

Feldmann, G., 1959. Une ustilaginale marine, parasite du *Ruppia maritima* L. *Rev. Gén. Bot.*, 66: 35—40.

Fontes, F.C., 1945. Algumas características fitossociológicas dos "salgados" de Sacavém. *Bol. Soc. Brot.*, 19 (serie 2): 789—813.

Frei, M., 1937. Studi fitosociologici su alcune associazioni littorali in Sicilia (Ammophiletalia e Salicornietalia). *Nuovo G. Bot. Ital., N.S.*, 44: 273—294.

Fries, N., 1944. Beobachtungen über die thamniscophage Mykorrhiza einiger Halophyten. *Bot. Not.*, 1944: 255—264.

Fröde, E. Th., 1958. Die Pflanzengesellschaften der Insel Hiddensee. *Wiss. Z. E.M. Arndt–Univ. Greifswald Math.-Naturw. Reihe*, 7: 277—305.

Géhu, J.M., 1960. Quelques observations sur la végétation et l'écologie d'une station réputée de l'Archipel des Chaussey: L'île aux Oiseaux. *Bull. Lab. Marit. Dinard*, 46: 78—92.

Géhu, J.M., 1968. Essai sur la position systématique des végétations vivaces halo–nitrophiles des côtes atlantiques françaises (Agropyretea pungentis cl. nov.). *Bull. Soc. Bot. Nord Fr.*, 21: 71—77.

Gillham, M.E., 1956. Ecology of the Pembrokeshire Islands. V. Manuring by the colonial seabirds and mammals, with a note on seed distribution by Gulls. *J. Ecol.*, 44: 429—454.

Gillham, M.E., 1957. Vegetation of the Exe Estuary in relation to water salinity. *J. Ecol.*, 45: 735—756.

Gillner, V., 1955. Strandängsvegetation i Nord–Norge. *Sven. Bot. Tidskr.*, 49: 217—228.

Gillner, V., 1960. Vegetations- und Standortsuntersuchungen in den Strandwiesen der Schwedischen Westküste. *Acta Phytogeogr. Suec.*, 43: 1—198.

Gimingham, C.H., 1964. Maritime and sub–maritime communities. In: J.H. Burnett (Editor), *The Vegetation of Scotland*. Oliver and Boyd, Edinburgh, pp. 67—142.

Harmsen, G.W., 1936. Systematische Beobachtungen der nordwesteuropäischen Seegrasformen. *Ned. Kruidkd., Arch.*, 46: 852—877.

Häyrén, E., 1956. Über die Algenvegetation des sandigen Geolitorals am Meere in Schweden und in Finnland. *Sven. Bot. Tidskr.*, 50: 257—269.

Heath, D.J., 1975. Factors affecting temperature and salinity conditions on a Scottish salt marsh, with notes on the ecology of *Sphaeroma rugicauda* (Leach). *Arch. Hydrobiol.*, 75: 76—89.

Heydemann, B., 1960. Verlauf und Abhängigkeit von Spinnensukzessionen im Neuland der Nordseeküste. *Verh. Dtsch. Zool. Ges. Bonn*, 1960: 431—457.

Heydemann, B., 1961. Vergleichend–ökologische Populationsanalysen an Micryphantiden (Aranae) von Nordseedeichen. *Verh. XI Int. Kongr. Entomol., Wien, 1960*, 1: 762—767.

Heydemann, B., 1962. Die biozönotische Entwicklung vom Vorland zum Koog. Vergleichend–ökologische Untersuchungen an der Nordseeküste. II. Teil. Käfer (Coleoptera). *Abh. Akad. Wiss., Lit. Math.–Naturwiss. Kl.*, 1962 (11): 767—964.

Heydemann, B., 1963. Deiche der Nordseeküste als besonderer Lebensraum. Ökologische Untersuchungen über die Arthropoden–Besiedlung. *Küste*, 11: 90—130.

Heydemann, B., 1964a. Die Carabiden der Kulturbiotope von Binnenland und Nordseeküste, ein ökologischer Vergleich (Coleopt. Carabidae). *Zool. Anz.*, 172: 49—86.

Heydemann, B., 1964b. Die Spinnenfauna des Naturschutzgebietes "Bottsand", der Kolberger Heide und des Schönberger Strandes (Araneae). *Faun. Mitt. Norddtschl.*, 2: 133—141.

Heydemann, B., 1967. *Die biologische Grenze Land—Meer im Bereich der Salzwiesen.* Steiner, Wiesbaden, 200 pp.

Høeg, O.A. and Lid, J., 1949. *Carex extensa*, ny for Norge. *Blyttia*, 7: 87—91.

Iversen, J., 1952—53. The zonation of the salt marsh vegetation of Skallingen in 1931—34 and in 1952. *Geogr. Tidsskr.*, 52: 113—118.

Jacquet, J., 1949. *Recherches écologiques sur le littoral de la Manche. Les Prés salés et la Spartine de Townsend. Les estuaires. La tangue. Encycl. Biogéogr. Ecol., V.* Lechevalier, Paris, 374 pp.

Jakobsen, B., 1954. The tidal area in South–Western Jutland and the process of the salt marsh formation. *Geogr. Tiddskr.*, 53: 49—61.

Jakobsen, B., 1964. Vadehavets morfologi. En geografisk analyse af vadelandskabets formudvikling med saerlig hensyntagen til Juvre Dybs Tidevandsområde. *Folia Geogr. Dan.*, 11: 1—176.

Jefferies, R.L., 1972. Aspects of salt–marsh ecology with particular reference to inorganic plant nutrition. In: R.S.K. Barnes and J. Green (Editors), *The estuarine environment.* London, Applied Sci. Publ., London, pp. 61—85.

Johnson, D.S. and York, H.H., 1915. The relations of plants to tide levels. A study of factors affecting the distribution of marine plants. *Carnegie Inst. Wash. Publ.*, No. 206: 1—162.

Jones, K., 1974. Nitrogen fixation in a salt marsh. *J. Ecol.*, 62: 553—565.

Jónsson, H., 1914. Strandengen i Sydvest–Island. *Mindeskr. J. Steenstrups Fødsel*, 12: 1—7.

Kalela, A., 1939. Über Wiesen und wiesenartige Pflanzengesellschaften auf der Fischerhalbinsel in Petsamo Lappland. *Acta For. Fenn.*, 48(2): 1—523 (1940).

Ketner, P., 1972. *Primary Production of Salt–Marsh Communities on the Island of Terschelling in The Netherlands.* Thesis, Catholic University of Nijmegen, 181 pp.

Koenders, J.W., 1968. Are tidal marshes suitable biotope for muskrats? *Eur. Plant Prot. Org. Publ., Ser. A*, Nr. 47: 67—69.

Korchagin, A.A., 1935. Die Vegetation der Wiesen und Moore des Meeralluviums der Mesener Bucht und Tscheschkaja Guba. *Acta Inst. Bot. Acad. Sci. U.R.S.S., Ser. III*, Fasc. 2: 223—344 (in Russian).

Kornaś, J., Pancer, E. and Brzyski, B., 1960. Studies on sea–bottom vegetation in the Bay of Gdańsk off Rewa. *Fragm. Florist. Geobot.*, 6: 3—92.

Lambert, J.M., Jennings, J.N., Smith, C.T., Green, C. and Hutchinson, J.N., 1960. *The Making of the Broads. A Reconsideration of Their Origin in the Light of New Evidence.* The Royal Geographical Society, London, 153 pp.

Lebret, T., 1965. The prospects for wild geese in the Netherlands. *Ann. Rep. Wildfowl Trust*, 16: 85—91.

Leiviskä, J., 1908. *Über die Vegetation an der Küste des Bottnischen Meerbusens zwischen Tornio und Kokkola.* Simelii, Helsinki, 209 pp.

Lemée, G., 1952. Végétation et écologie des tangues du havre de Portbail (Manche). *Bull. Soc. Bot. Fr. Mém.*, 1952: 156—165.

Libbert, W., 1940. Die Pflanzengesellschaften der Halbinsel Darsz. *Rep. Spec. Nov. Regni. Veg. Beih.*, 114: 1—95.

Lloyd, L.S. and Wilson, J.M., 1962. Development of the perithecium in *Lulworthia medusa* (Ell. et Ev.) Cribb et Cribb, a saprophyte on *Spartina townsendii. Trans. Br. Mycol. Soc.*, 45: 359—372.

Lorenzen, S., 1969. Freilebende Meeresnematoden aus dem Schlickwatt und den Salzwiesen der Nordseeküste. *Veröff. Inst. Meeresunters. Bremerhaven*, 11: 195—238.

Luther, H., 1951. Verbreitung und Ökologie der höheren Wasserpflanzen im Brackwasser der Ekenäs–Gegend in Südfinnland I—II. *Acta Bot. Fenn.*, 49/50: 1—231; 1—370.

Luxton, M., 1964. Some aspects of the biology of saltmarsh Acarina. *Acarologia C.R. 1er Congr. Int. Acarol., 1963*, pp. 172—182.

Luxton, M., 1967. The ecology of saltmarsh Acarina. *J. Anim. Ecol.*, 36: 257—277.

Marples, T.G., 1966. A radionuclide tracer study of Arthropod food chains in a *Spartina* salt marsh ecosystem. *Ecology*, 47: 270—277.

Martin, A.C., Zim, H.S. and Nelson, A.L., 1951. *American Wildlife and Plants, a Guide to Wildlife Food Habits.* Dover, New York, N.Y., 500 pp.

Mason, E., 1928. Note on the presence of mycorrhiza in the roots of salt marsh plants. *New Phytol.*, 27: 193—195.

McMillan, N.F., 1949. The brackish–water mollusca of Bromborough Pool, Cheshire. *J. Conchol.*, 23: 65—68.

McRoy, C.P., 1966. *The Standing Stock and Ecology of Eelgrass (Zostera marina L.) in Izembek Lagoon, Alaska.* Thesis, University of Washington, Seattle, 138 pp.

McRoy, C.P. and Barsdate, R.J., 1970. Phosphate absorption

in eelgrass. *Limnol. Oceanogr.*, 15: 6—13.

Meyer, K.O., 1955. Naturgeschichte der Strandschnecke *Ovatella myosotis* (Drapernaud). *Arch. Molluskenkd.*, 84: 1—43.

Moore, J.J., Fitzsimons, P., Lambe, E. and White, J., 1970. A comparison and evaluation of some phytosociological techniques. *Vegetatio*, 20: 1—20.

Morley, J.V., 1973. Tidal immersion of *Spartina* marsh at Bridgewater Bay, Somerset. *J. Ecol.*, 61: 383—386.

Mörzer Bruijns, M.F., Van Regteren Altena, C.O. and Butot, L.J.M., 1959. The Netherlands as an environment for land mollusca. *Basteria*, 23 (suppl.): 132—174.

Newell, R., 1962. Behavioural aspects of the ecology of *Peringia* (= *Hydrobia*) *ulvae* (Pennant) (Gastropoda, Prosobranchia). *Proc. Zool. Soc. Lond.*, 138: 49—75.

Newell, R., 1964. Some factors controlling the upstream distribution of *Hydrobia ulvae* (Pennant), (Gastropoda, Prosobranchia). *Proc. Zool. Soc. Lond.*, 142: 85—106.

Newell, R., 1965. The role of detritus in the nutrition of two marine deposit feeders, the prosobranch *Hydrobia ulvae* and the bivalve *Macoma balthica*. *Proc. Zool. Soc. Lond.*, 144: 25—45.

Nicol, E.A.T., 1935. The ecology of a salt–marsh. *J. Mar. Biol. Assoc. U.K.*, N.S., 20: 203—261.

Nienburg, W., 1927. Zur Ökologie der Flora des Wattenmeeres. 1. Teil Der Köningshafen bei List auf Sylt. *Wiss. Meeresunters. Abt. Kiel, N.F.*, 20: 145—196.

Nienhuis, P.H., 1970. The benthic algal communities of flats and salt marshes in the Grevelingen, a sea–arm in the south–western Netherlands. *Neth. J. Sea Res.*, 5: 20—49.

Nienhuis, P.H., 1972. The use of permanent sample plots in studying the quantitative ecology of algae in salt marshes. *Proc. 7th Int. Seaweed Symp. 1971*, Univ. of Tokyo Press, Tokyo, pp. 251—254.

Nienhuis, P.H., 1973. Salt–marsh and beach plain as a habitat for benthic algae. *Hydrobiol. Bull.*, 7: 15—24.

Nienhuis, P.H., 1975. *Biosystematics and Ecology of Rhizoclonium riparium (Roth) Harv. (Chlorophyceae: Cladophorales) in the Estuarine Area of the Rivers Rhine, Meuse and Scheldt.* Thesis, State University of Groningen, 240 pp.

Nienhuis, P.H. and Simons, J., 1971. *Vaucheria* species and some other algae on a Dutch salt marsh, with ecological notes on their periodicity. *Acta Bot. Neerl.*, 20: 107—118.

Nordhagen, R., 1923. Vegetationsstudien auf der Insel Utsire im westlichen Norwegen. *Bergen. Mus. Aarb.*, 1920—21: 1—149.

Nordhagen, R., 1940. Studien über die maritime Vegetation Norwegens, I. Die Pflanzengesellschaften der Tangwälle. *Bergen. Mus. Aarb., Naturv. Rekke*, 1(2): 1—123.

Nordhagen, R., 1954. Studies on the vegetation of salt and brackish marshes in Finmark (Norway). *Vegetatio*, 5/6: 381—394.

Oberdorfer, E., 1952. Beitrag zur Kenntnis der nordägäischen Küstenvegetation. *Vegetatio*, 3: 329—349.

Odum, E.P., 1971. *Fundamentals of Ecology*. Saunders, Philadelphia, Pa., 574 pp.

Odum, E.P. and Smalley, A.E., 1959. Comparison of population energy flow of a herbivorous and a deposit feeding invertebrate in a salt marsh ecosystem. *Proc. U.S. Natl. Acad. Sci.*, 45: 617—622.

Odum, W.E., 1970. Utilization of the direct grazing and plant detritus food chains by the striped mullet *Mugil cephalus*. In: J.H. Steele (Editor), *Marine Food Chains*. Oliver and Boyd, Edinburgh, pp. 222—240.

O'Reilly, H. and Pantin, G., 1957. Some observations on the salt marsh formation in Co. Dublin. *Proc. R. Ir. Acad. Sci.*, Sect. B, 58: 89—128.

Ostenfeld, C.H., 1908. On the ecology and distribution of the Grass–Wrack *(Zostera marina)* in Danish waters. *Rep. Dan. Biol. Station Board Agric.*, 16: 1—62.

Packham, J.R. and Liddle, M.J., 1970. Cefni salt marsh, Anglesey, and its recent development. *Field Stud.*, 3: 331—356.

Parke, M. and Dixon, P.S., 1968. Check–list of British marine algae. Second revision. *J. Mar. Biol. Assoc. U.K.*, 48: 783—832.

Patrick, W.H. and Khalid, R.A., 1974. Phosphate release and sorption by soils and sediments: Effect of aerobic and anaerobic conditions. *Science*, 186: 53—55.

Paviour–Smith, K., 1956. The biotic community of a salt meadow in New Zealand. *Trans. R. Soc. N.Z.*, 83: 525—554.

Payne, K.T., 1972. A survey of the *Spartina* feeding insects in Poole Harbour, Dorset. *Entomol. Mon. Mag.*, 108: 66—79.

Perring, F.H. and Walters, S.M., 1962. *Atlas of the British Flora*. Nelson, London, 432 + 12 pp.

Pethick, J.S., 1974. The distribution of salt pans on tidal salt marshes. *J. Biogeogr.*, 1: 57—62.

Philip, G., 1936. An enalid plant association in the Humber Estuary. *J. Ecol.*, 24: 205—219.

Phleger, F.B., 1970. Foraminiferal populations and marine marsh processes. *Limnol. Oceanogr.*, 15: 522—534.

Pignatti, S., 1966. La vegetazione alofila della laguna Veneta. *Ist. Veneto Sci. Lett. Arti Cl. Sci. Mat. Nat.*, 33: 1—174.

Pigott, C.D., 1969. Influence of mineral nutrition on the zonation of flowering plants in coastal salt–marshes. In: I.H. Rorison (Editor), *Ecological Aspects of the Mineral Nutrition of Plants*. Blackwell, Oxford, pp. 25—35.

Piotrowska, H., 1974. Maritime communities of halophytes in Poland and the problems of their protection. *Zakl. Ochr. Przyr. Pol. Akad. Nauk*, 39: 7—63 (in Polish).

Polderman, P.J.G., 1974. The algae of saline areas near Vlissingen (The Netherlands). *Acta. Bot. Neerl.*, 23: 65—79.

Polderman, P.J.G. and Prud'homme van Reine, W.F., 1973. *Chrysomeris ramosa* (Chrysophyceae) in Denmark and in the Netherlands. *Acta. Bot. Neerl.*, 22: 81—91.

Pomeroy, L.R., Johannes, R.E., Odum, E.P. and Roffman, B., 1969. The phosphorus and zinc cycles and productivity of a salt marsh. In: D.J. Nelson and F.C. Evans (Editors), *Proc. Natl. 2nd Symp. Radioecol. U.S. Energy Comm.*, pp. 412—419.

Pomeroy, L.R., Shenton, L.R., Jones, R.D.H. and Reimold, R.J., 1972. Nutrient flux in estuaries. In: G. Likens (Editor), *Nutrients and Eutrophication. Am. Soc. Limnol. Oceanogr., Spec. Symp.*, 1: 274—291.

Preuss, J., 1911—12. Die Vegetationsverhältnisse der deutschen Ostseeküste. *Schr. Naturforsch. Ges. Danzig*, 13: 45—258.

Pugh, G.J.F., 1974. Fungi in intertidal regions. *Veröff. Inst.*

Meeresforsch. Bremerhaven, Suppl., 5: 403—418.

Queen, W.H., 1974. Physiology of coastal halophytes. In: R.J. Reimold and W.H. Queen (Editors), Ecology of Halophytes. Academic Press, New York, N.Y., pp. 345—353.

Ragotzkie, R.A., 1959. Plankton productivity in estuarine waters of Georgia. Inst. Mar. Sci., 6: 146—158.

Ranwell, D.S., 1961. *Spartina* salt marshes in southern England. I. The effects of sheep grazing at the upper limits of *Spartina* marsh in Bridgewater Bay. J. Ecol., 49: 325—340.

Ranwell, D.S., 1964. *Spartina* salt marshes in southern England. III. Rates of establishment, succession and nutrient supply at Bridgwater Bay, Somerset. J. Ecol., 52: 95—105.

Ranwell, D.S., 1967. World resources of *Spartina townsendii* (sensu lato) and economic use of *Spartina* marshland. J. Appl. Ecol., 4: 239—256.

Ranwell, D.S., 1968. Coastal marshes in perspective. Reg. Stud. Group Bull. Strathclyde, No. 9: 1—26.

Ranwell, D.S., 1972. *Ecology of Salt Marshes and Sand Dunes.* Chapman and Hall, London, 258 pp.

Ranwell, D.S., 1974. The salt marsh to tidal woodland transition. Hydrobiol. Bull., 8: 139—151.

Ranwell, D.S. and Downing, B.M., 1959. Brentgoose *(Branta bernicla* (L.)): Winter feeding pattern and *Zostera* resources at Scolt Head Island, Norfolk. Anim. Behav., 7: 42—56.

Ranwell, D.S., Wyer, D.W., Boorman, L.A., Pizzey, J.M. and Waters, R.J., 1974. *Zostera* transplants in Norfolk and Suffolk, Great Britain. Aquaculture, 4: 185—198.

Rasmussen, E., 1973. Systematics and ecology of the Isefjord marine fauna (Denmark) with a survey of the eelgrass *(Zostera)* vegetation and its communities. Ophelia, 11: 1—507.

Rees, T.K., 1935. The marine algae of Lough Ine. J. Ecol., 23: 69—133.

Reimold, R.J., 1972. The movement of phosphorus through the salt marsh cord grass, *Spartina alterniflora* Loisel. Limnol. Oceanogr., 17: 606—611.

Remane, A., 1955. Die Brackwasser–Submergenz und die Umkomposition der Coenosen in Belt- und Ostsee. Kieler Meeresforsch., 11: 59—73.

Richards, F.J., 1934. The salt marshes of the Dovey estuary. IV. The rates of vertical accretion, horizontal extension and scarp erosion. Ann. Bot., 48: 225—259.

Ryther, J.H. and Dunstan, W.M., 1971. Nitrogen, phosphorus and eutrophication in the coastal marine environment. Science, 171: 1008—1012.

Schmeisky, H., 1974. *Vegetationskundliche und ökologische Untersuchungen in Strandrasen des Graswarders vor Heiligenhafen/Ostsee.* Thesis, University of Göttingen, 103 pp.

Serventy, D.L., 1960. The marine invertebrate fauna. In: J.A. Steers (Editor), *Scolt Head Island.* Heffer, Cambridge, pp. 216—245.

Shimwell, D.W., 1971. *The Description and Classification of Vegetation.* Sidgwick and Jackson, London, 322 pp.

Simons, J., 1974. *Vaucheria birostris* n. sp. and some further remarks on the genus *Vaucheria* in the Netherlands. Acta Bot. Neerl., 23: 399—413.

Simons, J. and Vroman, M., 1973. *Vaucheria* species from the Dutch brackish inland ponds "De Putten". Acta Bot. Neerl., 22: 177—192.

Stebbings, R.E., 1971. Some ecological observations on the fauna in a tidal marsh to woodland transition. Proc. Trans. Entomol. Soc., 4: 83—88.

Steinführer, A., 1955. Die Pflanzengesellschaften der Schleiufer und ihre Beziehungen zum Salzgehalt des Bodens. Jahrb. Heimatges. Kreis Eckernförde e.V., 13: 3—47.

Sterner, R., 1933. Vegetation och flora i Kalmarsunds skärgård. Acta Horti Gotob., 8: 189—280.

Stewart, G.R., Lee, J.A. and Orebamjo, T.O., 1972. Nitrogen metabolism of halophytes. I. Nitrate reductase activity in *Suaeda maritima.* New Phytol., 71: 263—268.

Strenzke, K., 1955. Verbreitung und Systematik der Collembolen der deutschen Nord- und Ostseeküste. Veröff. Inst. Meeresforsch. Bremerhaven, 3: 46—65.

Tansley, A.G., 1939. *The British Islands and Their Vegetation.* Cambridge University Press, London, 930 pp.

Teal, J.M., 1962. Energy flow in the salt marsh ecosystem of Georgia. Ecology, 43: 614—624.

Tutin, T.G., 1942. Biological flora of the British Isles. *Zostera* L. J. Ecol., 30: 217—226.

Tüxen, R., 1950. Grundrisz einer Systematik der nitrophilen Unkrautgesellschaften in der eurosibirischen Region Europas. Mitt. Flor.–Soziol. Arbeitsgem., N.F., 2: 94—175.

Tüxen, R. and Westhoff, V., 1963. Saginetea maritimae, eine Gesellschaftsgruppe im wechselhalinen Grenzbereich der europäischen Meeresküsten. Mitt. Flor.–Soziol. Arbeitsgem., N.F., 10: 116—129.

Tyler, G., 1967. On the effect of phosphorus and nitrogen, supplied to Baltic shore–meadow vegetation. Bot. Not., 120: 433—447.

Tyler, G., 1971. Distribution and turnover of organic matter and minerals in a shore meadow ecosystem. Studies in the ecology of Baltic sea–shore meadows. IV. Oikos, 22: 265—291.

Valiela, I., Teal, J.M. and Sass, W., 1973. Nutrient retention in salt marsh plots experimentally fertilized with sewage sludge. Estuarine Coastal Mar. Sci., 1: 261—269.

Van Goor, A.C.J., 1919. Het zeegras *(Zostera marina* L.) en zijn beteekenis voor het leven der visschen. Rapp. Verh. Rijksinst. Visscherijonderz., 1(4): 415—498.

Van Goor, A.C.J., 1921. Die *Zostera*–Assoziation des holländischen Wattenmeeres. Rec. Trav. Bot. Néerl., 18: 103—123.

Van Raalte, C.D., Valiela, I., Carpenter, E.J. and Teal, J.M., 1974. Inhibition of nitrogen fixation in salt marshes measured by acetylene reduction. Estuarine Coastal Mar. Sci., 2: 301—305.

Van Wijngaarden, A., 1961. De zoogdierfauna van Goeree–Overflakkee. Jaarb. 1961 Wet. Gen. Goeree–Overflakkee, pp. 93—99.

Van Wijngaarden, A., 1969. De Noorse Woelmuis, *Microtus oeconomus* Pall. in Nederland. Rep. R.I.V.O.N., 59 pp. (mimeograph).

Verhoeven, B., 1963. On the calciumcarbonate content of young marine sediments. Int. Inst. Land Reclam. Improv. Bull., 4: 27 pp.

Wagner, D.T., 1969. Ecological studies on *Leptosphaeria discors,* a graminicolous fungus of salt marshes. Nova Hedwigia, 18: 383—396.

Waisel, Y., 1972. *Biology of Halophytes.* Academic Press, New York, N.Y., 395 pp.

Wendelberger, G., 1950. Zur Soziologie der kontinentalen Halophyten–Vegetation Mitteleuropas. Unter besondere Berücksichtigung der Salzpflanzengesellschaften am Neusiedler See. *Denkschr. Österr. Akad. Wiss.*, 108(5): 1—180.

Westhoff, V., 1947. *The Vegetation of Dunes and Salt Marshes on the Dutch Islands of Terschelling, Vlieland and Texel.* Thesis, State University, Utrecht, 131 pp.

Westhoff, V. and Den Held, A.J., 1969. *Plantengemeenschappen in Nederland.* Thieme, Zutphen, 324 pp.

Westhoff, V. and Van der Maarel, E., 1973. The Braun–Blanquet approach. In: R.H. Whittaker (Editor), Ordination and Classification of Communities. Junk, The Hague, pp. 617—726.

Westhoff, V., Van Leeuwen, Chr. G. and Adriani, M.J., 1961. Enkele aspecten van vegetatie en bodem der duinen van Goeree, in het bijzonder de contactgordels tussen zout en zoet milieu. *Jaarb. 1961 Wet. Gen. Goeree–Overflakkee,* pp. 1—127.

Wiehe, P.O., 1935. A quantitative study of the influence of tide upon populations of *Salicornia europaea. J. Ecol.*, 23: 323—333.

Wolff, W.J., 1968. The halophilous vegetation of the lagoons of Mesolonghi, Greece. *Vegetatio,* 16: 95—134.

Wolff, W.J., De Koeijer, P., Sandee, A.J.J. and De Wolf, L., 1967. De verspreiding van Rotganzen in het Deltagebied in relatie tot de verspreiding van hun voedsel. *Limosa,* 40: 163—174.

Woodell, S.R.J., 1974. Anthill vegetation in a Norfolk saltmarsh. *Oecologia,* 16: 221—225.

Zonneveld, I.S., 1960. *De Brabantse Biesbosch. Een studie van bodem en vegetatie van een zoetwatergetijdendelta.* Pudoc, Wageningen, English summary, 210 pp.; Dutch text, 396 pp.

Chapter 7

MANGALS AND SALT MARSHES OF EASTERN UNITED STATES

ROBERT J. REIMOLD

INTRODUCTION

Along the Atlantic coast of the United States there are some 890 308 ha of wetlands which include 589 480 ha of salt marshes (Spinner, 1968). The distribution of salt marshes by state (Table 7.1, Fig. 7.1) reveals that the greatest expanses occur along the central and southern reaches of the east coast. South Carolina and Georgia alone contain over 62% of all the east coast salt marshes.

Management of coastal salt marshes of the eastern United States has typically included attempts to alter their value. Over the past two centuries, extensive efforts have been made to drain or fill these marshes for agricultural, industrial or residential expansion. Between 1954 and 1965, approximately 10% of the east coast salt marshes were destroyed or eliminated (Spinner, 1968) (see Chapter 14).

Duncan (1974) has listed the halophytic plants indigenous to eastern coast salt marshes. Several attempts have been made to classify eastern coast marshes by their fauna and flora (Wright, 1907; Martin et al., 1953; Shaw and Fredine, 1956;

TABLE 7.1

Geographic distribution of salt marshes and mangrove swamps along the eastern coast of United States (adapted from Spinner, 1968, and Shaw and Fredine, 1956)

State	Area (ha)	Percent of total
Maine	588	0.10
New Hampshire	151	0.03
Massachusetts	3213	0.55
Rhode Island	216	0.04
Connecticut	840	0.14
New York	4666	0.79
New Jersey	8445	1.43
Delaware	17 707	3.00
Maryland	27 899	4.73
Virginia	44 839	7.61
North Carolina	64 284	10.91
South Carolina	176 442	29.93
Georgia	192 508	32.66
Florida	47 631	8.08
Total	589 429	

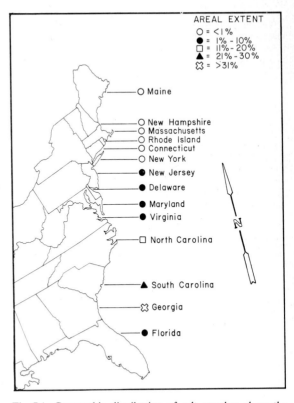

Fig. 7.1. Geographic distribution of salt marshes along the Atlantic coast of the United States (adapted from Spinner, 1968, and Shaw and Fredine, 1956).

Spinner, 1968). In general, divisions include coastal salt flats, coastal salt meadows, irregularly flooded salt marshes, regularly flooded marshes, and mangrove swamps. The coastal salt flats are areas which are irregularly inundated by tides. These areas are on the landward side of the salt meadows and marshes. The vegetation (Table 7.2) is patchy and the soil is waterlogged during the growing season, making these areas relatively unimportant as feeding grounds for waterfowl but providing a suitable resting habitat.

The coastal salt meadows generally on the landward side of salt marshes have soil that is waterlogged during the growing season but rarely covered by tides. The vegetation serves as food for waterfowl in both their production and wintering zones. There are usually open water areas (potholes) which increase use by waterfowl.

The irregularly flooded salt marshes do not produce food suitable for waterfowl. In these areas, the vegetation (*Juncus roemerianus*, needle rush, and *Ruppia maritima*, widgeon grass) grows in a rather well–drained sandy soil which is irregularly inundated by wind– and storm–generated tides during the plant growing season.

In the regularly flooded salt marshes, tides generally flood the marsh surface to a depth of 0.25 m or more on each cycle. These marshes occur along open ocean areas and are a major area for waterfowl feeding, resting and protection. The vegetation is predominately *Spartina alterniflora*, salt–marsh cord grass, one of the halophytic salt–marsh plants.

In mangrove swamps the soil is covered with 0.1—1.0 m of water (at high tide) during the year–round growing season. The tree growth and dense root system is limited to a very few species of plants including *Avicennia nitida*, black mangrove; *Rhizophora mangle*, red mangrove; and *Conocarpus erecta*, button mangrove (see p. 199).

The most serious threats to salt marshes result from man's activities. The National Estuary Study

TABLE 7.2

Vegetation typical of eastern United States coastal salt flats, salt meadows, irregularly flooded salt marshes, regularly flooded salt marshes, and mangrove swamps (adapted from Shaw and Fredine, 1956)

Scientific name	Common name	Occurrence					area of predominant occurrence
		salt flat	salt meadow	irregularly flooded salt marsh	regularly flooded salt marsh	mangrove swamp	
Avicennia nitida Jacq.	black mangrove					×	FL
Batis maritima L.	saltwort	×					MS to MD
Carex spp.	sedge		×				LA to MS
Conocarpus erectus L.	button mangrove					×	FL
Distichlis spicata (L.) Greene	salt grass	×	×				LA to FL
Juncus balticus Willd.	baltic rush		×				LA to NY
Juncus gerardi Loisel.	black rush		×				NF to NJ
J. roemerianus L.	needle rush			×			MD to FL
Monanthochloe littoralis Engelm.	salt–flat grass	×					FL
Plantago spp.	plantain		×				NS to DE
Pluchea camphorata (L.) DC	salt–marsh fleabane		×				ME to FL
Rhizophora mangle L.	red mangrove					×	FL
Ruppia maritima L.	widgeon grass			×			ME to FL
Salicornia bigelovii L.	dwarf saltwort	×					ME to FL
S. europaea L.	samphire	×					NS to FL
S. virginica L.	perennial saltwort	×					MS to FL
Scirpus olneyi L.	Olney threesquare		×				NS to FL
Spartina alterniflora Loisel.	salt–marsh cord grass				×		NF to FL
S. patens (Ait.) Muhl.	salt–marsh cord grass		×				NF to FL
Suaeda maritima (L.) Dum.	seablite	×					QU to VA

Abbreviations: DE = Delaware; FL = Florida; LA = Labrador; MD = Maryland; ME = Maine; MS = Massachusetts; NF = Newfoundland; NJ = New Jersey; NS = Nova Scotia; NY = New York; QU = Quebec; VA = Virginia.

(1970) lists sewage pollution, channel dredging and spoiling for transportation, bulkhead dredge and fill to create land, industrial waste, pesticide pollution, ditching and draining of wetlands, agricultural wastes, and river impoundment and flow control, as the most severe factors which have potentially adverse effects on the marshes (see p. 365). A greater in–depth view of the salt–marsh ecosystem along the Atlantic coast should help to put the entire natural system into a more comprehensible perspective.

The characteristic vegetation of the salt marshes from eastern Canada to south Florida (Fig. 7.1) is *Spartina alterniflora,* which dominates the intertidal zone. Many authors have attempted to document these growth zones (Table 7.3) and, in general, find that *S. alterniflora* occupies an area from near mean sea level to mean high water.

At the edge or near the high tide line, the next most abundant plant is *Juncus* spp. *J. gerardi* and *J. balticus* are found in New England and Canada; *J. gerardi* is common in the middle Atlantic region south to the Chesapeake Bay; *J. roemerianus* predominates south of Chesapeake Bay.

Several other plants which occur near and above the high water line include *Distichlis spicata, Spartina patens,* and several species of *Salicornia.* The rather cosmopolitan common reed, *Phragmites communis,* grows along the landward side of most salt marshes, especially where there is fresh water.

The algae of eastern coast salt marshes have been reviewed by Blum (1968). Pomeroy (1959) and Burkholder and Bornside (1957) considered the role of algae and bacteria respectively in salt–marsh ecosystems. Webber (1967) reviewed the blue–green algae of salt marshes.

Shanholtzer (1974) considered the relationship of vertebrates to eastern coast salt–marsh plants. Daiber (1974) reviewed the future of eastern coastal salt marshes in relation to animal management practices (see Chapter 5); Seneca (1974) also reviewed the importance of *Spartina alterniflora* in the stabilization of eastern coast dredged material. Reimold and Queen (1974) reviewed the ecology of the halophytic marsh plants and the interactions with animals including man.

The common avian fauna of the eastern U.S. salt marshes are nearly as widely distributed as the plants. The clapper rail, *Rallus longirostris,* is found in marshes from New Jersey southward throughout the year, and in more northern areas in the summer. It feeds on insects, snails, and small crustaceans but mostly on fiddler crabs (*Uca* sp.). Sharp–tailed sparrows, *Ammospiza caudacuta,* and seaside sparrows, *A. maritima,* are also common inhabitants relying on insects, spiders, snails, and sand fleas for 80% of their diet, and utilizing seeds of grasses and other plants for the remaining 20%. Common ducks include black ducks, *Anas rubripes,* and blue–winged teal, *Querquedula discors.* The former consume about three–fourths plant food such as pond weeds and seeds, and one–fourth

TABLE 7.3

Summary of tidal regimes in relation to growth of *Spartina alterniflora* (after Adams, 1963)

Tidal range (m)	Elevation range of *S. alterniflora* above mean sea level (m)	Geographic location of the investigation	Source
0.85	0.37	Beaufort, N.C.	Adams (1963)
1.22	0.52	Wrightsville Beach, N.C.	Adams (1963)
1.16	0.37	Delaware	Bourn and Cottam (1950)
1.10	0.61	Camp Lejeune, N.C.	Adams (1963)
1.22	0.64	Southport, N.C.	Adams (1963)
1.22	0.91	Oak Island, N.C.	Adams (1963)
1.52	1.52	South Carolina	Kurz and Wagner (1957)
1.89	1.77	Florida	Kurz and Wagner (1957)
2.23	1.46	Connecticut	Johnson and York (1915)
2.90	2.29	Massachusetts	Chapman (1940)

animal food such as snails, mussels, shrimps, etc. The teal feed largely on sedges, pond weeds and grasses. The American bittern, *Botaurus lentiginosus,* extends south into South Carolina in the summer and ranges north to Massachusetts in the winter. The willet, *Catotrophorus semipalmatus,* is found from New Jersey marshes southward. The marsh hawk, *Circus hudsonius,* occurs all along the eastern U.S. salt marshes, feeding on mice, other birds and carrion. The short–eared owl, *Asio flammeus,* also feeds on mice in the marsh. The red–winged blackbird, *Agelaius phoeniceus,* inhabits the entire coastal marsh areas during summer, but is found only from Massachusetts south during the winter. When these birds are scattered they feed on insects, but when congregated they consume seeds. The insectivorous meadow lark, *Sturnella magna,* occupies the coastal salt meadows from North Carolina northward. The marsh wren, *Telmatodytes palustris,* consumes insects and other small organisms and is found throughout the eastern marshes. Several conspicuous birds that frequently visit eastern U.S. salt marshes include the great blue heron, *Ardea herodias,* the little blue heron, *Florida caerulea,* the black–crowned night heron, *Nycticorax nycticorax,* the common egret, *Casmerodius albus,* and the snowy egret, *Egretta thula.*

The birds are one of the more predominant or visible faunal components of salt marshes. Noticeably absent are amphibians whose skins will not tolerate salt. One of the more noticeable vertebrates in southern marshes is the alligator, *Alligator mississippiensis,* which ranges from North Carolina southward. Diamondback terrapins, *Malaclemys centrata,* are found from New Jersey southward.

Common fish include several species of the killifish, *Fundulus* sp., which are found all along the coast, the sheepshead minnow, *Cyprinodon variegatus,* found from Cape Cod southward, and the silversides, *Menidia* sp., which extend along the entire coast.

No mammals exclusively inhabit salt marshes. The rice rat, *Oryzomus palustris,* is a common inhabitant that nests in *Spartina alterniflora.* The meadow mouse, *Microtus pennsylvanicus,* inhabits the coastal salt meadows. The muskrat, *Ondatra zibethica,* ventures into the marshes for food. Other occasional visitors include the otter, *Lutra canadensis,* the mink, *Mustela vison,* the opossum, *Didelphis virginiana,* and the white–tailed deer, *Odocoileus virginianus.* The raccoon, *Procyon lotor,* frequents the salt marshes to feed on fiddler crabs (*Uca*).

Davis and Gray (1966) summarize the insect fauna of a North Carolina salt marsh. Homoptera are the most predominant insects; Diptera are the next most abundant. Other orders represented include the Hemiptera, Orthoptera, Coleoptera and Hymenoptera. Generally the numbers of Homoptera decrease and other orders increase in abundance as the tidal elevation of the zone increases. Kraeuter and Wolf (1974) present a more comprehensive treatment of these arthropods in the salt marsh. Usinger (1957) reports that the best known of the salt–marsh insects are mosquitoes, *Aedes* sp. Other insects occurring near the high water line include the biting midges of the genera *Culicoides* and *Leptoconops.* In the mangrove swamps, the mosquito *Anopheles albimanus* is commonly found. Arndt (1914) reviews the insects and spiders commonly found between the tide zones. Barnes (1953) presents the best data available on the distribution of spiders in salt–marsh vegetation.

The remaining invertebrates found in salt marshes are not as widely distributed as the above–mentioned plants and animals. Kerwin (1972) relates the distribution of the salt–marsh snail, *Melampus bidentatus,* to the distribution of *Distichlis spicata, Spartina patens,* and the short form of *Spartina alterniflora.* Other common snails include *Littorina irrorata, Detracia floridana,* and *Littorina littorea. Littorina* sp. are usually most abundant in *Spartina alterniflora. Detracia* is found in the *Distichlis spicata* and *Spartina patens* community. Reviews of other faunal components of the marsh can be found in Miner (1950), Smith (1964), Arnold (1968), and Ursin (1972). The remainder of this chapter is devoted to describing the geographic differences in the ecosystem components of the eastern coast salt marshes.

NORTHERN MARSHES

Along the western shore of the Atlantic Ocean, the northermost marshes (see Chapter 1) are considered as the Bay of Fundy subgroup. Ganong (1903) produced a detailed survey of the vegetation of the Bay of Fundy salt marshes and considered their detailed formation in front of a rocky upland. The predominant vegetation is *Spartina alterniflora* and *Juncus balticus*. Other floral components include *Salicornia herbacea* agg. and *Suaeda linearis*.

NEW ENGLAND MARSHES

The New England subgroup of marshes is dominated by pure stands of *Spartina alterniflora* along the stream banks, and *Juncus balticus*, *Juncus gerardi* and *Spartina patens* in the higher salt marsh. Clear-cut patterns of zonation have been described by Chapman (1940), Miller and Egler (1950), Niering (1961), and Redfield (1965, 1972). In some instances, *Spartina alterniflora* fringes the stream (where there are steep sloping banks); in others it occurs over a large area where the slope is very gentle. In the intertidal zone above *Spartina alterniflora*, there is a large area dominated by a mixture of *Spartina patens* and *Distichlis spicata*. *S. patens* dominates in wetter, *D. spicata* in the dryer areas. Higher in the intertidal zone, *Juncus gerardi* and *Juncus balticus* form mixed stands (often in areas well above mean high water). In the upper marsh, there are many bare areas and salt pans. At the marsh fringe there is frequently a band of *Panicum virgatum* and *Spartina pectinata*.

The salt-marsh soils in the lower intertidal area contain large amounts of silt and plant remains. In the upper reaches of the intertidal zone, the substrate is almost entirely peat. Pilgrim (1973) has classified the tidal marsh soils of New Hampshire according to the latest principles.

A complete ecosystem study of a New England salt marsh (Nixon and Oviatt, 1973a) demonstrates system interactions and the dynamics of key components, *Spartina, Palaemonetes,* and *Fundulus*. Redfield (1972) provides a detailed account of the formation of these New England salt marshes.

Included in the New England subgroup are the marshes of Long Island (Conrad, 1924; Conrad and Galligar, 1929; Taylor, 1938), where *Spartina alterniflora* occupies the lower regularly flooded areas. The majority of these marshes are coastal salt meadows vegetated with *Spartina patens*. Man has modified much of the area by ditching, diking, or draining. The effects and extent of these alterations are considered in Chapter 17. The productivity and nutrient value of salt-marsh plants from Long Island (Udell et al., 1969) reveal that the *Spartina alterniflora—Distichlis spicata* mixture accounts for about 40% and the phytoplankton about 60% of the annual primary production of coastal marshes.

COASTAL PLAIN MARSHES

Further south we find the Coastal Plain subgroup of marshes. These marshes acquire a character more like the remainder of the marshes on the south Atlantic coastal plain. The vegetation of the marshes of New Jersey has been described by Harshberger (1900, 1902). These marshes have areas of *Spartina alterniflora,* mainly along the drainage channels. Since New Jersey and Delaware marshes have been extensively ditched (for mosquito control purposes), the water table in the upper marsh has been lowered. Bourn and Cottam (1950) demonstrated that ditching lowers the water table and that when this occurs the high marsh and marsh fringe plants quickly invade the area. There was a significant reduction in the numbers of invertebrates in the marsh due to ditching.

The majority of the New Jersey-Delaware salt marshes are vegetated with *Spartina patens* and *Distichlis spicata*. In some local areas there are cells of high-salinity interstitial water with *Salicornia* sp. being dominant. Along creeks draining into Delaware Bay there is a tendency for *Spartina alterniflora* to exhibit a tall growth form along the stream banks, contrasted with a short growth form in the high marsh (Morgan, 1961; Good, 1965; Squires and Good, 1974). In this area, too, we find the first general occurrence of *Spartina cynosuroides*. *Phragmites communis* borders the upland fringe of many of these areas. Recently Klemas et al. (1973) summarized the coastal vegetation of Delaware, whereas Anderson and Wobber (1973) have compiled a similar summary for New Jersey.

Further south, the marshes of Maryland and Virginia fringe the Chesapeake Bay. Chrysler (1910) showed that, along the eastern shore of Chesapeake Bay, *Spartina alterniflora* graded to *Spartina patens* and that *Juncus gerardi* was found on the upper marsh border. He also observed that, due to the influence of fresh water from streams draining from the Appalachian Mountain chain watershed, salt marshes along the western shore of the Chesapeake Bay often had *Spartina cynosuroides* bordering the streams. In the Chesapeake area at higher intertidal elevations (above *Spartina alterniflora* and *Spartina cynosuroides*), *Distichlis spicata* becomes far more abundant than *Spartina patens*. In the lower reaches (near the mouth of the Potomac River), *Juncus roemerianus* appears to replace *Juncus gerardi*. Metzgar (1973) has recently summarized the wetland vegetation for Maryland, whereas Wass and Wright (1969) have prepared a similar analysis of coastal wetlands of Virginia.

South of the Chesapeake Bay, the marshes of the Coastal Plain subgroup acquire a rather similar fauna and flora. These are the best developed and most extensive marshes (Table 7.1) along the eastern coast. Nearly three quarters of all the salt marshes of eastern United States (based on their areal extent) are found in North Carolina, South Carolina and Georgia. These marshes are formed in areas where major river watersheds, which all drain large areas of land, deposit heavy burdens of silt. In northern Florida, the marshes gradually give way to mangrove swamps (see Chapters 2 and 9).

There are extensive areas of brackish–water marshes in the southeast segment of the coast, especially from Cape Henry to Cape Hatteras. Limited tidal amplitude and large changes in the salinity of both the flood runoff and ground water result in a mixed stand of plants. The firm sand and peat substrate seldom support *Spartina alterniflora* except along the edge of rather straight tidal creeks near their mouths. In these areas, *Juncus roemerianus* is the most common pure stand of plant. In North Carolina there are nearly 40 000 ha of *Juncus roemerianus* marsh which in many places is just above mean high water. In slightly higher areas along the edge of head of the creeks, *Baccharis halimifolia, Borrichia frutescens,* and *Spartina cynosuroides* may occur on more sandy substrate.

Another plant community, *Spartina patens*, forms extensive stands which occupy a brownish peat. *Distichlis spicata, Scirpus robustus,* and *Pluchea purpurascens* are also scattered in this community. The latter species are usually subdominant to the *Spartina patens*. Frequently ponds, depressions, or "rotten spots" occur in *Spartina patens* meadows. Cooper and Waits (1973) considered the vegetation types of an irregularly flooded North Carolina salt marsh where *Juncus roemerianus* and *Spartina patens* were the two predominant forms of vegetation. In the regularly flooded salt marshes, *Spartina alterniflora* is predominant.

In the true salt marshes of the South Atlantic coast, the low marsh is consistent in that *Spartina alterniflora* dominates the lower intertidal range (where it grows to 3 m in height). Somewhat above mean high water, *Juncus roemerianus* occurs and *Spartina patens* is generally located above the *J. roemerianus*. In between the tall creekbank *Spartina alterniflora* and the *Juncus* is a short growth form of *S. alterniflora* (commonly reaching a height of 0.25 m). An intermediate–height form of *S. alterniflora* often occurs on the natural levees of tidal creeks. In the higher marsh areas where *Spartina patens* usually occurs, there are frequent ponds or "rotten spots". Salt water often intrudes into these ponds. When this happens, *S. alterniflora* (developing from seed) and *Salicornia* develop. The salinity regimes in these marshes range from the highest in the *S. alterniflora* areas to the lowest at the marsh fringe.

At the southern end of the eastern coast of the United States are mangrove swamps. In the lower portion of the intertidal zone red mangroves, *Rhizophora mangle,* are found (Walsh, 1974). The seedlings are established on shell fragments, sand and marl below the low tide level. As these mangroves grow, the roots become thickly entangled and act as natural weirs which trap debris. Mature *Rhizophora* plants are about 20 cm in diameter and 10—12 m tall. As the soil is built up, *Avicennia nitida,* the black mangrove, invades from the upper edge of the marsh and consocies occur at the upper edge of the red mangrove swamp where the soil is flooded only at irregular intervals. In areas where the black mangrove is not overly dense, the companion vegetative species are *Batis maritima, Distichlis spicata,* and *Spartina alterniflora*. Davis (1940) refers to this community as the *Avicennia*— salt marsh associes. Under optimum conditions, the black mangrove outcompetes and eliminates the salt–marsh plants. Severe cold can quickly kill or crop the black mangrove and thus allow the salt–marsh species to flourish. This is the case in northeastern Florida.

EASTERN UNITED STATES

TABLE 7.4

Comparison of biomass and primary production of salt-marsh plants from the eastern coast of North America (after Keefe, 1972)

Species	Biomass, aerial parts [g (dry) m^{-2}]	Net production, aerial parts [g (dry) m^{-2} yr^{-1}]	Locale	Source
Borrichia	785	—	Virginia	Wass and Wright (1969)
Carex stricta	1340	1699	New Jersey	Jervis (1964)
Distichlis spicata	359	—	Connecticut	Steever (1972)
	453	—	Delaware	Morgan (1961)
	648	—	New York	Udell et al. (1969)
	680	—	Rhode Island	Nixon and Oviatt (1973a,b)
	360	—	Virginia	Wass and Wright (1969)
Fimbristylis	605	—	Virginia	Wass and Wright (1969)
Juncus roemerianus	232	849	Florida	Heald (1969)
	—	560	North Carolina	Foster (1968)
	1173	796	North Carolina	Stroud and Cooper (1968)
	786	1360	North Carolina	Waits (1967)
	340	850	North Carolina	Williams and Murdock (1968)
	650	—	Virginia	Wass and Wright (1969)
Leersia oryzoides	1545	—	Virginia	Wass and Wright (1969)
Nuphar advena	245	—	Virginia	Wass and Wright (1969)
Phragmites communis	—	2695	Long Island	Harper (1918)
	800—1000	—	Rhode Island	Nixon and Oviatt (1973b)
Scirpus americanus	150	150	South Carolina	Boyd (1970)
Spartina alterniflora	717	—	Connecticut	Steever (1972)
	413	445	Delaware	Morgan (1961)
	3018	3990	Georgia	Odum and Fanning (1973)
	—	2000—3300	Georgia	Odum (1959, 1961)
	—	973	Georgia	Smalley (1959)
	1618	—	Georgia	Reimold et al. (1972)
	—	484	Maryland	Keefe and Boynton (1972)
	300	—	New Jersey	Good (1965)
	1592	592—1592	New Jersey	Squires and Good (1974)
	259—1320	329—1296	North Carolina	Stroud and Cooper (1968)
	545	650	North Carolina	Williams and Murdock (1969)
	250—2100	1000	North Carolina	Williams and Murdock (1966)
	—	757	Nova Scotia	Mann (1972)
	446—946	—	Rhode Island	Nixon and Oviatt (1973a,b)
	1332	—	Virginia	Wass and Wright (1969)
Spartina cynosuroides	724	1028	Georgia	Odum and Fanning (1973)
	1456	—	Virginia	Wass and Wright (1969)
Spartina patens	285	—	Connecticut	Steever (1972)
	—	993	Long Island	Harper (1918)
	503	—	New York	Udell et al. (1969)
	640	1296	North Carolina	Waits (1967)
	805	—	Virginia	Wass and Wright (1969)
Typha angustifolia	—	1733	Long Island	Harper (1918)
	930	—	Virginia	Wass and Wright (1969)
Typha latifolia	—	1358	Long Island	Harper (1918)
Typha angustifolia and *T. latifolia*[1]	1380	1905	New Jersey	Jervis (1964)
Zizania aquatica	1200	1547	New Jersey	Jervis (1964)
	560	—	Virginia	Wass and Wright (1969)

[1] For this community, underground biomass was also estimated. It amounted to 1800 g m^{-2}, giving a total biomass of 3180 g m^{-2}, and an annual net production of 3205 g m^{-2}.

On the landward fringe of the black mangroves, the button mangrove, *Conocarpus erectus*, lives in areas rarely or never flooded by the sea water. In all three areas, of red mangrove, black mangrove, and button mangrove, a fourth species, *Laguncularia racemosa*, occurs and may form small stands.

Along the low ridges of the brackish streams or on the shorelines of brackish lakes, the common vegetation includes *Baccharis halimifolia, Iva frutescens, Panicum virgatum*, and sometimes *Hibiscus moscheutos*.

In summary, the salt marshes of the eastern coast of the United States have faunal and floral components common to all. For comparative purposes, the marshes have been ranked according to primary production (Table 7.4). *Spartina alterniflora* has the greatest production in Georgia (2000—3000 g m^{-2} yr^{-1}), and there appears to be a latitudinal variation in primary production (based on the data summarized in Table 7.4). Since it is the energy that originates in these primary producers that actually nourishes higher trophic levels in the salt marsh, future quantitative comparisons of salt marshes might best be based on production measurements using standardized techniques. The marshes of eastern North America may be classified as *Spartina—Juncus* marshes for world–comparative purposes.

REFERENCES

Adams, D.A., 1963. Factors influencing vascular plant zonation in North Carolina salt marshes. *Ecology*, 44: 445–456.

Anderson, R.R. and Wobber, F.J., 1973. Wetlands mapping in New Jersey. *Photogramm. Eng.*, 39(4): 353—358.

Arndt, C.H., 1914. Some insects of between the tides zone. *Indiana Acad. Sci. Proc.*, 1914: 323—336.

Arnold, A.F., 1968. *The Sea Beach at Ebb Tide*. Dover, New York, N.Y., 490 pp.

Barnes, R.D., 1953. The ecological distribution of spiders in non–forest maritime communities at Beaufort, North Carolina. *Ecol. Monogr.*, 23: 315—337.

Blum, J.L., 1968. Salt marsh *Spartinas* and associated algae. *Ecol. Monogr.*, 38: 199—221.

Bourn, W.S. and Cottam, C., 1950. Some biological effects of ditching tidewater marshes. *U.S. Fish Wildl. Serv., Res. Rep.*, No. 19: 30 pp.

Boyd, C.E., 1970. Production, mineral accumulation and pigment concentrations in *Typha latifolia* and *Scirpus americanus. Ecology*, 51: 902—906.

Burkholder, P.R. and Bornside, G.H., 1957. Decomposition of marsh grass by anaerobic marine bacteria. *Bull. Torrey Bot. Club*, 84: 366—383.

Chapman, V.J., 1940. Succession on the New England salt marshes. *Ecology*, 21: 279–282.

Chrysler, M.A., 1910. The ecological plant geography of Maryland; Coastal Zone; Western Shore District. *Md. Weather Serv., Spec. Publ.*, No. 3: 149—197.

Conrad, H.S., 1924. Second survey of vegetation of a Long Island salt marsh. *Ecology*, 5: 379—388.

Conrad, H.S. and Galligar, G.C., 1929. Third survey of a Long Island salt marsh. *Ecology*, 10: 326—336.

Cooper, A.W. and Waits, E.D., 1973. Vegetation types in an irregularly flooded salt marsh on the North Carolina Outer Banks. *J. Elisha Mitchell Sci. Soc.*, 89: 78—91.

Daiber, F.C., 1974. Salt marsh plants and future coastal salt marshes in relation to animals. In: R.J. Reimold and W.H. Queen (Editors), *Ecology of Halophytes*. Academic Press, New York, N.Y., pp. 475—510.

Davis, J.H., 1940. The ecology and geologic role of mangroves in Florida. *Carnegie Inst. Wash. Publ.*, No. 517: 303—412.

Davis, L.V. and Gray, I.E., 1966. Zonal and seasonal distribution of insects in North Carolina salt marshes. *Ecol. Monogr.*, 36: 275—295.

Duncan, W.H., 1974. Vascular halophytes of the Atlantic and Gulf coasts of North America north of Mexico. In: R.J. Reimold and W.H. Queen (Editors), *Ecology of Halophytes*. Academic Press, New York, N.Y., pp. 23—50.

Foster, W.A., 1968. *Studies on the Distribution and Growth of Juncus roemerianus in Southeastern Brunswick County, North Carolina*. Thesis, North Carolina State University, Raleigh, N.C., 72 pp.

Ganong, W.F., 1903. The vegetation of the Bay of Fundy salt and diked marshes: An ecological study. *Bot. Gaz.*, 36(3): 161—186.

Good, R.E., 1965. Salt marsh vegetation, Cape May, New Jersey. *N. J. Acad. Sci. Bull.*, 10(1): 1—11.

Harper, R.M., 1918. Some dynamic studies of Long Island vegetation. *Plant World*, 21: 38—46.

Harshberger, J.W., 1900. An ecological study of the New Jersey strand flora. *Proc. Acad. Nat. Sci. Phila.*, 52: 623—671.

Harshberger, J.W., 1902. Additional observations on the strand flora of New Jersey. *Proc. Acad. Nat. Sci. Phila.*, 54: 642—669.

Heald, E.J., 1969. *The Production of Organic Detritus in a South Florida Estuary*. Thesis, University of Miami, Miami, Fla., 110 pp.

Jervis, R.A., 1964. *Primary Production in a Freshwater Marsh Ecosystem*. Thesis, Rutgers University, New Brunswick, N.J., 79 pp.

Johnson, D.S. and York, H.H., 1915. The relation of plants to tide-levels. *Carnegie Inst. Wash., Publ.*, No. 206: 1—162.

Keefe, C.W., 1972. Marsh production: A summary of the literature. *Contrib. Mar. Sci.*, 16: 165—181.

Keefe, C.W. and Boynton, W.R., 1972. Standing crop of salt marshes surrounding Chincoteague Bay, Maryland—Virginia. *Chesapeake Sci.*, 14(2): 117—123.

Kerwin, J.A., 1972. Distribution of the salt marsh snail (*Melampus bidentatus* Say) in relation to marsh plants in the Poropotank River area, Virginia. *Chesapeake Sci.*, 13(2): 150—152.

Klemas, V., Daiber, F.C., Bartlett, D.S., Crichton, O.W., and

Fornes, A.O., 1973. *Coastal Vegetation of Delaware: The Mapping of Delaware's Coastal Marshes.* College of Marine Studies, University of Delaware, Newark, Del., 29 pp.

Kraeuter, J.N. and Wolf, P.L., 1974. The relationship of marine macroinvertebrates to salt marsh plants. In: R.J. Reimold and W.H. Queen (Editors), *Ecology of Halophytes.* Academic Press, New York, N.Y., pp. 449—462.

Kurz, H. and Wagner, K., 1957. *Tidal Marshes of the Gulf and Atlantic Coasts of Northern Florida and Charleston, South Carolina.* Fla. State Univ. Studies, Tallahassee, Fla., No. 24, 168 pp.

Mann, K.H., 1972. Macrophyte production and detritus food chains in coastal waters. *Mem. Ist. Ital. Idrobiol.*, 29 (Suppl.): 353—383.

Martin, A.C., Hotchkiss, N., Uhler, F.M. and Bourn, W.S., 1953. Classification of wetlands of the United States. *U.S. Fish Wildl. Serv., Spec. Sci. Rep., Wildl.*, No. 20: 14 pp.

Metzgar, R.G., 1973. Wetlands in Maryland. *Md. Dep. State Planning, Publ.*, No. 157: 155 pp.

Miller, W.R. and Egler, F.E., 1950. Vegetation of the Westquetequock—Pawcatuck tidal-marshes, Connecticut. *Ecol. Monogr.*, 20: 143—172.

Miner, R.W., 1950. *Field Book of Seashore Life.* Putnam, New York, N.Y., 888 pp.

Morgan, M.H., 1961. *Annual Angiosperm Production of a Salt Marsh.* Thesis, University of Delaware, Newark, Del., 33 pp.

National Estuary Study, 1970. *U.S. Fish and Wildlife Service, Volumes 1—7.* U.S. Dep. Int., Washington, D.C.

Niering, W.A., 1961. Tidal marshes, their use in scientific research. *Conn. Arbor. Bull.*, 12: 3—7.

Nixon, S.W. and Oviatt, C.A., 1973a. Analysis of local variation in the standing crop of *Spartina alterniflora. Bot. Mar.*, XVI: 103—109.

Nixon, S.W. and Oviatt, C.A., 1973b. Ecology of a New England salt marsh. *Ecol. Monogr.*, 43(4): 463—498.

Odum, E.P., 1959. *Fundamentals of Ecology.* Saunders, Philadelphia, Pa., 546 pp.

Odum, E.P., 1961. The role of tidal marshes in estuarine production. *N. Y. State Conserv.*, 15(6): 12—15, 35.

Odum, E.P. and Fanning, M.E., 1973. Comparison of the productivity of *Spartina alterniflora* and *Spartina cynosuroides* in Georgia coastal marshes. *Bull. Ga. Acad. Sci.*, 31: 1—12.

Penfound, W.T., 1952. Southern swamps and marshes. *Bot. Rev.*, 18(6): 413—446.

Pilgrim, S.A.L., 1973. *Soil Survey of New Hampshire Tidal Marshes: A Progress Report.* Soil Conservation Service, Durham, N. H., 26 pp.

Pomeroy, L.R., 1959. Algal productivity in Georgia salt marshes. *Limnol. Oceanogr.*, 4: 386—397.

Redfield, A.C., 1965. The ontogeny of a salt marsh estuary. *Science*, 147: 50—55.

Redfield, A.C., 1972. Development of a New England salt marsh. *Ecol. Monogr.*, 42(2): 201—237.

Reimold, R.J. and Queen, W.H. (Editors), 1974. *Ecology of Halophytes.* Academic Press, New York, N.Y., 605 pp.

Reimold, R.J., Gallagher, J.L. and Thompson, D.E., 1972. Coastal mapping with remote sensors. *Proc. Coastal Mapping Symp., Am. Soc. Photogramm.*, pp. 99—112.

Seneca, E.D., 1974. Stabilization of coastal dredge spoil with *Spartina alterniflora.* In: R.J. Reimold and W.H. Queen (Editors), *Ecology of Halophytes.* Academic Press, New York, N.Y., pp. 525—530.

Shanholtzer, G.F., 1974. Relationship of vertebrates to salt marsh plants. In: R.J. Reimold and W.H. Queen (Editors), *Ecology of Halophytes.* Academic Press, New York, N.Y., pp. 463—474.

Shaw, S.P. and Fredine, C.G., 1956. Wetlands of the United States. *U.S. Fish Wildl. Serv., Circ.*, No. 39: 67 pp.

Smalley, A.E., 1959. *The Role of Two Invertebrate Populations, Littorina irrorata and Orchelimum fidicinium, in the Energy Flow of a Salt Marsh Ecosystem.* Dissertation, University of Georgia, Athens, Ga., 78 pp.

Smith, R.I., 1964. Keys to marine invertebrates of the Woods Hole region. *Syst.—Ecol. Progr., Mar. Biol. Lab., Woods Hole, Mass. Contrib.*, No. 11: 208 pp.

Spinner, G.P., 1968. The wildlife wetlands and shellfish areas of the Atlantic coastal zone. In: *Serial Atlas of the Marine Environment*, Folio 18. American Geographical Society, Washington D.C.

Squires, E.R. and Good, R.E., 1974. Seasonal changes in the productivity, caloric content, and chemical composition of a population of salt-marsh cord-grass *(Spartina alterniflora). Chesapeake Sci.*, 15(2): 1—32.

Steever, E.Z., 1972. *Productivity and Vegetation Studies of a Tidal Salt Marsh in Stonington, Connecticut: Cottrell Marsh.* Thesis, Connecticut College, New London, Conn., 56 pp.

Stroud, L.M. and Cooper, A.W., 1968. Color-infrared aerial photographic interpretation and net primary productivity of a regularly-flooded North Carolina salt marsh. *Univ. N. C., Water Resour. Res. Inst., Rep.*, No. 14: 86 pp.

Taylor, N., 1938. A preliminary report on the salt marsh vegetation of Long Island, New York. *N. Y. State Mus. Bull.*, 316: 21—84.

Udell, H.F., Zarudsky, J. and Doheny, T.E., 1969. Productivity and nutrient values of plants growing in the salt marshes of the town of Hempstead, Long Island. *Bull. Torrey Bot. Club*, 96(1): 42—51.

Ursin, M.J., 1972. *Life In and Around the Salt Marshes.* Crowell, New York, N.Y., 110 pp.

Usinger, R.L., 1957. In: J. Hedgpeth (Editor), *Treatise on Marine Ecology and Paleoecology. Ecology. Geol. Soc. Am. Mem.*, 67(1): 1177—1182.

Waits, E.D., 1967. *Net Primary Productivity of an Irregularly-Flooded North Carolina Salt Marsh.* Dissertation, North Carolina State University, Raleigh, N.C., 91 pp.

Walsh, G.E., 1974. Mangroves, a review. In: R.J. Reimold and W.H. Queen (Editors), *Ecology of Halophytes.* Academic Press, New York, N.Y., pp. 1—174.

Wass, M.L. and Wright, T.D., 1969. *Coastal Wetlands of Virginia. Interim Report of the Governor and General Assembly.* Special Report in Applied Marine Science and Ocean Engineering, Virginia Inst. Mar. Sci., Gloucester Point, Va., No. 10: 154 pp.

Webber, E.E., 1967. Bluegreen algae from a Massachusetts salt marsh. *Bull. Torrey Bot. Club*, 94: 99—106.

Williams, R.B. and Murdock, M.B., 1966. Annual production

of *Spartina alterniflora* and *Juncus roemerianus* in salt marshes near Beaufort, North Carolina. *Assoc. Southeastern Biol. Bull.*, 13: 49.

Williams, R.B. and Murdock, M.B., 1968. Compartmental analysis of production and decay of *Juncus roemerianus*. *Assoc. Southeastern Biol. Bull.*, 15: 59.

Williams, R.B. and Murdock, M.B., 1969. The potential importance of *Spartina alterniflora* in conveying zinc, manganese and iron into estuarine food chains. In: D.J. Nelson and F.C. Evans (Editors), *Proc. 2nd Natl. Symp. on Radiecology,* pp. 431—439.

Wright, J.O., 1907. Swamp and overflowed lands in the United States. *U.S. Dep. Agric., Circ.*, No. 825: 23 pp.

Chapter 8

PLANT AND ANIMAL COMMUNITIES OF PACIFIC NORTH AMERICAN SALT MARSHES

KEITH B. MACDONALD

INTRODUCTION

The objective of this chapter is to summarize present knowledge of the salt–marsh and mangal ecosystems developed along the North American Pacific coast between Point Barrow, Alaska (~ 71°N) and Cabo San Lucas, at the southern tip of Baja California (~ 23°N; Fig. 1). While the treatment is restricted to coastal settings lying below extreme high water, this still includes marshes developed under a wide variety of salinity regimes. The vegetation of most of these marshes is dominated by a rather small number of halophytes — often less than ten and only rarely as many as twenty species per site. As salinities decline, as local fresh-water sources appear, or as human disturbance increases, maritime, upland or weedy species appear and the diversity of the marsh flora rapidly increases.

This review emphasizes vegetation rather than floristics; vascular plants most characteristic of the salt marshes are stressed and exhaustive lists of "occasional" species are omitted. Species nomenclature follows Hultén (1968) for the northern coast, Munz (1973) for the central coast, and Shreve and Wiggins (1964) for the south coast. The five-volume *Vascular Plants of the Pacific Northwest* of Hitchcock et al. (1955—1969) was particularly useful for resolving problems of species synonymy. The bulk of the chapter consists of a region-by-region review of Pacific coast salt-marsh vegetation. Emphasis has been placed on major changes in species composition, descriptions of vertical zonation and succession, and impressions of local vegetational variability. Major literature sources, particularly recent ones, are listed and, where available, preliminary data on

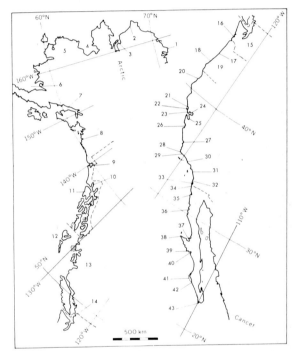

Fig. 8.1. The North American Pacific coast. Numbers refer to the following localities: *1* = Point Barrow; *2* = Cape Thompson; *3* = Kotzebue; *4* = Norton Sound; *5* = Yukon, Kuskokwim deltas; *6* = Bristol Bay; *7* = Cook Inlet; *8* = Copper River Delta; *9* = Yakutat Bay; *10* = Glacier Bay; *11* = Chicagof Island; *12* = Queen Charlotte Islands; *13* = Fjord coast, British Columbia; *14* = Strait of Georgia, Puget Sound; *15* = Washington; *16* = Willapa Bay; *17* = Yaquina Bay; *18* = Coos Bay; *19* = Oregon; *20* = Humboldt Bay; *21* = Bodega Bay; *22* = Tomales Bay; *23* = Drake's Estero, Bolinas Bay; *24* = San Pablo Bay, Suisun Bay; *25* = San Francisco Bay; *26* = Elkhorn Slough; *27* = Morro Bay; *28* = Point Conception; *29* = Goleta, Carpinteria; *30* = Mugu Lagoon; *31* = Newport Bay; *32* = Mission Bay; *33* = San Diego Bay; *34* = Tijuana Slough; *35* = Estero de Punta Banda; *36* = Bahia de San Quintin; *37* = Laguna Guerrero Negro, Ojo de Liebre; *38* = Punta Eugenia; *39* = Pond Lagoon; *40* = Laguna San Ignacio; *41* = Bahia de la Magdalena; *42* = Puerto Chale; *43* = Cabo San Lucas.

environment—species interactions are cited. This section is preceded by descriptions of the geologic setting and environmental variables characteristic of Pacific coast salt marshes, and followed by brief discussions of their phytogeography, introduced species and productivity. Finally, the animal communities of these salt marshes are described.

GEOLOGICAL SETTING

The Pacific coast is tectonically active and mountain building has left little room for coastal lowlands. Estuaries and lagoons make up only 10—20% of the shoreline, a marked contrast with the comparable figure of 80—90% for the Atlantic and Gulf coasts (Inman and Nordstrom, 1971; Gross, 1972).

From a geomorphic viewpoint, the Bering Sea and Arctic coasts of Alaska (Hartwell, 1972; Short and Wright, 1974), and the Baja California coast south of about 29°N (Phleger and Ewing, 1962; Phleger, 1965; Curray, 1970) provide the most suitable settings for extensive salt–marsh development. The continental shelf is broad and shallow in both regions and shoreline features usually synonymous with coastal marshes — barrier islands, shallow lagoons, spits, estuaries and deltas — are common. While the Baja California sites do support extensive marshlands, the extreme environmental conditions of northern Alaska severely restrict salt–marsh establishment. For up to nine months of the year sea ice is frozen fast to the coast, then, during the summer thaw, thermal, fluvial and marine processes continally erode and rework unstable nearshore sediments.

East and south, around the Gulf of Alaska, the continental shelf narrows and steepens. Recent glacial erosion has scoured the coast as far south as Puget Sound, creating an intricate system of deep fjords and archipelagoes. Most unconsolidated sediments have been scraped away and the lack of sediment moving along the coast precludes the formation of bars or spits which might otherwise shelter salt marshes. Tectonic uplift characterizes the region (Twenhofel, 1952; Reimnitz and Marshall, 1965; Reimnitz, 1972) and hinders marsh development.

South, from Washington into California, the continental shelf remains too steep and narrow for barrier beaches to form offshore. With the obvious exception of San Francisco Bay, salt–marsh formation is restricted to small bays, estuaries and river mouths. Many of these sites were created when rising postglacial sea levels flooded coastal river valleys, over–deepened during the preceding lower sea level. These estuaries have rapidly filled with marine and alluvial sediments and many are partially protected from open–ocean influences by bay–mouth sand spits (Upson, 1949; Johannessen, 1961; Daetwyler, 1965; Macdonald, 1971d; Pestrong, 1972).

Cores penetrating marsh deposits at Pony Slough (Coos Bay), Mugu Lagoon, Newport Bay, Mission Bay, Bahia de San Quintin and Laguna Guerrero Negro[1] (Macdonald, 1967 and including references) have all revealed sandy shoreline or bay–floor sediments beneath a thin veneer (0.2—1.8 m) of marsh clays or silts containing scattered plant remains. Thin, peaty sequences have been described from Oregon (Jefferson, 1974) and Alaska (Hanson, 1951), but the thick marine peats characteristic of some Atlantic coast marshes (Redfield, 1965, 1972) appear to be rare or absent (cf. p. 161). To date, no definitive records of the extension of marsh deposits below present mean sea level are known from the Pacific coast, although this may be the case at San Francisco Bay (Pestrong, 1972). Comparative studies of historical maps and photographs indicate that marshes at several Oregon and California sites have expanded rapidly over the last 100 years (Johannessen, 1961; Macdonald, 1967). Jefferson (1974) confirms that much of Oregon's marshland appears to be less than 200 years old. She further suggests that the mature high marsh represents a climax vegetation — rather than a sere — and as such may remain stable for much longer than previously believed. Supportive evidence for this comes from ^{14}C dates of 410 ± 135 and 770 ± 85 yr. B.P., obtained from *Triglochin* root material, buried at the base of a typical 1 m thick high–marsh sequence in Yaquina Bay.

Taken together, these data suggest that present–day Pacific coast salt marshes are of relatively recent origin and have short depositional histories. Older, buried lagoonal sequences and fossil salt–marsh molluscs suggest that comparable habitats existed along the coast during the Pleistocene.

---

[1] These and other localities are shown in the map (Fig. 8.1).

PACIFIC NORTH AMERICA

ENVIRONMENTAL VARIABLES

Climatic variations are considerable, but their severity is reduced by the maritime influence, and the fact that the warm North Pacific Current bathes southern Alaska and British Columbia whereas the cold California Current bathes the southern coastline.

Latitudinal gradients in January and July mean monthly temperatures, total days of frost, precipitation and evaporation are shown in Fig. 8.2. Mean monthly surface–water temperatures along the open coast increase southward (Point Barrow, 0—8°C; Anchorage, 4—13°C; San Francisco, 11—13°C; Cabo San Lucas, 20—28°C), although in spring and summer this general trend is interrupted by local areas of intense coastal upwelling. The combined effects of temperature, precipitation and evaporation are reflected in coastal salinity gradients. Salinities are substantially reduced along the Alaskan coast (Chuckhi Sea, 20—32‰; Prince William Sound, 11—29‰; Strait of Georgia, 10—20‰). Washington and Oregon are strongly influenced by the Columbia River plume and salinities frequently fall below 32‰. Near Cape Mendocino in northern California the long–term mean salinity is less than 33‰ but this value rises to 34.6‰ near Cabo San Lucas, Baja California.

Environmental conditions tend to be more extreme in the shallow estuaries and lagoons adjacent to salt–marsh habitats. At Bellingham Bay, Washington, for example, Heath (1969) recorded water temperatures from −1.0 to 35°C and salinities of 1—30‰. At Mission Bay, 17 degrees further south, Bradshaw (1968a) recorded water temperatures from −0.5 to 36°C and salinities of 30—50‰ while in nearby lagoons subject to seasonal closure salinities of up to 96‰ were noted (Carpelan, 1969).

Most of the Pacific coast is characterized by mixed tides. Distinctive "low marsh" and "high marsh" environments reflect abrupt increases in the duration of continuous exposure or submersion, produced by the inequality of successive tides (Doty, 1946; Chapman, 1960; Macdonald, 1969a; Jefferson, 1974). Tidal ranges on the open coast vary from less than 15 cm at Point Barrow to well over 2 m along the southern California coast. Ranges increase considerably in restricted bays and inlets: Copper River Delta, 4.7 m; Puget Sound, 1.5—3.3 m; San Francisco Bay, 2.6—4.3 m. Where tides are weak along the Alaskan coast, local storm surges may raise sea level as much as 2 m, causing extensive coastal flooding that can persist for several days.

Following Köppen's classification of world climates, significant changes in environmental conditions can be pinpointed along the Pacific coast as follows: at about 69°N, the transition from Polar Tundra (**ET**) or Arctic zone (see p. 63) to a Subarctic humid microthermal (**Dcf**) or Boreal climate is noted. In the Gulf of Alaska, at 60°N, rising winter temperatures cause a shift from microthermal to mesothermal (**Cfb**) climates. Rainfall increases ninefold between Anchorage and Yakutat; sea ice is no longer present and coastal salinities are low. At 46°N, precipitation patterns change from year–round, with accent on winter, to the summer drought—winter rain Mediterranean pattern (see Figs. 3.3B and 3.4B). Summer fogs are also common from here south. Peck (1961) notes a more abrupt change nearer the Oregon border (42°N) — winter gales become markedly

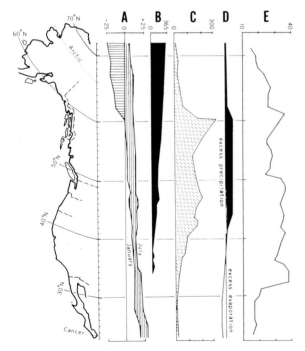

Fig. 8.2. North American Pacific coast: latitudinal variations in selected climatic variables and salt–marsh species diversity. A = mean monthly air temperatures, °C; B = day of frost; C = precipitation, cm; D = evaporation; E = potential number of salt–marsh species (see text).

less severe to the south, rainfall begins its steady southward decline, and coastal salinities rise again. Steadily rising temperatures and declining rainfall produce rapid climatic shifts further south: humid mesothermal (**Cs**) or warm temperate (see p. 61) to semiarid (**BS**) at about 37°N (evaporation now balances precipitation for the first time); semiarid to arid (**BW**) at about 31°N; middle latitude arid (**BWk**) to subtropical arid (**BWh**) at about 28°N, as mean annual temperatures rise above 18°C (see p. 61); and finally from arid to a tropical rainy climate with a winter dry season (**Aw**), near the southern tip of the Baja Peninsula 24°N) (see p. 61). As will be seen below, significant changes in the diversity and/or species composition of coastal salt–marsh vegetation accompany most of these environmental transitions (cf. Chapter 3).

REGIONAL REVIEW

Alaska and British Columbia

Major references include Cooper (1931), Hanson (1951, 1953), Crow (1966), Johnson et al. (1966), Stephens and Billings (1967), Eyerdam (1971) and Jefferies (1975). Supplementary data are provided by Porsild (1951), Polunin (1959), Spetzman (1959), Wiggins and Thomas (1962), Calder and Taylor (1968) and Hultén (1968).

Much of the Alaskan Arctic coastline is unstable and nearshore sediments are frequently re–worked by ice action. Well–developed salt–marsh communities are absent — rather, there is a mosaic of plant species, the extent and age of which is determined by the frequency of disruptive ice action. *Puccinellia phryganodes* is always the primary colonist on the open Arctic coast. This species is not known to set seed but undergoes extensive vegetative reproduction by means of surface stolons. Growth is particularly vigorous where the plants are partly covered by decaying algal mats, suggesting that nutritional deficiencies may otherwise restrict growth. *Stellaria humifusa* and *Cochlearia officinalis* agg. may be present at the seaward end of these marshes; *Primula borealis* and *Carex ursina* occur closer to shore, particularly where vegetational cover is poor. These open coast marshes, often only a few meters in extent, occur frequently between Cape Bathurst and Point Barrow (Jefferies, 1975).

Where conditions are more sheltered and the action of pack ice reduced, more extensive marsh communities are present. At the Mackenzie River Delta, for example, brackish conditions prevail and species characteristic of coastal fresh–water lagoons (*Arctophila fulva, Dupontia fischeri* and *Hippuris tetraphylla*) grow along the seashore. *Carex ramenskii* is the dominant species in these marshes and together with *Hippuris* accounts for most of the biomass; both species are inundated by the tide. *Carex subspathacea* occurs on bare mud where secondary erosion has occurred and *Carex glareosa* ssp. *glareosa* is found in the upper reaches of marshes above normal high–tide level. Additional species include *Calamagrostis neglecta, Plantago juncoides, Potentilla egedii, Puccinellia* spp. and *Stellaria humifusa*.

Puccinellia phryganodes remained the primary colonist in a sedge–grass, saline meadow community developed at Cape Thompson (Johnson et al., 1966). *Carex glareosa, Dupontia fischeri, Calamagrostis deschampsioides* and *Potentilla egedii* var. *groenlandica* were all abundant. Small brackish pools were common and frequently contained *Arctophila fulva, Hippuris vulgaris* and a rich algal flora. *Primula borealis* and *Stellaria humifusa* were still present but *Carex ramenskii* was rare and *Calamagrostis neglecta* and *Cochlearia officinalis* were restricted to shingle beach or tundra habitats. At least nine other species were represented. Clearly, overall species diversity was higher than along the shores of the Arctic Ocean and several widespread species exhibited marked changes in abundance.

A simpler assemblage was noted from the shores of the Kotzebue lowlands, further south (Hanson, 1951). Mosaic community patterns were characteristic and reflected the numerous ponds and poorly drained areas, shallow permafrost and severe frost action. *Carex subspathacea*, a salt–tolerant species, was the most abundant plant in low areas adjacent to tidal inlets, channels and ponds. *Potentilla pacifica, Puccinellia borealis* and *Stellaria humifusa* were all present, often as early invaders on saline mudflats. *Carex rariflora* and *C. glareosa* replaced *C. subspathacea* at higher elevations less subject to flooding by brackish water. *Deschampsia caespitosa, Chrysanthemum arcticum* and *Dupontia fischeri* were also represented in these less saline areas. Similar species lists are recorded from the more extensive marshes developed south into Norton Sound. The marshlands of the Yukon and Kuskokwim deltas, Bristol Bay and the Alaska Peninsula still await study.

Significant species composition changes are noted between the northwest Alaskan coastal marshes and those of Kodiak Island and Cook Inlet, south–central Alaska. *Carex lyngbyei* replaces *C. subspathacea* in similar habitats to the south; both species are present in estuaries around Nome. *Potentilla pacifica* and *Stellaria humifusa* remain prominent on the south coast and *Carex glareosa* continues to occur in the sequence, but *C. rariflora* drops out (Hanson, 1951, 1953).

East of the Alaska Peninsula, clear zonal succession replaces the irregular mosaic communities more typical of the Arctic and west Alaskan coasts. Hanson (1951) describes a simple succession from a gently sloping silt beach east of Anchorage: The first invaders — *Puccinellia phryganodes*, *Puccinellia triflora* and *Triglochin maritimum* — were subject to frequent submersion by high tides. All three species persisted at higher elevations but *Triglochin maritimum* became dominant and was joined by *Triglochin palustris*. *Scirpus maritimus* and *Carex lyngbyei* appeared at the inner edge of this zone and dense growth of the latter dominated higher levels. *Calamagrostis canadensis*, *Chrysanthemum arcticum* and *Potentilla pacifica* were scattered among the *Carex lyngbyei* in the driftwood zone. A more complex succession from Cook Inlet is outlined in Fig. 8.3.

In channel–bank communities of the Copper River Delta (Crow, 1966), pure stands of *Carex lyngbyei* colonized bare mud subject to regular tidal flooding. *Potentilla pacifica*, *Ranunculus cymbalaria* and *Salix arctica* appeared at higher elevations and were soon joined by *Eleocharis kamtschatica* and *Triglochin maritimum* to form a *Carex—Eleocharis* community. Ten additional species scattered within this zone included *Chrysanthemum arcticum*, *Deschampsia caespitosa*, *Festuca rubra*, *Poa eminens* and *Calamagrostis deschampsioides*. *D. caespitosa* increased in abundance on the channel levees to form a diverse *Deschampsia—Hedysarum alpinum* community containing over twenty species. This zone was only flooded by storm tides. Following regional uplift during the Alaska earthquake of 1964, the

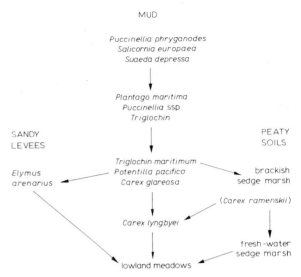

Fig. 8.3. Cook Inlet, Alaska: salt–marsh succession (after Hanson, 1951).

Deschampsia—Hedysarum marshes were replaced by species characteristic of more inland communities and *Puccinellia nutkaensis* and *Triglochin palustris* replaced *Carex lyngbyei* as primary mudflat colonists.

At Glacier Bay (Cooper, 1931) *Puccinellia pumila* was the primary colonist of intertidal mudflats. *Glaux maritima* appeared at slightly higher elevations. As plant cover became more continuous, *Plantago maritima* became dominant, co–occurring with *Hordeum brachyantherum* and *Triglochin maritimum*. A broad coastal meadow dominated by *Elymus arenarius* separated the brackish marsh from pine and spruce stands further inland.

In the Queen Charlotte Islands (Calder and Taylor, 1968), high–salinity marshes, fronted by shingle beaches or mudflats, are deeply dissected by tidal creeks. Between the creeks the marsh land forms a vegetative terrace that is only flooded by extreme tides or storms. The following species predominate in the frontal marsh zone: *Deschampsia caespitosa*, *Hordeum brachyantherum*, *Festuca rubra*, *Agrostis exarata*, *Carex lyngbyei*, *Plantago macrocarpa*, *Stellaria humifusa*, *Triglochin maritimum*, and *Trifolium wormskjoldii*. Between this zone and the adjacent closed forest, additional species such as *Apargidium boreale*, *Carex pluriflora*, *Galium trifidum* and *Calamagrostis nutkaensis* occur; *Aster subspicatus* and *Grindelia integrifolia* are occasionally found. These high–salinity marshes characteristically have few species and are very uniform throughout the islands. Low–salinity marshes are best developed on river deltas where the tidal waters that partially flood the marsh meadows are constantly diluted by fresh–water runoff. *Triglochin maritimum*, *Puccinellia pumila*, *Lilaeopsis occidentalis* and *Scirpus cernuus* occur along the margins of the muddy drainage courses, and grasses and sedges predominate on the vegetated terraces between them. More species are present here than on the higher–salinity, open coast marshes. The small tidal fringe and delta marshes of British Columbia's fjords may be similar to these Queen Charlotte Island marshes; however they still await study.

Washington and Oregon

The outstanding reference here is Jefferson (1973, 1974). Supplemental data are provided by House (1914), Jones (1936), Muenscher (1941), Peck (1961), Johannessen (1961, 1964), Macdonald (1967), and Macdonald and Barbour (1974).

Jefferson (1974) examined virtually all of Oregon's coastal marshes, as well as others in Washington and northern California. A flora of 6 macroalgae *(Cladophora gracilis, Enteromorpha compressa, E. intestinalis, Fucus distichus, Polysiphonia pacifica* and *Ulva linza)*, an intertidal moss *(Eurhynchium stokesii)* and 45 phanerogams, including one aquatic species *(Ruppia maritima)*, was identified. Quantitative distributional analysis

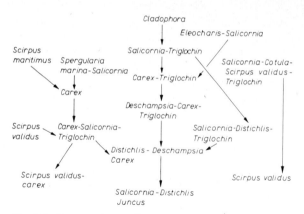

Fig. 8.5. Oregon salt marshes: succession on silt substrates (after Jefferson, 1974).

identified 6 salt–marsh vegetation types comprising 28 species associations. Plots of species cover versus distance were used to determine plant succession (Figs. 8.4, 8.5) and the results were verified by examining subsurface plant debris (Fig. 8.6). Comparisons with other less detailed studies from this region suggest that Jefferson's vegetation types and communities are characteristic of the Pacific coast from about 43°N to the Canadian border.

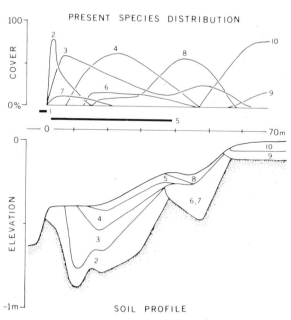

Fig. 8.6. Sand Lake, Oregon: present species distribution and buried plant debris along a 70–m sand marsh transect (after Jefferson, 1974). Species: *1 = Cladophora; 2 = Distichlis; 3 = Salicornia; 4 = Jaumea; 5 = Fucus; 6 = Triglochin; 7 = Scirpus; 8 = Plantago; 9 = Juncus; 10 = Deschampsia.*

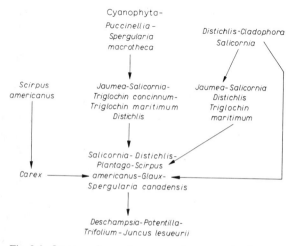

Fig. 8.4. Oregon salt marshes: succession on sand substrates (after Jefferson, 1974).

Low sandy marshes located on the inland side of bay–mouth sandspits or sandy bay islands were flooded by most high tides; tidal drainage was diffuse. The lowermost vegetation was dominated by *Salicornia virginica* or *Scirpus americanus* and the higher vegetation was mainly *Distichlis spicata*, *Jaumea carnosa* and *Plantago maritima*. *Spergularia canadensis*, *S. macrotheca*, *Puccinellia pumila*, *Carex lyngbyei* and *Glaux maritima* appeared frequently. Low silty marshes developed on silt or mud substrates wherever sedimentation was rapid. Circular islands of *Triglochin maritimum* colonized the mudflats and *Eleocharis parvula* and *Spergularia marina* were scattered between these colonies. Most high tides covered these marshes; tidal runoff was channeled between the plant colonies.

Sedge marshes dominated by almost pure stands of *Carex lyngbyei* developed on silty substrates between the low silty marshes and higher, more mature marshes. The nearly level marsh surface was often raised abruptly 30 cm or more above the surrounding tide flats. Most high tides flooded these marshes and while runoff was diffuse in the lower sedge marshes it was well contained in deep ditches on higher, older marshes. Bulrush and sedge marshes, characterized by *Scirpus validus* and *Carex lyngbyei*, occurred along tidal creeks and dikes or on islands where fresh–water influence was strong.

Immature high marshes developed on silty substrates rich in organic matter. They were located inland from the sedge and low sandy marshes and were elevated 5—10 cm above the surrounding low marshes and 60 cm or more above adjacent tideflats. *Deschampsia caespitosa* was often mixed with *Distichlis spicata* as co–dominant. Lesser quantities of *Salicornia virginica*, *Triglochin maritimum* and *Carex lyngbyei* were present. The mature high marshes were also developed on highly organic substrates, often overlying older clay sequences. Elevations of 90—100 cm above surrounding tideflats were typical. Only the higher high tides covered the marsh surface and runoff was restricted to well–defined tidal channels. *Deschampsia caespitosa*, *Juncus lesueurii* and *Agrostis alba* predominated; remnants of earlier plant communities were scattered across the surface and along ditches. The forbs *Grindelia integrifolia*, *Potentilla pacifica* and *Atriplex patula* (var. *hastata*, *littoralis* and *obtusa*) occurred on the highest elevations.

Extensive field observations at Yaquina Bay also led Jefferson (1974) to the following conclusions: Sand marshes began lowest in the intertidal zone, just above mean tide level. High marshes began near mean spring–tide high water, where periods of exposure increased sharply, and extended to extreme high water, where ecotone plants occurred. Both macro–algae and vascular plants were clearly sequenced across the intertidal zone. The lower elevational limits of various salt–marsh plants were different, indicating that there was no single critical environmental threshold for all species. Dates of new seedlings and sprouts varied each year, suggesting factors other than photoperiod stimulated new growth. Seedling success accounted for annual plant distributions; perennials reproduced vegetatively and may have extended beyond elevations suitable for initial establishment. Annuals sprouted last in the spring and lowest on the marsh; perennial plant growth began first furthest up the estuary and highest on the marsh. These patterns suggest springtime salinities and tidal exposure were important limiting factors. Maximum tidal heights were reached in winter, thus greatest flooding of the high marshes involved low–salinity water and occurred during a period of little or no growth (November—February salinities, 2—15‰; June—September, 12—36‰). During plant germination and sprouting both tide levels and salinities remained low. Despite the timing of spring growth, the plants reached maturity at consistent times. Termination of the growing season coincided with hot summer weather, high evaporation and peaks in tide water, soil water and soil salinities.

California

Despite a wealth of general descriptions (e.g., Filice, 1954; Howell, 1970; Pestrong, 1972), only Hinde (1954), Macdonald (1967), Mall (1969), Barbour et al. (1973) and Macdonald and Barbour (1974) provide quantitative data on northern California marshes. Major contributions on the better-known southern California marshes include Purer (1942), Stevenson and Emery (1958), Vogl (1966), Macdonald (1967), Bradshaw (1968b), Mudie (1969, 1970), Henrickson (1971) and Wayne (1971).

At Humboldt Bay, *Spartina foliosa* appears as a primary mudflat colonist for the first time. *Salicornia virginica* is a primary invader in sheltered situations and fucoid algae can be present. *S. virginica* and *Distichlis spicata* dominate the high–marsh vegetation. Additional high–marsh species include *Atriplex patula, Cordylanthus maritimus, Cuscuta salina, Limonium californicum, Myosurus minimus* and *Triglochin maritimum. Juncus lesueurii* and *Potentilla pacifica* appeared along the high–marsh margins.

The San Francisco Bay complex contains the most extensive salt marshes of California. Species compositions vary considerably between seasonally brackish sites towards the Sacramento–San Joaquin Delta and south San Francisco Bay, where salinities range from 27 to 29‰. Disturbed, brackish (7‰) leveed marshes at Suisun Bay (Mall, 1969) yielded 177 species, but only 15 were abundant. Diffuse vertical zonation reflected the artificially controlled seasonal submergence regime. *Salicornia virginica, Scirpus maritimus, Cotula coronopifolia* and *Typha angustifolia* dominated the low marsh; *Distichlis spicata, Juncus balticus* and *Atriplex patula* were abundant in the high marsh. Extensive, detailed studies of soil—water—salt relationships led Mall to conclude that the length of soil submergence had the greatest influence on plant distribution and competitive ability. Within the tolerances for submergence, root–zone salt concentrations, particularly in spring and early summer, determined the relative abundance of a given species.

In Castro Creek marsh (Filice, 1954), south San Pablo Bay, tidal creeks were bordered by *Spartina foliosa* and fresh–water rivulets marked by *Scirpus maritimus*. In both cases these species were replaced at higher elevations by *Salicornia virginica, Grindelia humilis* and *Atriplex patula* var. *hastata*. The highest parts of the marsh were covered by a diverse assemblage, dominated by *Distichlis spicata*, but also containing *Jaumea carnosa, Triglochin concinna, Triglochin maritimum, Frankenia grandifolia* and *Limonium californicum*. A similar succession is recorded from south San Francisco Bay (Hinde, 1954). *Spartina foliosa* was the conspicuous colonizer on the lowest mudflats, while *Salicornia virginica* dominated the high marsh. *Cuscuta salina, Distichlis spicata, Frankenia grandifolia, Jaumea carnosa*, and to a lesser extent *Limonium californicum* and *Triglochin concinna*, occurred scattered among the *Salicornia*. Other workers list *Salicornia europaea, Triglochin maritimum, Plantago maritima, Rumex occidentalis* and *Grindelia humilis* in the flora of these marshes (Macdonald and Barbour, 1974). Hinde (1954) noted that several native plants *(Grindelia, Limonium, Plantago, Triglochin)* had declined in abundance since the 1920's, while reclamation activity had allowed several new species *(Atriplex semibaccata, Chenopodium ambrosioides, Cotula coronopifolia* and *Spergularia marina)* to invade the salt–marsh fringes. *Atriplex hortensis*, an introduced species that occurs widely inland, is also well established in these salt marshes. Howell (1970) described a third successional zone intermediate between the *Salicornia* marsh and adjacent grasslands. *Distichlis spicata* and *Spergularia* were dominant and several additional native and introduced salt–tolerant plants were present: *Cotula coronopifolia, Monerma cylindrica, Parapholis incurva, Potentilla pacifica*.

Several small bays on the open coast contain salt marshes much like those described by Hinde (1954). Barbour et al. (1973) monitored soil–water salinities on a small sandy marsh in Bodega Bay, providing data still unavailable for most bay area marshes. Maximum soil–water salinities rose and fell with the tide but generally declined landward (June 1970: outer marsh, 10—23‰; inner marsh, 2—10‰). A tenfold seasonal drop in salinity recorded beneath the marsh center between October (11‰) and April (1‰) coincided with winter rains; by July, soil–water salinities had returned to their previous high values.

Among the outer coast marshes, *Spartina* is absent at Tomales and Bodega bays but present at Drake's Estero and Bolinas bay. *Carex lyngbyei*, so important in Oregon marshes, also occurs at Drake's Estero, but is rare. *Grindelia integrifolia*, another species common to the north, is present at Tomales and Bolinas bays but does not occur further south. Howell (1970), comparing the open coast and bay area marshes, notes that *Puccinellia lucida* and *Grindelia integrifolia* are present on the coast marshes but not in the bay, while *Grindelia humilis* and *Salicornia europaea* are present in the bay marshes but not along the coast. *Cordylanthus maritimus* is present in south San Francisco Bay, while at the less saline sites of the north bay *Cordylanthus mollis, Scirpus acutus, Glaux maritima, Rosa californica* and *Achillea borealis* are all represented. The occurrence of *Suaeda californica* in the bay marshes (Munz, 1973) should also be noted, for while the species is rare here at its northern limits, it becomes an important high–marsh component further south.

Only two major salt–marsh sites occur between San Francisco and Point Conception. At Elkhorn Slough *Spartina* is absent and *Salicornia* dominant; additional species included *Jaumea carnosa, Frankenia grandifolia, Distichlis spicata,* and less commonly *Cuscuta salina, Potentilla pacifica, Spergularia marina* and *Suaeda californica.* Morro Bay yielded a similar flora (Hoover, 1970); however, *Salicornia subterminalis* was an important new addition around the high–marsh fringes. Early records (Smith, 1952) indicate that sites just south of Point Conception (Goleta, Carpinteria) contained more diverse marsh assemblages: *Cordylanthus maritimus, Cuscuta salina, Distichlis spicata, Frankenia grandifolia, Jaumea carnosa, Lasthenia glabrata, Limonium californicum, Monanthochloe littoralis, Salicornia virginica, S. subterminalis, Suaeda californica, S. depressa* var. *erecta, Triglochin concinna* and *T. maritimum.* Still more species appeared on surrounding saline flats or near fresh–water sources: *Atriplex patula* var. *hastata, A. rosea, Cressa truxillensis, Cotula coronopifolia, Grindelia robusta, Juncus acutus* var. *sphaerocarpus* and others. These sites lacked *Spartina; Salicornia virginica* was the most widespread species. Urban encroachment has greatly modified these sites and many of the species listed are now rare or absent.

The larger tidal salt marshes developed southward to the Mexican border — at Mugu Lagoon, Newport Bay, Mission Bay, San Diego Bay and Tijuana Slough — form a homogeneous group with a succession similar to that shown in Fig. 8.7. Quantitative data from Newport Bay marshes (Vogl, 1966) supporting the sequence shown in Fig. 8.7, are given in Table 8.1. Noting that many species reach peak abundances in one "zone" but also occur in others, Vogl concluded that the salt–marsh plants followed a vegetation continuum — the floristic composition gradually changing along environmental gradients.

Mudie's (1970) extensive studies at San Diego Bay provide insights into the salinity regimes and nutrient flux of southern California salt marshes. Flood tides bring bay waters of 35—40‰ salinity into the marsh creeks; salinities rise sharply during ebb tide and can reach 55‰. Soil paste extracts yielded soil salinity values of 20—78‰ in low marshes, 25—31‰ in high marshes, and > 87‰ beneath salt flats. The soil salinities generally increased with moisture content and decreased with elevation. Soil moisture showed a positive correlation with peat and sand content, but generally decreased 50% between low and high marsh. The top 15 cm of marsh sediments evidenced seasonal salinity changes; soil salinities dropped off sharply at the littoral salt marsh—maritime sagebrush transition. Mudie concluded that there were few clear–cut relationships between plant distributions and soil salinities, but that the combined factors of tidal submergence, soil moisture and salinities probably control vegetation patterns (cf. Mall, 1969).

Nutrient analyses of tidal waters ebbing from the salt marshes showed three clear trends. Inorganic phosphorus levels were 1—9 times higher

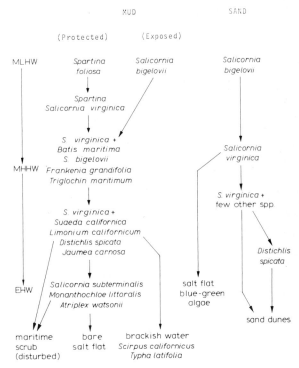

Fig. 8.7. Southern California: generalized salt–marsh succession (after Mudie, 1970).

TABLE 8.1

Upper Newport Bay, California: average percentage species cover, by zone (after Vogl, 1966)

Species	Littoral			Maritime bluffs
	lower	middle	upper	
Spartina foliosa	38(4—87)	1(0—2)	*	
Batis maritima	4(0—23)	15(3—42)	1(1—6)	
Salicornia virginica	4(1—10)	23(6—67)	40(2—65)	11(0—29)
Suaeda californica	*	*	2(1—7)	19(9—25)
Frankenia grandifolia		3(0—11)	2(2—12)	*
Distichlis spicata		2(0—10)	5(0—29)	
Triglochin maritimum		11(0—42)	1(0—8)	*
Limonium californicum		1(0—4)	1(0—10)	
Monanthochloe littoralis			15(0—66)	
Cuscuta salina		*	2(0—5)	
Juncus acutus			2(0—14)	
Scirpus californicus			14(0—90)	
Salicornia subterminalis				24(0—54)
Mesembryanthemum crystallinum				13(0—26)
Encelia californica				4(0—14)

Note: Values are rounded averages based on 240 quadrats; ranges for six separate stands bracketed. Asterisks indicate presence of species listed.

on the ebb than the flood tides; ammonia–nitrogen levels were 3—28 times higher in the marshes than the bay, and finally, nitrate–nitrogen levels were consistently high, while nitrite levels remained low. Clearly the ammonia was rapidly being oxidized to nitrates by an active nitrifying flora, while the turnover time of the nitrates was slow relative to the generation time. Mudie concluded that the San Diego Bay marshes acted as a significant reservoir of nutrients that were supplied to the adjacent bay waters by tidal flushing.

Baja California

Quantitative studies by Macdonald (1967), Neuenschwander (1972) and Thorsted (1972) now supplement general descriptions of Baja salt marshes provided by Dawson (1962), Phleger and Ewing (1962) and Shreve and Wiggins (1964). Mudie (in Macdonald and Barbour, 1974) has studied the temperate salt marsh—tropical mangal transition.

Estero de Punta Banda, south of Ensenada, and Bahia de San Quintin each contain *Spartina— Salicornia* marshes much like those of San Diego Bay. Thorsted (1972) lists only fifteen species from San Quintin marshes and of these only eight were common. Three distinctive vegetation types dominated by *Spartina foliosa, Salicornia virginica* and *Monanthochloe littoralis,* respectively, occupied the lower, middle and upper littoral. Comparisons with Vogl's Newport Bay data (Table 8.1) indicated that *Distichlis* was much rarer at San Quintin, while *Monanthochloe* and *Salicornia subterminalis* were more abundant. Thorsted supported Vogl's view that the marsh plants followed a vegetation continuum.

Salt–marsh vegetation extended inland 30 m beyond the high tide debris line (cf. maritime zone of Vogl, 1966). *Monanthochloe* and *Salicornia subterminalis* formed the majority of the cover in the lower part of the transition, while three upland succulent halophytes — *Frankenia palmeri, Lycium brevipes* and *Atriplex julacea* — accounted for most of the cover in the upper part. Neuenschwander (1972) concluded that the lower portion of the transition represented a capillary fringe around the bay, containing sub–irrigated salt–marsh species, while the upper part contained upland species controlled by adjacent desert conditions.

Laguna Guerrero Negro marshes (Macdonald, 1967; Macdonald and Barbour, 1974) are again much like those of San Quintin. *Spartina* and *Monanthochloe* dominated the lower and upper littoral zones, respectively; however, *Batis maritima* and *Salicornia bigelovii* dominated the middle littoral, rather than the *S. virginica—Batis—Frankenia* assemblages noted from San Quintin. Other species included: *Frankenia grandifolia, F. palmeri, Limonium californicum* var. *mexicanum, Sesuvium verrucosum, Salicornia subterminalis, Suaeda californica* and *Triglochin* sp. I. L. Wiggins (personal communication, 1972) notes that several of these species *(F. grandifolia, L. californicum* and *Triglochin maritimum)*, while common to the north, do not extend much below Punta Eugenia (∼ 28°N).

The salt marsh—mangrove transition occurs along the outer Baja coast between Bahia Asuncion (27°N) and Puerto Chale (24°30′N; Mudie, in Macdonald and Barbour, 1974). Several barrier beach—coastal lagoon complexes and estuaries are present; sandy sediments and hypersaline conditions predominate. Fresh-water runoff is limited to sporadic, but sometimes massive, river flooding. Low-marsh formation commences with the stabilization of sand bars by *Spartina*, followed by the establishment of *Laguncularia* or *Rhizophora* seedlings within the *Spartina* stands. The mangroves may later shade out the *Spartina* but it persists on aggrading channel banks in front of the mangroves. In the most wind-exposed northern lagoon (Pond Lagoon), dense scrubby *Laguncularia racemosa* (0.7—1.0 m high) comprises about 90% of the mangal association, with *Rhizophora mangle* (1.0—1.3 m high) occurring only as isolated stands within the *Laguncularia* scrub or on wind-sheltered channel banks. *Rhizophora* increases in height (as does *Laguncularia*) and frequency at sites southeast of San Ignacio Lagoon and becomes the dominant mangrove species from the northern end of Laguna Santo Domingo southwards. At Puerto Chale both species are 4—7 m high along the channels, becoming scrubbier with increasing elevation. *Avicennia germinans* joins the mangal association from Laguna Santo Domingo southwards. Colonizing active sand banks within the mangal formation it can become dominant, as a low-growing shrub, in poorly drained low areas around mean higher high water.

High-marsh vegetation is similar in structure and general appearance to that of northern Baja but is floristically simpler. *Frankenia grandifolia, Limonium californicum, Jaumea carnosa* and *Triglochin maritimum* are all absent and *Distichlis spicata* is rare. *Batis maritima, Salicornia virginica, S. bigelovii* or *Monanthochloe littoralis* are dominant. *Sesuvium verrucosum* occurs both in the high marsh, and on sandy low-marsh levees. Where the marshes are backed by dunes, extreme high water level is sharply demarcated by the presence of *Allenrolfea occidentalis, Salicornia subterminalis, Frankenia palmeri* and *Atriplex*. In many of the northerly areas, extreme high water is characterized by extensive, highly saline mudflats (Phleger, 1969), devoid of vegetation other than scattered *Allenrolfea occidentalis*. Southwards, low sandy plains occur behind the marshes; these are sparsely covered by a characteristic assemblage of *Salicornia subterminalis, Suaeda* cf. *taxifolia, Tricerma phyllanthoides, Haplopappus venetus* ssp. *furfurascens, Atriplex barclayana* and *Sporobolus contractus* (see p. 202).

Preliminary soil-water salinity studies (Mudie, in Macdonald and Barbour, 1974) indicated that low-marsh salinities remained close to those of sea water, while high-marsh values were higher by 50—75% of sea water values. The soils in poorly drained pans (characterized by *Batis* and *Salicornia bigelovii*) and on salt flats were 2.5—3 times more saline than the low marsh.

PHYTOGEOGRAPHY

Distributional ranges for the 100 or so plants that characterize Pacific coast salt-marsh vegetation (Macdonald and Barbour, 1974) were compiled from regional floras. The number and identity of all species *potentially* present at one-degree intervals along the Pacific Coast was then examined. The resulting diversity gradient (Fig. 8.2) and compositional changes are outlined below.

The lowest regional plant diversities were recorded from Baja California, the Arctic Slope and the west Alaskan coast (Fig. 8.2). The average diversity for the coastal segment between the northern Gulf of Alaska and Vancouver was 26 species per degree of latitude and for the Vancouver to Mexico segment, 33 species per degree of latitude. The diversity peak noted at 61°N (Anchorage) reflects the presence of both subarctic and temperate species within the ecotone between these two climate zones. The high diversity values from southern

California marshes (38 species per degree of latitude) partly reflect human disturbance that has permitted several introduced species to florish.

Major compositional changes in the salt–marsh vegetation developed at different latitudes have been reviewed above. Quantitatively, the most prominent shifts in species composition occurred as follows:

(1) Between 68 and 65°N, northern limits for 11 species.

(2) Between 61 and 58°N, northern limits for 10 species, southern limits for 14 species and 3 local endemics recorded.

(3) Between 55 and 54°N, northern limits for 5 species, southern limits for 2.

(4) Between 50 and 49°N, northern limits for ten species.

(5) At 41°N, northern limits for 3 species and a single southern limit recorded.

(6) At 38°N, northern limits for 5 species, southern limits for 6 species and 2 local endemics recorded.

(7) Between 35 and 32°N, northern limits for 14 species and southern limits for at least 20 species recorded.

(8) At 30°N, a single northern limit and 4 southern limits recorded.

(9) Between 28 and 27°N, northern limits for 3 species and southern limits for 4 species were recorded.

The general concordance of latitudinal environmental changes and regional differences in the salt–marsh vegetation — diversity, composition, dominant species, succession and floristic affinities — provide a basis for classifying Pacific coast salt marshes into a number of distinctive groups and subgroups. A tentative classification is outlined below. Group nomenclature closely follows that suggested by Breckon and Barbour (1974) in a recent review of Pacific coast beach vegetation. Obvious parallels with Chapman's classification of world salt marshes (see pp. 11—13) are also apparent.

Arctic Group

This group includes the marshes of west coast and Arctic Alaska; the environment is extreme and sea–ice action frequently disrupts succession. Mosaic communities of *Puccinellia* and *Carex* predominate.

(1) *Arctic Coast Subgroup* (71—69°N) — characterized by strikingly uniform, low diversity, *P. phryganodes—Stellaria humifusa* pioneer communities.

(2) *West Alaska Coast Subgroup* (69—60°N) — mosaic communities of *Carex subspathacea* succeeded by *C. glareosa* and *C. rariflora* are characteristic; diversity increases southward.

Subarctic Group

This group, developed between Anchorage and the Queen Charlotte Islands (60—54°N), coincides with the microthermal—mesothermal climatic shift. Warmer temperatures, high rainfall and low salinities are typical. *Puccinellia* and *Carex* species, more abundant to the north, overlap with the northern ranges of such genera as *Glaux, Salicornia, Spergularia* and *Suaeda*. Clear vertical zonation and succession are well documented; *Carex lyngbyei, Deschampsia caespitosa, Plantago maritima, Potentilla pacifica* and *Triglochin maritimum* are important salt–marsh components. Diversity declines southward as more Arctic species drop out.

Temperate Group

The marshes of this group are the most broadly distributed of the Pacific coast, extending between 54 and 35°N. Throughout most of this range a diverse general salt–marsh community is present. This includes: *Atriplex patula* var. *hastata, Cuscuta salina, Distichlis spicata, Jaumea carnosa, Salicornia virginica, Scirpus maritimus*, and *Triglochin maritimum*. Two introduced species, *Cotula coronopifolia* and *Spergularia marina*, are widespread. Latitudinal changes in co–occurring species suggest four distinct subgroups:

(1) *British Columbia Subgroup* (54—51°N) — most Arctic species have now dropped out; floral elements that first appeared in Subarctic marshes (*Carex lyngbyei, Deschampsia caespitosa, Festuca rubra*) gain in prominence but diversity remains low. Field data for this subgroup are scarce.

(2) *Washington—Oregon Subgroup* (51—43°N) — diversity increases sharply near the Canadian border as a large group of southerly floral elements joins the general community. While *Salicornia*

virginica is common, *Carex lyngbyei* and *Deschampsia caespitosa* occupy more central roles in succession.

(3) *Northern California Subgroup* (43—38°N) — diversity remains high. *Spartina foliosa* now joins the community as an important primary colonist and a dominant member of the low–marsh community. *Salicornia virginica* and *Distichlis spicata* are the most abundant species overall; *Limonium californicum* joins the high–marsh assemblage.

(4) *Central California Subgroup* (38—35°N) — evaporation now exceeds precipitation and salinities rise southward. *Salicornia virginica* predominates; several arid zone species join the high–marsh assemblage: *Frankenia grandifolia, Salicornia subterminalis, Sesuvium verrucosum* and *Suaeda californica. Spartina* is present around San Francisco, but locally absent south to Mugu Lagoon (34°N). This, and a southern diversity decline, may reflect a scarcity of suitable habitats.

Dry Mediterranean Group

This distinctive group of salt marshes is restricted to southern California (35—32°N). Plant diversity reaches the highest levels seen along the Pacific coast. Virtually all of the species important from Oregon northward have disappeared and several distinctly tropical floral elements join the general salt–marsh community: *Batis maritima, Frankenia palmeri, Salicornia bigelovii, Monanthochloe littoralis*. Several of these marshes are badly disturbed and introduced species are common.

Arid Group

This last group of Pacific coast salt marshes extends throughout Baja California. Environmental conditions become extreme as temperatures and salinities rise southward. Plant diversity is sharply lower and introduced species rare. Two subgroups can be distinguished:
(1) *Northern Baja Subgroup* (32—28°N) — low diversity and the absence of at least 15 natives and several introductions, seen to the north, characterize this subgroup. *Allenrolfea occidentalis* joins the community.
(2) *Southern Baja Subgroup* (28—23°N) — diversity remains low in this group. Further members of the general salt–marsh community (*Frankenia grandifolia, Juncus acutus, Limonium, Salicornia bigelovii* and *Triglochin*) drop out. Mangroves appear for the first time: *Avicennia germinans, Laguncularia racemosa* and *Rhizophora mangle*. This subgroup clearly represents the transition from Chapman's "Pacific American" salt marshes to the north to the "Gulf of California mangal subgroup".

NUMERICAL ANALYSIS

The objective reality of the proposed phytogeographic groups was further examined through numerical analysis. The species range data (above) were used to prepare "expected" species lists for *theoretical* sample sites located at one degree of latitude intervals along the coast. The resulting data matrix (88 species × 49 sample sites) was examined by Q– and R–mode unweighted pair-group cluster analysis (Sokal and Sneath, 1963; Bonham–Carter, 1967), using Jaccard coefficients (S_J) as a measure of sample similarity. The similarity relationships between salt–marsh floras expected from different coastal latitudes are summarized in the Q–mode dendrogram shown in Fig. 8.8. For comparative purposes, Fig. 8.8 also includes the tentative phytogeographic schema developed above. Groups of species showing similar presence–absence patterns along the coast were identified from an R–mode dendrogram (not shown), and were broadly similar to the floral groups already described in the regional review.

The theoretical patterns of floral similarity suggested by the Q–mode dendrogram (Fig. 8.8) are broadly similar to the phytogeographic groupings already proposed. The most pronounced floral break lies close to the Canadian border, all sites to the north and south, respectively, clustering into two separate groups. A further subdivision of sites at the $S_J = 0.5$ level, supports the separation of the northern marshes into Arctic and Subarctic groups. South of the Canadian border three additional clusters of localities are recognized. The largest of these (cf. Temperate Group) includes the marshes of Washington, Oregon and northern California (39—50°N), while the marshes of the remainder of California form a second cluster and those of Baja California (cf. Arid Group), a third.

The dendrogram also suggests an interesting

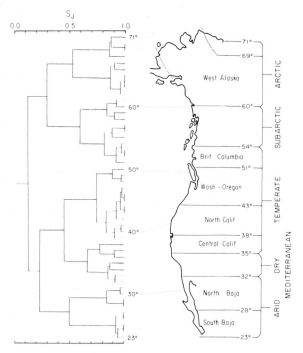

Fig. 8.8. North American Pacific coast. Left: *Q*–mode dendrogram summarizing floral similarity relationships (measured as Jaccard coefficients, S_J) among *theoretical* sample sites spaced at one–degree intervals along the coast. Right: Phytogeographic schema developed through qualitative data analysis. See text for additional explanation.

modification of the proposed Temperate Group of marshes, for British Columbia sites are clustered with those of the Subarctic Group and southern California's Dry Mediterranean marshes are included with those of the Temperate Group. Additional minor differences suggested by the dendrogram include several one–degree latitude changes in the boundaries between proposed salt–marsh groups.

INTRODUCED PLANT SPECIES

Human disturbance has played a significant role in modifying Pacific Coast salt–marsh vegetation. Hanson (1951), Wiggins and Thomas (1962), Hultén (1968) and Jefferies (1976), for example, note that harvesting of several Alaskan salt–marsh plants for food or basket making may have modified their geographic distributions. Jefferson (1974) notes that several introduced marsh plants in Oregon may have arrived with earthen ballast from nineteenth century ships, or with oyster spat imported from the Atlantic coast. The introduction of several weedy species to California with the livestock of European explorers and colonists is also well documented.

Many of the authors cited in the regional review list introduced species among salt–marsh floras. Examples include: *Agrostis alba, Atriplex hortensis, A. semibaccata, Cotula coronopifolia, Hordeum brachyantherum, H. jubatum, H. leporinum,* several *Mesembryanthemum* species, *Monerma cylindrica, Parapholis incurva, Polypogon monspeliensis,* and *Spergularia marina*. In most cases these species are restricted to disturbed areas around the salt–marsh fringes, or along artificial drainage ditches and dikes. Sometimes, however, they appear to fill empty niches in marsh succession. Jefferson (1974) notes that *Cotula coronopifolia* is common in Oregon marshes, growing among logs and bark chips — sites where most native species become badly stunted. *Spergularia marina* has become a primary invader with *Triglochin* on silt flats, mimicking and perhaps out–competing the native species, *S. canadensis* and *S. macrotheca. Spartina alterniflora,* introduced from the Atlantic coast, is actively colonizing sandflats in Willapa Bay, Washington, up to 80 m beyond the native marsh. *Spartina townsendii* introduced into Stillaguamish Bay, Washington, from Europe, as recently as 1960, is also expanding its range.

Severe environmental modifications, such as the restriction or exclusion of tidal influence from salt marshes, result in major floristic adjustments that often include invasion by opportunistic weed species. This is particularly common in southern California where most coastal lagoons have been altered by man. Several such sites studied by Bradshaw (1968b), Mudie (1969) and Henrickson (1971) yielded "salt marsh" floras of 65—168 species each; of these, more than 40% were naturalized. In marked contrast with these disturbed marshes, those at San Quintin Bay (Thorsted, 1972) contained only one introduced species: *Mesembryanthemum nodiflorum*. I.L. Wiggins (personal communication, 1972) notes that this species, along with *Spergularia marina, Phoenix dactylifera* and *P. canariensis* are all rather widespread introductions along the Baja coast.

TABLE 8.2

Annual net productivity of selected salt-marsh plants

Species Location (Reference)	Productivity[1] [kg (dry weight) m^{-2} yr^{-1}]
Carex ramenskii	
Arctic coast (Jefferies, 1975)	0.02—0.24[a]
Distichlis spicata	
Suisin marsh (Mall, 1969)	0.66
Puccinellia phryganodes	
Arctic coast (Jefferies, 1975)	0.05—0.14[b]
Salicornia virginica	
Suisin marsh (Mall, 1969)	0.65
San Pablo Bay (Cameron, 1972)	1.0—1.2
S. San Diego Bay (Mudie, 1970)	1.5—2.5
San Quintin Bay (Dawson, 1962)	~2.0[c]
Spartina foliosa	
San Pablo Bay (Cameron, 1972)	1.4—1.7
S. San Diego Bay (Mudie, 1970)	0.8
San Quintin Bay (Barnard, 1962)	~1.0

[1] Annual above–ground standing crop; includes ash weight.
[a,b,c] Above–ground below–ground biomass ratios of 1:1.9, 1:1 and 1:1, respectively.

PRODUCTIVITY

Salt–marsh productivity data from Pacific coast sites remain scarce. The values known to me, cited in Table 8.2, are all based on quadrat harvesting at the peak of the growing season for each species respectively. Since most salt–marsh plants have a restricted growth season, these standing–crop values approximate net annual above–ground productivity. These figures all include the ash weight, which can account for 20% or more of the dry weight (Mudie, 1970).

As expected, the marshes of California are far more productive than those of the Arctic coast. The tabulated values also compare favorably with those from North American Atlantic coast salt marshes. De la Cruz and Gabriel (1975), for example, cite the annual above–ground net production of Atlantic coast marshes as from 0.4 to 2.9 kg (dry weight) m^{-2} yr^{-1}. Values for individual species [all as kg (dry weight) m^{-2} yr^{-1}] include 0.9 for *Spartina patens* (Ranwell, 1967) and 0.6 and 0.84 for *Distichlis spicata* and *Spartina alterniflora*, respectively, from Long Island (~ 40°N; Udell, 1969).

Interestingly, the few data available suggest that the productivity of *Spartina foliosa* rises northward, between Baja California and San Francisco. Conversely, *Salicornia virginica* productivity values are lowest at brackish, northern sites and increase with higher salinities and lower latitudes.

ANIMAL COMMUNITIES

General

The animal communities of Pacific North American salt marshes are still very incompletely known. Comprehensive ecosystem syntheses such as those already published for the North American Atlantic coast (Dexter, 1947; Teal, 1962; Teal and Teal, 1969, among others), Europe (Green, 1968; Ranwell, 1972) and New Zealand (Paviour–Smith, 1956), are not yet available for any Pacific coast salt marsh. Instead, much of the literature is scattered, and both its geographic coverage and treatment of different faunal groups remains uneven. This contribution attempts to draw together some of this scattered material and present an overview of the marsh communities.

MacGinitie (1935), Warme (1971) and Barbour et al. (1973) provide an excellent introduction to the salt–marsh communities of California. I have also drawn extensively from my own studies (Macdonald, 1969a, b, 1971a—d) — particularly at Goleta Slough, a one-km² *Salicornia virginica* marsh adjacent to the University of California, Santa Barbara campus. Several recent reports of the California Department of Fish and Game (Bradshaw, 1968b; Mudie, 1970; Leach and Fisk, 1972; Jurek, 1973, among others) have proved particularly valuable. Taxonomic and ecological data on many of the species listed can be found in such excellent regional guides as Ricketts and Calvin (1952), Kozloff (1973) and Smith and Carlton (1975) on invertebrates; Roedel (1953), Fry (1973) and Hart (1973) on fish; Stebbins (1966) on reptiles; Martin et al. (1961) and Peterson (1961) on birds, and Ingles (1965) on mammals.

Taxonomic groups

Foraminifera

Phleger (1967) described three Pacific coast salt-marsh foraminiferal assemblages. A subarctic assemblage recorded north of Vancouver was dominated by a variant of *Trochammina inflata* and *Miliammina fusca* was sometimes present. An Oregonian assemblage yielded *T. inflata, M. fusca* and *Jadammina polystoma* in abundance, while *Ammobaculites exiguus* and *Haplophragmoides subinvolutum* were present but less common. Southern and Baja California salt marshes were characterized by a more diverse assemblage. The dominants were the same as in the Oregonian assemblage but associated species now included *Ammonia beccarii, Arenoparrella mexicana, Cellathus discoidale, Discorinopsis aguayoi, Glabratella* sp. and *Protoschista findens* (Scott et al., 1975).

Phleger (1967) found low-marsh and high-marsh environments were characterized by distinctive faunas, with the high marsh yielding the largest foraminiferal populations. Northern marsh faunas were almost entirely composed of arenaceous species, indicative of low-salinity areas with high runoff. Southern faunas contained more calcareous specimens, reflecting undiluted nearshore waters and a greater open-ocean influence.

Ostracoda

Warme (1971) found several ostracods represented at Mugu Lagoon but only one species, *Xestoleberis aurantia,* was abundant in the salt marsh.

This species is also common in upper Newport Bay and co-occurs with *Cyprideis (Goerlichia) castus* in tidal channel and salt-marsh habitats at Estero de Punta Banda, northern Baja California (Benson, 1959).

Nematoda

T. Dilcher and T. Yoshino (personal communication, 1970), working at Goleta Slough, confirmed that nematodes are an important component of the Pacific coast salt-marsh meiofauna (cf. Wieser and Kanwisher, 1961; Teal and Wieser, 1966). Analyses of 55 short sediment cores yielded nematode population densities of up to 9×10^6 m^{-2}. Some five or six species were represented, the individuals varying from 0.2 to 5 mm in length. The majority (50—95%) of the nematodes occurred in the top centimeter of sediment and very few were recovered from depths below 4 cm.

Annelida

Although recorded from other salt-marsh communities (Kraeuter and Wolf, 1974), virtually nothing is known of Pacific coast salt-marsh annelids. Reish and Barnard (1967) listed *Capitella capitata* and *Streblospio benedicti* — both recorded from Georgia salt marshes (Teal, 1962) — as abundant in marsh creeks at Morro Bay. Barnard (1970) noted *Exogone verugera* and *Fabricia limnicola* were abundant in fine silts bordering salt marshes at Bahia de San Quintin, Baja California. *Capitella capitata* is common in the wetter areas of the *Salicornia* marsh at Goleta Slough and frequently co-occurs with *Polydora nuchalis* in adjacent marsh creeks where some oligochaetes are also present.

Mollusca

Pacific coast salt-marsh mollusc assemblages (Macdonald, 1969a, b) consist almost entirely of epifaunal gastropods that eat algae and detritus. Population densities of several thousand gastropods per m² are not uncommon, yet over 90% of the individuals encountered at each site usually belong to only two or three species. The minute prosobranch *Assiminea translucens* is ubiquitous, being found on virtually every salt marsh between Vancouver Island and Baja California (25—48°N). *Littorina newcombiana* (41—47°N) and the pulmonate *Phytia myosotis* (35—49°N) overlap the northern portion of *Assiminea*'s range to form a three-species Oregonian assemblage. *Cerithidea californica* (27—38°N) and another pulmonate, *Melampus olivaceus* (23—37°N), co–occur with *A. translucens* in marshes within the California Province. Additional species include the rocky shore form *Littorina sitkana* found in Puget Sound salt marshes (Kozloff, 1973), and two introductions: *Batillaria zonalis* (36—49°N) from Japan (Whitlatch, 1974) and the horse mussel, *Modiolus demissus,* introduced into San Francisco Bay marshes from the Atlantic coast.

There is some tendency toward vertical zonation among the salt–marsh gastropods, the pulmonates *Melampus* and *Phythia* both reaching their maximum population densities at higher elevations than all other species. While the number and types of niches available in each marsh appear to remain the same, the standing crop of gastropods increases considerably from north to south (< 1 to > 10 g dry weight m^{-2}), suggesting that the resources available within each niche increase at lower latitudes.

Crustacea

Several species of crabs occur in the salt marshes. The mud crab, *Hemigrapsus oregonensis,* is common along much of the Pacific coast, while the fiddler crab, *Uca crenulata,* first appears near Los Angeles and is present in tremendous numbers in some of the sandier marshes of Baja California. Less frequently encountered species include *Hemigrapsus nudus, Pachygrapsus crassipes* and *Speocarnicus californiensis.*

Common salt–marsh amphipods include *Orchestia traskiana* and *Orchestoidea californica.* At Goleta Slough hundreds of *Orchestia* may be found beneath plant debris adjacent to tidal creeks. *Orchestia* declines in abundance away from the creeks, being replaced at higher, less frequently flooded elevations by several terrestrial isopods: *Armadillidium vulgare, Philoscia richardsonae, Porcellio laevis, P. dilatatus* and *P. scaber* (Filice, 1954; F. DiCarlo and K. Merrill, personal communications, 1970). The isopods *Ligia* sp. and *Portunion conformis* — a common endoparasite of the mud crab (Schultz, 1969) — as well as the ghost shrimp, *Callianassa californiensis,* have also been recorded from Californian salt marshes (Mudie, 1970; Warme, 1971; Smith and Carlton, 1975).

Insecta and Arachnida

Gustafson and Lane (1968) have reviewed the literature on Californian salt–marsh insects; most papers cited are taxonomic in character. Two notable ecological studies of insects from San Francisco Bay marshes have been published (Lane, 1969; Cameron, 1972) and a similar study is nearing completion at Goleta Slough. General observations from San Diego area marshes are described by Moore (1964), Bradshaw (1968b) and Stewart (1966, in Mudie, 1970).

Lane (1969), collecting from July through December, identified 134 insect species from a *Spartina—Salicornia* marsh in southern San Francisco Bay. Of several collecting methods tested, sweep netting yielded by far the greatest number and variety of insects. Further, 80% of all insects caught by sweeping were from the *Spartina foliosa* low marsh. Diptera (43 species) and Homoptera (12 species) were the prevalent orders throughout the marsh. *Prokelisia* sp. (Homoptera: Delphacidae) and *Oscinella* sp. (Diptera: Chloropidae) were the dominant species collected from *Spartina foliosa* and represented 85% of the low–marsh insect fauna. The bulk of the insects collected from the *Salicornia virginica* high marsh belonged to the dipteran families Ephydridae *(Ephydra cinerea,* and *E. riparia)* and Chironomidae *(Pseudosmittia* sp.), and the homopteran family Psyllidae *(Aphalara* sp.).

Cameron (1972) examined seasonal changes in insect diversity and trophic relationships in a closely similar marsh in northern San Francisco Bay. Unlike Lane, Cameron used a clip–quadrat sampling method and extracted insects from the clipped plants and associated detritus with Berlese funnels. Less than 10% of the 103 species of adult insects identified by Cameron were also collected by Lane. Apparently Cameron missed some of the more active flying insects, while Lane missed several sedentary ground–dwelling species.

Cameron categorized the adult marsh insects into three trophic levels: herbivores, saprovores and predators. He found a positive correlation between herbivore diversity and standing crop biomass of the marsh plants, and between saprovore diversity and litter accumulation. Herbivore diversity was highest in spring and lowest in winter. Saprovore diversity was highest in mid–winter.

The saprovore *Xenylla baconae* (Collembola:

Poduridae) was by far the most abundant species in *Spartina* stands. Other common saprovores included *Lachesilla pacifica* (Psocoptera: Lachesillidae) and *Corticaria* sp. (Coleoptera: Lathridiidae). Herbivore production was largely accounted for by *Haplothrips* sp. (Thysanoptera: Phloeothripidae) and *Anophothrips zeae* (Thysanoptera: Thripidae). *Xenylla baconae* remained the principal saprovore in the *Salicornia* community, while the scale insect, *Pseudococcus* sp. (Homoptera: Pseudococcidae) was the most abundant herbivore year–round. Important predators on the marsh insects included spiders, abundant year–round in both *Spartina* and *Salicornia* communities, shrews *(Sorex ornatus)*, particularly common in the *Salicornia*, and insectivorous birds.

Fish

While many fish occasionally enter marsh creeks, or feed among the flooded marsh plants at high tide, only a few species are truly characteristic of Pacific coast salt–marsh habitats. Foremost among these are euryhaline forms such as the arrow goby, *Clevelandia ios,* California killifish, *Fundulus parvipinnus* (Fritz, 1970), long–jawed mudsucker, *Gillichthys mirabilis* (Noble et al., 1963), Pacific staghorn sculpin, *Leptocottus armatus* (Jones, 1962) and topsmelt, *Atherinops affinis*.

Less frequently encountered species include marine forms such as the California halibut, *Paralichthys californicus,* Pacific herring, *Clopea harengus pallasi,* and starry flounder, *Platichthys stallatus,* that may enter estuarine shallows to spawn. The protected marshes and creeks of many northern estuaries also provide feeding, nursery and acclimation areas for the young of several anadromous species (Fry, 1973; Hart, 1973; Monroe, 1973) *en route* from spawning sites in tributary streams to their ocean feeding grounds. Salt–marsh ponds and streams characterized by low salinities often yield the tidewater goby, *Eucyclogobius newberryi,* and threespine stickleback, *Gasterosteus aculeatus*. The mosquito fish, *Gambusia affinis,* and fathead minnow, *Pimephales promelas,* have been widely introduced as a mosquito control measure.

Amphibians and reptiles

Field studies at Goleta Slough indicate that reptiles are well represented in more elevated, drier, salt–marsh habitats. The side–blotched lizard, *Uta standburiana,* southern alligator lizard, *Gerrhonotus multicarinatus,* and western fence lizard, *Sceloporus occidentalis,* are all common. The California king snake, *Lampropeltis getulus,* several species of garter snakes, *Thamnophis* sp., and the western rattlesnake, *Crotalus viridis,* have also all been collected. Except in areas of fresh–water influence, amphibians are rare at Goleta. The Pacific tree–frog, *Hyla regilla,* and western toad, *Bufo boreas,* are occasionally found in the salt marsh following heavy rains. These species are all representative of an assemblage once common along unpopulated and undisturbed areas of the California coast (Stebbins, 1966; Dingman, 1971). Similar faunas have been recorded from Bodega Bay (Barbour et al., 1973), and the Tijuana River Estuary (McIlwee, 1970).

Birds

Only a few species of birds are endemic to Pacific coast salt marshes. These include the secretive and increasingly rare black rail, *Laterallus jamaicensis,* and clapper rail, *Rallus longirostris,* both of which feed and nest in California *Salicornia* marshes (Moffitt, 1941; Zucca, 1954; Peterson, 1961; Leach and Fisk, 1972), and salt–marsh subspecies of both the savannah sparrow, *Passerculus sandwichensis beldingi,* and song sparrow, *Melospiza melodia* (Marshall, 1948). Despite the rarity of endemics, avian utilization of the salt marshes can be very high, particularly during major migratory periods.

The coastal wetlands that rim the Pacific between Alaska and Mexico make up an important segment of the North American Pacific flyway. Each spring, millions of migrant shorebirds and waterfowl follow the flyway north to breed and raise their young during the brief Arctic summer, when tundra vegetation and insect life reach their peak of development. As fall approaches and the Arctic freeze–up begins, the migrants fly south again to overwinter in the estuaries and coastal lagoons of California and Mexico. As many as a hundred different species utilize the flyway (Peterson, 1961; Linduska, 1964; Jurek, 1973). Shorebirds outnumber all other migrants along the coastal route. Western and least sandpiper, *Calidris mauri* and *C. minutilla,* dowichers, *Limnodromus* spp., willet, *Catoptrophorus semipalmatus,* and killdeer, *Chara-*

drius vociferus, are among the more abundant species encountered. Among the ducks, surface–feeding species, particular pintail, *Anas acuta,* American green–winged teal, *A. carolinensis,* mallard, *A. platyrhynchos,* American widgeon, *Mareca americana,* and shoveler, *Spatula clypeata,* predominate. The American coot, *Fulica americana,* is another common migrant. With the exception of the black brant, *Branta nigricans,* noted for its maritime distribution (Einarsen, 1965), geese are not commonly recorded along the Pacific coast (Bollman et al., 1970; Cowan, 1974).

Utilization of the marshes varies considerably between species; marsh plants, seeds, small marine invertebrates and insects provide an important food resource for some (Reeder, 1951; Recher, 1966), while others use the marshes as loafing and roosting areas when their intertidal feeding grounds are flooded, during periods of inclement weather, or at night. Still a third group of migrants nest in the marshes (Linduska, 1964; Jurek, 1973).

In addition to the endemics and migrants noted above, several other water–associated birds utilize the coastal marshes on a sporadic basis. Large flocks of gulls and terns roost on the marshes. Waders such as the great blue heron, *Ardea herodias,* green heron, *Butorides virescens,* common egret, *Casmerodius albus,* and snowy egret, *Leucophoyx thula,* take crabs, small fish and frogs from tidal creeks and brackish streams. Flocks of upland birds — particularly sparrows, finches, meadowlark, *Sturnella neglecta,* redwing blackbird, *Agelaius phoeniceus,* and swallows — move out onto the marshes to feed on seeds and insects, and raptors include the marshes in their hunting grounds, taking insects, small mammals and birds.

Mammals

Throughout much of California, salt marshes contain a mammalian fauna characteristic of other undisturbed coastal regions (Ingles, 1965; Bradshaw, 1968b; Gustafson, 1968; Dingman, 1971; Barbour et al., 1973). The more abundant species include the Californian meadow mouse, *Microtus californicus,* deer mouse, *Peromyscus maniculatus,* western harvest mouse, *Reithrodontomys megalotis,* and ornate shrew, *Sorex ornatus.* Several species of rabbits are also common — *Sylvilagus audubonii, S. bachmani* and *Lepus californicus.* The long–tailed weasel, *Mustela frenata,* racoon, *Procyon lotor,* and striped skunk, *Mephitis mephitis,* are the most frequently encountered carnivores. Additional species that reside in adjacent maritime or upland habitats (e.g., badger, *Taxidea taxus,* Beechey ground squirrel, *Otospermophilus beecheyi,* bobcat, *Lynx rufus,* Botta pocket gopher, *Thomomys bottae,* coyote, *Canis latrans,* gray fox, *Urocyon cinereoargenteus,* opossum, *Didelphis marsupialis,* and mule deer, *Odocoileus hemionus*) frequently range into the coastal marshes to hunt and feed.

San Francisco Bay area marshes contain two unique species: the salt–marsh harvest mouse, *Reithrodontomys raviventris,* and Suisun shrew, *Sorex sinuosus.* Water–associated mammals like the river otter, beaver and muskrat also occur in the brackish marshes of Suisun Bay (Leach and Fisk, 1972; Gill and Buckmann, 1974). Further north, the marsh shrew, *Sorex bendirii,* Oregon meadow mouse, *Microtus oregoni,* and Townsend meadow mouse, *M. townsendii,* all join the salt–marsh community. The mammalian faunas of salt marshes in Canada and Alaska have not been described. The smaller herbivores probably include hares and several species of voles and lemmings; mink, foxes and lynx are likely predators (Burt and Gorssenheider, 1964; Stonehouse, 1971).

SYNTHESIS

Consultation of a wide range of published sources has confirmed that Pacific North American salt–marsh communities in Alaska, British Columbia and Baja California have received little more than cursory attention. While the marsh communities of Washington, Oregon and California are somewhat better known, only those of San Francisco Bay (38°N) and metropolitan southern California (32—34°N) have been extensively studied. It is also apparent that data on several potentially significant groups of organisms — marsh microbiota (Zobell and Feltham, 1942; Lackey and Clendenning, 1965; Lackey, 1967), estuarine plankton (Kelley, 1966; Riley, 1967; MacGinitie and MacGinitie, 1969), nematodes, annelids and crustaceans, for example — remain woefully inadequate. In spite of these limitations, several conclusions of regional significance emerge from this review:

(1) Both the general organization and the faunal composition of middle latitude (30—50°N) Pacific coast salt–marsh communities are closely similar to those described for such communities in other parts of the world (cf. Dexter, 1947; Paviour–Smith, 1956; Teal, 1962; Green, 1968). Shared genera are common (e.g., the gastropods *Assiminea* sp., *Littorina* sp. and *Melampus* sp., the crustaceans *Orchestia* sp. and *Uca* sp., as well as several insects and vertebrates) and several examples of identical species can be cited (e.g., several foraminifera, the annelids *Capitella capitata* and *Streblospio benedicti,* the gastropod *Phytia myosotis* and several vertebrates).

(2) Only a few of the species encountered in Pacific coast salt marshes appear to be restricted to these habitats. Instead, many of the marine macroinvertebrates represent well–known estuarine genera commonly recorded from subtidal muds (e.g., *Xestoleberis* sp., *Capitella* sp., *Streblospio* sp.), rocky shores (e.g., *Ligia* sp., *Littorina* sp., *Pachygrapsus* sp.) or exposed ocean beaches (e.g., *Orchestia* sp., *Orchestoidea* sp.). These species rarely penetrate into the high marsh. Conversely, species such as the pulmonates *Melampus* sp. and *Phytia* sp. are clearly terrestrial in origin; while common in the high marsh these forms rarely occur in the more frequently submerged low marsh. Among more mobile salt–marsh species, terrestrial forms — the insects, arachnids, birds, reptiles and mammals — appear to be both more numerous and more diverse than mobile marine macro–invertebrates and fish. This general distribution of the salt–marsh fauna: terrestrial species > aquatic species > marsh species, is closely similar to that described by Teal (1962) for the salt–marsh communities of Georgia.

(3) A third community characteristic confirmed for several Pacific coast marsh taxa (e.g., foraminifera, ostracodes, gastropods, amphipods, insects) is the low species diversity—high–density population pattern widely regarded as typical of extreme, highly fluctuating physical environments.

While energy–flow studies (cf. Odum and Smalley, 1959; Teal, 1962; Marples, 1966) remain out of reach, sufficient data are available from the salt marshes of the San Francisco Bay area and southern California to outline some basic trophic relationships.

Several grazing food chains can be recognized.

The rooted marsh plants, dominated by *Spartina foliosa* in the low marsh and *Salicornia virginica* in the high marsh, support a diverse, largely terrestrial community. The rabbits *Sylvilagus* spp. and *Lepus* sp., as well as the California meadow mouse, graze extensively on leaves and shoots, while the western harvest mouse principally takes seeds. Coots, rails, several species of ducks and the deer mouse include salt–marsh plants in their more varied diet. Larval and adult insect herbivores and saprovores are abundant. Insect predators include other insects, spiders, lizards, the ornate shrew, opossum, and a variety of water birds (ducks, shorebirds) and land birds (swallows, sparrows, meadowlarks, redwings, shrikes and crows). Major predators upon the small mammals include the long–tailed weasel, snakes, and several birds — great blue heron, burrowing owl, white-tailed kite. Rabbits and coots are taken by larger raptors such as marsh and red–tailed hawks. Apex carnivores of this terrestrial salt–marsh community once included such wide–ranging species as the badger, grey fox, coyote and bobcat. More recently, however, coastal urbanization has isolated most of the marshes, effectively preventing these larger predators from fulfilling their important community role.

A second grazing food chain, that includes more aquatic species, is supported by benthic algae. Microscopic diatoms, flagellates and blue–green algae flourish on the marsh substrate and provide food for marsh ciliates, nematodes, foraminifera, ostracodes and gastropods. Larger benthic algae such as *Enteromorpha* and *Ulva* are grazed by the gastropods *Batillaria* and *Cerithidea,* as well as several fish (particularly *Atherinops affinis* and *Gillichthys mirabilis*). The larger algae also support a considerable fauna of aquatic insect larvae and amphipods (both groups feeding, in part, on organic detritus) that are eaten by fish and shorebirds. Bottom–feeding fish, several waterbirds and the crab *Pachygrapsus* are known to take the marsh gastropods. The smaller salt–marsh fish fall prey to larger fish, grebes, herons, gulls and terns.

During periods of tidal submergence, phytoplankton support yet a third grazing food chain: Herbivorous zooplankton are taken by carnivorous plankton, which in turn fall prey to *Fundulus* and *Gillichthys,* among others. Plankton are also utilized by the introduced, filter–feeding bivalve *Modiolus demissus*.

Teal (1962) found that detritus food chains accounted for greater community energy flow in Georgia salt marshes than did grazing food chains. *Spartina alterniflora* provided most of the detritus and much of it was exported from the marshes into estuarine waters. In San Francisco Bay, Cameron (1972) confirmed that half the *Spartina foliosa* detritus produced was carried offshore, while another 7% was moved landward, into the *Salicornia* community. From 400 to 600 g dry weight m^{-2} of plant detritus remained in the *Spartina* community year–round, while the dry–weight standing crop of live plants fluctuated from zero in January and February to almost 1800 g m^{-2} in July. Within the *Salicornia virginica* community the accumulated litter (300—600 g dry weight m^{-2} per month) represented an even higher percentage of the standing crop biomass (0—1100 g dry weight m^{-2} per month). Odum and Cruz (1967) have demonstrated that marsh detritus provides an excellent substrate for bacterial growth and promotes development of a florishing microbiota (Zobell and Feltham, 1942; Lackey, 1967). These microorganisms play important roles in the mineralization of bottom sediments, in water chemistry and nutrient recycling, and also provide food sources for larger organisms. Pacific coast salt–marsh organisms at least in part dependent upon detritus food chains include much of the meiofauna — nematodes and ostracodes, for example; a variety of terrestrial and aquatic insects and their larvae; and the majority of marine macroinvertebrates — polychaetes *(Capitella capitata, Streblospio benedicti)*, molluscs and crustaceans *(Hemigrapsus oregonensis, Uca crenulata, Orchestia traskiana)*.

Coastal salt marshes represent an interface between marine and terrestrial communities. As such the marsh communities are subject to continuous variations in composition and dynamics. In addition to changes that accompany tidal ebb and flow (Dexter, 1947), important seasonal variations are evident. The presence of thousands of migrant shorebirds and waterfowl in southern California between October and April, versus their virtual absence during the summer, is an obvious example. Less conspicuous is the change in dominance from insect–herbivores on the marshes during the spring, to insect–saprovores in mid–winter (Cameron, 1972). Additional seasonal changes reflect the reproductive cycles and hibernation behaviour of certain marsh animals (cf. Barbour et al., 1973).

ACKNOWLEDGEMENTS

Exchanges of preprints and unpublished data with J. Bradshaw, R.L. Jefferies, C.A. Jefferson, P.J. Mudie and I.L. Wiggins are gratefully acknowledged. This paper is dedicated to my wife, Barrie, for her continuing support and encouragement, and to E.W. Fager, who first kindled my interest in Pacific coast salt marshes.

REFERENCES

Barbour, M.A., Craig, R.B., Drysdale, F.R. and Ghiselin, M.T., 1973. *Coastal Ecology: Bodega Head.* University of California Press, Berkeley, Calif., 338 pp.

Barnard, J.L., 1962. Benthic marine exploration of Bahia de San Quintin, Baja California, 1960—61. *Pac. Nat.*, 3: 251—274.

Barnard, J.L., 1970. Benthic ecology of Bahia de San Quintin, Baja California. *Smithson. Contrib. Zool.*, 44: 1—60.

Benson, R.H., 1959. Ecology of recent ostracodes of the Todas Santos Bay region, Baja California, Mexico. *Contrib. Univ. Kansas Paleontol., Arthropoda,* Art. 1: 1—80.

Bollman, F.H., Thelin, P.K. and Forester, R.T., 1970. Bimonthly bird counts at selected observation points around San Francisco Bay, February 1964 to January 1966. *Calif. Fish Game*, 56: 224—239.

Bonham–Carter, G.F., 1967. Fortran IV program for Q–mode cluster analysis of non–quantitative data using IBM 7090/7075 computers. *Kansas Geol. Surv. Comput. Contrib.*, 17: 28 pp.

Bradshaw, J.S., 1968a. Environmental parameters and marsh foraminifera. *Limnol. Oceanogr.*, 13: 26—38.

Bradshaw, J.S., 1968b. Report on the biological and ecological relationships in the Los Penasquitos Lagoon and salt marsh area of the Torrey Pines State Reserve. *Calif. State Div. Beaches Parks Contr.*, 4–05094–033: 1—113.

Breckon, G.J. and Barbour, M.G., 1974. Review of North American Pacific Coast beach vegetation. *Madroño,* 22: 333—360.

Burt, W.H. and Grossenheider, R.P., 1964. *A Field Guide to the Mammals.* Houghton Mifflin, Boston, Mass., 2nd ed., 284 pp.

Calder, J.A. and Taylor, R.L., 1968. Flora of the Queen Charlotte Islands: (1) Systematics of the vascular plants. *Can. Dep. Agric. Monogr.*, 4: 1—659.

Cameron, G.N., 1972. Analysis of insect trophic diversity in two salt marsh communities. *Ecology,* 53: 58—73.

Carpelan, L.H., 1969. Physical characteristics of southern California coastal lagoons. In: A.A. Castañares and F.B. Phleger (Editors), *Coastal Lagoons, a Symposium (UNAM—UNESCO)*, Universidad Nacional Autónoma de México, Mexico City, pp. 319—334.

Chapman, V.J., 1960. *Salt Marshes and Salt Deserts of the World*. Hill, London, 392 pp.

Cooper, W.S., 1931. A third expedition to Glacier Bay, Alaska. *Ecology*, 12: 61—95.

Cowan, J.B., 1974. The fascination of migration. *Outdoor Calif.*, 35: 1—5.

Crow, J.H., 1966. *Plant Ecology of the Copper River Delta, Alaska*. Thesis, Washington State University, Seattle, Wash., 120 pp. (unpublished).

Curray, J.R., 1970. Quaternary influence, coast and continental shelf of the western U.S.A. and Mexico. *Quaternaria*, 12: 19—34.

Daetwyler, C.G., 1965. Marine geology of Tomales Bay, central California. *Pac. Mar. Station, Res. Rep.*, 6: 1—169.

Dawson, E.Y., 1962. Benthic marine exploration of Bahia de San Quintin, Baja California, 1960—61: Marine and marsh vegetation. *Pac. Nat.*, 3: 275—280.

Dexter, R.W., 1947. The marine communities of a tidal inlet at Cape Ann, Massachusetts: a study in bio-ecology. *Ecol. Monogr.*, 17: 263—292.

Dingman, R., 1971. Mammals. In: *An Environmental Evaluation of the Bolsa Chica Area*, II. Dillingham Environmental Co., La Jolla, Calif., pp. 1—14.

Doty, M.S., 1946. Critical tide factors that are correlated with the vertical distribution of marine algae and other organisms along the Pacific Coast. *Ecology*, 27: 315—328.

Einarsen, A.S., 1965. *Black Brant, Sea Goose of the Pacific Coast*. University of Washington Press, Seattle, Wash., 142 pp.

Eyerdam, W.J., 1971. Flowering plants found growing between pre- and postearthquake high-tide level lines during the summer of 1965 in Prince William Sound. In: *The Great Alaska Earthquake of 1964: Biology*. National Academy of Sciences, Washington, D.C., pp. 69–81.

Filice, F.P., 1954. An ecology survey of the Castro Creek area in San Pablo Bay. *Wasmann J. Biol.*, 12: 1—24.

Fritz, E.S., 1970. *The Life History of the California Killifish, Fundulus parvipinnis Girard, in Anaheim Bay, California*. Thesis, California State College, Long Beach, Calif., 97 pp. (unpublished).

Fry, D.H., 1973. *Anadromous Fishes of California*. Calif. Dep. Fish and Game, Sacramento, Calif., 111 pp.

Gabriel, B.C. and De la Cruz, A.A., 1974. Species composition standing stock, and net production of a salt-marsh community in Mississippi. *Chesapeake Sci.*, 15: 72—77.

Gill, R. and Buckmann, A.R., 1974. The natural resources of Suisun Marsh, their status and future. *Calif. Dep. Fish Game, Coastal Wetlands Ser.*, 8: 1—122.

Green, J., 1968. *The Biology of Estuarine Animals*. Sidgwick and Jackson, London, 401 pp.

Gross, M.G., 1972. *Oceanography, a View of the Earth*. Prentice-Hall, Englewood Cliffs, N.J., 581 pp.

Gustafson, J.F., 1968. *Ecological Study of Bolinas Lagoon, Marin County, California*. Resources and Ecology Projects, Mill Valley, Calif., 100 pp.

Gustafson, J.F. and Lane, R.S., 1968. An annotated bibliography of literature on salt marsh insects and related arthropods in California. *Pan–Pac. Entomol.*, 44: 327—331.

Hanson, H.C., 1951. Characteristics of some grassland, marsh and other plant communities in western Alaska. *Ecol. Monogr.*, 21: 317—375.

Hanson, H.C., 1953. Vegetation types in northwestern Alaska and comparisons with communities in other Arctic regions. *Ecology*, 34: 111—140.

Hart, J.L., 1973. Pacific fishes of Canada. *Fish. Res., Bull.*, 80: 1—740.

Hartwell, A.D., 1972. Coastal conditions of Arctic Northern Alaska. In: P.V. Sellmann, K.L. Carey, C. Keeler and A.D. Hartwell (Editors), *Terrain and Coastal Conditions on the Arctic Alaskan Coastal Plain*. U.S. Army Corps Eng., Cold Reg. Res. Eng. Lab., Spec. Rep., 165: 32—72.

Heath, W.G., 1969. Comparative osmotic regulation and temperature resistance in several Gulf of California and Puget Sound shallow-water fishes. In: A.A. Castañares and F.B. Phleger (Editors), *Coastal Lagoons, a Symposium (UNAM—UNESCO)*, Universidad Nacional Autónoma de México, Mexico City, pp. 671—678.

Henrickson, J., 1971. Vegetation. In: *An Environmental Evaluation of the Bolsa Chica Area*, V. Dillingham Environmental Co., La Jolla, Calif., pp. 1—64.

Hinde, H.P., 1954. Vertical distribution of salt marsh phanerogams in relation to tide levels. *Ecol. Monogr.*, 24: 209—225.

Hitchcock, C.L., Cronquist, A., Ownbey, M. and Thompson, J.W., 1955—1969. *Vascular Plants of the Pacific Northwest*. University of Washington Press, Seattle, Wash. (5 volumes), 2978 pp.

Hoover, R.F., 1970. *The Vascular Plants of San Luis Obispo County, California*. University of California Press, Berkeley, Calif., 350 pp.

House, H.D., 1914. Vegetation of the Coos Bay Region, Oregon. *Muhlenbergia*, 9: 81—100.

Howell, J.T., 1970. *Marin Flora*. University of California Press, Berkeley, Calif., 2nd ed., 366 pp.

Hultén, E., 1968. *Flora of Alaska and Neighboring Territories*. Stanford University Press, Stanford, Calif., 1008 pp.

Ingles, L.G., 1965. *Mammals of the Pacific States*. Stanford University Press, Stanford, Calif., 2nd ed., 506 pp.

Inman, D.L. and Nordstrom, C.E., 1971. On the tectonic and morphologic classification of coasts. *J. Geol.*, 79: 1—21.

Jefferies, R.L., 1976. Plant communities of muddy shores of Arctic North America. *Arctic* (in press).

Jefferson, C.A., 1973. Salt marsh mapping and description. In: G.J. Akins and C.A. Jefferson, *Coastal Wetlands of Oregon*. Oregon Coastal Conservation and Development Commission, Florence, Ore., pp. 14—108.

Jefferson, C.A., 1974. *Plant Communities and Succession in Oregon Coastal Salt Marshes*. Thesis, Oregon State University, Corvallis, Ore., 192 pp. (unpublished).

Johannessen, C.L., 1961. Some recent changes in the Oregon Coast: Shoreline and vegetation changes in the estuaries. *Dep. Geol., Univ. Ore., Final Rep.*, NR 338–062: 100—138.

Johannessen, C.L., 1964. Marshes prograding in Oregon: Aerial photographs. *Science*, 146: 1575—1578.

Johnson, A.W., Viereck, L.A., Johnson, R.E. and Melchior, H.,

1966. Vegetation and flora. In: N.J. Wilimovsky and J.N. Wolfe (Editors), *Environment of the Cape Thompson Region, Alaska*. U.S. Atomic Energy Commission, Division of Technical Information, Washington, D.C., pp. 277–354.

Jones, A.C., 1962. The biology of the euryhaline fish *Leptocottus armatus armatus* Girard (Cottidae). *Univ. Calif. Publ. Zool.*, 67: 321—365.

Jones, G.N., 1936. *A Botanical Survey of the Olympic Peninsula, Washington*. University of Washington Press, Seattle, Wash., 286 pp.

Jurek, R.M., 1973. *California Shorebird Study*. California Department of Fish and Game, Sacramento, Calif., 277 pp.

Kelley, D.W., 1966. Ecological studies of the Sacramento—San Joaquin estuary. *Calif. Dep. Fish Game, Fish Bull.*, 133: 1—133.

Köppen, W., 1923. *Die Klimate der Erde*. Bornträger, Berlin, 369 pp.

Kozloff, E.N., 1973. *Seashore Life of Puget Sound, the Strait of Georgia, and the San Juan Archipelago*. University of Washington Press, Seattle, Wash., 282 pp.

Kraeuter, J.N. and Wolf, P.L., 1974. The relationship of marine macroinvertebrates to salt marsh plants. In: R.J. Reimold and W.H. Queen (Editors), *Ecology of Halophytes*. Academic Press, New York, N.Y., pp. 449—462.

Lackey, J.B., 1967. The microbiota of estuaries and their roles. In: G.H. Lauff (Editor), *Estuaries. Am. Assoc. Adv. Sci., Publ.*, 83: 291—302.

Lackey, J.B. and Clendenning, K.A., 1965. Ecology of the microbiota of San Diego Bay, California. *Trans. San Diego Soc. Nat. Hist.*, 14: 9—40.

Lane, R.S., 1969. *The Insect Fauna of a Coastal Salt Marsh*. Thesis, California State College, San Francisco, Calif., 78 pp. (unpublished).

Leach, H.R. and Fisk, L.O., 1972. *At the Crossroads: a Report on California's Endangered and Rare Fish and Wildlife*. California Department of Fish and Game, Sacramento, Calif., 99 pp.

Linduska, J.P., 1964. *Waterfowl Tomorrow*. U.S. Department of the Interior, Washington, D.C., 770 pp.

Macdonald, K.B., 1967. *Quantitative Studies of Salt Marsh Mollusc Faunas From the North American Pacific Coast*. Thesis, University of California, San Diego, Calif. (No. 67-12907), 316 pp. (Univ. Microfilms, Ann Harbor, Mich.).

Macdonald, K.B., 1969a. Quantitative studies of salt marsh mollusc faunas from the North American Pacific Coast. *Ecol. Monogr.*, 39: 33—60.

Macdonald, K.B., 1969b. Molluscan faunas of Pacific Coast salt marshes and tidal creeks. *Veliger*, 11: 399—405.

Macdonald, K.B., 1971a. Variations in the physical environment of a coastal slough subject to seasonal closure. In: *Second Coastal and Shallow Water Research Conference*. University of Southern California, Los Angeles, Calif., p. 141 (abstract).

Macdonald, K.B., 1971b. Nutrient cycling in Goleta Slough, California. *Progr. Abstr., Am. Soc. Limnol. Oceanogr. (Pacific Section), San Diego, June 21—25*.

Macdonald, K.B., 1971c. Ecosystem studies in a southern California coastal slough. In: *Second Coastal and Shallow Water Research Conference*. University of Southern California, Los Angeles, Calif., p. 142 (abstract).

Macdonald, K.B., 1971d. The geologic setting of Bolsa Chica. In: *An Environmental Evaluation of the Bolsa Chica Area*, III. Dillingham Environmental Co., La Jolla, Calif., pp. 1—7.

Macdonald, K.B. and Barbour, M.G., 1974. Beach and salt marsh vegetation of the North American Pacific Coast. In: W.H. Queen and R.J. Reimold (Editors), *Ecology of Halophytes*. Academic Press, New York, N.Y., pp. 175—233.

Macdonald, K.B., Henrickson, J., Feldmeth, R., Collier, G., and Dingman, R., 1971. Changes in the ecology of a southern California wetland removed from tidal action. In: *Second Coastal and Shallow Water Research Conference*. University of Southern California, Los Angeles, Calif., p. 144 (abstract).

MacGinitie, G.E., 1935. Ecological aspects of a California marine estuary. *Am. Midl. Nat.*, 16: 629—765.

MacGinitie, G.E. and MacGinitie, N.L., 1969. A report on Mugu Lagoon. *Tabulata*, 2: 15—24.

Mall, R.E., 1969. Soil—water—salt relationships of waterfowl food plants in the Suisun Marsh of California. *Calif. Dep. Fish Game, Wildl. Bull.*, 1: 1—59.

Marples, T.G., 1966. A radionuclide tracer study of arthropod food chains in a *Spartina* salt marsh ecosystem. *Ecology*, 47: 270—277.

Marshall, J.T., 1948. Ecologic races of song sparrows in the San Francisco Bay region. *Condor*, 50: 193—215; 233—256.

Martin, A.C., Zim, H.S. and Nelson, A.L., 1961. *American Wildlife and Plants, a Guide to Wildlife Food Habits*. Dover, New York, N.Y., 500 pp.

McIlwee, W.R., 1970. *San Diego County Coastal Wetlands Inventory: Tijuana Slough*. California Department of Fish and Game, Sacramento, Calif., 62 pp.

Moffitt, J., 1941. Notes on the food of the California clapper rail. *Condor*, 43: 270—273.

Monroe, G.W., 1973. The natural resources of Humboldt Bay. *Calif. Dep. Fish Game, Coastal Wetlands Ser.*, 6: 1—160.

Moore, I., 1964. The Staphylinidae of the marine mud flats of southern California and northwestern Baja California (Coleoptera). *Trans. San Diego Soc. Nat. Hist.*, 13: 269—284.

Mudie, P.J., 1969. A survey of the coastal wetland vegetation of north San Diego County. *Calif. State Resour. Agency, Wildl. Manage. Admin. Rep.*, 70–4: 18 pp. (plus appendices).

Mudie, P.J., 1970. A survey of the coastal wetland vegetation of San Diego Bay. *Calif. Dep. Fish Game Contr.*, W26. D25-51: 79 pp. (plus appendices).

Muenscher, W.C., 1941. *The Flora of Whatcom County, State of Washington*. Published by author, Ithaca, N.Y., 139 pp.

Munz, P.A., 1973. *A California Flora and Supplement*. University of California Press, Berkeley, Calif., 1905 pp.

Neuenschwander, L.F., 1972. *A Phytosociological Study of the Transition Between Salt Marsh and Terrestrial Vegetation of Bahia de San Quintin*. California State College, Los Angeles, Calif., 60 pp. (unpublished).

Noble, E.R., King, R.E. and Jacobs, B.L., 1963. Ecology of

the gill parasites of *Gillichthys mirabilis* Cooper. *Ecology,* 44: 295—305.

Odum, E.P. and Cruz, A.A., 1967. Particulate organic detritus in a Georgia salt marsh—estuary ecosystem. In: G.H. Lauff (Editor), *Estuaries. Am. Assoc. Adv. Sci., Publ.,* 83: 383—388.

Odum, E.P. and Smalley, A.E., 1959. Comparison of population energy flow of a herbivorous and deposit–feeding invertebrate in a salt marsh ecosystem. *Proc. Natl. Acad. Sci., U.S.A.,* 45 : 617—622.

Paviour–Smith, K., 1956. The biotic community of a salt meadow in New Zealand. *Trans. R. Soc. N.Z.,* 83: 525—554.

Peck, M.E., 1961. *A Manual of the Higher Plants of Oregon.* Oregon State University Press, Corvallis, Ore., 2nd ed., 936 pp.

Pestrong, R., 1972. San Francisco Bay Tidelands. *Calif. Geol.,* 25: 27—40.

Peterson, R.T., 1961. *A Field Guide to Western Birds.* Houghton Mifflin, Boston, Mass., 2nd ed., 309 pp.

Phleger, F.B., 1965. Sedimentology of Guerrero Negro Lagoon, Baja California, Mexico. *Papers Colston Res. Soc.,* 17: 205—237.

Phleger, F.B., 1967. Marsh foraminiferal patterns, Pacific Coast of North America. *An. Inst. Biol., Univ. Nac. Autón. Méx., Ser. Cienc. Mar Limnol.,* 38(1): 11—38.

Phleger, F.B., 1969. A modern evaporite deposit in Mexico. *Am. Assoc. Pet. Geol. Bull.,* 53: 824—829.

Phleger, F.B. and Ewing, G.C., 1962. Sedimentology and oceanography of coastal lagoons in Baja California, Mexico. *Geol. Soc. Am. Bull.,* 73: 145—182.

Polunin, N., 1959. *Circumpolar Arctic Flora.* Oxford University Press, London, 542 pp.

Porsild, A.E., 1951. Plant life in the Arctic. *Can. Geogr. J.,* 42: 120—145.

Purer, E.A., 1942. Plant ecology of the coastal salt marshlands of San Diego County, California. *Ecol. Monogr.,* 12: 81—111.

Ranwell, D.S., 1967. World resources of *Spartina townsendii* (sensu lato) and economic use of *Spartina* marshland. *J. Appl. Ecol.,* 4: 239—255.

Ranwell, D.S., 1972. *Ecology of Salt Marshes and Sand Dunes.* Chapman and Hall, London, 258 pp.

Recher, H.F., 1966. Some aspects of the ecology of migrant shorebirds. *Ecology,* 47: 395—407.

Redfield, A.C., 1965. Ontogeny of a salt marsh estuary. *Science,* 147: 50—55.

Redfield, A.C., 1972. Development of a New England salt marsh. *Ecol. Monogr.,* 42: 201—237.

Reeder, W.A., 1951. Stomach analysis of a group of shorebirds. *Condor,* 53: 43—45.

Reimnitz, E., 1972. Effects in the Copper River Delta. In: *The Great Alaska Earthquake of 1964: Oceanography and Coastal Engineering.* National Academy of Sciences, Washington, D.C., pp. 290—302.

Reimnitz, E. and Marshall, N.F., 1965. Effects of the Alaska earthquake and tsunami on Recent deltaic sediments. *J. Geophys. Res.,* 70: 2363—2376.

Reish, D.J. and Barnard, J.L., 1967. The benthic Polychaeta and Amphipoda of Morro Bay, California. *Proc. U.S. Natl. Mus.,* 120: 1—25.

Ricketts, E.F. and Calvin, J., 1952. *Between Pacific Tides.* Stanford University Press, Stanford, Calif., 3rd ed., 502 pp.

Riley, G.A., 1967. The plankton of estuaries. In: G.H. Lauff (Editor), *Estuaries. Am. Assoc. Adv. Sci., Publ.,* 83: 316—326.

Roedel, P.M., 1953. Common ocean fishes of the California coast. *Calif. Dep. Fish Game, Fish Bull.,* 91: 1—184.

Schultz, G.A., 1969. *The Marine Isopod Crustaceans.* Brown, Dubuque, Iowa, 359 pp.

Scott, D.B., Mudie, P.J. and Bradshaw, J.S., 1976. Benthonic foraminifera of three southern California lagoons: modern ecological studies and interpretation of recent stratigraphy. *J. Foraminiferal Res.,* 6: 59—75.

Short, A.D. and Wright, L.D., 1974. Lineaments and coastal geomorphic patterns in the Alaskan Arctic. *Geol. Soc. Am. Bull.,* 85: 931—936.

Shreve, F. and Wiggins, I.L., 1964. *Vegetation and Flora of the Sonoran Desert.* Stanford University Press, Stanford, Calif. (2 volumes), 1669 pp.

Shuster, C.N., 1966. The nature of a tidal marsh. *N. Y. State State Conserv. Dep., Inform. Leafl.,* 8 pp.

Smith, C.F., 1952. *A Flora of Santa Barbara.* Santa Barbara Botanic Garden, Santa Barbara, Calif., 100 pp.

Smith, R.I. and Carlton, J.T., 1975. *Light's Manual: Intertidal Invertebrates of the Central California Coast.* University of California Press, Berkeley, Calif., 3rd ed., 716 pp.

Sokal, R.R. and Sneath, R.H.A., 1963. *Principles of Numerical Taxonomy.* Freeman, San Francisco, Calif., 359 pp.

Spetzman, L.A., 1959. Vegetation of the Arctic Slope of Alaska. *U.S. Geol. Surv., Prof. Pap.,* 302–B: 1—53.

Stebbins, R.C., 1966. *A Field Guide to Western Reptiles and Amphibians.* Houghton Mifflin, Boston, Mass., 279 pp.

Stephens, F.R. and Billings, R.F., 1967. Plant communities of a tide–influenced meadow on Chicagof Island, Alaska. *Northwest Sci.,* 41: 178—183.

Stevenson, R.E. and Emery, K.O., 1958. Marshlands at Newport Bay, California. *Occas. Pap. Allan Hancock Found.,* 20: 1—109.

Stonehouse, B., 1971. *Animals of the Arctic.* Holt, Rinehart and Winston, New York, N.Y., 172 pp.

Teal, J.M., 1962. Energy flow in the salt marsh ecosystem of Georgia. *Ecology,* 43: 614—624.

Teal, J.M. and Teal, M., 1969. *Life and Death of the Salt Marsh.* Little, Brown and Co., Boston, Mass., 278 pp.

Teal, J.M. and Wieser, W., 1966. The distribution and ecology of nematodes in a Georgia salt marsh. *Limnol. Oceanogr.,* 11: 217—223.

Thorsted, T.H., 1972. *The Salt Marsh Vegetation of Bahia de San Quintin, B.C., Mexico.* Thesis, California State College, Los Angeles, Calif., 64 pp. (unpublished).

Twenhofel, W.S., 1952. Recent shoreline changes along the Pacific Coast of Alaska. *Am. J. Sci.,* 250: 523—548.

Udell, H.F., 1969. Productivity and nutrient values of plants growing in the salt marshes of the town of Hempstead, Long Island. *Bull. Torrey Bot. Club,* 96: 42—51.

Upson, J.E., 1949. Late Pleistocene and Recent changes of sea level along the coast of Santa Barbara County, California. *Am. J. Sci.,* 247: 94—115.

Vogl, R.J., 1966. Salt—marsh vegetation of Upper Newport Bay, California. *Ecology*, 47: 80—87.
Warme, J.E., 1971. Paleoecological aspects of a modern coastal lagoon. *Univ. Calif. Publ., Geol. Sci.*, 87: 131 pp.
Whitlatch, R.B., 1974. Studies on the population ecology of the salt marsh gastropod *Batillaria zonalis*. *Veliger*, 17: 47—55.
Wieser, W. and Kanwisher, J., 1961. Ecological and physiological studies on marine nematodes from a small salt marsh near Woods Hole, Massachusetts. *Limnol. Oceanogr.*, 6: 262—270.
Wiggins, I.L. and Thomas, J.H., 1962. *A Flora of the Alaskan Arctic Slope*. University of Toronto Press, Toronto, Ont., 425 pp.
Zobell, C.E. and Feltham, C.B., 1942. The bacterial flora of a marine mud flat as an ecological factor. *Ecology*, 23: 69—78.
Zucca, J.J., 1954. A study of the California clapper rail. *Wasmann J. Biol.*, 12: 135—153.

Additional references

The following related references have been encountered since this chapter was prepared:

Atwater, B.F. and Hedel, C.W., 1976. Distribution of seed plants with respect to tide levels and water salinity in the natural tidal marshes of northern San Francisco Bay estuary, California. *U.S. Geol. Surv., Open File Rep.* 76—389: 41 pp.
Chapman, V.J., 1975. *Mangrove Vegetation*. Cramer, Lehre, 447 pp.
Eilers, H.P., 1975. *Plants, Plant Communities, Net Production and Tide Levels: the Ecological Biogeography of the Nehalem Salt Marshes, Tillamook County, Oregon*. Thesis, Oregon State University, Corvallis, Ore., 368 pp. (unpublished).
Macdonald, K.B., 1976. Coastal salt marsh. In: J. Major and M.G. Barbour (Editors), *Terrestrial Vegetation of California*. Wiley, New York, N.Y., in press.
Mahall, B.E. and Park, R.B., 1976. The ecotone between *Spartina foliosa* Trin. and *Salicornia virginica* L. in salt marshes of northern San Francisco Bay: I. Biomass and productivity; II. Soil water and salinity. *J. Ecol.*, in press.
Mudie, P.J., 1976. Ecological and taxonomic diversity in the modern salt marsh floras of southern California and western Baja California. In: W.W. Wright (Editor), *Plant Diversity in Aquatic Habitats. Second S. Calif. Botanists Symposium, Los Angeles, Calif., 1975*, in press.
Zedler, J.B., 1976. Salt marsh community structure in the Tijuana estuary, California. *Coastal Estuarine Mar. Sci.*, 4(6), in press.

Chapter 9

TIDAL SALT–MARSH AND MANGAL FORMATIONS OF MIDDLE AND SOUTH AMERICA

ROBERT C. WEST

INTRODUCTION

This chapter considers the wet coastal formations in the southern part of the Americas, including the Gulf coast of the United States and the coasts of Middle and South America. These formations occur mainly in depositional environments along low–energy coasts. In general, salt and brackish marshes predominate in the temperate areas, whereas mangrove woodland prevails in the tropical zone. Where temperature conditions are favorable, stunted mangal invades temperate salt marsh in subtropical areas, such as along the Gulf coast of the United States.

SALT–MARSH FORMATIONS

The Gulf of Mexico

Along the northern temperate shores of the Gulf of Mexico salt marshes are best developed on the fringes of the inactive Mississippi River delta and the chenier[1] coastal plain in the state of Louisiana. More than three–fourths of the total salt–marsh acreage along the Gulf coast (and about half of the total salt marsh in the United States) occurs there, in places extending inland for a distance of 40—50 km (Thorne, 1954, p. 197). Westward along the Texas coast only limited areas of salt marsh are found on the shores of bays enclosed by offshore bars. Eastward along the coasts of Mississippi, Alabama, and northern Florida areas of salt marsh are also small and disjunct, being limited to low alluvial pockets along protected bay shores. In both the northwestern and northeastern sectors of the Gulf coast, Pleistocene surfaces, 1.5—4.5 m above sea level, approach the shoreline, severely limiting the areas subject to tidal inundation. Moreover, the low tidal range within the Gulf of Mexico (45—60 cm) restricts the extent of salt marsh, except in the low deltaic plain of the Mississippi and portions of the west coast of Florida, which are barely above sea level. In southwestern Florida, however, salt marsh competes with mangrove swamp.

In terms of floristic composition and succession the salt–marsh formations of the northern Gulf coast are closely related to those of the southern Atlantic coast of the United States, with *Spartina, Juncus,* and *Distichlis* as dominants. In a recent survey, Chabreck (1972) indicated that *Spartina alterniflora* dominates the salt marsh that borders the Mississippi deltaic plain in areas where mean water salinity ranges from 14 to 19‰ (parts per thousand). Associates include *Distichlis spicata, Juncus roemerianus,* and other salt–tolerant grasses and sedges, as well as the shrub *Borrichia frutescens.* Near the coast patches of *Batis maritima* and *Salicornia* spp. frequently occupy small areas of highly saline soil. Landward from the salt marsh, in brackish–water areas (average water salinities between 7.5 and 11.5‰), the major species is *Spartina patens* with *Distichlis spicata, Juncus roemerianus,* and *Scirpus olneyi* as principal associates. In both salt and brackish marshes the shrubs *Iva frutescens* and *Baccharis halimifolia* occupy high ground, such as natural levees of tidal and fresh–water channels, beach ridges, and spoil banks along artificial channels. Behind the brackish sections lie extensive areas of fresh–water marsh

---

[1] Cheniers are wooded sandy ridges separating the depressions where the marshes occur.

composed of many grasses, sedges, and other herbs, among which various species of *Panicum, Eleocharis,* and *Sagittaria* dominate. Similar plant assemblages for the Louisiana marsh have been described by Penfound and Hathaway (1938) and O'Neil (1949) and were recently mapped in detail by Chabreck et al. (1968).

Stands of black rush, *Juncus roemerianus,* usually bordered on the seaward side by a narrow fringe of *Spartina alterniflora,* dominate the small, disjunct salt marshes of coastal Mississippi, Alabama, and northern Florida (Mohr, 1901; Kurz and Wagner, 1957; Eleuterius, 1972). A similar association continues along the coast of Florida as a narrow fringe, but south of Tampa Bay mangrove becomes dominant. According to Davis (1940, 1943), large patches of open salt marsh occur within and along the inland margins of the tall mangrove woodland in extreme southwestern Florida. Such marshes are dominated by *Spartina alterniflora, S. cynosuroides, Juncus roemerianus, Sporobolus virginicus,* and *Monanthochloe littoralis. Spartina alterniflora* grows best along the margins of tidal channels and may function as a pioneer plant, being effective in holding soil until mangrove plants, such as *Rhizophora,* can establish themselves (Davis, 1940, p. 344; Egler, 1948). The pioneer function of *Spartina* in mangrove communities has often been observed in the American tropics and will be considered in more detail further on (cf. 197). Once established, the mangrove woodland appears to shade out most of the salt-marsh grasses and sedges, although in Florida some persist as understory in mangrove forests composed of *Avicennia* (Thorne, 1954, p. 197).

Stands of *Distichlis spicata,* in places mixed with *Spartina alterniflora* and *Batis maritima,* characterize the salt marshes of southwestern Louisiana and continue into eastern Texas (Chabreck, 1972). Farther west, *Spartina alterniflora, Sporobolus virginicus, Salicornia perennis,* and *Batis maritima* become dominant in the small marshes that border the partially enclosed bays (Brown et al., 1971). Grassy salt marsh practically disappears around the margins of the Laguna Madre of southern Texas, due to hypersaline conditions (more than 40‰), caused by decreasing precipitation, increasing temperature and the closed nature of the lagoon (Hoese, 1967; Cooper, 1974). Low growths of *Salicornia* and *Batis* in places comprise the only vascular plants, and large areas are covered by algal mats. The latter are dominated by the blue-green alga *Microcoleus chthonoplastes,* and in extremely saline water within the lagoon a red alga, *Laurencia,* abounds, giving the water a rust–red color (Fisk, 1959, p. 115). Excepting the area immediately north of the Rio Grande delta, where stands of *Spartina spartineae* cover non–tidal saline soils (Clover, 1937), the hypersaline conditions of south Texas lagoons continue into those of northeastern Mexico (Ayala–Castañares, 1969). For example, on the shores of the Laguna Madre de Tamaulipas large tidal flats are completely bare of vegetation, but in places patches of *Batis maritima, Salicornia ambigua, Distichlis spicata,* and *Monanthochloe littoralis* surround algal flats of *Microcoleus chthonoplastes* (Hildebrand, 1957; Ayala–Castañares and Segura, 1968). In better drained areas on the landward side of the lagoons, *Spartina spartineae* forms dense growths on heavy clay soils.

Northwestern Mexico

Numerous salt marshes and tidal flats occur in deeply indented bays along the arid coasts of Baja California and Sonora, but few of these have been investigated (cf. pp. 176—177). Johnston (1924, p. 964) reported many salt marshes and flats along both shores of the Gulf of California, where *Salicornia pacifica, Allenrolfea occidentalis, Batis maritima* and *Monanthochloe littoralis,* dominate the tidal zones and the shrubs *Suaeda, Maytenus,* and *Frankenia* occupy higher, less saline areas. On the extensive mud flats of the Colorado River delta at the head of the Gulf, where the mean tidal range exceeds 7 m, vegetation is virtually non–existent. Scattered plants, such as *Distichlis* and *Salicornia,* are restricted to beach ridges and banks of tidal channels where tide waters are readily drained (Thompson, 1968, p. 107).

Temperate South America

In general the tidal marshes of temperate South America are inadequately known, but those of Argentina have received the most study. The dominance of cliffed coasts, extensive sand beaches, and dune fields along the Atlantic side from southern Brazil to Tierra del Fuego has limited tidal marshes to relatively small deltaic plains and river mouths. On the Pacific side tidal marshes are confined to small disjunct inlets along the mountainous high–energy coast of southern and central Chile, and the paucity of marshes along the arid coasts of northern Chile and Peru may result in part from the absence of lagoons and river deltas.

The floristic composition of the salt marshes is best known for the Atlantic side. South American species of *Spartina, Distichlis, Juncus,* and *Salicornia* dominate the tidal marshes of Uruguay and Argentina north of approximately 44°S (Ringuelet, 1938; Chebataroff, 1953). South thereof, in small disjunct marshes along the coast of Patagonia, low salt–tolerant shrubs (including the chenopods *Salicornia, Atriplex,* and *Suaeda,* the composites *Lepidophyllum* and *Statice,* and a species of *Frankenia*) become dominant in the formation, as *Spartina* and other salt grasses rapidly decline in abundance (Hauman, 1926).

Chebataroff (1952, 1953) has described the tidal marshes of Uruguay. These are limited mainly to the area east of Punta del Tigre near Montevideo, for westward thereof the shore is bathed by fresh and brackish waters of the La Plata Estuary, which receives the large discharge of the Paraná drainage basin. The Uruguayan marshes are confined chiefly to the lower stream courses, coastal lagoons, and inlets that receive tidal inflow. The shores of quiet muddy inlets washed by sea water carry an almost solid cover of *Spartina brasiliensis* (= *Spartina maritima* var. *brasiliensis*), in places accompanied by *Juncus acutus, Salicornia fruticosa, Chenopodium macrospermum,* and *Apium sellowianum*; *Spartina montevidensis* grows in the less saline spots. The country's largest tidal marsh, however, occurs along the banks of the lower Santa Lucía River, immediately west of Montevideo. There, the sedges *Scirpus californicus, S. olneyi,* and *Cyperus corymbosus* dominate the outer portions of the marsh, flooded daily by brackish river water. Farther inland on slightly higher ground, evaporation and capillarity, especially during the low–water period of summer, cause an increase in soil salinity. Thus, the vegetation of the inner marsh of the Santa Lucía is dominated by the halophytes listed above for the inlets, forming a zonation pattern that is reversed from the normal.

A similar "reversal" of marsh zonation occurs along the coast immediately south of Buenos Aires in Argentina. There, the shore is washed by the brackish water of the La Plata Estuary, and the low muddy areas are fringed by a narrow marsh formation, 100—300 m wide, briefly described by Parodi (1940) and shown in Fig. 9.1. An outer zone subject to daily tides is composed of fresh- and brackish–water grasses and sedges, including *Paspalum vaginatum, Scirpus riparius,* and *Eleocharis palustris.* Farther inland a narrow band of halophytes occupies slightly higher ground that only occasionally is inundated by tide, but whose substrate is markedly saline because of capillarity. Both *Spartina montevidensis* and *S. brasiliensis* are dominants, the latter in the wetter portions. Associates include low growths of *Salicornia gaudichaudiana, Atriplex hastata, Sesuvium portulacastrum* and *Apium sellowianum.*

In Argentina the largest tidal marshes occur along the muddy shores of Samborombón Bay (some 200 km south of Buenos Aires), where the Salado River has deposited quantities of fine sediment; and in the vicinity of Bahía Blanca, the

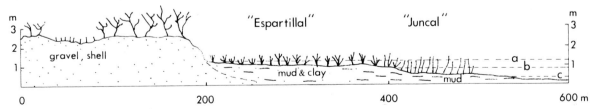

Fig. 9.1. Profile near Punta el Indio, La Plata Estuary (after Parodi, 1940), showing zonation of marsh vegetation. The *"juncal"* association is composed mainly of *Scirpus riparius, Eleocharis palustris, Paspalum vaginatum,* and *Cotula coronopifolia.* The *"espartillal"* association: *Spartina brasiliensis, S. montevidensis, Apium sellowianum, Atriplex hastata,* and *Salicornia gaudichaudiana. a* = mean spring tide; *b* = mean high tide; *c* = mean neap tide. Scales are approximate.

Colorado River delta, and San Blas Bay at the southern margin of the Pampas. Ringuelet (1938) has described in detail the northern part of the Samborombón marshes and Molfino (1921) those of Bahía Blanca. Both authors indicated that *Spartina brasiliensis* (called *S. alterniflora* by Molfino and *S. densiflora* by Cabrera, 1970, II, p. 387) forms dense stands along the sea margin, inundated twice daily by lunar tides (range 1 m or less), and acts as a pioneer species on mud flats. Landward a narrow band of *Spartina montevidensis*, associated with *Distichlis spicata, Scirpus maritimus, Juncus acutus,* and several species of salt–tolerant chenopods and other shrubs, occupies slightly higher elevations. Ringuelet specified the chenopods *Salicornia gaudichaudiana* and *Suaeda fruticosa* for the Samborombón area; but for Bahía Blanca, Molfino listed many more, including *Salicornia fruticosa, Suaeda fruticosa, Spirostachys (= Heterostachys) olivaceus, S. ritteriana,* and *Halopeplis patagonica,* suggesting an increase of salt–marsh shrubs poleward.

A high sea cliff extends along the coast of Patagonia (southern Argentina) from the Río Negro southward for 1500 km to the eastern entrance to the Strait of Magellan. In places the cliff is breached by streams flowing to the Atlantic, and at their mouths are found small tidal marshes. These have been described briefly by Hauman (1926, pp. 108—119), who indicated the increasing dominance of salt shrubs at the expense of grasses in the marsh formation as one goes poleward. For instance, *Spartina montevidensis* was recorded no farther south than Puerto Madryn (43°50′S) and *Spartina patagonica* (= *S. densiflora* subvar. *patagonica,* according to Saint–Yves, 1934) was poorly developed along the seaward margins of the more southern marshes. The salt shrubs as recorded by Hauman are given in Table 9.1. In the lower Gallegos River marsh (51°33′S) dense mats of chenopods *(Salicornia* sp., *Suaeda patagonica, Atriplex macrostyla),* with scattered bunches of *Spartina patagonica,* dominate the lower areas twice daily inundated by tides, while stands of the shrubs *Lepidophyllum cupressiforme* (Compositae), *Statice brasiliensis* (Plumbagineae) and *Frankenia microphylla* occupy higher land covered occasionally by the highest spring tides. Along this coast the large tidal range (10—15 m) causes the salt marshes to extend up the entrenched river valleys several kilometers from the coast.

Although salinity data are lacking, the present writer suggests that aridity along the Patagonian coast gives rise to hypersaline conditions on river-mouth mud flats that are only occasionally inundated by tides. High soil salinity on the flats may limit the extent of salt grasses, such as *Spartina,* and favor the growth of salt shrubs, such as *Lepidophyllum,* which also thrives on interior salt flats, 200 km from the coast (Hauman, 1926).

Few data on the tidal marshes of the Pacific coast of temperate South America are recorded in the literature. Salt shrubs, especially *Salicornia,* appear to dominate the small tidal marshes of southern Chile (Skottsberg, 1916, p. 42). Along the coast of central Chile, *Spartina densiflora* var.

TABLE 9.1

Occurrence of salt shrubs in five tidal marshes of Patagonia (compiled from Hauman, 1926)

Genus	Puerto Madryn (42°50′S)	Puerto Pirámides (43°40′S)	Comodoro Rivadavia (45°50′S)	San Julián (49°14′S)	Río Gallegos (51°33′S)
Salicornia	*S. corticosa*	*S. corticosa*	*S. fruticosa*	*S. corticosa*	*S.* sp.
Atriplex	*A. montevidense* *A. lampa*	*A. lampa*	*A. lampa*	*A. lampa* *A. sagittifolia*	*A. macrostyla*
Suaeda	*S. fruticosa* *S. maritima*	*S. maritima*	*S. fruticosa* *S. patagonica*	*S. fruticosa*	*S. patagonica*
Chenopodium					*C. rubrum*
Frankenia	*F. patagonica*	*F. patagonica* *F. microphylla*	*F. patagonica*		*F. microphylla*
Statice		*S. brasiliensis*		*S. brasiliensis*	*S.* sp.
Lepidophyllum				*L. cupressiforme*	*L. cupressiforme*
Polygonum					*P. maritimum*

typica has been reported for the tidal marshes near Puerto Montt and Corral and *S. densiflora* var. *patagonica* for those near Valdivia (Saint–Yves, 1934).

Tropical marsh formations

Salt–marsh formations in tropical latitudes are usually thought to be limited in extent because of competition with mangrove species (Chapman, 1960; Davies, 1972). In the American tropics these formations are probably more extensive than previously indicated. They occur under at least three different environmental situations, usually on the margins or within mangrove woodland:

(1) as a pioneer formation, colonizing recently formed mud flats along the open coast, the shores of estuaries or tidal channels fringing mangrove woodland;

(2) as a halophytic community occupying saline soils on the inner edge or within the mangrove woodland; and

(3) as a secondary formation on disturbed areas within the mangrove woodland, such as cutover or degraded sections.

The first environmental situation is found in many places on the Atlantic coast from Florida to southern Brazil, but is rare on the Pacific. The situation is dependent on the presence of extensive tidal mud flats exposed at least once daily and on the presence of *Spartina*, the main pioneer marsh species. The pioneer role of *Spartina alterniflora* in forming narrow bands of marsh along the margins of tidal channels in front of the mangrove woodland in southern Florida was described by Davis (1940) and Egler (1948), as mentioned earlier in this paper. Vázquez (1971) and Thom (1967) reported a similar example in the mangrove swamps of southern Veracruz and Tabasco, Mexico, respectively. Several investigators have commented on the role of *Spartina brasiliensis* as a colonizer of mud flats fronting mangrove woods along the coasts of the Guianas and Brazil. Martyn (1934) mentioned the extensive root system of *Spartina* as a device for catching fine mud particles during tidal ebb and flow, resulting in gradual rise of mud banks by deposition and subsequent trapping and growth of mangrove seedlings in the marsh area along the coast of British Guiana. This successional phenomenon was emphasized by Freyburg (1930, p. 109) in his study of the coast of northeastern Brazil (Fig. 9.2). He denied the importance of mangrove root systems as initial land builders along estuarine shores, but stressed the role of *Spartina brasiliensis* as a colonizer of mud flats, and the electrolytic precipitation of suspended material in river water on contact with salt water as the main depositional agent in estuaries. Both authors indicated that mangal eventually invades the *Spartina* marsh, and Martyn pointed out that mangrove trees eventually shade out the marsh grass, which is an obligate heliophyte. Other investigators who have reported the presence of pioneer *Spartina* marshes in front of mangrove woodland include Lindeman (1953, pp. 52, 107) for Surinam, Lamberti (1969, p. 50) and Luederwaldt (1919, p. 327) for São Paulo state, Brazil, Stellfeld (1949, p. 320) and Bigarella (1946) for Paraná state, Brazil, and Souza Sobrinho and Klein (1969) for Santa Catarina state, Brazil.

The second environmental situation is the most common for tidal marshes within the tropics. Along tropical coasts having a well–defined dry season, the slightly higher areas within or on the

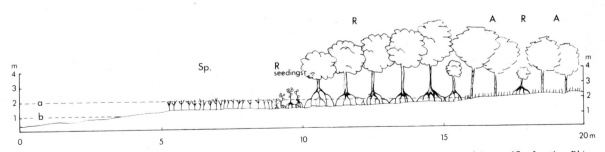

Fig. 9.2. Profile of a typical estuary in northeastern Brazil (after Freyburg, 1930), showing *Spartina* marsh on mudflat fronting *Rhizophora* forest. Sp = *Spartina brasiliensis*; R = *Rhizophora mangle*; A = *Avicennia*. a = mean high tide; b = mean low tide. Scales are approximate.

landward margins of the tidal mangal often have soil salinities greater than those of the mangrove woodland. Such areas are usually inundated only by spring tides, and the long dry season and high rate of evaporation induces strong capillarity and the concentration of salts near the surface. In the less saline spots, along the Brazilian coast for example, *Spartina brasiliensis, Sporobolus virginicus, Paspalum vaginatum, Scirpus maritima, Sesuvium portulacastrum, Batis maritima,* and *Iresine portulacoides* dominate the formation (Dansereau, 1947; Souza Sobrinho and Klein, 1969; Lamberti, 1969). Similar salt and brackish marshes back of and within mangrove woodland are found along the Gulf coast of Mexico (Vázquez, 1971), Yucatán (Miranda, 1959, p. 249) and Belize (Romney, 1959). On Ambergris Cay, Belize, the present writer visited several interior marshes, partially surrounded by mangal, where *Spartina* sp. and *Distichlis spicata* are dominant, but in the most saline spots algal mats are bordered by dense growths of *Salicornia* sp. (Fig. 9.3). In still drier areas, as on Puná Island in the Gulf of Guayaquil, Ecuador, Fosberg (1961) reported vegetation-free zones between mangrove-fringed tidal channels and suggested that hypersaline conditions were responsible. The present writer has observed similar situations on the southern shore of the Gulf of Fonseca, Pacific coast of Central America, where the tidal range exceeds 4 m and the dry season lasts six months.

Dansereau (1947, 1950) has described Brazilian examples of the third environmental situation, which involves the invasion of cutover or degraded mangal by marsh grasses. Large areas of mangrove woodland that once surrounded the shores of Guanabara Bay near Rio de Janeiro have been destroyed by cutting, and are now covered by dense stands of *Spartina brasiliensis*. Elsewhere in the New World tropics cutover mangrove forest is usually invaded by the pan-tropical fern *Acrostichum aureum*. A case of marsh succession in degraded mangal was reported by Boyé (1962) along the coast of French Guiana, where the tidal forest,

Fig. 9.3. Salt marsh, northwestern side of Ambergris Cay, Belize. *Distichlis spicata* occupies the center of the swale and is bordered by *Spartina spartineae*. Low trees of *Avicennia germinans* in background.

cut off from daily inundation of sea water, is slowly dying through soil compaction and increasing acidity, and is being replaced by stands of *Acrostichum aureum* and marsh grasses, such as *Paspalum vaginatum*. Børgesen (1909, p. 213) described a similar case for the Virgin Islands, West Indies.

MANGAL FORMATIONS

Floristic composition

The New World mangal formations are composed chiefly of four species of salt–tolerant trees: *Rhizophora mangle, Avicennia germinans, Laguncularia racemosa*, and *Conocarpus erectus*. All four also occur in West Africa. Although these species are dominant, in some areas others are recognized as part of the formation. Two additional species of *Rhizophora* are found in the New World: *R. harrisonii*, on both the Pacific and Atlantic coasts of Central and South America; and *R. racemosa*, confined to eastern Venezuela, the Guianas, and the Amazon mouth (Leechman, 1918; Hou, 1960; Pires, 1964; Breteler, 1969). Both these, as well as *R. mangle*, are also along the coast of West Africa (Keay, 1953). Salvoza (1936) and Cuatrecasas (1958a) claimed two other species, *R. brevistyla* and *R. samoensis*, for the Pacific coast of South and Central America, but the former is now regarded as a variety of *R. harrisonii*, and the latter a variety of *R. mangle* (Hou, 1960). According to Cuatrecasas (1958a, p. 94), *R. mangle* does not occur on the Pacific coast of South America. The New World species of *Rhizophora* have often been confused, and the distribution as given in Fig. 9.4 may be subject to correction (cf. Chapman, 1975).

According to Moldenke (1960), besides the dominant *Avicennia germinans*, three additional species of this genus occur in the New World mangal formations: *A. schaueriana*, found in the Lesser Antilles, the Guianas and Brazil; *A. bicolor*, on the Pacific coast of central America; and *A. tonduzii*, confined to the coasts of Costa Rica, Panama and northwestern Colombia. Another frequent associate of the mangal formations of the Pacific coast of Central and South America is the monospecific *Pelliciera rhizophorae* (Theaceae), an endemic species. The distribution of these species is shown in Fig. 9.5.

In terms of floristic composition several subgroups of the New World mangal formations can be recognized, as Chapman indicated in the introductory chapter of this volume (cf. p. 16). The richest of these in species are: (1) the Pacific coast of Central and South America, with two species of *Rhizophora*, three of *Avicennia*, one each of *Laguncularia* and *Conocarpus*, and the unique *Pelliciera rhizophorae;* and (2) the Atlantic coast of South America, including the Lesser Antilles, with three species of *Rhizophora*, two of *Avicennia*, and one each of *Laguncularia* and *Conocarpus*.

Salinity and shade reduce the number of plants that can be considered associates of the dominant mangal tree species. There is practically no ground cover, except in lightly shaded spots, where *Batis maritima* and low *Crinum* spp. may form an uneven carpet of green. Woody vines, in particular *Pavonia* spp. and *Rhabdadenia biflora*, are widespread in most of the New World mangal forests on both the Atlantic and Pacific sides, as are the leguminous shrubs *Dalbergia* spp. and *Machaerium lunatum*. On the landward side of the mangal forests, however, a great variety of herbs, shrubs, and trees form a transition between the saline and freshwater vegetation, but these cannot be considered a part of the mangal formations.

Latitudinal limits

Avicennia germinans, the mangrove species most tolerant of low temperatures, usually marks the poleward extent of mangal formations in the Americas, as it does elsewhere in the world (cf. Chapman, 1975). On the poleward margins the growth is shrubby, in most places mixed with salt–marsh plants. On the Atlantic coast of North America, it grows no farther north than St. Augustine, Florida (29°53′N), but with *Rhizophora* and other mangrove species it reaches the Bermuda Islands (32°20′N) in the north Atlantic. On the Gulf coast of the United States *Avicennia* occasionally forms dense thickets along the outer portion of the salt marsh in Lousiana (between 29° and 30°N), but periodically it is killed back by several days of continuous freezing temperatures in winter. For instance, during the winters of 1961—62 and 1962—63 the *Avicennia* thickets on the Louisiana coast were nearly exterminated by

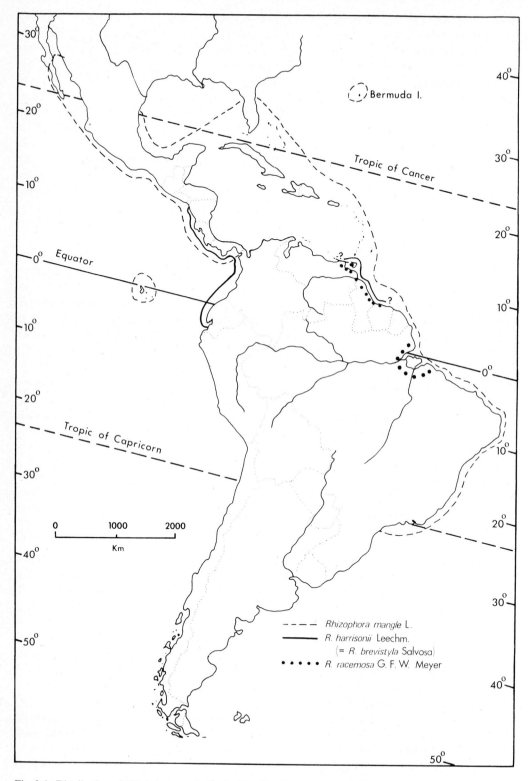

Fig. 9.4. Distribution of *Rhizophora* species in the New World. Based on Leechman (1918), Jonker (1959), Hou (1960) and Breteler (1969).

MIDDLE AND SOUTH AMERICA

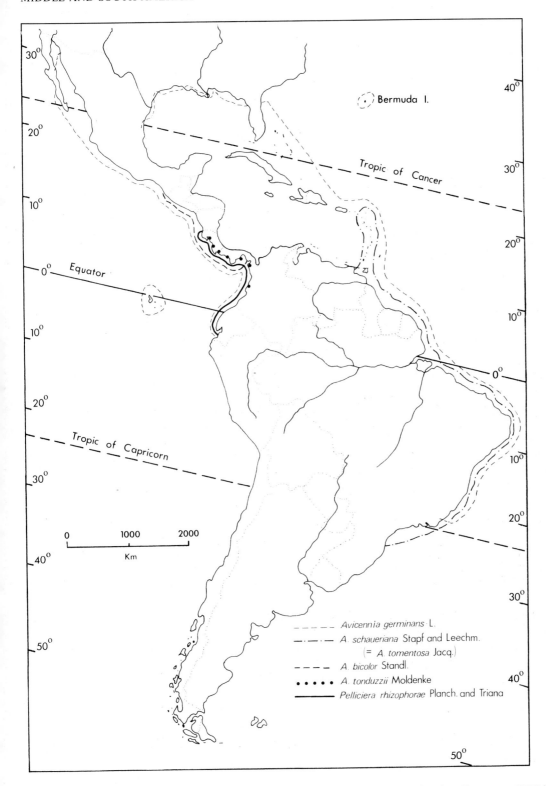

Fig. 9.5. Distribution of *Avicennia* species and *Pelliciera rhizophorae* in the New World. Based on Cuatrecasas (1958a) and Moldenke (1960).

hard freezes (−3 to −11°C). The absence of exceptionally cold winters along the coast since 1963 has permitted the plant to expand into many parts of the salt marsh at the expense of *Spartina alterniflora,* which it soon overshades; in 1974 some *Avicennia* shrubs in the Louisiana marsh measured 4 m high.

Avicennia shrubs have been reported only at a few places along the Texas coast, such as Harbor Island, near Corpus Christi (McMillan, 1971) and Clark Island in the Rio Grande delta near the Mexican border (Clover, 1937). In northeastern Mexico, scrubby *Avicennia* is found in the southern end of the Laguna Madre, but the northernmost occurrence of well-developed *Rhizophora* and *Laguncularia* woodland is around the margin of the Tamiahua Lagoon, 21°40′N (Sauer, 1967; Ayala–Castañares, 1969). The paucity of mangal formations along the northwestern Gulf coast stems largely from the freezing temperatures caused by the frequent invasion of cold polar air masses during the winter.

Along the Gulf coast of Florida the northern limit of dwarfed *Rhizophora* is Cedar Keys (29°N) (Graham, 1964). The association of all four New World mangal species begins to fringe the coast and lee sides of offshore reefs north of Tampa Bay and culminates in the magnificent tall forests of the Ten Thousand Islands district in southeastern Florida (25°45′N) which Davis (1940) has so well described. This forest, with trees reaching up to 30 m in height, represents the most poleward occurrence of a tall, mature *Rhizophoretum manglae* association in the New World. Its development is closely associated with the peculiar hydrographic and soil characteristics of the area, but more with the tropical climate engendered in part by the warm Gulf water at that latitude.

On the Atlantic side of South America, mangal formations extend to Santa Catarina state in southern Brazil. Low stands of both *Avicennia schaueriana* and *Rhizophora mangle* terminate at Florianopolis (27°30′S), but *Laguncularia racemosa* extends southward to the poleward limit of mangrove at the mouth of the Aranânguá River (29°S) (Lamberti, 1969; Souza Sobrinho and Klein, 1919).

On the Pacific side of the Americas the northern limit of *Avicennia* is near Puerto de Lobos (30°15′N) on the hot desert coast of Sonora, northwestern Mexico (personal observation). However, on the cool, foggy Pacific side of Baja California stunted forms of both *Rhizophora* and *Laguncularia* mark the poleward outpost of mangal at Ballenas Bay, some 320 km north of the nearest *Avicennia* recorded along that coast (Hastings et al., 1972). The southern limit of mangal formations (*Rhizophora* and *Laguncularia*) on the Pacific occurs at Punta Malpelo, Tumbes, near the Peruvian—Ecuadorian border at only 3°40′S of the equator (Peña, 1971). The absence of mangal as well as the paucity of salt marsh along the coasts of Peru and northern Chile has been attributed to extreme aridity and cool air temperatures generated by the Humboldt Current offshore (see Chapman, 1975). A more plausible explanation might be the lack of quiet bays and lagoons free from wave action and absence of river deltas with fine sediment for the development of tidal channels.

Zonation and succession

A characteristic of mangal formations that investigators have emphasized for both the Old and New World tropics is the banded or zonal arrangement of associations inland from the sea or tidal channel margins (Davis, 1940; Chapman, 1944; MacNae, 1968). Such zones are thought to be caused by a complex of factors, among which salinity gradients, tidal exposure, and substrate characteristics may dominate. The zones are also usually recognized as representing a succession. However, the zonal arrangement of mangal communities in the New World appear to be less pronounced than in the Old, possibly because of the lesser number of mangrove species in the former as compared with the latter area. Moreover, for both areas there exist few detailed studies entailing careful measurement of the vegetation and the environmental parameters that may explain the apparent zonation. Examples of such field studies for the Americas include Davis' (1940) monograph on the Florida mangals; Chapman's (1944) work on salinity and soil in the mangals of Jamaica; Thom's (1967) study of the influence of landform and substrate characteristics on the mangal formations of Tabasco, Mexico; and Lamberti's (1969) ecological analysis of the São Paulo mangals, Brazil.

Most general statements on mangal zonation in the New World place *Rhizophoretum manglae*

MIDDLE AND SOUTH AMERICA

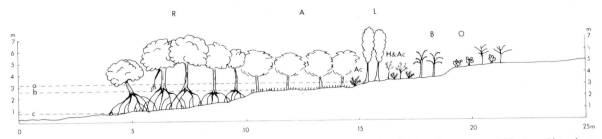

Fig. 9.6. Profile of mangal associations, Guanabara Bay, near Rio de Janeiro, Brazil (after Dansereau, 1947). R = *Rhizophora*; A = *Avicennia*; L = *Laguncularia*; Ac = *Acrostichum*; H = *Hibiscus tiliaceus*; B = *Bactris*; O = *Opuntia*. a = mean spring tide; b = mean high tide; c = mean low tide. Scales are approximate.

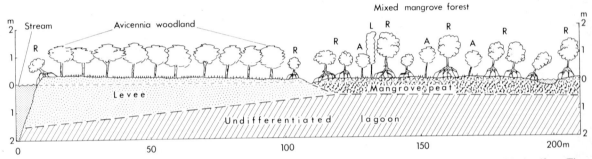

Fig. 9.7. Profile of mangal associations in inactive levee and interdistributary habitats, Grijalva delta, Tabasco, Mexico (from Thom, 1967, fig. 12). R = *Rhizophora*; A = *Avicennia*; L = *Laguncularia*.

as the pioneer association on the outer margins of lagoons and tidal channels. Ideally, this is followed inland by successive belts of the *Avicennietum*, *Laguncularietum racemosae*, and, in places above tidal influence, *Conocarpetum erectae* associations. In his report on the zonation of mangal formations in Guanabara Bay, near Rio de Janeiro, Dansereau (1947) described such a sequence, shown in Fig. 9.6. However, several investigators have revealed so many exceptions to this simple zonation that its utility as a model is doubtful.

Deltaic—lagoonal environments

Thom (1967) demonstrated that in dynamic deltaic—lagoonal environments, such as that along the Tabasco—Campeche coast of southeastern Mexico, a simple zonation of mangal communities does not prevail, and that there is a definite association of mangrove species with landform and substrate type (Fig. 9.7). Thus, the complexity of mangrove species distribution in a river delta reflects the active sedimentation and frequent shifts of distributary channels characteristic of such environments. In the Tabasco area, solid stands of *Avicennia germinans*, for example, although found on mud flats, occur chiefly on subsiding levees of abandoned distributaries (Fig. 9.8). The soils are composed of stiff clay, usually oxidized at depth. *Rhizophora*, on the other hand, thrives best on moist sites in deep, chemically reduced organic muck and peat, and usually lines the stable shores of lagoons and the edges of abandoned distributary channels where its prop roots aid in the slow deposition of fine alluvium and where salinities are higher than elsewhere in the delta (Fig. 9.9). In slowly subsiding interdistributary basins within the delta, however, a mixed forest of *Rhizophora mangle* and *Laguncularia racemosa* prevails. The latter species may form solid stands on sandy soils near the landward limit of tidal influence (Fig. 9.10), but *Conocarpus erectus* is relatively rare in the entire area. According to Thom, a definite pattern of succession is questionable, except on a short–term basis. "Long–range trends are dictated by physiographic processes in this deltaic plain, processes which are continually changing and which influence such phenomena as degree of water, soil type, and drainage of the surface." (1967, p. 340).

Fig. 9.8. *Avicennia germinans* forest, on inactive natural levee, Grijalva delta, Tabasco, Mexico. The pencil–like pneumatophores of this species occupy the ground between trees. Occasional specimens of *Rhizophora mangle* (middle background) occur among the dominant *Avicennia*.

Fig. 9.10. Thicket of young *Laguncularia racemosa* along banks of the brackish Verde River, inactive western edge of the Grijalva delta, Tabasco, Mexico.

Fig. 9.9. *Rhizophora mangle* fringe on tidal channel, Laguna de Carmen, Tabasco, Mexico, showing the characteristic tangle of prop roots of this species. The fringe is only three or four trees in width and is backed by *Avicennia* woodland on clay soils.

Fig. 9.11. Fluted buttresses characteristic of *Pelliciera rhizophorae*, in mangal forest north of San Juan River delta, Pacific coast of Columbia.

MIDDLE AND SOUTH AMERICA

An apparent absence of the classical zonation of mangrove species has also been observed in other deltaic environments, such as the Pacific coast of Colombia (West, 1956), the Atrato delta of northern Colombia (Vann, 1959a), and the Orinoco delta (Perales, 1952). Along the Pacific coast of Colombia, *Avicennia* may serve as the pioneer species on recently deposited mud flats along the strand more often than *Rhizophora;* but the latter species, almost without exception, occupies the outer fringe of abandoned distributaries and lower river courses where rich organic mud prevails. Moreover, it appears that *Avicennia* species tolerate soils with a high sand content better than do those of *Rhizophora,* although more experimental data are needed to substantiate this. Within the tall *Rhizophora* forests of Colombia and Ecuador, patches of *Pelliciera rhizophorae* occupy low elevations of stiff clay and shell (Cuatrecasas, 1958b; Acosta Solis, 1959), indicating an edaphic preference rather than a case of succession (Figs. 9.11, 9.12).

In the physiographically stable lagoonal environment along the coast of El Salvador, Central America, the distribution of mangrove species appears to be more the result of micro–landform and substrate type than the product of plant succession. Lötschert (1955) indicated that a narrow fringe of *Rhizophora* occupies the organic muck along the lagoon margins, but that the clay soils in the interior support a predominantly *Laguncularia racemosa* forest. A rough traverse made by the present writer through the mangal bordering the Estero de Jaltepeque in El Salvador (shown in Fig. 9.13) in part corroborates Lötschert's findings. However, it also reveals the presence of a mixed mangrove forest on the landward side of the lagoon, where *Rhizophora* usually occurs in mucky swales, while stands of *Laguncularia* and *Avicennia* (the former badly depleted

Fig. 9.12. Growth of tall *Pelliciera rhizophorae* on compact clay of point bar, distributary of the San Juan River, Pacific coast of Colombia. *Rhizophora harrisonii* occupies the low, muddy river bank in background.

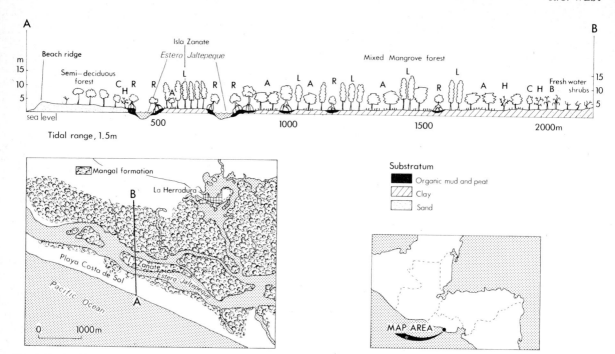

Fig. 9.13. Profile of mangal forest, Jaltepeque Lagoon, El Salvador. R = Rhizophora; A = Avicennia; L = Laguncularia; C = Conocarpus; B = Bactris subglobosa; H = Hibiscus tiliaceus.

by overcutting) occupy the slightly more elevated clay and sandy surfaces.

Since most of the New World mangal occurs along coasts where lagoons, bays, and river deltas are combined, future investigators of mangal ecology might consider with profit the relationships between landform, substrate, and plant life as exemplified by Thom's study.

Intertidal channel environment

In many deltaic, lagoonal, and quiet bayside environments of the American tropics where tidal range is large, the interior portions of the mangal cover between tidal channels consist of stunted growths of mangrove species (usually *Avicennia* and *Rhizophora*), in contrast to the tall forest along the channel banks and levees. An aerial view of this pattern is shown in Fig. 9.14 and a ground view of the interior scrub, in Fig. 9.15. This pattern has not been thoroughly studied, but has been mentioned for other mangal coasts of the world (Pynaert, 1933, and Rosevear, 1947, for West Africa; Kint, 1934, for Sumatra). Davis (1940) recognized three types of dwarf-form mangal communities in southwestern Florida. One, the scrub mangrove facies composed of mixed stands of *Avicennia, Laguncularia,* and *Rhizophora,* occurs in low basins inundated by stagnant water or on surfaces slightly above usual tide level, where extreme conditions of salinity and poor aeration of surface water prevail. The other two types (mainly dwarfed *Rhizophora*) occur on the inner margin of the mangal forest, in stagnant brackish to fresh water. In a brief visit to the mangal woodland along the Gulf of Nicoya, Costa Rica, the present writer observed some of the areas of dwarf mangrove shown in Fig. 9.14. Unfortunately, salinity was not measured, but the scrubby growth of *Avicennia* in the inter-channel zone appears to reflect stagnant and hypersaline water conditions. These areas appear to be slightly lower than the confining channel levees and may be inundated only by spring tides. Thus, increase of soil salinity through evaporation, plus soil compaction, probably are factors inimical to the growth of a mature mangal forest. Because of their widespread occurrence, the interior zones of dwarfed mangal formations are badly in need of further study.

Accreting strand environments

Another example of the influence of the substrate on mangal zonation occurs along the Guiana coast

MIDDLE AND SOUTH AMERICA

Fig. 9.14. Aerial view of a portion of the mangal woodland near Puerto Jesús along the shore of Nicoya Gulf, Costa Rica. The dark tones along the river and tidal channel banks indicate tall *Rhizophora* and *Avicennia* trees. The light tones in the inter–channel areas indicate low growth, mainly *Avicennia* scrub. (Photograph courtesy of the Instituto Geográfico Nacional de Costa Rica.)

of northern South America. For nearly 1600 km, from Amapá territory, Brazil (north of the Amazon mouth) to Guyana (British Guiana), a high forest of *Avicennia* fronts immediately on the low, muddy coast. Portions of this forest have been described briefly (French Guiana and Amapá by Vann, 1969; French Guiana by Boyé, 1962; Surinam by Lindeman, 1953, and Vann, 1959b; British Guiana by Martyn, 1934, and Fanshawe, 1952). Most of the forest floor is affected by semi–diurnal tides that have a range from 1 to 3 m, but only spring tides reach the inner forest margin (Pons, 1966). For most of its length the forest is from 0.5 to 4 km wide, with tree heights up to 30 m. Near river mouths in Amapá its width expands to 20 km, and for some distance west of Cayenne, French Guiana, it narrows to a fringe along rocky stretches of the seashore. Vann (1959b) reported that

Fig. 9.15. Ground view of the *Avicennia* scrub that occupies the inter–channel areas, mangal woodland, Gulf of Nicoya, Costa Rica. The scrub measures 0.75—1 m in height. In the background, a belt of higher *Avicennia* is followed by tall *Rhizophora* trees that fringe the tidal channel.

landward from the *Avicennia* forest in Surinam he found a narrow belt of *Rhizophora*, which practically disappears in French Guiana and Amapá.

The substrate that supports the *Avicennia* forest is composed of fine colloidal clays and silts with some sand. The clays consist of sediments that originate from the Amazon River. These are carried by littoral currents northwestward along the Guiana coast where they are deposited and washed onshore by waves and tides to form mud flats that are eventually occupied by *Avicennia* seedlings (Diephuis, 1966; Vann, 1969). Surface borings within the forest in Surinam show oxidation colors to a depth of 1.5—2 m (Pons, 1966), indicating lack of reduction that is usually associated with many mangrove soils.

Vann (1959b, 1969) and Boyé (1962) suggested various reasons for the presence of the unique *Avicennia* forest along the Guiana coast to the near–exclusion of other species. Vann mentioned salinity as a factor, indicating that *Avicennia* withstands higher concentrations of sea water than do other mangrove species. However, the salinity of the current off the shore of French Guiana has been measured at 20—25‰ (Boyé, 1962), less than that of normal sea water, and probably insufficiently high to exclude the growth of *Rhizophora* or *Laguncularia*.

Both authors emphasized the ability of *Avicennia* seedlings to establish themselves quickly on rapidly accreting shorelines, where mud has been deposited in sufficient quantity to raise flats above the water level at low tide. These conditions are characteristic of the Guiana coast. Because of the rapid development of its extensive shallow lateral root system from which extend upward its characteristic pencil–like pneumatophores, *Avicennia* may be able to adjust to rapid sedimentation better than

Rhizophora. Boyé noted a stepped profile of *Avicennia* growth along the French Guiana coast: a seaward belt of young seedlings on recently deposited flats backed by a belt of trees 10 m high, which is followed by a wide zone of mature trees 30 m high on compact clay soils. Such a pattern is indicative of rapid periodical colonization of depositional surfaces by *Avicennia,* a phenomenon noted elsewhere in the American mangal formations (Thom, 1967).

A more conclusive argument for the concentration of *Avicennia* forests along the Guiana shore may be the ability of this species to grow on the stiff clay soils that characterize the seaward margin of the Recent coastal plain. Again, the shallow lateral root system of *Avicennia* and its pneumatophore pattern may be basic adaptations to this type of substrate. As indicated previously, the association of *Avicennia* and heavy oxidized clay soils appears to be common in the American tropics. Further studies are needed to understand more thoroughly this apparent relationship.

According to Pons (1966), the soils under the *Rhizophora* belt behind the *Avicennia* forest in Surinam consists of reduced fine mud and peat to depths of several meters below the surface. Rich in organic matter, this kind of substrate appears to favor the growth of *Rhizophora*. The origin of this narrow zone behind the present belt of heavy clays nearer the strand line is unclear, but it may be a remnant of a much wider *Rhizophora* forest, which, according to Pons, existed along the Guiana coast during the time of rising sea level (Early Holocene), when rapid vertical sedimentation and reducing conditions prevailed immediately behind the strand line.

Along the lower river courses of the Guiana coast there occurs a zonation of mangrove species that is quite different from that of the strand. A fringe of *Rhizophora mangle* occupies the deep, reduced mud that has accumulated along the banks nearest the river mouths, where brackish water prevails. Farther upstream, *Laguncularia racemosa* mixed with *Rhizophora* begins to appear along the banks. But on the inner part of the natural levees, which are composed of compacted oxidized clays and are reached only by spring tides, *Avicennia* is dominant. According to Jonker (1959), both *Rhizophora harrisonii* and *R. racemosa* thrive in less saline water than *R. mangle,* and in Surinam the first two are found along the river banks upstream, as far as the tide is felt, in one case as much as 60 km from the coast in fresh to slightly brackish water[1].

Coral reef and island environments

Along the shores of the Caribbean mainland and in the West Indies, coral reefs and islands support mangal formations that are usually lower in height than those found in deltaic and lagoonal environments. Vermeer (1959, 1963) and Stoddart (1962) have described the formations on the coral cays off the coast of Belize (British Honduras). Dwarfed *Rhizophora mangle* occupies the lee side of the cays, growing in calcareous or marl mud deposited by tidal currents and by overwash during heavy storms. Hypersaline mud flats, in some cases poorly vegetated by *Salicornia,* in other cases completely bare of vegetation, occur in the interior of the low *Rhizophora* thickets (Fig. 9.16). Taller *Rhizophora* lines tidal channels on some of the large cays, such as Ambergris. Toward the windward side of the cays the back slope of the beach ridge supports low *Avicennia,* in places mixed with *Laguncularia* and *Conocarpus*. The prevalence of dwarfed mangal in these and other coral islands of the Caribbean may result from high salinity of surface and soil water. Moreover, the low content of organic matter in the calcareous mud may be a negative factor. On the mudflats of Ambergris Cay, *Rhizophora* seedlings abound and grow rapidly in the marl mud, but mature plants rarely attain heights over 2 m.

Mangal formations on the Florida Keys appear to be more luxuriant than those off Belize, despite similarity of substrate (Davis, 1940). Although the substrate is basically calcareous, the tall mangrove forests of the Ten Thousand Islands district of southwestern Florida probably owe their luxuriant growth to the abundance of peat in the substrate and to the occasional inundation of the tidal channels by fresh water from the Everglades.

Mangal is found on the lee shores of every island in the Bahama group, where most of the

---

[1] For West Africa, however, Savory (1953) noted that within the *Rhizophora* zone, *R. racemosa* is a pioneer along the sea margin and lower tidal channels, while *R. harrisonii* is dominant in the middle sections and *R. mangle* is found only in drier inner limits of the entire zone.

Fig. 9.16. Low *Rhizophora mangle* growing in calcareous mud around a salt flat on the lee (western) side of Ambergris Cay, Belize. Tall *Avicennia germinans* in background.

soils are derived from coral. Howard (1950) described the classical zonation of mangal associations for the Bimini Islands. *Rhizophora mangle* forms the outer zone, pioneering the calcareous mud flats. Inland this belt is followed by a low *Avicennia* woodland, which grades into scattered growths of *Laguncularia* and *Conocarpus*. Northrop (1902), however, reported extensive areas of low, scrubby *Avicennia* and *Conocarpus* growing on calcareous mud flats on the lee, or western, side of Andros Island in the Bahamas.

Stunted mangrove also has been reported growing on coral rock in various parts of the Caribbean (Børgesen, 1909; Chapman, 1944; Asprey and Robbins, 1953). Such occurrences are limited to scattered plants of *Rhizophora* and *Avicennia*, which anchor their roots in crevices of the coral and between coral boulders. Scattered clumps of low *Rhizophora* thriving on hard cemented coral sand at the sea edge can be seen on both the windward and leeward shores of the cays off Belize (personal observation). Such examples as given are not common and should be considered aberrant.

Brackish–water transitional zone

An inner zone transitional between the mangal and fresh–water formations has been frequently described for various parts of the New World tropics. This zone may be occasionally inundated by the highest spring tides or by storm tides, which bring in salt to brackish water, resulting in soil salinities that are greater than those normally found in fresh–water habitats. In areas of low or seasonal rainfall, the transitional zone may become quite saline or even hypersaline, supporting low, salt–tolerant herbaceous plants as previously mentioned in the section on tropical salt marsh. Scattered clumps of the malvaceous tree or shrub *Hibiscus tiliaceus* are also common in the less saline spots. In areas of abundant rainfall with little seasonal variation, brackish– to fresh–water swamp forest composed of non–mangrove species may characterize the transitional zone. In both cases the pan–tropical fern *Acrostichum aureum* forms extensive thickets within the transitional zone, and is also a colonizer of cutover or disturbed areas within the mangal itself.

Cuatrecasas (1958b) and Lamb (1959) described the large areas of tall *Mora megistosperma* forest followed inland by an equally extensive belt of *Campnosperma panamensis* in the traditional brackish–water zone on the inner edge of the mangal along the rainy Pacific coast of Colombia. The mora forest is also found on the inner side of the mangal in southern Ecuador (Acosta Solis, 1959). In Panama, Holdridge and Budowski (1956) indicated that *Campnosperma panemensis* and *Pterocarpus officinalis* are the dominants of the transitional zone, and that the fresh–water swamp forests farther inland are composed of *Prioria copaifera, Carapa guianensis,* and *Quararibaea* sp. The gnarled and buttressed *Pterocarpus officinalis* has been commonly reported as an important species in the transitional zone in most parts of the American tropics. Other constituents of the zone include various palms, such as *Bactris, Euterpe,* and *Raphia,* and (on the Atlantic coast from Central America to Brazil) the giant aroid *Montrichardia arborescens.* Like the mangrove species, the brackish– and fresh–water swamp plants tend to separate out in gregarious societies, in contrast to the multispecific nature of the normal tropical rain forest.

CONCLUSION

Despite a large literature on the subject, the wet coastal formations within the southern part of the Americas, excepting a few localities, are still poorly known. Most of the literature is descriptive rather than analytical. Quantitative data are few or lacking. Such data as soil and water salinity measurements, character of tidal movements and substrate analyses (moisture, texture, structure, chemistry) are fundamental in understanding the zonation and succession of wet coastal formations. A fruitful approach may be the study of the relationship between species distribution and micro–landform types within the intertidal zone. Far more detailed studies of this kind are necessary before meaningful generalizations or models can be made.

REFERENCES

Acosta Solis, M., 1959. *Los Manglares del Ecuador.* Instituto Ecuatoriana de Ciencias Naturales, Contribución no. 29, Quito, 82 pp.

Asprey, G.F. and Robbins, R.G., 1953. The vegetation of Jamaica. *Ecol. Monogr.,* 23: 359—412.

Ayala–Castañares, A., 1969. Datos comparativos de la geología marina de tres lagunas litorales del Golfo de México. *An. Inst. Biol., Univ. Nac. Auton. Mex.,* 40: 1—10.

Ayala–Castañares A. and Segura, L.R., 1968. Ecología y distribución de los foraminiferos recientes de la Laguna Madre, Tamaulipas, México. *Bol. Inst. Geol.,* 87: 1—89.

Bigarella, J.J., 1946. Contribução ao estudo da planície litorânea do Estado do Paraná. *Arq. Biol. Technol. (Inst. Biol. Pesq. Tecnol., Sec. Agr. Com. Paraná),* 1: 75—111.

Breteler, F.J., 1969. Las especies atlánticas de Rhizophora. *Bol. Inst. For. Lat. Invest. Capacitación,* no. 30—31: 3—13 (Venezuela).

Børgesen, F., 1909. Notes on the shore vegetation of the Danish West Indies. *Bot. Tidsskr.,* 29: 201—260.

Boyé, M., 1962. Les palétuviers du littoral de la Guyane Française; ressources et problèmes d'exploitation. *Cah. Outre–Mer,* 15: 271—290.

Brown, L.F., Fisher, W.L., Erxleben, A.W. and McGowan, J.H., 1971. Resource capability units, their utility in land– and water–use management with examples from the Texas coastal zone. *Geol. Circ. Bur. Econ. Geol., Univ. Texas, Austin,* No. 71–1: 22 pp.

Cabrera, A.L., 1970. *Flora de la Provincia de Buenos Aires. Parte 2, Gramíneas.* Colección Científica del I.N.T.A., Buenos Aires, 624 pp.

Chabreck, R.H., 1972. Vegetation, water and soil characteristics of the Louisiana coastal region. *La. State Univ., Agric. Exp. Station, Bull.,* No. 664: 72 pp.

Chabreck, R.H., Joanen, T. and Palmisano, A.W., 1968. *Vegative Type Maps of the Louisiana Coastal Marshes.* Louisiana Wildlife and Fisheries Commission, New Orleans, La.

Chapman, V.J., 1944. 1939 Cambridge Expedition to Jamaica. ... A study of the botanical processes concerned in the development of the Jamaican shoreline. *J. Linn. Soc. (Bot.),* 52: 407—533.

Chapman, V.J., 1960. *Salt Marshes and Salt Deserts of the World.* Hill, London, 392 pp.

Chapman, V.J., 1975. *Mangrove Vegetation.* Cramer, Lehre, 425 pp.

Chebataroff, J., 1952. Vegetación de los suelos salinos. *Rev. Urug. Geogr.,* 6: 71—100.

Chebataroff, J., 1953. Vegetación halófila de la costa uruguaya. *An. Assoc. Geogr. Bras.,* 4: 30—46.

Clover, E.U., 1937. Vegetational survey of the Lower Rio Grande Valley, Texas. *Madroño,* 4: 41—66; 77—100.

Cooper, A.W., 1974. Salt marshes. In: H.T. Odum, B.J. Copeland and E.A. McMahan (Editors), *Coastal Ecological Systems of the United States*. The Conservation Foundation, Washington, D.C., pp. 55—98.

Cuatrecasas, J., 1958a. Introducción al estudio de los manglares. *Bol. Soc. Bot. Mex.*, 23: 84—98.

Cuatrecasas, J., 1958b. Aspectos de la vegetación de Colombia. *Rev. Acad. Colomb. Cienc. Exactas, Fis. Nat.*, 10: 221—264.

Dansereau, P., 1947. Zonation et succession sur la restinga de Rio de Janeiro. I, Halosère. *Rev. Can. Biol.*, 6: 448—477.

Dansereau, P., 1950. Ecological problems in Southeastern Brazil. *Sci. Mon.*, 71: 71—84.

Davies, J.L., 1972. *Geographical Variation in Coastal Development*. Oliver and Boyd, Edinburgh, 204 pp.

Davis, J.H., 1940. The ecology and geologic role of mangroves in Florida. *Pap. Tortugas Lab.* 32: 303—412 *(Carnegie Inst. Wash., Publ.* No. 517*)*.

Davis, J.H., 1943. The natural features of Southern Florida, especially the vegetation, and the Everglades. *Fla. Geol. Surv., Geol. Bull.*, No. 25: 311 pp.

Diephuis, J.G.H.R., 1966. The Guiana coast. *Tijdschr. K. Ned. Aardrijkskd. Gen.*, 83: 145—152.

Egler, F.E., 1948. The dispersal and establishment of Red Mangrove, *Rhizophora*, in Florida. *Caribb. For.*, 9: 299—320.

Eleuterius, L.N., 1972. The marshes of Mississippi. *Castanea*, 37: 153—168.

Fanshawe, D.B., 1952. *The Vegetation of British Guiana*. Imperial Forestry Institute Paper no. 29. Oxford University Press, London, 96 pp.

Fisk, H.N., 1959. Padre Island and the Laguna Madre flats, coastal south Texas. In: R.J. Russell (Editor), *2nd Coastal Geography Conference, April 6—9, 1959, Coastal Studies Institute, Baton Rouge, La*. Office of Naval Research, Washington, D.C., pp. 103—151.

Fosberg, F.R., 1961. Vegetation–free zone on dry mangrove coasts. *U.S. Geol. Surv., Prof. Pap.*, 424–D: 216—218.

Freyburg, G.V., 1930. Zerstörung und Sedimentation an der Mangroveküste Brasiliens. *Leopoldina*, 6: 69—117.

Graham, S.A., 1964. The genera of Rhizophorae and Combretaceae in the Southeastern United States. *J. Arnold Arbor.*, 45: 285—301.

Hastings, J.R., Turner, R.M. and Warren, D.K., 1972. An atlas of some plant distributions in the Sonoran Desert. *Univ. Ariz., Inst. Atmos. Phys., Tech. Rep. Meteorol. Climatol. Arid Reg.*, No. 21: 255 pp.

Hauman, L., 1926. Étude phytogéographique de la Patagonie. *Bull. Soc. R. Bot. Belg.*, 58: 105—179.

Hildebrand, H.H., 1957. Estudios biológicos preliminares sobre la Laguna Madre de Tamaulipas. *Ciencias*, 17: 151—173.

Hoese, H.D., 1967. Effect of higher than normal salinities on salt marshes. *Contrib. Mar. Sci.*, 2: 249—261.

Holdridge, L.R. and Budowski, G., 1956. Report of an ecological survey of the Republic of Panama. *Caribb. For.*, 17: 92—110.

Hou, D., 1960. A review of the genus *Rhizophora* with special reference to the Pacific species. *Blumea*, 10: 625—634.

Howard, R.A., 1950. Vegetation of the Bimini Island Group. *Ecol. Monogr.*, 20: 317—349.

Johnston, I.M., 1924. Expedition of the California Academy of Sciences to the Gulf of California in 1921. The Botany (The Vascular Plants). *Proc. Calif. Acad. Sci., 4th Ser.*, 12: 951—1118.

Jonker, F.P., 1959. The genus *Rhizophora* in Surinam. *Acta Bot. Neerl.*, 8: 58—60.

Keay, R.W.J., 1953. *Rhizophora* in West Africa. *Kew Bull.*, 1953: 121—127.

Kint, A., 1934. De luchtfoto en de topografische terreingesteldheid in de mangrove. *Trop. Nat.*, 23: 173—189.

Kurz, H. and Wagner, D., 1957. Tidal marshes of the Gulf and Atlantic coasts of northern Florida and Charleston, S.C. *Fla. State Univ. Stud.*, No. 24: 168 pp.

Lamb, F.B., 1959. The coastal swamp forest of Nariño, Colombia. *Caribb. For.*, 20: 79—89.

Lamberti, A., 1969. Contribuição as conhecimento da ecología das plantas do manguezal de Ithanhaém. *Bol. Filos., Cienc. Letr. Univ. São Paulo*, No. 317, Botanica, 23: 7—217.

Leechman, A., 1918. The genus *Rhizophora* in British Guiana. *Kew Bull. Misc. Inf.*, 1918: 4—8.

Lindeman, J.C., 1953. The vegetation of the coastal region of Surinam. *Landbouw Proefstation, Paramaribo, Bull.* No. 63: 54 pp.

Lötschert, W., 1955. La vegetación de El Salvador. *Comun. Inst. Trop. Invest. Cient.*, 4: 65—79.

Luederwaldt, H., 1919. Os manguesaes de Santos. *Rev. Mus. Paul.*, 11: 309—408.

MacNae, W., 1968. A general account of the fauna and flora of mangrove swamps and forests in the Indo–West Pacific region. *Adv. Mar. Biol.*, 6: 73—270.

Martyn, E.B., 1934. A note on the foreshore vegetation near Georgetown. *J. Ecol.*, 22: 292—298.

Mason, H.L., 1957. *A Flora of the Marshes of California*. University of California Press, Berkeley, Calif., 878 pp.

McMillan, C., 1971. Environmental factors affecting seedling establishment of the black mangrove on the central Texas coast. *Ecology*, 52: 927—930.

Miranda, F., 1959. Estudios acerca de la vegetación. In: E. Beltran (Editor), *Los Recursos Naturales del Sureste y su Aprovechamiento*, 2. Inst. Mex. de Recursos Naturales Renovables, A. C., Mexico, D. F., pp. 215—271.

Mohr, C., 1901. *Plant Life of Alabama*. Contributions from the the U.S. National Herbarium, Vol. 6. U.S. Government Printing Office, Washington, D.C., 921 pp.

Moldenke, H.N., 1960. Materials toward a monograph of the genus *Avicennia*. *Phytologia*, 7: 123—168; 179—232; 259—293.

Molfino, J.F., 1921. Contribución a la flora de la región de Bahía Blanca. *Physis (Rev. Soc. Argent. Cienc. Nat.)*, 5: 1—27.

Northrop, A.R., 1902. Flora of New Providence and Andros. *Mem. Torrey Bot. Club*, 12: 1—98.

O'Neil, T., 1949. *The Muskrat in the Louisiana Coastal Marshes*. Louisiana Department of Wild Life and Fisheries, New Orleans, La., 152 pp.

Parodi, L.R., 1940. La distribución geográfica de los talares en la Provincia de Buenos Aires. *Darwiniana*, 4: 33—56.

Peña, G.M., 1971. Biocenosis de los manglares del Peru. *An. Cient.*, 9: 38—45.

Penfound, W.T. and Hathaway, E.S., 1938. Plant communities in the marshlands of southeastern Louisiana. *Ecol. Monogr.*, 8: 1—56.

Perales, P., 1952. La zona manglera oriental de Venezuela. *Rev. Fom. (Minist. Fom., Venezuela)*, 14(76): 179—226.

Phleger, F.B., 1967. Marsh foraminiferal patterns, Pacific Coast of North America. *An. Inst. Biol.*, 38: 11—33.

Phleger, F.B. and Ewing, G.C., 1962. Sedimentology and oceanography of coastal lagoons in Baja California. *Geol. Soc. Am., Bull.*, 73: 145—181.

Pires, J.M., 1964. The estuaries of the Amazon and Oiapoque rivers and their floras. In: *Humid Tropics Research. Scientific Problems of the Humid Tropical Zone Deltas and Their Implications. Proceedings of the Dacca Symposium.* UNESCO, Paris, pp. 211—217.

Pons, L.J., 1966. Geogenese en pedogenese in de Jong–Holocene Kustvlakte van de Drie Guyanas. *Tijdschr. K. Ned. Aardrijkskd. Gen.*, 83: 153—172.

Pynaert, L., 1933. La mangrove congolaise. *Bull. Agric. Congo Belge*, 23: 184—207.

Ringuelet, E.J., 1938. Estudio fitogeográfico del Rincón de Viedma (Bahía de Samborombón). *Rev. Fac. Agron., Univ. La Plata*, 3: 15—186.

Romney, D.N. (Editor), 1959. *Land in British Honduras.* Colonial Research Publ., No. 24. H.M. Stationery Office, London, 327 pp.

Rosevear, D.R., 1947. Mangrove swamps. *Farm For.*, 8: 23—30.

Saint–Yves, A., 1934. Monographia Spartinarum. *Candollea*, 5: 17—100.

Salvoza, F.M., 1936. Rhizophora. *Nat. Appl. Sci. Bull., Univ. Philippines*, 5: 179—237.

Sauer, J., 1967. Geographic reconnaisance of seashore vegetation along the Mexican Gulf Coast. *Coastal Stud. Inst., La. State Univ., Tech. Rep.* No. 56: 59 pp.

Savory, H.J., 1953. A note on the ecology of Rhizophora in Nigeria. *Kew Bull.*, 1953: 127—128.

Skottsberg, C., 1916. Die Vegetationsverhältnisse längs der Cordillera de los Andes, S. von 41° S. Br. (Botanische Ergebnisse der Schwedischen Expedition nach Patagonien und dem Feuerlande, 1907—09). *K. Sven. Vetenskapsakad. Handl.*, 56(5): 366 pp.

Souza Sobrinho, R.J. and Klein, R.M., 1969. *Os Manguezais na Ilha de Santa Catarina.* Universidade Federal Florinaopolis de Santa Catarina, 21 pp.

Stellfeld, C., 1949. Fitogeografia geral do Estado do Paraná. *Arq. Mus. Parana.*, 7: 309—349.

Stoddart, D.R., 1962. Three Caribbean atolls: Turneffe Islands, Lighthouse Reef, and Glover's Reef, British Honduras. *Atoll Res. Bull.*, No. 87: 151 pp.

Thom, B.G., 1967. Mangrove ecology and deltaic geomorphology: Tabasco, Mexico. *J. Ecol.*, 55: 301—343.

Thompson, R.W., 1968. Tidal flat sedimentation on the Colorado Delta, northwestern Gulf of California. *Geol. Soc. Am. Mem.*, 107: 137 pp.

Thorne, R.F., 1954. Flowering plants of the waters and shores of the Gulf of Mexico. In: P.S. Galtsoff (Editor), *The Gulf of Mexico, Its Origin, Waters, and Marine Life.* U.S. Fish Wildl. Serv., Fish. Bull., No. 55: 193—202.

Vann, J.H., 1959a. Landform—Vegetation relationships in the Atrato Delta. *Ann. Assoc. Am. Geogr.*, 49: 345—360.

Vann, J.H., 1959b. *The Physical Geography of the Lower Coastal Plain of the Guiana Coast.* Technical Report no. 1, Project NR 388–029, Contract Nonr–1575(02). Geography Branch, Office of Naval Research, Louisiana State University, Baton Rouge, La., 91 pp.

Vann, J.H., 1969. *Landforms, Vegetation, and Sea Level Change Along the Guiana Coast of South America.* Technical Report no. 3, Project NR 388–028, Contract Nonr 4501(00), Georgraphy Branch, Office of Naval Research. State University College, Buffalo, N.Y., 128 pp.

Vázquez, C., 1971. La vegetación de la Laguna de Mandinga, Veracruz. *An. Inst. Biol., Ser. Bot.*, 42: 49—94.

Vermeer, D.E., 1959. *The Cays of British Honduras.* University of California, Berkeley, Calif., 127 pp.

Vermeer, D.E., 1963. Effects of Hurricane Hattie, 1961, on the cays of British Honduras. *Z. Geomorphol.*, 7: 332—354.

West, R.C., 1956. Mangrove swamps of the Pacific coast of Colombia. *Ann. Assoc. Am. Geogr.*, 46: 98—121.

Chapter 10

AFRICA A. WET FORMATIONS OF THE AFRICAN RED SEA COAST

M.A. ZAHRAN

INTRODUCTION

The coastal salt marshes comprise areas of land bordering the sea, more or less covered with vegetation and subject to periodic inundation by tide (Chapman, 1974). They have certain qualities, related to the proximity of the sea, that distinguish them from inland salt marshes.

The problem of delineating the landward limit of littoral salt marshes is of ecological importance. Littoral salt marshes are really fringes of inland deserts, their landward area being defined by their desertic qualities. Ecological factors, such as terrain or climate, can be used to mark off the littoral marshes. Where there is a narrow belt of lowland along the coast that is shut off from the interior by a steep barrier of mountains, e.g., along the Red Sea coast, the limits are clear; but with a broad plain that stretches inland from the coast, there may be no clear physiographic limit to the littoral formation and we must depend upon other habitat features including vegetation. Vegetation characteristics, related to physiographic attributes reflecting both climatic and edaphic factors, provide the best single basis for delimiting littoral salt marshes. The width of littoral salt marshes may be a narrow strip of coast within the reach of salt spray, a few hundred metres wide, or it may extend inland for several kilometres (e.g., 100—150 km on the Somaliland coast; Meigs, 1966).

The East African littoral salt marshes are under the influence of two principal water bodies, namely: (1) Red Sea (in the north); and (2) Indian Ocean (in the south). The gulfs of Suez and Aden terminate the northern ends of the Red Sea and Indian Ocean, respectively.

The Red Sea extends between Africa and Asia, occupying the East African Rift valley (Fig. 10.1). "The deeps of the Red Sea are cut off from those of the Indian Ocean by a submarine sill at Bab el–Mandab, with the result that exchange circulation of water is impaired and the Red Sea has acquired its own distinctive feature" (Meigs, 1966). The widest part of the Red Sea (ca. 448 km) is in the region of Suakin (in the Sudan), while its narrowest part (ca. 33 km) is at the Bab el–Mandab Strait. "The water of the Sea is warm and evaporation has resulted in a high salt content" (Marshall, 1952). The Red Sea is bordered almost continuously by jagged coral reefs "the base of which extends downward for a depth more than 100 m. Such submerged coral reefs are exposed only at neap tides" (Ball, 1912).

In Egypt, the land adjacent to the Red Sea is generally mountainous, flanked on the western side by the range of coastal mountains (1705—2187 m above sea level; Kassas and Zahran, 1971). In the deep trough between the shoreline and the highlands extends a gently sloping plain. This coastal plain, which varies in width, is covered with sand, over which the drainage systems (wadis) meander with their shallow courses. In certain parts of the Gulf of Suez (e.g., Khashm El–Galala, ca. 60 km south of Suez), the coastal plain is practically non–existent and the mountains rise almost directly from the water of the gulf.

In the Sudan and Ethiopia, the Red Sea coastal plain ranges from 8 to 15 km wide throughout most of its length. It is often covered with gravelly alluvium mixed with drifted fine deposits. The coast is frequently fringed with a belt of uplifted coral about 10 to 20 m above sea level (about 1 km wide at Suakin).

French Somaliland is situated on the east coast

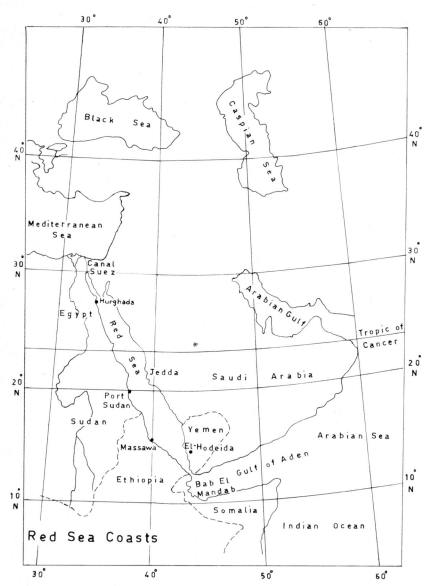

Fig. 10.1. Red Sea coastline.

of Africa around the Gulf of Tadjouva. The mountains press close to the sea, and ribbons of coastal lowland are lacking or are narrow and sandy except in a few small patches.

Climatically, the entire Red Sea and its margins are hot desert. The mean temperature of the hottest months are from 30°C to 35°C with mean daily maximum between 33°C and 43°C. The high humidity along the margins of the sea renders the summer heat especially oppressive, though a sea breeze often develops during the afternoon, and alleviates the heat. Even at night in summer the temperature ordinarily remains above 26°C. The northern part of the Red Sea, including the gulfs, is just cool enough in winter to fall outside the tropical category. Rainfall is deficient throughout. The northern third of the Red Sea is extreme desert with mean annual rainfall as low as 3 mm (in Hurghada, 400 km south of Suez, Egypt). The least dry stations are along the southern part, e.g., in Suakin (in the Sudan) and Massawa (in Ethiopia) with a rainfall more than 150 mm yr^{-1} (see also Chapter 3).

The erratic rainfall which characterises the arid

and semi–arid areas is obvious along the Red Sea coast. The mean annual rainfall fluctuates from 422 mm (in 1925) to 9 mm (in 1910) in Port Sudan, from 617 mm (in 1896) to 33 mm (in 1953) in Suakin (Kassas, 1957), from 2.00 mm (in 1949) to 56.8 mm (in 1952) in Suez (Kassas and Zahran, 1962). In Massawa rainfall fluctuates between 15% above and 59% below the annual mean, and the monthly variation is even greater — e.g., the December maximum is 188 mm and minimum is nil (Hemming, 1961). The annual pluviothermic coefficients (Emberger, 1951) for Suez, Port Sudan and Massawa are: 4, 17 and 27.5, respectively.

VEGETATION TYPES

The wet formations of the African Red Sea coast comprise: mangal, reed–swamp and salt–marsh vegetation types.

Mangal vegetation

The shoreline morphology and climate, as well as the local habitats of the African Red Sea coast, seem to favour the growth and domination of mangal vegetation. Most of the Red Sea coast consists of raised Quaternary coral reefs, along which there is a series of small bays that cut into the raised beach. These bays are partly land–locked by further coral reefs and the sea water in them is sheltered. Wave action is reduced to a minimum. Such protected and shallow lagoons and bays of the Red Sea coast provide a favourable habitat for the growth of mangal vegetation. Also, the southern part of the Red Sea (ca. three–fifths of its basin) is located south of the Tropic of Cancer, i.e., within the limits of the environmental conditions for the occurrence of mangroves.

Three mangrove species are recorded along the African Red Sea coast, namely: *Avicennia marina, Rhizophora mucronata* and *Bruguiera gymnorrhiza*. The cover, density, stratification and distribution of these species vary. They are absent from the shoreline of the coast of the Gulf of Suez but their presence at the mouth of the gulf at the Sinai side (lat. 27°40′N) was recorded by Ferrar (1914) and Zahran (1965, 1967). *A. marina* grows in a narrow lagoon (Marsa) with shallow water. Along the African Red Sea coast, the northern boundary of the mangal vegetation is represented by a few, scattered and depauperate individual bushes of *A. marina* in Myos Hormos Bay, about 22 km north of Hurghada (lat. 27°14′N). From Hurghada southwards to Marsa Halaib (lat. 22°N), the *A. marina* mangal is notable and is one of the common features of the vegetation of the littoral landscape of Egypt. Pure stands of *A. marina* mangal vary in extent from limited patches of a few individuals to continuous belts of dense growth extending for several kilometres. This distribution seems to depend upon the local conditions of the shoreline morphology (Kassas and Zahran, 1967). On Abu Minqar Island (offshore of Hurghada) and on Safaga Island (offshore of Safaga, ca. 60 km north of Hurghada) there are dense thickets of *A. marina* mangal.

In the most southern part of the Red Sea coast of Egypt, *Rhizophora mucronata* is recorded. In the shoreline between Marsa el–Madfa (lat. 23°N) (near Marsa Halaib on the Sudano–Egyptian border), *R. mucronata* forms an almost pure growth; in other places it is associated or co–dominant with *A. marina*.

A. marina thickets extend further southwards, dominating the shoreline of the Red Sea coast of the Sudan. In a localised area south of Suakin (lat. 19°15′—19°N) early reports (Brown and Massey, 1929; Andrews, 1950—1956) recorded that *R. mucronata* and/or *Bruguiera gymnorrhiza* dominate or co–dominate with *A. marina*. However, Professor M. Kassas (personal communication) has more recently visited Suakin (1966) but he did not find a single tree of *B. gymnorrhiza*. The disappearance of this species may be attributed to man's interference, because the inhabitants used to cut the bark for use in dyeing.

The shoreline of the Ethiopian coast and southward to the far end of the African Red Sea coast in French Somaliland is fringed by the growth of *A. marina* mangal. Other mangroves (*R. mucronata* and *B. gymnorrhiza*) are not recorded (Hemming, 1961; Verdcourt, 1968).

The structure of the mangal vegetation of the Red Sea coast is simple, usually one layer of *A. marina* (Zahran, 1976a). In the localities where *R. mucronata* and/or *B. gymnorrhiza* are included, these plants form a layer towering over that of *A. marina*. The ground layer is formed of associated marine phanerogams, e.g., *Cymodocea ciliata,*

C. rotundata, C. serrulata, Diplanthera uninervis, Halophila stipulacea and *H. ovalis*.

The soil associated with the Red Sea mangal vegetation generally consists of sandy mud, black in colour, rich in organic matter and decaying debris, and often foul-smelling. The great bulk of the soil (60—75%) is formed of two particle-size ingredients: 0.211—0.104 mm and 0.104—0.050 mm diameter, with a small proportion of coarse sand (10%, > 0.211 mm diameter). The very fine fraction (< 0.05 mm diameter) increases from 10% at the landward edge to 20% in the waterlogged mud. The soil reaction is alkaline (pH = 8 to 8.95) and its organic carbon content ranges between 0.32% and 2.22%. The total soluble salts of the soil range between 1.2% and 4.3%, on an oven-dry weight basis, mostly chlorides (0.5—1.75%) and partly sulphates (0.03—0.2%). The calcium carbonate (total carbonates) of the mangrove soils vary from 80% in the soil of *R. mucronata* to 4.5—19.5% in the soil of *A. marina*. "This edaphic distinction requires confirmation" (Kassas and Zahran, 1967).

It is not unusual to find *A. marina* on the terrestrial side of the shoreline. In the delta of Wadi Gimal (ca. 50 km south of Marsa Alam, Egypt), there is a stand of *A. marina* in a terrestrial habitat where the bushes are covered by sandy hillocks. This is apparently due to silting up of the shoreline zone, which was occupied originally by a lagoon with mangal growth.

Reed-swamp vegetation

The habitat of the reed-swamp vegetation of the African Red Sea coast is provided by the channels and creeks of the mouth of the big wadis, e.g., Wadi El-Ghweibba, and areas which represent the combined influences of the brackish-water springs, e.g., the Ain Sokhna area in Egypt (Kassas and Zahran, 1962).

The reed swamps are represented by *Phragmites australis* and *Typha domingensis*. *T. domingensis* usually inhabits areas where the soil is relatively less saline and the water is not too shallow, e.g., estuaries of wadis that collect the occasional surface drainage. *P. australis*, on the other hand, grows in swamps close to the dry land, often with higher soil salt content and drier than those of *T. domingensis* (Zahran, 1976b). The common associates of the reed-swamp vegetation include: *Berula erecta, Samolus valerandi, Cyperus articulatus, C. dives, C. mundtii, Scirpus mucronata, S. tuberosus, Lemna gibba, Spirodela polyrrhiza* and *Wolffia hyalina*.

Salt-marsh vegetation

The vegetation of the littoral salt marshes of the African Red Sea coast comprises twenty community types dominated by: *Halocnemon strobilaceum, Arthrocnemum glaucum, Salicornia fruticosa, Halopeplis perfoliata, Limonium pruinosum, L. axillare, Zygophyllum album, Aeluropus* spp., *Sporobolus spicatus, Nitraria retusa, Suaeda monoica, S. fruticosa, S. vermiculata, Tamarix mannifera, T. passerinoides, Cressa cretica, Imperata cylindrica, Alhagi maurorum* and *Juncus rigidus*. The ecological amplitudes of these dominants vary and this results in variations in their distribution, density, stratification, zonation and floristic composition. A brief account of each of these community types follows.

Halocnemon strobilaceum community type

This community occurs in the inland side of the shoreline. The ground is either a tidal flat, recurrently washed by tidal flow, or a shoreline bar of sand on rock detritus heaped up by wave and tidal action.

The growth of *Halocnemon strobilaceum* occurs in two forms: (1) circular patches on flat tidal mud ground (Fig. 10.2); or (2) sheets of irregular-shaped patches on the shoreline bars.

In Egypt, *H. strobilaceum* is common within

Fig. 10.2. Circular patches of *Halocnemon strobilaceum*, Myos Hormos coast, Gulf of Suez.

the littoral salt marshes of the Gulf of Suez (from Suez to Hurghada, a 400–km stretch) but not in the region farther south on the Red Sea coast. The plant cover is often pure stands of the dominant species but it may be associated with *Arthrocnemum glaucum, Zygophyllum album, Cressa cretica* and *Alhagi maurorum*.

H. strobilaceum is not recorded in the littoral salt marshes of the Sudan (Andrews, 1950—1956; Kassas, 1957). On the Red Sea coast of Ethiopia, Hemming (1961) does not refer to this community type, but *H. strobilaceum* is recorded among the species of the flora of Ethiopia (Cufodontis, 1961—1966).

Arthrocnemum glaucum community type

A. glaucum occupies the same shoreline zone as the *H. strobilaceum* community and shows a similar growth habit. In Egypt, *A. glaucum* occurs throughout the whole Red Sea coast but it is less common in the northern part (coast of the Gulf of Suez), where the *H. strobilaceum* community abounds. Its associated species are *Halocnemon strobilaceum, Zygophyllum album, Halopeplis perfoliata, Limonium axillare, Atriplex farinosa*, etc. (Kassas and Zahran, 1967).

On the Red Sea coast of the Sudan, *A. glaucum* covers long but narrow strips fringing the shoreline. The plant cover is dense on the sandy beach ridges and thins off a few metres inland. The second type of habitat consists of flat patches of newly formed littoral plains, where the land is just above the level of the high tide but may be washed by high seas (Kassas, 1957). *A. glaucum* forms circular patches that increase in size till they coalesce. In the meantime phytogenic mounds are gradually produced.

A. glaucum is very abundant on the Ethiopian Red Sea coast and is dominant in the salt flats lying about 20 m from the sea. "The areas dominated by *A. glaucum* did not appear to differ noticeably from that dominated by *Halopeplis perfoliata* except in that it is nearer to the sea" (Hemming, 1961). This is also repeated in the Egyptian and Sudanese Red Sea coast where *A. glaucum* is the first community inland from the mangal vegetation followed, in certain localities, by *H. perfoliata*.

Salicornia fruticosa community type

S. fruticosa and *A. glaucum* are difficult to distinguish from each other in their vegetative form; it is only at their time of flowering that they can be separated, *A. glaucum* flowering in April while *S. fruticosa* flowers much later (Tadros and Berlanta, 1958).

S. fruticosa is recorded in the floras of Egypt (Täckholm, 1956, 1974), the Sudan (Andrews, 1950—1956) and Ethiopia (Cufodontis, 1961—1966), but its dominance is recorded neither in the survey of the Sudan Red Sea coast (Kassas, 1957) nor in that of the Ethiopian coast (Hemming, 1961). On the Red Sea coast of Egypt, *S. fruticosa* is dominant over a very limited area (Mallaha, 260—280 km south of Suez). "There *S. fruticosa* forms patches of pure growth with plant cover ranges from 50—100%. The ground is covered with a thick crust of salts (total soluble salts = 80.05%)" (Kassas and Zahran, 1965).

Halopeplis perfoliata community type

H. perfoliata dominates a community type which occupies a zone that follows, on the landward side, the littoral zone of the *A. glaucum* community type. This littoral zone may be occupied also by sand mounds covered by *Zygophyllum album*. The ground zone of *H. perfoliata* is often lower in level than the littoral zone where the wave–heaped detritus or the wind–deposited sand may form slightly elevated bars. It is also lower in level than the higher ground of the zones further inland. This seems to cause conditions of poor drainage in most of the localities, so that the ground surface is usually wet and slippery. In a few localities the zone of *H. perfoliata* may be at the shoreline.

The biogeography of *H. perfoliata* along the African Red Sea coast is of ecological interest. Within the 950–km stretch from Suez to Marsa Kileis (in Egypt), *H. perfoliata* is recorded in one locality ca. 185 km south of Suez (Kassas and Zahran, 1962) but is otherwise very rare or absent. Southward of Marsa Kileis, *H. perfoliata* and its community is a common feature of the littoral salt marshes. The plant cover ranges from 5 to 40% mainly contributed by the dominant. *A. glaucum, Z. album, Cressa cretica, Suaeda vermiculata*, etc. are common associates. On the Sudan coast, the *H. perfoliata* community type occupies, as in Egypt, a zone inland to that of *A. glaucum*. The plant cover is thin (5—10%) and the associated species include *A. glaucum, Suaeda fruticosa,*

Aeluropus lagopoides, Suaeda monoica, S. vermiculata and *S. volkensii* (Kassas, 1957).

On the Ethiopian coast, *H. perfoliata* colonises the fringes of the salt flats which have no superficial sand mantle. Large areas of this habitat exist 16 km south of Marsa Cuba (lat. 16°15′N) and about 1 km from the sea. "*H. perfoliata* grows to about 23 cm and is able to tolerate more saline conditions than any other species found here. The plants are quite widely scattered" (Hemming, 1961).

Limonium pruinosum community type

L. pruinosum is a species with two distinctly different habitat types: littoral salt marsh (Kassas and Zahran, 1967) and desert limestone cliffs (Kassas and Girgis, 1964). It is very likely that it comprises two ecotypes.

The *L. pruinosum* community type occupies a salt-marsh zone bordering the shoreline zone of the *Halocnemon strobilaceum* community. It is common within the northern 100-km stretch of the Gulf of Suez (in Egypt). The plant cover ranges from 10 to 20% and the associated species include *Zygophyllum album, Nitraria retusa, H. strobilaceum, Suaeda calcarata, A. glaucum,* etc. The presence of *L. pruinosum* is not recorded southward in the littoral salt marshes of Egypt and the Sudan nor in Ethiopia.

Limonium axillare community type

L. axillare is a non-succulent salt-marsh half-shrub, which may extend inland to the fringes of the coastal desert plain. Its community abounds in the saline flats of the Red Sea littoral salt marshes.

Within the northern 650-km stretch of the Gulf of Suez and the Red Sea coasts (in Egypt), *L. axillare* is found in one or two localities that are widely spaced: 61—18 km south of Suez and 114 km south of Suez. In the former locality there are a few patches of *L. axillare* whilst in the latter locality there are a few individuals. Within the stretch extending southwards from 650 km south of Suez, *L. axillare* and the community type it dominates are among the common features of the salt-marsh formation. In this community the dominant species contributes the main part of the vegetation cover (5—50%). *Aeluropus* spp., *Z. album, Sevada schimperi, Salsola baryosma, H. perfoliata, Sporobolus spicatus, Salsola vermiculata* are among the common associate species of the stands of the *L. axillare* community in the Red Sea coast of Egypt (Zahran, 1964).

Kassas (1957) recorded *L. axillare* among the species commonly present in the coral limestone raised-beach habitat of the Red Sea coast of the Sudan. A surface crust of salts may render it sterile and the plant cover is usually 5—10%. The other species are *Suaeda vermiculata, Cyperus conglomeratus* var. *effusus* and *Aeluropus lagopoides*. Also, on the Ethiopian coast the raised-beach soil or the soil that overlies raised beaches is the favoured habitat for the growth and dominance of *L. axillare*. In the quartz sand over the raised-beach soil the vegetation is mixed. "*L. axillare* may extend on to the sand with diminutive species, *Salsola forskalei, Zygophyllum coccineum (Z. album), Indigofera argentea, I. semitrijuga, Eremopogon foveolatus,* etc." (Hemming, 1961).

Zygophyllum album community type

Z. album is a species that shows a wide ecological amplitude. Its growth and dominance are not restricted to the littoral and inland salt marshes since it is present in some wadis of the limestone desert, extending to the east of the Nile Valley, where it dominates a well-developed community (Kassas and Girgis, 1964); it may also grow in the sand dunes of the oases of the Western Desert of Egypt (Zahran, 1972).

Z. album is omnipresent in the Red Sea littoral salt marshes. On the Egyptian coast it may grow in the form of small individuals distantly spaced on the saline ground of the dried salt marsh. It may also build small phytogenic mounds of sand that stud the ground surface of the dried salt marsh. Being widely distributed, the community type dominated by *Z. album* includes numerous other species (44 species including 6 ephemerals) with differing ecological requirements (Kassas and Zahran, 1967). These include *Nitraria retusa, L. axillare, L. pruinosum, H. perfoliata, Aeluropus* spp., *Sporobolus spicatus,* etc., as salt-marsh species; *Panicum turgidum, Salsola baryosma, Launaea spinosa,* etc., as desert-plain species. The dominant contributes 5—50% of the plant cover.

On the Red Sea coast of the Sudan, *Z. album* is listed as among the most common associated species of the salt-marsh communities dominated by *Arthrocnemum glaucum, Aeluropus lagopoides,* etc. (Kassas, 1957). In his second survey (1966) of the Red Sea coast of the Sudan, Professor M. Kassas (personal communication) found that *Z. album* is widespread along the whole stretch southward to the Sudano–Ethiopian border. In Tokar (ca. 160 km south of Suakin and ca. 80 km north of the Ethiopian border), for example, the *Z. album* community covers vast areas of the salt marshes.

On the Ethiopian Red Sea coast, a *Z. coccineum* community has been included among the communities of the littoral salt marshes (Hemming, 1961). However, *Z. coccineum* is a xerophytic and salt–intolerant (0.3% maximum) plant dominating a widespread area in the Egyptian Desert (Täckholm, 1956, 1974, etc.; Kassas and El-Abyad, 1962) and the Sudanese Desert (Andrews, 1950—1956; and herbarium specimens, Botany Department, Faculty of Science, Cairo University). The *Z. coccineum* mentioned by Hemming (1961) in the Ethiopian littoral salt marshes has been confused, as far as identification is concerned, with *Z. album*, the true halophytic species. Cufodontis (1961–1966) recorded *Z. album* in the flora of Ethiopia. Hence, *Z. album (Z. coccineum)* is widely distributed on the Ethiopian Red Sea coast. It occurs in several habitats, e.g., in the coralline sand dunes and mantles, in the quartz sand over raised–beach soil, and in the normal raised–beach soils, etc., where it dominates a well–developed community type. The soil supporting the *Z. album* community (*Z. coccineum* of Hemming, 1961) is saline, the total soluble salts ranging between 3.1% in the superficial coral, 5.54% at 15 cm depth, and 2.91% at 30 cm depth. The associated species of the *Z. album* community on the Ethiopian coast include *Limonium axillare, Cornulaca ehrenbergii, Cyperus conglomeratus,* etc.

Aeluropus spp. community type

The grassland community type dominated by *Aeluropus* spp. *(A. lagopoides* and *A. massauensis)* occupies one of the inland zones of the Red Sea littoral salt marshes, the dominant grass forming patches or mats of dense growth. These are sometimes covered by spray–like crusts of salt which denote that the plants may be temporarily covered by saline water, which, on receding and drying, leaves the salt on the shoots of the plants. The growth form of the grass is usually that of the creeping type, but in one locality on the Red Sea coast of Egypt (Marsa Alam, 700 km south of Suez) it forms peculiar cone–like masses of interwoven roots, rhizomes and sand (Fig. 10.3). The plant cover (10%—80%) is mostly contributed by the dominant plant, and its associates are mainly halophytes, e.g., *Zygophyllum album, Sporobolus spicatus, Cyperus laevigatus, Limonium axillare, Halopeplis perfoliata, Sevada schimperi, A. glaucum* and *Tamarix mannifera*.

The community type dominated by *A. lagopoides* on the Red Sea coast of the Sudan follows the zone of *H. perfoliata*, occupying areas where the surface layer of the soil is apparently wind–borne material. The plant cover ranges from 50% to 70% and it is mainly of *A. lagopoides* plants associated with *Suaeda fruticosa, Sporobolus spicatus,* etc.

On the Ethiopian Red Sea coast, *A. lagopoides* prevails in impeded drainage sites, or often in areas behind coral sand dunes along the shore with alluvium deposits, where the surface of soil is generally a fine brown silty sand with a cracked surface. In the sand–flat areas which lie at the edge of the sands of the coastal plain, *A. lagopoides* is the common grass, where it is found on sand 6 cm deep and also on low hummocks of wind–blown sand.

Sporobolus spicatus community type

The *S. spicatus* community occupies a zone inland to that of *Aeluropus* spp. in which the sand deposits are much deeper than in the *Aeluropus* zone and the soil salinity is much less. In many localities one notes that the two grassland community types grow mixed in the same zone forming

Fig. 10.3. Close–up view of the peculiar cone–like masses of interwoven roots, rhizomes and sand of *Aeluropus* spp., Marsa Alam, Red Sea coast, Egypt.

a mosaic pattern, the *Sporobolus* grassland on the higher parts forming island–like patches amongst the sea of lower saline ground covered by *Aeluropus* grassland. In the *Aeluropus* grassland the ground–water level is usually shallow (40—100 cm) but in the *Sporobolus* grassland the ground–water level is usually deeper than 150 cm.

On the Red Sea coast of Egypt, the *S. spicatus* community is common in the southern part (south of 1030 km south of Suez) but not in the northern stretch. On the Sudan coast, the *S. spicatus* community is common and it follows the *A. lagopoides* community in succession sequences. The sand drift is deeper and shows no profile feature. Apart from the salt–marsh associate species, e.g., *A. lagopoides, Limonium axillare* and *Suaeda fruticosa*, there are about twelve species which are common in the inland plain and not typical salt–marsh plants, e.g., *Panicum turgidum, Lycium arabicum* and *Euphorbia monacantha*.

The *S. spicatus* community abounds in several habitats on the Ethiopian Red Sea coast, such as: (a) Sandy beaches, in areas where the soil is brown and more silty than the coral soil above. Here, *S. spicatus* is mixed with *Cyperus conglomeratus, Aeluropus lagopoides, Panicum turgidum, Dactyloctenium aristatum, Urochondra setulosa*, etc. (b) In alluvium over coralline soil, *S. spicatus* and *Eleusine compressa* are abundant between the clumps of *A. lagopoides*. (c) In salt flats with a superficial sand mantle and in a neighbouring area of the *Aeluropus* zone, the transitional zone between these salt flats is occupied by *S. spicatus* and *Eleusine compressa*.

Nitraria retusa community type

N. retusa is a salt–tolerant bush that grows in two types of habitat. In the first type *Nitraria* forms saline mounds or hillocks that stud the flat ground of the salt marsh (Fig. 10.4). Commonly *N. retusa* covers the north–facing part of the hillocks of this habitat type. The second type of habitat comprises sandy bars (actually chains of sandy hillocks) fringing the shoreline. These bars are less saline than the hillocks of the other type.

The *N. retusa* community type is wide–spread in the northern 600–km stretch of the Gulf of Suez and Red Sea littoral salt marshes of Egypt, but is absent in the coastal land south of Marsa Alam (700 km south of Suez). It is not recorded in the

Fig. 10.4. *Nitraria retusa* hillock, Hurghada coast, Egypt.

survey of the Sudan coast (Kassas, 1957) nor in the flora of the Sudan (Andrews, 1950—1956). In the survey of Hemming (1961) for the Ethiopian Red Sea coast, *N. retusa* is absent, though Cufodontis (1961—1966) mentioned *N. retusa* in the flora of Ethiopia.

The floristic composition of the *N. retusa* community includes *Z. album, Cressa cretica, Alhagi maurorum, Suaeda monoica, Tamarix mannifera*, etc.

Suaeda monoica community type

S. monoica is a frutescent plant of dry salt marshes (Fig. 10.5), comparable in habit and habitat to *Nitraria retusa*. The two species have an ecological range that extends beyond the limits of the salt marsh to the fringes of the coastal desert plain. "They may form and protect phytogenic

Fig. 10.5. *Suaeda monoica* community type, Red Sea coast, Sudano–Egyptian border.

mounds and hillocks of sand, though *Suaeda* hills may reach greater size" (Kassas and Zahran, 1967).

The biogeography of *S. monoica* and *N. retusa* differ. On the Red Sea coast of Egypt, *N. retusa* abounds in the northern 300–km stretch, *S. monoica* gradually replacing *N. retusa* within the 300—650–km stretch, until in the south (700 km south of Suez and southwards) *N. retusa* is absent and *S. monoica* is abundant. The stands of the *S. monoica* community type include 51 species (9 salt–marsh plants and 42 desert–plain plants) as well as 52 ephemerals. The diversity is attributed to the wide ecological range of this community.

On the Sudan Red Sea coast, the *S. monoica* community abounds in the inland margins, following in sequence of zonation the community type dominated by *Suaeda fruticosa* on the alluvial depressions. By virtue of its intermediate position between the salt marsh and the coastal plain, its floristic composition includes salt–marsh species, e.g., *Suaeda fruticosa, Cressa cretica, Eleusine compressa, Sporobolus spicatus* and *Cyperus conglomeratus*, and desert–plain species, e.g., *Panicum turgidum, Indigofera spinosa* and *Convolvulus hystrix*.

S. monoica is also present on the Red Sea coast of Ethiopia and of French Somaliland (Hemming, 1961; Verdcourt, 1968).

Suaeda fruticosa community type
S. fruticosa is recorded along the whole stretch of the African Red Sea coast. In Egypt and Ethiopia, *S. fruticosa* is a common associate species with many community types, but it is not a dominant (Hemming, 1961; Kassas and Zahran, 1962, 1965, 1967). On the Sudan Red Sea coast, *S. fruticosa* dominates a well–developed community type that follows, in sequence of zonation, the community type dominated by *Halopeplis perfoliata*. In certain localities vast stretches of land are covered by this community (plant cover 25—50%) and provide grazing ground for camels. *S. monoica, Aeluropus lagopoides* and *Eleusine compressa* are the common associates (Kassas, 1957). The ground covered by *S. fruticosa* is raised into low hummocks 5—10 cm above the general level of the sterile land. The salinity in soil layers exploited by roots of *S. fruticosa* is less (3.4%) than in deeper layers (9—20%).

Suaeda vermiculata community type
S. vermiculata is commonly recorded among the associate species of the Red Sea littoral community types. Its dominance is restricted to the delta of Wadi El–Ghweibba (Gulf of Suez, Egypt). The *S. vermiculata* community type occupies the zone of vegetation which lies in between the *Halocnemon strobilaceum* littoral zone and the *Nitraria retusa* inland zone of the dry salt marsh. It is associated with *Halocnemon strobilaceum, Zygophyllum album, Nitraria retusa, Limonium pruinosum, Cressa cretica, Salsola villosa*, etc. (salt–tolerant plants) and *Haloxylon salicornicum, Zilla spinosa, Farsetia aegyptiaca*, etc. (desert plants) (Hemming, 1961).

Juncus rigidus community type
J. rigidus is a halophyte very tolerant to increased soil salinity, soil water stresses and aridity of the climate (Zahran, 1976b). It is present in the littoral salt marshes of the Red Sea, but its densest growth and dominance are recorded in the salt marshes of the Gulf of Suez, e.g., the Ain Sokhna area (ca. 50 km south of Suez) where its cover is up to 90—100% (Kassas and Zahran, 1962). The associates are halophytes including *Halocnemon strobilaceum, Cressa cretica, Tamarix mannifera*, etc.

The *J. rigidus* community represents the sedge–meadow stage of the halosere succession of the Red Sea that follows the reed–swamp stage of *Phragmites australis* and *Typha domingensis*.

Tamarix mannifera community type
T. mannifera is one of the common bushes in the Red Sea littoral salt marshes from Suez southwards to French Somaliland (Andrews, 1950—1956; Cufodontis, 1961—1966; Kassas and Zahran, 1962, 1967; Verdcourt, 1968). It grows in a variety of habitats and in various forms. In many parts of the dried salt marsh it forms thickets; and, in the sand–choked deltaic parts of wadis that drain inland country and pour onto the shoreline, it forms sand hillocks that reach considerable size. The floristic composition of the *T. mannifera* community type include about 26 perennial species including 9 halophytes, e.g., *Z. album, N. retusa, J. rigidus, Cressa cretica* and *Suaeda monoica* and 17 xerophytes, e.g., *Zygophyllum coccineum, Haloxylon salicornicum, Reta-*

ma raetam, Tamarix aphylla, Acacia tortilis and Leptadenia pyrotechnica.

Tamarix passerinoides community type

T. passerinoides is morphologically and ecologically comparable to *T. mannifera*, but the former is rare along the Red Sea coast. It is recorded only once within the salina (El–Mallaha) 240 km south of Suez.

Apart from the previously mentioned communities of the Red Sea littoral salt marshes (African coast), there are certain halophytes present of ecological interest, e.g., *Cressa cretica*, *Alhagi maurorum* and *Imperata cylindrica*. *C. cretica*, *A. maurorum* and *I. cylindrica* are common associate species within a great number of the Red Sea littoral salt-marsh communities. They may also be dominant, but such communities are limited to the downstream part of Wadi Hommath that drains into the Gulf of Suez 30 km south of Suez (Kassas and Zahran, 1962). Their habitat is a littoral belt of sand extending parallel to the shoreline and varies in width from 150 to 300 m.

SOIL

The soil features are apparently one of the main factors influencing the plant growth of the littoral salt marshes, the plant cover and distribution and also the zonal pattern of the community types. It was found by many workers (Johnson and York, 1915; Harshberger, 1938; Penfound and Hathaway, 1938; Kassas and Zahran, 1967; Chapman, 1974) that there is a relationship between vegetation zones and soil salinity in maritime marshes. Harshberger (1938) considers that the distribution of the various salt–marsh species is dependent upon their resistance to salinity. Chapman (1974) states "It is doubtful whether salinity can be regarded as the major determining factor, tidal phenomena and soil may be equally important. It must, however, be remembered that salinity operating at a specific period or season may be the controlling factor as between zones." He also cites (p. 88) after Steiner (1934) that "the osmotic pressure of the soil solution varies over short distances horizontally". We may also quote Kassas and Zahran (1967): "These zones, though in a recognisable pattern, do not necessarily show a recognisable order of different salinity. This is due to the pattern of the distribution of the soluble salts within the successive layers of the soils of the salt marsh."

The climatic conditions of the African Red Sea coast have a pronounced effect on the edaphic characteristics of the salt marshes. Aridity of the climate (extremely arid in the stretch of the Egyptian coast) increases the rate of evaporation. As precipitation is low (very low in Egypt), there is insufficient leaching and salts accumulate in the form of surface crusts. The soil is generally alkaline with a pH of 7.0 to 8.9 (Zahran, 1976b).

Table 10.1 shows the total soluble salts in soil samples collected from different community types of the African Red Sea littoral salt marshes, namely: *Arthrocnemum glaucum*, *Halopeplis perfoliata*, *Limonium axillare*, *Zygophyllum album*, *Aeluropus* spp., *Suaeda monoica* and *Nitraria retusa*. "In general the main soil characteristics that seem to determine the vegetation are permeability and salinity" (Hemming, 1961).

TABLE 10.1

Total soluble salts (%) in soil samples collected from different community types of the African Red Sea littoral salt marshes (after Kassas, 1957; Hemming, 1961; and Kassas and Zahran, 1967)

Community type:	*Arthrocnemum glaucum*		*Halopeplis perfoliata*			*Limonium axillare*		*Zygophyllum album*		*Aeluropus* spp.			*Suaeda monoica*		*Nitraria retusa*
	Eg.	Sud.	Eg.	Sud.	Eth.	Eg.	Eth.	Eg.	Eth.	Eg.	Sud.	Eth.	Eg.	Sud.	Eg.
Surface layer	44.95	4.95	15.02	35.7	0.58	16.1	12.05	26.3	3.1	35.68	4.10	1.40	7.3	0.65	26.13
Subsurface layer	14.6	12.65	11.2	9.6	0.27	43.1	3.5	1.2	5.54	2.4	3.65	1.87	6.8	0.45	22.45
Third layer	1.36	11.45	3.02	8.45	0.49	2.7	3.6	1.9	2.9	1.1	0.95	1.18	0.8	0.6	0.84

Eg. = Egypt; Sud. = Sudan; Eth. = Ethiopia.

The data of Table 10.1 show the following features:

(1) The total amount of soluble salts is generally high.

(2) The surface layers usually contain the highest proportion of soluble salts (salt crusts) and the amount of soluble salts drops abruptly in the subsurface and lower layers — e.g., in the *Halopeplis perfoliata* community soil salinity drops from 35.7% in the surface layer to 9.6% and 8.5% in the subsurface and third layers, respectively. In the community types dominated by *Arthrocnemum glaucum* and *Zygophyllum album* the surface layers contain a lower amount of soluble salts than the subsurface layers. This is attributed to the accumulation of an aeolian sand covering the actual surface layers (represented by the subsurface saline layers). Here, the surface layer of sand has lower salt content (4.95% and 3.1%) than the subsurface (actual surface) layers (12.65% and 5.54%) in the *Arthrocnemum glaucum* and *Zygophyllum album* community types, respectively.

(3) As the mean annual rainfall of the African Red Sea coast increases southwards (see p. 216), the amount of total soluble salts of the soil associated with the comparable community types decreases southwards. This is well illustrated by the figures in Table 10.1 for the *Aeluropus* spp. community type.

CLASSIFICATION

Classification of the community types of the African Red Sea coast may be based on either the adaptability or the growth forms of their dominant species.

Walter (1961) classified the halophytes into: (1) facultative halophytes; and (2) euhalophytes. Facultative halophytes are plants that have an optimum development in non–saline soils, but may tolerate a certain amount of salt in the soil. The euhalophytes, on the other hand, are those plants which show optimum growth on soil with a specific salt content. They may grow also on soil that is poor in salts but they will not flourish.

The roots of the halophytes absorb the soil solution in a diluted form. As water is transpired and salt remains in the transpiring leaves, an accumulation of salts may eventually take place in leaves. The halophytes overcome this in different ways, viz.: with or without regulating mechanisms (Boughey, 1957; Chapman, 1974) as follows:

(1) With regulating mechanism. Here, the halophytes are either *excretives* or *succulents*. The excretives possess glandular cells capable of excreting excess salts. The succulents often lack this ability, but they thwart the increase in salt concentration by an increase in their water content, and they become more and more succulent during their development.

(2) Without regulating mechanism. In this type the salt concentration rises continuously during the growing season. When a certain level is reached the plant or organ dies. These are called *cumulative* halophytes.

Accordingly, the halophytic plants dominating the community types of the African Red Sea wet formations are classified into eight excretives (*Limonium axillare*, *L. pruinosum*, *Aeluropus* spp., *Sporobolus spicatus*, *Tamarix mannifera*, *T. passerinoides*, *Nitraria retusa* and *Cressa cretica*), eight succulents (*Halocnemon strobilaceum*, *Arthrocnemum glaucum*, *Salicornia fruticosa*, *Halopeplis perfoliata*, *Suaeda monoica*, *S. vermiculata*, *S. pruinosa* and *Zygophyllum album*) and four cumulatives (*Phragmites australis*, *Typha domingensis*, *Imperata cylindrica* and *Juncus rigidus*).

Scholander et al. (1962) classified the mangrove species into: salt–secreting and non–secreting plants. We may, accordingly, add *Avicennia marina* to the excretive halophytes and *Rhizophora mucronata* and *Bruguiera gymnorrhiza* to the cumulative halophytes[1].

Kassas and Zahran (1967) considered that *Alhagi maurorum* is a plant alien to the salt–marsh habitat. It has a long root system that may extend several metres deep, reaching soil layers that are less saline and permanently wet (Kassas, 1953). But, as *A. maurorum* is an abundant species dominating a characteristic salt–marsh community, both in the littoral zone (Kassas and Zahran, 1962) and inland (Zahran, 1972), we may consider it as a cumulative halophyte. The individual plants of *A. maurorum* show a range of morphological

---

[1] These are better classed as *excluding* halophytes since the roots exclude much of the salt from entering the plants (Editor) (see Chapman, 1975).

variations of leaf—spine relationships and may comprise a number of ecotypes (Zahran, 1976b).

According to the structure and growth forms of their dominant plants, the community types of the wet formations of the African Red Sea coast may be classified (Kassas and Zahran, 1967) into five main groups, namely: (1) grassland types; (2) succulent half-shrub types; (3) non-succulent half-shrub types; (4) succulent shrub types; and (5) non-succulent shrub types. We may also add (6) mat-shaped types; (7) reed types; and (8) rush types.

(1) *Grassland types*

Four community types are included in this group dominated by cryptophytes (geophytes; Raunkiaer, 1934): *Aeluropus* spp., *Sporobolus spicatus* and *Imperata cylindrica*. *Aeluropus* spp. usually form mats in a low layer (20 cm). A higher layer (40—50 cm) includes some of the associates, such as *Zygophyllum album*.

Sporobolus spicatus forms mounds that are higher than those of *Aeluropus* spp. The vegetation usually comprises a layer of the dominant (40—60 cm) and a lower one including such associates as *Aeluropus* spp.

Imperata cylindrica grows in the mouth of certain wadis and it prevails in the salt flats covered with sheets of sands. Its vegetation is often formed of one layer.

(2) *Succulent half-shrub types*

This group includes: *Halocnemon strobilaceum*, *Arthrocnemum glaucum*, *Salicornia fruticosa*, *Halopeplis perfoliata*, *Suaeda vermiculata*, *S. fruticosa* and *Zygophyllum album*. *H. strobilaceum*, *A. glaucum* and *Salicornia fruticosa* are similar with respect to their growth form and structure of vegetation. The dominant species form carpets or round patches of different density and simple organisation. The vegetation is often formed of a single layer 30—50 cm high; the higher (shrub) layer is usually absent or of negligible status. The vegetation may accumulate some sediments forming mounds of sand covered by the growth of the dominant species.

The vegetation in the *H. perfoliata* type is usually in the form of separate individuals that rarely exceed 40 cm in height. The plant cover is often thin (5—20%). In a few localities *H. perfoliata* forms sand mounds. The vegetation usually comprises a single layer.

The *Zygophyllum album* community is much more elaborately organised. The dominant species forms mounds and small hillocks of various heights. The areas between these mounds may be occupied by low-growing species (*Aeluropus* spp., *Cressa cretica*, etc.) forming a ground layer distinct from the layer of the dominant species. Occasional shrubs (*Tamarix mannifera*, *Nitraria retusa*) may form a frutescent layer which is distantly open.

(3) *Non-succulent half-shrub types*

This group includes the two *Limonium* community types, namely: *L. pruinosum* and *L. axillare*. *L. pruinosum* may form a much branched bushy growth with individuals up to 60 cm high, yet it does not usually form mounds. Individuals of *L. axillare* may reach equal size, but can build up mounds of considerable extent though not of considerable height. This community type rarely comprises more than one layer.

The *Alhagi maurorum* community type is included in this group. The vegetation is formed of the layer of the dominant (suffrutescent) and a ground layer may also be present represented by short plants such as *Cressa cretica*.

(4) *Succulent shrub types*

This group includes the *Nitraria retusa* and *Suaeda monoica* community types. The two dominant species are capable of building up sand hummocks, those of *N. retusa* reaching 1 to 1.5 m high, and those of *S. monoica* much higher and much more extensive. In the Tokar area (Sudan, M. Kassas, personal communication, 1974) *S. monoica* forms extensive tracts of forest-like growth, which is organised into three layers. The frutescent layer includes the dominant species. The suffrutescent layer includes the dwarfed perennials and, in the *S. monoica* community, many ephemerals.

(5) *Non-succulent shrub types*

In this group *Tamarix* thickets and mangrove thickets are included. The *Tamarix* (*T. mannifera* and *T. passerinoides*) thickets may be organised into several layers (ground, suffrutescent and frutescent of the dominant). The mangrove thickets are usually simple in structure and are formed of one layer of *Avicennia marina*. In the few localities where *Rhizophora mucronata* and/or *Bruguiera gymnorrhiza* are included in the vegetation, they form a layer towering over that of *A. marina*. The associated marine phanerogams, e.g., *Cymodocea* spp. and *Halophila* spp., may be considered as forming a ground layer.

(6) Mat–shaped type

The *Cressa cretica* community type is the sole representative of this type. It usually forms a pure growth of one low layer (up to 10 cm) of the dominant species. A few other species may be present, e.g., *Alhagi maurorum* and *Zygophyllum album*, and these form a second suffrutescent layer.

(7) Reed types

The *Phragmites australis* and *Typha domingensis* community types are included in this group. The vegetation is usually of one layer of the reed plants, but aquatic species may form the surface layer. Density of growth varies according to salinity of the habitat.

(8) Rush type

Juncus rigidus is a rush common in the littoral stretch. The vegetation is usually thick (plant cover ranges between 70 and 100%) and is formed of one layer of the dominant (suffrutescent). A frutescent layer of widely spaced individual shrubs of *Tamarix mannifera* and *T. passerinoides* may be present, as in the Wadi Ambagi and the Ain Sokhna area, Red Sea coast of Egypt.

BIOGEOGRAPHY

The geographical distribution of these community types of the African Red Sea coast is represented diagrammatically in Fig. 10.6. Except for the northern 400-km stretch — the coast of the Gulf of Suez — the mangal vegetation, dominated mainly by *Avicennia marina*, is well developed along the whole stretch (i.e., from Hurghada [lat. 17°14′N] southward). *Rhizophora mucronata* and *Bruguiera gymnorrhiza* are also recorded, but their distribution is limited to two localities:

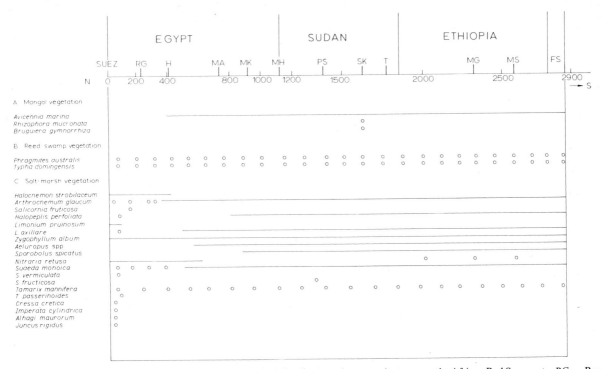

Fig. 10.6. Diagrammatic representation of the geographical distribution of community types on the African Red Sea coasts. *RG* = Ras Gharib; *H* = Hurghada; *MA* = Marsa Alam; *MK* = Marsa Kileis; *MH* = Marsa Halaib; *PS* = Port Sudan; *SK* = Suakin; *T* = Tokar; *MG* = Marsa Gulbub; *MS* = Massawa; *FS* = French Somaliland.

Marsa Halaib (at the Sudano–Egyptian border), where *R. mucronata* co–dominates with *A. marina*, and on the shoreline of Suakin in the Sudan, where *R. mucronata* and *B. gymnorrhiza* are both present with *A. marina*.

The reed–swamp vegetation dominated by *Phragmites australis* and *Typha domingensis* is distributed along the whole stretch of the Red Sea coast in limited patches related to local features of the habitat. The vegetation is not in continuous belts.

The communities of the salt marshes, even those which are ecologically and morphologically similar, are different in their geographical distribution. *Halocnemon strobilaceum* is abundant within the northern section of the Red Sea coast, from Suez to Hurghada (400 km), but is not present elsewhere (Kassas, 1957; Hemming, 1961; Kassas and Zahran, 1967). The ecologically comparable species, *Arthrocnemum glaucum*, on the other hand, is only occasionally found in the northern section (Suez—Marsa Alam, 670 km) but south of Marsa Alam it is a notable feature of the Red Sea littoral salt marsh. The *Salicornia fruticosa* community type abounds in only one locality, namely El–Mallaha (260—280 km south of Suez); elsewhere it is recorded among the associate species (Andrews, 1950—1956; Cunfondontis, 1953—1961; Täckholm, 1974).

Halopeplis perfoliata is rarely found (one locality) within the northern 960–km stretch of the Red Sea coast. Further south it is a very common salt–marsh species.

The *Zygophyllum album* community type extends all along the Egyptian and Sudanese Red Sea coasts (see p. 220). It is a most widespread plant.

Limonium pruinosum is confined to the northern 100–km stretch and is absent further southward. *L. axillare* is rarely recorded within the northern 500–km stretch, but within the southern stretch is a most common plant.

The two species of *Aeluropus* (*A. lagopoides* and *A. brevifolius*) are rare in the northern 700–km stretch. From Marsa Alam southward they are abundant species.

Sporobolus spicatus is absent in most of the Egyptian Red Sea coast (1100 km). Its presence starts at 1000 km south of Suez, and it is abundant southward.

Nitraria retusa has an interesting geographical distribution. It is a widespread plant within the 600–km stretch of the Egyptian Red Sea coast, but is absent within the coastal land south of Marsa Alam. It is not recorded in the littoral salt marshes of the Sudan (Andrews, 1950—1956; Kassas, 1957). Cufodontis (1961—1966) recorded *N. retusa* among the species of the flora of Ethiopia, but Hemming (1961) does not refer to it in his vegetation survey of the Ethiopian Red Sea coast. The ecologically comparable community dominated by *Suaeda monoica* is rarely recorded within the northern 250–km stretch of the Red Sea coast, but is widespread southward to Bab el–Mandab Strait (Kassas, 1957; Hemming, 1961; Kassas and Zahran, 1967; Verdcourt, 1968).

Suaeda fruticosa and *S. vermiculata* are common halophytes along the whole Red Sea coast, but the first is dominant only on the Sudanese coast (Kassas, 1957), while the second is dominant in a very limited area in the northern part of the Egyptian coast (Kassas and Zahran, 1962).

Tamarix mannifera is a widespread species in the Red Sea coast of Egypt where it may form thickets of dense growth. In the southern stretch it is also present but with thin cover. "Its scantiness within the southern stretch may be due to ecological not geographical causes" (Kassas and Zahran, 1967). The morphologically and ecologically comparable *T. passerinoides* only abounds within a limited inland salina (260—280 km south of Suez).

Cressa cretica, Imperata cylindrica, Alhagi maurorum and *Juncus rigidus* are common species over the whole stretch of the Red Sea coast but they are only dominant in the northern part of the Gulf of Suez (Kassas and Zahran, 1962).

ZONATION

Zonation is a universal characteristic feature of the littoral salt–marsh vegetation (Penfound and Hathaway, 1938; Stephenson and Stephenson, 1949; Tansley, 1949; Kassas, 1957; Hemming, 1961; Chapman, 1974). Several factors (tidal movement, relief of ground, sea–water spray, soil salinity, etc.) seem to play certain roles in determining the zonation pattern of the vegetation of the littoral salt marshes, but their individual effect varies. "The species on the lower marshes may be very different from those of the higher marshes, and in many instances only a few inches increase in level result in a profound change" (Chapman, 1974). Tansley (1949) states "The zones of the marshes are related only very approximately to tidal level." Southward (1958) mentioned "The range of the tide or amplitude is usually the major factor governing the vertical range of the intertidal zone, and hence the width or height of the various zones of plants and animals." In his work at Bank End Marsh (northwest of England), Zahran (1973) found that "the pattern of zonation of the littoral salt marsh is primarily controlled by tidal inundation and relief of land. Salinity is also an important factor that plays a role which is second in importance to the elevation–inundation factor." The results of studies carried out by Kassas and Zahran (1967) on the vegetation of the littoral salt marshes of the Egyptian Red Sea coast revealed that "Zonation is usually attributed to varying gradients in soil salinity... The development of salinity, amount of salts, pattern of salt content within the profile, and kind of salts associated with increase in ground level may differ in different zones. Again the dynamic processes of accretion, seem to produce different types of habitat within the different zones of the salt marsh."

Climate is also a very important factor. In humid countries, where the leaching and washing effects of rain is of great consequence, the zonation pattern follows the salt–marsh succession (halosere). Chapman (1974) states "...such zonation must imply that the vegetation on the salt marsh is seral in character and represents the salt–marsh succession." In a dry climate such as that of the Red Sea coast, on the other hand, where the leaching and washing effect of rain is of minor consequence, successional sequence may not be concomitant with the zonal sequence. In the littoral salt marshes of the Red Sea coast, and probably in all the littoral salt marshes of the arid climate, the salinity gradient does not form a regular pattern of decreasing salinity further from the shoreline. Areas that are nearer to the shoreline and that are subject to recurrent wash by sea water are less saline than adjacent areas that are only occasionally inundated by salt water. In the former, the salinity of the soil will remain that of sea water, while in the second capillary rise will result in an increase of the surface salinity.

Zonation sequence is complete only where the shore rises gently and gradually into the land, and this occurs rarely (MacNae and Kalk, 1962). The topography of the salt marsh comprises different types of shoreline bars, and its ground may be studded by mounds and hummocks of various dimensions. Again the ground of the littoral salt marsh may rise inland in such a fashion as can telescope salt–marsh zones, eliminate some, or the coastal plain may be practically non–existent.

Following Stephenson and Stephenson's (1949) classification of the shoreline zones, the wet formations of the Red Sea African coast comprise two zones: the whole stretch of the supralittoral and the supralittoral fringe with adjoining parts of the midlittoral zone. The former is referred to as the littoral salt marsh, the latter is the habitat of the mangrove vegetation.

There have been attempts to discover successional relationships between the various mangrove community types on the basis of zonation and water depth within the shoreline fringes (Dansereau, 1947; Schnell, 1952; Savory, 1953; Boughey, 1957). The whole mangal formation may be "regarded as initiating a primary succession leading up to the corresponding climax of the hinterland" (Boughey, 1957). It seems that the composition of the mangal is primarily controlled by the nature of the sediments of the substratum and by the salinity and depth of water (Zahran, 1964)[1] (see p. 218).

The initial salt–marsh vegetation (supralittoral zone) of the African Red Sea coast comprises 23 principal community types. These may be ecologically classified into two main groups: (1) communities of the saline flats, and (2) communities of the piled sand. The former comprises the community types growing on what may be described as the dried salt marsh: areas where the ground surface is higher than the levels of tides, or areas that are protected against tidal inundation by shoreline bars piled by waves or tides. The communities of the piled–up sand build and/or cover sand mounds or hillocks that are piled by wind (Zahran, 1964). The demarcation between the two groups is not sharp since a number of the salt–marsh species may grow in both types of habitats. "However, the community types of the saline flats are true salt–marsh vegetation, that is, they are confined to the supralittoral zone. The sand

---

[1] See also Chapman (1975).

community types grow in areas covering this zone and extending beyond to the coastal desert–plain" (Kassas and Zahran, 1967).

The communities of the saline flats occur in five main zones: the first or shoreline zone is the habitat of *Halocnemon strobilaceum* and *Arthrocnemum glaucum*. The second zone is occupied by the *Halopeplis perfoliata*, *Suaeda vermiculata* and *S. fruticosa* community types, the third zone is the habitat of *Limonium pruinosum* and *L. axillare*. The fourth and fifth zones are occupied by *Aeluropus* spp. and *Sporobolus spicatus*, respectively. *Salicornia fruticosa* occupies an inland salina with highly saline soil.

Saline flats covered by aeolian sand, forming littoral belts of sand extending parallel to the shoreline, constitute the habitats of the *Cressa cretica*, *Alhagi maurorum* and *Imperata cylindrica* community types. Such habitats are well represented in the delta of Wadi Hommath in the Gulf of Suez.

The communities of the piled sand include *Zygophyllum album*, *Suaeda monoica* and *Tamarix mannifera*, which may cover parts of the salt–marsh ecosystem.

The reed–swamp habitat may be dominated by *Typha domingensis* or *Phragmites australis*, and the habitat of the rush plants is dominated by *Juncus rigidus*. The reed–swamps are restricted to certain creeks associated with springs or mouth areas of wadis and their fringes that cross the salt–marsh zones — e.g., the creeks crossing *Halocnemon strobilaceum* community types in Ain Sokhna area (50 km south of Suez, Egypt).

From the discussion of the zonation of the African Red Sea littoral vegetation we may conclude that a consistent pattern of zonation of the salt–marsh communities may not exist in the field, and that any idealised pattern of zonation of a coastal stretch is only arbitrary. This is because (1) all the zones of the idealised pattern do not appear in one locality, or (2) representatives of certain communities may occupy zones different from their normal zones.

REFERENCES

Andrews, F.W., 1950—1956. *The Flowering Plants of the Sudan*. Volumes I—III. Sudan Government, Khartoum, 237 pp. (1950), 485 pp. (1952), 597 pp. (1956).

Ball, J., 1912. *Geography and Geology of South Eastern Egypt*. Ministry of Finance, Cairo, 394 pp.

Boughey, A.S., 1957. Ecological studies of tropical coastlines, I. The Gold Coast, W. Africa. *J. Ecol.*, 45: 665—687.

Brown, A.F. and Massey, R.E., 1929. *Flora of The Sudan*. Sudan Government, Khartoum, 502 pp.

Chapman, V.J., 1964. *Coastal Vegetation*. Pergamon Press, London, 262 pp.

Chapman, V.J., 1974. *Salt Marshes and Salt Deserts of the World*. Cramer, Lehre, 2nd ed., 392 pp. (complemented with 102 pp.).

Chapman, V.J., 1975. *Mangrove Vegetation*. Cramer, Lehre, 425 pp.

Cufodontis, G., 1961—1966. Enumeratio plantarum Aethiopiae. Spermatophyta (sequentiae). *Bull. Jardin Bot. Natl. Belg.*, 31: 709—772 (1961); 33: 879—924 (1963); 36: 1059—1114 (1966).

Dansereau, P., 1947. Zonation et succession sur la estinga de Rio de Janeiro. *Rev. Can. Biol.*, 6: 448.

Emberger, L., 1951. Rapport sur les régions arides et semiarides de l'Afrique du Nord. *UNESCO/NS/11/AZ/11*, No. 9: 50—61.

Ferrar, H.T., 1914. Note on a mangrove swamp at the mouth of the Gulf of Suez. *Cairo Sci. J.*, 8: 23—24.

Harshberger, J.W., 1938. The vegetation of the saltmarshes and the salt and fresh water ponds of northern coastal New Jersey. *Proc. Acad. Nat. Sci. Phila.*, 373—400.

Hemming, C.F., 1961. The ecology of the coastal area of northern Eritrea. *J. Ecol.*, 49: 55—78.

Johnson, D.S. and York, H.H., 1915. The relation of plants to tide levels. *Carnegie Inst. Wash. Publ.*, 206.

Kassas, M., 1953. On the distribution of *Alhagi maurorum* in Egypt. *Proc. Egypt. Acad. Sci.*, 3: 140—151.

Kassas, M., 1957. On the ecology of the Red Sea coastal land. *J. Ecol.*, 45: 187—203.

Kassas, M. and El–Abyad, M.S., 1962. On the phytosociology of the desert vegetation. *Ann. Arid Zone*, 1: 54—83.

Kassas, M. and Girgis, W.A., 1964. Habitat and plant communities of the Egyptian Desert. V. The limestone plateau. *J. Ecol.*, 52: 107—119.

Kassas, M. and Zahran, M.A., 1962. Studies on the ecology of the Red Sea coastal land. I. The district of Gebel Ataga and El–Galala El–Bahariya. *Bull. Soc. Géogr. Egypte*, 35: 129—175.

Kassas, M. and Zahran, M.A., 1965. Studies on the ecology of the Red Sea coastal land. II. The district from El–Galala El–Qibliya to Hurghada. *Bull. Soc. Géogr. Egypte*, 37: 155—193.

Kassas, M. and Zahran, M.A., 1967. On the ecology of the Red Sea littoral saltmarsh, Egypt. *Ecol. Monogr.*, 37: 297—316.

Kassas, M. and Zahran, M.A., 1971. Plant life on the coastal mountains of the Red Sea, Egypt. *J. Ind. Bot. Soc.*, 50A: 571—589.

MacNae, W., 1963. Mangrove swamps in South Africa. *J. Ecol.*, 51: 1—25.

MacNae, W. and Kalk, M., 1962. The ecology of mangrove swamps at Inhaca Island, Mozambique. *J. Ecol.*, 50: 19—34.

Marshall, N.B., 1952. Recent biological investigation in the

Red Sea. *World Fishing*, (Sept.): 201—205.

Meigs, P., 1966. Geography of coastal deserts. *Arid Zone Res. UNESCO*, 28: 140 pp.

Montasir, A.H. and Hassib, M., 1956. *Illustrated Manual Flora of Egypt*. Publ. Fac. Sci., Ain Shams University, Cairo, 615 pp.

Penfound, W.T. and Hathaway, E.S., 1938. Plant communities in the saltmarsh lands of southeastern Louisiana. *Ecol. Monogr.*, 5: 137—145.

Raunkiaer, C., 1934. *The Life Form of Plants and Statistical Plant Geography*. Oxford University Press, London, 632 pp. (English translation by Carter, Fausboll and Tansley).

Savory, H.J., 1953. A note on the ecology of *Rhizophora* in Nigeria. *Kew Bull.*, 1953: 127—128.

Schnell, R., 1952. Contribution à une étude phytosociologique et phytogéographique de l'Afrique occidentale. *Mem. Inst. Fr. Afr. Noire*, 18—43.

Scholander, P.E., Hammel, H.T., Hemmingsen, E. and Garey, W., 1962. Salt balance in mangroves. *Plant Physiol.*, 37: 722—729.

Southward, A.J., 1958. The zonation of plants and animals on rocky sea shores. *Biol. Rev.*, 33: 137.

Steiner, M., 1934. Zur Ökologie der Salzmarschen der nordöstlichen Vereinigten Staaten von Nordamerika. *J. Wiss. Bot.*, 81: 94—202.

Stephenson, T.A. and Stephenson, A., 1949. The universal features of zonation between tide marks on rocky coasts. *J. Ecol.*, 37: 389—305.

Täckholm, V., 1956. *Student's Flora of Egypt*. Anglo–Egyptian Bookshop, Cairo, 649 pp.

Täckholm, V., 1974. *Student's Flora of Egypt*. Cairo University, Giza, 2nd ed., 888 pp.

Tadros, T.M., 1953. A phytosociological study of halophilous communities from Mareotis (Egypt). *Vegetatio*, 4: 102—124.

Tadros, T.M. and Berlanta, A.A., 1958. Further contributions to the study of the sociology and ecology of the halophilous plant communities of Mareotis (Egypt). *Vegetatio*, 8: 136—160.

Tansley, A.G., 1949. *The British Islands and Their Vegetation*, I. Cambridge University Press, London, 484 pp.

Verdcourt, B., 1968. French Somaliland. *Conservation of Vegetation in Africa South of Sahara. Symp. 6th Plenary Meeting, Assoc. Etud. Taxon. Flora. Afr. Trop., Uppsala, Sept. 12th 1966*, pp. 140—141.

Walter, H., 1961. The adaptation of plants to saline soils. *Salinity Problems in the Arid Zone, Proc. Teheran Symp., UNESCO*, XIV: 129—134.

Zahran, M.A., 1964. *Contributions to the Study on the Ecology of the Red Sea Coast*. Thesis, Cairo University, Giza, 237 pp.

Zahran, M.A., 1965. Distribution of mangrove vegetation in UAR (Egypt). *Bull. Inst. Déserte Egypte*, 15: 6—11.

Zahran, M.A., 1967. On the ecology of the east coast of the Gulf of Suez. I. Littoral saltmarshes. *Bull. Inst. Déserte Egypte*, 17: 225—251.

Zahran, M.A., 1972. On the ecology of Siwa Oasis. *Egypt J. Bot.*, 25: 213—242.

Zahran, M.A., 1973. Comparative eco–physiological studies on *Puccinellia maritima* and *Festuca rubra*. *Bull. Fac. Sci. Mansoura Univ.*, 1: 43—66.

Zahran, M.A., 1976a. Biogeography of mangrove vegetation along the Red Sea coasts. *Proc. Int. Symp. for the Biology and Management of Mangrove Vegetation, Honolulu, 1974* (in press).

Zahran, M.A., 1976b. On the ecology of halophytic vegetation of Egypt. *Bull. Fac. Sci. Mansoura Univ.*, 4 (in press).

Chapter 11

AFRICA B. THE REMAINDER OF AFRICA[1]

V.J. CHAPMAN

NORTH AFRICA

Along the north African coasts salt marshes very similar to those in Egypt (p. 218) are to be found, and there is the same problem of determining where the salt marsh ends and salt desert commences. Along the northern coast (Tunisia and Algeria) the *Halocnemon strobilaceum* community is widespread as the pioneer. The plants collect sand around them and then the new soil may be colonised by *Cutandra memphitica, Bassia muricata* or *Traganum nudatum* (Killian and Lemée, 1948; Simmoneau, 1953). On the Solontchak soils the primary colonist is either *Arthrocnemum glaucum* or *A. macrostachyum*, the limits of the latter being set by frequency of submergence and rate of accretion (Chapman, 1974). In Tunisia the *Halocnemetum* is succeeded by a *Limoniastrum guyonianum* community in which *Zygophyllum album, Nitraria retusa* and *Suaeda vermiculata*, all plants of the Red Sea marshes, are to be found.

Information available from the Oran area shows that the pioneer coastal salt–marsh community is a *Salicornietum* dominated by *Salicornia arabica* and *S. herbacea* agg. At the mouths of rivers and streams this is replaced by a *Spartinetum maritimae* in which *Puccinellia palustris, P. distans* and *Crypsis aculeata* are to be found. This *Spartinetum* is essentially an eastern Atlantic community that spreads northwards into Europe (see Chapter 6) and appears southwards in South Africa. The Oran *Salicornietum* is succeeded by a *Limonietum* which covers much of the marshes and is quite rich floristically. The dominant species are *L. sinuatum* (which is pan–Mediterranean) and *L. sebkarum* (local). Farther east the former species is joined by *L. spathulatum*. These correspond to the *Limonium* communities of the Red Sea area (see p. 220). On the landward boundary of the salt marsh in Oran, adjacent to the *Arthrocnemetum*, one can find a *Salicornia fruticosa* community with either *Centaurium spicatum* (with submergence) or *Monerma cylindrica* (salt desert). Another Moroccan landward community is that of *Suaeda fruticosa* (see p. 223) which is rarely inundated. It exists in a number of facies which are said to be determined climatically (Chapman, 1974).

Elsewhere on the North African coast the *Halopeplis perfoliata* community of the east is replaced by an *H. amplexicaulis* community where there is little accretion and a high moisture content. In Tunisia a distinct *Zygophyllum album* community can be recognised and it presumably extends eastwards to Egypt. Where there is high soil moisture, salt and good drainage a *Juncus acutus* community frequently occurs. Species associated with it are *Inula crithmoides* and *Aeluropus littoralis*. Less saline areas on the landward side are occupied by *Tamarix africana* so that the genus is widespread along North and East Africa.

WEST AFRICA

Little information is available about the coastal vegetation from Oran to St. Louis in Senegal but there are probably isolated patches of salt marsh. In Senegal the average mean temperature is such that mangrove vegetation is able to grow and survive (Fig. 11.1). The species comprising this mangal are quite different to those on the east coast (see p. 217) and their relationships are with the mangroves of the western Atlantic (Chapman, 1975). At the present time six species of man-

---

[1] I had hoped this would have been written by the late Dr. W. MacNae (Ed.).

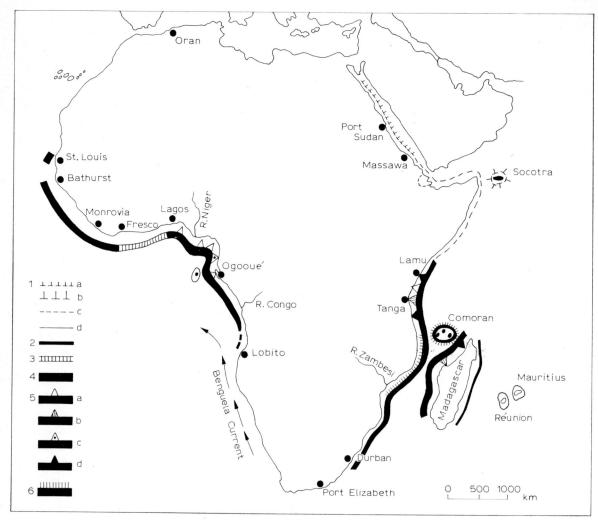

Fig. 11.1. Mangal distribution in Africa (after Grewe, 1941). *1* = *Avicennia:* (*a*) single and very small groups, (*b*) closed community, (*c*) with sporadic *Rhizophora*, (*d*) with better developed *Rhizophora*; *2* = most species present, well developed; *3* = as *2*, in lagoons; *4* = complete mangal, well developed; *5* = forest used for commerce: (*a*) not for export, (*b*) bark for export, (*c*) wood for export, (*d*) wood and bark for export; *6* = well-developed forest, bark used for export.

grove tree can be recognised on the West African coast.

Rhizophoraceae	*R. mangle*
	R. racemosa
	R. harrisonii
Avicenniaceae	*A. africana*
Combretaceae	*Laguncularia racemosa*
	Conocarpus erectus

The general pioneer species is *R. racemosa* (formerly termed "tall" *R. mangle*) whilst *R. mangle* ("short" trees) occupies drier and often more saline soils of the landward belt. Between these two species *R. harrisonii* occupies wetter areas of the main forest (Savory, 1953; Lawson, 1966).

In northern Senegal the long dry season and low mean winter temperature (10—15°C) provide a climate that is not suited to *Rhizophora* species so that *Avicennia africana* is the dominant mangrove. From Bathurst southwards well-developed mangal occurs, probably with all species represented, and this situation continues down to Angola where the mangal once more becomes sporadic. A feature of Ghana and Nigeria is the association of mangal with extensive lagoons (see also Chapter 9 for lagoon mangal). Grewe (1941) and

Ainslie (1926) state that mangal zonation in West Africa is essentially related to the physiography and the former recognises six types:

(1) *Avicennia*-dominated swamps of northern Senegal (Fig. 11.1).

(2) Bathurst to Monrovia in Liberia. The numerous estuaries, especially the Gambia, favour mangrove development with *Rhizophora racemosa* as the primary colonist. The main forest behind is dominated by *Avicennia africana* with *Paspalum vaginatum* as a frequent grass cover to the ground. The *Avicennia* areas often drain poorly and thus form an excellent area for malarial mosquitoes to breed. Drainage of such areas has been recommended as a means of control (Thomson, 1951) (Fig. 11.2).

(3) Monrovia to Lagos (Fig. 11.1). The cliff coast between Monrovia and Fresco reduces the mangal to small areas in the river estuaries. East of Fresco there are extensive lagoons where mangal is well developed. In those lagoons with open access to the sea *R. racemosa* is the principal colonist with *R. harrisonii* and some *Pandanus* behind, but in closed lagoons with a former sea opening there is a shrub zone of *Avicennia africana* at flood-water level. Between the strand and above the mangal there may be a grass zone of *Sporobolus virginicus* followed by *Fimbristylis obtusifolia* and *Paspalum vaginatum*. Next to the mangal *Sesuvium portulacastrum* and *Philoxerus vermicularis* can occur. In the open lagoons the prop roots and trunk bases can be clothed with the red algae *Bostrychia* and *Caloglossa* which have disappeared from closed lagoons. Some workers include *Drepanocarpus lunatus* in estuary mangal (Schnell, 1952; Jackson, 1964), whereas Boughey (1957) says it is typical of levels above flood waters.

Above the red algal *Bostrychia* turf on the prop roots the oyster *Ostrea tulipa* can form a distinct zone with an uppermost barnacle zone of *Chthamalus rhizophorae*.

(4) Niger delta to Ogooué (Gabon). This region provides the most extensive mangal in West Africa, though it is essentially a narrow belt extending 100 km up the Niger. The *Rhizophora* zones are similar to those in the Lagos lagoons with *R. mangle* and the fern *Acrostichum* on the landward, drier soils. The frontal *Rhizophora* may reach 46 m high but because of frequent felling they more usually attain to only 3 m. *Avicennia* trees of 30 m are reported from the Rio del Rey (Rosevear, 1947) where they may form small tree islands with *Rhizophora* at the mouth. Generally, however, *Avicennia* occurs between the frontal *Rhizophora* species and the *R. mangle*. The *Avicennia* is frequently accompanied by *Laguncularia racemosa*. The transition to fresh water is characterised by the invasion of the *R. mangle* zone by *Pandanus* and the *Raphia* palm. The animal life in this mangal is rich, the birds including pelicans, terns, sandpipers, crocodile bird and fish eagle: there are also two crocodiles, the Mona monkey and the mudskipper (a fish), *Periophthalmus koelreuteri*.

More comparable mangal occurs at the mouths of estuaries from Cameroun to Ogooué. On the island of Fernando Poó there is a sporadic mangrove community of *A. africana*, *Laguncularia* and the fern *Acrostichum aureum* (Adams, 1957). The generalised mainland zonation of the Cameroons (Chapman, 1975) is shown in Scheme A.

(5) Ogooué to the Congo River. Physiographically this is not very suited to mangal but narrow belts do occur along the banks of estuaries. The zonation is typical of the previous region.

(6) Angola. Mangal in this region is only sporadic and essentially forms fringes to rivers and estuaries. In the northern part *Rhizophora* is the primary colonist but in the southern part, south

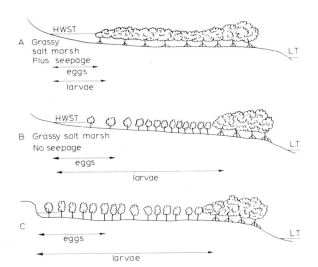

Fig. 11.2. Mangal in Sierra Leone showing three types of zonation and also the distribution of eggs and larvae of malarial mosquitoes (after Thomson, 1951).

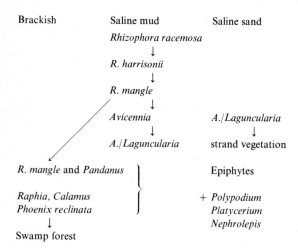

Scheme A. Generalised mainland zonation of Cameroun.

of Cuanza, *Avicennia* (on mud) or *Laguncularia* (on sand) are primary colonists associated with a ground flora of *Scaevola lobelia* and *Suaeda fruticosa* (Airy–Shaw, 1947).

SOUTH AFRICA

Little is known of the vegetation south of Angola until South Africa is reached. Temperatures here and around the Cape are too cool for mangal and the wet coastal formations are salt marsh. These have not been described in too much detail but the following account for Port Elizabeth is probably typical.

MacNae (1957) has described the marshes in the Zwartcops estuary near Port Elizabeth. The lowest marsh community, which is above a lower bed of *Zostera capensis*, is dominated by *Spartina maritima* with which is associated the red alga *Bostrychia*. The next zone is dominated by *Arthrocnemum perenne* and this in turn by a *Limonietum* (*L. linifolium*). This zonation is very reminiscent of that described earlier for around Oran. In wetter areas of both communities *Triglochin bulbosum* becomes a significant component of the flora. Other species common in the *Limonietum* are *Chenolea diffusa, Crassula maritima, Suaeda fruticosa* and *S. maritima*. In muddy areas the zone above the *Limonietum* is dominated by *Arthrocnemum africanum* or *A. pillansii*. Where the soil is sandy this zone is replaced by one dominated by *Sporobolus virginicus*.

In the *Spartinetum* the mud prawn *Upogebia africana* is the dominant animal but it is accompanied by the crabs *Sesarma catenata, Cyclograpsus punctata* and *Cleistostoma edwardsi*. The barnacles *Balanus elizabethae* and *B. amphitrite* clothe the lower stems of the *Spartina*. At higher levels the mangrove snail *Cerithidea decollata* with *Assiminea bifasciata* are quite abundant and both extend into the *Arthrocnemetum perenne* above. Fig. 11.3 shows the zonation of plants and animals in the estuary.

On the east coast the southern limit of the mangal is reached at the Kei River (Fig. 11.1) where only two mangrove species, *Rhizophora mucronata* and low *Bruguiera gymnorrhiza*, are to be found with an understorey of *Acrostichum aureum* and the occasional *Hibiscus tiliaceus*. *Avicennia marina* used to occur but now, rather surprisingly, reaches its southern limit at Kentani (Chapman, 1975). Farther north on the Bashee River *Avicennia* with *Bruguiera* are the pioneer species, and behind them there is a salt flat dominated by *Arthrocnemum natalensis* (West, 1945). Grewe (1941) included these Natal mangals in a geographic region which also included Moçambique. This region is interesting in that there are only the three species of mangrove present. Workers in this area (Day et al., 1954) report that the *Rhizophora* grows at a higher level than the other two species. For this

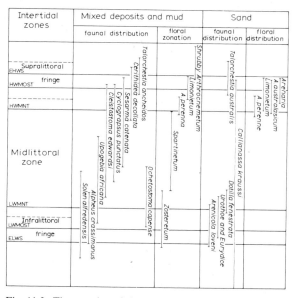

Fig. 11.3. The zonation of plants and animals in the Zwartkops Estuary (after MacNae, 1957).

reason the region would seem to be ideal for a detailed ecological study of the three species. The transition to brackish conditions is characterised by the advent of *Phragmites australis, Acrostichum aureum* and *Hibiscus tiliaceus*.

MacNae (1968) states that the dominant animals of these swamps are crabs, many of which occur farther north. The crabs include *Sesarma meinertii, S. ortmanni, S. eulimine, Uca annulipes* and *U. urvillei*. On the trunks and pneumatophores of the mangrove one finds the barnacle *Balanus amphitrite* var. *denticulata* and the mollusc *Assiminea*, and two polychaete worms, *Dendronereides zululandica* and *Lycastis indica*, have mud burrows. The fish are represented by a mudskipper, *Periophthalmus kalolo*, and a goby, *Gobius durbanensis*.

The other geographic region on the East African coast stretches from Lamu in the north to the Zambesi River and includes Moçambique and the island of Madagascar (Fig. 11.1). These mangal are luxuriant indeed, especially in the estuaries, and they also occur on offshore coral reefs.

EAST AFRICA

MacNae and Kalk (1962) have given a full account of the mangal on Inhaca Island (Moçambique). The salinity ranges from 19 to 42‰ in pools and channels, with the low values occurring in the rainy season. The mangrove zonation is thought to be related to soil salinity gradients, water table depths, pH and O_2 content of the soil atmosphere. The distribution of the mangal fauna, on the other hand, is determined by water table level, protection from the sun (i.e., canopy density), compactness of soil, food availability and the animal's own capacity to resist water loss.

The primary seaward colonist on the sandy soils is *Avicennia marina* with the slug–like *Peronia peroni* living among the pneumatophores. On the muddy soils up rivers and creeks *Rhizophora mucronata* replaces the *Avicennia*. On the pneumatophores and prop roots an algal turf of *Bostrychia, Catenella, Caloglossa, Murrayella* and the amphipod *Talorchestia malayensis* are to be found. The major portion of the mangal is comprised of *Ceriops tagal* and *Bruguiera gymnorrhiza*, the former in drier areas and the latter where it is wetter. A common parasite on the *Ceriops* is the mistletoe *Loranthus quinquinervis*. Towards the land *Avicennia* reappears and where the trees are not too crowded there is a ground cover of *Sesuvium portulacastrum* and/or *Arthrocnemum decumbens*. The former extends on to the sand flats behind where the salt content may reach high values. Here it is associated with *A. perenne* and a summer growth of *Salicornia pachystachya*. The landward fringe often contains trees of *Lumnitzera racemosa* with an occasional specimen of *Xylocarpus granatum*. All these mangrove species are Old World representatives and differ from those of the west coast or New World group (see Chapter 1). Along creeks the transition to brackish water is marked by the appearance of *Lumnitzera, Xylocarpus, Hibiscus tiliaceus* and *Thespesia acutiloba*. The final transition to fresh water is through a zone of *Juncus krausii* and *Phragmites*.

Along the creeks amid the *Rhizophora* roots there is a blue fiddler crab, *Uca urvillei*, with the allied *U. gaimardi* and *Ilyograpsus paludicola*. Flipping over the mud is the skipping goby *Periophthalmus sobrinus*. In the dense shade of the *Ceriops—Bruguiera* community *Uca* is replaced by species of *Sesarma*, whilst in damper areas and along creeks *Upogebia africana* and *Alpheus crassimanus* live and dwell in burrows. The molluscs *Littorina scabra* and *Cerithidea decollata* live on the lower mangrove branches with *Terebralia palustris* on the mud and *Cassidula labrella* on sandy soils. The general distribution of the fauna, which is more the fauna of mud soils than of mangroves, is shown in Table 11.1

TABLE 11.1

Distribution of mangrove fauna at Inhaca (Moçambique)

1. Landward fringe		*Sesarma eulimine, S. meinerti, S. ortmanni*
		Coenobia canpes, C. rugosa
		Cordiosoma carniflex
		Littorina scabra, Cerithidea decollata
2. Bare flat salinas		*Uca inversa* (very abundant)
3. *Avicennia* fringe		*Uca annulipes*
4. Channel		*Periophthalmus scobinus, Scylla serrata*
5. Shady mud banks		*Uca urvillei, U. gaimardi, Sesarma guttata,*
		Ilyograpsus paludicola, Macrophthalmus depressus
		Paracleistostoma fossula
		Upogebia africana, Terebralia palustris
		Cassidula labrella, Assiminea sp.
		Peronia peroni

Farther north at Inhambane and the mouth of the Zambesi the mangrove zonation is very similar except that a further pioneer species, *Sonneratia alba*, is very important. Its failure to extend southward appears to be due to cold winter temperatures (Chapman, 1975). For some reason which should be studied, mangal does not extend up the main Zambesi channel or the Rio Maputo (Lourenço Marques) to the limit of tidal influence. MacNae (1968) believes this may be related to high banks that stop tree roots from penetrating into the salt–water table. This clearly requires further study.

The east coast of Madagascar (Fig. 11.4) is not physiographically suited to mangal development, but there is abundant mangal on the drier west coast. *Sonneratia alba* and *Avicennia marina* are the primary colonists with *Rhizophora* and *Bruguiera* behind or up creeks. These latter do not reach a large size as they are heavily exploited (De la Bathie, 1921; Humbert, 1927; Jamet and De Balzac, 1930; Kiener, 1972). A species found here but not elsewhere on the mainland is *Xylocarpus moluccensis*. *Bruguiera*, *Ceriops tagal* and *Xylocarpus granatum* form the major part of the forest subject to tidal inundations, with the *Ceriops* extending landward onto higher ground. At the highest levels small trees of *Avicennia marina* are associated with *Heritiera littoralis*, *Xylocarpus* and *Lumnitzera racemosa* on creek banks. Where mounds of soil occur, *Calophyllum inophyllum* and *Barringtonia* sp. are to be found. According to Möbius (1921) and Voltzkow (1914) the communities described above all occur on the Seychelles and Comor Islands.

Some of the best East African mangal is to be found at the mouth of the Rufiji River and around Tanga (Walter and Steiner, 1936) (Fig. 11.5). *Sonneratia alba* is the pioneer on the open coast but at the mouths of rivers and upstream it is replaced by *Rhizophora mucronata*. On Ulenge Island the *Sonneratia* has a few frontal isolated trees of *Avicennia marina*, especially where the soil is rather sandy and firm. *Avicennia* also forms a zone behind *Sonneratia* on muddy soils where the mudskipper *Periophthalmus kalolo* and the crabs *Uca gaimardi* and *U. urvillei* are very common (MacNae, 1968). At Tanga the *Rhizophora* belt up the river is succeeded by a zone of *Ceriops tagal* (Fig. 11.6) with isolated trees of *Bruguiera*, which is not so abundant here as it is elsewhere.

The landward belt is dominated by *Avicennia*, the trees of which become progressively smaller and eventually disappear at the edge of a sandy salina. In this belt there is a ground flora of *Sesuvium*, *Arthrocnemum* and *Suaeda* with *Lumnitzera racemosa* in damper areas. The common crab is *Uca lactea* f. *annulipes*. The transition to fresh–water conditions is marked by the presence of the fern *Acrostichum*.

In the Rufiji Delta *Bruguiera* plays a more significant role in the zone behind the *Sonneratia*, the trees reaching up to 18 m high. Behind this zone is a major belt dominated by *Heritiera*

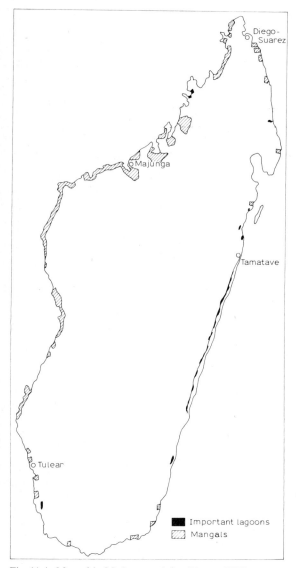

Fig. 11.4. Mangal in Madagascar (after Kiener, 1972).

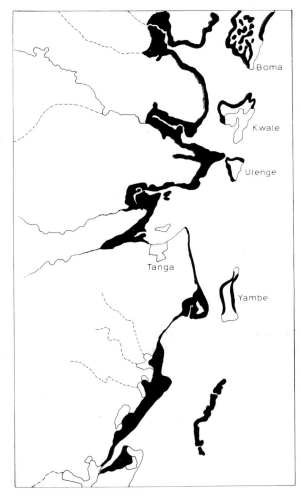

Fig. 11.5. Mangal in vicinity of Tanga (after Walter and Steiner, 1936).

littoralis with *Ceriops* and *Xylocarpus granatum*. This type of forest is a home for Sykes monkey *(Cercopithicus mitis)* and the otter *Lutra maculicollis* comes in to feed upon crabs.

On the coral reefs surrounding the offshore islands *Sonneratia* and *Avicennia* are both to be found, but from their distribution it would seem that chance determines which species occurs at any one locality.

Beyond the fact that mangal occurs northward to Lamu there are no detailed accounts of the swamps north of Tanga. North of Lamu mangrove does not occur again until the Somaliland mangroves are reached (see p. 217).

REFERENCES

Adams, C.D., 1957. Observations on the fern flora of Fernando Po. *J. Ecol.*, 45(2): 479—494.

Ainslie, J.R., 1926. The physiography of Southern Nigeria and its influence on the forest flora of the country. *Oxf. For. Mem.*, No. 5.

Airy–Shaw, H.K., 1947. The vegetation of Angola. *J. Ecol.*, 35: 23—48.

Boughey, A.S., 1957. Ecological studies of tropical coastlines. 1. The Gold Coast, West Africa. *J. Ecol.*, 45: 665—688.

Chapman, V.J., 1974. *Salt Marshes and Salt Deserts of the World*. Cramer, Lehre, 2nd ed., 392 pp. (complemented with 102 pp.).

Chapman, V.J., 1975. *Mangrove Vegetation*. Cramer, Lehre, 425 pp.

Day, J.H., Millard, N.A.H. and Brockhuysen, G.J., 1954. The ecology of South African estuaries. Part IV. The St. Lucia system. *Trans. R. Soc. S. Afr.*, 34(1): 129—156.

De la Bathie, M.H.P., 1921. *La végétation Malgache*. Paris.

Grewe, F., 1941. Afrikanische Mangrovelandschaften, Verbreitung und wirtschaftsgeographische Bedeutung. *Wiss. Veröff. Dtsch. Mus. Länderkd.*, N.F., 9: 105—177.

Humbert, H., 1927. Principaux aspects de la végétation à Madagascar. *Mem. Acad. Malgache*, 5: 78 pp.

Jackson, G., 1964. Notes on West African vegetation. 1. Mangrove vegetation at Ikorodu, Western Nigeria. *West Afr. Sci. Assoc. J.*, 9(2): 98—110.

Jamet, A. and De Balzac, H., 1930. Écorces de Palétuviers de Madagascar. Études Complémentaires. *Bull. Agence Gen. Colon.*, 23: 77.

Kiener, A., 1972. Écologie, biologie et possibilités de mise en valeur des mangroves malgaches. *Bull. Madagascar*, 308: 49—84.

Killian, C. and Lemée, G., 1948. Étude sociologique, morphologique et écologique de quelques halophytes Sahariens. *Rev. Gén. Bot.*, 55: 376—402.

Lawson, G.W., 1966. The littoral ecology of West Africa. *Oceanogr. Mar. Biol., Ann. Rev.*, 4: 405—448.

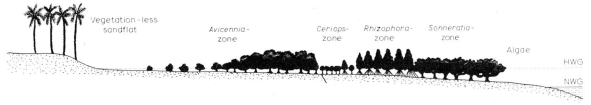

Fig. 11.6. Mangrove zonation at Tanga (after Walter and Steiner, 1936).

MacNae, W., 1957. The ecology of the plants and animals in the intertidal regions of the Zwartcops Estuary area, Port Elizabeth, S. Africa. *J. Ecol.*, 45(1): 113—131.

MacNae, W., 1968. A general account of the fauna and flora of mangrove swamps and forests in the Indo–West–Pacific region. *Adv. Mar. Biol.*, 6: 73—269.

MacNae, W. and Kalk, M., 1962. The ecology of mangrove swamps at Inhaca Island, Mozambique. *J. Ecol.*, 50: 19—35.

Möbius, K., 1921. *Eine Reise nach der Inseln Mauritius, im Jahre 1874/75.*

Rosevear, D.R., 1947. Mangrove swamps. *Farm For.*, 8: 23—30.

Savory, H.J., 1953. A note on the ecology of *Rhizophora* in Nigeria. *Kew Bull.*, 1953: 127—128.

Schnell, R., 1952. Contribution à une étude phytosociologique et géographique de l'Afrique occidentale: les groupements et les unités géobotaniques de la région guinéenne. *Mem. Inst. Afr. Noire,* No. 18: 14—234.

Simmoneau, P., 1953. Notes préliminaires sur la végétation des sols salés d'Oranie. *Ann. Inst. Cerc. Agron. Roman*, 3: 411—432.

Thomson, R.C.M., 1951. Studies on the salt water and fresh water *Anopheles gambiae* on the East African coast. *Bull. Entomol. Res.*, 41: 487—502.

Voltzkow, A., 1914. *Die Comoren.* Stuttgart, 380 pp.

Walter, H. and Steiner, M., 1936. Die Ökologie der Ostafrikanischen Mangroven. *Z. Bot.*, 30: 65—193.

West, O., 1945. Distribution of mangroves in the Eastern Cape province. *S. Afr. J. Sci.*, 41: 238—242.

Chapter 12

OUTLINES OF ECOLOGY, BOTANY AND FORESTRY OF THE MANGALS OF THE INDIAN SUBCONTINENT

F. BLASCO

INTRODUCTION

The Indian coasts fall within the bounds of the tropics, and their length is about 5100 km. On the eastern coast we find the gigantic deltas of the Ganges—Brahmaputra, Krishna—Godavari and Cauvery but there is no important delta on the western coast.

On a vegetation map of India (Fig. 12.1), it may be noted that, except in the Ganges and in the Andaman—Nicobar Islands, the mangals occupy a very small area. They are very often discontinuous and degraded. For a long time the vegetation of India has experienced many types of powerful degradation, particularly in the deltaic regions, where the population density generally exceeds 500 inhabitants per km^2.

GENERAL FEATURES

Area

The total area of the Indian mangals is estimated at about 356 500 ha as shown in Table 12.1.

The above figures bring out three noteworthy points:

(1) Nearly 85% of the Indian mangals are confined to West Bengal and to the islands of the Bay of Bengal.

(2) The mangals of Bangla Desh (600 000 ha) and those of West Bengal (India), covering about 200 000 ha, together represent one of the most important mangal areas in the world. For the sake of comparison, we may note that the mangals of the Mekong (South Vietnam) do not cover more than 250 000 ha (Vu Van Cuong, 1964).

(3) The present author's estimate of the total area of the Indian mangals is just one half of the

TABLE 12.1

Estimation of area of Indian mangals

Locality (see Fig. 12.1)	Approximate area under mangal forests (author's estimate)
Andaman and Nicobar Islands (Bay of Bengal)	100 000 ha
Gangetic Delta (West Bengal)	200 000
Mouth of Mahanadi	5000
Mouths of Godavari and Krishna	10 000
Cauvery Delta	1500
Bombay region	20 000
Saurashtra and Kutch coasts (Kathiawar)	20 000
Total	356 500

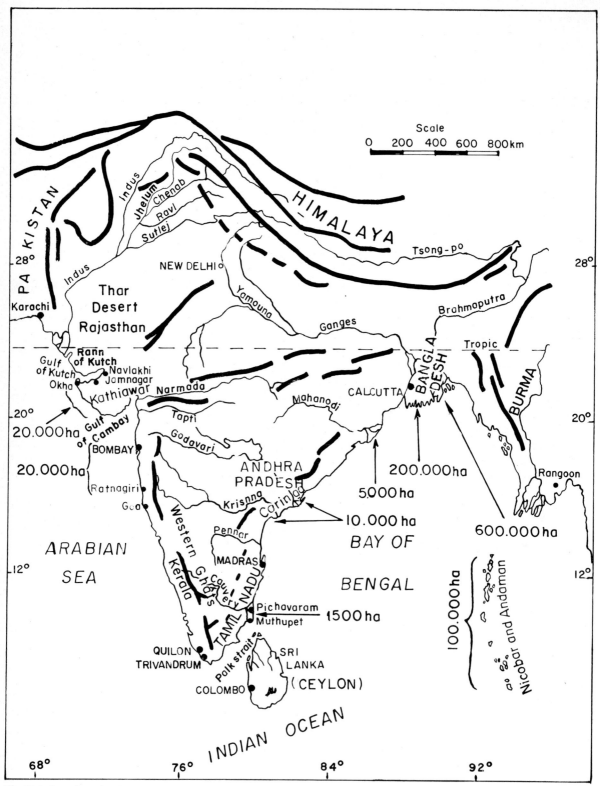

Fig. 12.1. Location of Indian mangroves.

official estimate (Sidhu, 1963). This difference may be attributed to the lack of a precise terminology in reference to the coastal formations of the subcontinent. In India, the term "mangrove forest" does not necessarily mean a forest cover. It may refer to various formations: arborescent, bushy, herbaceous and also regions which are devoid of any plant cover. As an example we may quote Qureshi (1959, p. 28): "Out of 97 sq. miles of *mangrove forests* in Saurashtra about 32 sq. miles are reported to be practically blank."

However, T.A. Rao and Sastry (1972) have suggested a new classification of Indian halophytes based on salt tolerance. This classification, which recalls that of Chapman (1942, 1954), has not yet been used by Indian botanists. Moreover, it is significant and surprising that the most complete book dealing with biogeography in India (Mani, 1974) devotes a few lines only to halophytes.

Our estimate of 356 500 ha is based mainly on our own field experience. Even this figure might be fairly high in view of the present very rapid regressive evolution encountered in certain regions. For example, in the Bombay region "reclamation for town planning and development of Greater Bombay and Twin Bombay is causing a rapid destruction of normal vegetation, especially the halophytes near the coasts from all sides" (Navalkar, 1974, p. 38).

Both anthropogenic and biotic influences lead to a noteworthy reduction in the floristic elements — so much so, only the very robust species are encountered in the secondary formations.

Flora

Thanks to the botanists of the past (Roxburgh, 1814; Griffith, 1836, 1851; Clarke, 1896; Prain, 1903; Blatter, 1905) and the present (T.A. Rao et al., 1963, 1972; T.A. Rao and Aggrawal, 1964), the foresters (V.S. Rao, 1950, 1959; Mathauda, 1959; Banerjee, 1964; Champion and Seth, 1968) and the present–day ecologists (Navalkar, 1940, 1951, 1956, 1959, 1974; Navalkar and Bharucha, 1949; Blasco and Caratini, 1973; Caratini et al., 1973; Blasco, 1975), we have a fairly good knowledge of Indian halophytes.

This flora, no doubt, belongs to the so–called "eastern" or "Indopacific" region. It is appreciably more diversified than that of the "western" or "Atlantic" region (see p. 233).

As shown in Table 12.2, there are 58 principal

TABLE 12.2

Distribution of principal halophytes in India

Species	West coast	East coast	Andaman and Nicobar Islands
Acanthaceae			
Acanthus ebracteatus Vahl			+
A. ilicifolius L.	+	+	+
A. volubilis Wall.			+
Aizoaceae			
Sesuvium portulacastrum L.	+	+	+
Apocynaceae			
Cerbera manghas L.	+	+	+
Asclepiadaceae			
Sarcolobus carinatus Wall.		+	+
S. globulus Wall.		+	+
Avicenniaceae			
Avicennia alba Blume	+	+	+
A. marina Vierh.	+	+	?
A. officinalis L.	+	+	+
Boraginaceae			
Heliotropum curassavicum L.		+	?

TABLE 12.2 *(continued)*

Species	West coast	East coast	Andaman and Nicobar Islands
Caesalpiniaceae			
Caesalpinia crista L.		+	+
Chenopodiaceae			
Arthrocnemum indicum (Willd.) Moq.	+	+	?
Atriplex stocksii Boiss.	+		
Salicornia brachiata Roxb.	+	+	?
Suaeda fruticosa Forsk.	+		
S. maritima (L.) Dum.	+	+	+
S. monoica Forsk.	+	+	
Combretaceae			
Lumnitzera racemosa (L.) Gaertn.	+	+	+
L. littorea (Jack.) Voigt			+
Convolvulaceae			
Stictocardia tiliaefolia Hallier f.		+	?
Cyperaceae			
Scirpus littoralis Schrad.		+	?
Euphorbiaceae			
Excoecaria agallocha L.	+	+	+
Ferns			
Acrostichum aureum L.	+	+	+
Stenochlaena palustre (Burm.) Bedd.	+	?	?
Gramineae			
Aeluropus lagopoides (L.) Trin.	+	+	?
Myriostachya wightiana HK. f.		+	?
Porteresia coarctata (Roxb.) Takeaka	?	+	+
Urochondra setulosa (Trin.) Hubbard	+		
Malvaceae			
Hibiscus tiliaceus L.	+	+	+
Meliaceae			
Xylocarpus granatum Koenig.		+	+
X. moluccensis (Lamk.) Roem.		+	+
Myrsinaceae			
Aegiceras corniculatum (L.) Bl.		+	+
Palmae			
Nypa fruticans Wurm.		+	+
Phoenix paludosa Roxb.		+	+
Pandanaceae			
Pandanus tectorius Soland.	+	+	+
Papilionaceae			
Derris trifoliata Lour.	+	+	+
Dalbergia spinosa Roxb.		+	+
Plumbaginaceae			
Aegialitis rotundifolia Roxb.		+	+
Rhizophoraceae			
Bruguiera cylindrica (L.) Bl.		+	+
B. gymnorrhiza (L.) Lamk.		+	+
B. parviflora (Roxb.) W. et A. ex Griff.	+	+	+

TABLE 12.2 (continued)

Species	West coast	East coast	Andaman and Nicobar Islands
B. sexangula (Lour.) Poir.		+	?
Ceriops decandra (Griff.) Ding Hou		+	?
C. tagal (Perr.) C.B. Robins	+	?	+
Kandelia candel (L.) Druce	+	+	?
Rhizophora apiculata Bl.	+	+	+
R. mucronata Lamk.	+	+	+
Rubiaceae			
Scyphiphora hydrophyllacea Gaertn.		+	+
Salsolaceae			
Salsola foetida Delile	+		
S. kali L.	+		
Salvadoraceae			
Salvadora oleoides Dcne.	+		
S. persica L.	+	+	
Sonneratiaceae			
Sonneratia alba Sm.	+		
S. apetala Buch.-Ham.	+	+	
Sterculiaceae			
Heritiera fomes Buch.-Ham.		+	
Tiliaceae			
Brownlowia lanceolata Benth.		+	+
Verbenaceae			
Clerodendrum inerme Gaertn.	+	+	+

halophilous non–parasitic species in India. Here, the genera *Bruguiera, Ceriops, Lumnitzera, Sonneratia* and *Xylocarpus (Carapa)*, which are absent in the "Atlantic mangals", play an important role.

The flora of the different deltas of India are affected considerably by the extent and the duration of the biotic and anthropogenic factors. *Rhizophora*, for example, is on the way to extinction, and has practically disappeared from the western coasts of India. It is rare in the deltas of the Krishna, Godavari and Mahanadi: even in the Gangetic Delta it is now less common.

The principal woody species of Indian mangals are those of *Avicennia* plus *Excoecaria agallocha*. These species are robust, with a very wide ecological amplitude and coppice readily. At present these are the best adapted to the considerable biotic pressures to which the Indian mangals are subjected.

Great ecological diversity

The following local factors are noteworthy:
(1) *Biotic factors*. These are very feeble or absent in the southern Gangetic Delta; contrariwise they are the main causes for the recent total disappearance of the Kerala mangals. In Godavari, cutting the mangals for firewood causes great damage and in Kutch over–grazing is excessive, particularly by camels.
(2) *Climate*. The climate varies considerably from region to region. Here, briefly, are the extremes: In Kathiawar, in the west, beyond 23°N, the climate is semi–arid, winter temperature is around 6°C. Violent winds are unknown. In the Gangetic Delta, in the east (22°N), the climate is humid, winter temperature is never less than 10°C, with extremely violent cyclones attacking every year.

Soils

Little has been written about them and no comprehensive study of the mangal soils of India is available. In a recent "Review of soil research in India" the mangal soils are grouped under the very vague heading: "Deltaic alluvium".

Our recent studies, with the help of personnel of the Office de la Recherche Scientifique et Technique Outre–Mer at Dakar, enable us to present the following provisional results:

(1) Granulometric analyses show variations from delta to delta. The texture is sandy to sandy-clayey in the basaltic region of Bombay (Elephanta Island), very fine and clayey in the eastern coasts, and loamy (silt types) in the Gangetic Delta.

(2) All the soils are very much saturated with exchangeable bases, those of the Ganges being less rich in this respect. Magnesium followed by sodium dominates in the absorbing complex.

(3) In general, the pH is neutral or slightly alkaline (between 7 and 8). However, in the Cauvery Delta, *Avicennia marina* formations grow on distinctly acidic soils, with pH between 3.5 and 4.5.

(4) Apart from the soils of the river banks, most of the Indian mangal soils are excessively saline, the conductivity exceeding 4000 μmho cm^{-1} (dilution 1:10).

Evolution of the deltas

In India, extensive areas of the forests along the river banks have been cleared, inducing a rapid evolution of the deltas due to the distinctly high rate of silting. The violence of the rains from June to November adds to this damage, so much so that most of the delta evolves at a great speed. On the eastern coast, the mangals of Muttupet (Cauvery) and of Coringa (Godavari) are now separated from the sea by a coastal sandy spit which makes the entry of sea water into the channels more difficult day by day. In these regions the tides are only of about 1 m twice a day.

These morphogenetic phenomena lead to a rapid ageing of the mangals.

Additionally, in the Gangetic plain, a tectonic movement leads to the tilting of the delta, which is higher in the western part (Indian territory) than in the eastern (Bangla Desh). Over the last four or five centuries, the Indian mangals of the Gangetic Delta have received practically no fresh water from the Ganges and this has caused profound floristic changes. Here spring tides are of about 6 m; neap tides about 2 m.

Present and past land uses

India is supposed to be the biggest shrimp- and prawn-producing country in the world. Yet there seems to be no methodical study for an organised scientific exploitation of mangals. However, one should make a distinction between the eastern and western coasts.

The mangals of the northwest region of the Indian Peninsula were under the jurisdiction of the Director of Marine Products (Pearl and Fisheries Department) until 1955, and there was no Forest Department. In these arid regions, the scarcity of wood and green fodder is so acute that it is very difficult to stop or to delay the destruction of the vegetation on which millions of the domestic ruminants feed. The degraded formations are transformed into discontinuous thickets with a total dominance of *Avicennia marina* var. *acutissima*.

By contrast, the foresters' working plans that have been carried out for the past fifty years on the eastern coasts have permitted a rational exploitation of the mangals. Nowadays, a clear felling every twenty years, on the average, is being practised. This technique is suitable to *Excoecaria agallocha* and *Avicennia,* which regenerate very well by vegetative means.

We may note that, in the Gangetic Delta, *Ceriops decandra* supports a felling every sixteen years very well.

Nowhere, other than for local use (see below), are the mangals now exploited commercially for tannin (see however pp. 353—355) which, for at least thirty years, has been extracted from Australian *Acacia* (chiefly *A. mearnsii*) introduced into the south Indian hills at the end of the last century.

However, the Indian mangals seem to have always been useful to the coastal people. The fishermen in particular, without any possible control, have been extracting from the mangals firewood *(Avicennia alba, A. marina, A. officinalis)*, planks and posts *(Sonneratia apetala),* wood for boat–building *(Heritiera fomes[1])*, floats for the nets *(Excoecaria agallocha)*, thatch material *(Myriostachia wightiana, Nypa fruticans)*, protective stains for the nets (bark of *Ceriops decandra*), wood for making toys *(Amoora cuculata)*, etc. Even now, the fishermen use a lot of bark from *Excoearia, Sonneratia* and *Kandelia candel*, which they boil to prepare an extract in which they soak their nets for two days. The nets then become rigid and resistant to salt water.

Apart from the honey, which is abundantly collected in the Gangetic Delta, the mangrove vegetation does not yield edible products. It must, however, be noted that the flowers of *Avicennia officinalis* are sometimes chewed with betel *(Piper betle)* and the pulpy fruits of *Sarcolobus carinatus* are edible. The leaves of *Suaeda* are also used as salad.

Natural dynamism

The Indian mangals are the victims of over–population. Practically nowhere do we notice the progressive dynamism of the woody elements. Everywhere there is a large regressive trend.

All the woody species of the mangals are light demanders. We note that presently most of the mangrove regions are monopolized by pioneers, mainly *Avicennia* spp. and *Excoecaria agallocha*.

Champion and Seth (1968, p. 163) and Chapman (1970, p. 11) have given a correct sketch of the evolution in the Godavari Delta. Further precise studies might explain the exact dynamic role of each of the three species of *Avicennia* found in India.

In any case, *Avicennia* and *Excoecaria* grow and remain established in the Indian mangals, thanks to their remarkable capacity to regenerate vegetatively, which permits them to subsist even when the mangal is destroyed entirely, without leaving any seed bearer.

Among examples of natural evolution of the flora, mention should be made of the practically total extinction of *Heritiera fomes* and of *Nypa fruticans* from the Gangetic Delta.

The dynamism of the herbaceous pioneer flora, on Indian coasts, has not yet been fully studied. We know, however, that in the Cauvery Delta the recent alluvial deposits are colonised by *Scirpus littoralis* (Cyperaceae); further north, in Godavari, it is *Myriostachia wightiana* which plays this role; in the Gangetic Delta, the pioneer is another grass: *Porteresia coarctata*.

Finally, the salinas bearing many hemixylous Chenopodiaceae *(Suaeda)*, which are a common feature in Kathiawar, do not generally show any sign of progressive evolution of the plant cover. Their denudation seems to be an irreversible phenomenon.

LOCAL DIVERSITY

One important point, set out in the previous pages, is that the Indian mangals, subjected to excessive anthropogenic and biotic aggression, are everywhere under regression and in some localities even on the way to extinction. The best–preserved formations are those of the Gangetic Delta, where the threat of their destruction increases day by day. In this section we will touch very briefly on each of the major mangal formations of the eastern and western coasts and those of the Andaman and Nicobar Islands.

Bengal mangals (Fig. 12.2)

Bengal belongs to two countries, India and Bangla Desh. It is located at the confluence of two big rivers: the Ganges and the Brahmaputra which together form the Gangetic Delta, supporting probably the most extensive mangrove forests of the world: 2000 km^2 in the Indian territory and 6000 km^2 in Bangla Desh. In both countries this deltaic region is known as *Sunderbans,* a Bengali term meaning beautiful *(Sunderi)* forests *(bans)*. In fact they are the most beautiful halophyte formations of India.

These extremely dense, not easily accessible forests, are the abode of the famous Bengal tigers, which are on the way to extinction.

Foresters have recognised the wealth of these forests from a very early date and prepared a cartographic survey accompanied by several rather precise enumerations (Curtis, 1933). However, botanists and ecologists have not yet paid much attention to these vegetation types.

---

[1] *H. fomes* Buch.–Ham., not *H. minor* Roxb., according to Kostermans (1959).

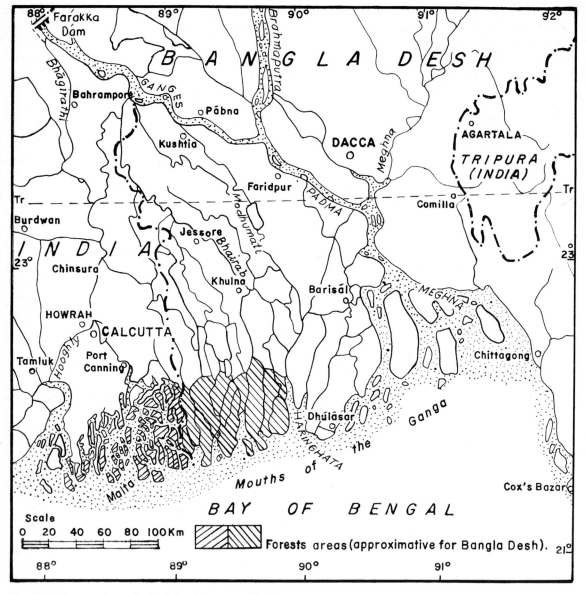

Fig. 12.2. Mangrove forests in the delta of Ganges and Brahmaputra.

Recently the present author visited these mangals and was able to obtain some first-hand knowledge of the ecology of the four or five major vegetation types recognised in these deltas.

Principal ecological factors

There are four important factors: lack of fresh water, violence of cyclones, alarming demographic situation, and soils.

Lack of fresh water in the Indian territory. The Ganges traverses the whole of northern India but its mouth is in Bangla Desh. Its ancient delta now suffers due to an extreme scarcity of fresh water, the only source of which is the monsoonal rain, from May—June to October—November. According to Gupta (1957) the natural modification of the lower course of the Ganges is a recent phenomenon which has taken place during the last five or six centuries. There seems to have been a tectonic and morphogenetic uplift of the western part and the river has naturally shifted its course towards the east.

The consequences on the mangals are many and are under investigation. West Bengal does not receive the fertile alluvium as before, and the terrain is not desalinated by the river floods. The most evident manifestation of these ecological changes is a floristic substitution (disappearance of *Heritiera* and *Nypa* which need leaching of the soil by fresh water). This point will be elaborated later on.

To cultivate the presently over–saline land and to save the threatened forest types, the Indian authorities have naturally thought of artificially restoring the ancient fluvial connections. It is with this aim, and also to avoid the rapid silting of Calcutta port, that the Farakka barrage across the Ganges has been constructed since 1971. Unfortunately, the commissioning of this dam now poses serious problems to India and Bangla Desh (see Fig. 12.2).

Humid climate characterised by violent cyclones. In these regions the total annual rainfall is between 1500 and 2500 mm. The dry season extends over five months from December to April.

At the relatively higher latitudes of these regions (22°N) the climate is still warm (tropical). The average temperature of the coolest month is 20°C (in January), the absolute minimum is above 10°C and annual average is about 26°C.

The most important climatic factor is the high frequency of the violent tropical cyclonic depressions (four to eight per year). Fosberg (1971) has aptly remarked: "The situation of the Gangetic Delta, at the head of the funnel shaped Bay of Bengal, poses perhaps the most serious threat from surges driven by storm waves to be found anywhere in the world (except on low coral atolls)." Cyclones are frequent, particularly from August to November. The marine invasion is then very important, the waves reaching 5—8 m high.

Explosive demographic situation. The population density of present Bengal is more than 1000 km^{-2} and the demographic increase is exceedingly high. In this impressive human concentration, which day by day reclaims more and more land for cultivation, one may foresee that the nearest mangals would not remain for very long, in spite of the desperate efforts of the Wild Life Conservation Services.

Some data on the regression of the Bengal mangals are available. The foresters estimate that the forest area of the delta has been reduced by half over the last two or three centuries. Banerjee (1964) estimates their regression during the last 100 years to be about 1500 km^2.

Mazumdar (1932) gives interesting data on cultivation in Gosaba Island, which at the beginning of the century was a virgin land. The impenetrable mangals were surrounded by channels rich with fish, but now the fishermen of Gosaba and the neighbouring villages have to go far south to get a good catch. In these regions, without any road connection to the neighbouring city, fish constitutes the principal source of animal protein. A scientific piscicultural scheme will be a great boon to these regions and will be rewarding to the country as well.

Soil. As far as the present author knows, the soils of this region have not yet been investigated. Here are some important general characteristics:

(1) In all the samples collected 40—60% of the soil is silt; the amount of coarse sand does not exceed 2%.

(2) The amount of organic matter remains very constant (between 1 and 1.5%).

(3) The pH of the fresh samples and the air–dried samples is always about 8.

(4) These soils are highly saturated with exchangeable bases (Ca^{2+}, Mg^{2+}, K^+, Na^+).

The measurements of salinity, to be significant, should be made several times a year during different seasons. Our results show that at the beginning of the dry season, there is a very high salinity with a conductivity of 1500—2000 μmho cm^{-1} (dilution 1:10). At the end of the dry season the salinity will be excessive, and this factor considerably limits the development, maintenance or regeneration of the less halophilous species.

Principal vegetation types

In the entire Gangetic Delta, we can distinguish four great groups of halophytes:

Curtis (1933)	Champion and Seth (1968)	Blasco (1975)
1 salt–water forests	mangrove scrub	back–mangals
2 mangrove forest	mangrove forest	dense mangals
3 moderately salt–water forests	salt–water mixed forest *(Heritiera)*	tall dense mangals with *Heritiera*
4 Fresh–water	brackish–water mixed forest *(Heritiera)*	
5	palm swamp	*Phoenix paludosa* belts

The principal physiognomic and floristic characteristics of each of these formations are given below.

Back–mangals. Here, as in the other Indian deltas, the back–mangals are bushy, discontinuous thickets of halophytes, often developed on soils situated at the equinoctial high–tide level. During the dry season, these lands become very saline at the surface where a thin crust of salt is deposited. As everywhere else in India, species of *Suaeda, Sesuvium, Heliotropium,* etc., are common.

A type with *Aegialitis rotundifolia,* the only member of the Plumbaginaceae in the Indian mangals, is confined to these regions, but this species is not very common.

From the geographical point of view, the back–mangals, as their name implies, are the most inland part of the mangal. They are therefore situated in that part of the delta where land reclamation is very acute. That is why these formations are found only on small areas in the north of the delta.

In spite of the high saline concentration, the land can be reclaimed for cultivation, but only during the rainy season, when the fresh rain water is enough for the cultivation of the local variety of rice, the one now cultivated commonly being *patnai*.

Dense mangals and Phoenix paludosa belts. These are the principal existing types in the Indian part of the Gangetic Delta. They entirely cover a multitude of islands and islets, 3—5 m above the water level.

The plant cover is very dense, there being on the average about 12 trees (of 10—18 m), 440 shrubs (of 4—8 m), 2200 under-shrubs (less than 4 m in height) per ha.

The floristic composition varies considerably from one island to another. The dominant species is sometimes *Excoecaria agallocha* or *Sonneratia apetala*. Each of these gregarious species may represent a practically pure stand. Sometimes, in the same island, a dense forest of 15 m dominated by *Sonneratia* may, suddenly, without any distinct transition, be followed by a thicket about 4 m high, wherein the species of *Ceriops* with a few *Aegiceras corniculatus* are the principal woody plants.

The dense mangals of Indian Bengal are the richest in species. None of the tree species of the other Indian mangals are absent here. The important arborescent and shrubby species are given here, in descending order of their abundance: *Excoecaria agallocha, Ceriops decandra, Sonneratia apetala, Avicennia* spp., *Bruguiera gymnorrhiza, Xylocarpus granatum (Carapa obovata* Bl.*), X. moluccensis (Carapa moluccensis* Lamk.*), Aegiceras corniculatus, Rhizophora mucronata*.

Mention must be made of the "sea date", *Phoenix paludosa,* which is found all over the delta, but always at the edge of the water. This elegant palm, of 3—5 m height, constitutes a narrow band along the banks and so serves as a "fixing" agent. *Phoenix paludosa* survives even if the stipe is submerged for a long period. It also resists high concentrations of salt at the surface of the soil. Being very gregarious its stands are extremely conspicuous between the water's edge and the mangal proper. For this reason we feel that its communities must be separated from the other mangal formations. A *Phoenix paludosa* belt may be found over a stretch of 60 km or even more inland, on the embankments where there is heavy accumulation of salts during the dry season. This palm is unknown in the other mangals of the Indian Peninsula, except in the neighbouring delta of Mahanadi.

Some autecological remarks:

Excoecaria agallocha, Xylocarpus granatum and *X. moluccensis* seem to prefer the areas immediately behind the *Phoenix paludosa* belt; these terrains, raised to 3—5 m above the water level, are submerged by sea water mostly during exceptional tides.

Rhizophora mucronata and specially *Bruguiera gymnorrhiza* are rather characteristic of the borders of the local drainage channels (called *Khals* in Bengali) where the soil is submerged daily by the high tides.

Ceriops decandra, on the other hand, occupies the highest terrain.

Sonneratia apetala is noticed mainly along the gently sloping borders where the soil is constantly soaked in water.

Porteresia coarctata (Gramineae) fixes the recent alluvial deposits. Very abundant in the north of the delta, it becomes rare in the south.

Avicennia is distinctly more numerous near the inhabited zones where the biotic and anthropogenic interferences are frequent.

Nypa fruticans is nearing total extinction in West Bengal. The rare small stands observed, are located at the edge of the water. This species, which is less competitive, needs a constantly humid soil and frequent supply of fresh water. It is no longer in ecological equilibrium, since the Ganges has stopped flowing into the Indian part of the Gangetic Delta. The cause of its disappearance is not only ecological but also anthropogenic and biotic. This palm is over-exploited for the leaves which are highly valued for thatching, and it is observed that the deer *(Axix axix* and *Muntiacus muntjak)* uproot the *Nypa* palms for unknown reasons.

Tall dense mangals with Heritiera. These forests, the most beautiful of the delta, are known only in the eastern part of the delta, i.e., in the territory of Bangla Desh. They may be characterised by the abundance or the dominance of *Heritiera fomes.* This member of the Sterculiaceae attains a height of 30 m and forms almost pure stands, particularly in Burma (Arakan coast and Tenasserim). It is a gregariously developing element, producing an extraordinary network of conical pneumatophores. Very often *Bruguiera gymnorrhiza* is found growing with *Heritiera.* Sometimes, however, *Acrostichum aureum,* a pantropical fern, is able to accomodate and flourish in the neighbourhood of *Heritiera.*

Heritiera fomes (*Sundri* in Bengali) is one of the most useful trees of India. Its timber is very durable; in an inundated medium the posts of this species can resist for almost 15 years. Its wood is strong but easy to work on and takes a beautiful polish. It resists well to insect attacks and to a certain extent even marine borers. It is used for boat building. This tree is very much valued and over-exploited. This should explain the considerable reduction of these forests since the beginning of this century.

The most beautiful mangals of *Heritiera* are those which develop on the young fluvial deposits where the influence of the marine water is significant but not preponderant. They are therefore localised inland and absent towards the south of the delta. The western part of the delta (i.e., West Bengal of India), at present not receiving the Gangetic floods, does not contain *Heritiera,* though 5000 years ago the arborescent vegetation of the Calcutta region was dominated by *Heritiera* (Mukherjee, 1972).

These notes enable us to understand that there are in fact two types of *Heritiera fomes* communities: one at the top of the delta where the influence of the fresh water from the rivers is constant and dominant ("brackish–water mixed forest") and the other, towards the middle of the delta ("salt–water mixed forest"), which is only a complex transition between the communities of the less saline soils and the dense mangals studied previously. In the "brackish–water mixed forests" of Champion and Seth (1968), *Heritiera* finds its ecological optimum. Further south, its growth is delayed, the trees are smaller and the ecology is suitable for Rhizophoraceae.

Such are the principal characters of the mangals of Bengal, the most important and the least disturbed in India.

Mangals of the Krishna and Godavari deltas

These two large deltas are in contact with one another at about latitude 17°N. Their mangals, which cover a total of about 10 000 ha, differ considerably from the other Indian mangals because of their dominant trees: *Sonneratia apetala* and several *Avicennia* spp. In addition there is a grass, *Myriostachia wightiana,* which is very common here, but practically unknown elsewhere.

From the ecological viewpoint, in the region of Coringa particularly, there is the development of a sandy offshore bar which daily diminishes the marine influence in the mangals. On the contrary, during the heavy flooding of the rivers, mostly in October and November, fresh water invades large areas. The seasonal variation of the water salinity is thus very high, changing from 20—30‰ during the hot, dry season (March—May) to less than 10‰ during the rainy season. The Indian arborescent genera which are most adapted to support such variations of water salinity are *Avicennia, Sonneratia* and especially *Excoecaria.*

It is not necessary to give here data on the soils, as they resemble very much those of Muttupet (Cauvery Delta) which we will discuss later on. Besides, T.A. Rao et al. (1972) have published some figures concerning the mouth of the Coringa, where there are still some important mangals.

In these regions there are two main types of halophytic populations: those growing on the river banks, and those which grow on the ancient alluvial soil in the interior of the delta.

Mangals of river banks
About 9 km inland, the bank of the Coringa support tall, dense, closed thickets composed of halophytic trees, climbing shrubs and herbs. At such a distance from the ocean, the banks are low, about a metre above the water level. The tides are practically imperceptible.

The physionomy of these riparian thickets is simple: it consists of a band 4—5 m wide and 1.5 m high of *Myriostachia wightiana* (Gramineae) and *Cyperus* spp. The rhizomes of these species are constantly under water.

Behind this band the thicket is very compact, with a good representation of *Avicennia officinalis* and *Hibiscus tiliaceus*. Both these species attain a height of 3—6 m. In the present author's opinion nowhere else in India is this member of the Malvaceae represented in such great numbers. In these thickets of *Avicennia—Hibiscus*, we also find some scattered *Sonneratia apetala* measuring 8—12 m high.

These tall thickets also comprise a number of very thorny, climbing shrubs which render them not easily penetrable. Here, we find particularly *Caesalpinia crista (C. nuga)* and *Dalbergia spinosa*. Further, *Stictocardia tiliaefolia* (Convolvulaceae), rare or absent in other mangals, is very common here.

In the vicinity of these thickets under-shrubs such as *Acanthus ilicifolius* and *Clerodendrum inerme*, which snatch the place of the cut trees, can be observed.

Towards the ocean, the marine influence increases and the flora is modified. *Avicennia* persists but the dominance is that of *Sonneratia apetala*. Some rare Rhizophoraceae appear, whereas *Hibiscus tiliaceus* and *Stictocardia tiliaefolia* disappear completely.

Low bushy mangals on old alluvial soils
Towards the interior of the delta, the physiognomy of the mangal is that of a discontinuous thicket about 0.5 to 2 m high. We note everywhere a profound mark of anthropogenic and biotic interferences. It is observed that all the present trees have a rapid but distorted vegetative regeneration, which gives them an irregular bushy appearance. These trees are mostly *Avicennia (A. officinalis, A. alba, A. marina), Excoecaria agàllocha* (Euphorbiaceae without pneumatophores) and also some *Prosopis spicigera*.

The structure of this formation comprises a discontinuous bushy stratum and also a herbaceous stratum which tends to cover the soil, and which includes plants of *Suaeda maritima, Heliotropium curassavicum, Fimbristylis spathacea, Aeluropus lagopoides, Cressa cretica,* etc., species commonly found in the back-mangals.

Since the characteristic species of the mangals, particularly those of Rhizophoraceae, are absent, we may say that the actual thickets of the Godavari and Krishna represent, to a greater extent, a stage of ageing of the primitive mangal. In addition, in a good number of secondary thickets, the daily rhythm of the tides is no longer perceptible.

Regional flora
When we cross the delta, the general impression is one of a great floristic simplification. All halophytic vegetation types are poor in woody species. This probably results from an ancient method of forest exploitation, i.e., a clear felling of the trees every 15—25 years, depending on the region. The new population is composed of species which coppice abundantly, whereas the other species become rare or disappear. Mathauda (1959, p. 74) who studied the regeneration, states that "*Avicennia officinalis, A. marina, Excoecaria agallocha* and *Lumnitzera racemosa* form nearly 90% of the stock in mangroves of the Godavari and Krishna deltas. The first two species occur in lower ground. They regenerate satisfactorily from seeds... Seedling regeneration in the case of *Excoecaria* and *Lumnitzera*, which together occur above the *Avicennia* belt, is deficient."

For the whole of the two deltas, the floristic

inventories of Venkateshwarlu (1944) and of R.S. Rao (1959), show 26 species, which seems to be relatively rich. In the field, however, it is to be noted that almost all the species of Rhizophoraceae and Meliaceae are rare and on the way to extinction.

In this region the floristic influence of the Ganges is still felt with the presence of *Porteresia coarctata* and *Myriostachia wightiana* (Gramineae), mentioned, and also *Sarcolobus carinatus* (Asclepiadaceae). This last species is important further east, particularly in the mangals of Burma.

The evolutionary tendency of most of the mangals of these deltas is towards a reduction, as more and more trees are exploited for firewood. The discontinuous thickets become more and more open and give place to Chenopodiaceae which are an index of evolution towards over-saline scalds (salinas).

Mangals of the Cauvery Delta

In this large delta, in the south of the peninsula, the actually wooded surface of the mangals is probably not more than 1500 ha. Nevertheless, the mangals found here are among the more interesting ones of India for two reasons:

(1) A more or less complete zonation is seen in Pichavaram; it is here that we have studied the pollen content of the surface sediments.

(2) The forest management methods for the Muttupet region have resulted in beautiful, pure formations of *Avicennia marina*. These regions, situated in Tamil Nadu, have been studied by foresters (Venkatesan, 1966), botanists (Rajagopalan, 1952), ecologists and palynologists (Blasco and Caratini, 1973; Caratini et al., 1973).

Characteristic and complete zonation of Pichavaram (Fig. 12.3)

(1) A narrow *Rhizophora* belt on clayey, constantly wet or soaked terrain.

(2) A wide *Avicennia* belt on terrain submerged mostly during equinoctial tides.

(3) A back-mangal on generally very saline sandy terrain. Only exceptional tides reach this belt.

Rhizophora belt. This is well developed along the channels.

In this part of the delta the tides are feeble, about 0.5—1.5 m, twice a day. The vegetation belt of the "slikke" (a very fluid bed with an undifferentiated profile, submerged at each tide) supports essentially Rhizophoraceae 7—8 m high. *Rhizophora apiculata* is the most common, but *R. mucronata* is also abundant.

Under the *Rhizophora* spp., which constitute the dominant stratum, we may recognise two more macrophyllous Rhizophoraceae: *Bruguiera cylindrica* and *Ceriops decandra*. A few scattered *Sonneratia apetala* can also be observed. It is one of the tallest trees of the Cauvery (12—15 m tall, 50 cm girth), but, being selectively exploited, its extinction by anthropogenic factors is almost complete.

This belt of Rhizophoraceae (5—10 m is width)

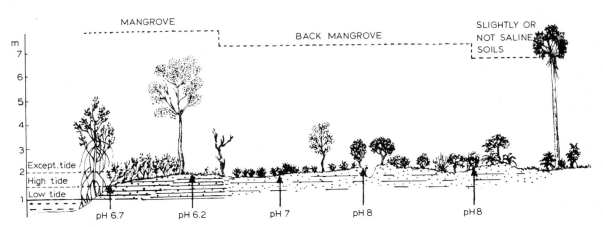

Fig. 12.3. Mangrove zonation in the Cauvery Delta.

also consists of the rather common *Lumnitzera racemosa* (a member of the Combretaceae with lanceolate leaves) and *Aegiceras corniculatum* (a member of the Myrsinaceae with thick leaves). Two climbing shrubs, *Dalbergia spinosa* and *Derris trifoliata*, are common among evergreen trees and shrubs. This vegetation zone is a thick, dense curtain, hardly penetrable.

There are no epiphytic angiosperms but *Dendrophthoe falcata* is a common parasite on *Rhizophora*.

Avicennia belt. Behind the very dense formation of Rhizophoraceae, there is an entirely different vegetation type, invaded mostly by equinoctial tides, in which *Avicennia marina* is conspicuous ("white mangrove tree"). The physiognomy is not dense, 3—6 m high, and small trees or shrubs are often mutilated to a great extent. Under them occur dwarf under–shrubs *(Suaeda maritima,* stunted *Excoecaria agallocha)*, which are rather dense in some places, but never closed.

Back–mangal. The back–mangal is the third vegetation belt. It does not consist of *Avicennia*. The flora includes mostly herbs and hemixylous species: *Suaeda maritima, S. monoica, Sesuvium portulacastrum, Heliotropium curassavicum, Salicornia brachiata, Aeluropus lagopoides,* etc. These species growing on firm muddy soils are highly tolerant of salt. This is a classical example of back–mangal, with its flora, accumulation of salt over the soil surface, and the irregular immersion régime.

The environment of the back–mangal favours growth of *Acanthus ilicifolius,* a bush with spinescent leaves and large blue flowers blossoming from June to August. It is also suited to *Excoecaria agallocha,* a highly tortuous arborescent member of the Euphorbiaceae, which is the only deciduous species in the Indian mangals. In Pichavaram, the proportion of sand in the back–mangal may reach 60% and the pH normally ranges from 7.5 to 8.5.

Pollen spectrum of recent sediments

The study of fossil pollen grains of these regions has given us some provisional results (Blasco and Caratini, 1973; Caratini et al., 1973) which may be summarized as follows:

A comparison of the average actual pollen spectrum (regrouping the results of all the surface samples examined) with the floristic composition of these regions shows that the pollen spectrum does not give a faithful image of the floristic composition. *Rhizophora* and *Sonneratia* are over-represented while the Chenopodiaceae are represented in normal proportion in the pollen spectrum. *Avicennia* and *Excoecaria* are very poorly represented in the sediments, which besides contain no pollen grains of *Derris trifoliata, Aegiceras corniculatum, Azima tetracantha* or *Clerodendrum inerme*.

It must be noted that, on the average, 25% of the pollen grains found in the sediments are allochthonous and do not belong to mangroves. *Thus, there is a strong distortion between the mangal vegetation of Pichavaram and the representation of its pollen in the actual surface sediments*[1].

Forest management of the Muttupet—Chatram mangal is the most characteristic and the most efficient one in the southern part of the delta. The natural zonation has been completely wiped out and there are only pure stands of *Avicennia marina*. There are no longer any members of the Rhizophoraceae or of *Sonneratia*.

The forest operations comprise a clear felling every 25 years, leaving *in situ* about 50 seed bearers per ha. Recently this cycle has been reduced to 20 years and no seed–bearing stock is left, and regeneration is mainly through coppices. However, the foresters take care to collect in nets the floating seedlings of *Avicennia* and plant them during low tides in sites for reforestation. The seedlings are preferably pressed slightly by foot into the mud.

The effect of this policy is to produce a pure forest stand of *Avicennia marina,* composed of trees with multiple stems developed vegetatively. At 15 years, the trees measure 10—15 m in height, 35—45 cm in girth.

Our soil studies bring out three important points:

(1) Among all the mangal soils which we have examined, they are the only ones which are distinctly acidic (pH less than 4 at 30 cm depth).

(2) They are very saline (conductivity more than 4000 μmho cm^{-1}, dilution 1:10) and highly saturated with exchangeable bases. Magnesium dominates the complex.

(3) As in the soils of Godavari, the texture is very fine.

Finally, we may note that the recent clayey

---

[1] Studies conducted also by G. Thanikaimoni.

sediments are colonised by *Scirpus littoralis* (Cyperaceae) which does not have such great importance anywhere else in India.

Mangals of the basaltic coasts of Bombay Region

In these western coasts there are no big watercourses. The narrow and very rainy littoral plains (annual rainfall more than 2500 mm from June to October) support small stretches of mangals on the basaltic creeks and on small islands. The most fully explored are those of the Bombay region (thanks to the works of Navalkar and Bharucha, 1949), Ratnagiri (Bhosale and Joshi, 1973) and the Goa region (Shinde and Mustafa, 1974).

The unity and the distinctness of the geological substratum is its Trappean basalt of the Cretaceous, Paleocene and Eocene periods which is very resistant to erosion. This type of rock is totally unknown on the other coasts.

Practically everywhere the anthropogenic pressures are so strong that there remain mostly stunted formations of more or less dense *Avicennia* and *Acanthus ilicifolius*.

It has been shown that during the last few years there has been a constant regression of the mangals of these regions and some of the species have become extinct due to human interference. Navalkar (1951, p. 157) remarked that *Lumnitzera racemosa* was present in the Bandra region in 1934 but now it is no longer to be seen.

Under such conditions, an understanding of the natural zonation is very difficult, although in Ghodbunder (north of Bombay) there are a few transects where the original plant succession remains visible. The belt nearest the sea consists of several Rhizophoraceae *(R. mucronata, Ceriops tagal, Bruguiera cylindrica)* and some scattered *Sonneratia apetala*. Further inland, the dominant shrubs are *Avicennia*[1] and *Excoecaria agallocha*. Beyond the reach of the daily tides, there appears a lawn in which *Aeluropus lagopoides* (*A. villosus* Trin.), a graminaceous creeper, is very common.

In the volcanic Elephanta Island, situated in the Bay of Bombay, a complete zonation is no longer found. There are only three arborescent species: *Avicennia alba*, *Sonneratia apetala* and *S. alba*, a species which the present writer has not seen elsewhere in the Indian Peninsula. Its abundance is quite surprising, and it is tempting to suggest that it might be a recent introduction in this important tourist centre.

In these formations, where all the shrubs are very much mutilated and traversed constantly by cattle, we do not find any other species. *Avicennia alba* occurs towards the sea, whereas *Sonneratia alba* occurs inland. The soil is covered by a dense mat of small pneumatophores of both species. The soil, a sandy–clayey deposit with the pH ranging between 7 and 8, is very shallow (depth less than 30 cm).

Floristic remarks

In the whole western coast, the halophytic flora does not contain any palm, Sterculiaceae *(Heritiera)*, or Meliaceae *(Xylocarpus)*. The present author believes that even *Kandelia candel* (= *K. rheedii* W. et A.), reported by Navalkar (1956) and Cooke (1901—1908) should now be excluded.

Another floristic feature is the presence of two *Salvadora* species *(S. persica* and *S. oleoides)*, both of which are quite characteristic of ancient mangals. *S. persica* is also found on the eastern coast.

Concerning the Rhizophoraceae, it should be emphasised that they are on the verge of extinction, especially on the whole of the western coast and in practically all the Indian mangals. In some places, we know the precise cause of their recent disappearance. At the beginning of the present century, Bourdillon (1908, as cited by Troup, 1921, Vol. 2) noted *Bruguiera gymnorrhiza* and two *Rhizophora* species to be very common in the Quilon region. They have disappeared today, being over–exploited during a long time for firewood and for tannin. The increasing rarity of these species in some mangals is also due to the fact that their seedlings are consumed by herds of cattle, and sexual reproduction has become quite insufficient. Since the species of *Bruguiera*, *Ceriops* and *Rhizophora* do not coppice, their survival in the present mangals is rather difficult and often impossible.

Mangals of the arid zone of Kathiawar

This coastal region in the northwest of India (lat. 23°N) is the driest in the country; the bio-

---

[1] In these regions mostly *A. alba* has been reported, but Moldenke (1960, p. 226) feels that it is *A. marina* var. *acutissima*.

climates near Jamnagar are sub–desertic: nine dry months with annual rainfall of about 400 mm.

There are two kinds of halophytic formations: One is immense and found in the interior of the peninsula; it is of no great economic value to the local population. The other is coastal, not very extensive, but it is of regional economic importance.

Salt marshes of the interior

The better known ones are those of "Little" and "Great Rann of Kutch". The origin of the salt in this peninsula is still disputed (Auden, 1952; Ghose, 1964). Several factors, more or less complementary, should be taken into account. They are geological or tectonic (progressive silting up and uplift of the Gulf of Cambay), topographic (poorly drained plains, underground water level near the surface), climatic (the aridity accentuates a more or less pronounced halomorphism), etc.

Fig. 12.4 gives an idea of the extraordinary extent of these vegetation types. Their physiognomy is that of an under–shrubby discontinuous thicket, in which the soil appears almost naked, as though sterile, over vast areas. Locally, the vegetation becomes more dense but never closed.

Some small trees *(Salvadora persica* and *Prosopis juliflora)* may be quite common in some saline depressions. The latter has been introduced from Mexico. The sporadic presence of some other trees like *Tamarix dioica* and *Acacia nilotica* subsp. *indica,* is mostly a question of chance. Even among the dwarf under–shrubs the flora is poor; it consists practically of *Suaeda fruticosa,* which is the commonest, *Salsola baryosma* and *Haloxylon salicornicum.*

On the other hand, Gramineae and Cyperaceae seem to be better adapted to the hydromorphic conditions. Their flora is more diversified, including especially *Aeluropus lagopoides,* species of *Cenchrus (C. setigerus, C. ciliaris), Sporobolus (S. marginatus, S. helveticus)* and among the sedges, *Cyperus rotundus.* Complete lists of plants are given by Blatter et al. (1927, 1928) and Saxena and Gupta (1973).

A thing rare in India, the country people and the animals are not mainly responsible for this vast, more or less empty, area. Here, it is the very special soil, with the following principal characteristics of a profile as observed during dry periods (October to May): (1) a cracked, saline crust, 1—5 mm thick; (2) a clayey or loamy–clayey horizon, varying thickness, dark grey colour; (3) a bed of gypsum and sodium chloride crystals; and (4) a very clayey, compact horizon, bearing stains of rust.

In general, at about one metre depth, the soil is very hydromorphic with a predominance of clay with a clear, grey colour.

As emphasised by Chapman (1974) these saline and alkali soils are distinct from the maritime mangal soils.

The mangals and derived formations

Although the foresters' official estimate of the mangals of these regions is 52 000 ha, the present author does not think that it exceeds 20 000 ha.

The conditions of maintenance for these mangals are particularly unfavourable because of the scarcity of wood and fodder in this arid region, which nevertheless has a population density of 80—100 inhabitants km^{-2}. Further, winters are cold. It is in these mangals that we note the lowest winter temperatures of 5—8°C.

For the numerous coastal ports (Salaya, Navlakhi, Kandla, Mundra, etc.) the mangals are the only source of firewood and also represent the main source of green fodder. It seems that the leaves as well as the fruits of *Avicennia* are particularly appreciated by milch cattle. Kulkarni and Junagad (1959) have stated that "over 2500 camels belonging to Rabanis are estimated to be dependent on browsing in the mangrove forests in this tract".

This explains why the physiognomy of these mangals is generally an open scrubby type forming narrow, discontinuous, coastal fringes.

The most important woody species is *Avicennia marina* var. *acutissima,* most valued by the local population. It has a bushy and mutilated appearance with multiple vegetative shoots attaining a height of about 2 m. Rhizophoraceae are rare and seem to have practically disappeared everywhere.

Thus, the zonation and flora are very simplified. The *Avicennia* zone is followed inland by a sort of back–mangal consisting essentially of *Salicornia brachiata, Suaeda fruticosa, Atriplex stocksii* and a caespitose grass, *Urochondra setulosa,* which is here at the extreme eastern limit of its geographical distribution.

THE INDIAN SUBCONTINENT

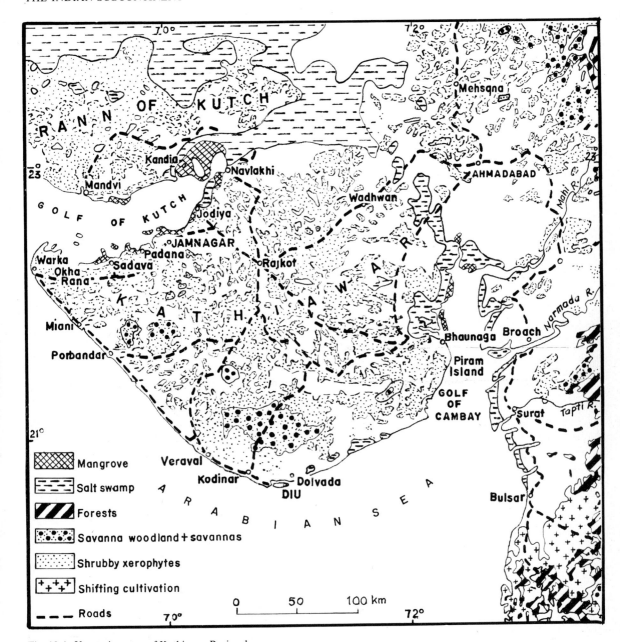

Fig. 12.4. Vegetation map of Kathiawar Peninsula.

Mangals of the Andaman—Nicobar Archipelago

The archipelago is highly diversified, composed of more than 200 islands and islets, situated in the Bay of Bengal. Compared to peninsular India, the most important difference is that these islands remained almost uninhabited until the middle of the nineteenth century.

The Andaman and Nicobar Islands have not been explored sufficiently, either from the botanical or ecological points of view. The works of Kloss (1902), Parkinson (1923), Chengapa (1944) and Thothathri (1960a, b, 1962) give the best information.

Even today the mangals remain as a wealth of these islands. In the Andaman group of islands, which covers about 6400 km², the mangals cover about 1150 km². The difficulties in exploiting them explain their good conservation in the protected creeks and on the banks of the water courses.

This zonation is simple and resembles greatly that of Pichavaram in the Cauvery Delta.

A dense curtain of *Rhizophora* (*R. mucronata*, the commonest, and some *R. apiculata*) 10 m high (on average), rises on either side of the channels.

A little towards the interior, two species of *Bruguiera* are abundant (*B. parviflora* and especially *B. gymnorrhiza*, the tallest trees of these mangals exceeding 25 m). Quite commonly there is an undergrowth of *Ceriops tagal*. These five members of the Rhizophoraceae generally form a very distinct coastal fringe, in which *Aegiceras corniculatum* and *Xylocarpus granatum* may also be found.

In its composition, the zonation gets complicated from one bay to another or from one islet to another. This is the only formation, in India, where *Nypa fruticans* is common, but *Sonneratia apetala* seems curiously to be absent.

The foresters distinguish "tidal forests" from "riverian forests" or "low–level evergreen forests" which follows them up–stream. Among the transitional species mention may be made of: *Cerbera manghas, Heritiera littoralis, Brownlowia lanceolata* and *Scyphiphora hydrophyllacea*.

CONCLUSIONS

In the vast Indian territory, the mangals cover about 350 000 ha. Of this wooded area 85% lies in the Gangetic Delta and in the islands of the Bay of Bengal as shown in Fig. 12.1. In all the other deltas, the mangals, victims of over–population and of excessive grazing, are highly degraded and are not of economic importance unless scientific forest management and pisciculture are undertaken.

Practically every delta has a forest with distinct ecological and botanical traits.

The total halophytic flora is rich; 58 major species have been enumerated.

At present the more common species of the Indian mangals are an arborescent member of the Euphorbiaceae, *Excoecaria agallocha*, and diverse species of *Avicennia*. These species have a great ecological amplitude and a remarkable faculty for vegetative regeneration.

On the contrary, the *Rhizophora* species, which show a poor resistance against the strong biotic pressures, become rare and have disappeared or tend to disappear, except in the Cauvery Delta and in the Andaman Islands.

The almost complete disappearance of *Nypa* and *Heritiera fomes* is mostly due to the ecological factors which we have discussed with reference to the mangals of the Ganges.

REFERENCES

Anonymous, 1971. *Review of Soil Research in India.* Indian Society for Soil Science, Indian Agricultural Research Institute, New Delhi, 229 pp.

Auden, J.B., 1952. Some geological and chemical aspects of the Rajasthan salt problem. In: *Symposium on Rajasthan Desert.* Natl. Inst. Sci. India, New Delhi, pp. 53—67.

Banerjee, A.K., 1964. Forests of Sunderbans. In: *Centenary Commemoration Volume. West Bengal Forests.* D.F.O. Planning and Statistical Cell, Calcutta, pp. 166—175.

Bhosale, L.J. and Joshi, G.V., 1973. Ecological studies in mangroves at Ganpatipule. *Proc. 60th Session Indian Science Congress,* III: 373.

Blasco, F., 1975. Mangroves of India. *Inst. Fr. Pondichéry, Trav. Sect. Sci. Tech.,* 14: 180.

Blasco, F. and Caratini, C., 1973. Mangrove de Pichavaram: Phytogéographie et palynologie. *Trav. Doc. Géogr. Trop.,* 8: 164—185.

Blatter, E., 1905. The mangrove of the Bombay Presidency and its biology. *J. Bombay Nat. Hist. Soc.,* 16: 644.

Blatter, E. et al., 1927. The flora of the Indus delta. *J. Indian Bot. Soc.,* 6: 31—47; 57—78; 115—132.

Blatter, E. et al., 1928. The flora of the Indus delta. *J. Indian Bot. Soc.,* 7: 22—43; 70—96.

Bourdillon, T.F., 1908. *The Forest Trees of Travancore.* Travancore Government Press, Trivandrum, 250 pp.

Caratini, C., Blasco, F. and Thanikaimoni, G., 1973. Relation between the pollen spectra and the vegetation of a South India mangrove. *Pollen Spores,* 15: 281—292.

Champion, H.G. and Seth, S.K., 1968. *A Revised Survey of the Forest Type of India.* Manager of Publications, Delhi, 404 pp.

Chapman, V.J., 1942. The new perspective in the halophytes. *Q. Rev. Biol.,* 17: 291—311.

Chapman, V.J., 1954. The influence of salt upon the terrestrial halophytes. *7th Int. Bot. Congr., Paris,* pp. 194—200.

Chapman, V.J., 1970. Mangrove phytosociology. *Trop. Ecol.,* 11: 1—19.

Chapman, V.J., 1974. *Salt Marshes and Salt Deserts of the World.* Cramer, Lehre, 2nd ed., 392 pp. (complemented with 102 pp.).

Chengapa, B.S., 1944. Andaman forests and their regeneration. *Indian For.,* 70: 297—304.

Clarke, C.B., 1896. Presidential address to the Linnean Society on the Sunderbans of Bengal. *Proc. Linn. Soc. Lond.*, 32: 14—29.

Cooke, T., 1901—1908. *Flora of Bombay.* Taylor and Francis, London, 1946 pp. (3 vols.).

Cornwell, R.B., 1937. *Working Plan for the Godavari Lower Division (1934—1944).* Forest Department, Andhra Pradesh.

Curtis, S.J., 1933. *Working Plan for the Sunderbans Division (1931—51).* Forest Department, Calcutta.

Deb, S.C., 1956. Paleoclimatology and geophysics of the Gangetic delta. *Geol. Rev. India,* 18: 11—18.

Ding Hou, 1955—1958. Rhizophoraceae. In: *Flora Malesiana.* Noordhoff—Kolff, Djakarta, pp. 429—493.

Fosberg, F.R., 1971. Mangroves v. tidal waves. *Biol. Conserv.,* 4: 38—39.

Gamble, J.S., 1916—1935. *Flora of the Presidency of Madras.* Botanical Survey of India, Calcutta, 1389 pp. (3 vols.).

Ghose, B., 1964. Geomorphological aspects of the formation of salt basin in lower Luni basin. *UNESCO General Symposium on Problems of Indian Arid Zone,* pp. 169—178.

Griffith, W., 1836. On the family of Rhizophoraceae. *Trans. Med. Phys. Soc., Calcutta,* 8.

Griffith, W., 1851. On the development of the ovulum in *Avicennia. Trans. Linn. Soc. Lond.,* 20: 1—6.

Gupta, A.C., 1957. The Sunderbans, its problems, its possibilities. *Indian For.,* 83: 481—487.

Heald, E.J., 1970. The Everglades estuary: an example of seriously reduced inflow of fresh water. *Trans. Am. Fish. Soc.,* 99: 847—850.

Kloss, C.B., 1902. *The Andamans and Nicobars.* Vivek Publishing House, Kamla Nagar, Delhi, 373 pp.

Kostermans, A.J.G.H., 1959. Revision of the genus *Heritiera. Reinwardtia,* 4: 465—583.

Kulkarni, D.H. and Junagad, C.F., 1959. Utilisation of mangrove forests in Saurashtra and Kutch. *Proc. Mangrove Symposium, Calcutta, 1957,* pp. 30—35.

Mani, M.S., 1974. *Ecology and Biogeography in India.* Junk, The Hague, 773 pp.

Mathauda, G.S., 1959. The mangroves of India. *Proc. Mangrove Symposium, Calcutta, 1957,* pp. 66—87.

Mazumdar, S.P., 1932. *Gosaba Cooperative Commonwealth.* West Bengal Cooperative Press, Calcutta, 2nd ed., 23 pp.

Moldenke, H.N., 1960. Materials toward a monograph of the genus *Avicennia. Phytologia,* 7: 123—293.

Mukherjee, B.B., 1972. Quaternary pollen analysis as a possible indicator of prehistoric agriculture in deltaic part of West Bengal. *J. Palynol.,* 8: 144—151.

Muller, J. and Hou–Liu, S.Y., 1966. Hybrids and chromosomes in the genus *Sonneratia* (Sonneratiaceae). *Blumea,* 14: 337—343.

Navalkar, B.S., 1940. Studies in the ecology of Mangrove. *J. Univ. Bombay,* N.S., 8(5): 58—73; 9(5): 78—91.

Navalkar, B.S., 1951. Succession of mangrove vegetation in Bombay and Salsette Islands. *J. Bombay Nat. Hist. Soc.,* 50: 157—160.

Navalkar, B.S., 1956. Geographical distribution of the halophytic plants of Bombay and Salsette Islands. *J. Bombay Nat. Hist. Soc.,* 53: 335—345.

Navalkar, B.S., 1959. Studies in the ecology of the mangroves. *J. Univ. Bombay,* N.S., 28(3): 6—10.

Navalkar, B.S., 1974. Ecological study and geographical distribution of the Halophytes of Bombay and Maharashtra. *Indian Sci. Congr. Assoc.,* 61st session: 38.

Navalkar, B.S. and Bharucha, F.R., 1949. Studies in the ecology of the mangroves. *J. Univ. Bombay,* N.S., 18(3): 17—35.

Parkinson, C.E., 1923. *A Forest Flora of the Andaman Islands.* Bishen Singh, Dehra Dun, 325 pp.

Prain, D., 1903. Flora of the Sunderbans. *Rec. Bot. Surv. India,* 2: 231—370.

Qureshi, I.M., 1959. Botanical, silvicultural features of mangrove forests of Bombay State. *Proc. Mangrove Symposium, Calcutta, 1957,* pp. 20—26.

Rajagopalan, V.R., 1952. *Ecological Adaptations of the Mangrove Vegetation at Pichavaram.* Thesis, Annamalainagar University, India.

Rao, R.S., 1959. Observations on the mangrove vegetation of the Godavari Estuary. *Proc. Mangrove Symposium, Calcutta, 1957,* pp. 36—44.

Rao, T.A., 1973. An ecological approach towards classification of coastal vegetation in India. *Indian Sci. Congr. Assoc.,* 60th session: 396.

Rao, T.A. and Aggrawal, K.R., 1964. Ecological studies of Saurashtra Coast and neighbouring Islands. *Bull. Bot. Surv. India,* 6: 173—183.

Rao, T.A. and Mukherjee, A.K., 1972. Ecological aspects along the shores of the Burabalanga tidal estuary, Balasore District. *Proc. Indian Acad. Sci.,* 76(5): 201—206.

Rao, T.A. and Sastry, A.R.K., 1972. An ecological approach towards classification of coastal vegetation of India. *Indian For.,* 98: 594—607.

Rao, T.A., Aggrawal, K.R. and Mukherjee, A.K., 1963. Biological account of the vegetation of Rameswaram Island. *Bull. Bot. Surv. India,* 5: 301—323.

Rao, T.A., Sastry, A.R.K. and Shanware, P.C., 1972. Analysis of the basic patterns of an estuarine shore in the vicinity of Coringa Bay, Andhra Pradesh. *Proc. Indian Acad. Sci.* 75(1): 40—50.

Rao, V.S., 1950. Afforestation of waste lands as practised in West Bengal in combating the soil erosion problem and meeting the scarcity of fuel and fodder. *Indian For.,* 76: 323—326.

Rao, V.S., 1959. Mangrove forests and the problem of reclaiming saline blanks. *Proc. Mangrove Symposium, Calcutta, 1957,* pp. 58—65.

Roxburgh, W., 1814. *Hortus bengalensis or a Catalogue of Plants Growing in the Honourable East India Company's Botanical Garden at Calcutta.* Mission Press, Serampore, 105 pp.

Sahni, K.C., 1953. Botanical exploration in the Great Nicobar. *Indian For.,* 79: 3—16.

Saxena, S.K. and Gupta, R.K., 1973. Vegetation of Pachpadra salt basin in Western Rajasthan. *J. Bombay Nat. Hist. Soc.,* 70: 104—127.

Shinde, S.D. and Mustafa, F.R., 1974. A geographical analysis on the floral distribution and composition in the Terekhol Inlet. *Geol. Rev. India,* 36: 23—30.

Sidhu, S.S., 1963. Studies on the mangroves of India. I — East Godavari region. *Indian For.,* 89(5): 337—351.

Thothathri, K., 1960a. Botanical exploration in Car Nicobar and Nancouri Islands. *Bull. Bot. Surv. India,* 2: 341—346.

Thothathri, K., 1960b. Studies on the flora of Andaman Islands. *Bull. Bot. Surv. India,* 2: 357—373.

Thothathri, K., 1962. Contribution to the flora of the Andaman and Nicobar Islands. *Bull. Bot. Surv. India,* 4: 281—296.

Troup, R.S., 1921. *The Silviculture of Indian Trees.* Oxford University Press, London, 1195 pp. (3 vols.).

Venkatesan, K.R., 1966. The mangrove of Madras State. *Indian For.,* 92: 27—34.

Venkateshwarlu, V., 1944. The estuarial flora of the Godavary. *J. Bombay Nat. Hist. Soc.,* 44: 431—435.

Vu Van Cuong, 1964. *Flore et végétation de la mangrove de la région de Saïgon. Cap Saint Jacques.* Thesis, Université de Paris, Paris, 189 pp.

Chapter 13

WET COASTAL FORMATIONS OF INDO-MALESIA AND PAPUA-NEW GUINEA

V.J. CHAPMAN[1]

INTRODUCTION

Since this is the area in which the greatest number of mangal species are to be found it is here also that the largest number of communities occur. In Malaya the most extensive forests with numerous communities are to be found on the west coast and in the Riau Islands south of Singapore (Van Bodegom, 1929) where there are 22 600 ha. Fig. 13.1 shows extensive mangal on the eastern shore of Sumatra and mainly on the southern and western shores of Borneo. Papua–New Guinea also possesses extensive forests especially in the areas around Port Moresby, Kairuku and Sepik. A number of rivers in the region are substantial and in their lower reaches flow through an extensive lowland. Tidal influence thus extends a long way up the estuaries and so do mangals. In Borneo mangal reaches up the Baram River for 130 km and up the Kapuas River for 240 km, whilst in Papua–New Guinea 140 km marks the limit up the Kikori River and 320 km limit up the Fly River (Van Steenis, 1958).

INDO-MALESIA

The distribution of the various communities depends very largely on physiography, e.g., open

---

[1] I wish to express thanks to Prof. C.G.G.J. van Steenis for helpful suggestions on this chapter.

Fig. 13.1. Map of mangal in Sumatra, Java and Borneo (after Chapman, 1975).

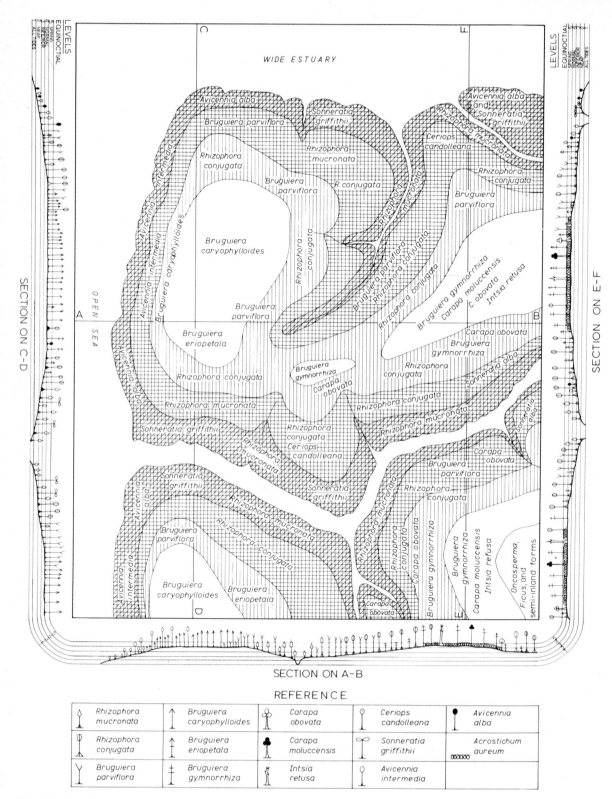

Fig. 13.2. Zonation of mangrove in Malaya (after Watson, 1928).

protected coast, large estuary or small river (Kint, 1934). Watson (1928) in his classic account of Malayan mangrove forests illustrated this variation very well (Fig. 13.2). The chief determining factors for the communities seem to be salinity, soil type, degree of exposure and drainage (Van Bodegom, 1929; Kint, 1934; Bünning, 1947; Chapman, 1975) with frequency of inundation, age of swamp and light as less significant.

On open protected coast the primary community in the region is dominated by either *Avicennia marina* or *Sonneratia alba* or a combination of the two. At the edges of large estuaries *Avicennia alba* may also be a dominant. The *Avicennia* tends to be associated with firmer, more sandy soils and *Sonneratia* with softer muds. This is an example of soil composition affecting the distribution of communities (Kint, 1934). Whilst Watson (1928) and Richards (1964) regard the *Avicennia* species and *Sonneratia alba* as forming a single community, Chapman (1975) considers that since the species are segregated by soil requirements it is better to regard them as forming two separate communities, the *Avicennia* pioneer community and the *Sonneratia alba* pioneer community. The plants range from 2 m or so in front to 15 m or more in the rear.

The pioneer species occur normally on accrescent shores and in such places where the mud in front of the vegetation zone is semi-fluid the mudskipper *Scartelaeos viridis* is predominant with the crab *Macrophthalmus latreillei* (MacNae, 1968). The advent of trees with their underground and aerial roots produces a less fluid and hence firmer substrate. The predominant animals change and major species comprise another mudskipper, *Boleophthalmus boddaerti*, the mud crab *Metaplax crenulatus*, and two fiddler crabs, *Uca coarctata* and *U. dussumieri* (Fig. 13.3), which extend well into the *Avicennia* and/or *Sonneratia* pioneer belt. Associated more specifically with *Avicennia* are two other mudskippers, *Periophthalmus chrysopilos*, which climbs up the trees as the tide rises, and *Periophthalmodon schlosseri*, which extends throughout the various communities up to high water mark.

Around the pioneer bushes several small crabs, *Ilyoplax*, *Metaplax elegans* and *Leiopecten sordidulum* are common. In the higher and older parts of *Avicennia—Sonneratia* communities another

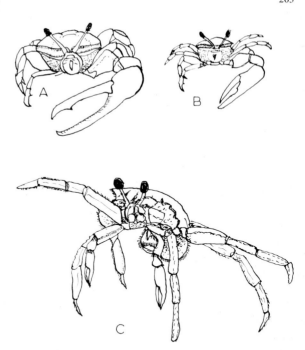

Fig. 13.3. Crabs of the mangal. A. *Uca dussumieri*. B. *Uca lactea*. C. *Metaplax crenulatus*. (After MacNae, 1968).

fiddler crab, *Uca rosea*, is widespread. In the Malayan pioneer mangal *U. triangularis* also occurs, whilst in Borneo pioneer mangal its variety *variabilis* is found along with *U. rhizophorae*. Water snakes *(Cerberus rhynchops)* chase after crabs in their burrows and at low tide lie on the mud in the shade of trees. On the pneumatophores two smaller crabs, *Nannosesarma minuta* and *Clistocoeloma merguiense* are frequent. Several species of snail wander over the mud between the trees. *Syncera brevicula* and *Haminea* are found where there is shade, *Telescopium telescopium* and *T. mauritsi* in wetter runnels, and *Cassidula aurisfelis*, with *C. mustellina*, not only on the mud but also on the bole bases.

Up smaller creeks *Rhizophora mucronata* replaces *Avicennia* and *Sonneratia* as the pioneer community. The trees can grow up to 16 m but are generally not so tall. With increasing height of the ground *R. apiculata* enters the community and a mixed *R. apiculata* community develops. In Indo-Malesia these two communities of *Rhizophora* cover a very large area indeed, with the *R. apiculata* community being rather more significant. In Papua-New Guinea *R. mucronata* is dominant on rich, sandy-clay soils whilst *R. api-*

culata is associated with more muddy soils. In these native communities the trees range from 25 to 45 m tall. On the eastern more sandy shores of Malaya *R. stylosa* is the pioneer and it continues to occupy this position in Australia (see p. 302) and in parts of Indonesia such as Bangka Island (Kint, 1934) and the Celebes (Steup, 1936). *Bruguiera parviflora, B. gymnorrhiza* and *Xylocarpus granatum* can be associated with the *Rhizophora* communities whilst older stands of both *Rhizophora* communities have an understorey of *Ceriops tagal*. On the east coast of Malaya the principal pioneer species is *Rhizophora stylosa*, but it is doubtful if it forms a recognisable community. There are areas in Papua–New Guinea and Indonesia where it can be regarded as forming a community, especially on more sandy shores. Animals associated with the *Rhizophora* forest communities include snapping alpheid prawns, *Macrophthalmus* spp. and *Metapograpsus latifrons* (both crabs), all where the soil is relatively unconsolidated. On firmer soils a wider range of crab species are to be found, most of which extend into the *Bruguiera* communities behind. Some of these, e.g., *Metaplax elegans, Leiopecten sordiculum* and *Cleistocoeloma merguiense* can be regarded as survivors from the *Sonneratia—Avicennia* communities. Other crabs, e.g., *Ilyograpsus* spp., *Utica borneensis* and *Paracleistostoma*, reach their lower limit here and extend higher up on the shore into the *Bruguiera* communities.

A *Bruguiera parviflora* community is a characteristic feature of wetter areas in the *Rhizophora* belt and also behind it. It establishes quickly, probably because the small seedling can be transported both by tide and rainwater. Particularly at the higher levels the mud lobster, *Thalassina anomala*, forms large mounds in which large species of *Sesarma* crabs, e.g., *S. indica, S. mederi*, make their burrows. On the *Thalassina* mounds the fern *Acrostichum aureum* finds a very suitable habitat and its growth may be so dense that mangal regeneration is no longer possible. The *Bruguiera parviflora* community appears to be somewhat opportunist with a vertical range comparable to that of *Rhizophora*, so that if the *Rhizophora* is felled in a wet area *B. parviflora* is likely to replace it and so make it difficult for *Rhizophora* to re–enter the area.

At higher levels and on stiff–clay soils *Bruguiera cylindrica* (= *B. caryophylloides*) forms well–marked communities. It is generally associated with active accretion forests and is widespread along the west Malayan coast. The trees are usually not more than 20 m high. *Ceriops tagal* is a common understorey shrub and a wide range of animals are to be found, including small *Sesarma* spp., *Metapograpsus frontalis*, species of *Uca*, etc. (Fig. 13.4). In fact this community has more animals associated with it than any other. This community is not typical of river forests and where sites are drier *Lumnitzera littorea* enters as an associated species. It is reported that because of the stiff–clay soils *B. cylindrica* is very dependent upon its pneumatophores and if waterlogged the trees die (cf. also p. 2). MacNae (1968) reported a species of the nudibranch genus *Elysia* in brackish pools as well as tadpoles of *?Rana cancrivora*.

On the landward side of the swamps there are generally extensive areas of *Bruguiera gymnorrhiza* with large trees up to 36 m tall (Chapman, 1975). This community is regarded as the final stage in the mangrove succession. Another species, *B. hainesii*, grows in similar situations but its ecological requirements have not yet been worked out. At the higher levels the fern *Acrostichum aureum* can form a dense undergrowth and so inhibit regeneration. In Bangka Island this community also includes *Xylocarpus granatum* (Kint, 1934). MacNae (1968) reports that this type of mangal has a greater variety of ground–dwelling animals than elsewhere. Crabs are present in abundance, e.g., species of *Sarmatium, Helice, Ilyograpsus* and *Sesarma*. The representatives of the last–named genus are generally small, e.g., *S. eumolpe, S. dussumieri, S. onchophora* and *S. indiarum*. A characteristic denizen is the mud lobster *Thalassina anomala* whose presence is marked by the heaps of soil at the entrance to the burrows. Large *Sesarma* spp. burrow into these mounds, and because of the change of level the fern *Acrostichum aureum* or *A. speciosum* colonise them and add to the problems of natural regeneration.

Thalassina extends up to the land fringe of the mangal where the mounds are penetrated by the burrows of a number of large *Sesarma* species, e.g., *S. indica, S. mederi, S. palawanensis, S. singaporensis, S. tetragona* and *S. versicolor*.

B. gymnorrhiza forest can contain shallow mud pools which are occupied by small crabs of the

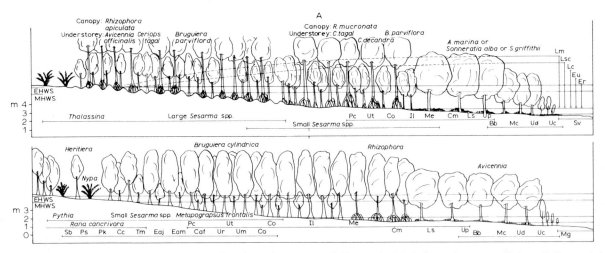

Fig. 13.4. Plant and animal zonation in mangal as shown in transects. A. Pattern in deltaic forests in western Malaya. B. Pattern in western seaboard of Malaya south of Kuda Selangor. (After MacNae, 1968.)

Key to symbols: Bb = *Boleophthalmus boddaerti*; Caf = *Cassidula aurisfelis*; Cc = *Cerithidea cingulata*; Cm = *Clistocoeloma merguiense*; Co = *Cerithidea obtusa*; Eaj = *Ellobium aurisjudae*; Eam = *Ellobium aurismidae*; Er = *Enigmonia rosea*; Eu = *Eulimine*; Il = *Ilyograpsus* sp. or spp.; Lc = *Littorina carinifera*; Lm = *Littorina melanostoma*; Ls = *Leiopecten sordidulus*; Lsc = *Littorina scabra*; Ma = *Macrophthalmus* sp.; Mc = *Metaplax crenulatus*; Me = *Metaplax elegans*; Pc = *Periophthalmus chrysospilos*; Pk = *Periophthalmus kalolo*; Ps = *Periophthalmodon schlosseri*; Sb = *Syncera brevicula*; Sv = *Scartelaos viridis*; Tm = *Telescopium* spp.; Uc = *Uca coarctata*; Ud = *Uca dussumieri*; Um = *Uca manii*; Up = *Upogebia* sp.; Ur = *Uca rosea*; Ut = *Uca triangularis*.

genus *Ilyoplax* which, according to Tweedie (1950) are restricted in their distribution. Other small crabs found in this type of habitat are *Paracleistostoma depressum*, *P. microcheirum*, *P. longimanum*, *Tylodiplax tetratylophorus* and *Utica borneensis*. In clearings *Uca manii* and *U. rosea* are typical crab inhabitants. If there are shallow creeks with sandy bottoms in this forest than one may find *Macrophthalmus depressus* and *M. pacificus*.

Among the molluscs, species of *Haminea*, *Terebralia palustus*, *T. sulcata* and *Syncera brevicula* all occur on the firm mud, but where the mud is softer they are replaced by *T. telescopium* and *T. mauritsi*. Six species of *Cerethidea* are common and climb up to form clumps on the bases of the tree trunks.

On the landward edge of these mangal swamps there can be a scrub of *Ceriops tagal* which is generally so heavily exploited that the plants are dense and thin-stemmed. The fauna of this scrub is similar to that of the *B. gymnorrhiza* forest. Where conditions become more brackish a community dominated by *Lumnitzera littorea* is found associated with creek banks. This species is typical of loamy soils whereas on clay soils it is replaced by *L. racemosa* (see later, and Fig. 13.5).

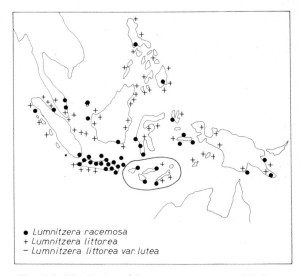

Fig. 13.5. Distribution of *Lumnitzera racemosa* and *L. littorea* in Indonesia (after Chapman, 1975).

Another brackish-water community is represented by a mixture of *Sonneratia caseolaris*, *Nypa fruticans* and *Heritiera littoralis* (Kint, 1934; Bünning, 1947). In this type of community, especially where there is *Nypa*, *Sesarma moeschii*, *S. crassimana* and *S. sediliensis* are common crab species. In the ever-wet mangal of Malaya this landward belt contains a number of prosobranch and pulmonate snails such as *Cassidula mustellina, Ellobium aurisjudae, Nerita birmanica, Pythia scarabeaus* and *Cerithidea obtusa.*

Very similar forest communities occur extensively on the north (Bünning, 1947) and east coasts of Sumatra where there is quite some protection from wave action associated with large areas of low–lying country (Bünning, 1944). The Javan forests are less extensive, being mostly confined to the northern coast (see Fig. 13.1). A large area in the Kindzee (Segara Anakan) has been described by Karsten (1891) and De Haan (1931). The open coast pioneers are *Avicennia marina* and *Sonneratia alba* with an undergrowth of *Acanthus* and *Aegiceras corniculatum*. *Avicennia* tends to be predominant on more sandy soils and *Sonneratia ovata* on coral limestone (Bünning, 1944). *Rhizophora* forms the next belt but in the lagoon area *R. mucronata* and *R. apiculata* are the pioneer species with *Ceriops tagal, Sonneratia alba, Xylocarpus moluccensis, Bruguiera cylindrica* and *Aegiceras* mixed in. Other communities are similar to those already described for Malaya but *Xylocarpus moluccensis* when present in a dominant form represents an additional community. Brackish water is characterised by *Nypa, Sonneratia caseolaris, Acanthus ilicifolius* and *Dolichandrone spathacea*. The presence of *Heritiera littoralis* and *Lumnitzera littorea* also mark a transition to fresh–water conditions. In the landward portion of these communities De Haan (1931) recognised two additional communities, one dominated by *Bruguiera sexangula* with the palm *Oncosperma filamentosa* and the other a mixed *Nypa, Xylocarpus* complex with *Cerbera odollam*.

Physically the mangal of Borneo is more extensive than that of Malaya but the communities are essentially the same with comparable faunal elements. An early account was given by Foxworthy (1910) and a more recent one by Chai (1973). The last named author has recognised eight types of forest in Indonesian Borneo. These are respectively *Sonneratia* forest, *Avicennia* forest, *Rhizophora* forest, *Xylocarpus granatum* forest, *Nypa fruticans* swamp, two types of *Bruguiera* forest and *Excoecaria* forest.

The *Sonneratia* forest with trees up to 15—25 m high is the pioneer on mud and sandy mud flats along sheltered river coasts and protected coastlines. *Avicennia* forest occurs in two forms, one on muddy coastlines and up rivers and dominated by *A. alba;* the other on sandy coastlines and dominated by *A. marina*. The former has an understorey of *Kandelia candel, Aegiceras corniculatum* and young *Rhizophora apiculata,* whilst the latter has various species of *Bruguiera* and *Ceriops* representing the invaders of the next community in the succession.

Bruguiera parviflora forest occurs immediately behind the *Avicennia* forest with trees up to 15 m tall, though on higher and drier ground they may grow to 40 m. A transition to brackish water conditions is marked by the advent of *Heritiera littoralis* and *Xylocarpus granatum*. Inland regions may also contain *Excoecaria agallocha, Oncosperma tigillaria* and *Myrsine umbellata*, though the last two are not strict mangal species but rather upland invaders. The presence of the mud lobster and *Acrostichum* on its mounds inhibits regeneration.

Rhizophora apiculata forest forms an excellent rich mangrove community with trees up to 30—40 m high. Chai (1973) reports very little mud lobster and hence little *Acrostichum*, so that regeneration is good. On higher ground *Xylocarpus granatum* enters as a co–dominant and perhaps represents a transitional phase to a still drier *Excoecaria* or *Bruguiera sexangula* type. *B. gymnorrhiza* only occasionally forms pure stands, one of the best being the Kenalian forest reserve in Lawas (Chai, 1973). It contains lobster mounds and *Acrostichum* fern and in such places there is no regeneration.

Forest dominated by *Bruguiera sexangula* marks a transition to fresh–water peat–swamp forest so that the soil is dark and peaty and the water tea–coloured. *Pandanus affinis* and *Caesalpinia nuga* are common understorey species but cannot be regarded as true mangal denizens. There are also the inevitable lobster mounds. Inland on stiff consolidated clay soils *Excoecaria agallocha* is the dominant forest. The raising of the soil level is caused by the coalescing of the numerous lobster mounds so that small pits are left in which water persists. Near channels or on softer soils *Ceriops decandra* and *C. tagal* are associated species whilst inland *Heritiera littoralis* can become very prominent. Still further inland *Dolichandrone spatha-*

cea and *Myrsine umbellata* form the understorey. *Nypa* swamp is a typical transition to fresh–water conditions, as also is *Oncosperma tigillaria* forest. The former has been described by Foxworthy and Matthews (1917). This latter palm forest contains other marginal species such as *Heritiera littoralis, Dolichandrone spathacea, Planchonella obovata* and *Pandanus affinis*. The dominant species may have stems reaching up to 40 m in height.

The above represent the major plant communities in Borneo but there are several minor types. These include river–fringing *Rhizophora mucronata* forest replaced inland by *Sonneratia caseolaris*. Also near rivers but on higher ground there is a *R. apiculata—Bruguiera* spp. forest and in some places *Xylocarpus granatum* is added as an additional co–dominant. Further inland the same species occur together but *Excoecaria* also becomes a co–dominant. A final inland mangrove community is represented by a combination of *Heritiera littoralis—Bruguiera sexangula—Excoecaria agallocha*. In Ambon Island (the Moluccas) *Sonneratia ovata* occurs on raised reefs with the main forest composed of mixed species, mostly of *Bruguiera*. The presence of *Camptostemon schultzii* is of considerable interest because this can be regarded as marking an affinity with some north Australian forests (see Chapter 15) and forests of the Philippines (see Chapter 14).

In northern Celebes many of the mangal species are absent and such mangal as there is occurs primarily on coral reefs (Steup, 1936). The zonation is essentially the same as that described for Malaya and Indonesia. An exception is the presence of *Camptostemon philippinensis* which elsewhere in this general region occurs only in one or two places in Borneo.

PAPUA–NEW GUINEA

In New Guinea accounts of the mangrove swamps have been given by Lam (1934), Cavanaugh (1950), Womersley and McAdam (1957), Taylor (1959), Heyligers (1966), Paijmans (1967) and Robbins (1968).

In Papua–New Guinea the pioneer community is typical of the Indo–Malesian region as well as of North Australia (see p. 305). It is dominated by either *Sonneratia alba* or *Avicennia marina* or both.

A. marina occurs either as var. *alba* (generally close to open water) or var. *resinifera* on higher ground. The pioneer community occurs on the open coast and also at the mouths of creeks. Further up the creeks *Ceriops* enters the community of pioneers and then as conditions become brackish these pioneer species are replaced by *Sonneratia caseolaris* with a frontal fringe of *Nypa fruticans*. Other plants include *Myristica* and *Brownlowia argentata*. In these upper reaches there may also be a ground cover of *Cyperus* and *Leersia* but these are invaders from fresh water.

In the Watam and Murik lagoons of the lower Sepik area Robbins (1968) states that the pioneer fringe is composed of low trees (8 m high) of *Aegiceras corniculatum*. The main forest both here and elsewhere in the Papua–New Guinea area is composed of a mixture of *Rhizophora* and *Bruguiera* species with a mixed crab fauna and with the large bivalve *Mactua eximia* also common. *Rhizophora mucronata* predominates on rich sandy–clay soils and *R. apiculata* on the rich mud soils. In other places *Bruguiera gymnorrhiza* is the predominant rather than the *Rhizophora* species. The crab population is regarded as being responsible for the relatively small number of seedlings though sufficient survive to give adequate regeneration. In the Port Moresby area *B. gymnorrhiza* is joined with *B. cylindrica, B. parviflora* and *B. sexangula* and occasional emergent trees of *Avicennia marina* and *A. officinalis. B. parviflora* occurs on firmer rich muds and *B. sexangula* on higher landward areas with *B. cylindrica* on the fringes (Womersley and McAdam, 1957). Interesting associates of *B. parviflora* on the firmer muds include *Camptostemon schultzii* and *Xylocarpus moluccensis*. The occurrence of the former and its distribution in this region as a whole is of considerable interest (Fig. 13.6). Undergrowth in this type of *Bruguiera—Rhizophora* forest consists of *Acanthus* and *Acrostichum* and there may be epiphytic orchids and the ant plant *Myrmecodia. Acanthus* and *Acrostichum* are generally associated with the mounds of the mud lobster and here they inhibit regeneration of the main species.

Where salinity is still high but flooding is less bushes of *Ceriops tagal* and *Xylocarpus granatum* occur as undergrowth. The latter persists into the next belt which is often a forest of *Bruguiera* species and *Heritiera littoralis*. In the Fly R. Delta region

Fig. 13.6. Distribution of *Camptostemon schultzii* and *C. philippinensis* (after Chapman, 1975).

very large areas are dominated by *Heritiera* (Cavanaugh, 1950). In quite a few areas *Heritiera* marks the landward fringe of the mangal.

On higher ground with less frequent flooding *Lumnitzera racemosa* or *L. littorea* occupy the more saline areas with *Avicennia marina* and *Ceriops tagal* in the less saline and better drained areas. The two species of *Lumnitzera* are typical of the Indo–Malesian region and their distribution is shown in Fig. 13.5. *L. racemosa* can be regarded as having a predominantly southerly distribution, whereas *L. littorea* is more frequent to the north. In such places *Sesuvium portulacastrum*, *Chloris* sp. and *Sporobolus virginicus* may provide a ground cover. An alternative community is a savannah of *Avicennia* with *Acrostichum aureum* and *A. speciosum*. These are essentially beach invaders.

At still higher levels with reduced salinity but poor drainage *Ceriops* is replaced by *Excoecaria agallocha* as the dominant with spaced trees of *Avicennia*. A shrub cover of *Aegiceras*, *Thespesia populnea*, *Clerodendron inerme*, *Hibiscus tiliaceus* and the fern *Acrostichum speciosum* is also present. The final landward communities of the mangal are either a forest of tall *Excoecaria* with *Hibiscus* as a second storey or a forest of tall *Melaleuca* with *Excoecaria* as the second storey. On the landward tidal flats, where conditions are brackish, *Sonneratia caseolaris* is the dominant but it can have a whole range of associated plants such as the *Hibiscus*, *Nypa*, *Avicennia marina*, *Bruguiera cylindrica* and *Aegiceras corniculatum*. *Dolichandrone spathacea*, *Brownlowia lanceolata* and *Myristica* sp. mark the transition to fresh–water swamp forest. Chapman (1975) has summarised the succession in this region as shown in Fig. 13.7.

VIETNAM

On the Vietnamese coast extensive mangrove swamps are to be found and are estimated to occupy some 2800 km^2. *Avicennia marina* is the pioneer species and in the belt behind the dominants are *Rhizophora apiculata*, *Bruguiera parviflora* and *Ceriops tagal* (Ho, 1963). On land that is high and dry *Bruguiera gymnorrhiza* predominates,

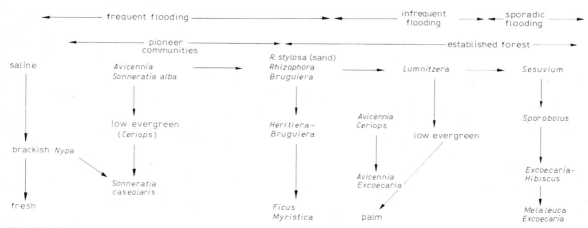

Fig. 13.7. Succession in mangals, Papua–New Guinea (after Chapman, 1975).

and there is the inevitable *Nypa fruticans* along streams where conditions are brackish. Similar conditions in the actual swamps are characterised by the presence of *Sonneratia caseolaris*, though this species does not play any major role. The final landward transition community is dominated by *Melaleuca leucodendron*. Other species present in these swamps, but not acting as dominants, are *Lumnitzera littorea*, *Xylocarpus granatum* and *Excoecaria agallocha*.

The fauna appears to be dominated by crabs which are present in such numbers that they affect regeneration (see below). With increase in latitude species gradually disappear and ultimately only *Kandelia candel* remains in southern Japan (see p. 286).

The extensive swamps between North and South Vietnam have in recent years secured notoriety because of the use made of herbicides to defoliate them for military purposes. Walsh (1974) has reported on the effect of various herbicides as a means of controlling mangroves. Prior to the use in Vietnam 2,4–D, 2,4–T and MCPA (2–methyl–4–chlorophenoxyacetic acid) had been used in Sierra Leone to eradicate *Rhizophora racemosa* and *Avicennia africana*. Extensive defoliation occurred after seven months by which time many trees were dead. Application of 3–(4–chlorophenyl–1)–1,1–dimethyl urea to pneumatophores of *Avicennia* killed all trees, as did Dalapon at 45 kg ha^{-1}. Tshirley (1969), Orians and Pfeiffer (1970), Aaronson (1971), Boffey (1971) and Westing (1971a, b, c) have reported on defoliation in Vietnam where the various mangrove species all proved highly susceptible to spraying by a mixture of 2,4–D and Picloram. The sprayed areas remained uncolonised for at least six years. This could be due to a variety of causes: crab predators of seedlings washed in (see above), compaction of soil after death of trees making establishment difficult, or actual lack of seedling invaders.

LAND ANIMALS

At this point some account may be given of the larger animals to be encountered in these Indo–Malesian swamps. For full details reference should be made to MacNae (1968). Birds, apart from "mangrove kingfishers" are mostly visitors, nesting in the trees and seeking their food from the fish and crustacea in the creeks. Two cormorants, *Phalacrocorax melanogaster* (New Guinea) and *P. niger* (Malaya to Borneo) occur in the region with herons using the banks as fishing grounds. The white heron, *Egretta alba*, and the little egret, *E. garzetta* and *E. eulophotes*, are all found throughout the region. The night heron, *Nycticorax caledonicus*, also occurs in Indonesia together with the large heron, *Ardea sumatrana*. From Malaysia to Indonesia and Australia the principal scavenger is the Brahminy kite, *Haliastur indus*, which can be found in the vicinity of all mangrove villages. A troublesome poacher of fish and prawns in Indonesia is represented by the "sea eagle", *Ichthyophaga ichthyaetus*.

The principal mangal kingfisher is represented by *Halcyon chloris* together with *Pelargopsis capensis* in Southeast Asia. Waders are represented by the whimbrel *Numenius phaeopus*, the redshank, *Tringa toanus*, and the Terek sandpiper, *T. terek*. These birds perch on the mangrove branches and then scatter over the mud flats as the tide recedes. Smythies (1960) reported a whistler, *Pachycephala cinerea*, as common in mangal of Borneo, and also *Zosterops chloris*. MacNae (1968) refers to a number of passerine birds in the mangrove swamps of Malaya. These include the grey tit, *Parus major*, the flycatcher *Muscicapa rufigastra* (both restricted to mangal), another flycatcher, *Rhipidura javanica*, and two bulbuls, *Pycnonotus goiaver* and *P. plumosus*.

Among the mammals a few monkeys have been reported as visitors. Macaque monkeys, *Macaca irus*, are common from Malaya to Borneo whilst the leaf monkey *Presbytis cristatus* is restricted to Malayan mangal and the proboscis monkey *Nasalio larvatus* is restricted to Bornean mangal. The otters *Ambonyx cinerea* and *Lutra perspicillata* both occur in Southeast Asia but are not seen too often. Wild and mouse deer are said to be common in the *Nypa* swamps (MacNae, 1968).

The edges of rivers and creeks are favourite hunting grounds for the large crocodile, *Crocodylus porosus*, and one may also encounter a water monitor, *Varanus salvator*. In the Malayan and Sumatran mangal there are two pit vipers, *Trimeresurus wagleri* and *T. purpureomaculatus*. The Malayan mangal is also a home to at least three other snakes and a frog, *Rana cancrivora*.

Insects are either visitors or have larval stages living in the soil (e.g., midges). The "visitors" come for flower honey, or are parasites, or prey on other insects. Nests of species of the weaver or tailor ant *Oecophylla* spp. are very evident on trees of *Bruguiera*, *Sonneratia caseolaris* or *Ceriops*. The principal anopheline mosquito is *Anopheles sundaicus* that breeds in water of salinity up to 13‰. Other anopheline mosquitoes occurring in New Guinea include *A. faranti* and *A. amictus* subsp. *hilli*. A number of culicine mosquitoes also occur throughout the region. These include *Aedes fumidus*, breeding in pools on the ground, *A. niveus* and *A. littoreus*, breeding in water holes in trees, *A. butleri*, breeding in *Nypa* communities, and *A. amesii*, breeding in rot holes and said to be the commonest biter of man. MacNae (1968) draws special attention to spectacular phenomenon of synchronously flashing fireflies. The insects responsible are the males of *Pteroptyx malaccae* and are generally restricted to certain trees of *Sonneratia caseolaris*, *Avicennia officinalis*, *Acanthus ilicifolius* and *Rhizophora apiculata*. In some places specific trees of other mangrove genera may occasionally be selected. It has been suggested that the frequency may be temperature–controlled. It is evident that this phenomenon is worthy of further study.

REFERENCES

Aaronson, T., 1971. A tour of Vietnam. *Environment*, 13: 34—43.
Boffey, P.M., 1971. Herbicides in Vietnam: AAAS study finds widespread devastation. *Science*, 171: 43—47.
Bünning, E., 1944. Botanische Beobachtungen in Sumatra. *Flora, N.S.*, 37: 334—344.
Bünning, E., 1947. *In die Wäldern Nordsumatras. Flora, N.S.*, 40: 148—157.
Cavanaugh, L.G., 1950. Mangrove–delta area. *For. Resour. Rep., Malay*.
Chai, P.K., 1973. *The Types of Mangrove in Sarawak*. Forest Department of Sarawak, 34 pp.
Chapman, V.J., 1975. *Mangrove Vegetation*. Cramer, Lehre, 425 pp.
De Haan, J.H., 1931. Het een en ander over de Tjilatjapsche vloedbosschen. *Tectona*, 24: 39—75.
Foxworthy, F.W., 1910. Distribution and utilisation of mangrove swamps in Malay. *Ann. Jard. Bot. Buitenzorg*, 2nd Ser., Suppl. II, Part I: 319—344.
Foxworthy, F.W. and Matthews, D.M., 1917. Mangrove and nipah swamps of British North Borneo. *B.N.B. Dep. For. Bull.*, 3: 1—67.
Heyligers, P.L., 1966. Vegetation and ecology of the Port Moresby—Kairuku area. *CSIRO, Pap. Land Res.*, Ser. 14: 146—173.
Ho, Pham–Hoang, 1963. Écologie et phytogéographie de la mangrove. *Bull. Soc. Biol. Vietnam*, 1963: 39—54.
Karsten, G., 1891. Über die Mangrovevegetation in Malayischen Archipel. *Bibl. Bot.*, 22: 71.
Kint, A., 1934. De luchtfoto en de topographische gesteldheid in de mangrove. *Trop. Nat.*, 23: 173—189.
Lam, H.J., 1934. Materials towards a study of the flora of the Island of New Guinea. *Blumea*, 1: 115—159.
MacNae, W., 1968. A general account of the fauna and flora of mangrove swamps and forests in the Indo–West–Pacific region. *Adv. Mar. Biol.*, 6: 73—269.
Orians, G.H. and Pfeiffer, E.W., 1970. Ecological effects of the war in Vietnam. *Science*, 168: 544—554.
Paijmans, K., 1967. Vegetation of the Safia—Pongani area. *CSIRO, Pap. Land Res.*, Ser. 17: 142—167.
Paijmans, K., 1969. Vegetation and ecology of the Karema—Vailala area. *CSIRO, Pap. Land Res.*, Ser. 23: 95—116.
Rand, A.L. and Brass, L.J., 1940. Results of the Archbold Expeditions No. 29. Summary of the 1936—37 New Guinea Expedition. *Bull. Am. Mus. Nat. Hist.*, 77: 341—380.
Richards, P.W., 1964. *The Tropical Rain Forest: an Ecological Study*. Cambridge University Press, London, 450 pp.
Robbins, R.G., 1968. Vegetation of the Wewak—Lower Sepik area. *CSIRO, Pap. Land Res.*, Ser. 22: 109—124.
Smythies, B.E., 1960. *The Birds of Borneo*. Oliver and Boyd, Edinburgh, 562 pp.
Steup, F.K.M., 1936. Botanische aanteekeningen over Noord Celebes. IV. *Trop. Nat.*, 25: 29—31.
Steup, F.K.M., 1946. Boschbeheer in de vloedbosschen van Riouw. *Tectona*, 36: 289—298.
Taylor, B.W., 1959. The classification of lowland swamp communities in Northeastern Papua. *Ecology*, 40: 703—711.
Tshirley, F.H., 1969. Defoliation in Vietnam. *Science*, 163: 779—786.
Tweedie, M.W.F., 1950. Grapsoid crabs from Labuan and Sarawak. *Sarawak Mus. J.*, 4: 338—369.
Van Bodegom, A.H., 1929. De vloedbosschen in het gewest Riouw en onderhoorigheden. *Tectona*, 22: 1302—1332.
Van Steenis, C.G.G., 1958. Ecology of mangroves. In: *Flora malesiana*, 1. Noordhoff—Kolff, Djakarta, pp. 431—444.
Walsh, G.E., 1974. Mangroves: a Review. In: R.J. Reimold and W.H. Queen (Editors), *Ecology of Halophytes*. Academic Press, New York, N.Y., pp. 51—174.
Watson, J.G., 1928. Mangrove forests of the Malay Peninsula. *Malay. For. Rec.*, 6: 275 pp.
Westing, A.H., 1971a. Ecological effects of military defoliation on the forests of South Vietnam. *Bioscience*, 21: 893—898.
Westing, A.H., 1971b. Forestry and the war in South Vietnam. *J. For.*, 69: 773—784.
Westing, A.H., 1971c. Herbicides as weapons in South Vietnam, a bibliography. *Bioscience*, 21: 1225—1227.
Womersley, J.S. and McAdam, J.B., 1957. Forest and forest conditions in the territories of Papua and New Guinea. Government Printer, Port Moresby.

Chapter 14

MANGALS OF MICRONESIA, TAIWAN, JAPAN, THE PHILIPPINES AND OCEANIA

T. HOSOKAWA, H. TAGAWA and V.J. CHAPMAN

MICRONESIA (T. Hosokawa)

Introduction

Climate

Based on average air temperatures the Micronesian Islands are situated in the tropics. From the viewpoint of climatic conditions, we can recognise four major zones in the tropical Pacific from the north to the south; Northeast Trade–wind Zone, Equatorial Humid Zone, Equatorial Arid Zone, and Southeast Trade–wind Zone. The second zone lies to the north of the equator, and the equator runs through the third zone. Considering the average monthly and annual rainfall in the Micronesian Islands, the Marianne Islands, which are situated in the Northeast Trade–wind Zone, are regarded as in Köppen's (1923) **Amwi** climatic type. Palau and the Caroline Islands and the Marshall Islands, all situated in the Equatorial Humid Zone, are in the **Afi** type, and those islands of the Equatorial Arid Zone may lie in either **BSwi** or **Awi** of Köppen's climatic types (Table 14.1).

Status

Mangal vegetation occurs as an edaphic climax characteristically in the coastal flooded areas of tropical and subtropical shores, being specially well developed in Indo–Malesia and Papua–New Guinea (Lam, 1934) (cf. Chapter 13) and in the Philippines (Dickenson et al., 1928) (cf. p. 287). The Bonins (Hosokawa, 1938), the northern Marianas (Hosokawa, 1934), Tinian, Rota (Fosberg, 1960) and most of the Polynesian Islands are, however, entirely or almost devoid of mangal vegetation. Samoa delimits the eastern border of its range in the eastern hemisphere (Rechinger, 1908; Chapman, 1954). It seems that the development and geographical distribution of mangal vegetation in Oceania is much more influenced by edaphic and orographic factors than by the climate.

In Micronesia, the occurrence of some mangrove stands of *Bruguiera gymnorrhiza*, *Sonneratia caseolaris* and *Lumnitzera littorea* was reported from the Marshall Islands by Hatheway (1953) and Fosberg (1953, 1955), but no mangal vegetation in a state of nature has ever developed in any island of the northern Marianas. They flourish well on the major high islands of Micronesia, such as Yap, Palau, Truk (Hosokawa, 1937), Ponape (Glassman, 1952) and Kusaie (Hosokawa, 1954c). In the southern Marianas, Hosokawa (1934) found forest of the *Bruguiera gymnorrhiza* type in Saipan and Guam, with forest of the *Rhizophora mucronata* type associated with it in Guam (Merrill, 1914; Fosberg, 1960). In the Palaus and Carolines (Hosokawa, 1937), where the mangrove forests are well developed, the strand forests are restricted to limited areas of headlands or capes, which are exposed directly to the prevailing wind. Few or no strand forests grow on the coastal areas surrounded by the thick mangrove belt. In such cases, instead of strand forests, either the swamp forests (Hosokawa, 1952a) or the plantation forests or groves of coconut palm or breadfruit trees are developed there in contact with the mangal areas. The mangrove forests, therefore, occur as synecological parallels of swamp forests; with few exceptions swamp forests are not found in the Marianas, though various types thrive quite well in the Palaus and Carolines (Table 14.2).

Fosberg (1960) states that mangrove swamps in Micronesia are not so extensive or well developed as those in the subtropical region farther west. Considerable areas around all the high islands in the Carolines are in the form of narrow fringes

TABLE 14.1

Rainfall averages in Micronesia (Hosokawa, 1967)

Zone		NE Trade-Wind Zone							Equatorial Humid Zone									Eq. Mod. Zone		Eq. Arid. Zone
		Bonin Islands	Midway	Mariana Islands			Saipan		Ujelang Marshall Is.	Caroline Islands				Ponape		Kusaie		Jaluit (Marshall Is.)	Fanning	Gilbert			Christmas
Station		Titi Is. (Chichi)		Guam (Sumay)	Rota	Tinian	Garapan			Korror	Palau	Yap	Truk	Eten	Colonia	Nipit	Lelahafen	Mission			Nauru	Banaba	Malden	
Elevation (m)		4.3	6	19	<10	<10	10	207	9	<10	32	35	32		<10	450	<10	±100	3	3	5	28	8	<3
Years (total)		1907–1940 (34)	1921–1930 (10)	1906–1922 (17)	1926–1934 (9)	1926–1934 (9)	1901–1913 (13)	1927–1929 (3)	1894–1912 (9)	1905–1913 (9)	1924–1929 (6)	1900–1930 (31)	(3)	(8)	1901–1913 (13)	1934–1940 (7)	1904–1931 (10)	(5)	1892–1913 (22)	1903–1930 (28)	1894–1913 (20)	1904–1930 (27)	1890–1925 (36)	1916–1919 (4)
											Precipitation (mm)													
Jan.		89	110	61	132	112	51	104	53	219	400	177	160	178	281	414	391	483	244	257	301	300	86	**176**
Febr.		82	97	80	135	62	93	147	56	196	210	*173*	102	232	223	314	340	380	224	*278*	239	234	54	136
March		108	85	81	68	69	102	96	65	*196*	*172*	124	66	241	359	494	403	692	362	268	*175*	*190*	114	127
April		138	111	57	49	47	73	94	157	*177*	294	131	161	297	496	613	489	542	417	349	143	154	117	117
May		**204**	79	102	135	125	76	168	160	293	438	245	263	340	507	724	452	730	407	313	*160*	113	*107*	141
June		149	58	145	139	120	143	146	*189*	310	332	273	352	260	348	530	433	573	381	260	112	115	54	48
July		92	*103*	**357**	**297**	**302**	**224**	**402**	**215**	**456**	619	420	254	333	419	511	313	608	**385**	213	246	*159*	*49*	68
Aug.		153	92	**393**	**267**	**254**	**352**	**252**	**222**	**387**	**388**	413	223	339	408	504	312	**513**	**298**	114	**191**	118	39	30
Sept.		150	132	**415**	**308**	**331**	**328**	**454**	**274**	**285**	**328**	339	173	**339**	**391**	376	332	**632**	**340**	*82*	140	110	*21*	27
Oct.		159	64	**315**	**297**	**294**	**303**	**361**	*257*	244	**388**	**298**	*157*	261	**393**	464	267	**395**	**302**	*92*	*133*	109	*24*	*14*
Nov.		150	67	186	**204**	141	**211**	*127*	*277*	**338**	253	258	232	290	**419**	587	364	492	306	*80*	*107*	150	*19*	*8*
Dec.		*140*	*72*	120	145	134	141	135	125	**382**	**347**	**227**	126	261	**407**	669	339	**424**	**334**	212	**243**	**229**	*21*	*57*
Year		1614	*1070*	**2312**	**2176**	1990	**2103**	**2486**	**2050**	**3433**	**4078**	**3080**	**2270**	**3371**	**4651**	6200	4495	**6472**	**4000**	2518	2200	1981	705	949
Ep.–Q		8.1		10.4	10.4	4.7	7.8			9.0		5.1	9.2		16.0		15.2		6.7	3.1			0	
Climatic formula		Cfah	Awa	Amwi	Amwi	Amwi	Amwi	Afwi	Amwi	Afi	Afi	Afi	Afi	Afi	Afi	Afi	Afi	Afi	Afi	Afi	Afi	Afi	BSwi	Awi

Italic: less than 100 mm/month and less than 1200 mm/year. Boldface: more than 170 mm/month and more than 2000 mm/year. For Ep.-Q see footnote[2] on p. 277.

TABLE 14.2

The geographical distribution and fidelity of the species growing in the mangrove forests of Micronesia (Hosokawa, 1957)

Life-form[1]	Species	Saipan	Guam	Yap	Palau	Truk	Ponape	Kusaie	Arno, Marshalls	Fidelity
HH	*Acrostichum aureum* L.	+	+							4
HH	*Acrostichum speciosum* Willd.			+	+	+				5
M	*Pandanus kanehirae* Martelli				+					1
HH	*Nypa fruticans* Wurmb.		+	+	+	+	+	+		5
MM	*Crudia cynometroides* Hosok.				+					3
MM	(?) *Cynometra bijuga* Spanoghe				+					1
M–L	*Dalbergia candenatensis* Prain	+	+	+	+	+	+	+		5
M–L	*Derris trifoliata* Lour.		+	+	+	+	+	+		3
M	*Xylocarpus granatum* Koenig		+	+	+	+	+	+		5
M	*Excoecaria agallocha* L.		+	+	+	+	+	+		3
M	*Samadera indica* Gaertn.				+					1
M–L	*Hippocratea macrantha* Korth.				+					1
M	*Stemonurus ellipticus* Sleumer				+					3
MM	*Heritiera littoralis* Dryand.	+	+	+	+	+	+	+		3
MM	*Calophyllum cholobtaches* Laut.				+					3
MM	*Sonneratia caseolaris* Engl.			+	+	+	+	+	+	5
MM	*Bruguiera conjugata* Merr.	+	+	+	+	+	+	+	+	5
M	*Ceriops roxburghiana* Arn.			+	+					5
M	*Rhizophora apiculata* Bl.		+	+	+	+	+	+		5
M	*Rhizophora mucronata* Lamk.		+	+	+	+	+	+		5
M	*Lumnitzera littorea* Volgt.	+	+	+	+	+	+	+	+	5
M	*Barringtonia racemosa* Bl.		+	+	+	+	+	+		1
M–L	*Finlaysonia maritima* Backer				+					5
M–L	*Sarcolobus sulphureus* Schltr.				+					1
M–L	*Tylophora polyantha* Volk.			+						1
M	*Thespesia populnea* Soland.			+	+	+	+	+		1
MM	*Avicennia marina* Vierh.			+	+					5
N	*Clerodendron inerme* Gaertn.	+	+	+	+	+	+	+	+	1
N	*Acanthus ebracteatus* Vahl.				+					5
MM	*Scyphiphora hydrophyllacea* Gaertn.			+	+					5
HH	*Paspalum vaginatum* Sw.	+	+	+	+	+	+	+		3
	Total no. of species	7	14	20	29	16	15	15	4	

[1] Raunkiaer's life-form system. M–L indicates the microphanerophytic lianes.

along many stretches of coastline, with large swamps in estuaries and filled lagoons (as in Kusaie). They are a much less conspicuous element in the vegetation of the Marianas and Marshalls. In the Marianas, only around the southern half of Guam there is any important development of mangal, and the creation of the large naval installations around Apra Harbour has destroyed the most extensive areas formerly existing on this island. Fosberg (1960) commenting further on the lack of mangal noted that in the Marshall Islands there are few examples of mangrove swamps with muddy or silty bottoms, but that a modification, which has been termed "mangrove depressions" (Fosberg, 1947), did occur here and in the atolls of the Carolines. In these, several of the species of trees characteristic of mangrove swamps, especially *Bruguiera gymnorrhiza* (= *Bruguiera conjugata* Merr.) and *Lumnitzera littorea,* as well as *Pemphis acidula* and *Intsia bijuga* found elsewhere on dry limestone rock, grow in clear water in rock–lined depressions. Other depressions are muddy but separated from the sea. The water in the different depressions is of varying salinity, and the floristic composition seems to differ with salinity. Some, at least, of the mangal depressions in the Marshall Islands are abandoned taro pits into which the natives have thrown *Bruguiera* seedlings.

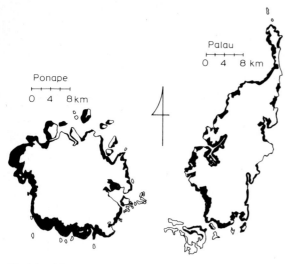

Fig. 14.1. Map showing the mangal areas (blackened) of Ponape and Palau, which are well developed on the leeward coast of the prevailing wind (see Fig. 14.2) (Hosokawa, 1957).

The development of mangrove forests in Oceania is strictly influenced by the direction of the prevailing wind. It is clear that in every main island of Micronesia, such as Ponape, Kusaie, Yap and Palau the widest belt of mangrove forests is found on the leeward coast and the narrowest on the windward coast of the island (Figs. 14.1, 14.2).

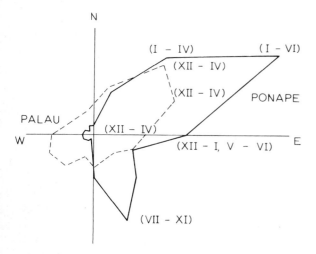

Fig. 14.2. Diagrams showing frequency of each wind direction, which are expressed in a broken line by the annual mean of the observation number (4-hourly values) in Palau (1926—1930) and in a continuous line by that of Ponape (1931). The letters in parentheses indicate months (Hosokawa, 1957).

Forest types

We can recognise three types of forest in Micronesian mangal vegetation:

(1) The *Sonneratia caseolaris* type of forests, which are developed in the islands of Yap, Palau, Truk, Ponape and Kusaie. The forests of this type grow occasionally near the outer or seaward side of the mangal area. Even if the flat bottom of the shallow sea in lagoons is of coral reef, *Sonneratia caseolaris* is able to grow there and resist the winds and waves, occurring usually on the outermost edge of the mangal area.

(2) The *Rhizophora mucronata* type of forests, which are developed in the islands of Guam, Yap, Palau, Truk, Ponape and Kusaie. These forests thrive well along the banks of lower courses of streams and near their mouths and the shallow seas of inlets, where *Rhizophora mucronata* is mixed sometimes with *Rhizophora apiculata,* or both of them grow respectively in pure stands.

(3) The *Bruguiera gymnorrhiza* type of forests, which are developed in the islands of Saipan, Guam, Yap, Palau, Truk, Ponape and Kusaie. This type occurs in the inner parts of mangal areas, especially in Palau and Ponape where it grows along the banks of middle courses of streams and is frequently continuous with either the swamp forests or lowland rain forests.

The islands (e.g., Ponape)

Glassman (1952), who studied the flora of Ponape, described the mangrove forest as encircling the entire main island, except in places where the trees have been cut to clear the way for building harbours, and also extending for some distance up a number of tidal streams. It is further found around the inner shores of some of the basaltic islands, and it constitutes the entire vegetation of a number of the inshore deposit islets. Along the seaward side of the mangrove swamp, *Sonneratia caseolaris* (kotoh), is generally the dominant species. At the mouth of the larger rivers and around bay indentations, *Rhizophora mucronata* and *R. apiculata* form almost pure stands, or are mixed with *Sonneratia caseolaris* and *Bruguiera gymnorrhiza*.

Imanisi and others had earlier carried out ecological studies on Ponape in 1941, and described

(Imanisi and Kira, 1944) in detail the mangal vegetation on Ponape. They were interested especially in the ecological succession of Ponapean mangal vegetation. In the case of vegetation extending over submerged reef flats the pioneer is *Rhizophora mucronata,* and the next zone is *R. apiculata.* Shrubby species of mangrove usually occupy the seaward edge of the mangal vegetation, and this is probably a pioneer feature. Small masses of mangal growing here and there in the open shallow sea of a submerged reef flat are always shrubby in growth. The sandy or muddy bottom of a shallow sea, in addition to the slowness of tidal flow, may enable seedlings of mangrove species to root there and gradually to form a colony far apart from massive areas of mangal vegetation. Sand deposits at the sea bottom may be stabilised with the assistance of growing submerged communities of *Enhalus acoroides,* so that mangal succession really commences with the *Enhalus acoroides* communities.

The succession diagrams (Fig. 14.3) illustrate the seral stages in Ponapean mangal vegetation as shown by Imanisi and Kira (1944).

The reason why they recognised the *Xylocarpus—Sonneratia—Bruguiera* forest to be the climax forest on Ponape is that it covers an extensive area and they also considered the occurrence of *Xylocarpus granatum* to be a leading indication of the climax.

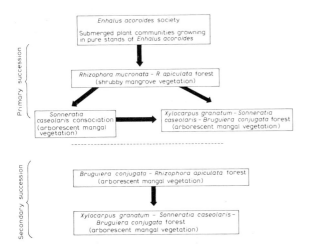

Fig. 14.3. A diagram showing the seral stages in the ecological succession of Ponapean mangal vegetation (Imanisi and Kira, 1944).

Synusiae

According to the present writer (Hosokawa, 1957), forests of the *Sonneratia caseolaris* and *Rhizophora mucronata* types consist of only a single synusia, while the *Bruguiera gymnorrhiza* type consists of two strata, viz. it has the *Acrostichum aureum* society (Du Rietz, 1936) on the forest floors in Saipan and Guam (?) and the *Acrostichum speciosum* society on those of Yap and Palau. We are able to recognise also vascular epiphyte communities growing only in sunny and dry environments in the mangal areas, the communities being developed in crowns and on crown bases of trees as the main niches throughout the mangrove forests of Micronesia. The following single aerosynusia have been recognised (Hosokawa, 1951): (a) the *Phymatodes scolopendria* epies (Hosokawa et al., 1954) in the forests of Saipan; (b) the *Dischidia hahliana—Bulbophyllum volkensii* epilia (Hosokawa et al., 1954) in those of Yap; (c) the *Dendrobium elongaticolle—Bulbophyllum volkensii* epilia in those of Palau; (d) the *Davallia solida* epies in those of Truk.

The epilia communities in Yap and Palau are also found growing as an aerosynusia in sunny and dry environments on trees in the *Campnosperma* rain forests which stand inland in both islands (Hosokawa, 1954d, 1957). The three types of mangrove forests in Ponape and Kusaie have only one aerosynusia, the *Nephrolepis hirsutula* epies, which is found chiefly in tree crowns and crown bases of sunny and dry environments.

Although the structure and the floristic composition of the arborescent vegetation of lowland rain forests and mangrove forests are quite different from each other, we can, however, recognise a similarity of epiphyte communities between them (Table 14.3) (Hosokawa, 1954c, 1955, 1957). Thus, the *Dischidia hahliana—Bulbophyllum volkensii* epilia of Yap (Hosokawa, 1954d, 1955, 1957) is found ranging over sunny and drier places in crowns and crown bases of the trees standing not only in the mangal area but also in the area of the *Semecarpus venenosa—Pandanus japensis* association (Hosokawa, 1954c), which is composed of lowland rain forests. Likewise in Palau the *Dendrobium elongaticolle—Bulbophyllum volkensii* epilia (Hosokawa, 1954b, 1955) is developed on the sunny and drier places not only in trees growing

TABLE 14.3

The floristic composition of the vascular epiphyte communities growing in the mangrove forests of Micronesia (Hosowaka, 1957)

Island	Social unit of epiphyte communities		Major component species
Saipan	*Phymatodes scolopendria* epies	(Rr)	*Phymatodes scolopendria* Ching
Yap	*Dischidia hahliana—Bulbophyllum volkensii* epilia	(Rr, Se)	*Dischidia hahliana* Volkens
		(Rr, Se)	*Bulbophyllum volkensii* Schltr.
		(Mc, Se)	*Bulbophyllum profusum* Ames
		(Rr)	*Davallia solida* Sw.
		(Rr)	*Phymatodes scolopendria* Ching
		(Rr)	*Nephrolepis hirsutula* Presl
		(Rd)	*Vittaria elongata* Sw.
Palau	*Dendrobium elongaticolle—Bulbophyllum volkensii* epilia	(C, Se)	*Dendrobium elongaticolle* Schltr.
		(Rr, Se)'	*Bulbophyllum volkensii* Schltr.
		(Rr, Se)	*Aglossorhyncha micronesiaca* Schltr.
		(C)	*Dendrobium implicatum* Fukuyama
		(Mc)	*Phreatia palawensis* Tuyama
		(Rr)	*Nephrolepis hirsutula* Presl
		(Rr)	*Humata ophioglossa* Cav.
		(Fi)	*Mecodium polyanthos* Copel.
		(Rr)	*Humata gaimardiana* J. Sm.
		(Rr)	*Davallia solida* Sw.
		(Rr, Se)	*Bulbophyllum gibbonianum* Schltr.
Truk	*Davallia solida* epies	(Rr)	*Davallia solida* Sw.
		(Rr)	*Humata trukensis* H. Ito
		(Rr)	*Nephrolepis hirsutula* Presl
		(Rr)	*Phymatodes scolopendria* Ching
		(Rd)	*Vittaria elongata* Sw.
Ponape	*Nephrolepis hirsutula* epies	(Rr)	*Nephrolepis hirsutula* Presl
		(C)	*Dendrobium carolinense* Schltr.
		(Rr)	*Davallia solida* Sw.
		(Rr)	*Humata ophioglossa* Cav.
		(Rr)	*Phymatodes scolopendria* Ching
Kusaie	*Nephrolepis hirsutula* epies	(Rr)	*Nephrolepis hirsutula* Presl
		(C)	*Dendrobium carolinense* Schltr.
		(Rr)	*Davallia solida* Sw.
		(Rd)	*Vittaria elongata* Sw.
		(F)	*Neottopteris nidus* J. Sm.

The epilias and epies of epiphyte communities growing in the mangrove forests are all found in sunny and drier places in crowns and on crown bases, and moreover sometimes on the upper part of trunks.

The letters in parentheses indicate the life-forms of vascular epiphytes, which are original with the author (1943a, 1949, 1955) (see footnote, p. 277).

Asterisk indicates the characteristic species.

in the mangal area but also in the area of the *Campnosperma* rain forests (Hosokawa, 1954b). The reason why these epilias occur in these quite different forest types may be explained by the fact that the habitats of sun epiphytes are almost similar to each other, so far as microclimatic conditions are concerned, in spite of different types of arborescent host vegetation.

Life form

Another synecological feature that can be examined is that of life form (Raunkiaer, 1934). In mangrove forest communities the reason why the value of E is rather high, more than 50%, as compared with that of woody species, MM, M or N, may be attributed to the fact that the woody species which grow in mangrove swamps are confined to those adapted to such severe conditions, while the environmental conditions of vascular epiphytes on mangrove trees are not so different in general from those of the adjacent lowland rain forest (Hosokawa, 1957)[1]. In Palau, among the inland tropical rain forests of the *Campnosperma brevipetiolata—Pandanus aimiriikensis* association and the *Planchonella obovata* association, and the swamp rain forests of the *Horsfieldia amklaal—Donax canniformis* association (Hosokawa, 1954b), the percentage values of E, MM, M and N in the life–form spectra scarcely differ from each other. Because the last–named association occurs in swampy places the values of Ch, H and G are almost nil, while the value of HH amounts to 7%, a situation that is not found in the non–swampy forests (Table 14.4). In the mangrove forest communities the values of most life forms (E excepted) are similar to those in swamp forests (Hosokawa, 1952a). The comparatively high value for E in mangrove forest is considered to arise from the small number of woody species capable of growing in muddy shallow sea. Summarising, there is not any fundamental difference between the life–form spectra of epiphyte societies in four different kinds of forests of Palau (Table 14.5).

One of the physiognomic characteristics of mangrove forests in every island of Micronesia is the relatively high values for Ep–Q[2] (Table 14.6) (Hosokawa, 1950, 1957). These values are probably in almost direct proportion to the annual amount

---

[1] The meanings of the symbols used are as follows:
General life–form symbols (Raunkiaer, 1934): Ch = chamaephytes; E = epiphytes; G = geophytes; H = hemicryptophytes; HH = helo– and hydrophytes; M = microphanerophytes; ML = microphanerophytic lianes; MM = mega– and mesophanerophytes; N = nanophanerophytes.
Symbols for epiphyte types (Hosokawa, 1949): C = caespitose epiphytes; D = dendro–epiphytes; Eph = ephemeral epiphytes; F = fasciculate epiphytes; Fi = filmy–type epiphytes; He = hemi–epiphytes; Mc = monocauli–epiphytes; O = occasional epiphytes; Rd = creeping–tufted epiphytes; Rr = creeping–shoot epiphytes; Rt = root–tuber epiphytes; Se = stem/leaf succulent epiphytes; SV = climbing epiphytes.
[2] "Ep–Q" is the epiphyte quotient, i.e., the number of epiphytic species in the vegetation, expressed as a percentage of the total number of vascular species.

TABLE 14.4

Palau: life-form spectrum of vascular plants (Hosokawa, 1967)

Forest community	Life form	S	E	MM	M	N	Ch	H	G	HH	Th	Number of species
Campnosperma—Pandanus aimiriikensis assoc. (<150 m)			41	12	26	10	5	4	3			190
Planchonella obovata assoc. (<150 m)			38	16	22	6	9	8	2			120
Horsfieldia amklaal—Donax canniformis assoc. (± 10 m)			42	12	32	6	1			7		118
Mangrove forest (± 0 m)			59	9	25	4				3		55

TABLE 14.5

Palau: life-form spectrum of vascular epiphytes (Hosokawa, 1967)

Forest community	Life form	D	Mc	SV	Rr	Rd	C	F	Rt	Fi	(Se)	He	O	Eph	Number of species
Campnosperma—Pandanus aimiriikensis assoc. (<150 m)		9	2	20	8	22	17			9	(24)	6	5	3	71
Planchonella obovata assoc. (<150 m)		6	2	30	12	18	18			8	(28)	2	2		49
Horsfieldia amklaal—Donax canniformis assoc. (±10 m)		8	2	22	10	22	18			14	(23)	2	2		50
Mangrove forest (±0 m)		7	2	27	12	24	10			2	(27)	2	5	7	41

TABLE 14.6

The life-form[1] spectrum of vascular epiphytes in mangrove forests of Micronesia (%) (Hosokawa, 1957)

Island	EP–Q	Fi	D	SV	Mc	Rr	Rd	C	F	Rt	(Se)	He	O	Eph	Sp. No.
Saipan	16.7					50								50	2
Guam	(?)														?
Yap	34.6				2	50	10		10		(30)			10	10
Palau	51.5	2		2	7	27	12	24	10		(27)	2	5	7	41
Truk	40.0					63	25	13							8
Ponape	45.8					54	8	15	8		(15)	8	8		13
Kusaie	52.2		8	8		23	8	23	23		(15)	8			13
Arno, Marshalls	(?)					50			50						2

[1] Hosokawa (1943a, 1949, 1955, 1957 and 1968).
Since Se, succulent, is a life form which is generalised with regard to succulency, and not to the state of a persistent axis and arrangement of leaves and shoots, the species numbers of Se are shown with parentheses in percentage which corresponds to the same species-number of a given one of the other life forms.

of precipitation for each island. Every stand of mangal is inundated with brackish water at high tide up to the middle level of trunks or near the crown bases, so that no heliophobous or sciophilous communities are able to grow on the lower parts of trunks. In point of fact only heliophilous communities can develop in these forests. In the life–form spectrum of vascular epiphytes (Hosokawa, 1949) in communities growing only in sunny and dry places on trees in the three types of mangrove forests throughout Micronesia Rr (see p. 278), one of the life forms of vascular epiphytes (Hosokawa, 1943a, 1949, 1955, 1957) is predominant with C and Se next (Table 14.6). Nevertheless, the low values of F and Fi in mangrove forest, and by contrast, the high values of F and Fi in swamp forest, may be the result of the influence of water level as an important edaphic factor in mangal and swamp forest areas (Table 14.5) (Hosokawa, 1967).

The effect of forest microclimatic conditions on plant life form should be studied. Among the ecologically important microclimatic factors in forest and in the Micronesian mangrove forests light intensity and the conditions of atmospheric humidity and evaporation stand out. According to the present writer's investigation on the spatial distribution of the life forms of vascular epiphytes on trees within the forests (Hosokawa, 1967, 1968), we can group those life forms into 4 ecological types, viz. the sun type of direct sunshine (Rr and C), the stem and/or leaf succulent xerophilous type (Se), the shade-tolerant type of diffused light (Rd and F), and the hygrophilous type (Fi and Rt). Accordingly one may recognise an adaptation of every epiphyte life form to the microclimatic conditions in the forests.

TAIWAN (T. Hosokawa)

Climatic relations

In Taiwan, at altitudes less than about 500 m having an **Afw** or **Cfaw** climate (Köppen, 1923), is found the Machilion kusanoi (the *Machilus* forest areas) consisting of the *Ctenideto–Ficetum cuspidatocaudatae* and *Elatostemeto–Machiletum kusanoi* of tropical and subtropical rain forests. Areas of **Awa** climate in the southwestern part and of **Cwah** climate in the western part of Taiwan, which correspond climatically to the lowland areas of Kuang–chou and Hai–nan Tao of China, were covered originally by thin and sparse rain–green woods, but nowadays most of those areas of seasonally dry and natural wasteland woody vegetation have been brought under cultivation, e.g., rice fields, sugar–cane fields and so on (Hosokawa, 1954a).

However, there still remain, in several **Awa** areas between Kao–hsiung City and Heng–ch'un Town in the southwestern part of Taiwan, rain–green forests of the *Bridelieto–Albizzietum procerae* and *Imperateto–Bombacetum ceibae* (Hosokawa, 1952b, 1954a). Rain–green thorn scrubs, consisting predominantly or mainly of *Pistacia chinensis*,

Sapindus mukorossi, Drypetes formosana, Gymnosporia diversifolia, Scolopia oldhami, Elaeagnus oldhami, Breynia officinalis, Cudrania cochinchinensis var. *gerontogea* and *Zanthoxyllum setosum* (Hibino and Shimada, 1937), which have been conserved, are situated near the mangal vegetation at Xian–ke–shi, Hsin–chu prefecture, in the **Cwah** climatic region of the northwestern part of Formosa. Comparable mangal vegetation has developed also in a **Awa** climate in Kao–hsiung Bay.

Generally, Taiwan mangal vegetation has developed in localities of quite seasonally dry climate (**Awa** and **Cwah**), but then none of the vascular epiphytes can be found there. We cannot call it tropical rain forest or rain scrub, and we should perhaps term it an edaphic–climax littoral woody vegetation (Fig. 14.4).

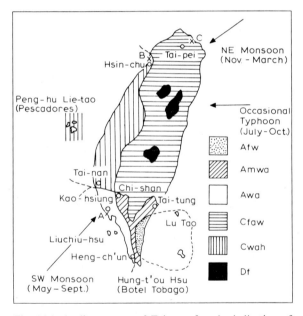

Fig. 14.4. A climate map of Taiwan after the indication of Köppen's method. A (Kao–hsiung Bay), B (Hong–mao) and C (Chi–lung) indicate the localities of mangal vegetation.

Distribution

Kao–hsiung

Mangal vegetation occurs in Taiwan at three localities along the northern and southwestern parts of the coast. One of the three localities includes the Kao–hsiung Bay and the Kao–hsiung River, where the most massive mangrove forests in Taiwan occur. The climatic condition of the Kao–hsiung Bay is closely allied to that of Tai–nan City, on the west coast of Taiwan (lat. 23°N, and about 45 km north of Kao–hsiung City). The annual rainfall and rainy days of Tai–nan City are set out in Table 14.7. According to Kudo (1932), five species of mangroves are to be encountered, namely, *Rhizophora mucronata, Bruguiera gymnorrhiza, Ceriops tagal, Lumnitzera racemosa*, and *Avicennia officinalis*. In every place around the bay this type of mangal vegetation is well developed. The principal factor against the development of mangal vegetation is violent wave action, so that it is difficult for young plants to establish themselves and they are liable to be carried bodily away from the muddy bottom soil of shallow seas. This is the case with the mangal vegetation on the side of the Chi–hou Peninsula, where one more often finds *Avicennia officinalis* and *Lumnitzera racemosa*, both of which are able to withstand the prevalent wind that blows directly in toward the coast. *Rhizophora mucronata* and *Bruguiera gymnorrhiza* only occur in sheltered places, and then but rarely.

On the opposite side of the peninsula, in the vicinity of innermost Chien–chen, rich mangals have developed over long stretches, though in some cases they may be largely due to man's interference. *Rhizophora mucronata* is very common, with excellent stilt roots. *Bruguiera gymnorrhiza, Avicennia officinalis* and *Ceriops tagal* are also found there. According to Kudo (1932), who

TABLE 14.7

The amounts of rainfall (in mm) and rainy days observed in Tai–nan City

	Length of record (year)	Elevation (m)	Jan.	Febr.	March	April	May	June	July	Aug.	Sept.	Oct.	Nov.	Dec.	annual
Rainfall	26	14	25	34	43	58	182	347	331	433	159	37	21	12	1676
Rainy days	26	14	5	5	6	7	10	15	16	19	11	5	4	4	107

made a careful investigation of the mangal vegetation in Kao–hsiung Bay, there are two places, each with fine mangal vegetation, both situated near the innermost part of the bay.

In one place, the mangal vegetation stretches along the shore in a belt and is composed of large trees of *Avicennia officinalis* and shrubby *Rhizophora mucronata*. In the other, the banks of a stream are occupied by a dense growth of *Rhizophora mucronata*, among which are found isolated trees of *Bruguiera gymnorrhiza*, *Avicennia officinalis* and *Ceriops tagal*.

Almost the same type of mangal vegetation occurs along the Kao–hsiung River and in Kao–hsiung, *Bruguiera gymnorrhiza* being the common species with *Avicennia officinalis* as well. On the left bank of the stream, the latter species mostly grows to tree size and forms almost a pure society in a small area. The canopy of this pure stand is well closed, and there is no undergrowth except for some shrubby *Bruguiera gymnorrhiza*.

Hong–mao

As the second locality of Taiwan mangal vegetation, the mouth of the Hong–mao River, Hsin–chu prefecture (west coast), may be cited. Differing distinctly from the mangal vegetation of Kao–hsiung Bay, this mangal may be developed from an artificial origin, since some 180 years ago a few plants of *Kandelia candel* were introduced here from the southwest coast of continental China. Nowadays, however, it seems to be quite natural.

The climatic conditions of the locality are probably allied to those of the Agricultural Experiment Station of Hsin–chu, about 6 km to the southeast of the mangal vegetation (Table 14.8).

The annual mean air temperature is about 20°C, but in the most muggy summer July period it rises to 32°C, with an absolute maximum of 37°C. The coldest month of the year is February (12°C) when sometimes a thin frost may occur on the ground. The annual rainfall amounts to about 1690 m. The distribution of rainfall in the year is not uniform, lesser monthly amounts than the average are observed from October to mid–January, these months being relatively dry. Abundant rainfall is observed in March and April every year, but this decreases in May, and again increases from June to September under the influence of frequent typhoons. Apart from the typhoons, the SW Monsoon blowing from May to September generally provides calm weather. From October to February the NE—NNE Monsoon is usually predominant and in January and February wind velocities of about 10 m sec^{-1} continue sometimes for several days.

TABLE 14.8

Several climatic conditions of average monthly and annual amounts, which were observed at the Agricultural Experiment Station of Hsing–chu (Taiwan)

	Air temperature (°C)			Wind		Rainfall (mm)	Rainy days
	mean	mean max.	mean min.	direction	daily velocity (m sec^{-1})		
Jan.	16.3	?19.2	11.9	NE	7.1	88.4	8.0
Febr.	16.1	19.5	12.1	NNE	6.4	200.4	14.4
March	18.4	?20.4	12.9	NNE	5.9	136.3	9.2
April	21.5	24.3	16.7	NNE	5.8	294.1	14 2
May	25.6	28.7	20.0	SW	4.2	161.3	8.2
June	27.6	30.0	23.1	SW	4.2	283.9	11.4
July	29.8	32.1	24.3	SSW	3.7	77.4	7.6
Aug.	30.7	31.5	24.4	SW	3.9	243.4	9.8
Sept.	28.4	30.1	23.2	NE	4.6	104.8	6.8
Oct.	25.2	26.9	19.6	NE	6.5	44.5	6.8
Nov.	18.4	25.1	16.3	NNE	6.4	14.6	4.6
Dec.	18.4	12.0	12.8	NE	6.4	41.6	6.8
Annual	19.6	25.0	18.2			1690.7	107.8

Chi–lung

A locality near the mouth of the river at Chi–lung Bay provides the third site of Formosan mangal. Here *Kandelia candel* is the only species, forming a very poor mangal vegetation.

Dispersal

Fruits and seeds produced from mangal vegetation have been observed on the sandy beach of Chi–lung in northern Taiwan, on the stony shore in the island of Yaku–jima, about 140 km south of Kagoshima City, and on the beach of Sado situated in the Sea of Japan (Nihon–Kai).

At Chi–lung (lat. 25°N) *Nypa* phalanges landing on the beaches usually display the cotyledon and root there, but the soil condition is not muddy and the climatic condition is not good enough. The winter climate of Chi–lung is too cold for *Nypa* palms to grow there vigorously against such environmental conditions, although *Nypa fruticans, Sonneratia caseolaris* and *Acrostichum aureum,* none of which are found in Formosa, grow in the mangrove swamp areas of Iriomote–shima Island, situated in lat. 24°20′N.

On the island of Yaku–jima (Fig. 14.5) phalanges of *Pandanus odoratissimum* commonly occur on sandy beaches. Another example of phalanges of *Nypa fruticans* drifting inshore comes from Sado Island in the Sea of Japan, about 50 km apart from the coast of the Niigata prefecture in the Kokuriku district. Such phalanges supposedly drift north by means of the Tsushima warm current from the south (Kitami, 1970).

Epiphytes

The mangrove vegetation of the Kao–hsiung district is located in the area which bears the characteristics of Köppen's **Aw** climate, i.e., rain–green forest climate. Compared with the physiognomy of mangrove forest in the Malay Archipelago, Taiwanese examples of mangal vegetation are not so luxuriantly developed, not only in the Kao–hsiung Bay but also in the other two sites (pp. 279—281). Mangal vegetation developed under such climatic conditions does not contain any vascular epiphytes. Hence it appears that mangal vegetation is not only developed under the climatic condition of Köppen's **Afi** climate, i.e., tropical rain–forest climate, but also under **Aw** or **Cw** climates. Those mangrove forests containing numerous vascular epiphytes occur usually in the tropical rain–forest region of Micronesia (cf. p. 271ff). In such areas one can, with Rübel (1930), refer to the mangal as a kind of tropical rain forest or tropical rain scrub. However mangal is not always a tropical rain forest or rain scrub, because it can be developed luxuriantly in the rain–green forest climatic region (**Aw**), too. Mangal vegetation is usually produced under particular edaphic conditions, so that it is fairly defined rather as an edaphic climax than as a climatic climax developed under the influence of certain climatic conditions.

MANGALS IN SOUTHWEST JAPAN (H. Tagawa)

Introduction

The islands of Nansei–shoto, southwest Japan, form stepping stones for mangroves from tropic Asia to the warm temperate region of Kyushu in Japan. The areas occupied by the mangrove forests and the diversity of the species composition decrease with increasing latitude. At the northern

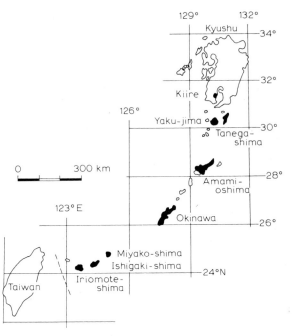

Fig. 14.5. The islands of Nansei–shoto, southwest Japan. The dark areas indicate mangal vegetation.

limit and in adjacent areas mangal is represented by stands of one species, *Kandelia candel*.

Localities where mangal vegetation is found are Iriomote–shima, Ishigaki–shima, Miyako–shima, Okinawa, Amami–oshima, Yaku–jima, Tanega–shima, and Kiire (Fig. 14.5). The following account gives the general trend of simplification in the specific diversity of mangals and their ecological characteristics with increasing latitude.

Distribution

Iriomote–shima
Iriomote–shima is the only island of Ryukyu Retto where the vegetation has been very well preserved in the natural condition. Forested low mountains formed of sandstone and shale of Tertiary origin and abundant precipitation (2630 mm annually) have produced rivers of low head with consequent development of alluvial deposits on both sides of rivers and estuaries.

Every bank of the rivers is lined with mangal vegetation, the extent and species composition of which differ from place to place. Odani (1964) recognised the following four different stands from the lowest reaches of the river to the upper on the Nakara River: (a) mixed stand of *Kandelia candel* and *Rhizophora mucronata*; (b) pure stand of *R. mucronata*; (c) mixed stand of *R. mucronata* and *Bruguiera gymnorrhiza*; (d) pure stand of *B. gymnorrhiza*. The last extends to the middle and upper reaches. According to Miyata and Odani (1963), *B. gymnorrhiza* and *Pandanus tectorius* are segregated. *B. gymnorrhiza* prefers a habitat on the undercut slope of the meander in the middle reaches, while *P. tectorius* is limited in its distribution to the slip–off slope.

Nypa fruticans, which is a component of the mangals in tropical Asia, is only found on the river Yashiminato of this island. On the right bank of Yashiminato River, 150 individuals of *N. fruticans* have been recorded in the mangal, and have been assigned as a natural monument since 1959. A few plants appear also in the mangal on the Shiira River. From the above account, it may safely be said that the mangals on the river banks are simple in their species combinations. Fig. 14.6 shows the distribution of diameter and height of *B. gymnorrhiza* from the pure stand on the Nakara River.

Compared with the poverty of species in the

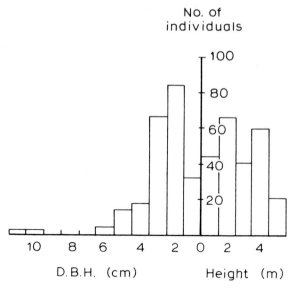

Fig. 14.6. Frequency distribution of diameter (D. B. H.) and stature of *Bruguiera gymnorrhiza* in 10 × 10 m quadrats (modified from fig. 1.5 and table 1.5 of Miyata and Odani, 1963).

riverside mangals, the species composition becomes more complex in the estuarine mangals. From extensive observation of many estuaries Miyata and Odani (1963) arrived at the conclusion that there were four zones from low water to high–water level. They are successive zones of *Sonneratia alba*, *R. mucronata*, *K. candel*, and *B. gymnorrhiza* (Fig. 14.7). The most inland zone of *B. gymnorrhiza* is adjacent to a belt of luxuriant *Pandanus tectorius*. *Pandanus* scrub forms the frontal zonation of swamp forest where *Barringtonia racemosa* occurs as dominant, with *Heritiera littoralis* and many species of banyans (*Ficus septica*, *F. ampelas*, *F. benguetensis*, *F. microcarpa*, etc.).

Islet colonies of *Avicennia marina* represent pioneer mangrove in the Maera River (Sekizuka and Shimizu, 1970), but this species does not always appear in such a place. *Lumnitzera racemosa* is a very rare species in the mangals of Iriomote–shima, but its ecological performance has not been sufficiently studied here.

Intensive study of *B. gymnorrhiza* carried out by Miyata et al. (1963) proved that there are large and small clumps composed of a similarly aged population, and the saplings and seedlings grow up so far as to fill spaces not occupied by the mother trees. They consider that a number of muddy mounds

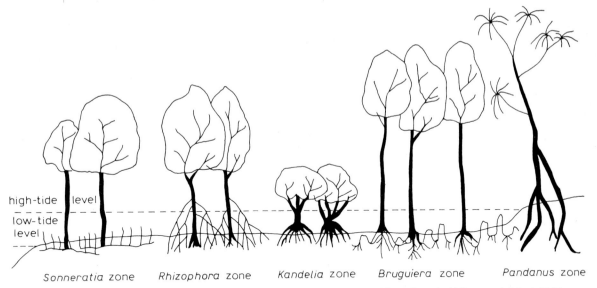

Fig. 14.7. Zonation of mangrove species in the estuaries of Iriomote–shima (revised from fig. 1.4 of Miyata and Odani, 1963).

(50 cm high in average) made by a crab species, *Sesarma* sp., on the forest floor have a great effect on the distribution of seedlings and on regeneration of the community. The top surface of the mounds is usually dried up so that seedlings cannot become established. A similar phenomenon was observed in Amami–oshima by the present author.

Ishigaki–shima

This island has a central Palaeozoic plateau and irrupted granite which produces a mountain range (maximum height is 526 m on Mt. Omoto) running from north to southwest. Some rivers, therefore, flow into the Pacific Ocean, and others into the East China Sea. Precipitation averages 2700 mm in a year.

Mangal on the Tsuro River, which flows eastwards, is a nearly pure community of *B. gymnorrhiza*, but on the rivers flowing westwards, such as the Nagura and Jiba and their estuaries, *R. mucronata* and a few individuals of *B. gymnorrhiza* are the constituent species (Sekizuka and Shimizu, 1970).

No adequate reason has been found to explain why these two species exhibit a habitat segregation. In this connection it may be noted that there is a mixed population of *R. mucronata* and *B. gymnorrhiza* on the banks of the Miyara River which flows southwards into the sea. This mangal has been assigned as a natural monument since 1959.

Okinawa

Okinawa is divided geologically into two districts. The northern half is characterised by the mountainous topography called Yambaru, which is on Palaeozoic rock. The southern half is on the flat Ryukyu limestone which covers a Tertiary bed.

Mangal in the Kesaji River estuary is the most representative. The trees are comparatively young, and reach only 3 m in height, because people had frequently cut down the mature trees of *R. mucronata* and *B. gymnorrhiza* until this forest was assigned as a natural monument in 1959. The former species is dominant. *B. gymnorrhiza* is, however, taller than *R. mucronata,* and is emergent from the *Rhizophora* canopy in places. There is even less *K. candel* than *B. gymnorrhiza*.

The landward boundary of this mangal is adjacent to a *Pandanus tectorius* zone, but to date no intensive study of mangals in Okinawa has ever been published.

Amami–oshima

Amami–oshima is composed of Palaeozoic shales, clay slate and green tuff, though a limited area of the northwestern end is occupied by Cenozoic strata.

Large–scale mangal is found in the western estuary of Sumiyo Bay, but is limited to only two species; *Kandelia candel* and *Bruguiera gymnorrhiza*. In general, *K. candel* occupies the seaward

TABLE 14.9

The number of individuals in 10 × 10 m quadrats placed in the mangal of Sumiyo Bay, Amami–oshima, in 1974

Species \ Quadrat	OS-3	OS-2	OS-1	OS-6	OS-7	OS-5	OS-4	OS-8	OG-1
Kandelia candel									
seedlings	967	5	54	12	.	.	1	.	632
saplings	922	2	.	4	.	.	1	.	.
adults	100	34	28	64	44	12	8	.	28
Bruguiera gymnorrhiza									
seedlings	.	.	.	76	60	.	23	4	.
saplings	.	.	.	68	8	12	40	1	.
adults	.	.	.	44	108	96	156	21	.
Clerodendron inerme	.	.	.	.	24	24	.	14	.
Heritiera littoralis	.	.	.	.	.	.	.	13	.
Drypetes karapinensis	.	.	.	.	.	.	.	24	.
Quercus glauca var. *amamiana*	.	.	.	.	.	.	.	2	.

zone of the mangal. A *B. gymnorrhiza* stand succeeds it gradually, and is backed by the terrestrial forest of which the dominant is *Quercus glauca* var. *amamiana*. *Drypetes karapinensis, Heritiera littoralis,* and lianes of *Caesalpinia crista* and *Clerodendron inerme* are also abundant in the *Bruguiera* forest. The gradual change of species composition of the mangal is illustrated in Table 14.9.

At Gusuku, on the eastern end of Sumiyo Bay, mangrove again occurs (OG–1 in Table 14.9) on a small scale, with a *K. candel* population principally composing the community with an occasional rare *B. gymnorrhiza*. In this tall community (4 m high), many seedlings of a few years old grow under the parent canopy. *Kandelia* seedlings are hardly found under the denser canopy of *B. gymnorrhiza* (OS–4 to OS–8), but *Bruguiera* seedlings do occur in such places. From this fact, it may be thought that *B. gymnorrhiza* is more resistant to weak light than *K. candel*. If this is correct it could have an important implication on the ecological succession. However, no experiments on the photosynthetic and respiratory rates of these mangrove species have been carried out.

As already described, crab mounds have an effect on regeneration of mangrove in Iriomote–shima. Here, in Amami–oshima, there are many more and larger mounds (about 1 m high) than those in Iriomote–shima. These mounds are built by the mangrove shrimp, *Thalassina anomala*. The upper half of a mound is usually emergent from the tidal water, and on this part grow terrestrial plants such as *Miscanthus sinensis, Gardenia jasminoides, Heritiera littoralis, Caesalpinia crista,* etc. (Fig. 14.8). The number of the mounds is greater on the landward side of the mangal. Mounds built side by side become connected with each other. The present author would recognise the ecological significance of this type of mound in producing land and, therefore, on lotic succession.

Morphological variation of the roots in *K. candel* is an outstanding feature. The basal part of the root is buttress–like on plants near creeks with flowing water and adjacent areas, but away from the creeks or in estuaries with comparatively still water the basal part forms a typical prop–root and several stems arise from near the base. Furthermore, "landed" individuals on the river terrace in Yaku–jima produce their roots just like those of ordinary trees growing on land (Fig. 14.9).

Yaku–jima

Uplifted granite built up the present Yaku–jima in the Tertiary (the highest peak is 1935 m in altitude), and many jagged mountains fall steeply into the sea. Streams flowing down in the deep valley abruptly subside into still waters in lower reaches but there is no developed estuary with a thick deposit of silt. Annual precipitation at Isso reaches 4314 mm on average, and at Arakawa (740 m in altitude) it rises to 9345 mm.

Mangal in Yaku–jima is restricted to the emergent river terrace of the Kurio River in the south-

Fig. 14.8. Muddy mounds built by mangrove shrimp in Amami–oshima. A and B. The apical parts were colonised by *Miscanthus sinensis*. C. Mound is built on the base of *Kandelia candel*.

Fig. 14.9. Morphological variation in roots of *Kandelia candel*. A. Small scale of buttresses in lotic estuary (Amami–oshima). B. Prop–root in landward estuary (Amami–oshima). C. Terrestrial type of root on the river terrace (Yaku–jima).

west. It is never submerged at ordinary high tide, but is submerged at high spring tides. The area around is cultivated for vegetables.

Only one species, *Kandelia candel* (maximum height is 4 m), occurs in the community (Table 14.10). Since the plants are almost permanently out of the water, prop–roots are not conspicuous, and the roots look like the ordinary roots of terrestrial trees. Stems are scarcely divided, and usually one stem grows upright (Fig. 14.9C).

The morphological changes in *Kandelia* roots may arise from the large difference in habitat in passing from submerged estuary to mesic terrestrial humid soil and not from any difference in latitude. Thus, *K. candel* growing in a lagoon in Tanega–shima at approximately the same latitude has prop–roots and branched stems.

The mangal in Yaku–jima can have a ground vegetation composed of one or more of the following species: *Rumex japonicus, Lactuca lanciniata, L. raddeana, Polygonum blumei, Commelina communis, Bothriochloa parviflora* and *Phragmites japonica*.

The crab *Sesarma dehaani* constructs nesting pits that open from place to place on the forest floor, but this species does not make a mound around the pit.

The proportion of seedlings and saplings to the total trees of the *Kandelia* community is 58.6% and 3.7%, respectively, and the percentage of saplings is very much smaller than the corresponding value in Amami–oshima (47.1% and 42.6%, respectively in OS–1 to 3 in Table 14.9). A *Kandelia* community with a high regenerative ability is effective in maintaining and extending itself to the seaward front in an estuary, while on land the

TABLE 14.10

The number of individuals in each quadrat (10 × 10 m) set in the mangals of Kurio, Yaku–jima (YM), Minato, Tanega–shima (TM) and Nukumi, Kiire (KM)

Species \ Quadrat	YM–1	YM–2	YM–3	YM–4	TM–2	TM–3	TM–4	KM–1	KM–2	KM–3	KM–4
Kandelia candel											
seedlings; alive	100	50	400	50	1825	275	.	168	136	132	20
seedlings; dead	.	.	.	25	275	88	.	4	16	184	4
saplings; alive	24	1	7	6	.	.	.	.	.	.	.
saplings; dead	4	.	.	.	.	.	.	.	.	.	.
adults	128	58	70	131	.	44	.	752	464	260	316
Hibiscus hamabo	.	.	.	.	.	13	.	.	.	.	.
Paliurus ramossissimus	.	.	.	.	.	5	5	.	.	.	.
Machilus thunbergii	.	.	.	.	.	.	3	.	.	.	.
Prunus jamasakura	.	.	.	.	.	.	1	.	.	.	.
Quercus glauca var. *amamiana*	.	.	.	.	.	.	2	.	.	.	.
Ilex integra	.	.	.	.	.	.	1	.	.	.	.
Eurya japonica	.	.	.	.	.	.	4	.	.	.	.
Neolitsea sericea	.	.	.	.	.	.	1	.	.	.	.
Syzygium buxifolium	.	.	.	.	.	.	1	.	.	.	.
Rapanaea neriifolia	.	.	.	.	.	.	1	.	.	.	.
Raphiolepis umbellata	1	.	.	.	.	.	8	.	.	.	.
Ligustrum japonicum	.	.	.	.	.	.	3	.	.	.	.
Distylium racemosum	.	.	.	.	.	.	3	.	.	.	.
Cinnamomum japonicum	.	.	.	.	.	.	2	.	.	.	.
Ternstroemia japonica	.	.	.	.	.	.	6	.	.	.	.
Meliosma rigida	.	.	.	.	.	.	1	.	.	.	.
Gardenia jasminoides f. *grandiflora*	.	.	.	.	.	.	11	.	.	.	.
Lagerstroemia subcostata	.	.	.	.	.	.	4	.	.	.	.
Elaeocarpus sylvestris var. *ellipticus*	.	.	.	.	.	.	1	.	.	.	.
Clerodendron trichotomum	1	.	.	.	.	.	.	.	.	.	.

viviparous seedlings lose their ordinary means of dispersion and they only drop on the forest floor.

Another stand of *K. candel* on a small scale is found on the opposite and right bank of the Kurio River. The community is surrounded by *Zoysia macrostachya* and *Phragmites japonicus* along the frontal water creek. The landward area behind the mangal is covered by a scrub of *Hibiscus hamabo*.

Tanega–shima

Tanega–shima exhibits a striking contrast to Yaku–jima because of the gentle topography. The whole island is hilly, but the highest peak is only 237 m in altitude. The annual rainfall, therefore, is much lighter (2674 mm at Nishino'omote) than on Yaku–jima.

On both sides of the river near Minato, there is a closed and well–developed but narrow *Kandelia* community (maximum tree height is 8 m) on the undercut and slip–off slopes of the meander.

A characteristic feature of this community is a narrow belt of young plants of *K. candel* in front of the adult population (TM–2 of Table 14.10). The mortality of seedlings and saplings is higher under the adult population (TM–3) than in the young frontal zone. The *Kandelia* community on the gentle slip–off slope changes gradually into evergreen broad–leaved forest (TM–4) through an intermediate zone of deciduous trees such as *Hibiscus hamabo* and *Paliurus ramossissimus*.

Many trees of *Kandelia* have been partially damaged since 1969 by a scale insect, *Ceroplastes rubens*. The Board of Education of Nishino'omote City tried in 1972 to stamp out this noxious insect by using a parasitic chalcid fly, *Anicetus beneficus*, but it was not successful. All foliage with the scale insects was removed in 1973, and a number of new branches are now arising from the cut portions.

Clerodendron inerme and *Myoporum bontioides* are companions of the present mangal. *C. inerme* is frequently observed in the mangal on Atake River (Ohno, 1973).

Young and open–scrub type mangal can be seen in the lagoon developed on the lower reaches of the Oura River. This mangal occupies the largest area of any in Japan. According to the intensive study of this community over ten years from 1952 to 1961 carried out by Yamashiro (1954, unpubl.), growth of the plumule of *K. candel* is greatly influenced by the NaCl concentration. The plumule grows favourably between fresh water and 1/2 diluted sea water, but the growth is considerably depressed in the sea water of 2/3 dilution or more concentrated. Yamashiro also examined rooting and root elongation of *K. candel* and showed that the rooting is limited under dark conditions, and that fresh water and much diluted sea water are more favourable for the elongation of root than sea water.

Kiire

A vast area of southern Kyushu is covered with thick volcanic ash and pumice from the volcano Aira, which became depressed in the Early Tertiary to make a big caldera, Kinto Bay. Kiire beach is located on the western side of the bay. Mangal in Kiire has developed on the estuary of a small stream which flows near the colony of Nukumi. It is said that this mangal was transplanted artificially in old times, but this report is of dubious authenticity. In 1952 this northernmost community was assigned as a natural monument.

Viviparous seedlings of *K. candel* fall from the middle of June to July during the rainy season. They establish successfully and root, but there are no saplings to be found in the community (see Table 14.10). Observations made by the present author show that the *Kandelia* trees produce more seeds in the forest margin than in the central area.

THE PHILIPPINES (V.J. Chapman)

On the muddy banks of creeks, especially in the areas of Pagbilao Bay, Calapan and Cotabato (Fig. 14.10), *Rhizophora mucronata* represents the principal pioneer community with an *R. apiculata* community behind. On the seaward front a mixed community of *Sonneratia alba* and *Avicennia officinalis* represents the frontal zone. *Bruguiera gymnorrhiza* dominates the mangal on high ground whilst in the wetter or more brackish areas a mixed community of *Xylocarpus moluccensis*, *X. granatum*, *Lumnitzera littorea* and *Aegiceras corniculatum* is to be found (Brown and Fischer, 1920; Dickenson et al., 1928; Chapman, 1975). The most landward community is dominated by *Heritiera littoralis* together with *Hibiscus tiliaceus*, *Thespesia populnea* and *Glochidion littorale*. Epiphytes, such as species of *Cymbidium*, *Dendrobium* (orchids),

Fig. 14.10. Some mangal areas in the Philippines (after Chapman, 1975).

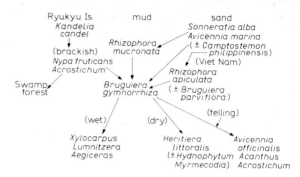

Fig. 14.11. Zonation of mangal in the Philippines.

Myrmecodia tuberosa, Dischidia saccata (ant plants), *Drynaria quercifolia, Polypodium sinuatum* and *Asplenium nidus,* represent invaders from the mesophytic rain forest, as well as lianes such as *Derris trifoliata, Dalbergia candenatensis* and *Caesalpinia* spp. (Merrill, 1945). Tree felling allows *Avicennia officinalis* to invade such an area as well as permitting a dense ground cover of *Acrostichum aureum, Acanthus ilicifolius* and *A. ebracteatus* to develop which discourages tree-seedling establishment and regeneration of the original forest. Up creeks the transition to fresh water is marked by the inevitable *Nypa fruticans.*

It is clear that mangal in the Philippines shows a close affinity to the patterns found in Indo–Malesia and there is not the paucity of species that occurs north into Japan or eastwards into Melanesia. This will apply not only to the plants but also to the fauna, though the latter has not received serious attention. The zonation is represented in Fig. 14.11.

NEW CALEDONIA AND OCEANIA (V.J. Chapman)

Baltzer (1969) has provided a detailed account of the mangal in the Dumbéa River delta of New Caledonia. Here *Bruguiera gymnorrhiza, Sonneratia* and *Avicennia marina* are either absent, or present only as isolated individuals. *Rhizophora mucronata* is the principal pioneer community and is generally succeeded by a *Rhizophora—B. sexangula* zone. With increasingly brackish conditions the *Rhizophora* gradually disappears and *Acrostichum aureum* and Cyperaceae form a ground cover to the *Bruguiera*. At higher levels, with fewer inundations, *Avicennia officinalis* can replace *Bruguiera* and this leads to a sward of *Salicornia australis*. At extreme high-water mark there can be a belt of *Lumnitzera racemosa* (Fig. 14.12). The fauna of these communities is in need of study and it may not be so rich as that of mangal in Papua–New Guinea (see p. 268).

In the Fiji Islands, mangrove species of east and west meet. The principal species are *Rhizophora mangle, R. stylosa* and *Bruguiera gymnorrhiza*. *R. mucronata* is also present but has been confused with *R. stylosa* by earlier writers (Guppy, 1906; Parham, 1964), and there is also a putative hybrid between the two major *Rhizophora* species described by Salvosa (1936) as var. *selala* of *R. mucronata*. The physiography of the islands does not encourage the growth of extensive mangrove swamps, and the major one is found associated with the Rewa River near Suva on Viti Levu. *R. mangle* is the pioneer species up the rivers, whilst *R. stylosa* or the "selala" form are the coastal pioneers, often only forming a broken fringe. Behind the *Rhizophora, Bruguiera* forms a distinct community in

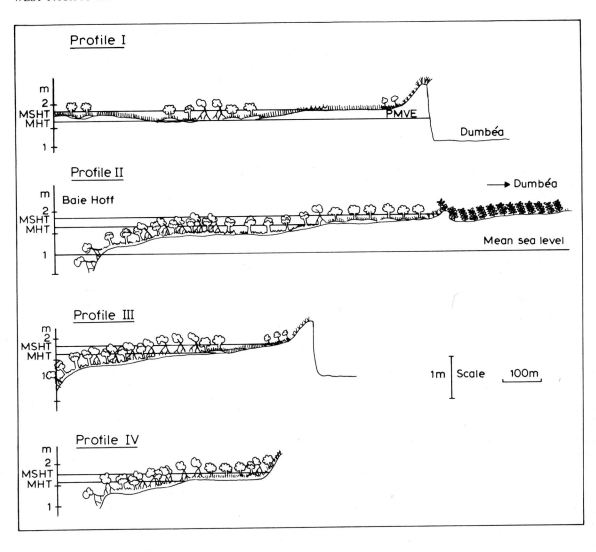
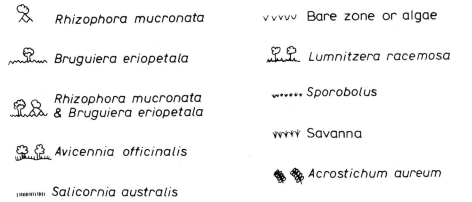

Fig. 14.12. Mangrove zonation in New Caledonia (after Baltzer, 1969). MSHT = mean spring high tide; MHT = mean high tide.

the Rewa River, but elsewhere occurs only sporadically (Chapman, 1975). In those few areas where there are extensive swamps the most landward community is mixed with *Heritiera littoralis, Excoecaria agallocha, Barringtonia racemosa* and *Hibiscus tiliaceus,* with lianes such as *Derris heterophylla, Entada scandens* and *Mucuna gigantea.* In very wet brackish areas there is a giant sedge, *Scirpodendron costatum.*

In the Friendly Islands, *Rhizophora mangle, R. mucronata* (this needs re-examination to see if it is not *R. stylosa), Bruguiera gymnorrhiza, Lumnitzera littorea* and *Xylocarpus granatum* have all been recorded (Hemsley, 1893; Guppy, 1906) but there is no real account of the communities. *Hydnophytum* and *Myrmecodia* occur as epiphytes on trees of landward mangal. *Rhizophora* and *Bruguiera* also occur in the sparse mangal of Samoa (Reinecke, 1898). In both groups of islands the landward zone is occupied by a mixed community of *Heritiera littoralis, Excoecaria agallocha* and *Chlorodendron inerme.* In Samoa this belt also contains *Barringtonia racemosa* and *Scirpodendron costatum.*

It should be evident from this brief account (as there are no recent intensive studies, either of flora or fauna) that the mangal ecosystems in Polynesia are essentially similar, but in no case do they achieve very extensive status.

REFERENCES

Baltzer, F., 1969. Les formations végétales associées au delta de la Dumbéa. *Cah. ORSTOM, Sér. Géol.,* 1(1): 59—84.
Braun–Blanquet, J., 1951. *Pflanzensoziologie. Grundzüge der Vegetationskunde.* Springer, Berlin, 865 pp.
Brown, W.H. and Fischer, A.F., 1920. Philippine mangrove swamps. In: W.H. Brown (Editor), *Minor Products of Philippine Forests, 1. Bur. For. Bull.,* 2.
Burger, Hzn., D., 1972. *Seedlings of Some Tropical Trees and Shrubs Mainly of South East Asia.* Centre for Agricultural Publishing and Documentation, Wageningen, 399 pp.
Busgen, M., Jensen, H. and Busse, W., 1905. Vegetationsbilder aus Mittel– und Ost–Java. *Vegetationsbilder,* 3(3).
Chapman, V.J., 1975. *Mangrove Vegetation.* Cramer, Lehre, 425 pp.
Dickenson, R.E. et al., 1928. *Distribution of Life in the Philippines.*
Du Rietz, G.E., 1936. Classification and nomenclature of vegetation units, 1930—1935. *Sven. Bot. Tidskr.,* 30: 580—589.
Fosberg, F.R., 1947. Micronesian mangrove. *J. N. Y. Bot. Gard.,* 48: 128—138.
Fosberg, F.R., 1949. Atoll vegetation and salinity. *Pac. Sci.,* 3: 89—92.
Fosberg, F.R., 1953. Vegetation of Central Pacific Atolls: A brief summary. *Atoll Res. Bull.,* 23: 17.
Fosberg, F.R., 1955. Northern Marshall Islands expedition, 1951—1952. Land biota: vascular plants. *Atoll Res. Bull.,* 39: 16.
Fosberg, F.R., 1960. The vegetation of Micronesia. 1. General descriptions, the vegetation of the Marianas Islands, and a detailed consideration of the vegetation of Guam. *Bull. Am. Mus. Nat. Hist.,* 119(1): 1—76.
Glassman, S.F., 1952. The flora of Ponape. *Bernice P. Bishop Mus. Bull.,* 209: 1—152.
Guppy, H.B., 1906. *Observations of a Naturalist in the Pacific Between 1896 and 1899, II. Plant Dispersal.* Macmillan, London.
Hatheway, W.H., 1953. The land vegetation of Arno Atoll, Marshall Islands. *Atoll Res. Bull.,* 16: 38—44.
Hemsley, W.B., 1894. The flora of Tonga or Friendly Islands. *J. Linn. Soc. (Lond.) Bot.,* 30: 158—217.
Hibino, S. and Shimada, Y., 1937. *Tennen Kinenbutu Chosa Hokoku.* (A report of Natural Monument Research.) Government of Taiwan, Taipeh, 4: 1—26.
Hosokawa, T., 1934. Preliminary account of the vegetation of the Marianne Islands group. *Bull. Biogeogr. Soc. Jap.,* 5: 124—172.
Hosokawa, T., 1935. *Nypa* phalanges drifting ashore in Formosa and adjacent islands. *Kagaku (Tokyo),* 5: 14—15.
Hosokawa, T., 1937. A preliminary account of the phytogeographical study on Truk, Caroline. *Bull. Biogeogr. Soc. Jap.,* 7: 171—255.
Hosokawa, T., 1938. On phytogeographical considerations of the Bonin Islands. *Shokubutsu Oyobi Doobutu,* 6(3), (4) and (5).
Hosokawa, T., 1943a. Studies on the life forms of vascular epiphytes and the epiphyte flora of Ponape, Micronesia. *Trans. Nat. Hist. Soc. Formosa,* 33: 234—236.
Hosokawa, T., 1943b. *A General Consideration of the Vegetation of the Asiatic Tropics,* pp. 70—83 (in Japanese).
Hosokawa, T., 1949. Studies on the life form of vascular epiphytes and the spectrum of their life forms. *J. Jap. Bot.,* 24: 41—45.
Hosokawa, T., 1950. Epiphyte–quotient. *Bot. Mag. Tokyo,* 63: 18—20.
Hosokawa, T., 1951. On the nomenclature of aerosynusia. *Bot. Mag. Tokyo,* 64: 107—111.
Hosokawa, T., 1952a. A synchorological study of the swamp forests in the Micronesian islands. *Mem. Fac. Sci. Kyushu Univ., Ser. E. (Biol.),* 1: 101—123.
Hosokawa, T., 1952b. On the relationship of climate to vegetation in the southern part of Formosa; in special reference to the raingreen forest. *Bull. Soc. Plant. Ecol. (Jap.),* 2: 1—9.
Hosokawa, T., 1954a. Outline of the vegetation of Formosa together with the floristic characteristics. *Angew. Pflanzensoziol., Festschr. Aichinger,* 1: 504—511.
Hosokawa, T., 1954b. On the structure and composition of the *Campnosperma* forests in Palau, Micronesia. *Mem. Fac. Sci. Kyushu Univ., Ser. E. (Biol.)*, 1: 199—218.
Hosokawa, T., 1954c. On the *Campnosperma* forests of Kusaie

in Micronesia, with special reference to the community units of epiphytes. *Vegetatio*, 5—6: 351—360.

Hosokawa, T., 1954d. On the *Campnosperma* forests of Yap, Ponape and Kusaie in Micronesia. *Mem. Fac. Sci. Kyushu Univ.*, *Ser. E (Biol.)*, 1: 210—243.

Hosokawa, T., 1955. On the vascular–epiphyte communities in tropical rainforests of Micronesia. *Mem. Fac. Sci. Kyushu Univ.*, *Ser. E. (Biol.)*, 2: 31—44.

Hosokawa, T., 1957. Outline of the mangrove and strand forests of the Micronesian islands. *Mem. Fac. Sci. Kyushu Univ.*, *Ser. E. (Biol.)*, 2: 101—118.

Hosokawa, T., 1967. Life form of vascular plants and the climatic conditions of the Micronesian islands. *Micronesia*, 3: 19—30.

Hosokawa, T., 1968. Ecological studies of tropical epiphytes in forest ecosystem. *Proc. Symp. Recent Adv. Trop. Ecol.*, pp. 482—501.

Hosokawa, T., Omura, M. and Nishihara, Y., 1954. Social units of epiphyte communities in forests. *VIIIe Congr. Int. Bot., Paris, Rapp. Commun., Sect. 7*, pp. 109—191.

Imanisi, K. and Kira, T., 1944. In: K. Imanisi, *The Ponape Island*, pp. 62—91 (in Japanese).

Kagoshimaken Kyoiku Iinkai, 1973. Kagoshimaken no Bunkazai, pp. 246.

Kanehira, R., 1933. *Flora Micronesia* (in Japanese).

Kanehira, R., 1935. An enumeration of Micronesian plants. *Dep. Agric. Kyushu Imp. Univ.*, 4: 237—464.

Kanehira, R., 1936. *Formosan Trees*.

Kariyone, T., 1927. The mangroves of the south sea islands. *J. Jap. Bot.*, 4: 116—120.

Kira, T., 1967. Ecology of mangrove. *Nettai Ringyo*, 5(9): 1—16.

Kitami, H., 1970. Phalanges of *Nypa fruticans* Warmb. drifting inshore of Sado Island in the Japan Sea. *J. Jap. Bot.*, 45: 64.

Koidzumi, G., 1915. The vegetation of Jaluit Island. *Bot. Mag. Tokyo*, 29: 242—257.

Köppen, W., 1923. *Die Klimate der Erde*. Bornträger, Berlin.

Kudo, Y., 1930. *Formosan Plants* (in Japanese).

Kudo, Y., 1932. The mangrove of Formosa. *Bot. Mag. Tokyo*, 46: 147—156.

Lam, H.J., 1934. Materials towards a study of the flora of the island of New Guinea. *Blumea*, 1: 115—159.

Merrill, E.D., 1914. An enumeration of the plants of Guam. *Philipp. J. Sci., Ser. C., Bot.*, 9: 17—155.

Merrill, E.D., 1945. *Plant Life in the Pacific World*. Macmillan, London.

Meteorological Observatory of South Seas Bureau, 1928. *Data of Tropical Climate. 1. Rainfall*.

Meteorological Observatory of South Seas Bureau, 1930. *Data of Tropical Climate. 2. Air—Temperature, Relative Humidity, Wind Direction and Velocity, Amount of Cloud and Sunshine*.

Miyata, I. and Odani, N., 1963. The vegetation of Iriomote-jima, Yaeyama Group, the Ryukyus. *Rep. Comm. Foreign Sci. Res., Kyushu Univ.*, 1: 23—42.

Miyata, I., Odani, N. and Ono, Y., 1963. An analysis by I-method of dispersion on the population of *Bruguiera conjugata* (L.) Merrill. *Rep. Comm. Foreign Sci. Res., Kyushu Univ.*, 1: 43—48.

Odani, N., 1964. An analysis of dispersion on the mangrove communities of Iriomote-jima, Yaeyama Group, the Ryukyus. *Rep. Comm. Foreign Sci. Res. Kyushu Univ.*, 2: 181—246.

Ohno, T., 1973. Mangrove forests in Kagoshima Prefecture. *Kagoshimaken Bunkazai Chosa Hokokushu*, 20: 1—6 (in Japanese).

Parham, J.W., 1964. *Plants of the Fiji Islands*. Government Printer, Fiji.

Raunkiaer, C., 1934. *The Life Forms of Plants and Statistical Plant Geography*. Oxford University Press, London (English translation by Carter, Fausboll and Tansley).

Rechinger, K., 1908. Samoa. *Vegetationsbilder*, 6(1).

Reinecke, F., 1898. Die Flora der Samoa–Inseln. *Engl. Bot. Jahrb.*, 25(5).

Rübel, E., 1930. *Pflanzengesellschaften der Erde*, pp. 58—60.

Ryukyu Seifu Bunkazai Hogo–Iinkai, 1966. Bunkazai Yoran 1965—Nen Ban. pp. 92.

Salvosa, F.M., 1936. Rhizophoraceae. *U.P. Natl. Appl. Sci. Bull.*, 5: 179—255.

Sekizuka, R. and Shimizu, K., 1970. Mangroves of Yaeyama-gunto. *Iden*, 7: 37—43.

Yamashiro, M., 1954. Viviparity of *Kandelia candel* (L.) Druce, preliminary report, *Kagoshima Hakubutsugaku Kaiho*, 1(1): 22—27 (in Japanese).

Yamashiro, M., (unpublished). Ecological study in *Kandelia candel* (L.) Druce. Thesis, Hiroshima University, 99 pp.

Chapter 15

MANGAL AND COASTAL SALT-MARSH COMMUNITIES IN AUSTRALASIA

PETER SAENGER, MARION M. SPECHT, RAYMOND L. SPECHT and V.J. CHAPMAN

AUSTRALIA (P. Saenger, M.M. Specht and R.L. Specht)

Introduction

Mangal vegetation

The edaphic complex, known as mangal vegetation, is found around the coasts of all mainland states of Australia. In both species–diversity and extent, the mangal complex shows greatest development in northern tropical Australia where 27 species of tree, belonging to 14 families of angiosperms have been recorded:

Arecaceae: *Nypa* (1 sp.)
Avicenniaceae: *Avicennia* (2 spp.)
Bombacaceae: *Camptostemon* (1 sp.)
Caesalpiniaceae: *Cynometra* (1 sp.)
Combretaceae: *Lumnitzera* (2 spp.)
Euphorbiaceae: *Excoecaria* (1 sp.)
Meliaceae: *Xylocarpus* (2 spp.)
Myrsinaceae: *Aegiceras* (1 sp.)
Myrtaceae: *Osbornia* (1 sp.)
Plumbaginaceae: *Aegialitis* (1 sp.)
Rhizophoraceae: *Bruguiera* (5 spp.)
 Ceriops (2 spp.)
 Rhizophora (3 spp.)
Rubiaceae: *Scyphiphora* (1 sp.)
Sonneratiaceae: *Sonneratia* (2 spp.)
Sterculiaceae: *Heritiera* (1 sp.)

Ten extra species from eight families have been noted as associated liane, epiphyte, or understorey species.

Progressively fewer species, occupying smaller and smaller areas of the landscape, are found in the subtropical region, while only one species (*Avicennia marina*) extends in small pockets into temperate Australia.

Mangal vegetation in New Guinea, to the north of Australia, shows an even richer flora (probably 30 tree species) than that recorded in the Cape York Peninsula region of Queensland. Dense mangrove forests (canopy cover 70—100%), often reaching 30 m or more in height, are common around these coasts and have been classified as *closed–forest* to *tall closed–forest* in Specht et al. (1974) (see Chapter 13).

There is a definite progression not only in number of species but also in structural characteristics in the mangal vegetation from tropical to temperate Australia. The mangals commonly form *closed–forests* (10—30 m tall) in the tropical north. They are not as tall in the subtropical region where they usually form a *low closed–forest* (5—10 m tall).

In temperate Australia, the mangrove trees tend to be more widely spaced than in the north — a *low open–forest* (canopy cover 30—70%) or *low woodland* (canopy cover 10—30%). In some cases the mangrove species tend to branch near the base to form gnarled, shrub–like trees in an *open–scrub* (canopy cover 30—70%) to *tall shrubland* (canopy cover 10—30%) formation.

However, variations in micro–habitat are apparent in all regions from the tropical to the temperate part of Australia. The structural development of the mangroves is often considerably reduced from the optimal described above. Even stunted mangal communities, sometimes as low as 2 m tall, may be seen in New Guinea and Cape York Peninsula — areas where *closed–forests* tend to predominate. These stunted communities tend to be somewhat more open in structure than that observed in the *closed–forests,* but not necessarily so. In temperate Australia the canopy cover of stunted communities is almost invariably open.

The Australian mangal vegetation thus shows a gradual reduction in species–number, structure, and area occupied from tropical to temperate regions (Fig. 15.1). In Tasmania mangals are completely replaced by salt marshes.

Tidal salt–marsh vegetation

Tidal salt–marsh vegetation, in contrast to the mangal vegetation, becomes increasingly more complex in structure and contains many more species as latitude increases from tropical to temperate Australia.

Only seven species (belonging to four angiosperm families) have been recorded in salt marshes of tropical Australia and some of these are relatively rare. One of the seven species, *Tecticornia australasica,* behaves as an annual, germinating in April at the end of the monsoonal rainy season, and completing its life cycle by November (Van Royen, 1956); the salt–marsh vegetation, dominated by this species, may be devoid of vegetation for the rest of the year. Salt flats, lacking even seasonal vegetation, are a frequent occurrence on tidal salt marshes in the drier parts of the tropical Australian coastline (MacNae, 1966).

The salt marshes increase progressively in species–richness towards temperate Australia where up to 34 species (belonging to 16 angiosperm families) have been recorded. Some of these species may be typical samphire bushes (0.5—2.0 m tall) of the family Chenopodiaceae and form a *low shrubland* (canopy cover 10—30%) to *low open–shrubland* (canopy cover 0—10%) formation (Specht et al., 1974). Salt–tolerant grasses and herbs may occur with the succulent, rhizomatous chenopod *Salicornia quinqueflora* to form either a carpet to the *low shrubland* formation or alone as a *closed–herbland* or *herbland* formation. Coastal salt marshes in southern Australia rarely have a vegetation–free zone.

Inland salt–marsh and mangal vegetation

Burbidge (1960) suggested that the Eremaean flora of inland Australia was developed from the littoral and sand–dune flora of the more extensive coasts of the Cretaceous. A considerable portion of the southern part of the continent was submerged by sea during the Tertiary (see Specht, 1972b). Remnants of the salt–marsh flora along the estuarine parts of the Tertiary coastline were stranded far inland when the sea retreated to the present–day coastline.

On both sides of the Flinders Range in South Australia, and extending 500 or more kilometres inland from the nearest coastline, salt lakes, devoid of vegetation, and salt–marsh flats are a common feature of the landscape (Murray, 1931; Carrodus et al., 1965; Specht, 1972b). Many salt–tolerant species, particularly samphires, which are common in coastal salt marshes, extend into the inland salt marshes — viz. *Arthrocnemum halocnemoides, A. leiostachyum, Frankenia pauciflora, Maireana brevifolia, Nitraria schoberi, Salicornia quinqueflora, Salsola kali.* The samphire *Pachycornia tenuis* is apparently one of the few species which is confined to inland salt marshes and not found in coastal areas.

Inland salt marshes are by no means an unusual observation; but it is surprising to find a report (Beard, 1967a) of a salt–water creek lined with low mangrove trees *(Avicennia marina)* stranded some 40 km from the coast of the Eighty Mile Beach of Western Australia (19°45′S 121°20′E). The seaward exit of this creek, fed by water seeping from springs in the arid countryside, has been buried by an alluvial coastal plain built–up in relatively recent times to a distance of over 30 km from the coast.

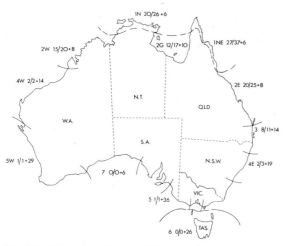

Fig. 15.1. Mangal—salt–marsh biogeographic zones in Australia are labelled 1NE, 1N, 2W, 2G, 2E, 3, 4W, 4E, 5W, 5E (subdivided 5Ea for eastern Victoria, 5Eb for Port Phillip—Westernport Bay, 5Ec for western Victoria—southeastern S. Australia, and 5Ed for the peninsulas and gulfs of S. Australia). 6, and 7. The numbers of species of mangrove trees, of mangrove trees plus understorey, and of salt–marsh plants found in each zone are shown as follows: Zone 1NE 27 tree species/37 tree plus understorey species + 6 salt–marsh species, i.e., 1NE 27/37 + 6.

The environment

Geomorphology

Australia, though the smallest of the continents, covers a latitudinal range from 11° to 44°S. The climate thus varies markedly with geographical position: summer rainfall predominates in the north, winter rainfall is characteristic of the south, while the eastern seaboard shows a more general distribution of rainfall throughout the year (see Chapter 3).

The total coastline, 19,650 km in length, shows a wide range of geomorphological features described by Bird (1972a): cliffs; beaches, spits and barriers; coastal dunes; estuaries and lagoons; deltas; coral reefs and atolls. Each of these features provides a habitat, albeit modified, for the development of mangal vegetation, salt marsh or salt flat. Least development is seen on coastal cliffs, though salt spray communities with strong affinities to salt–marsh vegetation have developed in crevices and cliff–tops, especially on offshore islands (T.G.B. Osborn, 1922, 1923, 1925; Willis, 1953; Gillham, 1961; Sauer, 1965; Parsons and Gill, 1968; Specht, 1969, 1972b; Symon, 1971).

A multiple barrier system, formed by barrier beaches (or occasionally by barrier islands) with intervening tracts of lagoon or swamp, is a common feature of the coastline of the southern half of Australia. The inner barriers are generally of Pleistocene age, whereas the outer barrier is much younger, often of Recent age. The calcareous sands, forming the barrier dunes of southern and western Australia, have usually been lithified to form a relatively durable calcarenite; the siliceous beach sands of eastern Victoria, New South Wales and southeastern Queensland show little evidence of consolidation. The swamp deposits stranded between these barrier beaches are often saline and provide a habitat for salt–marsh and occasionally mangal vegetation (T.G.B. Osborn, 1914a; Fenner and Cleland, 1935; Jessup, 1946; Blake, 1947; Blackburn, 1952; Coaldrake, 1961; Turner et al., 1962, 1968; Bird, 1972a; Clifford and Specht, 1976).

The greatest development of salt–marsh and mangal vegetation is found in the northern half of Australia on saline muds deposited in estuaries, deltas (such as the mouth of the Burdekin River, Queensland) and quiet bays where a relatively stable land surface has been exposed by a eustatic fall in sea level during Recent times (Gill and Hopley, 1972; Hopley, 1974).

South of Mackay, Queensland, at least as far south as the Victorian border, tectonic movements of the land have tended to neutralise any eustatic fall in sea level (Thom et al., 1969, 1972). Only small areas of saline muds have developed around the tidal estuaries of rivers and in the quiet waters, protected by long sandy islands, on the eastern sides of Moreton Bay and Hervey Bay, Queensland.

Southeastern and southern Australia also show evidence of a eustatic fall in sea level in Recent times (Gill, 1971; Gill and Hopley, 1972). Exposed coastal flats, subjected to strong wave action, have developed beach barriers described above. Shallow areas exposed in quiet parts of bays, inlets and gulfs have developed as mud–flats with mangal and salt–marsh vegetation (e.g., in Corner Inlet, Westernport Bay, and Port Phillip Bay in Victoria; the upper waters of the Gulfs in South Australia; protected inlets on Kangaroo Island and Eyre Peninsula, South Australia; and at Bunbury, Western Australia). Small estuarine mud–flats have developed near outlets of rivers in Tasmania, Victoria, and near the outlet of the River Murray in South Australia.

Mangals are also found on the "low wooded islands", developed on coral reef platforms in the lagoon–like sea waters north of Cairns, Queensland. Apparently in the protection of the nearby Great Barrier Reef, a sand cay develops on the leeward (NW) side of the reef platform while a shingle embankment forms on the windward (SE) side; between them, a depression or shallow lagoon, in which mangroves become established, is found (Steers, 1929, 1937; T.A. Stephenson et al., 1931; Fairbridge and Teichert, 1947; W. Stephenson et al., 1958). Wooded Island in the Abrolhos Group off Western Australia shows a similar mangal habitat in the centre of the island (see also Chapter 2).

Soils (see also Chapter 4)

The dominant soils of salt pans, tidal flats and mangal areas are highly saline and often gypseous clays usually with a very soft loose surface, not showing seasonal cracking; little change in texture occurs with depth. Northcote et al. (1960—68) in the *Atlas of Australian Soils* have classified uniform–textured profiles (U) composed of fine–textured material (f) as Uf soils. Profile variations have been observed as follows:

Uf1   clayey soil materials darkened by organic material in the surface soil but without other pedological changes
    Uf1.41   pale colours below the surface soil (Kimberley Region of Western Australia to Blue Mud Bay in Arnhem Land, Northern Territory)
Uf6   clayey soils that are plastic throughout the profile; organic material (peat) may be present in relatively large amounts in some surface soils
    Uf6.51   grey colour below the surface soil (Gulf of Carpentaria, Cape York Peninsula)
    Uf6.61   mottled grey colour below the surface soil which is usually dark (Gulf of Carpentaria, Cape York Peninsula, East Coast of Queensland, pipe–clay flats and swamps at the southern end of the Coorong, South Australia, coastal plain between Port Pirie and Port Broughton, South Australia)
    Uf6.62   mottled yellow–brown colour below the surface soil (Gulf of Carpentaria, Cape York Peninsula, Northeastern Queensland)

Climate (see also Chapter 3)

In this review, the coastline of Australia has been subdivided into zones as shown in Fig. 15.1, and listed in Table 15.1 in clockwise sequence beginning at the Kimberley Region of Western Australia (Zone 2W). The climatic characteristics of selected coastal towns are tabulated for each zone (Table 15.1).

Monthly values of solar radiation have been estimated by Hounam (1964). Data on air temperature and precipitation were extracted from a 30–year summary of Australian climatic records (Comm. Aust., Bur. Meteorol., 1956). Pan evaporation has been estimated by means of the formulae developed by Fitzpatrick (1963). Mean run–off values have been computed by Specht using data collated by the Australian Water Resources Council (1965). Foley (1945) provided statistics on the number of days per annum on which frost has been recorded throughout Australia. The frequency, tracks and intensities of tropical cyclones in the Australian region have been summarised by F. Coleman (1972).

Köppen's system of classifying climate provides a broad summary of the climatic types around the Australian coastline (Dick, 1972). The zones, shown in Fig. 15.1, are listed below against the relevant Köppen climatic type and with the equivalents in Walter's classification (Chapter 3, above):

A hot climate with no month below 18°C
    Af climate with uniform rain (= I Walter, p. 61): zone 1NE (Tully—Innisfail, Qld.)
    Am climate with very short dry season (= I Walter, p. 61): zone 1NE (Ingham to Cairns, Qld.)
    Aw climate with a dry winter (= II Walter, p. 61): zones 2W, 2G, 2E (northern half)
    Awi climate with small annual temperature range (below 5°C): zones 1N, 1NE
B dry climate (= III Walter, pp. 61, 63)
    BS semi–arid climate
        BSh mean annual temperature warm (above 18°C)
            BShw climate with dry winter: zones 2W (southwestern); 2G (southern)
            BShs climate with dry summer (= IV Walter, p. 61): zone 5W (Shark Bay)
        BSk mean annual temperature cool (below 18°C): zones 5Ed (northern half of both Gulfs), 7 (Great Australian Bight)
    BW arid climate
        BWh mean annual temperature warm (above 18°C)
            BWhw climate with dry winter: zone 4W
C warm climate (at least one month below 18°C) (= V Walter, pp. 61, 62)
    Cf climate with uniform rain
        Cfa climate with hot summer (hottest month above 22°C): zones 3, 4E
        Cfb climate with long mild summer (hottest month below 22°C and at least four months above 10°C); zones 5Ea—c, 6
    Cs climate with dry summer (= VII Walter, p. 61)
        Csa climate with hot summer (hottest month above 22°C); zone 5W (western coast)
        Csb climate with long mild summer (hottest month below 22°C and at least four months above 10°C): zones 5Ec, 5Ed (except northern half of both Gulfs), 5W (southern coast)
    Cw climate with dry winter (= VI Walter, pp. 61, 63)
        Cwa climate with hot summer (hottest month above 22°C): zone 2E (southern half)

Coastal hydrology

Over the last two decades increasing attention has been given to the marine hydrology of Australian coastlines (Halligan, 1921; Rochford, 1951; Spencer, 1956; Hamon, 1961; Newell, 1961, 1966, 1971, 1973; Easton, 1970; King, 1970; Albani, 1973; Kenny, 1974; Bullock, 1975). Hydrological studies on currents, sea temperatures, salinity, oxygen levels, etc., of large bodies of water, can give only a limited guide to the environment within the intertidal ecosystems dominated by mangal and salt–marsh vegetation. Tidal amplitude and rhythm exert a strong influence on water supply, aeration, salinity and temperature of the saline muds of these ecosystems, but must be considered in relation to the evaporative power of the atmosphere, and the local precipitation plus surface run–off or seepage of fresh water from the surrounding terrestrial landscape.

In tropical monsoonal waters, with a marked dry season, a sheltered intertidal area subjected to regular inundation by sea water, alternating with periods of aeration, but not desiccation, when the tide falls, is usually colonised by a dense stand of mangroves. Similarly, a supra–littoral fringe, only occasionally inundated by sea water, may receive continual fresh–water seepage — a dense mangrove stand results. In between these two extreme habitats, the environment may be less favourable — water deficits and high salinities result in depauperate, more open, mangal vegetation; a salt marsh or salt flat devoid of vegetation is often observed in the central zone of dry coastlines (Fosberg, 1961; MacNae, 1966; Saenger and Hopkins, 1975).

Intertidal areas in tropical regions receiving rainfall throughout the year will show little change in community structure with tidal zonation across the ecosystem and will tend to merge with dense tropical rain forest.

In temperate areas mangrove trees may colonise the coastal fringe of soft intertidal muds, within daily tidal ranges. Salt-marsh vegetation is found in areas less frequently flooded and thus subjected to increasing periods of physiological drought.

Thus, general statistics on coastal hydrology must be qualified for each mangal—salt-marsh complex. Table 15.1 presents statistics on tidal range (Osborn, 1972) and sea–surface temperature (R. Neth. Meteorol. Inst., 1949) for selected coastal towns in the zones delineated on Fig. 15.1. Tidal ranges of mean high to mean–low water springs of 0.4—9.8 m are recorded around the Australian coastline. Sea–surface temperatures range from 15° to 30°C in summer and from 12° to 26°C in winter.

Mangal—salt–marsh complexes are only developed around coastlines exposed to low–energy wave action. High–energy wave action prevents the sedimentation of mud — coastal and barrier dunes result instead (see Chapter 2).

In northern Australia, tidal surges associated with tropical cyclones (Table 15.1) may inundate low–lying coastal land not usually affected by tides. Propagules of mangroves, washed inland by these surges, may germinate and become established in less salt–tolerant, but well–watered, coastal communities (Specht et al., 1976).

Biogeography

Vascular plants

Avicennia marina var. *australasica* is the most widespread mangrove in Australia and it is the only species occurring south of Merimbula Estuary (36°50'S) on the east coast and south of Carnarvon (25°S) on the west coast. Along the coastline south of these localities, the distribution of this species is sporadic with isolated occurrences on the Abrolhos Islands (28°40'S 113°35'E), at Bunbury (33°20'S 115°40'E) and around Ceduna (32°08'S 133°35'E); along the upper shores of Spencer and St. Vincent Gulfs the distribution of this species is more or less continuous. Further east the distribution again becomes sporadic with occurrences at Barwon Heads (38°14'S 144°30'E), Limeburners Creek (38°05'S 144°28'E) within Port Phillip Bay, Westernport Bay (38°45'S 145°20'E) and Corner Inlet (38°45'S 146°30'E) — the most southerly known occurrence of a mangrove in Australia (see p. 3).

Salt-marsh plants show a similar, but inverted, distribution with the greatest number of species recorded from southeastern and southwestern Australia (Patton, 1942; Black, 1943—57; Curtis and Sommerville, 1947; Willis, 1962, 1972; Smith, 1973). The distribution pattern of the salt-marsh flora is somewhat complicated by the unclear distinction between coastal salt-marsh plants and

TABLE 15.1

Climate and coastal hydrology of the Australian coastline

Zone	Locality	Solar radiation (cal. cm^{-2} day^{-1})			Mean air temperature (°C)			Mean precipitation (mm)			Mean pan evaporation (mm)			Mean run-off (mm)			Days of frost per year		Cyclones per 10 year (crossing 5° lat. × 5° long.)	Tidal range (m)			Sea-surf. temperat. (°C)	
		annual	January	July	annual	January	July	annual	January	July	annual	January	July	annual	January	July	0–2°C	<0°C		(HAT–LAT)[1]	(MHWS–MLWS)[1]	(MHWN–MLWN)[1]	summer	winter
2W	Broome	525	555	427	27	30	21	581	187	4	2102	156	155	33	20	0	0	0	12.8	11.0	8.5	1.8	29	23
	Derby	517	520	432	28	30	22	609	194	7	2386	176	168	38	23	0	0	0	12.2	?	9.8	4.5	29	24
	Port George IV (Port Warrender)	505	475	438	26	29	21	1280	384	8	1944	125	159	600	230	0	0	0	7.8	?	6.4	0.9	29	24
	Wyndham	506	495	440	29	31	24	639	172	2	2846	235	199	44	18	0	0	0	6.4	8.0	6.5	2.8	28	23
1N	Darwin	481	440	460	28	29	25	1491	411	0	2073	146	184	599	247	0	0	0	6.8	7.8	5.5	1.8	28	25
	Cape Don	468	430	470	26	28	24	1291	290	2							0	0	5.7	?	1.5	0.9	28	25
	Millingimbi	469	430	470	26	28	24	1080	239	1							0	0	6.2	?	3.6	1.9	28	25
	Gove	468	430	470	26	28	24										0	0	5.9	?	2.0	1.2	29	25
2G	Port Langdon, Groote Eylandt	480	450	455	26	29	23	1647	232	0									5.8	?	0.8	0.2	29	23
	Port McArthur	491	490	435	26	29	22										0	0	4.9	?	1.5	0.3	28	21
	Normanton (Karumba)	492	510	415	27	30	22	954	295	3	2469	187	163	213	133	0	0	0	7.0	?	1.9	?	30	23
	Mapoon (Weipa)	452	440	400	28	29	25	1579	458	2	1881	169	121	600	275	0	0	0	5.0	3.3	1.6	0.3	30	26
1NE	Thursday Is.	433	435	370	27	28	25	1688	441	13	1581	154	103	599	265	0	0	0	4.0	3.7	1.9	0.5	29	24
	Cooktown	440	485	360	26	28	22	1724	365	26	1591	165	102	597	186	0	0	0	10.8	?	1.8	0.4	28	23
	Cairns	452	515	350	25	28	21	2193	419	40	1553	171	95	747	156	0	0	0	12.8	3.0	1.8	0.6	28	22
	Cardwell (Lucinda)	465	525	350	23	27	19	1956	421	30	1289	136	75	680	235	0	0	0	12.8	3.7	2.4	0.9	27	22
2E	Townsville	477	525	370	24	28	20	1094	278	19	1680	166	114	363	167	0	0	0	11.8	3.8	2.5	0.8	27	22
	Mackay	466	525	365	22	27	17	1604	344	40	1313	144	78	600	207	1	0.8	0	10.1	6.4	4.9	2.3	27	21
	Gladstone	460	545	340	22	26	17	972	155	51	1416	147	90	212	52	1	0	0	9.2	4.1	3.1	1.5	26	20

AUSTRALASIA

	Location																							
3	Bundaberg	455	550	320	26	16	1076	216	39	1364	150	83	336	130	1	0	0	9.2	2.8	2.0	1.1	26	20
	Brisbane	436	535	300	25	15	1018	145	48	1472	168	79	259	87	2	0	0	5.8	2.6	1.8	1.1	25	19
	Clarence Heads (Coffs Harb.)	442	572	285	23	14	1402	142	122	1133	116	70	597	85	54	3.5	0.4	3.3	1.8	1.2	0.8	23	18
4E	Newcastle	412	560	246	22	13	1051	77	113	1187	139	59	300	9	52	0.2	0	0	2.0	1.2	0.8	22	17
	Sydney	403	548	237	22	12	1138	98	124	1282	161	66	438	36	75	0.2	0	0	2.1	1.3	0.9	22	17
	Wollongong (Port Kembla)	402	549	232	21	12	1119	106	92	1123	135	56	400	45	22	0	0	0	1.7	1.2	0.9	21	16
5E	Bega (Eden)	404	575	223	21	9	912	103	65	1212	150	62	160	34	5	2.3	0	0	2.3	1.1	0.6	20	15
a–c	Westernport (Stony Point)	347	525	165	18	10	741	41	74	925	132	35	72	0	9	6.5	1.2	0	?	2.3	1.5	17	14
	Melbourne (Williamstown)	366	575	170	20	10	658	48	49	1205	175	45	48	3	3	11.1	0.9	0	0.9	0.6	0.4	17	14
	Portland	357	548	170	17	10	850	30	104	920	128	41	120	1	23	9.4	3.2	0	?	0.7	0.2	17	14
6	Stanley	343	553	155	16	9	925	42	109	847	122	31	169	1	37	2.6	0.1	0	?	2.5	2.1	16	13
	Hobart	331	535	135	16	8	636	46	54	1046	147	41	43	2	4	12.8	1.7	0	?	0.9	0.3	15	12
5E	Adelaide (Outer Harbour)	402	610	205	23	12	536	19	63	1606	240	48	26	0	4	0.3	0	0	2.5	1.9	0.3	19	13
d	Port Pirie	433	640	225	24	12	330	19	32	1603	235	48	8	0	1	1.9	0	0	2.8	1.6	0.4	20	13
	Port Lincoln	384	580	200	20	12	463	10	74	1136	160	45	17	0	5	0	0	0	2.0	1.0	0.4	19	14
7	Fowlers Bay (Thevenard)	427	590	240	21	12	296	8	43	1210	140	64	6	0	1	2.9	0.2	0	1.6	0.9	0.4	19	15
	Eucla	434	598	260	21	12	253	15	23	1297	152	67	5	0	1	3.2	0.6	0	?	0.9	?	20	16
	Eyre	431	598	250	21	12	289	14	32	1247	152	61	6	0	1	12.0	5.0	0	?	?	?	20	16
5W	Esperance	408	585	226	20	12	679	20	107	1098	142	50	53	0	15	1.8	0.3	0	?	?	?	19	15
	Albany	361	525	180	19	12	1008	35	152	1017	130	49	246	1	78	0.3	0	0	?	0.4	?	19	16
	Bunbury	412	630	190	21	13	844	7	172	1158	176	45	149	0	43	2.5	0	0	?	0.5	?	20	17
	Fremantle	434	640	215	23	14	791	8	158	1293	182	56	92	0	27	0	0	0	?	0.4	?	21	18
	Geraldton	479	670	276	24	15	472	8	96	1506	196	68	18	0	4	0	0	3.6	?	0.5	?	22	19
4W	Carnarvon (Learmonth)	517	680	337	27	17	229	13	34	1666	186	89	4	0	1	0	0	8.4	3.0	2.1	0.6	26	21
	Onslow	530	635	370	30	18	238	21	11	2174	247	112	4	0	0	0	0	10.0	2.5	1.8	0.5	27	21
	Port Hedland	533	610	396	31	20	280	47	6	2318	202	143	6	1	0	0	0	10.3	7.7	5.8	1.4	29	22

[1] HAT and LAT = highest and lowest astronomical tides; MHWS and MLWS = mean high and low water springs; MHWN and MLWN = mean high and low water neaps. Place names between parentheses are nearest ports recording coastal hydrological data.

arid–zone salt–tolerant species constituting part of the Eremaean floristic element (Burbidge, 1960). Nevertheless, the generalized temperate and tropical nature of the salt–marsh and mangal floras respectively (Appendix, Tables 15.5 and 15.6), can be seen in their inverse relationship (Fig. 15.2) in terms of species number for the various coastal regions around Australia.

The factors responsible for present–day mangrove distributions are undoubtedly complex and various suggestions have been stressed by different authors (e.g., MacNae, 1966; Clarke and Hannon, 1967, 1969, 1970, 1971; Gill, 1975; Saenger and Hopkins, 1975; Thom et al., 1975). Broadly speaking, however, these factors can be classified into four interrelated categories: historical, environmental, biological, and fortuitous. It is not proposed to discuss each of these groups of factors in detail but rather to summarise information on those factors likely to determine large–scale distribution patterns.

From present species distribution (Van Steenis, 1962) it appears that the Malaysian region was the centre of dispersal of modern–day mangrove floras (MacNae, 1966, 1968a). Muller (1964) has demonstrated an unbroken succession of tropical mangal vegetation from the Lower Tertiary to the Recent in northwestern Borneo, thereby lending support to the suggestion that modern mangal floras originated in the Malaysian area with a thermal régime similar to that existing today (see pp. 19—22).

Recent fossil pollen and wood evidence (Churchill, 1973) from the southern margins of Western Australia indicates that tropical coastal waters extended along these shores during the Middle to Late Eocene. The late Eocene fossil occurrence of *Nypa, Sonneratia, Avicennia* and the Rhizophoraceae mangroves in southwestern Australia (Churchill, 1973) indicates that tropical mangal vegetation existed at that time and that there has been a loss of these elements from southern Australia since the Eocene. In view of the more restricted present distribution of these elements, past changes in climate and coastal conditions appear to have had a sifting effect on Australian mangal vegetation. This sifting effect has undoubtably contributed to the existing species gradients not only of the mangroves themselves, but also of associated plants and animals of these communities.

The large concentration of mangrove species and associated plants in the northeastern area of Australia (Appendix, Table 15.5) can thus be attributed to three main factors:

(1) This region was the origin of dispersal for mangroves (as well as other floristic elements — Burbidge, 1960) into Australia by virtue of its land connections with South East Asia (Walker, 1972) during the various changes in paleo–sea levels.

(2) The climatic régime of this area approximates to that under which mangal vegetation first developed — consequently little or no sifting of species has occurred.

(3) Coastline configuration in this region, with its numerous estuaries generally sheltered by the offshore Barrier Reef, provided large areas suitable for mangrove colonisation and development.

Apart from mangrove and salt–marsh species distribution *per se*, long–term climatic and coastal changes have influenced the physiognomy of the mangal communities around the Australian coastline.

Various coastal regions are delineated in Fig. 15.1, based primarily on floristic and faunistic distribution records. These regions correlate closely with certain environmental (particularly meteorological and tidal) as well as physiognomic features of the mangal and salt–marsh vegetation. Variation

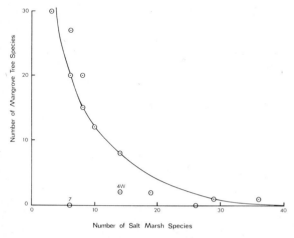

Fig. 15.2. An inverse relationship is apparent between the number of species of mangrove trees and the number of species of salt–marsh plants found in any biogeographic zone (see Fig. 15.1). The cliffs of the Great Australian Bight (Zone 7) and the long sandy beaches of Zone 4W provide few habitats for the development of mangal—salt–marsh vegetation.

exists not only in the basic zonation pattern but also in the relative width, height and densities of the various zones, and their species composition (Figs. 15.3—15.14).

The basic zonation pattern of Australian mangal coasts has been discussed by MacNae (1966, 1967, 1968a) and Saenger and Hopkins (1975), and the following zones have been identified:
1 the landward fringe (including salt marsh)
1a the landward *Avicennia* zone (*Avicennia* parkland)
2 *Ceriops* thickets
2a vegetation-free high tidal flats
3 *Bruguiera* forests
4 *Rhizophora* forests
5 the seaward fringe

Some variation in zonation occurs with the topography of the shoreline and only topographically similar mangal communities can be validly compared.

The following summary of zonation and physiognomy of the various regions (Fig. 15.1) is restricted to "fringe forests", described by Lugo

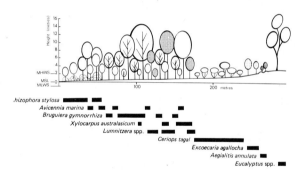

Fig. 15.3. Zone 1NE. Cross-section through mangrove vegetation in the Daintree River Area, northern Queensland.

and Snedaker (1974) as the vegetation type occurring along the fringes of relatively protected shorelines and islands.

Mangals of regions 1N, 1NE, 2W, 2G, 2E are distinctly zoned while those of region 3 are less clearly zoned. The zonation of species in regions 4W, 5W and 5E is erratic, not clearly recognisable or absent where *Avicennia marina* is the sole mangrove.

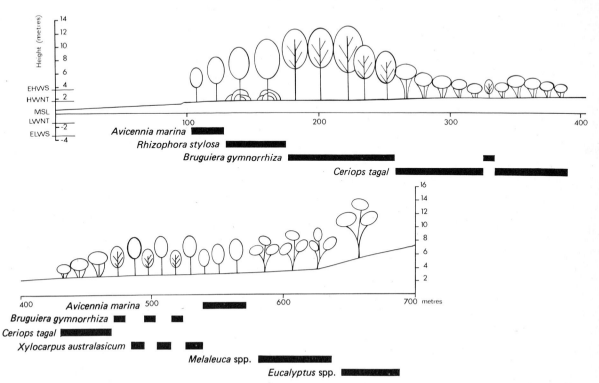

Fig. 15.4. Zone 2E. Cross-section through mangroves along an open shoreline near Townsville, northern Queensland (after MacNae, 1966).

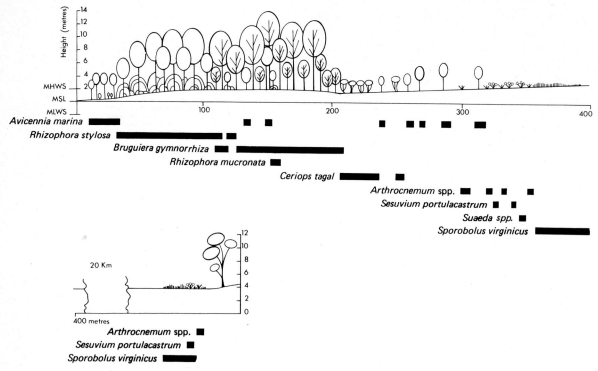

Fig. 15.5. Zone 2G. Cross-section through mangroves along gulf shoreline at Tarrant Point, Gulf of Carpentaria (after Saenger and Hopkins, 1975).

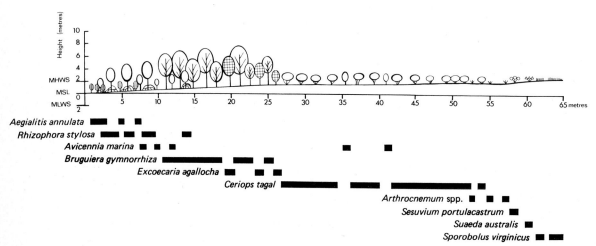

Fig. 15.6. Zone 2G. Cross-section through mangroves on a convex meander along Channon Creek, Gulf of Carpentaria (after Saenger and Hopkins, 1975).

AUSTRALASIA

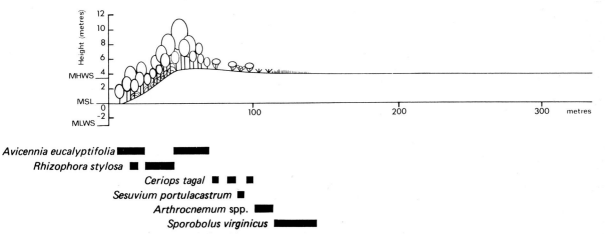

Fig. 15.7. Zone 2W. Cross–section through mangroves along gulf shoreline of Cambridge Gulf, Western Australia (after Thom et al., 1975).

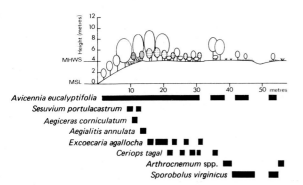

Fig. 15.8. Zone 2W. Cross–section through mangroves along the King River, Cambridge Gulf, Western Australia (after Thom et al., 1975).

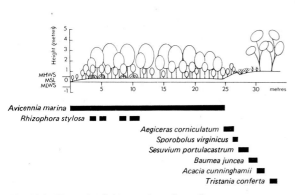

Fig. 15.9. Zone 3. Cross–section through mangroves near Amity Point, North Stradbroke Island, southeastern Queensland.

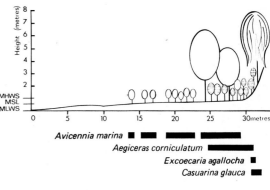

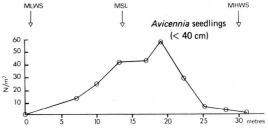

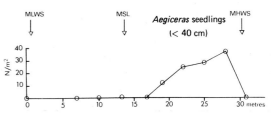

Fig. 15.10. Zone 3. Cross–section through mangroves at Hayes Inlet, southeastern Queensland.

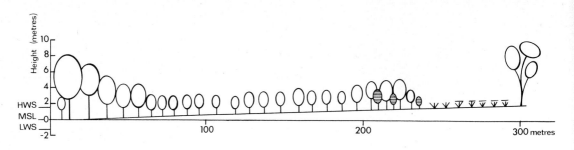

Fig. 15.11. Zone 4E. Cross–section through mangroves at Careel Bay, New South Wales (based on Hutchings and Recher, 1974).

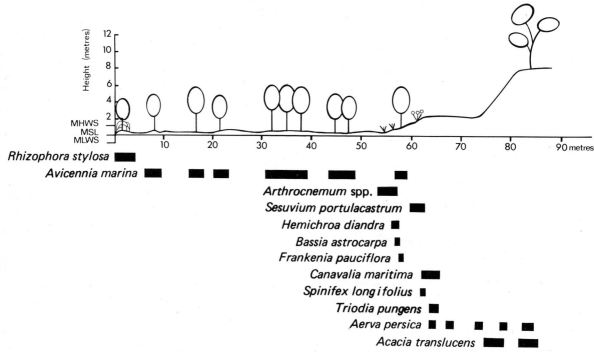

Fig. 15.12. Zone 4W. Cross–section through mangroves in a tidal inlet at Port Hedland, Western Australia (after Sauer, 1965).

AUSTRALASIA

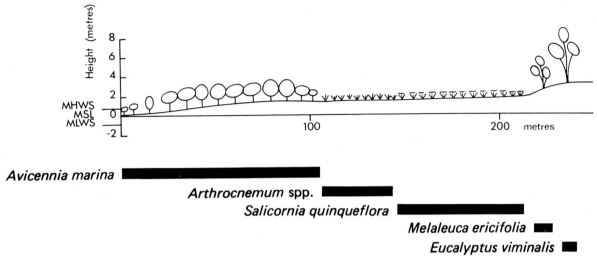

Fig. 15.13. Zone 5Eb. Cross-section through mangroves and salt marsh at Yaringa, Westernport Bay, Victoria (after Bird, 1971a).

Sonneratia generally constitutes the seaward fringe in regions 1NE and 1N (MacNae, 1966; Jones, 1971a, b) although locally *Rhizophora* may occupy this position (Specht, 1958). In regions 2W, 2G, 2E and 3, *A. marina* forms a narrow fringe seaward of the *Rhizophora* zone (MacNae, 1967; Saenger and Hopkins, 1975; Thom et al., 1975). In contrast, the seaward fringe in New Guinea consists of *Sonneratia* followed by a distinct *Nypa* zone (MacNae, 1968a; Percival and Womersley, 1975). Other changes in the general zonation have been summarized in Figs. 15.3—15.14. These include the virtual absence of *Ceriops* thickets, vegetation-free high tidal flats, *Avicennia* parklands and landward salt-marsh fringe in regions 1N and 1NE which possess the highest rainfall; the poor development (in extent, height and density) of the *Rhizophora* and *Bruguiera* forests and the good development of the *Ceriops* thickets in regions 2E, 2G, 2W and to a lesser extent in 3; the relatively simple zonation along southern shorelines where a short (1—5 m high) *Avicennia* seaward fringe merges into an extensive *Avicennia* parkland which abuts a well-developed salt-marsh vegetation; occasionally a narrow bare zone separates the *Avicennia* parkland from the salt-marsh vegetation (Patton, 1942; Kratochvil et al., 1973).

The flora of tidal salt marshes shows an increase in number of species from tropical to temperate Australia (Figs. 15.1 and 15.2; Appendix, Table 15.6). Of the species found in the most frequently inundated parts of Australian salt marshes, *Batis argillicola* and *Tecticornia australasica* are the only ones confined to the tropics; a few extra species (*Sesuvium portulacastrum, Xerochloa barbata*, and, in New Guinea, *Eriachne pallescens*), which are found at the landward edge of the salt marsh, may also be considered as typically tropical. Most dominant Australian salt-marsh species have a wide geographical range, while a number, usually understorey species, are confined to temperate Australia. Many extra species have been reported in the landward fringe of salt marshes (and mangal vegetation) or on sandy ridges, rarely if ever inundated by salt water, within the marsh. These species have been listed in the Appendix (Table 15.7).

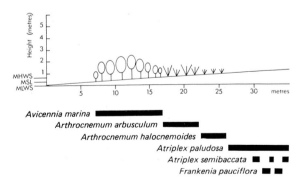

Fig. 15.14. Zone 5Ed. Cross-section through salt marsh at Port Wakefield, South Australia (based on Osborn and Wood, 1923).

The frequency of tidal inundation decreases across the salt marsh from seaward to landward edge. Often a distinct ecotonal zonation may be observed; in other areas the zonation is blurred. The zonal patterns observed from north to south in Australia are summarised in Table 15.2. This table clearly illustrates the intergrading tropical and temperate floras. The zones labelled 1 to 5 contain species which are progressively less tolerant both to waterlogging and to salt. All appear to be intolerant of shade (Clarke and Hannon, 1971).

Algae

Information on the distribution of algae in mangal—salt-marsh vegetation of Australia is fragmentary (see Appendix, Table 15.8). Almost no algologist has worked on the algae of the extensive salt flats of northern Australia; in the south, little attention has been paid, until recently, to the algae of the disjunct pockets of mangal and salt-marsh vegetation (Womersley and Edmonds, 1958; Post, 1963, 1964). The algae of the mangal—salt-marsh complex of southeastern Queensland are probably reasonably well-known (McLeod, 1969; Cribb, in Clifford and Specht, 1976).

The muddy substrate under mangal vegetation is not a favourable habitat for algal growth. Only a few species of *Bostrychia, Caloglossa, Catenella, Monostroma* and *Rhizoclonium* form stunted tufts and mats attached to pneumatophores and the bases of mangrove trees — and these survive the daily tidal agitation of the mangal mud. Occasional mats of *Enteromorpha* may be found (see p. 18).

The salt marshes of southeastern Queensland are characterised by low shrubby samphires and wide expanses of bare mud. Nevertheless, McLeod (1969) notes that a great abundance of blue–green algae forms dense, unbroken mats on the mud between the phanerogams, particularly in the upper *Suaeda australis* and *Salicornia quinqueflora* zones and in the *Salicornia—Sporobolus* ecotone. A short period of rain is sufficient to stain the ground with bright green patches of the green alga *Hormidium subtile* and the blue–green *Chroococcus turgidus* in particular.

McLeod (1969) examined cross-sections of thirteen salt marshes in southeastern Queensland and was forced to conclude that, with the exception of a salt marsh near Gladstone, only a small percentage of algal species were confined to one level of the marsh. *Phormidium angustissimum*, ubiquitous through all salt marshes, is best

TABLE 15.2

Zonation across Australian salt marsh from the mangal to terrestrial fringing communities

Zone	Northern Australia	Southeast Queensland	Sydney District, N.S.W.	Victoria S. Australia
Mangrove zone	well developed	well developed	present	disjunct areas
Zone 1	*Arthrocnemum leiostachyum*	*Arthrocnemum leiostachyum* (± *Suaeda arbusculoides*)	—	*Arthrocnemum arbusculum*
Zone 2	*Tecticornia australasica* (an annual species)	*Arthrocnemum halocnemoides*	—	*Arthrocnemum halocnemoides*
Zone 3	*Batis argillicola*	*Suaeda australis*	(± *Suaeda australis*)	*Suaeda australis*
Zone 4	salt-flat	*Salicornia quinqueflora*	*Salicornia quinqueflora*	*Salicornia quinqueflora*
Zone 5	*Sporobolus virginicus*	*Sporobolus virginicus*	*Sporobolus virginicus*	*Distichlis distichophylla* (± *Sporobolus virginicus*)
References	Specht (1958); Van Royen (1956b); Specht et al. (1976)	McLeod (1969)	Collins (1921); Clarke and Hannon (1967); Kratochvil et al. (1973)	Osborn and Wood (1923); Patton (1942); Specht (1972b); Bridgewater (1975)

developed in the *Suaeda* level through to the fringing forest. *Calothrix crustacea* is a predominantly lower marsh alga but persists in low frequencies through to the fringing forest, as do *Anabaena torulosa, Anacystis marina,* and *Gloeocapsa alpicola*. The Cyanophyta, many nitrogen-fixing, tend to show a maximum in the *Salicornia* and *Sporobolus* zones.

Womersley and Edmonds (1958) report mats of *Gelidium pusillum, Bostrychia simpliciuscula* and occasionally *Chaetomorpha capillaris* under samphire bushes in salt marshes around the South Australian coast. A few mats of *Enteromorpha* may be found in the mangals.

Very few algae have been reported from mangal areas in Westernport Bay, Port Phillip Bay and Corner Inlet of Victoria (Post, 1963, 1964; King et al., 1971). The red algae *(Bostrychia moritziana, B. tenuis,* and *Caloglossa leprieurii),* commonly found in the supra-littoral zone along rocky shorelines, where they form the *Bostrychia—Caloglossa* association (Post, 1963), extend into mangal areas.

Most of the algae found in the mangal—saltmarsh complex of Australia have a much wider distribution in other coastal ecosystems.

Lichens

Lichens found in mangals are almost entirely species common on nearby terrestrial vegetation. A considerable lichen flora — crustose, foliose and fruticose — is found on mangroves of southeastern Queensland. Preliminary studies by Dr. R.W. Rogers and associates have yielded 23 species from the mangrove stands around Moreton Bay near Brisbane; 16 species have been recorded from Gladstone (see Appendix, Table 15.9).

Lichens are apparently rare in terrestrial vegetation in the monsoonal part of Australia according to studies made by Specht (1958) in Arnhem Land, Northern Territory, at Weipa in northern Queensland (Specht et al., 1976), and at Mt. Isa, Queensland (Specht and Rogers, personal communication, 1975). Mangals and rain forests appear to be the only habitats in which lichens can develop. Three species have been recorded from the mangals of Arnhem Land, eight species from Weipa — a flora by no means as rich as that found in southern Queensland.

The lichen flora of the mangals of southern Australia is sparse and depauperate if the collections made by Dr. E.M. Wollaston from the mangals at Port Adelaide, South Australia, are representative.

Marine invertebrates

Mangal faunas, except for those animals (birds, reptiles, mammals) of land origin, are disseminated largely by ocean currents. Many marine animals found in mangal are also found on open shorelines where they attach themselves to rocky substrata, or on sand or mud where mangroves cannot become established. The only gastropod whose occurrence is virtually restricted to mangal habitats, *Melarapha scabra,* occurs also in Japan, the Cocos-Keeling Islands, East Africa and on islands in the Pacific; it occurs among mangroves around the Australian coastline, except in Victoria.

The richness and species composition of the fauna vary around the Australian coast. Definite distribution patterns can be delineated and the range of any species is generally discrete.

Intertidal zonation of the faunas down the mangrove trees and across a stand is clearly marked (Table 15.3).

TABLE 15.3

Zonation of Mollusca in mangals in Moreton Bay, southeastern Queensland

	On trees	On mud
Supralittoral fringe	*Melarapha* sp. *Cerithidea* sp.	*Salinator* sp. *Ophiocardelus* sp. *Haminoea* sp. *Onchidium* sp.
Midlittoral zone	*Bembicium* sp. *Austrocochlea* sp. *Nerita* sp. *Saccostrea* sp.	*Bembicium* sp. *Austrocochlea* sp. *Pyrazus* sp. *Velacumantis* sp.

It is possible to make some observations about the molluscan fauna. Molluscs of the mangals are inhabitants of the upper mid–littoral zone and the supra–littoral fringe of the shoreline. Some molluscs are exclusively "mud" fauna; they inhabit the mud on the floor of the mangrove forest. Others are exclusively "epi–fauna" of the trees, e.g., *Melarapha scabra;* a few constitute "in–fauna", e.g., shipworm (family Teredinidae). Of the epifauna, some species may be found both on the mud or on the trunks and roots of mangrove trees, e.g., *Bembicium* sp.

Of the molluscan fauna of the mangal flats, only a few species are found on nearby salt marshes, or on the upper limit of mangals. Members of the families Pythiidae, Ellobiidae and Amphibolidae may be found generally in salt marshes.

Zonation of molluscs has been treated for convenience. It should be noted that Crustacea are similarly distributed across the mud, with species diversity declining markedly from Moreton Bay to Sydney. In southern Queensland some eight species are common in the supra–littoral fringe. These species extend their distribution into the mid–littoral zone, where other species increase the number of Crustacea present. With the exception of isopods, which are found in decaying wood, and of barnacles, Crustacea are not found on the trees.

From bay shores, where species diversity is high, a decrease in species numbers occurs as the distance up a tidal estuary increases (Fig. 15.15). The decrease in species number follows the change in salinity of the water. A similar decrease is seen in the number of species of mangrove.

Species distribution as represented by the accompanying lists (see Appendix) indicates marked change in composition of fauna with latitude and longitude. It is not possible to draw conclusions from these lists alone; many more published records exist from, for example, Moreton Bay, Queensland, and the Sydney District, New South Wales, than from certain other areas.

The size of mollusc species which occur in northern areas is generally larger than that of those species in southern mangal areas. The number of species represented decreases as the latitude rises.

The pattern of distribution of Mollusca and Crustacea may be seen from the accompanying lists (see Appendix, Tables 15.10 and 15.11).

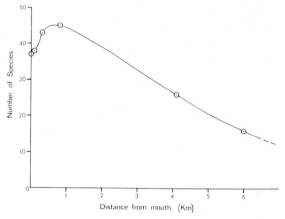

Fig. 15.15. In southeastern Queensland (the Serpentine Creek, for example), the number of invertebrate fauna (particularly Mollusca and Crustacea) recorded in mangals are high near the mouth of rivers, tend to reach a peak within a kilometre of the mouth and then decrease in number as the waters become less tidal.

It would seem that two major faunal groups intermingle at the extremes of their ranges; one group is temperate, the other tropical. In temperate waters different species occupy the niches filled further north by tropical species, and one–sixth of the mollusc species recorded from northern Queensland are also present in Victorian waters. Of the species whose distribution is no farther north than Moreton Bay, one–third also occur in Victoria.

A list of the annelid fauna is appended (see Appendix, Table 15.12). Though most recordings are from Moreton Bay, it is perhaps a lack of records that makes the figures for northern Queensland so sparse.

It may be of interest that the sipunculid *Phascolosoma lurco* recorded here from Moreton Bay, also occurs in Malaya (Berry, 1963).

Insects

Certain insects are common in mangals; mosquitoes and biting midges are frequently very plentiful — generally the visitor may expect them. They have been listed for the Tallebudgera area in south Queensland by Dr. E.J. Reye (Shine et al., 1973). Spiders have been listed in the same publication, and also in that on Careel Bay, New South Wales (Hutchings and Recher, 1974). Colonies of the weaver ant *Oecophylla smaragdina* are frequently seen in tropical mangal forests.

Mention should be made of certain animals whose presence is determined by the presence in the community of their host plant. Listed among the plants are the ant–house plants, *Myrmecodia antoinii* and *Hydnophytum formicarum*, found in the mangals of northeastern Queensland (Jones, 1971a). The gouty, tuberous stems of these plants are often hollow and ants of several kinds take advantage of this feature for housing. The ant *Phedale myrmecodiae* is confined to these plants. A species of butterfly, *Hypochrysops apollo*, breeds in these plants (Monteith, personal communication, 1975); the ants take care of the butterfly's larvae.

Fish

Mangal areas contain many species of fish. Many ocean species spend part of their juvenile stage in mangals. Other species of fresh–water fish may come down into estuaries for feeding (Shine et al., 1973). The species list (see Appendix, Table 15.13) is long; there would appear to be a disjunct distribution from the Northern Territory to southern Queensland, perhaps due to lack of published records. In listing animals of mangal habitats, it is clear that some species are not listed from regions in which their occurrence is known, but, as no record of their occurrence in the mangals of that region has been found, they have not been included for such a region. It is known that research by post–graduate students into fishes of estuaries in northern Queensland is at present under way.

Fish particularly associated with the mangal habitat are animals of the genus *Periopthalmus*, the mud skippers, to be seen on the surface of the mud in mangrove swamps at low tide in tropical areas. These fish, with their bulging eyes and ability to crawl on the mud, can also survive for quite long periods out of water.

Birds

A list of birds which may be found in mangals is appended (see Appendix, Table 15.14). Of those listed, some are found only in mangals, and many others utilise the habitat for feeding purposes. Some are seasonal residents, some vagrants. These last are classified as visitors.

It would appear that in the northeast the variety of avian fauna is richest, with 125 visitors, 54 associated and 7 birds exclusive to the mangals. The north and northwest has the same number of visitors as has the southeast, but more birds in the north are associated with or are exclusive to mangals than in the southeast (Table 15.4). Over the whole range of coastline being reviewed, only 13 species are exclusive to the mangal habitat.

The mistletoe bird, *Dicaeum hirundinaceum*, feeds on the fruits of mistletoe, by eating the fruit–wall and discarding the seed itself. The seed, which has a sticky coat, is by that coat attached to the branch on which it falls from the bird's mouth, and there germinates. A mangrove mistletoe *(Amyema mackayense)* has been recorded in mangals from Port Curtis northward; the mistletoe bird is common in these northern and northeastern mangals. Although the bird is distributed south to Tasmania, it does not frequent mangals of the southeast.

This absence of a species from mangal areas, though it is known to occur in that geographical region, is not confined to *Dicaeum*. Other species of birds are differently associated with mangals in different localities. This may be a reflection of changing flora composition of different areas, providing a changing source of food. That a bird is not listed for an area means only that it does not visit mangals there. It does not follow that it is absent from the geographical region.

TABLE 15.4

Distribution of birds in Australian mangals

Region	Visitor	Associated	Exclusive
North and northwest	104	48	13
Northeast	125	54	7
Southeast	106	20	4

Reptiles and mammals

Snakes and goannas *(Varanus* spp.*)* are seen from time to time in the habitat, but it is the crocodile, *Crocodylus porosus,* present in tropical waters, which is the best-known reptile in the ecosystem.

Of the mammals, the agile wallaby, *Wallabia agilis,* and the flying fox or fruit bat, *Pteropus alecto,* are commonly seen. The fruit bat is an inhabitant of mangals from northwestern Australia to southern Queensland, forming colonies of hundreds of animals in large trees in the swamps. From their "batteries" the large bats emerge at twilight to feed in the neighbouring countryside.

In northern Australia the mammalian population of rats (native and introduced), of mice, and of flying squirrels (Phalangeridae) have from time to time been seen in mangal forests.

NEW ZEALAND (V.J. Chapman)

From North Cape to Tauranga the principal ecosystem is dominated by the mangrove *Avicennia marina* var. *resinifera* (Forst. f.) Bakh. (this is the same as *A. marina* var. *australasica;* see p. 297), with the crab *Helice crassa,* the oyster *Saccostrea cucullata,* barnacle *Elminius modestus* and mussel *Modiolus neozelanicus* on the pneumatophores, and molluscs *(Amphibola crenata, Zeacumantus lutulentus)* crawling over the mud. A recent arrival in the far north mangrove swamps is the mangrove swimming crab *(Scylla serrata)* from the tropics. The ubiquitous algal turf of *Catenella, Bostrychia* and *Caloglossa* clothes the lower pneumatophores and the surface mud can harbour the red alga *Gracilaria secundata* var. *pseudoflagellifera* and a free-living form of *Hormosira banksii,* the southern equivalent of the Northern Hemisphere salt-marsh fucoids.

Behind the mangal there is an ecotone with *Salicornia australis* mixed in, leading generally to a pure *Salicornietum* still with *Helice* and *Amphibola* and the burrowing worm *Nicon aestuariensis.* South of Tauranga (Fig. 15.16) the *Salicornietum* is commonly the pioneer, especially where seawater salinity predominates. In better drained areas, e.g., shingle admixed, *Salicornia* is often replaced by *Samolus repens* and where there is fresh-water influence *Triglochin striata* var. *filifor-*

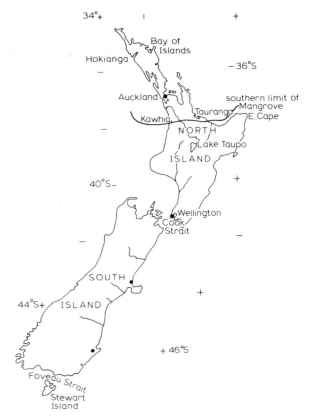

Fig. 15.16. Map of New Zealand.

mis, Scirpus cernuus and *S. nodosus* replace *Salicornia* as the pioneers.

Where normal tidal flooding occurs a *Juncetum maritimi australiensis* can cover large areas and then give way to a *Leptocarpus* zone, though these two zones can be interchanged and chance may well determine which species occupies the lower zone. Fauna here is scarce but the red alga *Bostrychia harveyi* can be found around the bases of the plants. Beetles, Collembola, Hemiptera and Lepidoptera larvae, mites, and spiders are the main faunal components. In some areas, e.g., Hokianga, the *Salicornia* zone may be replaced by a zone of *Puccinellia distans* (Fig. 15.16).

Where there is a fresh-water inflow the rush belt is gradually replaced by *Typha angustifolia* which eventually leads to a swamp association of *Leptospermum scoparium* and the New Zealand flax *Phormium tenax.* On drier ground the rushes are replaced by the shrubby *Plagianthus divaricatus* and the tufted grass *Stipa teretifolia.*

Brackish-water impact is marked in the north

AUSTRALASIA

by the appearance of dense swards of the composite *Cotula coronopifolia*. In the Kokianga, where sea walls have impounded water that becomes increasingly more brackish the mangroves die and a sward of *Cotula* covers the soil (Chapman, 1975).

In the South Island the *Salicornietum* (with the green alga *Rhizoclonium* and some Myxophyceae) generally gives way to a general salt–marsh community with *Selliera radicans*, *Puccinellia stricta*, *Parapholis incurva*, *Cotula dioica* (replacing *C. coronopifolius*) and *Plantago coronopus* (Chapman,

1974). In better drained areas *Samolus* continues to replace *Salicornia*. The transition to brackish–water conditions is marked in the South Island by *Scirpus lacustris*, *S. americanus* var. *polyphyllus* and *S. caldwellii*.

On all the marshes there may be found a wide range of birds but none specifically restricted to them. Important members are waders, especially the pied stilt (*Himantopus leucocephalus*) and oyster–catchers (Haematopodidae).

APPENDIX: SPECIES LISTS

TABLE 15.5
Australian mangal flora

Region (see Fig. 15.1):	1N	1NE	2E	2G	2W	3	4E	4W	5E a	5E b	5E c	5E d	5W	6	7
MANGROVE TREES															
Arecaceae															
Nypa fruticans Wurmb.	×	×	—	—	—	—	—	—	—	—	—	—	—	—	—
Bombacaceae															
Camptostemon schultzii Masters	×	×[1]	—	—	×	—	—	—	—	—	—	—	—	—	—
Caesalpiniaceae															
Cynometra ramiflora var. *bijuga* (Spanoghe) Benth.	—	×	×	—	—	—	—	—	—	—	—	—	—	—	—
Combretaceae															
Lumnitzera littorea (Jack) Voigt	×	×	—	×	—	—	—	—	—	—	—	—	—	—	—
L. racemosa Willd.	×	×	×	—	×	×	—	—	—	—	—	—	—	—	—
Euphorbiaceae															
Excoecaria agallocha L.	×	×	×	×	×	×	—	—	—	—	—	—	—	—	—
Meliaceae															
Xylocarpus australasicum Ridl.	×	×	×	—	×	—	—	—	—	—	—	—	—	—	—
X. granatum Koenig	×	×	×	×	×	×	—	—	—	—	—	—	—	—	—
Myrsinaceae															
Aegiceras corniculatum (L.) Blanco	×	×	×	×	×	×	×	×	—	—	—	—	—	—	—
Myrtaceae															
Osbornia octodonta F. Muell.	×	×	×	—	—	—	—	—	—	—	—	—	—	—	—
Plumbaginaceae															
Aegialitis annulata R. Br.	×	×	×	×	×	—	—	—	—	—	—	—	—	—	—
Rhizophoraceae															
Bruguiera cylindrica (L.) Bl.	—	×[1]	—	—	—	—	—	—	—	—	—	—	—	—	—
B. exaristata Ding Hou	—	×	×	—	—	?	—	—	—	—	—	—	—	—	—
B. gymnorrhiza (L.) Lam. [syn. *B. conjugata* Meir., *B. rheedii* Bl.]	×	×	×	×	×	—	—	—	—	—	—	—	—	—	—
B. parviflora (Roxb.) Wight and Arn.	×	×	×	×	—	—	—	—	—	—	—	—	—	—	—
B. sexangula (Lour.) Poir.	×	—	—	—	—	—	—	—	—	—	—	—	—	—	—
Ceriops decandra (Griff.) Ding Hou	—	×	—	—	—	—	—	—	—	—	—	—	—	—	—
C. tagal var. *australis* C.T. White	×	×	×	×	×	×	—	—	—	—	—	—	—	—	—

TABLE 15.5 (continued)

Region (see Fig. 15.1):	1N	1NE	2E	2G	2W	3	4E	4W	5E a	5E b	5E c	5E d	5W	6	7
C. tagal var. *tagal* (Perr.) C.B. Rob.	×	×	×	—	—	—	—	—	—	—	—	—	—	—	—
Rhizophora apiculata Bl.	—	×	×	—	—	—	—	—	—	—	—	—	—	—	—
R. mucronata Lam.	×	×	×	×	×	—	—	—	—	—	—	—	—	—	—
R. stylosa Griff.	×	×	×	×	×	×	—	—	—	—	—	—	—	—	—
Rubiaceae															
Scyphiphora hydrophyllacea Gaertn. f.	×	×	—	—	—	—	—	—	—	—	—	—	—	—	—
Sonneratiaceae															
Sonneratia alba J. Sm.	×	×	×	—	×	—	—	—	—	—	—	—	—	—	—
S. caseolaris (L.) Engl.	—	×	—	—	—	—	—	—	—	—	—	—	—	—	—
Sterculiaceae															
Heritiera littoralis (Dryand.) Ait.	—	×	×	—	×	—	—	—	—	—	—	—	—	—	—
Verbenaceae (Avicenniaceae)															
Avicennia eucalyptifolia Zip. ex Miq.	—	×	×	×	×	—	—	—	—	—	—	—	—	—	—
A. marina (Forsk.) Vierh. var. *australasica* (Walp.) Moldenke	×	×	×	×	×	×	×	×	×	×	—	×	×	—	—
MANGAL UNDERSTOREY															
Acanthaceae															
Acanthus ilicifolius L.	×	×	×	×	—	—	—	—	—	—	—	—	—	—	—
Amaryllidaceae															
Crinum penduculatum R.Br.	—	—	—	—	—	×	×	—	—	—	—	—	—	—	—
Pteridaceae															
Acrostichum speciosum Willd.	×	×	×	×	×	×	—	—	—	—	—	—	—	—	—
MANGAL LIANAS															
Asclepiadaceae															
Ischnostemma carnosum (R.Br.) Merr. and Rolfe [syn. *Cynanchum carnosum* (R.Br.) Domin]	×	×	×	×	×	—	—	—	—	—	—	—	—	—	—
Gymnanthera nitida R.Br.	×	×	—	×	×	×	—	—	—	—	—	—	—	—	—
Fabaceae															
Derris trifoliata Lour.	×	×	—	×	—	—	—	—	—	—	—	—	—	—	—
MANGAL EPIPHYTES															
Asclepiadaceae															
Dischidia nummularia R.Br.	—	×	—	—	—	—	—	—	—	—	—	—	—	—	—
Loranthaceae															
Amyema mackayense (Blakely) Dans. ssp. *mackayense*	—	—	×	—	—	—	—	—	—	—	—	—	—	—	—
A. mackayense (Blakely) Dans. ssp. *cycnei–sinus* (Blakely) Barlow	×	—	—	—	×	—	—	—	—	—	—	—	—	—	—
Orchidaceae															
Cymbidium canaliculatum R. Br.	?	×	—	—	×	—	—	—	—	—	—	—	—	—	—
Dendrobium discolor Lindl.	—	×	×	—	—	—	—	—	—	—	—	—	—	—	—
Rubiaceae															
Hydnophytum formicarum Jack	—	×	—	—	—	—	—	—	—	—	—	—	—	—	—
Myrmecodia antoinii Becc.	—	×	—	—	—	—	—	—	—	—	—	—	—	—	—

[1] Restricted to coastline north of Daintree River, Queensland.

These data are derived from the following sources: New South Wales: Collins (1921); Northern Territory: Specht (1958); Chippendale (1972); Queensland: Jones (1971a,b), Hopkins (1974), Specht et al. (1976), Saenger and Hopkins (1975); South Australia: Wood (1937), Womersley and Edmonds (1958), Specht (1972b); Victoria: Willis (1962, 1972), Monash Westernport Bay survey; Western Australia: Beard (1967a), Thom et al. (1975).

AUSTRALASIA

TABLE 15.6

Plants recorded in tidal salt marshes in Australia

Region (see Fig. 15.1):	1N	1NE	2E	2G	2W	3	4E	4W	5E				5W	6	7
									a	b	c	d			
Aizoaceae															
Disphyma blackii R.J. Chinnock	—	—	—	—	—	?	—	—	×	×	×	×	—	×	×
Lampranthus tegens[1] (F. Muell.) N.E. Br.	—	—	—	—	—	×	—	—	×	—	—	—	—	—	—
Sesuvium portulacastrum L.	×	×	×	×	×	×	—	×	—	—	—	—	—	—	—
Trianthema turgidifolia F. Muell.	—	—	—	×	—	—	—	—	—	—	—	—	—	—	—
Apiaceae															
Apium prostratum Labill. ex Vent.	—	—	—	—	—	×	×	—	×	×	×	×	×	×	—
Asteraceae															
Angianthus preissianus (Steetz) Benth.	—	—	—	—	—	—	—	—	×	×	×	×	×	×	—
Cotula coronopifolia L.	—	—	—	—	×	×	—	×	×	×	×	×	×	×	—
Batidaceae															
Batis argillicola van Royen	×	×	—	—	×	—	—	—	—	—	—	—	—	—	—
Caryophyllaceae															
Spergularia media (L.) C. Presl [syn. *S. marginata* (DC) Kittel]	—	—	—	—	—	—	—	—	×	×	×	×	—	×	—
S. rubra[1] (L.) J. and C. Presl	—	—	×	—	—	×	×	—	×	×	×	×	×	×	—
Chenopodiaceae															
Arthrocnemum arbusculum (R.Br.) Moq.	—	—	—	—	—	?	×	×	×	×	×	×	×	×	×
A. bidens Nees	—	—	—	—	—	—	—	—	—	—	—	—	×	—	—
A. halocnemoides Nees var. *pergranulatum* J.M. Black	—	—	×	×	—	×	—	×	×	×	×	×	—	—	×
A. leiostachyum Benth. (Paulsen)	×	×	×	×	×	×	—	×	—	—	—	—	×	—	—
Atriplex paladosa R.Br.	—	—	—	—	—	—	—	—	—	×	×	×	×	×	×
A. patula[1] L.	—	—	—	—	—	—	×	—	×	×	×	×	×	—	—
Bassia astrocarpa F. Muell.	—	—	—	—	—	—	×	—	—	—	—	—	—	—	—
Enchylaena tomentosa R.Br.	—	—	×	—	—	×	×	×	×	×	×	×	—	—	×
Hemichroa diandra R.Br.	—	—	—	—	—	—	—	×	—	×	×	×	×	—	—
H. pentandra R.Br.	—	—	—	—	—	—	—	—	×	×	×	×	×	×	—
Maireana brevifolia (R.Br.) P.G. Wilson	—	—	—	—	—	—	—	—	—	—	—	×	—	—	—
Maireana oppositifolia (F. Muell.) P.G. Wilson	—	—	—	—	—	—	—	—	—	×	×	×	—	—	—
Rhagodia baccata (Labill.) Moq.	—	—	—	—	—	×	×	×	×	×	×	×	×	×	×
Salicornia blackiana Ulbrich	—	—	—	—	—	—	—	—	×	×	×	×	—	—	—
S. quinqueflora Bunge ex Ung. Sternb.	—	—	×	×	—	×	×	—	×	×	×	×	×	×	—
Salsola kali L.	×	×	×	×	×	×	×	×	×	×	×	×	—	—	—
Suaeda arbusculoides L.S. Smith	—	—	—	—	×	—	—	—	—	—	—	—	—	—	—
S. australis (R.Br.) Moq. [syn. *S. maritima* var. *australis* (R.Br.) Domin]	—	—	×	×	×	×	×	×	×	×	×	×	×	×	—
Tecticornia australasica (Moq.) P.G. Wilson [syn. *T. cinerea* (F. Muell.) Hook. f. and Jackson]	×	×	—	×	×	—	—	—	—	—	—	—	—	—	—
Theleophyton billardieri (Moq.) Moq. [syn. *Atriplex billardieri* (Moq.) Hook.f.]	—	—	—	—	—	—	—	—	×	×	—	—	—	×	—
Threlkeldia diffusa R.Br.	—	—	—	?	—	?	—	—	×	×	×	×	×	×	—
Convolvulaceae															
Wilsonia backhousei Hook.f.	—	—	—	—	—	×	×	×	×	×	×	×	×	—	—
W. humilis R.Br.	—	—	—	—	—	—	—	—	×	×	×	×	×	—	—
W. rotundifolia Hook.	—	—	—	—	—	—	—	—	×	×	×	×	×	—	—

TABLE 15.6 *(continued)*

Region (see Fig. 15.1):	1N	1NE	2E	2G	2W	3	4E	4W	5E a	b	c	d	5W	6	7
Cyperaceae															
Baumea juncea (R.Br.) Palla [syn. *Cladium junceum* R.Br.]	—	—	—	—	—	?	×	—	×	×	×	×	×	×	—
Scirpus maritimus L.	—	—	—	—	—	×	×	—	×	×	×	×	×	×	—
Frankeniaceae															
Frankenia pauciflora DC.	—	—	—	—	—	—	—	×	×	×	×	×	×	×	—
Goodeniaceae															
Selliera radicans Cav.	—	—	—	—	—	—	×	—	×	×	×	×	—	×	—
Juncaginaceae															
Triglochin striata Ruiz and Pav.	—	—	—	—	—	×	×	—	×	×	×	×	×	×	—
Malvaceae															
Lawrencia spicata Hook. [syn. *Plagianthus spicatus* (Hook.) Benth.]	—	—	—	—	—	—	—	—	×	×	×	×	×	—	—
Plantaginaceae															
Plantago coronopus[1] L.	—	—	—	—	—	—	×	—	×	×	×	×	×	×	—
Plumbaginaceae															
Limonium australe (R.Br.) O. Kuntze	—	—	—	—	—	×	×	—	—	×	—	—	—	×	—
L. binervosum[1] (G.E. Smith) C.E. Salmon	—	—	—	—	—	—	—	—	—	—	—	×	—	—	—
L. psilocladon[1] (Boiss.) O. Kuntze	—	—	—	—	—	—	—	—	—	—	—	×	—	—	—
L. salicorneacea (F. Muell.) O. Kuntze	—	—	—	—	—	—	×	—	—	—	—	—	—	—	—
Poaceae															
Distichlis distichophylla (Labill.) Fassett	—	—	—	—	—	—	—	—	×	×	×	×	—	×	—
Hordeum marinum[1] Huds.	—	—	—	—	—	?	×	—	×	×	×	×	×	×	—
Parapholis incurva[1] (L.) C.E. Hubbard	—	—	—	—	—	?	×	—	×	×	×	×	×	×	—
Puccinellia stricta (Hook.f.) C. Blom	—	—	—	—	—	—	×	—	×	×	×	×	×	×	—
Sporobolus virginicus (L.) Kunth	×	×	×	×	×	×	×	×	×	×	×	×	×	×	—
Spartina anglica[1] C.E. Hubbard	—	—	—	—	—	—	—	—	×	—	—	×	—	—	—
Xerochloa barbata R.Br.	×	?	—	×	×	—	—	—	—	—	—	—	—	—	—
Primulaceae															
Samolus junceus R.Br.	—	—	—	—	—	—	×	—	—	—	—	—	×	—	—
S. repens (Forst. and Forst.f.) Pers.	—	—	—	—	—	×	×	—	×	×	×	×	×	×	—
Scrophulariaceae															
Mimulus repens R.Br.	—	—	—	—	—	—	×	—	×	×	×	×	×	×	—

[1] Introduced plants.

These data are derived from the following sources: New South Wales: Collins (1921), Clarke and Hannon (1967, 1969, 1970, 1971); Northern Territory: Specht (1958), Chippendale (1972); Queensland: McLeod (1969), Specht et al. (1976), Saenger and Hopkins (1975); South Australia: Osborn and Wood (1923), Wood (1937), Specht (1972b); Tasmania: Curtis and Sommerville (1947); Victoria: Patton (1942), Willis (1962, 1972), Monash Westernport Bay survey; Western Australia: Beard (1970), Smith (1973).

AUSTRALASIA

TABLE 15.7

Plants found at the fringe of tidal salt–marsh or mangal vegetation in Australia

Region (see Fig. 15.1):	1N	1NE	2E	2G	2W	3	4E	4W	5E a	5E b	5E c	5E d	5W	6	7
Aizoaceae															
Carpobrotus rossii (Haw.) Schwantes	—	—	—	—	—	—	—	×	×	×	×	×	—	×	×
Gasoul crystallinum[1] (L.) Rothmaler	—	—	—	—	—	—	—	×	×	—	×	×	×	×	—
Apiaceae															
Hydrocotyle capillaris F. Muell.	—	—	—	—	—	—	—	—	×	×	×	×	×	×	—
Asteraceae															
Baccharis halimifolia[1] L.	—	—	—	—	—	×	—	—	—	—	—	—	—	—	—
Senecio lautus Forst. f. ex Willd.	—	—	—	—	—	×	×	×	×	×	×	×	×	×	×
Caryophyllaceae															
Sagina maritima G. Don	—	—	—	—	—	—	×	—	—	×	—	×	—	×	—
Casuarinaceae															
Casuarina glauca Sieber ex Spreng.	—	—	—	—	—	×	×	—	—	—	—	—	—	—	—
Centrolepidaceae															
Centrolepis polygyna (R.Br.) Hieron	—	—	—	—	—	×	—	—	×	×	×	×	×	×	—
Chenopodiaceae															
Atriplex cinerea Poir.	—	—	—	—	—	?	×	—	×	×	×	×	×	×	—
A. semibaccata R.Br.	—	—	—	—	—	—	—	—	—	×	—	—	—	—	—
A. stipitata Benth.	—	—	—	—	—	—	—	—	—	—	—	—	×	—	×
Chenopodium glaucum ssp. *ambiguum* (R.Br.) Murr. and Thell. ex Thell.	—	—	—	—	—	×	—	—	×	×	×	×	?	×	—
Cyperaceae															
Cladium filum (Labill.) R.Br. [syn. *Gahnia filum* (Labill.) F. Muell.]	—	—	—	—	—	—	—	—	×	×	×	×	—	×	—
Cyperus polystachyos Rottb.	×	×	×	×	×	×	—	—	×	—	—	—	×	—	—
C. stoloniferus Retz. [syn. *C. littoralis* R.Br.]	×	×	—	—	—	—	—	—	—	—	—	—	—	—	—
C. vaginatus R.Br.	—	—	—	×	—	—	×	—	—	—	—	—	×	×	—
Eleocharis geniculata (L.) R. and S.	×	×	×	×	×	×	—	×	—	—	—	—	—	—	—
Fimbristylis densa S.T. Blake	×	×	—	×	—	—	—	—	—	—	—	—	—	—	—
F. ferruginea (L.) Vahl	×	×	×	×	×	—	×	—	—	—	—	—	—	—	—
F. punctata R.Br.	×	—	×	—	—	×	—	—	—	—	—	—	—	—	—
F. rara R.Br.	×	—	—	×	—	—	—	—	—	—	—	—	—	—	—
Rhynchospora pterochaeta F. Muell.	×	—	×	—	—	—	—	—	—	—	—	—	—	—	—
Schoenus nitens (R.Br.) Poir.	—	—	—	—	—	×	—	—	×	×	×	×	×	×	—
Scirpus antarcticus L.	—	—	—	—	—	×	—	—	×	×	×	×	×	×	—
S. litoralis Schrad.	—	—	×	×	—	×	—	—	—	—	—	—	×	—	—
S. nodosus Rottb.	—	—	—	—	—	×	—	×	×	×	×	×	×	×	—
Fabaceae															
Canavalia maritima (Aubl.) Thouars	×	×	×	?	×	×	×	×	—	—	—	—	—	—	—
Melilotus indica[1] (L.) All.	—	—	—	—	—	—	×	—	×	×	×	×	×	×	—
Gentianaceae															
Centaurium pulchellum[1] (Swartz) Druce	—	—	—	—	—	×	—	—	×	×	×	—	—	×	—
C. spicatum (L.) Fritsch	—	—	—	—	×	×	×	—	—	×	—	—	×	×	—
Sebaea albidiflora F. Muell.	—	—	—	—	—	—	—	—	×	×	×	×	—	×	—
Juncaceae															
Juncus maritimus Lam. var. *australiensis* Buch.	—	—	—	—	—	×	×	×	×	×	×	×	×	×	—

TABLE 15.7 *(continued)*

Region (see Fig. 15.1):	1N	1NE	2E	2G	2W	3	4E	4W	5E a	5E b	5E c	5E d	5W	6	7
Juncaginaceae															
Triglochin mucronata R.Br.	—	—	—	—	—	—	—	—	×	—	×	—	×	×	—
Lythraceae															
Pemphis acidula Forst. and Forst.f.	×	×	—	—	—	—	—	—	—	—	—	—	—	—	—
Malvaceae															
Hibiscus tiliaceus L.	×	×	×	×	—	×	—	—	—	—	—	—	—	—	—
Thespesia lampas (Cav.) Dalz. ex Dalz. and Gibs. var. *thespesioides* (R.Br. ex Benth.) P.A. Fryxell	×	—	—	—	×	—	—	—	—	—	—	—	—	—	—
T. populnea (L.) Sol. ex Corr.	—	—	×	—	—	—	—	—	—	—	—	—	—	—	—
T. populneoides (Roxb.) Kostel.	×	×	—	×	×	—	—	—	—	—	—	—	—	—	—
Myoporaceae															
Myoporum acuminatum R.Br.	—	—	×	—	×	×	×	×	—	—	—	—	—	—	—
Myoporum insulare R.Br.	—	—	—	—	—	—	×	—	×	×	×	×	×	×	—
Myrtaceae															
Melaleuca acacioides F. Muell.	×	×	—	—	×	—	—	—	—	—	—	—	—	—	—
M. ericifolia Sm.	—	—	—	—	—	?	×	—	×	×	—	—	—	×	—
M. halmaturorum F. Muell. ex Miq.	—	—	—	—	—	—	—	—	—	—	×	×	—	—	—
M. quinquenervia (Cav.) S.T. Blake	—	×	×	—	—	×	—	—	—	—	—	—	—	—	—
M. squarrosa Donn ex Sm.	—	—	—	—	—	—	×	—	×	×	×	×	—	×	—
Poaceae															
Agrostis billardieri R.Br.	—	—	—	—	—	—	×	—	×	×	×	×	—	×	—
Cynodon dactylon (L.) Pers.	×	×	×	×	×	×	×	×	×	×	×	×	×	×	—
Lepturus repens (Forst. f.) R.Br.	×	×	×	×	—	×	—	—	—	—	—	—	—	—	—
Lolium loliaceum[1] (Bory and Chaub.) Hand.	—	—	—	—	—	?	×	—	—	—	—	—	×	×	×
Paspalum distichum L. var. *littorale* (R.Br.)	—	—	—	—	—	×	×	—	—	—	—	—	—	—	—
Periballia minuta[1] (L.) Ashers and Graebn.	—	—	—	—	—	—	—	—	—	—	—	×	—	—	—
Phragmites australis (Cav.) Trin. ex Steud.	—	—	—	—	×	×	—	—	×	×	×	×	—	×	—
P. karka (Retz.) Trin. ex Steud.	×	×	×	×	×	—	—	×	—	—	—	—	—	—	—
Stipa teretifolia Steud.	—	—	—	—	—	—	—	—	×	×	×	×	?	×	—
Vetiveria elongata (R.Br.) Stapf ex C.E. Hubbard	×	×	—	×	—	—	—	—	—	—	—	—	—	—	—
Zoysia macrantha Desv.	—	—	—	—	×	×	—	—	×	×	—	—	—	—	—
Verbenaceae															
Clerodendrum inerme (L.) Gaertn.	×	×	×	×	—	×	—	—	—	—	—	—	—	—	—
Zygophyllaceae															
Nitraria schoberi L.	—	—	—	—	—	—	—	—	—	—	—	—	×	×	×

[1] Introduced plants.

AUSTRALASIA

TABLE 15.8

Algae recorded in mangal and tidal salt–marsh vegetation in Australia
(substrate: m = mud surfaces; A–p = *Avicennia* pneumatophores; R–sr = *Rhizophora* stilt–roots)

Species	Substrate	Localities
CHLOROPHYTA		
Caulerpa fastigiata Mont.	A–p	North Stradbroke Is., Qld.
Chaetomorpha capillaris (Kütz.) Boerg.	m	South Australia
Chlorococcum humicola (Naeg.) Rabenhorst	m	SE Qld.
Cladophora sp.	A–p	North Stradbroke Is., Qld.
Cladophorella calcicola Fritsch	m	SE Qld.
Enteromorpha clathrata (Roth) Grev.	m	SE Qld.
Enteromorpha sp.	m	South Australia
Hormidium subtile (Kütz.) Heering	m	SE Qld.
Monostroma crepidinum Farlow	A–p	Tallebudgera Creek to North Pine River, Qld.
Monostroma sp.	m	Gladstone, Qld.
Protococcus viride Agardh	m	SE Qld.
Pseudendoclonium submarinum Wille	m	SE Qld.
Rhizoclonium capillare Kütz.	m	Gladstone, Qld.
	m	Noosa, Qld.
Rhizoclonium implexum (Dillwyn) Kütz.	A–p, m	Gladstone, Qld.
	A–p, m	SE Qld.
Stichococcus bacillaris Naegeli	m	SE Qld.
Ulva lactuca L.	A–p	North Stradbroke Is., Qld.
PHAEOPHYTA		
Colpomenia sinuosa (Roth) Derbes and Solier	A–p	North Pine River, Qld.
RHODOPHYTA		
Bostrychia binderi Harv.	A–p	Gulf of Carpentaria, Qld.
	A–p	Redcliffe, Qld.
Bostrychia flagellifera Post	A–p, m	Parramatta R., N.S.W.
	A–p	Wellington Point, Qld.
Bostrychia kelanensis Grun. ex Post	m	Brisbane River, Qld.
	A–p	Noosa to Southport, Qld.
	A–p	Gladstone, Qld.
Bostrychia mixta J.D. Hook. and Harv.	A–p, m	Brisbane River, Qld.
	m	Kangaroo Is., S.A.
Bostrychia moritziana (Sond.) J.Ag.	A–p	Westernport Bay, Vic.
	A–p, m	Georges River, N.S.W.
	A–p	Nerang River, Qld.
	A–p, m	Brisbane River, Qld.
Bostrychia radicans Mont.	A–p	Gulf of Carpentaria, Qld.
	A–p	Gladstone, Qld.
	m	Noosa, Qld.
Bostrychia scorpioides Mont.	?	"Australia"
Bostrychia simpliciuscula Harv. ex J. Ag.	m	Noosa, Qld.
	m	South Australia
Bostrychia tenella (Vahl) J. Ag.	A–p	Gladstone, Qld.
	m	Noosa to Southport, Qld.
Bostrychia tenuis Post	A–p	Westernport Bay, Vic.
	m	Kangaroo Is., S.A.
	A–p, m	Georges River, N.S.W.
Bostrychia vaga J.D. Hook. and Harv.	m	Kangaroo Is., S.A.
Bostrychia sp.	m	SE Qld.

TABLE 15.8 *(continued)*

Species	Substrate	Localities
Caloglossa adnata (Zanard.) de Toni	A–p, m	Brisbane River, Qld.
	A–p	Noosa to Southport, Qld.
	A–p	Gladstone, Qld.
Caloglossa bombayensis Boerg.	m	Brisbane River, Qld.
		"east coast of Australia"
Caloglossa leprieurii (Mont.) J. Ag.	A–p	Gladstone, Qld.
	m	Cooktown—Rockhampton, Qld.
	m	Noosa to Southport, Qld.
	m	Brisbane River, Qld.
	A–p	Westernport Bay, Vic.
	A–p, m	Port Phillip Bay, Vic.
Caloglossa ogasawaraensis Okam.	A–p	Deception Bay to Southport, Qld.
	A–p	Gladstone, Qld.
	A–p	Cooktown, Qld.
Catenella nipae Zanard.	A–p	Cooktown, Qld.
	A–p	Gladstone, Qld.
	A–p, m	SE Qld.
	A–p	North Pine River, Qld.
	A–p	Lane Cove, Sydney, N.S.W.
	m	Brisbane River, Qld.
	?	Kissing Pt., N Qld.
Gelidium pusillum (Stackhouse) Le Jolis	m	South Australia
Gelidium sp.	A–p	North Stradbroke Is., Qld.
Gracilaria confervoides (L.) Grev.		
f. *gracilis*	A–p	North Stradbroke Is., Qld.
f. *tumida*	m	Qld. and N.S.W.
Gracilaria lichenoides (L.) Harv.		
f. *lemanea*	m	Qld. and N.S.W.
Hypnea sp.	A–p	North Stradbroke Is., Qld.
Laurencia nidifica J. Ag.	A–p	Gladstone to Wynnum, SE Qld.
Polysiphonia macrocarpa Harv.	A–p, R–Sr	SE Qld.
Spyridia filamentosa (Wulf.) Harv. in Hook.	A–p	North Stradbroke Is., Qld.

CYANOPHYTA

1. Chroococcaceae

Anacystis marina (Hansg.) Dr. and Daily	m	SE Qld.
Chroococcus turgidus (Kütz.) Näg.	m	SE Qld.
Entophysalis deusta (Meneg.) Dr. and Daily	m	SE Qld.
Gloeocapsa alpicola (Lyngb.) Born.	m	SE Qld.
Gloeocapsa atrata (Turp.) Kütz.	m	SE Qld.

2. Nostocaceae

Anabaena fertilissima C.B. Rao	m	SE Qld.
Anabaena torulosa (Carm.) Lagerh. ex Born. and Flah.	m	SE Qld.
Calothrix crustacea Thuret	m	SE Qld.
Calothrix pilosa Harv.	m	SE Qld.
Fischerella muscicola (Thuret) Gomont	m	SE Qld.
Nostoc commune Vaucher	m	SE Qld.
Nostoc entophytum Born. and Flah.	m	SE Qld.
Nostoc linckia (Roth) Bornet ex Born. and Flah.	m	SE Qld.
Nostoc muscorum Ag. ex Born. and Flah.	m	SE Qld.
Scytonema rhizophorae Zeller	m	SE Qld.
Tolypothrix tenuis (Kütz.) Johs. Schmidt	m	SE Qld.

3. Oscillatoriaceae

Lyngbya aesturii (Mart.) Liebmann.	m	SE Qld.

TABLE 15.8 (continued)

Species	Substrate	Localities
Lyngbya epiphytica Hieron	m	SE Qld.
Lyngbya lutea (Ag.) Gomont	m	SE Qld.
Lyngbya majuscula (Dillwyn) Harv.	m	Gulf of Carpentaria, Qld.
Microcoleus chthonoplastes Thuret ex Gomont	m	SE Qld.
Microcoleus lyngbyaceus (Kütz.) Crouan	m	Gladstone, Qld.
Microcoleus tenerrimus Gomont	m	SE Qld.
Oscillatoria corallinae (Kütz.) Gom.	m	SE Qld.
Oscillatoria earlei Gardner	m	SE Qld.
Oscillatoria laete-virens (Crouan) Gomont	m	SE Qld.
Oscillatoria nigro-viridis Thwaites ex Gomont	m	SE Qld.
Phormidium angustissimum W. and G.S. West	m	SE Qld.
Phormidium autumnale (Ag.) Gomont	m	SE Qld.
Phormidium corium (Ag.) Gomont	m	SE Qld.
Phormidium foveolarum (Mont.) Gomont	m	SE Qld.
Phormidium luridum (Kütz.) Gomont	m	SE Qld.
Phormidium molle (Kütz.) Gomont	m	SE Qld.
Phormidium tenue (Menegh.) Gomont	m	SE Qld.
Porphyrosiphon kurzii (Zeller) Dr.	m	Gladstone, Qld.
Spirulina subtilissima Kütz. ex Gomont	m	SE Qld.
Symploca laete–viridis Gomont	m	SE Qld.

CHRYSOPHYTA

Species	Substrate	Localities
Vanheurckia lewisiana Breb.	m	Gladstone, Qld.

Biogeographical data were derived from the following references: Fischer (1940), May (1948), Cribb (1956, 1958, and unpublished records), Womersley and Edmonds (1958), Post (1963), McLeod (1969), King et al. (1971), Cribb, in Clifford and Specht (1976), and Saenger (unpublished records).

TABLE 15.9

Lichens recorded on mangrove trees in Australia

Species	Localities
Crustose lichens	
Arthonia cf. *fissurina* Nyl.	Gladstone, central Qld.
Arthopyrenia cf. *cinereopruinosa* (Schaer.) Korb.	Gladstone, central Qld.
Buellia callispora (Kn.) Stirton	Tallebudgera, southern Qld.; North Stradbroke Is., southern Qld.
Buellia punctata (Hoffm.) Mass.	Gladstone, central Qld.
Calicium cf. *pallidellum* Willey	Gladstone, central Qld.
Caloplaca byrsonimae Melme	Tallebudgera, southern Qld.
Caloplaca pyracea (Ach.) Th.Fr.	Port Adelaide, S.A.
Graphis anguiliformis Tayl.	Weipa, northern Qld.
Graphis cf. *caesiella* Wain.	Weipa, northern Qld.
Graphis cf. *longula* Kremplh.	Weipa, northern Qld.
Haematomma sp.	North Stradbroke Is., southern Qld.
Lecanora cf. *hageni* (Ach.) Ach.	Gladstone, central Qld.
Lecanora pallida (Schreb.) Rabenh.	Tallebudgera, southern Qld.; North Stradbroke Is., southern Qld.; Gladstone, central Qld.
Lecanora subfusca (L.) Ach.	Gladstone, central Qld.
Lecanora varia (Hoffm.) Ach.	Port Adelaide, S.A.
Leptotrema sp.	Weipa, northern Qld.
Ochrolechia subpallescens Vers.	Gladstone, central Qld.
Pertusaria bispora (Farlow) Linder	Tallebudgera, southern Qld.; North Stradbroke Is., southern Qld.
Pertusaria multipuncta (Turn.) Nyl.	Gladstone, central Qld.

TABLE 15.9 *(continued)*

Species	Localities
Pertusaria cf. *oahuensis* Magn.	Tallebudgera, southern Qld.; North Stradbroke Is., southern Qld.
Pertusaria thiospoda Knight	Gladstone, central Qld.
Pertusaria sp.	Weipa, northern Qld.
Pyrenula sp.	Weipa, northern Qld.
Foliose lichens	
Anaptychia tremulans (Müll. Arg.) Kurok.	Tallebudgera, southern Qld.
Coccocarpia sp.	Tallebudgera, southern Qld.
Collema glaucophthalmum Nyl.	North Stradbroke Is., southern Qld.
Dirinaria applanata (Fee) Awasthi	North Stradbroke Is., southern Qld.
Dirinaria aspera (Magn.) Awasthi	Gladstone, central Qld.
Leptogium sp.	Tallebudgera, southern Qld.
Parmelia dilatata Wain.	North Stradbroke Is., southern Qld.
	Gladstone, central Qld.
Parmelia maura Pers.	Tallebudgera, southern Qld.
Parmelia sydneyensis Gyelnik	North Stradbroke Is., southern Qld.
Parmelia tinctorum Nyl.	Tallebudgera, southern Qld.
Physma byrsinum (Ach.) Müll. Arg.	Tallebudgera, southern Qld.
Pyxine berterana (Fee) Imshaug	Tallebudgera, southern Qld.
	North Stradbroke Is., southern Qld.
Pyxine coceos (Sw.) Nyl.	North Stradbroke Is., southern Qld.
Pyxine physciiformis (Malme) Imshaug	North Stradbroke Is., southern Qld.
Pyxine subcinerea Stirton	Tallebudgera, southern Qld.
	North Stradbroke Is., southern Qld.
Pyxine sp.	Weipa, northern Qld.
Xanthoria ectanea (Ach.) Raes. ex R. Fils	Port Adelaide, S.A.
Fruticose lichens	
Ramalina calicaris (L.) Röhl.	Groote Eylandt, N.T.
Ramalina leiodea Nyl.	Weipa, northern Qld.
	Melville Bay, N.T.
Ramalina reagens (B. de Lesd.) Culb.	Gladstone, central Qld.
Ramalina pusilla Le Prev.	Gladstone, central Qld.
Ramalina cf. *sideriza* Zahlbr.	North Stradbroke Is., southern Qld.
	Gladstone, central Qld.
Roccella montagnei Bel.	Groote Eylandt, N.T.
Teloschistes flavicans (Sw.) Norm.	North Stradbroke Is., southern Qld.
	Gladstone, central Qld.
Usnea cf. *scabrida* Tayl.	Tallebudgera, southern Qld.
	North Stradbroke Is., southern Qld.

This preliminary list of lichens, found on mangrove trees, was identified by Dr. R.W. Rogers, Botany Department, University of Queensland. It is based on the following collections: Port Adelaide, S.A.: collector E.M. Wollaston; Tallebudgera Creek, southern Qld.: collector R.W. Rogers (in Shine et al., 1973); North Stradbroke Island, southern Qld.: collector R.W. Rogers (in Clifford and Specht, 1976); Gladstone, central Qld.: collector P. Saenger; Weipa, northern Qld.: collector R.L. Specht; Groote Eylandt and Melville Bay, N.T.: collector R.L. Specht (in Bibby, 1958).

AUSTRALASIA

TABLE 15.10

Mollusca recorded in mangal and tidal salt–marsh vegetation in Australia

Species	N.T.	Queensland			N.S.W.	Vic.	Tas.	S.A.	W.A. (S.W.)
		N.	C.	S.					
POLYPLACOPHORA									
Chitonidae									
Acanthozostera gemmata (Blainville)	×	×	—	—	—	—	—	—	—
GASTROPODA									
Fissurellidae									
Montfortula rugosa Quoy and Gaimard	—	—	—	×	×	×	—	—	×
Patellidae									
Cellana tramoserica (Sowerby)	—	—	—	×	×	×	—	×	—
Trochidae									
Austrocochlea adelaidae (Philippi)	—	—	—	—	—	×	—	×	×
Austrocochlea concamerata (Wood)	—	—	—	×	×	×	×	×	×
Austrocochlea obtusa Dillwyn or *Austrocochlea constricta* (Lamarck)	×	×	×	×	×	×	×	×	×
Euchelus atratus (Gmelin)	×	×	×	×	—	—	—	—	—
Monodonta labio (Linné)	×	×	—	—	—	—	—	—	—
Neritidae									
Melanerita melanotragus Smith or *Nerita atramentosa* Reeve	—	—	—	×	×	×	—	×	×
Nerita albicilla Linné	—	—	—	×	—	—	—	—	—
Nerita chamaeleon Linné	×	×	×	×	—	—	—	—	—
Nerita lineata Gmelin	×	×	×	×	—	—	—	—	—
Nerita planospira Anton	×	×	—	—	—	—	—	—	—
Nerita plicata Linné	×	×	—	—	—	—	—	—	—
Nerita polita Linné	×	×	—	×	—	—	—	—	—
Littorinidae									
Bembicium auratum Quoy and Gaimard	×	×	×	×	×	×	×	×	—
Bembicium melanostomum (Gmelin)	—	—	—	×	×	×	—	×	—
Melarapha infans Smith	—	—	—	×	×	—	—	—	—
Melarapha scabra L.	×	×	×	×	×	—	—	×	—
Melarapha undulata Gray	×	×	×	×	—	—	—	—	—
Peasiella tantilla (Gould)	×	×	—	—	—	—	—	—	—
Turritellidae									
Turritella cerea Reeve or *Turritella terebra* L.	×	×	—	—	—	—	—	—	—
Planaxidae									
Planaxis sulcatus (Born)	×	×	×	×	—	—	—	—	—
Quoyia decollata (Quoy and Gaimard)	×	×	—	—	—	—	—	—	—
Potamididae (= Telescopiidae)									
Pyrazus ebeninus Bruguière	×	×	×	×	×	×	—	—	—
Telescopium telescopium (L.)	×	×	×	—	—	—	—	—	—
Terebralia palustris (L.)	×	×	—	—	—	—	—	—	—
Terebralia sulcata Born	×	×	—	—	—	—	—	—	—
Velacumantus australis (Quoy and Gaimard)	×	×	×	×	×	×	×	×	—

TABLE 15.10 (continued)

Species	N.T.	Queensland			N.S.W.	Vic.	Tas.	S.A.	W.A. (S.W.)
		N.	C.	S.					
Cerithiidae									
Clypeomorus carbonarius Sowerby	—	—	—	×	×	—	—	—	—
Cerithium dorsuosum Mencke	×	×	—	—	—	—	—	—	—
Cerithium jannelli Hombron and Jacquinot	×	×	—	—	—	—	—	—	—
Cerithidea kieneri Hombron and Jacquinot	—	—	—	×	—	—	—	—	—
Cerithidea largillierti (Phillippi)	—	—	—	×	—	—	—	—	—
Cerithidea obtusa (Lamarck)	×	×	—	×	—	—	—	—	—
Diala rufilabris (Adams)	—	—	—	—	×	×	×	—	—
Strombidae									
Lambis lambis (L.)	×	×	—	—	—	—	—	—	—
Strombus luhuanus L.	×	×	—	—	—	—	—	—	—
Naticidae									
Conuber melanostoma (Swainson)	—	—	—	×	×	×	—	—	—
Conuber sordida (Swainson) (syn. *C. strangei*)	—	—	—	×	×	×	—	—	—
Natica violacea Bruguière	×	×	—	—	—	—	—	—	—
Polinices conicus (Lamarck)	×	×	×	×	×	×	—	×	—
Thaisidae									
Agnewia pseudamygdala (Reeve)	—	—	—	×	×	—	—	—	—
Muricidae									
Bedeva hanleyi (Angas)	×	×	×	×	×	×	—	×	—
Morula marginalba (Blainville)	×	×	×	×	×	—	—	—	—
Nassidae									
Nassarius burchardi (Philippi)	—	—	×	—	×	×	—	×	—
Mitridae									
Strigatella retusa Lamarck	—	—	—	×	—	—	—	—	—
Hydatinidae									
Bullina lineata Gray	×	×	—	—	—	—	—	—	—
Atydidae									
Atys naucum L.	×	×	—	—	—	—	—	—	—
Haminoea sp.	—	—	×	×	—	—	—	—	—
Ellobiidae									
Cassidula augulifera Petit	×	×	×	×	—	—	—	—	—
Cassidula rugata Menke	×	×	×	—	—	—	—	—	—
Melosidula nucleus Gmelin	×	×	—	—	—	—	—	—	—
Melosidula zonata (H. and A. Adams)	—	—	—	×	×	—	—	—	—
Ophiocardelus ornatus (Férussac)	—	—	—	×	×	×	×	—	—
Ophiocardelus quoyi H. and A. Adams	—	—	—	×	×	×	—	—	—
Ophiocardelus sulcatus H. and A. Adams	×	×	×	×	×	×	—	—	—
Pythia scarabeus L.	×	×	—	—	—	—	—	—	—
Ellobium aurisjudae L.	×	×	×	—	—	—	—	—	—
Amphibolidae									
Salinator fragilis (Lamarck)	×	×	×	×	×	×	×	×	×
Salinator solida von Martens	×	×	×	×	×	×	×	×	×

AUSTRALASIA

TABLE 15.10 (continued)

Species	N.T.	Queensland			N.S.W.	Vic.	Tas.	S.A.	W.A. (S.W.)
		N.	C.	S.					
Siphonariidae									
Siphonaria bifurcata Reeve	—	—	—	×	—	—	—	—	—
Siphonaria denticulata Quoy and Gaimard	×	×	—	—	—	—	—	—	—
Siphonaria marza (Iredale)	×	×	—	—	—	—	—	—	—
Onchidiidae									
Onchidina australis Semper	×	×	—	×	×	×	—	—	—
Onchidium buetschli Stantshinsky	×	×	—	—	—	—	—	—	—
Onchidium damellii Semper	—	—	×	×	×	—	—	—	—
Onchidium verruculatum Cuvier	×	×	—	×	×	—	—	—	—
BIVALVIA									
Arcidae									
Anadara trapezia Deshayes	—	—	—	×	×	—	—	—	×
Mytilidae									
Modiolus inconstans (Dunker)	—	—	—	—	—	×	—	×	—
Modiolus pulex (Lamarck)	×	×	—	×	×	×	×	×	×
Isognomontidae									
Isognomon isognomon (L.)	—	—	—	×	—	—	—	—	—
Perna nucleus Lamarck	×	×	—	—	—	—	—	—	—
Pinnidae									
Atrina gouldii (Reeve)	×	×	—	—	×	—	—	—	—
Lucinidae									
Austrosarepta sp.	×	×	—	—	—	—	—	—	—
Geloinidae									
Geloina coaxans (Gmelin)	×	×	—	—	—	—	—	—	—
Batissa triquetra Deshayes	×	×	—	—	—	—	—	—	—
Pteriidae									
Pinctada margaritifera (L.)	×	×	—	—	—	—	—	—	—
Anomiidae									
Patro australis (Gray)	×	—	—	—	—	—	—	—	—
Ostreidae									
Ostrea nomades Iredale	—	—	—	×	—	—	—	—	—
Saccostrea amasa (Iredale)	×	×	—	×	—	—	—	—	—
Saccostrea commercialis (Iredale and Roughley)	—	—	—	×	×	×	—	—	—
Saccostrea cucullata (Quoy and Gaim.)	—	—	—	—	×	—	—	—	—
Mactridae									
Notospisula parva (Angas)	—	—	×	×	×	×	×	×	—
Glaucomyidae									
Glaucomya virens Linné	—	—	—	×	—	—	—	—	—
Glaucomya rugosa Hanley	×	×	—	—	—	—	—	—	—

TABLE 15.10 (continued)

Species	N.T.	Queensland			N.S.W.	Vic.	Tas.	S.A.	W.A. (S.W.)
		N.	C.	S.					
Teredinidae									
Bactronophorus subaustralis Iredale	×	×	—	—	—	—	—	—	—
Dicyathifer caroli Iredale	×	×	—	—	—	—	—	—	—
Teredo tristi Iredale	—	—	—	×	—	—	—	—	—
Laternulidae									
Laternula creccina Reeve	—	—	—	—	—	×	×	×	—
Laternula tasmanica (Reeve)	—	—	—	×	—	—	—	—	—
Laternula vagina Reeve	×	×	—	—	—	—	—	—	—
Veneridae									
Circe tumidum (Bolten) (syn. *Gafrarium tumidum*)	×	×	—	—	—	—	—	—	—
Paphia hiantina (Lamarck)	×	×	×	—	—	—	—	—	—
Tapes watlingi Iredale	—	—	—	×	×	—	—	—	—

The biogeographic data presented above were extracted from the following references: T.A. Stephenson et al. (1931), Gillies (1951), W. Stephenson et al. (1958), Womersley and Edmonds (1958), Allan (1959), Rippingdale and McMichael (1961), Iredale and McMichael (1962), Macpherson and Gabriel (1962), Cotton (1964), MacNae (1966, 1967), Hegerl and Timmins (1973), Hutchings (1973), Shine et al. (1973), Campbell et al. (1974), and Hutchings and Recher (1974).

TABLE 15.11

Crustacea recorded in mangal vegetation in Australia

Species	Far north Qld.	Central Qld.	Southern Qld.	Sydney District, N.S.W.
ISOPODA				
Ligidae				
Ligia australiensis (Dana)	—	—	—	×
Ligia sp.	×	—	—	—
Sphaeromidae				
Exosphaeroma sp.	—	—	×	—
Sphaeroma sp.	—	—	×	—
Indet. spp.	×	—	—	—
AMPHIPODA				
Talitridae				
Indet. spp.	×	—	—	—
Talorchestia sp.	—	—	×	—
CIRRIPEDIA				
Balanidae				
Balanus amphitrite Darwin	×	×	×	—
Tetraclita caerulescens (Spengler)	×	—	—	—
Tetraclita squamosa (Bruguière)	×	—	—	—
Tetraclita vitiata Darwin	×	—	—	—

AUSTRALASIA

TABLE 15.11 *(continued)*

Species	Far north Qld.	Central Qld.	Southern Qld.	Sydney District, N.S.W.
Chthamalidae				
Chthamalus caudatus Pilsbry	×	—	—	—
Chthamalus malayensis Pilsbry	×	—	—	—
Lapididae				
Ibla cumingi Darwin	×	—	—	—
ANOMURA				
Laomedia healyi Yaldwyn and Wear	—	—	—	×
DECAPODA				
Alpheidae				
Alpheus edwardsi (Andouin)	×	—	—	×
Callianassinae				
Callianassa australiensis Dana	—	×	×	×
Grapsidae				
Clistocoeloma merguinense de Man	×	×	×	—
Grapsus strigosus (Herbst)	×	—	—	—
Helice leachii Hess	×	×	×	×
Helograpsus haswellianus (Whitelegge)	×	×	×	×
Ilyograpsus paludicola (Rathbun)	—	—	×	—
Metapograpsus frontalis Miers [syn. *M. gracilipes* (de Man)]	×	×	×	×
Metapograpsus latifrons (White)	—	×	—	—
Metapograpsus messor (Forskål)	×	—	×	—
Paragrapsus laevis (Dana)	—	—	×	×
Sesarma bidens (de Haan)	×	—	—	—
Sesarma borneensis Tweedie	×	×	—	—
Sesarma erythrodactyla Hess	—	×	×	×
Sesarma guttata Milne Edwards	—	×	—	—
Sesarma meinerti de Man	×	×	×	—
Sesarma messa Campbell	×	×	×	—
Sesarma moluccensis de Man	×	×	—	—
Sesarma semperi longicristatum Campbell	×	×	×	—
Sesarma smithii Milne Edwards	×	×	×	—
Sesarma villosa (Milne Edwards)	×	—	—	—
Mictyridae				
Mictyris longicarpus (Latreille)	×	×	×	×
Ocypodidae				
Cleistostoma wardi Rathbun	×	×	×	—
Euplax tridentata Milne Edwards	×	×	×	×
Heloecius cordiformis Milne Edwards	×	×	×	×
Leipocten sordidulum Latreille	—	—	×	—
Ocypode ceratopthalmus (Pallas)	—	×	×	—
Macrophthalmus crassipes Milne Edwards	—	—	×	×
Macrophthalmus depressus (Rüppel)	—	×	—	—
Macrophthalmus latreillei (Desmarest)	×	×	—	—
Macrophthalmus pacificus Dana	×	×	—	—

TABLE 15.11 (continued)

Species	Far north Qld.	Central Qld.	Southern Qld.	Sydney District, N.S.W.
Macrophthalmus punctulatus Miers	—	—	×	—
Macrophthalmus setosus Milne Edwards	—	—	×	×
Uca arcuata (de Haan)	—	—	×	—
Uca bellator (White)	×	×	—	—
Uca coarctata Milne Edwards	×	×	—	—
Uca dussumieri Milne Edwards	×	×	×	—
Uca lactea (de Haan)	×	×	—	—
Uca longidigitum Kingsley	×	×	×	—
Uca marionis (Desmarest)	—	—	×	—
Uca signata Hess	×	×	—	—
Paguridae				
Clibanarius virescens Krauss	×	—	—	—
Paguristes sp.	—	×	×	—
Penaeidae				
Penaeus plebejus Hess	—	—	×	×
Penaeus sp.	×	×	—	—
Portunidae				
Portunus pelagicus (L.)	—	×	×	—
Scylla serrata (Forskål)	×	×	×	×
Thalamita danae Stimpson	—	—	×	—
Thalassinidae				
Thalassina anomala (Herbst)	×	×	—	—
Xanthidae				
Eurycarcinus integrifrons de Man	×	×	×	—
Eurycarcinus maculatus (Milne Edwards)	—	—	×	—
Pilumnus sp.	—	×	×	—

The biogeographic data presented above were extracted from the following references: Campbell et al. (1974): Moreton Bay, southern Qld.; B. Campbell (personal communication, 1975): Port Curtis, central Qld.; Gillies (1951): North Stradbroke Is., southern Qld.; Hegerl and Timmins (1973): Noosa River, southern Qld.; Hutchings and Recher (1974): Careel Bay, N.S.W.; MacNae (1966): northern Qld.; Shine et al. (1973): Tallebudgera Creek, southern Qld.

TABLE 15.12

Annelida, Sipuncula, and Echiura recorded in mangal vegetation in Australia

Species	Far north Qld.	Moreton Bay, Qld.	Sydney District, N.S.W.
PHYLUM ANNELIDA			
Class Polychaeta			
Nereidae			
Australonereis ehlersi (Augener)	—	×	×
Ceratonereis erythraensis Fauvel	—	×	×
Namalycastis abiuma Mueller	—	×	—
Neanthes vaallii Kinberg	—	×	×
Nereis uncinula (Russel)	—	×	—
Platynereis cf. *dumerilii antipoda* Hartman	—	×	—

TABLE 15.12 *(continued)*

Species	Far north Qld.	Moreton Bay, Qld.	Sydney District, N.S.W.
Amphinomidae			
Eurythoë complanata (Pallas)	×	—	—
Eunicidae			
Diopatra dentata Kinberg	—	×	—
Eunice antennata Savigny	×	—	—
Marphysa sanguinea (Montagu)	—	×	×
Capitellidae			
Capitella capitata (Fabricius)	—	—	×
Notomastus hemipodus Hartman	—	×	×
Nephtyidae			
Nephtys australiensis Fauchald	—	×	×
Spionidae			
Polydora sp.	—	—	×
Scolecolepis indica Fauvel	×	—	—
Orbiniidae			
Haploscoloplos n.sp.	—	—	×
Oweniidae			
Owenia fusiformis Delle Chiaje	—	×	×
Phyllodocidae			
Phyllodoce malmgreni Gravier	×	×	—
Phyllodoce novaehollandiae Kinberg	—	—	×
Chaetopteridae			
Chaetopterus sp.	×	×	×
Mesochaetopterus minuta Potts	×	—	—
Sabellidae			
Laonome sp.	—	×	—
Salmacina sp.	×	—	—
Class Oligochaeta			
Pontodrilus bermudensis Beddard	×	—	—
Pontoscolex corethrurus (Bahl)	—	×	—
PHYLUM SIPUNCULA[1]			
Phascolosoma dunwichi Edmonds	—	×	—
Phascolosoma lurco Selenko and de Man	—	×	—
Themista lageniformis Baird (syn. *Dendrostomum signifer* Selenko and de Man)	—	×	—
PHYLUM ECHIURA[1]			
Ochetostoma australiense Edmonds	—	×	—

[1] Information supplied by Dr. S.J. Edmonds, Zoology Department, University of Adelaide.
The biogeographic data presented above were extracted from the following references: Campbell et al. (1974): Moreton Bay, Qld; Gillies (1951): North Stradbroke Is., Qld.; Hutchings and Recher (1974): Careel Bay, N.S.W.; W. Stephenson et al. (1958): Low Isles, Qld.

TABLE 15.13

Fish recorded in mangal vegetation in Australia

Species	N.T.	Far north Qld.	Southern Qld.	Sydney District, N.S.W.
Ambassidae				
Ambassis macleayi (Castelnau)	×	—	×	—
Ambassis marianus Günther	—	—	×	—
Atherinidae				
Melanotaenia nigrans (Richardson)	×	—	—	—
Batrachoididae				
Halophryne diemensis (Le Sueur)	×	—	—	—
Belonidae				
Tylosaurus macleayanus (Ogilby)	—	—	×	—
Blennidae				
Dasson variabilis (Cantor)	—	—	×	—
Petroscirtes anolius (Cuvier and Valenciennes)	—	—	×	—
Carangidae				
Caranx sexfasciatus Quoy and Gaimard	×	—	×	—
Chorinemus tolooparah (Rüppel)	—	—	×	—
Chaetodontidae				
Chaetodon flavirostris Günther	—	—	×	—
Chelmon rostratus (L.)	×	—	×	—
Heniochus acuminates (L.)	×	—	×	—
Clupeidae				
Harengula castelnau (Ogilby)	—	—	×	—
Hyperlophus vittatus (Castelnau)	—	—	×	—
Diodontidae				
Dicotylichthys myersi Ogilby	—	—	×	—
Diodon holocanthus L.	×	—	—	—
Diodon punctulatus (Kaup)	—	—	—	×
Eleotridae				
Parioglossus rainfordi McCulloch	—	×	—	—
Gerridae				
Gerres oyena (Forskål)	×	—	×	—
Girellidae				
Girella tricuspidata (Quoy and Gaimard)	—	—	×	—
Gobiidae				
Arenigobius frenatus (Günther)	—	—	—	×
Bathygobius kreffti (Steindachner)	—	—	—	×
Nesogobius pulchellus (Castelnau)	—	—	—	×
Hemirhamphidae				
Hemirhamphus regularis Günther	—	—	×	—
Zenarchopterus buffonis (Cuvier and Valenciennes)	×	—	—	—
Zenarchopterus gilli Smith	×	—	—	—

AUSTRALASIA

TABLE 15.13 *(continued)*

Species	N.T.	Far north Qld.	Southern Qld.	Sydney District, N.S.W.
Lutjanidae				
Lutjanus argentimaculatus (Forskål)	×	×	×	—
Lutjanus carponotatus (Richardson)	×	—	×	—
Lutjanus fulviflamma (Forskål)	×	×	—	—
Monodactylidae				
Monodactylus argenteus (L.)	×	—	×	—
Mugilidae				
Liza vaigiensis (Quoy and Gaimard)	×	×	×	—
Mugil cephalus L.	×	—	×	—
Mugil georgii Ogilby	—	—	×	—
Mullidae				
Parupeneus jeffi (Ogilby)	×	—	—	—
Pseudupeneus indicus (Shaw)	—	—	×	—
Pseudupeneus signatus (Gunther)	—	—	×	—
Upeneus tragula Richardson	×	—	—	—
Orectolobidae				
Hemiscyllium ocellatum (Bonaterre)	×	×	—	—
Periophthalmidae				
Periophthalmus australis Castelnau	×	×	—	—
Periophthalmus kalolo Lesson	×	×	—	—
Periophthalmus koelreuteri (Pallas)	—	×	—	—
Periophthalmus schlosseri (Pallas)	—	×	—	—
Plotosidae				
Cnidoglanis macrocephalus (Cuvier and Valenciennes)	—	—	×	—
Plotosus lineatus (Thunberg)	×	—	×	—
Pomacentridae				
Abudefduf sexfasciatus (Lacépède)	×	—	×	—
Pomadasyidae				
Plectorhinchus multivittatus (Macleay)	×	×	—	—
Plectorhinchus pictus (Thunberg)	×	×	—	—
Plectorhinchus sordidum (Klunzinger)	—	×	—	—
Plectorhinchus unicolor (Macleay)	×	×	—	—
Pomatomidae				
Pomatomus saltator (L.)	—	—	×	—
Scatophagidae				
Selenotoca multifasciata (Richardson)	×	—	—	—
Scorpaenidae				
Centropogon australis (Shaw)	—	—	×	—
Liocranium praepositum Ogilby	×	—	×	—
Notesthes robusta (Gunther)	—	—	×	—
Microcanthus strigatus (Cuvier and Valenciennes)	—	—	×	—
Serranidae				
Epinephelus gilberti (Richardson)	×	×	—	—

TABLE 15.13 (continued)

Species	N.T.	Far north Qld.	Southern Qld.	Sydney District, N.S.W.
Sillaginidae				
Sillago ciliata (Cuvier and Valenciennes)	×	—	×	—
Sillago maculata Quoy and Gaimard	×	—	×	—
Sparidae				
Acanthopagrus (= *Mylio*) *australis* (Gunther)	—	—	×	—
Rhabdosargus sarba (Forskål)	—	—	×	—
Sphyraenidae				
Sphyraena barracuda (Walbaum)	×	—	—	—
Sphyraena obtusata Cuvier and Valenciennes	—	—	×	—
Synodontidae				
Saurida gracilis (Quoy and Gaimard)	—	—	×	—
Teraponiidae				
Pelates quadrilineatus (Bloch)	—	—	×	—
Terapon jarbua (Forskål)	×	—	×	—
Tetraodontidae				
Arothron hispidus (L.)	—	—	×	—
Sphaeroides hamiltoni (Richardson)	×	—	×	×
Sphaeroides pleurostictus (Günther)	×	—	×	—

The biogeographic data presented above were extracted from the following references: Campbell et al. (1974): Moreton Bay, southern Qld.; Hutchings and Recher (1974): Careel Bay, N.S.W.; Shine et al. (1973): Tallebudgera Creek, southern Qld.; W. Stephenson et al. (1958): Low Isles, far north Qld.; Taylor (1964): Arnhem Land, N.T.

TABLE 15.14

Birds recorded in mangal and tidal salt–marsh vegetation in Australia

The biogeographic data presented below were extracted with the generous help of Dr. Jiro Kikkawa, of the Zoology Department, University of Queensland.
The continent has been divided into three major zones for the purpose of listing the avian fauna:
southeast (S.E.)   southeast Australia, north to the vicinity of Noosa, Qld.
northeast (N.E.)   from the vicinity of Noosa, Qld. to Cape York, Qld.
north and northwest (N. and N.W.)   Gulf of Carpentaria, Northern Territory, and Kimberley, W.A.
No specific attention has been given to birds in mangroves and salt marshes of the South Australian and southwest Western Australian coasts. In making these divisions it was recognised that birds, unlike intertidal animals, are a land fauna. Factors affecting their distribution are different from those affecting distribution of molluscs, crustacea, etc.
Those birds which are found only in mangals or salt marshes are listed as *exclusive* to their habitat. Those birds which utilize mangals or salt marshes as an integral part of their habitat, but which are not exclusively confined to mangals are listed as *associated*. Those birds which may be seen in mangals or salt marshes as migrants or visitors are listed as *visitors*. These birds are commonly known in other habitats.
The distribution of birds is tied up with the floristic composition of the mangal vegetation; the extensive mangrove forests of the north and northwest offer good food sources.
References consulted include Deignan (1964), Slater (1970, 1974), Hegerl and Timmins (1973), Hutchings (1973), Macdonald (1973), Hutchings and Recher (1974). Nomenclature follows Macdonald (1973).

Species	S.E.		N.E.		N. and N.W.	
	mangrove	salt marsh	mangrove	salt marsh	mangrove	salt marsh
Pelecanidae						
Pelecanus conspicillatus Temminck (pelican)	visitor		visitor		visitor	

AUSTRALASIA

TABLE 15.14 *(continued)*

Species	S.E.		N.E.		N. and N.W.	
	mangrove	salt marsh	mangrove	salt marsh	mangrove	salt marsh
Anhingidae						
Anhinga rufex (Daudin) (darter)	visitor		visitor		visitor	
Phalacrocoracidae						
Phalacrocorax carbo (L.) (black cormorant)	visitor		visitor			
Phalacrocorax melanoleucos (Vieillot) (little pied cormorant)	visitor		visitor		visitor	
Phalacrocorax sulcirostris (Brandt) (little black cormorant)	visitor		visitor		visitor	
Phalacrocorax varius (Gmelin) (pied cormorant)	visitor		visitor		visitor	
Ardeidae						
Ardea novaehollandiae Latham (white–faced heron)	associated		associated	associated	associated	
Ardea pacifica Latham (white–necked heron)	visitor		visitor	associated	visitor	
Ardea sumatrana Raffles (great–billed heron)			exclusive	associated	exclusive	
Ardeola ibis (L.) (cattle egret)	visitor		visitor	associated	visitor	
Botaurus poiciloptilus (Wagler) (brown bittern)	visitor					
Butorides striatus (L.) (mangrove heron)	exclusive		exclusive		exclusive	
Egretta alba (L.) (white egret)	visitor		associated	associated	associated	
Egretta garzetta (L.) (little egret)	visitor		associated	associated	associated	
Egretta intermedia (Wagler) (plumed egret)	visitor		associated	associated	associated	
Egretta sacra (Gmelin) (reef heron)	visitor		associated	associated	associated	
Hydranassa picata (Gould) (pied heron)			associated	associated	associated	
Ixobrychus flavicollis (Latham) (black bittern)	visitor		visitor	associated	visitor	
Ixobrychus minutus (L.) (little bittern)	visitor		visitor	associated		
Nycticorax caledonicus (Gmelin) (nankeen night–heron)	visitor		associated	associated	associated	
Ciconiidae						
Xenorhynchus asiaticus (Latham) (jabiru)	visitor		associated	associated	associated	
Threskiornithidae						
Platalea flavipes (Gould) (yellow–billed spoonbill)	visitor			associated	visitor	
Platalea regia Gould (royal spoonbill)	visitor		visitor	associated	visitor	
Plegadis falcinellus (L.) (glossy ibis)	associated			associated	associated	
Threskiornis molucca (Cuvier) (white ibis)	associated		associated	associated	associated	
Threskiornis spinicollis (Jameson) (straw–necked ibis)	visitor		visitor	associated	visitor	
Anatidae						
Anas gibberifrons Muller (grey teal)	visitor		visitor	associated	visitor	
Anas superciliosa (Gmelin) (black duck)	visitor		visitor	associated	visitor	
Anseranus semipalmata (Latham) (magpie goose)			visitor	associated	visitor	
Dendrocygna arcuata (Horsfield) (water whistle duck)			visitor	associated	visitor	
Tadorna radjah (Garnet) (Burdekin duck)			associated	associated	associated	
Accipitridae						
Accipiter cirrocephalus (Vieillot) (collared sparrowhawk)	visitor		visitor		visitor	
Accipiter fasciatus (Vig. and Horsf.) (brown goshawk)	visitor		visitor		visitor	

TABLE 15.14 *(continued)*

Species	S.E.		N.E.		N. and N.W.	
	mangrove	salt marsh	mangrove	salt marsh	mangrove	salt marsh
Accipiter novaehollandiae (Gmelin) (grey goshawk)			visitor		visitor	
Circus approximans Peale (swamp harrier)		associated		associated		associated
Elanus notatus Gould (black–shouldered kite)	visitor		visitor		visitor	
Elanus scriptus Gould (letter–winged kite)					visitor (Karumba)	associated
Haliaetus leucogaster (Gmelin) (white–breasted sea eagle)	associated		associated		associated	
Haliastur indus (Boddaert) (Brahminy kite)	associated		associated		associated	
Haliastur sphenurus (Vieillot) (whistling kite)	visitor		visitor		visitor	
Milvus migrans (Boddaert) (black kite)	visitor		visitor	associated	visitor	
Pandionidae						
Pandion haliaetus (L.) (osprey)	associated		associated		associated	
Falconidae						
Falco cenchroides (Vig. and Horsf.) (nankeen kestrel)	visitor		visitor	associated	visitor	
Falco peregrinus Tunstall (peregrine falcon)	visitor		visitor	associated	visitor	
Megapodiidae						
Megapodius freycinet Gaimard (scrub fowl)			associated		associated	
Gruidae						
Grus antigone (L.) (Sarus crane)				associated	visitor	
Grus rubicunda (Perry) (brolga)	visitor		visitor	associated	visitor	
Rallidae						
Eulabeornis castaneoventris Gould (chesnut rail)				associated	exclusive	
Fulica atra L. (coot)		visitor		visitor		visitor
Gallinula porphyrio (L.) (swamphen)	visitor		visitor	associated	visitor	
Gallinula tenebrosa Gould (dusky moorhen)	visitor		visitor	associated		
Porzana cinerea (Vieillot) (whitebrowed crake)			visitor	associated	visitor	
Rallus phillipensis L. (banded landrail)	associated		associated	associated	associated	
Haematopodidae						
Haematopus fuliginosus Gould (sooty oyster–catcher)	associated		associated	associated	associated	
Haematopus ostralegus L. (pied oyster–catcher)			associated	associated	associated	
Charadriidae						
Charadrius ruficapillus Temminck (red–capped dotterel)	visitor	associated	visitor		visitor	
Charadrius asiaticus veredus Gould (oriental dotterel)	visitor	visitor	visitor		visitor	
Charadrius bicinctus Jard. and Selb. (double–banded dotterel)	visitor	visitor				
Charadrius cinctus (Gould) (red–kneed dotterel)	visitor	associated	visitor		visitor	
Charadrius leschenaultii Lesson (large dotterel)	visitor	visitor	visitor		visitor	
Charadrius melanops Vieillot (black–fronted dotterel)	visitor	associated	visitor		visitor	
Charadrius mongolus Pallas (Mongolian dotterel)	visitor	visitor	visitor		visitor	
Pluvialis dominica (Muller) (golden plover)	visitor	visitor	visitor		visitor	

AUSTRALASIA

TABLE 15.14 *(continued)*

Species	S.E.		N.E.		N. and N.W.	
	mangrove	salt marsh	mangrove	salt marsh	mangrove	salt marsh
Pluvialis squatarola (L.) (grey plover)	visitor	associated	visitor		visitor	
Vanellus miles (Boddaert) (masked plover)	associated	visitor	associated		visitor	
Vanellus novaehollandiae Stephens (spur–winged plover)	visitor	associated				
Arenariidae						
Arenaria interpres (L.) (ruddy turnstone)	visitor	visitor	visitor		visitor	
Scolopacidae						
Calidris acuminata (Horsfield) (sharp–tailed sandpiper)	visitor	associated	visitor		visitor	
Calidris canutus (L.) (knot)		associated	visitor		visitor	
Calidris ferruginea (Brunnich) (curlew sandpiper)	visitor	visitor	visitor		visitor	
Calidris ruficollis (Pallas) (red–necked stint)	visitor	visitor	visitor		visitor	
Calidris tenuirostris (Horsfield) (grey knot)		visitor	visitor		visitor	
Gallinago hardwickii (Gray) (Japanese snipe)	visitor	associated				
Limicola falcinellus (Pontoppidan) (broad–billed sandpiper)		visitor	visitor		visitor	
Limosa lapponica (L.) (bartailed godwit)	associated	visitor	associated		associated	
Numenius madagascariensis (L.) (eastern curlew)	visitor	associated	visitor		visitor	
Numenius minutus Gould (little whimbrel)	visitor	associated	associated		associated	
Numenius phaeopus (L.) (whimbrel)	associated	associated	associated		associated	
Tringa brevipes (Vieillot) (grey–tailed tattler)	visitor	visitor	visitor		visitor	
Tringa cinerea (Guldenstaedt) (Terek sandpiper)	visitor	associated	visitor		visitor	
Tringa glareola L. (wood sandpiper)		associated			visitor	
Tringa hypoleucos L. (common sandpiper)	visitor	associated	visitor		visitor	
Tringa nebularia (Gunnerus) (greenshank)	visitor	associated	visitor		visitor	
Recurvirostridae						
Himantopus himantopus (L.) (black–winged stilt–pied stilt)	visitor	associated	visitor		visitor	
Recurvirostra novaehollandiae Vieillot (red–necked avocet)	visitor	associated				
Burhinidae						
Burhinus magnirostris (Latham) (southern stone–curlew)	visitor	associated	visitor		visitor	
Esacus magnirostris (Vieillot) (beach stone–curlew)	associated	associated	associated		associated	
Laridae						
Anous tenuirostris (Temminck) (lesser noddy)					nests in mangroves in Albrolhos Is.	
Chlidonias (*Sterna hybrida*) Pallas (whiskered tern)	visitor	associated	visitor		visitor	
Larus novaehollandiae Stephens (silver gull)	visitor	visitor	visitor		visitor	
Sterna albifrons Pallas (little tern)	visitor		visitor		visitor	
Sterna bergii Lichtenstein (crested tern)	visitor		visitor		visitor	
Sterna beugalensis Lesson (lesser–crested tern)	uncommon visitor		visitor		visitor	
Sterna caspia Pallas (Caspian tern)	visitor		visitor		visitor	
Sterna hirundo L. (common tern)	visitor					

TABLE 15.14 (continued)

Species	S.E.		N.E.		N. and N.W.	
	mangrove	salt marsh	mangrove	salt marsh	mangrove	salt marsh
Sterna leucoptera Temminck (white–winged black tern)	visitor	associated	visitor		visitor	
Sterna nilotica Gmelin (gull–billed tern)	visitor	associated	visitor		visitor	
Columbidae						
Chalcophaps indica (L.) (green–winged pigeon)			visitor		visitor	
Ducula spilorrhoa (Gray) (Torres Strait pigeon)			associated		associated	
Geopelia humeralis (Temminck) (bar–shouldered dove)	associated		associated		associated	
Geopelia striata (L.) (peaceful dove)			visitor		visitor	
Macropygia phasianella (Temminck) (brown pigeon)			visitor		visitor	
Ptilinopus regina Swainson (red–crowned pigeon)			associated		associated	
Ptilinopus superbus (Temminck) (purple–crowned pigeon)			associated		associated	
Psittacidae						
Aprosmictus erythropterus (Gmelin) (red–winged parrot)					visitor	
Cacatua galerita (Latham) (sulphur–crested cockatoo)			visitor		visitor	
Pezoporus wallicus (Kerr) (ground parrot)		associated				
Platycercus adscitus (Latham) (pale–headed rosella)	visitor		visitor			
Platycercus venustus (Kuhl) (northern rosella)					visitor	
Probosciger aterrimus (Gmelin) (palm cockatoo)			visitor			
Psephotus chrysopterygius Gould (golden–shouldered parrot)					visitor	
Trichoglossus moluccanus (Gmelin) (rainbow lorikeet)	visitor		visitor			
Cuculidae						
Cacomantis castaneiventris (Gould) (chesnut–breasted cuckoo)			visitor			
Cacomantis pyrrophanus (Vieillot) (fan–tailed cuckoo)	visitor		visitor			
Cacomantis variolosus (Vig. and Horsf.) (brush cuckoo)			associated		associated	
Centropus phasianus (Latham) (pheasant coucal)	visitor	visitor	visitor		visitor	
Chrysococcyx lucidus (Gmelin) *plagosus* (golden bronze cuckoo)	associated		associated			
Chrysococcyx malayanus (Raffles) *russatus* (rufous breasted bronze cuckoo)			associated			
Eudynamys scolopacea L. (koel)	visitor		visitor		visitor	
Caprimulgidae						
Caprimulgus macrurus Horsfield (white–tailed nightjar)			visitor		visitor	
Eurostopodus guttatus (Vig. and Horsf.) (spotted nightjar)					visitor	

AUSTRALASIA

TABLE 15.14 *(continued)*

Species	S.E. mangrove	S.E. salt marsh	N.E. mangrove	N.E. salt marsh	N. and N.W. mangrove	N. and N.W. salt marsh
Strigidae						
Ninox connivens (Latham) (barking owl)			visitor		visitor	
Podargidae						
Podargus papuensis Quoy and Gaim. (Papuan frogmouth)			visitor			
Tytonidae						
Tyto longimembris (Jerdon) (grass owl)				visitor		
Apopidae						
Apus pacificus (Latham) (fork–tailed swift)	visitor	visitor	visitor		visitor	
Chaetura caudacutus (Latham) (spine–tailed swift)		visitor	visitor			
Collocalia spodiopygia (Peale) (grey swiftlet)		visitor	visitor			
Alcedinidae						
Ceryx azureus (Latham) (azure kingfisher)	visitor		associated		associated	
Ceryx pusillus (Temminck) (little kingfisher)			associated		associated	
Dacelo gigas (Boddaert) (laughing kookaburra)	visitor					
Dacelo leachii Vig. and Horsf. (blue–winged kookaburra)			visitor		visitor	
Halcyon chloris (Boddaert) (mangrove kingfisher)	exclusive		exclusive		exclusive	
Halcyon macleayi Jard. and Selb. (forest kingfisher)	visitor		visitor		visitor	
Halcyon sancta Vig. and Horsf. (sacred kingfisher)	visitor		visitor		visitor	
Halcyon torotoro (Lesson) (yellow–billed kingfisher)			visitor			
Tanysiptera sylvia Gould (white–tailed kingfisher)			visitor			
Meropidae						
Meropus ornatus Latham (rainbow bird)	visitor	visitor	visitor		visitor	
Hirundinidae						
Hirundo neoxena Gould (welcome swallow)	visitor	visitor	visitor			
Petrochelidon ariel (Gould) (fairy martin)	visitor	visitor	visitor		visitor	
Petrochelidon nigricans (Vieillot) (tree martin)	visitor		visitor		visitor	
Ptittidae						
Pitta iris Gould (rainbow pitta)			visitor			
Motacillidae						
Anthus novaeseelandiae (Gmelin) (Australian pipit)		associated		associated		associated
Motacilla cinerea (Tunstall) (grey wagtail)				visitor		
Motacilla flava L. (yellow wagtail)				visitor		
Grallinidae						
Coracina lineata (Swainson) (barred cuckoo shrike)			visitor			
Coracina novaehollandiae (Gmelin) (black–faced cuckoo shrike)	visitor		visitor		visitor	

TABLE 15.14 *(continued)*

Species	S.E.		N.E.		N. and N.W.	
	mangrove	salt marsh	mangrove	salt marsh	mangrove	salt marsh
Coracina papuensis (Gmelin) (little cuckoo shrike)	visitor		visitor		visitor	
Coracina tenuirostris (Jardine) (Cicada bird — Jardine triller)			visitor		visitor	
Grallina cyanoleuca (Latham) (magpie lark)	visitor	visitor	visitor		visitor	
Lalage leucomela (Vig. and Horsf.) (varied triller)	associated		associated		associated	
Lalage sueurii (Vieillot) (white–winged triller)			visitor		visitor	
Timaliidae						
Psophodes olivaceus Latham (eastern whipbird)	visitor					
Maluridae						
Malurus amabilis Gould (lovely wren)		visitor	associated			
Malurus cyaneus (Latham) (superb blue wren)	visitor					
Malurus melanocephalus (Latham) (red–backed wren)	visitor	associated	visitor		visitor	
Stipiturus malachurus (Shaw) (southern emu–wren)	visitor	associated				
Sylviidae						
Acrocephalus australis (Gould) (reed warbler)		associated		associated		associated
Cisticola exilis (Vig. and Horsf.) (golden–headed cisticola)		associated		associated		associated
Megalurus timoriensis Wallace (tawny grassbird)		visitor		visitor		visitor
Acanthizidae						
Gerygone chloronata Gould (green–backed warbler)					exclusive	
Gerygone flavida Ramsay (fairy warbler)			associated			
Gerygone laevigaster Gould (mangrove warbler)	exclusive				exclusive	
Gerygone magnirostris Gould (large–billed warbler)			associated		associated	
Gerygone olivacea (Gould) (white–throated warbler)	visitor		visitor		visitor	
Gerygone palpebrosa Wallace (black–throated warbler)			associated			
Gerygone tenebrosa (Hall) (dusky warbler)					associated	
Muscicapidae						
Arses telescopthalmus lorealis Devis (frill–necked flycatcher)			visitor			
Machaerirhynchus flaviventer Gould (boat–billed flycatcher)			visitor			
Microeca flavigaster Gould (lemon–breasted flycatcher)			associated		associated	
Microeca griseoceps Devis (yellow flycatcher)			visitor			
Monarcha frater Sclater (black–winged flycatcher)			visitor			
Monarcha leucotis Gould (pied flycatcher)			visitor			
Monarcha melanopsis (Vieillot) (black–faced flycatcher)	visitor		visitor			
Monarcha trivirgata (Temminck) (spectacled flycatcher)	visitor		associated			

AUSTRALASIA

TABLE 15.14 *(continued)*

Species	S.E. mangrove	S.E. salt marsh	N.E. mangrove	N.E. salt marsh	N. and N.W. mangrove	N. and N.W. salt marsh
Myiagra alecto (Temm. and Lang.) (shining flycatcher)			associated	visitor	associated	visitor
Myiagra rubecula (Latham) (leaden flycatcher)	associated		associated		associated	
Myiagra ruficollis (Vieillot) (broad–billed flycatcher)			associated		associated	
Peneoenanthe pulverulenta (Bonaparte) (mangrove robin)			exclusive		exclusive	
Poecilodryas cerviniventris (Gould) (buff–sided robin)					exclusive	
Poecilodryas superciliosa (Gould) (white–browed robin)			associated			
Rhipidura fuliginosa (Sparrman) (grey fantail)	visitor		visitor		visitor	
Rhipidura leucophrys (Latham) (willy wagtail)		associated		associated		associated
Rhipidura rufifrons (Latham) (rufous fantail)	visitor		visitor		visitor	
Rhipidura rufiventris (Vieillot) (northern fantail)			visitor		visitor	
Pachycephalidae						
Colluricincla brunnea Gould (brown shrike thrush)					visitor	
Colluricincla harmonica Latham (grey shrike thrush)	visitor		visitor		visitor	
Colluricincla megarhyncha Quoy and Gaim. (rufous shrike thrush)			visitor			
Colluricincla parvula Gould (little shrike thrush)					associated	
Dicaeum hirundinaceum (Shaw and Nodder) ((mistletoe bird)			associated		associated	
Pachycephala griseiceps Gray (grey whistler)			visitor			
Pachycephala lanioides Gould (white–breasted whistler)					exclusive	
Pachycephala melanura Gould (mangrove golden whistler)			exclusive		exclusive	
Pachycephala rufiventris (Latham) (rufous whistler)	visitor		visitor		visitor	
Pachycephala simplex Gould (brown whistler)					exclusive	
Nectariniidae						
Nectarinia jugularis (L.) (yellow–breasted sunbird)			associated			
Zosteropidae						
Zosterops lateralis (Latham) (grey–breasted silvereye)	visitor					
Zosterops lutea Gould (yellow silvereye)			exclusive (Barrata Creek only)		exclusive	
Meliphagidae						
Acanthorhynchus tenuirostris (Latham) (eastern spinebill)	visitor					
Acridotheres tristis L. (Indian mynah)			visitor			
Aegintha temporalis (Latham) (red–browed finch)	associated	associated	associated			
Aplonis metallica (Temm. and Laug.) (shining starling)			visitor			

TABLE 15.14 *(continued)*

Species	S.E.		N.E.		N. and N.W.	
	mangrove	salt marsh	mangrove	salt marsh	mangrove	salt marsh
Conopophila albogularis (Gould) (rufous–banded honey–eater)					exclusive	
Entomyzon cyanotis (Latham) (blue–faced honey–eater)	visitor		visitor		visitor	
Lichmera indistincta (Vig. and Horsf.) (brown honey–eater)	associated	visitor	associated		associated	
Lonchura castaneothorax (Gould) (chestnut–breasted finch)		visitor		visitor		visitor
Lonchura flaviprymna (Gould) (yellow rumped finch)				visitor		visitor
Meliphaga fasciogularis (Gould) (mangrove honey–eater)	exclusive					
Meliphaga flava (Gould) (yellow honey–eater)			visitor			
Meliphaga flaviventer (Lesson) (tawny–breasted honey–eater)			visitor			
Meliphaga gracilis (Gould) (graceful honey–eater)			associated			
Meliphaga lewinii (Swainson) (Lewin honey–eater)	visitor					
Meliphaga notata (Gould) (lesser Lewin honey–eater)			visitor			
Meliphaga versicolor (Gould) (varied honey–eater)			exclusive			
Melithreptus albogularis Gould (white–throated honey–eater)	visitor		visitor		visitor	
Melithreptus laetior (Gould) (golden–backed honey–eater)					visitor	
Myzomela erythrocephala Gould (red–headed honey–eater)				visitor	associated	
Myzomela obscura Gould (dusky honey –eater)			associated	visitor	associated	
Philemon argenticeps (Gould) (silver–crowned friar bird)					associated	
Philemon citreogularis (Gould) (little friar bird)	visitor	visitor	visitor		visitor	
Philemon corniculatus (Latham) (noisy friar bird)	visitor					
Philemon gordoni Mathews (Melville Island friar bird)					associated	
Philemon novaeguineae (Muller) (helmeted friar bird)			associated			
Ramsayornis fasciatus (Gould) (bar–breasted honey–eater)					visitor	
Ramsayornis modestus (Gray) (brown–backed honey–eater)		visitor	associated			
Oriolidae						
Oriolus flavocinctus (Vigors) (yellow oriole)			associated		associated	
Oriolus sagittatus (Latham) (olive–backed oriole)	associated		visitor		visitor	
Sphecotheres flaviventris Gould (yellow figbird)			visitor		visitor	
Sphecotheres vieilloti Vig. and Horsf. (southern figbird)	visitor					
Dicruridae						
Dicrurus bracteatus Gould (spangled drongo)	visitor		associated	visitor	associated	visitor

AUSTRALASIA

TABLE 15.14 *(continued)*

Species	S.E.		N.E.		N. and N.W.	
	mangrove	salt marsh	mangrove	salt marsh	mangrove	salt marsh
Artamidae						
Artamus leucorhynchus (L.) (white–breasted wood swallow)	associated	associated	associated		associated	
Cracticidae						
Cracticus mentalis Salv. and d'Alb. (black–backed butcher bird)		visitor	visitor		visitor	
Cracticus nigrogularis (Gould) (pied butcher bird)	visitor	visitor	visitor		visitor	
Cracticus quoyi (Lesson) (black butcher bird)		visitor	associated		associated	
Cracticus torquatus (Latham) (grey butcher bird)	visitor	visitor			visitor	
Streptera graculina (White) (pied currawong)	visitor					
Ptilonorhynchidae						
Ailuroedus melanotus (Gray) (spotted catbird)			visitor			
Chlamydera cerviniventris Gould (fawn–breasted bower bird)			visitor			
Chlamydera nuchalis (Jard. and Selb.) (great bower bird)		visitor	visitor		visitor	
Paradisaeidae						
Phonygammus keraudrenii (Less. and Garn.) (manucode)			visitor			
Ptiloris magnificus (Vieillot) (magnificent rifle bird)			visitor			
Ptiloris victoriae Gould (Victoria rifle bird)			visitor			
Corvidae						
Corvus cecilae Mathews (Australian crow)	visitor	visitor	visitor		visitor	

REFERENCES

A comprehensive bibliography of the subject follows, including all publications cited in the text, but not confined to them.

Abe, N., 1942. Ecological observations on *Melaraphe (Littorinopsis) scabra* (Linnaeus) inhabiting the mangrove tree. *Palaeotrop. Biol. Station Stud.*, 2: 391—435.

Albani, A.D., 1973. A hydrological investigation of Brisbane Water, N.S.W. *Aust. J. Mar. Freshwater Res.*, 24: 127—141.

Allan, J., 1959. *Australian Shells*. Georgian House, Melbourne, 487 pp.

Anonymous, 1972. Mangroves and Man. *Aust. Conserv. Found. Viewpoint*, No. 7: 12 pp.

Atkinson, M.R., Findlay, G.P., Hope, A.B., Pitman, M.G., Saddler, H.D. and West, K.R., 1967. Salt regulation in the mangroves *Rhizophora mucronata* Lam. and *Aegialitis annulata* R.Br. *Aust. J. Biol. Sci.*, 20: 589—599.

Attiwill, P.M. and Clough, B.F., 1974. *The Role of the Mangrove and Sea–Grass Communities in Nutrient Cycling in Westernport Bay*. Unpublished manuscript.

Australian Water Resources Council, 1965. *Review of Australia's Water Resources (Stream Flow and Underground Resources) 1963*. Department of National Development, Canberra, A.C.T., 107 pp.

Baas Becking, L.G.M., 1959. Some aspects of the ecology of Lake Macquarie, N.S.W., with regard to an alleged depletion of fish. III. Characteristics of water and mud. *Aust. J. Mar. Freshwater Res.*, 10: 297—303.

Baas Becking, L.G.M. and Wood, E.J.F., 1955. Biological processes in the estuarine environment. *Proc. Aust. Acad. Sci.*, B, 58: 160; 59: 398.

Baas Becking, L.G.M., Thomson, J.M. and Wood, E.J.F., 1959. Some aspects of the ecology of Lake Macquarie, N.S.W., with regard to an alleged depletion of fish. I. General introduction. *Aust. J. Mar. Freshwater Res.*, 10: 269—270.

Baker, R.T., 1915. The Australian grey mangrove. *J. Proc. R. Soc. N.S.W.*, 49: 257—288.

Baldwin, J.G. and Crocker, R.L., 1941. The soils and vegetation of portion of Kangaroo Island, South Australia. *Trans. R. Soc. S. Aust.*, 65: 263—275.

Beard, J.S., 1967a. An inland occurrence of mangroves. *West. Aust. Nat.*, 10: 112—115.

Beard, J.S., 1967b. Some vegetation types of tropical Australia in relation to those of Africa and America. *J. Ecol.*, 55: 271—290.

Beard, J.S. (Editor), 1970. *West Australian Plants*. Society for Growing Australian Plants and King's Park Board, Perth, 2nd ed., 142 pp.

Bennett, I., 1968. The mud lobster. *Aust. Nat. Hist.*, 16(1): 22—25.

Bennett, I. and Pope, E.C., 1953. Intertidal zonation of the exposed rocky shores of Victoria, together with a rearrangement of the biogeographical provinces of temperate Australian shores. *Aust. J. Mar. Freshwater Res.*, 4: 105—159.

Berry, A.J., 1963. Faunal zonation in mangrove swamps. *Singapore Natl. Mus. Bull.*, 32: 90—98.

Bibby, P., 1958. Some lichens collected in Arnhem Land. In: R.L. Specht and C.P. Mountford (Editors), *Records of the American—Australian Scientific Expedition to Arnhem Land. 3. Botany and Plant Ecology*. Melbourne University Press, Melbourne, Vic., p. 169.

Billard, B., 1972. *Microelectrode and Current Clamping Investigations Into Salt and Water Flow Through the Salt Glands of Aegiceras corniculatum*. Thesis, University of Queensland, St. Lucia, Qld., 134 pp.

Billard, B. and Field, C.D., 1974. Electrical properties of the salt gland of *Aegiceras*. *Planta*, 115: 285—296.

Bird, E.C.F., 1971a. Mangroves as land—builders. *Vic. Nat.*, 88: 189—197.

Bird, E.C.F., 1971b. The conservation of mangroves in North Queensland. *Qld. Nat.*, 39(156): 6—8.

Bird, E.C.F., 1972a. Mangroves on the Australian coast. *Aust. Nat. Hist.*, 17: 167—171.

Bird, E.C.F., 1972b. *An Introduction to Systematic Geomorphology, 4. Coasts*. Australian National University Press, Canberra, A.C.T., 246 pp.

Bird, E.C.F., 1973. Mangroves and coastal morphology in Cairns Bay, North Queensland. *J. Trop. Geogr.*, 35: 11—16.

Black, J.M., 1943—57. *Flora of South Australia, I—IV*. Government Printer, Adelaide, S.A., 2nd ed., 1008 pp.

Blackburn, G., 1952. The soils of the Kingston–Avenue Drainage Area, South Australia. *CSIRO, Div. Soils, Soils Land Use Ser.*, No. 7: 36 pp.

Blake, S.T., 1940. The vegetation of Goat Island and Bird Island in Moreton Bay. *Qld. Nat.*, 11: 94—101.

Blake, S.T., 1947. The vegetation of Noosa. *Qld. Nat.*, 13: 47—50.

Blake, S.T., 1968. The plants and plant communities of Fraser, Moreton and Stradbroke Islands. *Qld. Nat.*, 19: 23—30.

Blake, S.T., 1969. A revision of *Carpobrotus* and *Sarcozona* in Australia, genera allied to *Mesembryanthemum* (Aizoaceae). *Contrib. Qld. Herb.*, No. 7: 65 pp.

Boomsma, C.D., 1946. The vegetation of the Southern Flinders Ranges, South Australia. *Trans. R. Soc. S. Aust.*, 70: 259—276.

Bostrom, T.E. and Field, C.D., 1973. Electrical potentials in the salt gland of *Aegiceras*. In: W.P. Anderson (Editor), *Ion Transport in Plants*. Academic Press, London, pp. 385—392.

Bridgewater, P.B., 1975. Peripheral vegetation of Westernport Bay. *Proc. R. Soc. Vic.*, 87: 69—78.

Brünnich, J.C. and Smith, F., 1911. Some Queensland mangrove barks and other tanning materials. *Qld. Agric. J.*, 27: 86—94.

Bullock, D.A., 1975. The general water circulation of Spencer Gulf, South Australia, in the period February to May. *Trans. R. Soc. S. Aust.*, 99: 43—53.

Burbidge, N.T., 1960. The phytogeography of the Australian region. *Aust. J. Bot.*, 8: 75—211.

Butler, A.J., Depers, A.M., McKillup, S.C. and Thomas, D.P., 1975. *The Conservation of Mangrove Swamps in South Australia*. A report to the Nature Conservation Society of S. Aust. Dep. Zool., University of Adelaide, S.A., 76 pp.

Cameron, A.M., 1966. Some aspects of behaviour of the soldier crab *Mictyris longicarpus*. *Pac. Sci.*, 20: 224—234.

Campbell, B., Wallace, C. and King, H., 1974. *Field Study of Marine Littoral Invertebrate Macrofauna From the Proposed Brisbane Airport Extension Area*. Queensland Museum, Brisbane, Qld., 49 pp.

Cardale, S. and Field, C.D., 1971. The structure of the salt gland of *Aegiceras corniculatum*. *Planta*, 99: 183—191.

Cardale, S. and Field, C.D., 1975. Ion transport in the leaf of *Aegiceras corniculatum*. *Proc. Int. Symp. Biol. Manage. Mangroves, Hawaii*, 2: 608—614.

Carey, G. and Fraser, L., 1932. The embryology and seedling development of *Aegiceras majus* Gaertn. *Proc. Linn. Soc. N.S.W.*, 57: 341—360.

Carrodus, B.B., Specht, R.L. and Jackman, M.E., 1965. The vegetation of Koonamore Station, South Australia. *Trans. R. Soc. S. Aust.*, 89: 41—57.

Chapman, V.J., 1952. Problems in ecological terminology. *Rep. Aust. Assoc. Adv. Sci.*, 29: 259—279.

Chapman, V.J., 1970. Mangrove phytosociology. *Trop. Ecol.*, 11: 1—19.

Chapman, V.J., 1974. *Salt Marshes and Salt Deserts of the World*. Cramer, Lehre, 2nd ed., 392 pp. (complemented with 102 pp.).

Chapman, V.J., 1975. *Mangrove Vegetation*. Cramer, Lehre, 425 pp.

Chippendale, G.M., 1972. Check list of Northern Territory plants. *Proc. Linn. Soc. N.S.W.*, 96: 207—267.

Christian, C.S. and Stewart, G.A., 1952. General report on a survey of Katherine–Darwin region, 1946. *CSIRO, Land Res. Ser.*, No. 1: 156 pp.

Churchill, D.M., 1973. The ecological significance of tropical mangroves in the Early Tertiary floras of southern Australia. *Geol. Soc. Aust., Spec. Publ.*, 4: 79—86.

Churchill, D.M. and De Corona, A., 1972. *The Distribution of Victorian Plants*. The Dominion Press, North Blackburn, Vic., 130 pp.

Clarke, L.D. and Hannon, N.J., 1967. The mangrove swamp and salt–marsh communities of the Sydney District. I. Vegetation, soils and climate. *J. Ecol.*, 55: 753—771.

Clarke, L.D. and Hannon, N.J., 1969. The mangrove swamp and salt–marsh communities of the Sydney District. II. The holocoenotic complex with particular reference to physiography. *J. Ecol.*, 57: 213—234.

Clarke, L.D. and Hannon, N.J., 1970. The mangrove swamp and salt–marsh communities of the Sydney District.

III. Plant growth in relation to salinity and waterlogging. *J. Ecol.*, 58: 351—369.

Clarke, L.D. and Hannon, N.J., 1971. The mangrove swamp and salt–marsh communities of the Sydney District. IV. The significance of species interaction. *J. Ecol.*, 59: 535—553.

Clifford, H.T. and Specht, R.L., 1976. *The Vegetation of North Stradbroke Island, Queensland (With Notes on the Fauna of Mangrove and Marine Meadow Ecosystems by M.M. Specht)*. University of Queensland Press, St. Lucia, Qld., in press.

Clough, B.F. and Attiwill, P.A., 1976. Nutrient cycling in a community of *Avicennia marina* in a temperate region of Australia. *Proc. Int. Symp. Biol. Manage. Mangroves, Hawaii*, 1: 137—146.

Coaldrake, J.E., 1961. The ecosystem of the coastal lowlands ("wallum") of southern Queensland. *CSIRO Bull.*, No. 283: 138 pp.

Coleman, F., 1972. Frequencies, tracks and intensities of tropical cyclones in the Australian region 1909—1969. *Commonw. Aust., Bur. Meteorol., Meteorol. Summ.*, 42 pp.

Coleman, J., Gagliano, S.M. and Smith, W.G., 1966. Chemical and physical weathering on saline high tidal flats, North Queensland, Australia. *Bull. Geol. Soc. Am.*, 77: 205—206.

Collins, M.I., 1921. On the mangrove and salt–marsh vegetation near Sydney, N.S.W. with special reference to Cabbage Tree Creek, Port Hacking. *Proc. Linn. Soc. N.S.W.*, 46: 376—392.

Commonwealth Australia, Bureau of Meteorology, 1956. *Climatic Averages in Australia. Temperature, Relative Humidity, Rainfall*. Department of Supply, Maribyrnong, Vic., 107 pp.

Connor, D.J., 1969. Growth of grey mangrove (*Avicennia marina*) in nutrient culture. *Biotropica*, 1: 36—40.

Cotton, B.C., 1950. An old mangrove mud–flat exposed by wave scouring at Glenelg, South Australia. *Trans. R. Soc. S. Aust.*, 73: 59—61.

Cotton, B.C., 1964. Molluscs of Arnhem Land. In: R.L. Specht (Editor), *Records of the American—Australian Scientific Expedition to Arnhem Land, 4. Zoology*. Melbourne University Press, Melbourne, Vic., pp. 9—43.

Cribb, A.B., 1956. Records of marine algae from south–eastern Queensland. II. *Polysiphonia* and *Lophosiphonia*. *Univ. Qld. Pap., Dep. Bot.*, 3(16): 131—147.

Cribb, A.B., 1958. Records of marine algae from south–eastern Queensland. III. *Laurencia* Lamx. *Univ. Qld. Pap., Dep. Bot.*, 3(19): 159—191.

Curtis, W.M. and Somerville, J., 1947. Boomer Marsh — a preliminary botanical and historical survey. *Proc. R. Soc. Tasm.*, 1947: 151—157.

Dakin, W.J., 1960. *Australian Seashores*. Angus and Robertson, Sydney, N.S.W., 372 pp.

Deignan, H.G., 1964. Birds of the Arnhem Land Expedition. In: R.L. Specht (Editor), *Records of the American—Australian Scientific Expedition to Arnhem Land, 4. Zoology*. Melbourne University Press, Melbourne, Vic., pp. 345—425.

Dick, R.S., 1971. Vegetation mapping at large scales. *Qld. Geogr. J.*, 3rd Ser., 1: 65—79.

Dick, R.S., 1972. A map of the climates of Australia: according to Köppen's principles of definition. *Qld. Geogr. J.*, 2(3) (in press).

Ding Hou, 1957. A conspectus of the genus *Bruguiera* (Rhizophoraceae). *Nova Guinea*, 8: 163—171.

Easton, A.K., 1970. Tides of the continent of Australia. *Horace Lamb Centre, Flinders Univ., S. Aust., Res. Pap.*, 37: 1—326.

Eichler, H., 1965. *Supplement to J.M. Black's Flora of South Australia*. Government Printer, Adelaide, S.A., 385 pp.

Enright, J., 1969. *Processes and Patterns of Coastal Change in Westernport Bay*. Thesis, University of Melbourne, Melbourne, Vic.

Enright, J., 1973. Mangrove shores in Westernport Bay. *Vic. Resourc.*, 15: 12—15.

Fairbridge, R.W. and Teichert, C., 1947. The rampart system of Low Isles 1928—45. *Rep. Great Barrier Reef Comm.*, 6: 1—16.

Fenner, C. and Cleland, J.B., 1935. The geography and botany of the Adelaide coast. *Field Nat. Sect. R. Soc. S. Aust. Publ.*, No. 3.

Field, C.D., 1969. Ion transport in mangroves. *Third Int. Biophys. Congr., Abstr.*, p. 91.

Fischer, P.H., 1940. Notes sur les peuplements littoraux d'Australie. III. Sur la faune de la mangrove australienne. *Mém. Soc. Biogéogr., Paris*, 7: 317—329.

Fitzpatrick, E.A., 1963. Estimates of pan evaporation from mean maximum temperature and vapor pressure. *J. Appl. Meteorol.*, 2: 780—792.

Foley, J.C., 1945. Frost in the Australian region. *Commonw. Aust., Bur. Meteorol. Bull.*, No. 32: 142 pp.

Fosberg, F.R., 1961. Vegetation–free zone on dry mangrove coasts. *U.S. Geol. Surv., Prof. Pap.*, 424D: 216—218.

Frankenberg, J., 1971. *Nature Conservation in Victoria*. Victorian National Parks Association and Halstead Press, Sydney, N.S.W., 145 pp.

Gardner, C.A., 1923. Botanical Notes. Kimberley Division of Western Australia. *W. Aust. For. Dep. Bull.*, No. 32: 105 pp.

Gardner, C.A., 1942. The vegetation of Western Australia with special reference to the climate and soils. *J. R. Soc. W. Aust.*, 28: 12—87.

Gentilli, J., 1956. Tropical cyclones as bioclimatic activators. *W. Aust. Nat.*, 5: 82—86; 107—117.

Gill, A.M., 1975. Australia's mangrove enclaves: A coastal resource. *Proc. Ecol. Soc. Aust.*, 8: 129—146.

Gill, E.D., 1971. The far–reaching effects of Quaternary sealevel changes on the flat continent of Australia. *Proc. R. Soc. Vic.*, 84: 189—205.

Gill, E.D. and Hopley, D., 1972. Holocene sea levels in eastern Australia — a discussion. *Mar. Geol.*, 12: 223—233.

Gillham, M.E., 1961. Plants and seabirds of granite islands in southeast Victoria. *Proc. R. Soc. Vic.*, 74: 21—35.

Gillies, M.M., 1951. *The Ecology of Mangrove Faunas*. Thesis, University of Queensland, St. Lucia, Qld., 38 pp.

Goodall, D.W., 1961. Objective methods for the classification of vegetation. IV. Pattern and minimal area. *Aust. J. Bot.*, 9: 162—196.

Goodrick, G.N., 1970. A survey of wetlands of coastal New South Wales. *CSIRO Div. Wildl. Res., Tech. Mem.*, No. 5: 36 pp.

Halligan, G.H., 1921. The ocean currents around Australia. *J. R. Soc. N.S.W.*, 55: 188—195.

Hamilton, A.A., 1919. An ecological study of the salt–marsh vegetation of the Port Jackson District. *Proc. Linn. Soc. N.S.W.*, 44: 463—513.

Hamon, B.V., 1961. The structure of the east Australian current. *CSIRO, Div. Fish. Oceanogr., Tech. Pap.*, No. 11: 11 pp.

Harrison, G.G.T., 1967. Conservation of marine life of mangrove swamps, estuaries and coastal streams. In: *Caring for Queensland.* Symp. Aust. Conserv. Found., Brisbane, Qld., pp. 27—30.

Hattersley, R.T., Hutchings, P.A. and Recher, H.F., 1973. Careel Bay, Pittwater N.S.W. Development proposals. Environmental Studies. *Univ. N.S.W., Water Res. Lab., Tech. Rep.*, No. 73/6: 35 pp.

Hedley, C., 1910. The mangrove swamp. *Aust. Nat.*, 12(2): 21.

Hedley, C., 1915. An ecological sketch of the Sydney beaches. *J. Proc. R. Soc. N.S.W.*, 49: 15—77.

Hegerl, E.J. and Timmins, R.D., 1973. The Noosa River tidal swamps. A preliminary report on the flora and fauna. *Operculum*, 3(4): 38—43.

Herbert, D.A., 1951. The vegetation of south–eastern Queensland. In: *Handbook of Queensland.* Government Printer, Brisbane, Qld., pp. 25—35.

Hindwood, R.A., 1956. The mangrove honeyeater in N.S.W. *Emu*, 56: 353—356.

Hopkins, A.J.M., 1974. *The Population Ecology, Reproduction and Rehabilitation of the Mangrove Avicennia marina.* Thesis, University of Queensland, St. Lucia, Qld., 56 pp.

Hopley, D., 1974. Investigations of sea level changes along the coast of the Great Barrier Reef. In: *Proc. 2nd Int. Coral Reef Sym.*, 2. Great Barrier Reef Committee, Brisbane, Qld., pp. 551—562.

Hounam, C.E., 1964. *Australia: Average Monthly Total Radiation.* Commonwealth of Australia, Bureau of Meteorology, Melbourne, Vic., 12 maps.

Hutchings, P., 1973. The mangroves of Jervis Bay. *Operculum*, 3(1—2): 39—42.

Hutchings, P.A., 1974. The polychaeta of Wallis Lake, N.S.W. *Proc. Linn. Soc. N.S.W.*, 98: 175—195.

Hutchings, P.A. and Recher, H.F., 1974. The fauna of Careel Bay with comments on the ecology of mangrove and sea–grass communities. *Aust. Zool.*, 18: 99—128.

Ingwersen, F., 1973. Vegetation of the Jervis Bay District with special reference to land use and conservation. *Operculum*, 3(1—2): 31—38.

Iredale, T. and McMichael, D.F., 1962. A reference list of the marine Mollusca of New South Wales. *Aust. Mus. Mem.*, No. 11: 109 pp.

Jardine, F., 1925. The physiography of the Port Curtis district. *Rep. Great Barrier Reef Comm.*, 1: 73—110.

Jardine, F., 1928a. The topography of the Townsville littoral. *Rep. Great Barrier Reef Comm.*, 2: 70—87.

Jardine, F., 1928b. The Broad Sound drainage in relation to the Fitzroy River. *Rep. Great Barrier Reef Comm.*, 2: 88—92.

Jennings, J.N. and Bird, E.C.F., 1967. Regional geomorphological characteristics of some Australian estuaries. In: G.H. Lauff (Editor), *Estuaries. Am. Assoc. Adv. Sci.*, 83: 121—128.

Jessup, R.W., 1946. The ecology of the area adjacent to Lakes Alexandrina and Albert. *Trans. R. Soc. S. Aust.*, 70: 3—34.

Johnson, D.H., 1964. Mammals of the Arnhem Land Expedition. In: R.L. Specht (Editor), *Records of the American—Australian Scientific Expedition to Arnhem Land, 4. Zoology.* Melbourne University Press, Melbourne, Vic., pp. 427—515.

Johnston, T.H. and Mawson, P.M., 1946. A zoological survey of Adelaide beaches. In: *Handbook of South Australia, 25th Meeting of ANZAAS.* Government Printer, Adelaide, S.A., pp. 42—47.

Jones, D., 1973. The tidal flats of the Cairns Esplanade with reference to other marine ecosystems. *North Qld. Nat.*, 40(160): 1—7.

Jones, W.T., 1971a. The field identification and distribution of mangroves in Eastern Australia. *Qld. Nat.*, 20: 35—51.

Jones, W.T., 1971b. Mangroves. In: L. Pedley and R.F. Isbell (Editors), *Plant Communities of Cape York Peninsula. Proc. R. Soc. Qld.*, 82: 51—74.

Kenny, R., 1974. Inshore surface sea temperatures at Townsville. *Aust. J. Mar. Freshwater Res.*, 25: 1—5.

King, R.J., 1970. Surface sea–water temperatures at Port Phillip Bay Heads, Victoria. *Aust. J. Mar. Freshwater Res.*, 21: 47—50.

King, R.J., Hope Black, J. and Ducker, S.C., 1971. Intertidal ecology of Port Phillip Bay with systematic lists of plants and animals. *Mem. Natl. Mus. Vic.*, 32: 93—128.

Kratochvil, M., Hannon, N.J. and Clarke, L.D., 1973. Mangrove swamp and salt–marsh communities in Southern Australia. *Proc. Linn. Soc. N.S.W.*, 97: 262—274.

Loder, J.W. and Russell, G.B., 1969. Tumor inhibitory plants. The alkaloids of *Bruguiera sexangula* and *Bruguiera exaristata* (Rhizophoraceae). *Aust. J. Chem.*, 22: 1271—1275.

Lugo, A.E. and Snedaker, S.C., 1974. The ecology of mangroves. *Ann. Rev. Ecol. Syst.*, 5: 39—64.

Macdonald, J.D., 1973. *Birds of Australia.* Reed, Sydney, N.S.W., 552 pp.

MacNae, W., 1966. Mangroves in eastern and southern Australia. *Aust. J. Bot.*, 14: 67—104.

MacNae, W., 1967. Zonation within mangroves associated with estuaries in North Queensland. In: G.H. Lauff (Editor), *Estuaries. Am. Assoc. Adv. Sci.*, 83: 432—441.

MacNae, W., 1968a. A general account of the fauna and flora of mangrove swamps and forests in the Indo–West–Pacific region. *Adv. Mar. Biol.*, 6: 73—270.

MacNae, W., 1968b. Mangroves and their fauna. *Aust. Nat. Hist.*, 16(1): 17—21.

Macpherson, J.H. and Gabriel, C.J., 1962. *Marine Molluscs of Victoria.* Melbourne University Press, Melbourne, Vic., 475 pp.

Malcolm, C.V., 1964. Effect of salt, temperature, and seed scarification on germination of two varieties of *Arthrocnemum halocnemoides. J. R. Soc. W. Aust.*, 47: 72—74.

May, V., 1948. The algal genus *Gracilaria* in Australia. *Counc. Sci. Ind. Res. Aust. Bull.*, No. 235: 64 pp.

McArthur, W.M., 1957. Plant ecology of the coastal islands near Fremantle, W.A. *J. R. Soc. W. Aust.*, 40: 46—64.

McLeod, J., 1969. *Tidal Marshes of South–Eastern Queensland and Their Associated Algal Flora.* Thesis, University of Queensland, St. Lucia, Qld., 224 pp.

McMillan, C., 1975. Adaptive differentiation to chilling in mangrove populations. *Proc. Int. Symp. Biol. Manage. Mangroves, Hawaii.*

Montgomery, S.K., 1931. Report on the Crustacea *Brachyura* of the Percy Sladen Trust Expedition to the Abrolhos Islands (under the leadership of Prof. W.J. Dakin) in 1913 along with other crabs from W.A. *J. Linn. Soc. Lond. Zool.*, 37: 405—465.

Morris, D., 1971. *Field Manual for Tidal Swamps, 1971. Subtropical Eastern Australia.* Qld. Litt. Soc., Wetlands Committee, Brisbane, Qld., 43 pp.

Muller, J., 1964. A palynological contribution to the history of the mangrove vegetation in Borneo. In: L.M. Cranwell (Editor), *Ancient Pacific Floras.* University of Hawaii Press, Honolulu, Hawaii, pp. 33—42.

Muller, J. and Van Steenis, C.G.G.J., 1968. The genus *Sonneratia* in Australia with notes on hybridization of its two species. *North Qld. Nat.*, 35(147): 6—8.

Murray, B.J., 1931. A study of the vegetation of the Lake Torrens Plateau, South Australia. *Trans. R. Soc. S. Aust.*, 55: 90—112.

Musgrave, A., 1929a. The lure of a mangrove swamp. *Aust. Mus. Mag.*, 3(9): 318—324.

Musgrave, A., 1929b. Life in a mangrove swamp. *Aust. Mus. Mag.*, 3(10): 341—347.

Newell, B.S., 1961. Hydrology of south—east Australian waters: Bass Strait and New South Wales tuna fishing area. *CSIRO, Div. Fish. Oceanogr., Tech. Pap.*, No. 10: 37 pp.

Newell, B.S., 1966. Seasonal changes in the hydrological and biological environments off Port Hacking, Sydney. *Aust. J. Mar. Freshwater Res.*, 17: 77—91.

Newell, B.S., 1971. The hydrological environment of Moreton Bay, Queensland, 1967—68. *CSIRO, Div. Fish. Oceanogr., Tech. Pap.*, No. 30: 35 pp.

Newell, B.S., 1973. Hydrology of the Gulf of Carpentaria, 1970—71. *CSIRO, Div. Fish. Oceanogr., Tech. Pap.*, No. 35: 29 pp.

Northcote, K.H. et al., 1960—68. *Atlas of Australian Soils.* CSIRO, Melbourne University Press, Melbourne, Vic., sheets 1—10.

Orr, A.P. and Moorhouse, F.W., 1933. Physical and chemical conditions in mangrove swamps. *Great Barrier Reef Exped. 1928—29, Sci. Rep.*, 2: 102—110.

Osborn, J.H.S., 1972. *Australian National Tide Tables 1973.* Aust. Hydrogr. Publ. No. 11. Aust. Gov. Publ. Service, Canberra, A.C.T., 254 pp.

Osborn, T.G.B., 1914a. Sketches of vegetation at home and abroad. VIII. Notes on the flora around Adelaide, South Australia. *New Phytol.*, 13: 109—121.

Osborn, T.B.G., 1914b. Types of vegetation on the coast in the neighbourhood of Adelaide, South Australia. *Rep. Br. Assoc. Adv. Sci.*, 1914: 584—586.

Osborn, T.G.B., 1922. Flora and fauna of Nuyt's Archipelago. 3. A sketch of the ecology of the Franklin Islands. *Trans. R. Soc. S. Aust.*, 46: 194—206.

Osborn, T.G.B., 1923. The flora and fauna of Nuyt's Archipelago and the Investigator Group. No. 8. The ecology of Pearson Islands. *Trans. R. Soc. S. Aust.*, 47: 97—118.

Osborn, T.G.B., 1925. The flora and fauna of Nuyt's Archipelago and the Investigator Group. No. 18. Notes on the vegetation of Flinders Island. *Trans. R. Soc. S. Aust.*, 49: 276—289.

Osborn, T.G.B. and Wood, J.G., 1923. On the zonation of the vegetation in the Port Wakefield district, with reference to the salinity of the soil. *Trans. R. Soc. S. Aust.*, 57: 244—256.

Ostenfeld, C.H., 1916. Stray notes from tropical west Australia. *Dan. Vidensk. Biol. Medd.*, 2(8): 1—29.

Parsons, R.F., 1966. The soils and vegetation of Tidal River, Wilson's Promontory. *Proc. R. Soc. Vic.*, 79: 319—354.

Parsons, R.F. and Gill, A.M., 1968. The effects of salt spray on coastal vegetation at Wilson's Promontory, Victoria, Australia. *Proc. R. Soc. Vic.*, 81: 1—10.

Patton, R.T., 1942. Ecological studies in Victoria. VI. Saltmarsh. *Proc. R. Soc. Vic.*, 54: 131—144.

Paulsen, O., 1918. Chenopodiaceae from Western Australia. *Dan. Bot. Arb.*, 2: 56—66.

Percival, M. and Womersley, J.S., 1975. *Floristics and Ecology of the Mangrove Vegetation of Papua New Guinea.* Department Forests, Lae, 96 pp.

Perry, R.A., 1953. The vegetation communities of the Townsville—Bowen Region. *CSIRO, Land Res. Ser.*, No. 2: 44—54.

Perry, R.A., 1970. Vegetation of the Ord—Victoria area. *CSIRO, Land Res. Ser.*, No. 28: 104—119.

Perry, R.A. and Christian, C.S., 1954. Vegetation of the Barkly Region. *CSIRO, Land Res. Ser.*, No. 3: 78—112.

Perry, R.A. and Lazarides, M., 1962. Vegetation of the Alice Springs area. *CSIRO, Land Res. Ser.*, No. 6: 208—236.

Perry, R.A. and Lazarides, M., 1964. Vegetation of the Leichhardt—Gilbert area. *CSIRO, Land Res. Ser.*, No. 11: 152—191.

Pidgeon, I.M., 1940. The ecology of the central coastal area of New South Wales. III. Types of primary succession. *Proc. Linn. Soc. N.S.W.*, 65: 221—249.

Post, E., 1963. Zur Verbreitung und Ökologie der *Bostrychia—Caloglossa* Assoziation. *Int. Rev. Ges. Hydrobiol.*, 48: 47—152.

Post, E., 1964. *Bostrychietum* aus dem National Park von Melbourne. *Rev. Algol.*, 3: 242—255.

Rains, D.W. and Epstein, E., 1967. Preferential absorption of potassium by leaf tissue of the mangrove, *Avicennia marina:* an aspect of halophytic competence in coping with salt. *Aust. J. Biol. Sci.*, 20: 847—857.

Reye, E.J. and Lee, D.J., 1962. The influence of the tide cycle on certain species of *Culicoides* (Diptera, Ceratopogonidae). *Proc. Linn. Soc. N.S.W.*, 87: 377—387.

Riggert, T., 1966. *Wetlands of Western Australia — Study of Wetlands of Swan Coast — Particularly Water Fowl 1964—66.* Fish. and Fauna, Western Australia, Perth, W.A., 36 pp.

Rippingale, O.H. and McMichael, D.F., 1961. *Queensland and Great Barrier Reef Shells.* Jacaranda Press, Brisbane, Qld., 210 pp.

Rochford, D.J., 1951. Studies in Australian estuarine hydrology. I. Introductory and comparative features. *Aust. J. Mar. Freshwater Res.*, 2: 1—116.

Royal Netherlands Meteorological Institute, 1949. Sea areas around Australia. *Oceanogr. Meteorol. Data*, No. 124: 176 pp.

Saenger, P. and Hopkins, M.S., 1975. Observations on the mangroves of the south–eastern Gulf of Carpentaria, Australia. *Proc. Int. Symp. Biol. Manage. Mangroves, Hawaii.*

Sauer, J.D., 1965. Geographic reconnaissance of Western Australian seashore vegetation. *Aust. J. Bot.*, 13: 39—69.

Shapiro, M.A., 1975. The philosophy and application of environmental studies. *Proc. R. Soc. Vic.*, 87: 89—93.

Shapiro, M.A. and Connell, D.W., 1975. The Westernport Bay environmental study. *Proc. R. Soc. Vic.*, 87: 1—10.

Shine, R., Ellway, C.P. and Hegerl, E.J., 1973. A biological survey of the Tallebudgera Creek estuary. *Operculum*, 3(5—6): 59—83.

Slater, P., 1970. *A Field Guide to Australian Birds. Non–passerines.* Rigby, Adelaide, S.A., 428 pp.

Slater, P., 1974. *A Field Guide to Australian Birds. Passerines.* Rigby, Adelaide, S.A., 309 pp.

Smith, G.G., 1973. *A Guide to the Coastal Flora of South–Western Australia.* W. Aust. Nat. Club Handbook No. 10, Perth, 60 pp.

Smith, L.S., 1945. *Mangroves and Adjacent Communities on the North Pine River.* University of Queensland, St. Lucia, Qld.

Snelling, B., 1959. The distribution of intertidal crabs in the Brisbane River. *Aust. J. Mar. Freshwater Res.*, 10: 67—83.

Specht, R.L., 1958. The climate, geology, soils and plant ecology of the northern portion of Arnhem Land. In: R.L. Specht and C.P. Mountford (Editors), *Records of the American—Australian Scientific Expedition to Arnhem Land, 3. Botany and Plant Ecology.* Melbourne University Press, Melbourne, Vic., pp. 333—414.

Specht, R.L., 1969. The vegetation of the Pearson Islands, South Australia: A re–examination, February 1960. *Trans. R. Soc. S. Aust.*, 93: 143—152.

Specht, R.L., 1970. Vegetation. In: G.W. Leeper (Editor), *The Australian Environment.* CSIRO, Melbourne University Press, Melbourne, Vic., 4th ed., pp. 44—67.

Specht, R.L., 1972a. Water use by perennial evergreen plant communities in Australia and Papua New Guinea. *Aust. J. Bot.*, 20: 273—299.

Specht, R.L., 1972b. *The Vegetation of South Australia.* Government Printer, Adelaide, S.A., 2nd ed., 328 pp.

Specht, R.L., 1975. Stradbroke Island: A place for teaching biology. *Proc. R. Soc. Qld.*, 86: 81—83.

Specht, R.L. and Cleland, J.B., 1961. Flora conservation in South Australia. 1. The preservation of plant formations and associations in South Australia. *Trans. R. Soc. S. Aust.*, 85: 177—196.

Specht, R.L. and Perry, R.A., 1948. The plant ecology of part of the Mount Lofty Ranges (1). *Trans. R. Soc. S. Aust.*, 72: 91—132.

Specht, R.L., Roe, E.M. and Boughton, V.H. (Editors), 1974. Conservation of major plant communities in Australia and Papua New Guinea. *Aust. J. Bot. Suppl.*, No. 7: 667 pp.

Specht, R.L., Salt, R.B. and Reynolds, S., 1976. Vegetation in the vicinity of Weipa, North Queensland. *Proc. R. Soc. Qld.*, in press.

Speck, N.H., 1960. Vegetation of the North Kimberley area, W.A. *CSIRO, Land Res. Ser.*, No. 4: 41—63.

Speck, N.H., 1963. Vegetation of the Wiluna—Meekatharra area. *CSIRO, Land Res. Ser.*, No. 7: 143—161.

Speck, N.H. and Lazarides, M., 1964. Vegetation and pastures of the West Kimberley area. *CSIRO, Land Res. Ser.*, No. 9: 140—174.

Spencer, R.S., 1956. Studies in Australian estuarine hydrology. II. The Swan River. *Aust. J. Mar. Freshwater Res.*, 7: 193—253.

Steenson, J. and Barratt, T., 1970. Field survey of saline mud–flats, Smith's Creek, Ku–ring–gai Chase. *J. Nat. Parks*, September: 11—14.

Steers, J.A., 1929. The Queensland Coast and the Great Barrier Reef. *Geogr. J.*, 74: 232—257; 341—370.

Steers, J.A., 1937. The coral islands and associated features of the Great Barrier Reef. *Geogr. J.*, 89: 1—28; 119—146.

Stephenson, T.A., Stephenson, A., Tandy, C. and Spender, M.A., 1931. The structure and ecology of Low Isles and other reefs. *Sci. Rep. Great Barrier Reef Exped. (1928—29)*, 3: 17—112.

Stephenson, W., Endean, R. and Bennett, I., 1958. An ecological survey of the marine fauna of Low Isles, Qld. *Aust. J. Mar. Freshwater Res.*, 9: 261—318.

Stick, M.M. and Lergessner, D.A., 1969. Conservation of plant communities on the near North Coast, Queensland. *Capricornia*, 5: 13—19.

Story, R., 1963. Vegetation of the Hunter Valley. *CSIRO, Land Res. Ser.*, No. 8: 136—150.

Story, R., 1969. Vegetation of the Adelaide—Alligator Area. *CSIRO, Land Res. Ser.*, No. 25: 114—130.

Story, R., 1970. Vegetation of the Mitchell—Normanby area. *CSIRO, Land Res. Ser.*, No. 26: 75—88.

Symon, D.E., 1971. Pearson Island Expedition 1969. 3. Contributions to the land flora. *Trans. R. Soc. S. Aust.*, 95: 131—142.

Taylor, B.W., 1959. The classification of lowland swamp communities in north–eastern Papua. *Ecology*, 40: 703—711.

Taylor, W.R., 1964. Fishes of Arnhem Land. In: R.L. Specht (Editor), *Records of the American—Australian Scientific Expedition to Arnhem Land, 4. Zoology.* Melbourne University Press, Melbourne, Vic., pp. 45—307.

Thom, B.G., Hails, J.R. and Martin, A.R.H., 1969. Radiocarbon evidence against higher postglacial sea levels in eastern Australia. *Mar. Geol.*, 7: 161—168.

Thom, B.G., Hails, J.R., Martin, A.R.H. and Phipps, C.V.G., 1972. Postglacial sea levels in eastern Australia — a reply. *Mar. Geol.*, 12: 233—242.

Thom, B.G., Wright, L.D. and Coleman, J.M., 1975. Mangrove ecology and deltaic–estuarine geomorphology; Cambridge Gulf—Ord River, Western Australia. *J. Ecol.*, 63: 203—232.

Thomson, J.M., 1972. Mangroves, heathlands, waterfowl and life in the intertidal zone. In: *Conservation of the Australian coast. Aust. Conserv. Found., Spec. Publ.*, 7: 29—36.

Troll, W., 1933. *Camptostemon schultzii* Mart. und *C. philippensis* Becc. als neue Vertreter der australasiatischen Mangrove–Vegetation. *Flora, N.S.*, 28: 348—352.

Truman, R., 1961. The eradication of mangroves. *Aust. J. Sci.*, 24: 198—199.

Turner, J.S., Carr, S.G.M. and Bird, E.C.F., 1962. The dune succession at Corner Inlet, Victoria. *Proc. R. Soc. Vic.*, 75: 17—33.

Turner, J.S., Ashton, D.H. and Bird, E.C.F., 1968. The plant ecology of the coast. *Vic. Year Book,* 1968: 1—7.

Valentin, H., 1959. Geomorphological reconnaissance of the north–west of Cape York Peninsula (Northern Australia). *2nd Coastal Geogr. Conf., Louisiana,* pp. 213—231.

Valentin, H., 1961. The central coast of Cape York Peninsula. *Aust. Geogr.,* 8: 65—72.

Van Royen, P., 1956. Notes on a vegetation of clay-plains in southern New Guinea. *Nova Guinea,* 7: 175—180.

Van Royen, P., 1956a. Notes on *Tecticornia cinerea* (F.v.M.) Bailey (Chenopodiaceae). *Nova Guinea,* 7: 180—186.

Van Royen, P., 1956b. A new Batidacea, *Batis argillicola. Nova Guinea,* 7: 187—196.

Van Royen, P., 1960. *Sertulum Papuanum.* 3. The vegetation of some parts of Waigeo Island. *Nova Guinea,* 10(5): 25—62.

Van Royen, P., 1963. *Sertulum Papuanum.* 7. Notes on the vegetation of south New Guinea. *Nova Guinea,* 10(13): 195—241.

Van Steenis, C.G.G.J., 1962. The distribution of mangrove plant genera and its significance for palaeogeography. *Verh. K. Ned. Akad. Wet. Ser. C.,* 65(2): 164—169.

Van Steenis, C.G.G.J., 1968. Do *Sonneratia caseolaris* and *S. ovata* occur in Queensland or the Northern Territory? *North Qld. Nat.,* 35(145): 3—6.

Walker, D. (Editor), 1972. *Bridge and Barrier — The Natural and Cultural History of Torres Strait.* Australian National University, Research School of Pacific Studies, Canberra, A.C.T., 437 pp.

Walsh, G.E., 1974. Mangroves: A review. In: T.J. Reimold and W.H. Queen (Editors), *Ecology of Halophytes.* Academic Press, New York, N.Y., pp. 51—174.

Webb, L.J., 1966. The identification and conservation of habitat–types in the wet tropical lowlands of North Queensland. *Proc. R. Soc. Qld.,* 78: 59—86.

Webb, L.J., 1969. Mangroves. *Wildl. Austr.,* 6: 38—39.

Wester, L., 1967. *Distribution of Mangroves in South Australia.* Thesis, University of Adelaide, Adelaide, S. A., 109 pp.

White, C.T., 1926. A variety of *Ceriops tagal* C.B. Rob. *(= C. candolleana). J. Bot. (Lond.),* 64: 220—221.

Whitley, G.P., 1954. Goggle-eyed mangrove fish. *Aust. Mus. Mag.,* 11(6): 187—188.

Whitlock, F.L., 1947. Animal life in mangroves. *W. Aust. Nat.,* 1: 53—56.

Willis, J.H., 1944. Excursion to Seaholm salt marsh flora and mangroves. *Vic. Nat.,* 61: 40—41.

Willis, J.H., 1953. The Archipelago of the Recherche. III. Plants 3(a) Land flora. *Aust. Geogr. Soc. Rep.,* 1: 3—35.

Willis, J.H., 1962, 1972. *A Handbook to Plants in Victoria, I and II.* Melbourne University Press, Melbourne, Vic., 448 and 832 pp.

Wilson, P.G., 1972. A taxonomic revision of the genus *Tecticornia* (Chenopodiaceae). *Nuytsia,* 1: 277—288.

Womersley, H.B.S., 1956. The marine algae of Kangaroo Island. IV. The algal ecology of American River Inlet. *Aust. J. Mar. Freshwater Res.,* 7: 64—87.

Womersley, H.B.S. and Edmonds, S.J., 1952. Marine coastal zonation in southern Australia in relation to a general scheme of classification. *J. Ecol.,* 40: 84—90.

Womersley, H.B.S. and Edmonds, S.J., 1958. A general account of the intertidal ecology of South Australian coasts. *Aust. J. Mar. Freshwater Res.,* 9: 217—260.

Womersley, J.S. and McAdam, J.B., 1957. *The Forests and Forest Conditions in the Territories of Papua and New Guinea.* Government Printer, Port Moresby, 62 pp.

Wood, E.J.F., 1959a. Some aspects of the ecology of Lake Macquarie, N.S.W. with respect to an alleged depletion of fish. VI. Plant communities and their significance. *Aust. J. Mar. Freshwater Res.,* 10: 322—340.

Wood, E.J.F., 1959b. Some eastern Australian sea-grass communities. *Proc. Linn. Soc. N.S.W.,* 84: 218—226.

Wood, J.G., 1937. *Vegetation of South Australia.* Government Printer, Adelaide, S.A., 164 pp.

Wright, L.D., Coleman, J.M. and Thom, B.G., 1972. Emerged tidal flats in the Ord River estuary, Western Australia. *Search,* 3: 339—341.

Wright, L.D., Coleman, J.M. and Thom, B.G., 1973. Processes of channel development in a high–tide–range environment: Cambridge Gulf—Ord River Delta, Western Australia. *J. Geol.,* 81: 15—41.

Chapter 16

EXPLOITATION OF MANGAL[1]

GERALD E. WALSH

INTRODUCTION

Species that comprise the mangrove formation are useful to man. The discussion that follows is the result of communication with many people throughout the world and my own library researches. The consensus of opinion among my correspondents is that knowledge of uses of mangrove is difficult to obtain, out–of–date, tenuous at best, and often frustratingly lacking because most usage is local and never reported. I concur with that opinion.

EARLY USES

Little is known about early uses of mangroves. The earliest references to use that I have found are cited in Bowman (1917) who stated that Nearchus (325 B.C.) and Theophrastus (305 B.C.) referred to seedlings of *Rhizophora* as having an aphrodisiac effect when ingested and of their use in philters in Arabia. Abou'l Abbas en–Nebaty (1230), a Moorish botanist, wrote that mangroves were used as food, fuel, medicine for curing sore mouth, and tanning leather. Oviedo (1526) and Clusius (1601) wrote that natives of the West Indies used the hypocotyl of *Rhizophora* seedlings as food in times of famine. Sloane (1725) described the oysters that grew on mangrove roots in Trinidad as "well tasted" and mentioned that seedlings were used as food by humans, the bark as a source of tannin and dye, and that bark, when mixed with milk or fresh butter, aided in curing diseases of the liver. Arnott (1869) and Le Maut and Decaisne (1876) described wine made from mangrove propagules in the West Indies. Bowman (1917) described use of mangroves for retention of ballast along the shores of the Florida Keys and in Venezuela.

MacNae (1968) suggested that sea–going vessels from the shores of the Gulf of Oman and the Persian Gulf in prehistoric and early historic times had keels of mangrove wood. He also reported that poles of *Rhizophora mucronata* and *Bruguiera gymnorrhiza* were used for construction of buildings in Arabian cities. Crossland (1903) was the first to describe use of mangrove wood by Arabs for house and furniture construction. Wood from *R. mucronata* and *B. gymnorrhiza* was especially desirable because it is resistant to attack by termites.

Chapman (1970) speculated that early man carried mangrove propagules from South America to southern Pacific Ocean islands. They were to serve as seed material for trees that produced tannin and wood.

Canestri and Ruiz (1973) reported that when Amerigo Vespucci visited Maracaibo, he was impressed by native buildings that were supported over water by stilts, presumably of mangrove origin. He named the area Venezuela ("Little Venice") because the stilt–supported homes over water reminded him of Venice.

A.V.N. Sarma (personal communication, 1973) demonstrated that mangals have existed on Ecuadorean coasts since 6500 B.C. and gave archaeological evidence that primitive man exploited the shellfish *Anadara tuberculosa,* peccaries, deer, and other animals of the mangal.

---

[1] Contribution No. 228 from the Gulf Breeze Environmental Research Laboratory.

Today, mangals are of great importance to many people who live along tropical shorelines, but few data are available with regard to production of wood and its by-products or to monetary value. Extensive local use for production of such things as posts, poles, dye, tannin, boats, railroad ties, matchboxes, pencils, salt, sodium carbonate, incense wood, cigarette wrappers, fishing floats, fish poison, thatching material, and alcohol precludes the keeping of accurate records.

FOREST MANAGEMENT AND PRODUCTS

Kuenzler (1969) stated that mangrove forests are among the most productive of estuarine ecosystems, fixing as much energy as other estuarine systems, eutrophic ponds, evergreen forests, or good farmland (cf. p. 26). Watson (1928) was the first to describe the importance of mangal as a manageable forest resource subject to rules of agriculture and silviculture, and Graham (1929) described a replanting schedule by the Forest Department of Kenya in Africa. Watson did not give production figures, but described uses of mangroves in Malaysia, some of which are given in Table 16.1. Silviculture of mangroves in Malaysia was described briefly by MacNae (1968) and the Ministry of Primary Industries of West Malaysia (1970) reported that production of poles, firewood, and charcoal was 958 600 m^3 and that most came from mangal.

Walker (1937) recommended that in Thailand, *Rhizophora apiculata (= R. conjugata* Lamk.*)* and

TABLE 16.1

Uses of mangroves in Malaysia (modified from Watson, 1928)

Species	Use
Acanthus ebracteatus	fruit used as a blood purifier and dressing for boils; leaves used to relieve rheumatism; fruit and roots applied to snakebite and arrow poisoning
Acrostichum aureum	litter for cattle and roofing for homes
Aegiceras majus	bark extract used as fish poison
Avicennia alba *Avicennia intermedia* *Avicennia lanata* *Avicennia officinalis*	inferior firewood and timber; bark a source of tannin and astringent; ointment made from seeds applied to tumors and smallpox ulcerations; sap used as contraceptive; wood used to make water pipes
Bruguiera caryophylloides *Bruguiera gymnorrhiza*	firewood; timber; young radicles eaten; medicine for sore eyes from fruit; scent from pneumatophores; condiment from bark
Bruguiera parviflora	firewood; timber
Ceriops tagal (= *C. candolleana*)	firewood; timber; dyebark; tanbark
Lumnitzera littorea (= *L. coccinea*) *Lumnitzera racemosa*	timber; fence posts
Nypa fruticans	leaves for thatch; fruit eaten; leaves as cigarette wrappers; sugar from sap; vinegar; alcohol
Rhizophora conjugata (= *R. apiculata*)	firewood; timber; charcoal; tanbark
Rhizophora mucronata	firewood; timber; charcoal; tanbark; wine from fruit
Sonneratia acida *Sonneratia alba* (= *S. ovata*)[1]	firewood; fruit eaten
Sonneratia griffithii (= *S. alba*)[1]	firewood; timber; pneumatophores used as fishing floats and as substitute for cork

[1] Cf. Chapman (1975).

R. mucronata be managed for firewood production in sub–coupes of 40 ha or less. He discussed a 20–cm felling girth and suggested planting of seedlings two years after exploitation of the sub–coupes. He later (1938) stated that, in addition to *R. conjugata* and *R. mucronata*, *Bruguiera caryophylloides*, *B. gymnorrhiza*, and *Ceriops candolleana* could also be cultured for firewood. Species of *Avicennia* and *Sonneratia* were of poor quality as firewood.

Unfortunately, neither Watson nor Walker gave production figures for wood or mangrove products.

Durant (1941) was the first to report growth rates of mangroves in Malaya. He gave rates of growth in girth of several mangrove species over a 25–year period. The volume of wood per hectare in mixed *Rhizophora* forests, including only timber 20 cm in girth and greater, increased from 39 m^3 when the sub–coupe was 10 years old to 160 m^3 at 50 years. Durant recommended harvesting at 22—23 years when the median annual volume increment was at its maximum of 147.7 and the volume of wood approximately 232 m^3 ha^{-1}. Durant pointed out, however, that approximately 20% of the volume was composed of bark and approximately 20% was unuseable. Therefore, only about 60% of the total volume could be considered as commercial wood output.

Noakes (1950, 1951, 1955, 1957, 1958) stated that the total area of mangrove–type forest in Malaya was about 1500 km^2 of which approximately 1200 km^2 was under sustained yield management in forest reserves. Fuel was the main product, but poles for fish traps were also produced. *Rhizophora apiculata* (= *R. conjugata*) and *R. mucronata* were by far the most abundant and economically important species, although *Ceriops tagal* was locally abundant and produced the best quality fuel and tanbark.

In studies by Noakes, natural regeneration was normally obtained over the annual coupe. When planting of seedlings was necessary, it was usually successful. The *Rhizophora* grew slowly however, reaching felling size of 15—18 m in height and 46—76 cm in girth at 20—30 years. Trees were harvested by clear–felling following intermediate felling to induce regeneration of the forest. Rotation of 30 years appeared proper for mixed *Rhizophora* forest. Yield from the above practice was approximately 210 m^3 ha^{-1}. Yield from an unmanaged forest in Selangor was only 96 m^3 ha^{-1}.

Mangrove forest in Thailand is of considerable importance, not only as a source of firewood, timber, and tanbark, but also as a shoreline stabilizer (Banijbatana, 1957), a function that it performs elsewhere (Savage, 1972). There are approximately 133 400 ha of mangrove forests available for silviculture in Thailand. Banijbatana recommended: (1) clearing or thinning for young forest that contains trees of 20 cm girth and under; (2) heavier thinning for middle–aged forests in which the majority of trees are between 20 and 50 cm in girth, and seedling felling for removal of *Ceriops decandra* (= *C. roxburghiana*) and *Bruguiera cylindrica*; and (3) heavy seedling felling in older mangrove forests composed of trees greater than 50 cm in girth. At the final felling, three or four seed bearers were retained at distances of 20 m apart for natural regeneration.

Species of choice for mangrove management in Thailand were *R. mucronata* and *R. apiculata* (= *R. candelaria*). These trees reached 65—70 cm in girth between the ages of 39 and 43 years. A rotation period of 40 years was adopted, with yield of 125—150 m^3 ha^{-1}. Most of the yield was used as firewood.

Banerji (1958a, b) stressed the value of *B. gymnorrhiza* for poles in the Andamans. He estimated that approximately 17 300 transmission and telegraph poles were produced annually from that species. When treated with a creosote:earth–oil mixture, poles from *B. gymnorrhiza* remained useable for at least 12 years. The poles had the high modulus of rupture value of 1000 kg^{-1} cm^{-1} and scored high in criteria set by the American Standards and Canadian Standards Associations. The same was true of poles made from *Rhizophora mangle* and *Avicennia germinans* (= *A. nitida* Jacq.) in Venezuela (Arroyo P., 1971). Arroyo P. gave an extensive account of the physical and mechanical properties of mangrove poles. Graham (1929) stated that more than 500 000 poles from mangroves were used in one year in Kenya. Poles from *Sonneratia alba* and *Casuarina equisetifolia* were especially good for boat masts.

Clear–felling and planting of *B. gymnorrhiza* was recommended by Banerji (1958a, b) as the best silvicultural method. The species grows slowly and a rotation period of 100 years was recommended for an exploitable girth of 70 cm. On some plantations, *B. gymnorrhiza* attained a height of 9—12 m and a girth of 23—30 cm in 15 years and could be used at that time.

Banerji also recommended mangement of *R. mucronata* and *R. apiculata (= R. conjugata)* for use as firewood. The genus grows faster than *Bruguiera,* yielding 180 m length of wood per ha, whereas *B. gymnorrhiza* yielded only 65 m length of wood per ha.

Chatterjee (1958), quoting Sahni, stated that the Andaman and Nicobar Islands might yield, with proper management, over 20 000 kg of *Bruguiera* bark each year. *Bruguiera* yielded approximately 185 poles per ha, and about half a million timber poles could be expected per year from the Nicobars. He also suggested possible use of *Nypa* as a source of alcohol. Approximately seven million liters of alcohol were distilled from *Nypa* juice each year, and the Andamans could probably support an alcohol industry.

Five classes of wood, based on their value as firewood in relation to specific gravity, were described by Cox (1911):

Class	Description	Specific gravity
I	excellent	> 0.90
II	very good	0.75—0.90
III	good	0.60—0.75
IV	fair	0.45—0.60
V	poor	0.30—0.45

Becking et al. (1922), using the classification of Cox, rated species of the Malayan region as follows:

Cox class	Species
I	*Rhizophora* spp., *Bruguiera* spp., *Ceriops* spp., *Heritiera littoralis,* and *Cynometra ramiflora*
II	*Lumnitzera* spp., *Sonneratia alba, Pithecolobium umbellatum*
III	*Xylocarpus* spp., *Aegiceras corniculatum*
IV	*Dolichandrone longissima*
V	*Avicennia* spp., *Sonneratia acida, Excoecaria agallocha, Cerbera manghas*

Becking et al. (1922) also calculated the yields of fuel wood and timber from *Rhizophora* and *Bruguiera* of various diameters in Indonesia (Tables 16.2 and 16.3).

Schneider (1916) reported that *Rhizophora, Bruguiera, Ceriops,* and *Kandelia* (all of the family Rhizophoraceae) were used in the Philippines for foundation pilings, mine timbers, house posts, furniture, cabinets, musical instruments, and native implements. According to Marco (1935) inland genera of the Rhizophoraceae *(Carallia, Anisophyllea, Compretocarpus, Anopyxis,* and *Gynotroches)* show a handsome silvery grain when quartered and are suitable for furniture, cabinets, planks, posts, and flooring. Unwin (1920) recommended *Poga* as a substitute for cedar and mahogany and judged the flavor of its nut as better than the Brazil nut. None of the authors commented on potential economic value of the products.

Table 16.1 shows that fruit from several species is edible. In Oceania, the fruit of *Bruguiera eriopetala* are peeled, sliced, and soaked in water for several hours. They may then be stored for several months after air–drying, or they may be steamed or boiled and eaten with coconut cream (Barrau, 1959).

Jones (1971) gave uses of mangroves in Australia and his list is summarized in Table 16.4. There is, however, no commercial industry in Australia that exploits mangroves at this time.

Dugros (1937) mentioned many of the above uses of mangroves in Indo–China but decried the fact that poor management led to destruction of large areas.

Womersley (1975) reviewed the management of mangrove forests in Papua–New Guinea. Keith (1935) reported export of over five million m^3 of mangrove firewood from Sabah in the early part of this century. Since the development of petroleum fuels, decline in export has been rapid, and Liew That Chin (1970) reported export as only 17 m^3 in 1968.

Uses of mangroves in Puerto Rico are similar to those in Malaya (Wadsworth, 1959). *Laguncularia racemosa,* the most common mangrove, was used for timber production. *Avicennia germinans, L. racemosa,* and *R. mangle* were used in production of fence posts whose service life was estimated to be over ten years.

In Puerto Rico, large saplings in undisturbed stands had an average girth of 4.3 cm and volume of 111 m^3 ha^{-1}. Thinning of the sapling stand tripled the girth in four years, and clear–cutting was the harvest method of choice.

TABLE 16.2

Yield, in m³, of mangrove fuel wood in Indonesia (from Becking et al., 1922)

Species	Diameter (cm)						
	15—20	20—25	25—30	30—35	35—40	40—45	45—50
Rhizophora apiculata (= *R. conjugata*)	0.23	0.44	0.75	1.15	1.66	2.3	—
Bruguiera gymnorrhiza	0.17	0.32	0.52	0.78	1.11	1.48	1.97
B. parviflora	0.23	0.50	0.97	—	—	—	—

TABLE 16.3

Yield, in m³, of mangrove timber in Indonesia (from Becking et al., 1922)

Species	Diameter (cm)						
	15—20	20—25	25—30	30—35	35—40	40—45	45—50
Rhizophora apiculata (= *R. conjugata*)	—	0.28	0.41	0.57	0.79	1.07	—
Bruguiera gymnorrhiza	—	—	0.34	0.52	0.71	0.90	1.08
B. parviflora	0.16	0.29	0.42	—	—	—	—
B. caryophylloides	0.11	0.18	0.26	0.33	—	—	—

TABLE 16.4

Uses of mangrove in Australia (after Jones, 1971)

Species	Use
Acanthus ilicifolius	ornamental shrub
Aegialitis annulata	cultivated
Aegiceras corniculatum	tanbark for fishnets; honey
Avicennia eucalyptifolia	boat keels and elbows
Avicennia marina	boat keels and elbows; mallets
Bruguiera parviflora	tanbark; poles; fish traps
Bruguiera exaristata	tanbark; poles; rafters; fish traps
Bruguiera gymnorrhiza	tanbark fishnets
Ceriops tagal	tanbark fishnets; bleached roots used as ornaments
Excoecaria agallocha	interior carpentry
Heritiera littoralis	boat–building, furniture, interior carpentry
Lumnitzera littorea	beautiful flowers
Nypa fruticans	"toddy" from fruit
Osbornea octodanta	insect repellent
Rhizophora apiculata *Rhizophora mucronata*	timber; net racks; rough building; firewood; tanbark
Rhizophora stylosa	timber; tanbark
Sonneratia alba	canoe building; ornamental shrub
Xylocarpus granatum	cabinet wood; boat keels

In Nigeria, Rosevear (1947) stated that the chief useful product of mangrove swamps was wood from *R. mangle*. It was useful for poles, and withstood the hot, damp atmosphere of coal mines better than any other wood. The poles lasted for over twelve years, whereas those from teak lasted approximately six years. The wood was also used extensively for firewood. Boyé (1962) described the potential use of *Avicennia nitida* for production of paper pulp. A minimum rotation period of

25 years for production of *R. mangle* was recommended by Dale (1938).

In the Americas, Morton (1965) reviewed the potential use of *R. mangle* as a source of tannin, plywood adhesive, dyebark, fuel, cattle feed, and human dietary supplement. She concluded that mangroves are an untapped resource in the United States and that red mangrove leaves may warrant investigation as an abundant source of protein.

Potential use of *R. mangle, A. germinans,* and *Laguncularia racemosa* in Mexico was reviewed by Vazquez–Yanes et al. (1972). The authors stressed the economic potential of mangroves and recommended silviculture of forests and development of industries related to them. Similar recommendations were made for Venezuela by Flores (1968), who cited the extensive work of Budowski (1951, 1952). *Rhizophora mangle, A. nitida (germinans?), Laguncularia racemosa* and *Conocarpus erectus* were all considered exploitable species.

A recent development in utilization of mangroves has been production of high–alpha (dissolving) pulps for the manufacture of rayon, cellophane, lacquers, cellulose acetate, and other cellulose derivatives. Pulpwood of *R. mucronata* has been exported from the Philippines to Japan for such manufacture (Table 16.5). Methods for production of high–alpha pulps were described by Nicolas and Bawagan (1970).

A portable hogger that operates within the mangrove swamp and transports wood chips to ships offshore was described by Womersley (1975). He felt that this machine, when coupled with clear–cutting of mangroves, would revolutionize harvesting. He suggested that *R. mucronata, R. apiculata, C. tagal,* and *Bruguiera* spp. would be suitable for such harvesting.

There is some question at this time in the Philippines as to whether mangal should be utilized as a forest resource or whether it should be removed and the land used for mariculture. De la Cruz and Banaag (1967) and De la Cruz (1969) warned against clear–cutting of swamps for mariculture and recommended laws for establishment of mangrove forest reserves. J.M. Lawas et al. (personal communication, 1973) stated that the justifications for maintaining *Rhizophora* spp., *Bruguiera parviflora,* and *Nypa fruticans* are: (1) the wood is exportable and therefore a source of foreign exchange income; (2) the wood is a source of raw material for manufacture of rayon and other products; (3) the wood is used for firewood, posts, furniture, charcoal, tannin, and dye, and *Nypa* is used for thatch, alcohol, and sugar; and (4) charcoal from *Rhizophora* spp. and *B. parviflora* is a substitute for petroleum coke that is used in the manufacture of calcium carbide and ferro–alloys, which, in turn, are used in the chemical, plastics, and metal industries. Statistics related to use of mangal as a forest resource in the Philippines are given in Table 16.6.

TABLE 16.6

Potential production of wood and income from Philippine mangrove forests when utilized as a forest resource (modified from J.M. Lawas et al., personal communication, 1973)

Total mangrove area, ha	448 310
Total volume of trees 20 cm in diameter, m^3	17 289 000
Wood volume per ha, m^3	39
Age of trees at cutting, years	20
Annual cutting area, ha	22 415
Annual wood production, m^3	865 250
Gross income, pesos	32 144 000
Net income, pesos	11 265 500

TABLE 16.5

Quantity and value of pulpwood exported from the Philippines to Japan (personal communication from A.A. de la Cruz, 1974)

Year	Quantity (m^3)	F.O.B. value U.S. dollars	F.O.B. value/m^3 U.S. dollars
1971	117 510	1 120 782	9.50
1972	57 588	535 445	9.30
1973	28 302	288 374	10.20
1974 (Jan.—Apr.)	1 798	26 972	15.00

MARICULTURE

Mariculture in mangrove swamps has been described by Le Mare (1950), Tung–Pai Chen (1952), Herre (1953), Hofstede et al. (1953), Ancona (1954), Schuster et al. (1954), Huet (1956), Korringa (1956), Angelis (1959), Schuster (1960), Tampi (1960), Thomson (1960), Hora and Pillay (1962), Pillay (1962, 1965), Fischer (1963), Pakrasi (1964), Summers (1964), Bigot (1968), Lin (1968), Hickling (1970), Pagan–Font (1971), and Kiener (1972).

According to MacNae (1968) mariculture in brackish–water ponds created by clearing mangrove forests has been practiced for many years in Java, Sumatra, the Philippines, and Taiwan. Production of fish, crabs, and prawns in Java was described by Schuster (1952). Schuster suggested that the milkfish, *Chanos chanos,* was well–suited for cultivation in ponds. Overall production from 57 000 ha of fishponds in Indonesia was 8 962 000 kg of *C. chanos* and 2 947 000 kg of prawns. One pond of 12 ha yielded approximately 3200 *Scylla serrata* crabs. Schuster showed that production of fish was related to soil type (Table 16.7), and it is clear from the work of Watts (1969) that production is related to soil and water conditions. Watts reported very poor yield from fishponds constructed in sulphidic soils.

TABLE 16.7

Average annual yield of *Chanos chanos* per ha in mangrove–lined ponds in Java (from Schuster, 1952)

Soil type	Annual yield (kg ha^{-1})
Juvenile volcanic soil	225—425
Colloidal clay	168—314
Juvenile lateritic soil	133—269
Calcareous clay	84—133
Senile lateritic soil	57—111
Rocky or sandy soil	44—84

J.M. Lawas et al. (personal communication, 1973) gave the following statistics for potential annual yield of *C. chanos* from fishponds in the Philippines:

Total area of ponds (ha)	359 000
Total production (kg)	365 500
Gross income (pesos)	539 440 000
Net income (pesos)	57 539 300

Estimated net income from utilization of mangals for fish production was about five times greater than for wood production. J.M. Lawas et al. concluded that fishpond development of swamps should be given priority over wood production.

Other fishes can be cultivated in mangal ponds. MacNae (1968) suggested cultivation of the mullet *(Mugil* spp.*)* and *Tilapia mossambica.* Tan Lee Wah (1972) reported yield of four species of carp in fishponds of mangrove swamps in Singapore. Annual yields from individual ponds varied between 590 and 1100 kg, depending on carp species and pond location. The author stated that the less accessible and therefore less valuable parts of Singapore could be cleared for profitable production of fish.

Oysters are common inhabitants of mangal swamps and may be cultivated for food. Mattox (1949) recommended mangal swamps for cultivation of *Ostrea rhizophorae* (= *Crassostrea rhizophorae)* in Puerto Rico. Flores (1968) stated that *C. rhizophorae* was the most important bivalve of Venezuela in terms of abundance in mangals and nutritive value. According to Bacon (1971), commercial exploitation of swamps is currently being investigated in Venezuela, the Dominican Republic, and Jamaica. Using 16 stakes made of mangrove wood in a mangal lagoon in Trinidad, he obtained the following yield of *C. rhizophorae:* number of oysters — 809; mean shell length — 25 mm; total weight of meat — 331 g. Bacon concluded that the oyster can be cultivated comparatively easily and cheaply in mangal lagoons. The same conclusions were reached by Nikolic and Melendez (1968) using stakes of artificial fiber in Cuba and by Hunter (1970), who discussed cultivation of oysters in mangal swamps of Sierra Leone.

TANNIN PRODUCTION

Howeson (1804) was probably the first to de-describe use of mangroves by the tannin industry in India. The method was insufficient because extraction was made without the use of heat and the liquor was concentrated by evaporation under sunlight. This process caused deterioration of the product. Hamilton (1846) described R. mangle and mentioned its possible use as a source of tannin in the Americas. The chemical properties and composition of tannin from R. mangle bark were described by Trimble (1892). He gave the empirical formula $C_{25}H_{25}O_{11}$, and tannin comprised 24% of the weight of air–dried bark.

A report (Anonymous, 1904) in the Bulletin of the Imperial Institute stated that bark of R. mucronata was regularly imported into Zanzibar and that it was also in demand as a tanning material in Marseilles. Tannin comprised approximately 35% of air–dried bark. The author concluded that leathers produced with mangrove tannin were of poor commercial quality but could be used for production of cheap boots.

Sack (cited in Drabble and Nierenstein, 1907) reported that, in R. mangle, older plants contained more tannin than younger, with an average yield from air–dried bark of 24.5%. Sack gave the empirical formula $C_{24}H_{20}O_{12}$. Drabble and Nierenstein stated that both Laguncularia racemosa and Avicennia germinans from the Americas produced commercially suitable tannin. Drabble (1908) described the bark of R. mangle and Laguncularia racemosa in West Africa and suggested that the latter was the better source of tannin. Baillaud (1912) recommended that the bark of Rhizophora and Bruguiera be collected at the end of each growing season because their tannin imparts less color to leather at that time.

Becking et al. (1922) compared the tannin contents of dried mangrove bark from the Philippines and Indonesia (Table 16.8). There were often large differences between samples, but all, except B. parviflora, B. caryophylloides, and Sonneratia alba, had relatively high average tannin contents. Becking et al. stated that over ten million kilograms of bark were produced annually in Penang and over five million kilograms produced annually in Singapore.

Buckley (1929) gave similar figures for tannin contents of bark from Malaysia. He noted that, although thick bark from old trees contained the greatest amount, the tanning value of extracts varied greatly with position on the tree; no two strips of bark from any single tree contained the same concentration of tannin. The differences among trees of a single species were often as great as differences among species. Also, tanning value was related to method of preparation of bark before extraction. Buckley also demonstrated that the tannin contents of leaves and fruit were generally much lower than those of bark. The same was true of African (Pynaert, 1933) and Indian (Mudaliar and Kamath, 1952; Venkatesan, 1966) mangroves. In Sarawak, tannin from bark of R. apiculata, B. gymnorrhiza, and B. eriopetala was uniformly of very poor quality (Bennett and Coveney, 1959).

Dey (quoted in Chatterjee, 1958) stated that up to 5000 metric tons of mangrove tannin extracts were exported annually from Borneo and that dry

TABLE 16.8

Percent tannin concentrations in dried mangrove barks from the Philippines and Indonesia (from Becking et al., 1922)

Species	Philippines			Indonesia		
	minimum	maximum	average	minimum	maximum	average
R. mucronata	18	28	25	—	—	32
R. apiculata (= R. conjugata)	18	39	27	24	36	30
B. gymnorrhiza	24	32	31	21	37	29
B. eriopetala	27	32	32	—	—	—
B. parviflora	9	15	10	—	—	—
B. caryophylloides	—	—	—	6	9	8
C. candolleana	17	31	27	25	30	27
X. granatum	22	25	23	20	21	21
X. moluccensis	23	23	23	24	27	26
S. alba	12	12	12	—	—	9

bark from *C. decandra* (= *C. roxburghiana*) was the major source.

In the United States, Bowman (1917) reported that several tannin factories operated in the Florida Everglades. The commercial ventures were unsuccessful however, and Rogers (1950—1951) explained that *R. mangle* bark in Florida is nearly inaccessible and costly to collect. He estimated that it would yield no more than one million kilograms of pure tannin annually for five years. In addition, tannin from red mangrove bark did not prove useful in Venezuela (Pustelnik, 1954), and degraded when stored (Acosta–Solis, 1947, 1959; Behrens Motta, 1960; Pearman, 1957).

According to Pearman (1957) tannin from red mangrove is undesirable because: (1) it must be mixed with other tannins to reduce astringency that causes cracky grain in leather; (2) the high salt content in mangrove extract limits its use in tannin mixtures; and (3) its intense dark red color has little trade appeal. A method for reduction of color was given by Del Sagrario (1957).

Conversely, collection and sale of red mangrove bark was reported by West (1956) to be a lucrative business on a local scale in Colombia. Rodriguez et al. (1968) reported that 2 617 000 kg of tannin were produced from mangroves in Colombia in 1967. In contradiction to Pustelnik (1954), Arroyo P. (1971) stated that tannin from *R. mangle* in Venezuela is of excellent quality. Leaves of *Laguncularia racemosa* were processed by tanneries in Brazil in 1951 (Howes, 1953).

Russell (1943) recommended utilization of *C. erectus* as a source of tannin in the Americas. He stated that tannin content of bark was approximately 20% dry weight, that extraction is simple, and that extracts have a "good" color and contain soluble material that disperses readily. The extract gave a firm, well–filled, strong piece of leather of a light pink color. Although Russell stated that leather tanned with *C. erectus* extracts was of good quality, I know of no commercial exploitation of the species in the United States.

Joshi (personal communication, 1973) stated that *Ceriops candolleana* is a valuable source of tannin in India (see however p. 246). Tannin percentage of dry weight is 16 in leaves, 26 in twig bark, and 41 in bole bark. It imparts a good red color to leather and required only six weeks to complete the tanning process. In contrast, oak tannin requires six months. *Kandelia candel* is also a good source of tannin for heavy leather tannage in India. Joshi commented that leather prepared from tannin of *R. mucronata* is inferior in color and quality. Navalkar (1962) listed *Excoecaria agallocha* as a good source of tannin in India.

The bark of *Rhizophora* and *Bruguiera* has been used for tanning in Australia, but it is now used mainly to render fishing nets more resistant to water and destructive organisms (MacNae, 1966). It is also used to toughen ropes and sails in Australia (Australian Conservation Foundation, 1972).

AGRICULTURAL USE OF MANGAL SOIL

High productivity of mangals is well known (Heald and Odum, 1970; Heald, 1971; Odum, 1971). Schuster (1952) stressed the production potential of mangal soil. MacNae (1968) stated that the quality of mangal soil depends upon the source of alluvium; rivers that drain quartzitic and granitic areas carry silt of poor quality, whereas rivers that drain areas of recent or moderately recent volcanic soils produce alluvium of high quality.

Use of high–quality mangal soils for growing rice has been practiced for many years in Sierra Leone, West Africa. Due and Karr (1973) described the need for additional rice–land in that country and, among other methods, recommended clearing and reclamation of mangal forest to increase production. The authors suggested varieties of rice that should be planted in relation to soil salinity and stated that rice yields from mangal swamps were second only to inland valley swamps. Their potential rice productivity, with improved technology, is equal to or greater than inland valley swamps (Table 16.9).

The literature with regard to reclamation of mangal land and its use for rice production in Sierra Leone is voluminous. For an overview of the subject, see the following reports: Doyne and Glanville (1933), Dent (1947), Macluskie (1952), Jordan (1954, 1959, 1964), Tomlinson (1957a, b), Hart (1959, 1962, 1963), Hesse (1961a, b, 1962), Hart et al. (1963), Hesse and Jeffrey (1963), Das Gupta (1969, 1970, 1971, 1972a, b), and Terry and Das Gupta (1973).

Hesse (1961a) reviewed the use of mangrove soils for growing rice in Sierre Leone. He wrote

TABLE 16.9

Current (1973) and possible yields of rice from six farming situations in Sierra Leone (from Due and Karr, 1973)

	Approximate yield (kg ha^{-1})					
	uplands	river terraces	bolilands[1]	riverain	inland swamp	mangrove swamp
Current	1000	1700	1300	1800	2000	1800
Expected with improved technology	1600	2000	1700	1800	2700	2700—4000

[1] Areas of low, swamp grasslands in the central and northern part of Sierra Leone in the region of Makani, Pedemu and Kamala. Most of these large areas do not have natural water outlets and are flooded to various depths during the monsoon season by runoff from surrounding higher country. No means are available to control water which makes cropping hazardous.

that land reclaimed from *Rhizophora* forest, when empoldered, developed soil conditions that were adverse to growth of rice. Land reclaimed from *Avicennia* forest could be used with impunity. This is because empoldering caused desiccation of topsoil. Upon desiccation, pH of the highly sulphidic and fibrous *Rhizophora* soils was reduced due to production of sulphuric acid with concomitant liberation of aluminum ions. Rorison (1973) pointed out that aluminum toxicity and depression of phosphorus uptake by aluminum in acid solution are important in reducing production. Hesse (1961a) also showed that there was a high carbon: nitrogen ratio in *Rhizophora* swamps and suggested that land reclaimed from them should give a pronounced response to nitrogen fertilizers. He cited an unpublished report that described a twofold increase in rice production after application of sulphate of ammonia to empoldered *Rhizophora* land.

Attempts to grow field and fruit crops on mangal soil in Thailand were described by Van Breeman et al. (1973). Only rice grown during the wet season produced a crop, and that was of high yield (1940 kg ha^{-1}). The authors ascribed crop failure to soil acidity (pH 2—3) that resulted from pyrite oxidation.

Other uses of mangal soil have been reported. In Puerto Rico, mangal swamps have been reclaimed for cultivation of sugar cane (Holdridge, 1940). In Australia, the slightly saline soils of the landward fringes of mangal swamps have been used for the same purpose (MacNae, 1966; Australian Conservation Foundation, 1972). Walsh (1967) reported that the American Sugar Company introduced seedlings of *R. mangle* from Florida into Hawaii to stabilize mud flats of Molokai. *Avicennia resinifera* is of little economic importance in New Zealand, but its land can be reclaimed as good pasture (Kuchler, 1972).

In Cambodia, the coconut palm grows well on reclaimed mangal swamp, but the land must be drained (Briolle, 1969).

For detailed studies on chemistry of mangal soils as it relates to use and fertility, see the following publications of Freise (1935), Doyne (1937), Navalkar (1941, 1959, 1962), Navalkar and Bharucha (1949, 1950), Hart (1959), Hesse (1961b, 1962, 1963), Hart et al. (1963), Hesse and Jeffrey (1963), Cate and Sukhai (1964), Pons (1964), Thornton and Giglioli (1965), Vieillefon (1969, 1973), Augustinius and Slager (1971), Allbrook (1973), Andriesse et al. (1973), Beye (1973), Bloomfield and Coulter (1973), Coulter (1973), Driessen and Ismangun (1973), Frink (1973), Grant (1973), and Rorison (1973).

DISCUSSION

It is clear that mangal is potentially of great commercial value. This is not often recognized, and it is unfortunate that, in many parts of the tropical world, mangrove forests are "reclaimed" indiscriminately or used as dumping grounds for garbage or sewage sludge (Wadsworth, 1959; Bird, 1971) in the belief that the forest is good for nothing else. Canestri et al. (1973) described changes in species composition and diversity of a mangal ecosystem in Venezuela, where 32 ha of swamp were deforested, filled, and dredged for human habitation and recreation. The authors concluded that biological and aesthetic values were lost because of indiscriminate manipulation without proper planning.

Mangrove stands serve as nurseries and sources of food for many game and food fish (Austin, 1971; Austin and Austin, 1971a, b). In Florida, Heald and Odum (1970), Sastrakusumah (1971), and Clark (1971) described the reliance of twelve commercially important species of fish and shellfish on mangal swamps. Robas (1970) gave the following weights and values for commercial landings of mangal–dependent species in Florida in 1968: striped mullet (605 100 kg, $100 000); spotted seatrout (1 679 700 kg, $132 000); red drum (396 500 kg, $132 000); blue crab (6 805 100 kg, $1 241 000); and shrimp (14 550 100 kg, $15 719 000).

Robas estimated that one acre (0.405 ha) of undisturbed estuary yields $7980 worth of commercial fish products in twenty years. She also estimated that for every acre of estuary filled or dredged, two additional acres are lost to fish production, yielding a total capitalized loss of $23 940 during the same period. She also cited a report by Wharton (1970) who estimated that the value of the Alcony River swamp in Georgia, 930 ha in area, exceeds seven million dollars per year in fisheries resources.

Heald and Odum (1970) stated the problem succinctly: "If mangrove estuarine systems are destroyed, a valuable protective habitat for juvenile fishes will be lost. More importantly, it will cut off the input of mangrove organic material which is largely contributed in leaf fall. The leaf fall supports the detritus based food webs, a large population of detritus feeders, and their predators. The production of phytopankton and benthic and epiphytic algae in such an area is so much less than the production of detritus of a mangrove origin that the yield of gamefish must decline in proportion to mangrove destruction." This comment on importance of mangals to fisheries was echoed by Bird (1971) in Australia.

Human needs will best be served by recognition that mangal can be managed as a commercial resource and used for recreation. The Australian Conservation Foundation, whose goal is "the wisest possible use, over a long term, of all our natural resources, applied for the benefit of man" published a thoughtful article entitled *Mangroves and Man* (1972). I conclude this review with its list of recommendations for use of mangal, edited to permit world–wide application.

Recommendations of the Australian Conservation Foundation:

(1) That a very careful examination should be made by relevant state fisheries and/or fauna authorities and by plant ecologists before any area of mangroves is allowed to be altered in any way or allocated for any form of development. Its potential contribution to commercial and recreational fisheries as well as long–term environmental stability should be weighed against the proposed alternative land–use values for nearby centres of population. Applications to clear mangrove areas should be made public and state the intended purpose.

(2) The unknown and remote mangrove swamps ... should be surveyed intensively as a matter of urgency. This work should be carried out by a competent land–use authority or, where such does not exist, by the co–operative efforts of the appropriate existing state instrumentalities.

(3) Mangrove islands in the bays and estuaries adjacent to developing areas should be reserved both for aesthetic reasons and for their value to fisheries.

(4) More extensive mangrove areas, in which the habitat will be protected but fishing allowed in adjacent waters, should be reserved. Total protection should be given to areas of special interest by creating national parks or similarly restricted areas.

(5) Research should be intensified by bodies including universities ... and state departments concerned with land and water resources. We need to know far more about our mangrove woodlands — about the animals and plants that live in them, about the operation of the food chains, about the specialized saline environment, about the chemical processes and role of the micro–organisms that produce nutrients in the mud, and about the total productivity of mangrove vegetation in representative localities. Present studies need to be extended beyond the marginal mangroves of the temperate and subtropical areas to the more typical mangrove associations ... The effect of removal of mangrove woodlands on fisheries and on the coastal environment should be assessed ... where extensive development has already occurred and by "before–and–after" studies ... The effects of pollutants, including sewage, mining effluents, and pesticides

on the mangrove complex should also be assessed.

(6) Integrated educational campaigns should be launched by government departments, conservation organizations and local tourist and progress associations, particularly in towns and cities near mangrove areas, to inform the public about their commercial value and their important role in preserving the environment. Such educational campaigns should be designed to counter the common view that the intertidal zone is a wasteland and to achieve recognition of its true value as one of the most prolific biotic regions on earth.

REFERENCES

Abou'l Abbas en–Nebaty, 1230. *Introduction to "Ibu el–Beithar"* (Leclerq), V. Notices des Manuscrits, 23. (Cited from Bowman, 1917.)

Acosta–Solis, M., 1947. Commercial possibilities of the forests of Ecuador — mainly Esmeraldas Provence. *Trop. Woods Yale Univ. School For.*, 89: 1—47.

Acosta–Solis, M., 1959. Los manglares del Equador. *Contrib. Inst. Ecuat. Cienc. Nat.*, No. 29: 82 pp.

Allbrook, R.F., 1973. The identification of acid sulphate soils in northwest Malaysia. In: H. Dost (Editor), *Acid Sulphate Soils, II*. Int. Inst. Land Reclam. Improv., Wageningen, pp. 131—139.

Ancona, U. d', 1954. La pêche et la pisciculture dans les lagunes d'eaux saumâtres. *FAO Fish. Bull.*, 7: 165—194.

Andriesse, J.P., Van Breeman, N. and Blokhuis, W.A., 1973. The influence of mud lobsters *(Thalassina anomala)* on the development of acid sulphate soils in mangrove swamps in Sarawak (East Malaysia). In: H. Dost (Editor), *Acid Sulphate Soils, II*. Int. Inst. Land Reclam. Improv., Wageningen, pp. 11—32.

Angelis, R. de, 1959. La technique des pêcheries dans les lagunes saumâtres. *FAO Gen. Fish. Counc. Méditerr., Stud. Rev.*, No. 7: 16 pp.

Anonymous, 1904. Mangrove barks, and leather tanned with these barks, from Pemba and Zanzibar. *Bull. Imp. Inst., Lond.*, September: 163—166.

Arnott, G.A.W., 1869. *Proc. Linn. Soc., Lond.*, pp. 101—102. (Cited from Canestri et al., 1973.)

Arroyo, P.J., 1971. Propiedades y usos posibles de los mangles de la region del Rio San Juan en la reserva forestal de Guarapiche (Estado Monagas). *Inst. For. Lat.-Am. Invest. Capac., Merida, Venezuela. Bol.*, Nos. 33—34: 53—76.

Augustinius, P.G.E.F. and Slager, S., 1971. Soil formation in swamp soils of the coastal fringe of Surinam. *Geoderma*, 6: 203—211.

Austin, H.M., 1971. A survey of the ichthyofauna of the mangroves of western Puerto Rico during December 1967—August 1968. *Caribb. J. Sci.*, 11: 27—39.

Austin, H.M. and Austin, S.E., 1971a. Juvenile fish in two Puerto Rican Mangroves. *Underwater Nat.*, 7: 26—30.

Austin, H.M. and Austin, S.E., 1971b. The feeding habits of some juvenile marine fishes from the mangroves in western Puerto Rico. *Caribb. J. Sci.*, 11: 171—178.

Australian Conservation Foundation, 1972. *Mangroves and Man*. Carlton, Vic., 12 pp.

Bacon, P.R., 1971. Studies on the biology and cultivation of the mangrove oyster in Trinidad with notes on other shellfish resources. *Trop. Sci.*, 12: 265—278.

Baillaud, M.E., 1912. La situation et la production des matières tannantes tropicales. *J. Agric. Trop.*, 12: 105—107.

Banerji, J., 1958a. The mangrove forests of the Andamans. *Trop. Silvic.*, 20: 319—324.

Banerji, J., 1958b. The mangrove forests of the Andamans. *World For. Congr.*, 3: 425—430.

Banijbatana, D., 1957. Mangrove forest in Thailand. *Proc. 9th. Pac. Sci. Congr., Bangkok*, pp. 22—34.

Barrau, J., 1959. The sago palms and other food plants of marsh dwellers in South Pacific islands. *Econ. Bot.*, 13: 151—162.

Becking, J.H., Den Berger, L.G. and Meindersma, H.W., 1922. Vloed– of mangrovebosschen in Ned.–Indie. *Tectona*, 15: 561—611.

Behrens Motta, A., 1960. Estudios preliminares para la producción de extracto curtiente de corteza de mangle *(Rhizophora mangle)*. *Esc. Farm.*, 20: 28—31.

Bennett, H. and Coveney, R.D., 1959. Mangrove bark from Sarawak. *Trop. Sci.*, 1: 116—130.

Beye, G., 1973. Acidification of mangrove soils after empoldering in Lower Casamance. Effects on the type of reclamation system used. In: H. Dost (Editor), *Acid Sulphate Soils, II*. Int. Inst. Land Reclam. Improv., Wageningen, pp. 359—371.

Bigot, L., 1968. Contribution à l'étude écologique des peuplements halophiles de la région Tuléar. 1 — La mangrove. *Fac. Sci. Ann. Univ. Madagascar*, 6: 237—247.

Bird, E.C.F., 1971. The conservation of mangroves in North Queensland. *North Qld. Nat.*, 39: 6—8.

Bloomfield, C. and Coulter, J.K., 1973. Genesis and management of acid sulfate soils. *Adv. Agron.*, 25: 265—326.

Bowman, H.H.M., 1917. Ecology and physiology of the red mangrove. *Proc. Am. Philos. Soc.*, 56: 589—672.

Boyé, M., 1962. Les palétuviers du littoral de la Guyane française. Resources et problèmes d'exploitation. *Cah. Outre-Mer*, 15: 271—290.

Briolle, C.E., 1969. Le cocotier sur les terres de mangrove au Cambodge: aménagement de plantations familiales. *Oleagineaux*, 24: 545—549.

Buckley, T.A., 1929. Mangrove bark as a tanning material. *Malay. For. Rec.*, 7: 40 pp.

Budowski, G., 1951. *Información sobre manglares*. Preparado por la Oficina Técnica, Dirección Forestal, M.A.C. Mecanografiado. Biblioteca del Ministerio de Agricultura y Criá, Caracas. (Cited in Canestri et al., 1973.)

Budowski, G., 1952. Los manglares de la costa Atlántica de Venezuela. *Ser. For., Min. Agric. Criá, Caracas*, No. 47. (Cited in Bowman, 1917.)

Canestri, V. and Ruiz, O., 1973. The destruction of mangroves. *Mar. Pollut.*, 4: 183—185.

Canestri, V., Ruiz, S.O., Rhode, F.A. and Saavedra, B.L., 1973. Diagnostico de la destruccion de los ecosistemas de manglares en las areas Tucacas—Chichiriviche (Edo. Falcon) y

Carenero (Edo. Miranda). *Inf. Tecn., Rep. Venez., Min. Agric. Criá, Of. Nacl. Pes., Caracas,* No. 61: 33 pp.

Cate, R.B. and Sukhai, A.P., 1964. A study of aluminum in rice soils. *Soil Sci.,* 98: 85—93.

Chapman, V.J., 1970. Mangrove phytosociology. *Trop. Ecol.,* 11: 1—19.

Chapman, V.J., 1975. *Mangrove Vegetation.* Cramer, Lehre, 425 pp.

Chatterjee, D., 1958. Symposium on mangrove vegetation. *Sci. Cult.,* 23: 329—335.

Clark, S.H., 1971. Factors affecting the distribution of fishes in Whitewater Bay, Everglades National Park, Florida. *Univ. Miami, Sea Grant Tech. Bull.,* No. 8: 100 pp.

Clusius, C., 1601. *Rariorum Plantarum Historia.* (Cited in Bowman, 1917.)

Coulter, J.K., 1973. The management of acid sulphate and pseudo–acid sulphate soils for agriculture and other uses. In: H. Dost (Editor), *Acid Sulphate Soils, I.* Int. Inst. Land Reclam. Improv., Wageningen, pp. 255—269.

Cox, A.J., 1911. Philippine firewood. *Philip. J. Sci., Sect. A,* 6. (Cited from Becking et al. , 1922.)

Crossland, C., 1903. Note on dispersal of mangrove seedlings. *Ann. Bot.,* 17 : 267. (Cited in Bowman, 1917.)

Dale, I.R., 1938. Kenya mangroves. *Z. Weltforstwirtsch.,* 5: 413—421.

Das Gupta, D.K., 1969. The response of rice varieties to applied nitrogen under swamp cultivation in Sierra Leone. I. Wet season. *Trop. Agric. Trinidad,* 46: 311—323.

Das Gupta, D.K., 1970. Effects of Cycocel on lodging and grain yield of upland and swamp rice in Sierra Leone. *Exp. Agric.,* 7: 157—160.

Das Gupta, D.K., 1971. Effects of levels and time of nitrogen application and interaction between phosphorus and nitrogen on grain yield of rice varieties under tidal mangrove swamp cultivation in Sierra Leone. *Afr. Soils,* 16: 59—67.

Das Gupta, D.K., 1972a. Effect of methods of nitrogen application on yield of rice varieties under tidal mangrove swamp and boliland cultivation in Sierra Leone. *Ghana J. Agric. Sci.,* 5: 121—126.

Das Gupta, D.K., 1972b. Effects of nitrogen application on nitrogen content of grains of swamp rice in Sierra Leone. *Exp. Agric.,* 8: 155—160.

De la Cruz, A.A., 1969. Mangroves — tidal swamps of the tropics. *Sci. Rev.,* 11: 9—16.

De la Cruz, A.A. and Banaag, J.F., 1967. The ecology of a small mangrove patch in Matabungkay Beach, Batangas Province. *Nat. Appl. Sci. Bull.,* 4: 486—494.

Del Sagrario, C.G., 1957. Recovery of tannin from mangrove bark and its conversion into a colorless tannin extract. *Chem. Abstr.,* 53: 4792D.

Dent, J.M., 1947. Some soil problems in empoldered rice lands in Sierra Leone. *Emp. J. Exp. Agric.,* 15: 206—212.

Doyne, H.C., 1937. A note on the acidity of mangrove swamp soils. *Trop. Agric. Trinidad,* 14: 236—237.

Doyne, H.C. and Glanville, R.R., 1933. Some rice growing soils of Sierra Leone. *Trop. Agric. Trinidad,* 10: 132—138.

Drabble, E., 1908. The bark of the red and white mangroves. *Q.J. Inst. Comm. Res. Trop.,* 3: 33—39.

Drabble, E. and Nierenstein, M., 1907. A note on the West-African mangroves. *Q.J. Inst. Comm. Res. Trop.,* 2.

Driessen, P.M. and Ismangun, 1973. Pyrite–containing sediments of southern Kalimantan, Indonesia. Their soils, management and reclamation. In: H. Dost (Editor), *Acid Sulphate Soils, II.* Int. Inst. Land Reclam. Improv., Wageningen, pp. 345—356.

Due, J.M. and Karr, G.L., 1973. Strategies for increasing rice production in Sierra Leone. *Afr. Stud. Rev.,* 16: 23—71.

Dugros, M., 1937. Le domaine forestier inondé de la Cochine. *Bull. Econ. Indochine,* 40: 283—314.

Durant, C.C.L., 1941. The growth of mangrove species in Malaya. *Malay. For. Rec.,* 10: 3—15.

Fischer, W., 1963. Die Fische des Brackwassergebietes Lenga bei Concepcion (Chile). *Int. Rev. Ges. Hydrobiol.,* 48: 419—511.

Flores, C., 1968. Anotaciones sobre los manglares Venezolanos, su importancia ecologica y economica. *Lagena,* 19: 21—31.

Freise, F.W., 1935. A importancia de conversazao dos mangues como viveiros de peixes. *An. 1° Congr. Nacl. Pesca, Rio de Janeiro, 1934,* pp. 315—319.

Frink, C.R., 1973. Aluminum chemistry in acid sulphate soils. In: H. Dost (Editor), *Acid Sulphate Soils, I.* Int. Inst. Land Reclam. Improv., Wageningen, pp. 131—168.

Graham, R.M., 1929. Notes on the mangrove swamps of Kenya. *J. East Afr. Nat. Hist. Soc., Nairobi,* 36: 157—164.

Grant, C.J., 1973. Acid sulphate soils in Hong Kong. In: H. Dost (Editor), *Acid Sulphate Soils, II.* Int. Inst. Land Reclam. Improv., Wageningen, pp. 215—228.

Hamilton, W., 1846. On the medical and economic properties of the *Rhizophora mangle* or mangrove tree. *Pharmacol. J. Trans.,* 6: 11.

Hart, M.G.R., 1959. Sulphur oxidation in tidal mangrove soils of Sierra Leone. *Plant Soil,* 11: 215—236.

Hart, M.G.R., 1962. Observations on the source of acid in empoldered mangrove soils. I. Formation of elemental sulphur. *Plant Soil,* 17: 87—98.

Hart, M.G.R., 1963. Observations on the source of acid in empoldered mangrove soils. II. Oxidation of soil polysulphides. *Plant Soil,* 19: 106—114.

Hart, M.G.R., Carpenter, A.J. and Jeffrey, J.W.O., 1963. Problems in reclaiming saline mangrove soils in Sierra Leone. *Agron. Trop.,* 18: 800—802.

Heald, E., 1971. The production of organic detritus in a south Florida estuary. *Univ. Miami, Sea Grant Tech. Bull.,* No. 6: 110 pp.

Heald, E. and Odum, W.E., 1970. The contribution of mangrove swamps to Florida fisheries. *Proc. Gulf Caribb. Fish. Inst.,* 22nd Ann. Session, pp. 130—135.

Herre, A., 1953. The pond culture of mullet in the New Territory, Hong-Kong, China. *Proc. 7th Pac. Sci. Congr.,* pp. 457—460.

Hesse, P.R., 1961a. Some differences between the soils of *Rhizophora* and *Avicennia* mangrove swamps in Sierra Leone. *Plant Soil,* 14: 335—346.

Hesse, P.R., 1961b. The decomposition of organic matter in a mangrove swamp soil. *Plant Soil,* 14: 249—263.

Hesse, P.R., 1962. Phosphorus fixation in mangrove swamp muds. *Nature,* 193: 295—296.

Hesse, P.R., 1963. Phosphorus in mangrove swamp mud with particular reference to aluminum toxicity. *Plant Soil,* 19: 205—218.

Hesse, P.R. and Jeffrey, J.W.O., 1963. Some properties of Sierra Leone soils. *Agron. Trop.*, 18: 803—805.

Hickling, C.F., 1970. Estuaries fish farming. In: F.S. Russell (Editor), *Advances in Marine Biology, 8*. Academic Press, New York, N.Y., pp. 119—213.

Hofstede, A.E., Ardiwinata, R.O. and Botke, F., 1953. Fish-culture in Indonesia. *Indo–Pac. Fish. Counc., Spec. Publ.*, No. 2: 129 pp.

Holdridge, L.R., 1940. Some notes on the mangrove swamps of Puerto Rico. *Caribb. For.*, 1: 19—29.

Hora, S.L. and Pillay, T.V.R., 1962. Handbook on fish culture in the Indo–Pacific region. *FAO Fish Biol. Tech. Pap.*, No. 14: 204 pp.

Howes, F.N., 1953. *Vegetable Tanning Materials*. Butterworth, London, pp. 77—79.

Howeson, V., 1804. Preparation of tan made in the East Indies from the bark of the mangrove tree. *Trans. Soc. Arts*, 22: 201.

Huet, M., 1956. Aperçu de la pisciculture en Indonésie. *Bull. Agric. Congo Belge*, 47: 901—958.

Hunter, J.B., 1970. A survey of the oyster population of the Freetown estuary, Sierra Leone, with notes on the ecology, cultivation, and possible utilization of mangrove oysters. *Trop. Sci.*, 11: 276—285.

Jones, W.T., 1971. The field identification and distribution of mangroves in eastern Australia. *Qld. Nat.*, 20: 35—51.

Jordan, H.D., 1954. The development of rice research in Sierra Leone. *Trop. Agric. Trinidad*, 31: 27—32.

Jordan, H.D., 1959. The utilization of saline mangrove soils for rice growing. *Trans. 3rd Int. Afr. Soils Conf., Dalaba*, pp. 327—331.

Jordan, H.D., 1964. The relation of vegetation and soil to the development of mangrove swamps for rice growing in Sierra Leone. *J. Appl. Ecol.*, 1: 209—212.

Keith, H.G., 1935. Timber supply consumption and marketing in North Borneo. *Fourth Br. Emp. For. Congr.* (Cited from Womersley, 1975.)

Kiener, A., 1972. Écologie, biologie et possibilités de mise en valeur des mangroves malgaches. *Bull. Madagascar*, 308: 49—84.

Korringa, P., 1956. Oyster culture in South Africa. Hydrographical, biological and osteological observations in the Knysna lagoons, with notes on conditions in other South African waters. *Div. Fish. Invest., Rep.*, No. 20: 86 pp.

Kuchler, A.W., 1972. The mangrove in New Zealand. *N.Z. Geogr.*, 28: 113—129.

Kuenzler, E.J., 1969. Mangrove swamp systems. In: H.T. Odum, B.J. Copeland and E.A. McMahon (Editors), *Coastal Ecological Systems of the United States, 1*. Institute of Marine Sciences, University of North Carolina, Chapel Hill, N.C., pp. 353—383.

Le Mare, D.W., 1950. The application of the principles of fish culture to estuarine conditions in Singapore. *Proc. Indo–Pac. Fish. Counc.*, 2: 175—180.

Le Maut, E. and Decaisne, J., 1876. *Traité de Botanie Générale*, p. 419. (Cited in Bowman, 1917.)

Liew That Chin, 1970. Research on mangrove forest in Sabah. *Sabah For. Dep. Rep.*, 80 pp. (Cited from Womersley, 1975.)

Lin, S.Y., 1968. Milkfish farming in Taiwan. *Taiwan Fish. Res. Inst., Fish Cult. Rep.*, No. 3: 63 pp.

Macluskie, H., 1952. The reclamation of mangrove swamp areas for rice cultivation. *World Crops*, 2: 129—132.

MacNae, W., 1966. Mangroves in eastern and southern Australia. *Aust. J. Bot.*, 14: 67—104.

MacNae, W., 1968. A general account of the fauna and flora of mangrove swamps and forests in the Indo–West–Pacific region. *Adv. Mar. Biol.*, 6: 73—270.

Marco, H.F., 1935. Systematic anatomy of the woods of the Rhizophoraceae. *Trop. Woods*, 44: 1—26.

Mattox, N.T., 1949. Studies on the biology of the edible oyster, *Ostrea rhizophorae* Guilding, in Puerto Rico. *Ecol. Monogr.*, 19: 339—356.

Ministry of Primary Industries, 1970. *Annual Report on Forestry in West Malaysia (Malaya)*. Kuala Lumpur, pp. 65 66.

Morton, J.F., 1965. Can the red mangrove provide food, feed and fertilizer? *Econ. Bot.*, 19: 113—123.

Mudaliar, C.R. and Kamath, H.S., 1952. Distribution of *Rhizophora mucronata* Lam., in the "back-water" of the west coast and its economic importance. *Madras Agric. J.*, 39: 610—615.

Navalkar, B.S., 1941. Studies in the ecology of mangroves. II. Physical factors of the mangrove soil. *J. Univ. Bombay*, 9: 78—92.

Navalkar, B.S., 1959. Studies in the ecology of mangroves. 1959. VII. Humus content of mangrove soils of Bombay and Salsette Islands. *J. Univ. Bombay*, 28: 6—10.

Navalkar, B.S., 1961. Importance of mangroves. *Trop. Ecol.*, 2: 89—93.

Navalkar, B.S., 1962. Mangrove habitat and its economic exploitation. *Symposium on Evaluation of Tropical Habitat for Production of Food, Fodder, Fuel and Fertilizers*. International Society for Tropical Ecology, Cuttack, India, p. 16.

Navalkar, B.S. and Bharucha, F.R., 1949. Studies in the ecology of mangroves. V. Chemical factors of the mangrove soil. *J. Univ. Bombay*, 18: 17—35.

Navalkar, B.S. and Bharucha, F.R., 1950. Studies in the ecology of mangroves. VI. Exchangeable bases of mangrove soils. *J. Univ. Bombay*, 18: 7—16.

Nearchus, 325 B.C. *Arr Anab VI*. (Cited from Bowman, 1917.)

Nicolas, P.M. and Bawagan, B.O., 1970. Production of high-alpha (dissolving) pulps from bakuan–babae (*Rhizophora mucronata* Lam.). *Philipp. Lumberman*, 16: 4 pp.

Nikolić, M. and Melendez, S.A., 1968. El ostión del mangle, *Crassostrea rhizophorae* Guilding, 1828. Experimentos iniciales en el cultivo. *Nota Invest. Cent. Invest. Pesq. Bauta*, 7: 1—30.

Noakes, D.S.P., 1950. The mangrove charcoal industry. *Malay. For.*, 24: 201—203.

Noakes, D.S.P., 1951. Notes on the silviculture of the mangrove forests of Matang, Perak. *Malay. For.*, 14: 183—196.

Noakes, D.S.P., 1955. Methods of increasing growth and obtaining natural regeneration of the mangrove type in Malaya. *Malay. For.*, 18: 23—30.

Naokes, D.S.P., 1957. Mangrove. *Trop. Silvic.*, 2: 309—318.

Noakes, D.S.P., 1958. Mangrove. *World For. Congr.*, 3: 415—424.

Odum, W.E., 1971. Pathways of energy flow in a south Florida estuary. *Univ. Miami, Sea Grant Tech. Bull.*, No. 7: 162 pp.

Oviedo, G.F., 1526. *Primera parte de la historia natural general de las Indias.* (Cited from Bowman, 1917.)

Pagan–Font, F.A., 1971. Utilization of aquatic resources: specifically the development of aquaculture. *Caribb. Conserv. Assoc. Envir. Newsl.*, 2: 33—38.

Pakrasi, B.B., 1964. Culture of brackish–water fishes in impoundments in west Bengal. *Indo–Pac. Fish. Comm., Bull.*, No. 11.

Pearman, R.W., 1957. Mangrove bark — its value as a tanning material. *Leather Trades Rev.*, 125: 315—316.

Pillay, T.V.R., 1962. Fish farming methods in the Philippines, Indonesia, and Hong–Kong. *FAO Biol. Tech. Pap.*, No. 18: 75 pp.

Pillay, T.V.R., 1965. A bibliography of brackish–water fish culture. *FAO Fish. Circ.*, No. 21: 20 pp.

Pons, L.J., 1964. Pyrites as a factor controlling chemical "ripening" and formation of "cat clay" with special reference to the coastal plain of Surinam. *Agric. Exp. Stat. Paramaribo Bull.*, 82: 141—161.

Pustelnik, W., 1954. Análisis de la corteza del mangle colorado (*Rhizophora mangle*) del oriente de Venezuela y su importancia en la elaboración de extractos curtientes sólidos. *Rev. Soc. Venez. Quím.*, 5: 19—31.

Pynaert, L., 1933. La mangrove Congolaise. *Bull. Agric. Congo Belge*, 24: 185—207.

Robas, A.K., 1970. South Florida's mangrove–bordered estuaries: their role in sport and commercial fish production. *Univ. Miami Sea Grant Inf. Bull.*, No. 4: 28 pp.

Rodriguez, M., Luis, A. and Gil, A.J., 1968. *Informe (a la C.V.F.) sobre la industria manglera colombiana.* Caracas, 25 pp. (Cited from Arroyo, P. 1970.)

Rogers, J.S., 1950—51. Native sources of tanning materials. *U.S.D.A. Yearb.*, pp. 709—715.

Rorison, I.H., 1973. The effect of extreme soil acidity on the nutrient uptake and physiology of plants. In: H. Dost (Editor), *Acid Sulphate Soils, I.* Int. Inst. Land Reclam. Improv., Wageningen, pp. 223—254.

Rosevear, D.R., 1947. Mangrove swamps. *Farm. For.*, 8: 23—30.

Russell, A., 1943. *Conocarpus erecta* (Buttonwood, Zaragosa-mangrove), a new domestic source of tannin. *Chemurgic Dig.*, 2: 27—30.

Sack, *Inspec. Landb. W. Ind., Bull.*, No. 5. (Cited from Drabble and Nierenstein, 1970.)

Sastrakusumah, S., 1971. A study of the food of juvenile migrating pink shrimp, *Penaeus duorarum* Burkenroad. *Univ. Miami, Sea Grant Tech. Bull.*, No. 9.

Savage, T., 1972. Florida mangroves as shoreline stabilizers. *Fla. Dep. Nat. Res., Prof. Pap.*, No. 19: 46 pp.

Schneider, E.E., 1916. Commercial woods of the Philippines: their preparation and uses. *Philipp. Bur. For., Bull.*, No. 14: 179—182.

Schuster, W.H., 1952. Fish culture in brackish water ponds of Java. *Indo–Pac. Fish. Counc., Spec. Publ.*, No. 1: 143 pp.

Schuster, W.H., 1960. Synopsis of biological data on milkfish *Chanos chanos* (Forsk.). *FAO Fish. Biol. Tech. Pap.*, No. 4: 64 pp.

Schuster, W.H., Kesteven, G.L. and Collins, G.E.P., 1954. Fish farming and inland fishery management in rural economy. *FAO Fish Stud.*, No. 3: 64 pp.

Sloane, H., 1725. *A Voyage to the Islands Madeira, Barbados, Jamaica, etc.* (Cited from Bowman, 1917.)

Summers, C.C., 1964. Hawaiian fishponds. *Bernice P. Bishop Mus., Spec. Pub.*, No. 52: 26 pp.

Tampi, P.R., 1960. Utilization of saline mud flats for fish culture. An experiment in marine fish farming. *Indian J. Fish.*, 7: 137—146.

Tan Lee Wah, 1972. Carp culture in Singapore: A case study. *J. Trop. Geogr.*, 35: 67—74.

Terry, E.R. and Das Gupta, D.K., 1973. The effects of nitrogen phosphorus on nematode (*Aphelencoides basseyi*) infestation and grain yields of two rice varieties under tidal mangrove swamp cultivation in Sierra Leone. *W. Afr. Rice Dev. Assoc. Spec. Publ.*, 7 pp.

Theophrastus, 305 B.C. *Historia Plantarum IV.* (Cited from Bowman, 1917.)

Thomson, J.M., 1960. Brackish water fish farming. *Fish. Newsl.*, 19: 17—20.

Thornton, I. and Giglioli, M.E.C., 1965. The mangrove swamps of Keneba, Lower Gambia River Basin. II. Sulphur and pH in the profiles of swamp soils. *J. Appl. Ecol.*, 2: 257—269.

Tomlinson, T.E., 1957a. Changes in a sulphide–containing mangrove soil on drying and their effect upon the suitability of the soil for the growth of rice. *Emp. J. Exp. Agric.*, 25: 108—118.

Tomlinson, T.E., 1957b. Relationship between mangrove vegetation, soil texture and reaction of surface soil after empoldering saline swamps in Sierra Leone. *Trop. Agric. Trinidad*, 34: 41—50.

Trimble, H., 1892. Mangrove tannin. *Contrib. Bot. Lab. Univ. Penn.*, 1: 50—55.

Tung–Pai Chen, 1952. Milkfish culture in Taiwan. *Joint Comm. Rural Reconstr., Fish. Ser.*, No. 1: 17 pp.

Unwin, A.H., 1920. *West African Forests and Forestry.* Dutton, New York, N.Y., 416 pp.

Van Breeman, N., Tandatemiya, M. and Chanchareonsook, S., 1973. A detailed survey on the actual and potential soil acidity at the Bang Pakong Land Development Centre, Thailand. In: H. Dost (Editor), *Acid Sulphate Soils, II.* Int. Inst. Land Reclam. Improv., Wageningen, pp. 159—168.

Vazquez–Yanes, C. et al., 1972. Problemas ecologicos de la explotacion del manglar. In: *Problemas Biológicos de la Región de los Tuxtlas, Veracruz.* UNAM, Guadarrama Impresores, Mexico City, pp. 135—163.

Venkatesan, K.R., 1966. The mangroves of Madras State. *Indian For.*, 92: 27—34.

Vieillefon, J., 1969. La pédogénèse dans les mangroves tropicales. Un example de chronoséquence. *Sci. Sol*, 2: 115—148.

Vieillefon, J., 1973. Sur quelques transformations sédimentologiques et minéralogiques dans les sols sulphates acides du Sénégal. In: H. Dost (Editor), *Acid Sulphate Soils, II.* Int. Inst. Land Reclam. Improv., Wageningen, pp. 99—113.

Wadsworth, F.H., 1959. Growth and regeneration of white mangrove in Puerto Rico. *Caribb. For.*, 20: 59—71.

Walker, F.S., 1937. The management and exploitation of the Klang mangrove forest. *Malay. For.*, 6: 71—78.

Walker, F.S., 1938. Regeneration of Klang mangroves. *Malay. For.*, 7: 71—76.

Walsh, G.E., 1967. An ecological study of a Hawaiian mangrove swamp. In: G.H. Lauff (Editor), *Estuaries. Am. Assoc. Adv. Sci. Publ.*, No. 83: 420—431.

Watson, J.G., 1928. Mangrove forests of the Malay Peninsula. *Malay. For. Rec.*, 6: 1—275.

Watts, J.C.D., 1969. Phosphate retention in acid sulphate pond muds from the Malacca area. *Malay. Agric. J.*, 47: 187—202.

West, R.C., 1956. Mangrove swamps of the Pacific coast of Colombia. *Ann. Assoc. Am. Geogr.*, 46: 98—121.

Wharton, C.H., 1970. *The Southern River Swamp — a Multiple-Use Environment*. University of Georgia, Athens, Ga. (Cited from Robas, 1970.)

Womersley, H.B.S., 1975. Management of mangrove forests: Utilization versus conservation with special reference to the forests of the Papuan Gulf. *Int. Symp. Biol. Manage. Mangroves, Proc.*, 2: 732—741.

Chapter 17

HUMAN USES OF SALT MARSHES

WILLIAM H. QUEEN

INTRODUCTION

People have always been attracted to the coastal zone. In prehistoric times, the abundance of food in marine waters was the primary motivating factor. Later, food–gathering activities were augmented by maritime commerce. More recently, the amenity of a residence on or near the water has been a major attractant for individuals. And the availability of the water itself, as a coolant and as a waste disposal system, has been an attractant for industries. Until the present century, adequate acreage of fastland (areas within the coastal zone not routinely affected by tidal waters) has been available in most regions of the world to meet the spatial needs for residential, commercial, and industrial sites. However, during the past several decades the situation has changed rapidly. Fastland has become scarce — and expensive. Strong pressures have developed in many areas to meet the demand for shore zone sites by the creation of artificial land areas by fill and by the dredging for new lagoons. Frequently, the fill and dredging operations involve tidal marshes. The rate and scale of these operations has been such that substantial marsh acreage has been lost over the past several decades. Data compiled by the Departments of Agriculture and Interior (U.S. Government) indicate that over two million acres (800 000 ha) of salt marsh along the American coast were destroyed between 1932 and 1954. The two million acres represented approximately one quarter of the 1932 total acreage (Teal and Teal, 1969).

Until recently, loss of marsh was not something that many people worried much about. In fact, more often than not, marshes were considered to be worthless wastelands inhabited primarily by unwanted mosquitoes and flies. Almost all marsh alterations, even those resulting in complete marsh loss, were viewed as beneficial. Largely as the result of the work of a few ecologists over the past two decades, this view of marsh value has changed. Results of ecological studies which have been important in establishing the role of marshes as producers of nutrients for coastal waters, as an important habitat for many animal species, and as a potential purifier of domestic waste water have become widely known. These studies will be reviewed in the following section. However, emphasis in this chapter will be placed not on marsh uses but on marsh value. Salt–marsh processes such as organic production and waste–water purification are widely acknowledged but the value, or potential value, to society of such processes is vigorously disputed. Factors contributing to the strong differences of opinion concerning marsh value will also be considered, as will efforts to place monetary values on marsh uses.

MARSH USES

In several areas of the world, Holland being the best known, marshes have been diked and drained for the purpose of creating additional agricultural land (Chapman, 1974a). Once a marsh conversion process has been completed, the diked and drained area is no longer marsh, and the benefits resulting from the subsequent agricultural activity should not be thought of as coming from marsh use. Similarly, economic benefits resulting from residential, commercial and industrial activity occurring on filled marsh sites is in no way dependent on the natural characteristics and processes of marshes and should not be considered as marsh uses or values. Only those uses and values that can be linked, either directly or indirectly, with natural characteristics or processes of marshes should be considered in calculations of marsh values. This criterion is followed in almost all assessments of marsh value and will be used in this chapter.

Even though salt marshes have long been considered to be worthless land, various useful products have been harvested from them over a prolonged period of time. From colonial days in America until the present century, *Spartina patens* marshes along the New England coast were used for grazing. Also, these same marshes were frequently cut for the purpose of obtaining hay. Shoots from the coarser *Spartina alterniflora* plants of the lower marshes were used as thatching for farm–house roofs (Teal and Teal, 1969). Similar uses of products from marshes along the English and northern European coasts have been recorded by Chapman (1974a). Commercial exploitation of marsh products has been more limited than private use. The harvesting of fish and shellfish from the marsh–estuarine system for commercial purposes has occurred on a rather large scale. Teal and Teal (1969) have graphically described the exploitation of waterfowl populations in New England marshes for profit. The use of dried plants as packing material is a less well known commercial use of a marsh product.

With the exception of fish and shellfish production, current efforts to assess the values of undeveloped marshes discount the earlier uses and products described above as being of only marginal value. Present calculations of marsh value emphasize natural characteristics and processess that have been revealed by ecological studies over the past several decades. For the following review, these characteristics and processes have been grouped into four categories: (1) biological production; (2) aquaculture; (3) waste–water assimilation; and (4) other uses.

Biological production

The term "biological production" is frequently applied to a variety of marsh processes. Primary productivity of marsh plants is a type of production, as is the production of shellfish. However, the discussion of biological production in the next several paragraphs will focus on those marsh organisms of immediate utility to man, since this article is concerned with human uses of salt marshes. Immediate utility means that the organisms are harvested, either commercially or recreationally, by man. Hay, thatch and packing material could have been considered at length in any discussion of marsh production during earlier periods. The value of these products is negligible however, in relation to the value of animals harvested from the marsh–estuarine system at the present time and for this reason they will not be discussed.

The extensive use of salt marshes by waterfowl is immediately obvious to even the most casual marsh visitors. These general impressions of marsh use by waterfowl are amply supported by observations of ornithologists for decades. Substantial data has been collected on the various bird species that inhabit salt marshes, the type of marsh (vegetation, salinity, etc.) preferred by each species, and the use (feeding, reproductive and roosting) of the different marsh types by the various bird species (Shaw and Fredine, 1956; Daiber, 1974).

Several fur–bearing species of importance to man inhabit salt marshes. Of these, muskrats, racoons, minks and nutria are among the more important. The type of information available on the use of marshes by fur–bearers is similar to that for waterfowl. Species inhabiting marshes have been identified and their preferences for marsh types has been studied (Stearns et al., 1939; Shaw and Fredine, 1956; Shanholtzer, 1974).

Unlike waterfowl and fur–bearers which physically occupy marshes, finfish and most shellfish depend on salt marshes in only indirect ways. Oysters and clams are harvested from tidal creeks that meander through salt marshes and from estuaries bordered by marshes. Shrimp, during their life cycle, migrate from open ocean to salt marsh and then back to open ocean. Considerable growth occurs while the shrimp are in the marshes. Estuaries bordered by marshes, and coastal waters in the vicinity of extensive marsh–estuarine systems, are among the world's most productive waters in terms of both commercial and sports fisheries (Rounsefell, 1963). Both estuarine shellfish and many of the coastal–water finfish are doubly dependent on coastal salt marshes. The marshes, being an integral part of the estuarine system, are a vital component of the habitat of these estuarine animals. Some estuarine organisms, such as oysters, occupy the marsh–estuarine system throughout their entire life cycle. Other animals, some anadromous fish species for example, utilize limited areas within the system only for brief periods of time. In both cases, however, the marsh–estuarine area is being used as a habitat by the organism. These habitat uses are well known (Palmisano, 1972). Of equal importance with the habitat role, is the function of marshes as producers of nutrient material — especially organic material. From primary productivity studies on *Spartina alterniflora* marshes at Sapelo Island, Georgia, E.P. Odum (1961) concluded that salt marshes were among the most productive ecosystems in the world. Odum's conclusion about salt–marsh productivity was later verified by other investigators (Keefe, 1972). Interest in the primary production of salt marshes was again stimulated with Teal's report (1962) that 45% of the organic material produced in Georgia marshes is flushed into the estuary and eventually into coastal waters. Comparable data on organic export has been presented for the Louisiana marshes (Stowe et al., 1971). Thus, coastal waters bordering extensive salt marshes should possess quantities of organic nutrients sufficient to support large fish and shellfish populations.

Aquaculture

Economic demand has stimulated interest in increasing fish and shellfish production in estuarine waters by the use of aquaculture techniques. Recently developed oyster culture methods in Japan have resulted in yields as high as 4500 pounds of meat per acre (5000 kg ha^{-1}) per year (H.T. Odum, 1971, chapter 13). High shrimp yields have also been obtained in Japan with aquaculture techniques. Culture methods developed by Lunz have been successfully employed with fish in South Carolina (E.P. Odum, 1968). Although all these culture techniques represent commercial exploitation of marshes, they are dependent on a natural marsh process — the production of organic nutrients by marsh plants.

Waste–water assimilation

Grant and Patrick (1970) reported that water flowing out of Tinicum marsh (Pennsylvania, U.S.A.) contained much less phosphorus and nitrogen than the highly polluted water that had flowed into the marsh earlier. Calculations based on differences in ion content of water entering and leaving the marsh indicated that 7.2 kg of phosphorus and 14.7 kg of nitrogen were removed from the polluted water by each hectare of marsh covered. In additional studies, Patrick et al. (1971) found a substantial loss of inorganic nitrogen by denitrification in Delaware Bay marshes. And Sweet (1971) reported that mid–Atlantic marshes along the American coast were able to remove up to 21.8 kg of BOD (biological oxygen demand) per hectare per day. Other investigators have also reported improvements in water quality as a result of the flow of polluted water across marshes (Valiela and Teal, 1972). The magnitude of these purification activities per unit area of marsh has led to considerable speculation that marshes can be used to partially purify domestic waste water, thereby reducing the need for tertiary sewage treatment (Gosselink et al., 1974). Such an activity, if proven to be feasible, would be based on natural processes of marshes.

Other uses

A variety of other marsh characteristics and activities are mentioned with considerable frequency as being useful to man. Foremost among these are control of erosion along estuarine shorelines, prevention of damage resulting from tidal flooding, and the purported role of marshes in global cycles of nitrogen and sulfur (Deevey, 1970). Marshes are thought to reduce the rate of shoreline erosion because marsh plants develop a thick mat of interlacing roots which is a very effective binder of coastal sediments (Redfield, 1972). Removal of shoreline vegetation has been found almost invariably to result in an increased rate of erosion. Conversely, the planting of marsh grasses along an eroding unvegetated shoreline has been shown to substantially reduce the rate of erosion (Wass and Wright, 1969). Arguments for a marsh role in the prevention of damage from tidal floods is related to the fact that residential, commercial and industrial facilities located on filled marshlands suffer inordinate damage from flooding (Gosselink et al., 1974).

Recycling of nitrogen and sulfur on a worldwide basis has recently become a matter of concern. Nitrogen has been accumulating in the earth's surface as a result of the increased use of nitrogenous fertilizers (Delwiche, 1970). Present industrial activities release large quantities of sulfur into the atmosphere which is later converted by nature into sulfuric acid. Microbial processes responsible for denitrification and sulfuric acid reduction require oxidized and reduced zones in close proximity. This important condition is met in tidal marshes but not in many other habitats. Some ecologists are of the opinion that this "life support" activity is an important marsh use (Gosselink et al., 1974).

MARSH VALUE

Generally, the above described characteristics and processes of marshes have not been challenged. Most coastal–zone property owners and developers, as well as conservationists and environmentalists, agree that salt marshes play a role in supporting waterfowl, shellfish and finfish, and that marsh plants along the shoreline reduce the rate of erosion. Also, most academic investigators and government officials who have reviewed data pertaining to aquaculture and marsh effects on waste water agree that there is a potential role for marshes in these areas. Thus, the question confronting society today is not, "What are the uses of marshes in their natural state?" Rather, the important question is, "How should marshes be used?": as nutrient producers for shellfish and finfish, nesting areas for waterfowl, aquaculture, waste-water treatment — uses related to natural characteristics of marshes?; or, should marsh property owners and developers be allowed to freely alter them for residential, commercial and industrial uses? Agreement among the interested parties mentioned above — concerning marsh roles — does not extend to these questions. At least four factors contribute to the lack of agreement. First, most of the economic benefits that are derived from natural marsh processes and characteristics do not accrue to the owners of marsh property (e.g., marsh owners do not profit from commercial fish catches in coastal waters even though the fish are dependent on the privately owned marshes). Second, scientific information concerning many marsh processes and characteristics is very limited — the ability to assimilate waste from polluted waters is based on only a half dozen or so experiments. Third, even for those marsh characteristics and processes that are based on rather extensive scientific data bases, significant differences of opinion exist as to their value to humans. For example, H.T. Odum (1967) attributes considerable importance to the primary productivity of marshes while Rounsefell (1963) minimizes the value of this particular activity. Wass and Wright (1969) cite erosion control as a beneficial human value of marshes while Walker (1973) vigorously disputes this conclusion. Fourth and finally, benefit/cost analysis that is routinely used with traditional market–place items cannot be employed with common–property items such as marsh processes and characteristics.

Of these four factors, only the first — the problem of private ownership — even gives the appearance of being amenable to resolution within the immediate future. A variety of mechanisms for dealing with this problem have been advocated with some frequency over the past several years and the widespread awareness of them has undoubtedly contributed to the optimism. Among the more frequently mentioned mechanisms are: (1) government purchase of privately owned marshes; (2) purchase of scenic easements; (3) restrictive zoning; and (4) property–tax relief in exchange for agreements not to develop. These mechanisms are attractive in that they would, if employed, protect natural–marsh values without unfairly penalizing private citizens by the seizure of their property. However, application of any of these mechanisms with respect to specific cases has proven to be difficult. Invariably, almost unanswerable questions arise. Could the public benefit more by expending the funds for other purposes? Could other marsh acreage of equal value be purchased at less cost? These questions pertain to marsh purchase, of course. However, comparable questions arise with respect to the other options. In almost all instances, the inadequacy of the response to the questions posed above can be attributed to insufficient scientific data, strong differences of opinion concerning the importance (to man) of a marsh characteristic or process, and the inability to place monetary values on the uses of marshes in their natural state — the other three factors contributing to the disagreement over, "How should marshes be used?"

The adequacy of existing scientific data with respect to questions concerning marsh value is a matter of considerable controversy. Some interested parties, especially government officials having responsibility for coastal–zone activities, cite the large number of marsh studies that have already been undertaken over the past several decades and then ask, "Why is more scientific data needed in order to place values on marshes?" Conversely, scientists continually express an urgent need for additional data before expressing opinions on specific management policies concerning marsh use or value. Both positions give the appearance of being easily justified. Lengthy bibliographies can readily be prepared to document the large number of salt–marsh studies. However, a review of the papers themselves will reveal that most of the studies have focused on marsh flora and fauna, life cycles of marsh animals, and descriptive ecology. Relatively few studies of nutrient cycling, food webs and population dynamics are reported. Information on such topics is essential if the routinely asked questions concerning marsh values are to be answered.

Because benefit/cost analysis cannot be used with common–property items, attempts have been made to develop new methodologies for placing monetary values on salt marshes. Gosselink et al. (1974) considered a "component" approach and H.T. Odum (1971) proposed an approach based on marsh energy flow. The component approach involves assigning monetary values separately for each marsh use (biological production, waste–water assimilation, aquaculture, etc.) and then summing the individual values to obtain a total marsh value. Individual values are calculated from data on the monetary value in the market place of products dependent on marsh characteristics and processes. For example, a biological production value was calculated (in part) from data on the monetary value of the yearly fisheries catch along the Georgia coast (Gosselink et al., 1974). Various criticisms have been voiced against the component approach; e.g., the scientific data for some marsh processes is so weak that the monetary values that have been assigned to marsh uses dependent on these processes may be unrealistic (e.g., only limited scientific data have been collected on the capacity of marshes to assimilate waste–water, yet considerable value is attributed to marshes for this role). Also, several marsh uses are conflicting. A marsh could not be used simultaneously for aquaculture and waste–water purification. Nevertheless, the separate values for the two uses are added, along with others, to obtain a total marsh value. Similar criticisms are not raised against the energy flow approach which assumes a relationship between national energy consumption per year and the Gross National Product. However, the energy flow approach has not been widely accepted as a method for placing monetary values on marshes.

MARSH PROTECTION

Deficiencies in scientific data on marshes, strong differences of opinion concerning the value to humans of natural marsh characteristics and processes, and difficulties in placing monetary values on marshes have not proved to be insurmountable obstacles to conservationists and environmentalists concerned with marsh protection and preservation. During the past several years, severe restrictions have been placed on further marsh development by many regional and national governments (Bradley and Armstrong, 1972; Ponder, 1974; Chapman, 1974b). Undoubtedly, enactment of legislation and imposition of policies restricting marsh development were influenced by present concepts of marsh use.

CONCLUSION

Salt marshes have been used by people for centuries — for hunting, fishing, grazing and shellfish harvesting. However, our present concepts of use emphasize marsh characteristics and processes (organic production, waste assimilation, etc.) that have been brought to our attention over the past several decades. Efforts have recently been made to place monetary values on specific marsh characteristics and processes, and on marshes in general. The results of these efforts have not been widely accepted. Hence, agreement has not been achieved on the important question of, "How should marshes be used?" Nonetheless, many coastal areas (governments) have enacted legislation or adopted policies over the past several years that severely restrict further marsh development.

ACKNOWLEDGEMENT

The support of the Research Applied to National Needs Program of the National Science Foundation (Grant No. ERT 74—22179) is gratefully acknowledged.

REFERENCES

Bradley Jr., E.H. and Armstrong, J.M., 1972. A description and analysis of coastal zone and shoreland management programs in the United States. *Univ. Mich. Sea Grant Tech. Rep.*, No. 20: 426 pp.

Chapman, V.J., 1974a. *Salt Marshes and Salt Deserts of the World.* Cramer, Lehre, 2nd ed., 392 pp. (complemented with 102 pp.).

Chapman, V.J., 1974b. Coastal zone management in New Zealand. *Coastal Zone Manage. J.*, 1: 333—345.

Daiber, F.C., 1974. Salt marsh plants and future coastal salt marshes in relation to animals. In: R.J. Reimold and W.H. Queen (Editors), *Ecology of Halophytes.* Academic Press, New York, N.Y., 605 pp.

Deevey, E.S., 1970. In defense of mud. *Bull. Ecol. Soc Am.*, 51(1): 5—8.

Delwiche, C.C., 1970. The nitrogen cycle. *Sci. Am.*, 223(3): 136—146.

Gosselink, J.G., Odum, E.P. and Pope, R.M., 1974. *The Value of the Tidal Marsh.* Center for Wetland Resources, Louisiana State University, Baton Rouge, La., Publication No. LSU—SG—74—03, 30 pp.

Grant, R.R. and Patrick, R., 1970. Tinicum marsh as a water purifier. In: *Two Studies of Tinicum Marsh.* The Conservation Foundation, Washington, D.C., pp. 105—123.

Keefe, C.W., 1972. Marsh production: A summary of the literature. In: *Contributions to Marine Science, 16.* University of Texas, Austin, pp. 163—181.

Odum, E.P., 1961. The role of tidal marshes in estuarine production. *Conservationist*, 15(6): 12—15.

Odum, E.P., 1968. Energy flow in ecosystems: A historical review. *Am. Zool.*, 8: 11—18.

Odum, E.P., 1971. *Fundamentals of Ecology.* Saunders, Philadelphia, Pa., 3rd ed., 574 pp.

Odum, H.T., 1967. Biological circuits and the marine systems of Texas. In: T.A. Olson and F.J. Burgess (Editors), *Pollution and Marine Ecology.* Wiley, New York, N.Y., pp. 99—157.

Odum, H.T., 1971. *Environment, Power and Society.* Wiley, New York, N.Y., 331 pp.

Palmisano, A.W., 1972. Habitat preference of waterfowl and fur bearers in the northern Gulf coast marshes. In: R.H. Chabreck (Editor), *Coastal Marsh and Estuary Management.* Louisiana State University, Baton Rouge, La., pp. 163—190.

Patrick, W.H., Delaune, R.D., Antie, D.A. and Engler, R.M., 1971. Nitrate removal from water at the water—soil interface in swamps, marshes and flooded soils. *Ann. Progr. Rep. PFWOA*, EPA (Project 1605 FJR, LSU).

Ponder, H., 1974. *Survey of State Coastal Management Laws. Special Report.* Chesapeake Research Consortium, Inc. The Johns Hopkins University, Baltimore, Md., 50 pp.

Redfield, A.C., 1972. Development of a New England salt marsh. *Ecol. Monogr.*, 42(2): 201—237.

Rounsefell, G.A., 1963. Realism in the management of estuaries. *Ala. Mar. Resour. Lab., Dauphin Island, Alab., Mar. Resour. Bull.*, No. 1: 13 pp.

Shanholtzer, G.F., 1974. Relationship of vertebrates to salt marsh plants. In: R.J. Reimold and W.H. Queen (Editors), *Ecology of Halophytes.* Academic Press, New York, N.Y., 605 pp.

Shaw, S.P. and Fredine, C.G., 1956. Wetlands of the United States. *U.S. Fish Wildl. Serv., Circ.*, No. 39: 67 pp.

Stearns, L.A., MacCreary, D. and Daigh, F.C., 1939. Water and plant requirements of muskrat on a Delaware tidewater marsh. *Proc. N. J. Mosq. Exterm. Assoc.*, 36: 212—221.

Stowe, W.C., Kirby, C., Brkich, S. and Gosselink, J.G., 1971. Primary production in a small saline lake in Barantaria, Louisiana. *La. State Univ. Coastal Stud. Bull.*, No. 6: 27—37.

Sweet, D.C., 1971. *The Economic and Social Importance of Estuaries.* Environmental Protection Agency, Water Quality Office, Washington, D.C., pp. 49—58.

Teal, J.M., 1962. Energy flow in the salt marsh ecosystem of Georgia. *Ecology*, 43: 614.

Teal, J. and Teal, M., 1969. *The Life and Death of the Salt Marsh.* Little, Brown and Co., Boston, Mass., 274 pp.

Valiela, I. and Teal, J.M., 1972. Nutrient and sewage sludge enrichment experiments in a salt marsh ecosystem. *Int. Symp. on Physiological Ecology of Plants and Animals in Extreme Environments, Dubrovnik.*

Walker, R., 1973. Wetlands preservation and management of Chesapeake Bay: the role of science in natural resource policy. *Coastal Zone Manage. J.*, 1(1): 75—101.

Wass, M.L. and Wright, T.D., 1969. *Coastal Wetlands of Virginia — Interim Report.* Special Report No. 10, Virginia Institute of Marine Science, Gloucester Point, Va., 154 pp.

AUTHOR INDEX [1]

Aaronson, T., 269, *270*
Abderhalden, E., *151*
Abe, N., *339*
Abou'l Abbas en-Nebaty, 347, *358*
Abraham, R., 146, *149*
Acostas Solis, M., 205, 211, 355, *358*
Adams, C.D., 235, *239*
Adams, D.A., 12, 27, 159, *164*
Adriani, M.J., 114, 138, *149, 155*
Agadzhanov, S.D., 12, *27*
Aggrawal, K.R., 243, *259*
Ainslie, J.R., 235, *239*
Airy-Shaw, H.K., 236, *239*
Akins, G.J., *188*
Albani, A.D., 297, *339*
Aldrich, J.W., *106, 108*
Allan, J., 324, *339*
Allbrook, R.F., 356, *358*
Allen, E.A., 72, *76*
Allen, J.R.L., 72, *76*
Allorge, P., 135, *149*
Almquist, E., 135, *149*
Amanieu, M., 144, 146, *150*
Ancona, U.d'., 353, *358*
Anderson, R.R., 161, *164*
Anderson, W.P., *340*
Andrews, F.W., 217, 219, 221–3, 228, *230*
Andriesse, J.P., 356, *358*
Angelis, R. de., 353, *358*
Antie, D.A., *368*
Apley, M.L., 86, *106, 108*
Ardiwinata, R.O., *360*
Armstrong, J.M., 367, *368*
Arndt, C.H., 160, *164*
Arnold, A.F., 160, *164*
Arnott, G.A.W., 347, *358*
Arroyo, P., 349, 355, *358*
Ashton, D.H., *345*
Asprey, G.F., 210, *211*
Athern, W.D., 82, *108*
Atkinson, M.R., *339*
Attiwell, P.M., *339, 341*

Atwater, B.F., *191*
Auden, J.B., 256, *258*
Augustinius, P.G.E.F., 356, *358*
Austin, H.M., 357, *358*
Austin, S.E., 357, *358*
Australian Conservation Foundation, 356, 357, *358*
Australian Water Resources Council, 296, *339*
Avolizi, R.J., *106*
Ayala-Castañares, A., 69, 74, 76–7, *188*, 194, 202, *211*

Baas Becking, L.G.M., *339*
Bacon, P.R., 353, *358*
Baillaud, M.E., 354, *358*
Baker, R.T., *339*
Baker, S.D., 18, *27*
Baker, S.M., 141, *150*
Bakker, C., 116, *150*
Baldwin, J.G., *339*
Ball, J., 215, *230*
Baltzer, F., 288–9, *290*
Banaag, J.F., 352, *359*
Banerjee, A.K., 243, 249, *258, 259*
Banerji, J., 349–50, *358*
Banijbatana, D., 349, *358*
Barbour, M.G., 172–4, 176–8, 181, 184–5, 187, *189, 191*
Barnard, J.L., 181–2, *187, 190*
Barnes, R.D., 90, 105, *106*, 160, *164*
Barnes, R.S.K., *150, 152*
Barratt, T., *344*
Barrau, J., 350, *358*
Barrett, R., *29*
Barsdate, R.J., 148, *152*
Bartholomew, G.A., 102, *106, 108*
Bartlett, D.S., *164*
Bascand, L., 9, *27*
Bawagan, B.O., 352, *360*
Beard, J.S., 294, 312, 314, *340*
Becking, J.H., 350–1, 354, *358*
Beeftink, W.G., 4, 8, 10–11, *27*, 40, 59, 109–10, 112–5, 120–1, 126–35, 137–9, 145, 147–9, *150*

Behrens Motta, A., 355, *358*
Beltran, E., *212*
Benda, W.K., 74, *76*
Bennett, H., 354, *358*
Bennett, I., *340, 344*
Benson, R.H., 76, 182, *187*
Bent, A.C., 96–7, *106*
Berlanta, A.A., 219, *231*
Berry, A.J., 308, *340*
Beye, G., 356, *358*
Bharucha, F.R., 243, 255, *259*, 356, *360*
Bhosale, L.J., 255, *258*
Bibby, P., 320, *340*
Bigarella, J.J., 197, *211*
Bigot, L., 353, *358*
Bilio, M., 19, *27*, 113, 115, 141, 146–7, 149, *150*
Billard, B., *340*
Billings, R.F., 170, *190*
Bird, E.C.F., 11, *28*, 110, *150*, 295, 305, *340, 342, 344*, 356–7, *358*
Black, J.M., 297, *340*
Blackburn, G., 295, *340*
Blake, S.T., 295, *340*
Blasco, F., 241, 243, 248, 250, 252–4, *258*
Blatter, E., 243, 256, *258*
Blokhuis, W.A., *358*
Bloomfield, C., 356, *358*
Blum, J.L., *28*, 159, *164*
Boffey, P.M., 269, *270*
Bohling, M.H., 141, *150*
Bollman, F.H., 185, *187*
Bongiorno, S.F., 100, *106*
Bonham-Carter, G.F., 179, *187*
Boomsma, C.D., *340*
Boorman, L.A., 126, *154*
Børgesen, F., 199, 210, *211*
Bornside, G.H., 118, *150*, 159, *164*
Borror, A.C., 81, *106*
Bostrum, T.E., *340*
Botke, F., *360*
Boughey, A.S., 225, 229, *230*, 235, *239*

---
[1] Page references to text are in roman type, to bibliographical entries in italics.

Boughton, V.H., *344*
Bourdillon, T.F., 255, *258*
Bourn, W.S., 79, *106*, 159, 161, *164, 165*
Bowman, H.H.M., 347, 355, *358*
Boyd, C.E., 163, *164*
Boyé, M., 198, 207–9, *211*, 351, *358*
Boynton, W.R., 163, *164*
Bradley Jr., E.H., 367, *368*
Bradshaw, J.S., 72–3, 75–6, 77, 81–2, *108*, 169, 173, 180, 182–3, 185, *187, 190*
Bradshaw, R., 77
Brass, L.J., *270*
Braun-Blanquet, J., 109, 120, 124, *150, 290*
Breckon, G.J., 178, *187*
Brereton, A.J., 140–1, *150*
Breteler, F.J., 199–200, *211*
Bridgewater, P.G., 306, *340*
Brinsdon, R., 42
Briolle, C.E., 356, *358*
Brkich, S., *368*
Brockhuysen, G.J., *239*
Brown, A.F., 217, *230*
Brown, L.F., 194, *211*
Brown, W.H., 24, 287, *290*
Brümmer, G., 112, *150*
Brünnich, J.C., *340*
Brzyski, B., *152*
Buckley, T.A., 354, *358*
Buckmann, A.R., 185, *188*
Budowski, G., 211, *212*, 352, *358*
Bullock, D.A., 297, *340*
Bünning, E., 263, 265–6, *270*
Burbidge, N.T., 294, 300, *340*
Burger Hzn, D., *290*
Burgess, F.J., *368*
Burkholder, L.M., 118, *150*
Burkholder, P.R., 118, *150*, 159, *164*
Burnett, J.H., *28, 151*
Burrows, E.M., 11, *28*
Burt, W.H., 185, *187*
Burton, P.J.K., 144, *150*
Busgen, M., *290*
Busse, W., *290*
Butler, A.J., *340*
Butot, L.J.M., 146, *153*
Byrne, J.V., 72, *76*

Cabrera, A.L., 196, *211*
Cade, T.J., 102, *106*
Calder, J.A., 170–1, *187*
Calvin, J., 182, *190*
Cameron, A.M., *340*
Cameron, G.N., 181, 183, *187*
Campbell, B., 324, 326–7, 330, *340*
Campbell, J.W., 144, *150*
Canestri, V., 347, 356, *358*

Caratini, C., 243, 253–4, *258*
Cardale, S., *340*
Carey, G., *340*
Carey, K.L., *188*
Carlton, J.T., 182–3, *190*
Carpelan, L.H., 169, *188*
Carpenter, A.J., *359*
Carpenter, E.J., 147–8, *154*
Carr, S.G.M., 344
Carrodus, B.B., 294, *340*
Carter, N., 141–2, *150*
Cate, R.B., 356, *359*
Cavanaugh, L.G., 267–8, *270*
Chabreck, R.H., 193–4, *211, 368*
Chai, P.K., 266, *270*
Champion, H.G., 243, 247, 250–1, *258*
Chanchareonsook, S., *361*
Chapman, V.J., 1–4, 11, 17–9, 21–7, *28*, 33–4, 38, 39, 46, 54, 58, *59*, 67, 79, *106*, 111, 118, 128, 139–42, *150*, 159, 161, *164*, 169, 178, *188, 191*, 197, 199, 202, 210, *211*, 215, 224–5, 228–9, *230*, 233, 235–6, 238, *239*, 243, 247, 256, *258*, 261, 263, 265, 268, *270*, 271, 287–8, 290, 293, 310–1, *340*, 347, *359*, 363–4, 367, *368*
Chatterjee, D., 350, 354, *359*
Chebataroff, J., 195, *211*
Chengapa, B.S., 257, *258*
Chippendale, G.M., 312, 314, *340*
Christian, C.S., *340, 343*
Christiansen, W., 127, *150*
Chrysler, M.A., 161, *164*
Churchill, D.M., 300, *340*
Clapham, A.R., 109, *150*
Clark, S.H., 357, *359*
Clarke, C.B., 243, *258*
Clarke, L.D., 13, *28*, 300, 306, 314, *340, 341, 342*
Cleland, J.B., 295, *341, 344*
Clendenning, K.A., 185, *189*
Clifford, H.T., 295, 306, 319–20, *341*
Clough, B.F., *339, 341*
Clover, E.U., 194, 202, *211*
Clusius, C., 347, *359*
Coaldrake, J.E., 295, *341*
Coleman, F., 296, *341*
Coleman, J.M., 72, *77, 344, 345*
Collier, G., *189*
Collins, G.E.P., *361*
Collins, M.I., 306, 312, 314, *341*
Commonwealth of Australia, Bureau of Meteorology, 296, *341*
Connell, D.W., 344
Connell, W.A., 88, *106*
Connor, D.J., *341*
Conrad, H.S., 161, *164*

Cook, G.H., 29
Cooke, T., 255, *258*
Cooper, A.W., 162–3, *164, 165*, 194, *212*
Cooper, W.S., 170–1, *188*
Copeland, B.J., *212, 360*
Corillion, R., 135, 137, *150*
Cornwell, R.B., *259*
Corré, J.J., 12, *28*
Cottam, C., 79, *106*, 126, *150*, 159, 161, *164*
Cotton, A.D., 141–2, *151*
Cotton, B.C., 324, *341*
Coulter, J.K., 356, *358*
Coveney, R.D., 354, *358*
Coventry, R.J., 54, *59*
Cowan, J.B., 185, *188*
Cox, A.J., 350, *359*
Craig, R.B., *187*
Cranwell, L.M., *343*
Cribb, A.B., 306, 319, *341*
Crichton, O.W., 87–8, *106, 164*
Crocker, R.L., *339*
Cronquist, A., *188*
Crossland, C., 347, *359*
Crow, J.H., 170–1, *188*
Cruz, A.A., *187, 190*
Cuatrecasas, J., 199, 201, 205–6, *212*
Cufodontis, G., 219, 221–3, 228, *230*
Curran, H.A., 72, *76*
Curray, J.R., 168, *188*
Curtis, D.M., 74, *77*
Curtis, S.J., 247, 250, *259*
Curtis, W.M., 297, 314, *341*

Daane, 145
Daetwyler, C.G., 168, *188*
Dahl, E., 130, *151*
Daiber, F.C., 79, 88, 92, 95, *106*, 118, *151*, 159, *164*, 364, *368*
Daigh, F.C., *368*
Daigh, L.V., 102–3, *108*
Dakin, W.J., *341*
Dalby, D.H., 11, *28, 151*
Dalby, R., *151*
Dale, I.R., 352, *359*
Dansereau, P., 198, 203, *212*, 229, *230*
Das Gupta, D.K., 355, *359*
Davies, J.L., 197, *212*
Davis, J.H., 46, 48, *59*, 162, *164*, 194, 197, 202, 209, *212*
Davis, L.V., 12, 24, *28*, 90, 105–6, 160, 162, *164*
Dawson, E.Y., 176, 181, *188*
Day, J.H., 236, *239*
Deb, S.C., *259*
De Balzac, H., 238, *239*
De Corona, A., *340*
Deevey, E.S., 365, *368*

AUTHOR INDEX

De Haan, J.H., 266, *270*
Deignan, H.G., 330, *341*
De Koeijer, P., *155*
De la Bathie, M.H.P., 238, *239*
De la Cruz, A.A., 187, *188*, 352, *359*
Delaune, R.D., *368*
Del Sagrario, C.G., 355, *359*
Delwiche, C.C., 366, *368*
De Molenaar, J.G., 131, 136–7, *151*
Den Berger, L.G., *358*
Den Hartog, C., 109, 119–20, 126–7, 146, *151*
Den Held, A.J., 135, 137, *155*
Dent, J.M., 355, *359*
De Pauw, N., 116, *150*
Depers, A.M., *340*
Descaisne, J., 347, *360*
Devaux, J.P., *28*
DeWitt, P., 95, *106*
De Wolf, L., 144, *155*
Dexter, R.M., 86, *106*, 181, 186–7, *188*
Dey, A.C., *354*
Di Carlo, F., *183*
Dick, R.S., 296, *341*
Dickenson, R.E., 271, 287, *290*
Dickinson, C.H., 118, *151*
Diels, L., 24
Diephuis, J.G.H.R., 208, *212*
Dilcher, T., *182*
Ding Hou, 21, *28*, 259, *341*
Dingman, R., 184–5, *188*
Dixon, P.S., 109, *153*
Doheny, T.E., *165*
Doing Kraft, H., 126, *151*
Dost, H., *358, 359, 361*
Doty, M.S., 169, *188*
Downing, B.M., 144, *154*
Doyne, H.C., 355–6, *359*
Dozier, H.L., 102–3, *106*
Drabble, E., 354, *359*
Driessen, P.M., 356, *359*
Drummond, D., 144, *151*
Drysdale, F.R., *187*
Ducker, S.C., *342*
Due, J.M., 355–6, *359*
Duffey, E., 146, *151*
Dugros, M., 350, *359*
Duncan, W.H., 157, *164*
Dunstan, W.M., *154*
Durant, C.C.L., 349, *359*
Du Rietz, G.E., 109, 142, *151*, 275, *290*

Easton, A.K., 297, *341*
Edmonds, S.J., 306–7, 312, 319, 324, 327, *345*
Egler, F.E., 25, *28*, 161, *165*, 194, 197, *212*
Eichler, H., *341*

Eilers, H.P., *191*
Einarsen, A.S., 185, *188*
El-Abyad, M.S., 221, *230*
Eleuterius, L.N., 194, *212*
Elliott, J.S.B., *151*
Ellis, E.A., 119, 146, *151*
Ellway, C.P., *344*
Emberger, L., 217, *230*
Emery, K.O., 173, *190*
Endean, R., *344*
Engel, H., 146, *151*
Engler, R.M., *368*
Englund, B., 135, *151*
Enright, J., *341*
Epstein, E., 118, *151, 343*
Erxleben, A.W., *211*
Evans, F.C., *153, 166*
Evans, G., 45, *59*, 72, *77*
Ewing, G.C., 74, *77*, 168, 176, *190*
Eyerdam, W.J., 170, *188*

Fairbridge, R.W., 53, *59*, 295, *341*
Fanning, M.E., 163, *165*
Fanshawe, D.B., 207, *212*
Feekes, W., 117, *151*
Fevejee, J.Ch.L., 69, *77*
Feldmann, G., 119, *151*
Feldmeth, R., *189*
Feltham, C.B., 185, 187, *191*
Fenner, C., 295, *341*
Ferrar, H.T., 217, *230*
Ferrigno, F., 88–9, 95–6, *106*
Field, C.D., *340, 341*
Filice, F.P., 173–4, 183, *188*
Findlay, G.P., *339*
Fischer, A.F., 287, *290*
Fischer, P.H., 319, *341*
Fischer, W., 353, *359*
Fisher, W.L., *211*
Fisk, H.N., 194, *212*
Fisk, L.O., 182, 184–5, *189*
Fisler, G.F., 104–5, *107*
Fitzpatrick, E.A., 296, *341*
Fitzsimmons, P., 115, *152*
Flores, C., 353, *359*
Foley, J.C., 296, *341*
Fontes, F.C., 131, *151*
Forester, R.T., *187*
Fornes, A.O., *165*
Fosberg, F.R., 198, *212*, 249, *259*, 271, 273, *290*, 297, *341*
Foster, W.A., 163, *164*
Foxworthy, F.W., 266–7, *270*
Frankenberg, J., *341*
Fraser, L., *340*
Fredine, C.G., 157–8, *165*, 364, *368*
Frei, M., 127, *151*
Freise, F.W., 356, *359*
Frey, R.W., 72, *77*
Freyburg, G.V., 197, *212*

Fries, N., 118, *151*
Frink, C.R., 356, *359*
Fritz, E.S., 184, *188*
Fröde, E.Th., 135–6, *151*
Fry, D.H., 182, 184, *188*

Gabriel, B.C., 181, *188*
Gabriel, C.J., 324, *342*
Gagliano, S.M., *341*
Gallagher, J.L., *165*
Galligar, G.C., 161, *164*
Galtsoff, P.S., *213*
Gamble, J.S., *259*
Ganong, W.F., 160, *164*
Gardner, C.A., *341*
Garey, W., *231*
Geerts, S.J., 146, *151*
Géhu, J.M., 4, 8, *27*, 127–9, 138, 150, *151*
Gentilli, J., *341*
Gerry, B.I., 89–90, *107*
Getz, L.L., 104, *107*
Ghiselin, M.T., *187*
Ghose, B., 256, *259*
Giglioli, M.E.C., 356, *361*
Gil, A.J., *361*
Gill, A.M., 295, 300, *341, 343*
Gill, E.D., 295, *341*
Gill, R., 185, *188*
Gillham, M.E., 115, 144, *151*, 295, *341*
Gillies, M.M., 324, 326–7, *341*
Gillner, V., 11, *28*, 39, *59*, 66, *67*, 110, 112, 121, 124, 127–8, 130–1, 134–6, *151*
Gimingham, C.H., 11, *28*, 112, *151*
Girgis, W.A., 220, *230*
Glanville, R.R., 355, *359*
Glassman, S.F., 271, 274, *290*
Godwin, H., 56, *59*
Godwin, M.E., 56, *59*
Golley, F., 19, 26, *28*
Good, R.E., 161, 163, *164, 165*
Goodall, D.W., *341*
Goodrick, G.N., *341*
Gorodkov, B.N., *67*
Grossenheider, R.P., 185, *187*
Gosselink, J.G., 26, *28*, 365–7, *368*
Graham, R.M., 348–9, *359*
Graham, S.A., 202, *212*
Grant, C.J., 356, *359*
Grant, R.R., 365, *368*
Gray, E.H., 87–8, *107*
Gray, I.E., 12, *28*, 90, 105–6, 160, *164*
Green, C., 38, *59*, *152*
Green, J., *152*, 181, 186, *188*
Grewe, F., 234, 236, *239*
Griffith, W., 243, *259*
Gross, M.G., 168, *188*

Grünwaldt, H.S., 112, *150*
Guppy, H.B., 288, 290
Gupta, A.C., 248, 256, *259*
Gustafson, J.F., 183, 185, *188*

Hackney, A.G., 86, *107*
Hadač, E., 130, *151*
Hails, J.R., *344*
Haines, H., 104, *107*
Halligan, G.H., 297, *342*
Hamilton, A.A., *342*
Hamilton, W., 354, *359*
Hammel, H.T., *231*
Hamon, B.V., 297, *342*
Handley, C.O., 102, *108*
Hannon, N.J., 13, *28*, 300, 306, 314, *340, 341, 342*
Hansens, E.J., 91–2, *107*
Hanson, H.C., 168, 170–1, 180, *188*
Harmsen, G.W., 120, *152*
Harnickell, E., *67*
Harper, R.M., 163, *164*
Harrington, E.S., 91–2, *107*
Harrington, R.W., 91–2, *107*
Harris, E.S., 102, *107*
Harrison, G.G.T., *342*
Harshberger, J.W., 161, *164*, 224, *230*
Hart, J.L., 182, 184, *188*
Hart, M.G.R., 355–6, *359*
Hartwell, A.D., 168, *188*
Hassib, M., *231*
Hastings, J.R., 202, *212*
Hathaway, E.S., 194, *212*, 224, 228, *231*
Hatheway, W.H., 271, *290*
Hattersley, R.T., *342*
Hauman, L., 195–6, *212*
Hauseman, S.A., 86, *107*
Häyrén, E., 142, *152*
Heald, E.J., 163, *164, 259,* 355, 357, *359*
Heath, D.J., 146, *152*
Heath, W.G., 169, *188*
Hedel, C.W., *191*
Hedgpeth, J., *165*
Hedley, C., *342*
Hegerl, E.J., 324, 326, 330, *342, 344*
Hemming, C.F., 217, 219, 221–4, 228, *230*
Hemmingsen, E., *231*
Hemsley, W.B., 290
Henrickson, J., 173, 180, *188, 189*
Herbert, D.A., *342*
Herre, A., 353, *359*
Hesse, P.R., 355–6, *359, 360*
Heydemann, B., 144, 146, *152*
Heyligers, P.L., 267, *270*
Hibino, S., 279, *290*
Hickling, C.F., 353, *360*

Hildebrand, H.H., 194, *212*
Hinde, H.P., 74, 77, 173–4, *188*
Hindwood, R.A., *342*
Hitchcock, C.L., 167, *188*
Ho, Pham-Hoang, 268, *270*
Høeg, O.A., 135, *152*
Hoese, H.D., 194, *212*
Hofmann, W., 67
Hofstede, A.E., 353, *360*
Holdridge, L.R., 211, *212*, 356, *360*
Holle, P.A., 86, *107*
Hollister, T.A., *29*
Hopkins, A.J.M., 312, *342*
Hopkins, M.S., 297, 300–2, 305, 312, 314, *344*
Hoover, R.F., 175, *188*
Hopley, D., 295, *341, 342*
Hope, A.B., *339*
Hope Black, J., *342*
Hora, S.L., 353, *360*
Hosokawa, T., 271–8, *290, 291*
Hotchkiss, N., *165*
Hou, D., 199–200, *212*
Hou-Liu, S.Y., *259*
Hounam, C.E., 296, *342*
House, H.D., 172, *188*
Howard, J.D., 72, 77
Howard, R.A., 210, *212*
Howell, J.T., 173–4, *188*
Howes, F.N., 355, *360*
Howeson, V., 353, *360*
Huet, M., 353, *360*
Hult, R., 124
Hultén, E., 167, 170, 180, *188*
Humbert, H., 238, *239*
Hunter, J.B., 353, *360*
Hunter, P.D., 86, 106, *108*
Hutchings, P.A., 18–9, *28*, 304, 308, 324, 326–7, 330, *342*
Hutchinson, J.N., 39, *59, 152*

Imanisi, K., 274–5, *291*
Ingles, L.G., 182, 185, *188*
Ingwersen, F., *342*
Inman, D.L., 168, *188*
Iredale, T., 324, *342*
Isbell, R.F., *342*
Ismangun, 356, *359*
Ito, K., 12, *28*
Iversen, J., 131, *152*

Jaccard, 179
Jackman, M.E., *340*
Jackson, G., 235, *239*
Jacobs, B.L., *189*
Jacquet, J., 112, *152*
Jakobsen, B., 111, *152*
Jamet, A., 238, *239*
Jamnback, H., 90, *107*

Jardine, F., *342*
Jefferies, R.L., 118, 149, *152*, 170, 180–1, *188*
Jefferson, C.A., 168–9, 172–3, 180, *188*
Jeffrey, D.W., 117
Jeffrey, J.W.O., 355, *359*
Jennings, J.N., 54, *59*, 112, *152, 342*
Jensen, H., *290*
Jervis, R.A., 163, *164*
Jessup, R.W., 295, *342*
Joanen, J., *211*
Johannes, R.E., 148, *153*
Johannessen, C.L., 168, 172, *188*
Johns, D., *60*
Johnson, A.W., 170, *188*
Johnson, D., 56
Johnson, D.H., *342*
Johnson, D.S., 147, *152*, 159, *164*, 224, *230*
Johnson, R.E., *188*
Johnston, I.M., 194, *212*
Johnston, R.F., 100–1, 104–6, *107*
Johnston, T.H., *342*
Jones, A.C., 184, *189*
Jones, D., *342*
Jones, G.N., 172, *189*
Jones, K., 116, 118, *152*
Jones, O.T., *60*
Jones, R.D.H., 148, *153*
Jones, W.T., 305, 309, 312, *342*, 350–1, *360*
Jonker, F.P., 200, 209, *212*
Jónsson, H., 131, *152*
Jordan, H.D., 355, *360*
Joseph, St., 44
Joshi, G.V., 255, *258*, 355
Junagad, C.F., 256, *259*
Jurek, R.M., 182, 184–5, *189*

Kagoshimaken Kyoiku Iinkai, *291*
Kale, H.W., 102, 106, *107*
Kalela, A., 135, *152*
Kalk, M., 229, *230*, 237, *240*
Kamath, H.S., 354, *360*
Kanehira, R., *291*
Kanwisher, J., 83, 182, *191*
Kariyone, T., *291*
Karr, G.L., 355–6, *359*
Karsten, G., 266, *270*
Kassas, M., 215, 217–26, 228–30, *230*
Keay, R.W.J., 199, *212*
Keefe, C.W., 163, *164*, 365, *368*
Keeler, C., *188*
Kenny, R., 297, *342*
Keith, H.G., 350, *360*
Kelley, D.W., 185, *188*
Kerwin, J.A., 86–8, *107*, 160, *164*

AUTHOR INDEX

Kesteven, G.L., *361*
Ketner, P., 149, *152*
Khalid, R.A., 147, *153*
Kiener, A., 238, *239*, 353, *360*
Kikkawa, J., 330
Killian, C., 233, *239*
King, H., *340*
King, R.E., *189*
King, R.J., 297, 307, 319, *342*
Kint, A., 206, *212*, 263–5, *270*
Kira, T., 275, *291*
Kirby, C., *368*
Kitami, H., 281, *291*
Klein, R.M., 197–8, 202, *213*
Klemas, V., 161, *164*
Kloss, C.B., 257, *259*
Knoerr, A., 12, *28*
Koenders, J.W., 144, *152*
Koidzumi, G., *291*
Kolehmainen, S., 1, *28*
Köppen, W., 169, *189*, 271, 279, 281, *291*
Korchagin, A.A., 135, *152*
Kornaś, J., 121, *152*
Korringa, P., 353, *360*
Koster, J.T.L., 11, *28*
Kostermans, A.J.G.H., 247, *259*
Kozloff, E.N., 182–3, *189*
Kraeuter, J.N., 79, 105, *107*, 160, *165*, 182, *189*
Kratochvil, M., 305–6, *342*
Küchler, A.W., 356, *360*
Kudo, Y., 279, *291*
Kuenen, Ph.H., 70, *77*
Kuenzler, E.J., 84–5, *107*, 348, *360*
Kulkarni, D.H., 256, *259*
Kurz, H., 159, *165*, 194, *212*

Lackey, J.B., 185, 187, *189*
Lam, H.J., 267, *270*, 271, *291*
Lamb, F.B., 211, *212*
Lambe, E., 115, *153*
Lambert, J.M., *59*, 112, *152*
Lamberti, A., 197–8, 202, *212*
Lane, R.W., 183, *188, 189*
Lankford, R.R., 69, 74, *77*
Lauff, G.H., *59, 60, 189, 190, 342, 362*
Lawas, J.M., 352–3
Lawson, G.W., 234, *239*
Lazarides, M., *343, 344*
Leach, H.R., 182, 184–5, *189*
Lebret, T., 144, *152*
Lee, D.J., *343*
Lee, J.A., 117, *154*
Lee, J.J., 81–2, *107*
Leechman, A., 199–200, *212*
Leeper, G.W., *344*
Le Forestier, *28*

Leiviskä, J., 127, 135, *152*
Le Mare, D.W., 353, *360*
Le Maut, E., 347, *360*
Lemée, G., 138, 233, *152, 233, 239*
Lent, C.M., 84–5, 106, *107*
Lergessner, D.A., *344*
Le Roy, D.O., 72, *76*
Leu, T., 12, *28*
Libbert, W., 135, *152*
Lid, J., 135, *152*
Liddle, M.J., 11, *28*, 33, *59*, 140, *153*
Lieth, H., 61, *67*
Liew That Chin, 350, *360*
Likens, G., *153*
Lin, S.Y., 353, *360*
Lindeman, J.C., 197, 207, *212*
Linduska, J.P., 184–5, *189*
Llewellyn, L.M., 102, *106*
Lloyd, L.S., 119, *152*
Loder, J.W., *342*
Lorenzen, S., 146, *152*
Lötschert, W., 205, *212*
Luederwaldt, H., 197, *212*
Lugo, A.E., 25–7, *28*, 301, *342*
Luis, A., *361*
Luther, H., 126, *152*
Luxton, M., 90–1, 105–6, *107*, 146, *152*

McAdam, J.B., 267, *270*, 345
McArthur, W.M., *342*
McCann, C., *258*
MacCreary, D., 102–3, *108, 368*
Macdonald, J.D., 330, *342*
Macdonald, K.B., 11–12, 81, *107*, 168–9, 172–4, 176–7, 182, *189, 191*
MacGinitie, G.E., 181, 185, *189*
MacGinitie, N.L., 185, *189*
McGowan, J.H., *211*
McIlwee, W.R., 184, *189*
McKillup, S.C., *340*
McLeod, J., 306, 314, 319, *342*
Macluskie, H., 355, *360*
McMahan, E.A., *212, 360*
McMichael, D.F., 324, *342, 343*
McMillan, C., 202, *212, 343*
McMillan, N.F., 146, *152*
MacMillen, R.E., 104–6, *107*
MacNae, W., 1, 18–9, 22, *28*, 51, 53, *59*, 105, *107*, 202, *212*, 229, *230*, 236–8, *240*, 263–5, 268–9, *270*, 294, 297, 300–1, 305, 324, 326, *342*, 347–8, 353, 355–6, *360*
Macpherson, J.H., 324, *342*
McRoy, C.P., 127, 148, *152*
Mahal, B.E., *191*
Major, J., *191*
Malcolm, C.V., *342*

Mall, R.E., 173–5, 181, *188*
Mani, M.S., 243, *259*
Mann, K.H., 163, *165*
Marco, H.F., 350, *360*
Markley, M.H., 102, *106*
Marker, M.E., 57–8, *59*
Marples, T.G., 149, *152*, 186, *189*
Marsden, I.D., 86, *107*
Marshall, D.R., 39, *59*
Marshall, J.T., 100–1, 105, *107*, 168, 184, *189*
Marshall, N.B., 215, *230*
Marshall, N.F., 168, *190*
Martin, A.C., 144, *152*, 157, *165*, 182, *189*
Martin, A.R.H., *344*
Martyn, E.B., 197, 207, *212*
Mason, E., 118, *152*
Massey, R.E., 217, *230*
Matera, N.J., 81–2, *107*
Mathauda, G.S., 252, *259*
Matthews, D.M., 267, *270*
Mattox, N.T., 353, *360*
Maurer, D., 87, *107*
Mawson, P.M., *342*
May, M.S., 86, *107*
May, R.C., 95, *107*
May, V., 319, *342*
Mazumdar, S.P., 249, *259*
Meade, R.H., 69, *77*
Meanley, B., 96–7, 106, *107*
Meigs, P., 215, *231*
Meindersma, H.W., *358*
Melchior, H., *188*
Melendez, S.A., 353, *360*
Merrill, E.D., 271, 288, *291*
Merrill, K., 183
Meteorological Observatory of the South Seas Bureau, *291*
Metzgar, R.G., 162, *165*
Meyer, K.O., 146, *153*
Millard, N.A.H., *239*
Miller, K.G., 87, *107*
Miller, W.R., 161, *165*
Miner, R.W., 160, *165*
Ministry of Primary Industries [Malaysia], 348, *360*
Miranda, F., 198, *212*
Miyata, I., 282–3, *291*
Miyawaki, A., 12, *28*
Möbius, K., 238, *240*
Moffitt, J., 184, *189*
Mohr, C., 194, *212*
Moldenke, H.N., 199, 201, *212*, 255, *259*
Molfino, J.F., 196, *212*
Molinier, R., 12, *28*
Monroe, G.W., 184, *189*
Montasir, A.H., *231*

Monteith, G.B., 309
Montgomery, S.K., *343*
Moore, C.J., 95, *108*
Moore, I., 183, *189*
Moore, J.J., 115, *153*
Moorehouse, F.W., *343*
Morley, J.V., 147, *153*
Morgan, M.H., 161, 163, *165*
Morgan, T., 1, *28*
Morris, D., *342*
Morton, J.F., 352, *360*
Mörzer Bruijns, M.F., 146, *153*
Mountford, C.P., *340, 344*
Mudaliar, C.R., 354, *360*
Mudie, P.J., 173, 175–7, 180–3, *189, 190, 191*
Mueller-Dombois, D., *67*
Muenscher, W.C., 172, *189*
Mukherjee, A.K., *259*
Mukherjee, B.B., 251, *259*
Muller, J., *259*, 300, *343*
Munz, P.A., 167, 175, *189*
Murdock, M.B., 163, *165, 166*
Murray, B.J., 294, *342*
Murray, J.W., 73, 77, 82, 105, *108*
Musgrave, A., *343*
Mustafa, F.R., 255, *259*

National Estuary Study [U.S.A.], 158, *165*
Navalkar, B.S., 243, 255, *259*, 355–6, *360*
Nearchus, 347, *358*
Nelson, A.L., *152, 189*
Nelson, D.J., *153, 166*
Nelson-Smith, A., 1, *28*
Neuenschwander, L.F., 176, *189*
Newell, B.S., 297, *343*
Newell, R., 146, *153*
Nicol, E.A.T., 79, 86, 105, *108*, 146, *153*
Nicolas, P.M., 352, *360*
Nielsen, N., 37, 40, *59*
Nienburg, W., 141, *153*
Nienhuis, P.H., 141–4, *153*
Nierenstein, M., 354, *359*
Niering, W.A., 161, *165*
Nikolić, M., 353, *360*
Nishihara, Y., *291*
Nixon, S.W., 161, 163, *165*
Noakes, D.S.P., 349, *360*
Noble, E.R., 184, *189*
Nordhagen, R., 130, 134–6, 138, *153*
Nordstrom, C.E., 168, *188*
Northcote, K.H., 295, *343*
Northrop, A.R., 210, *212*

Oberdorfer, E., 127, *153*

Odani, N., 282–3, *291*
Odum, E.P., 25–6, *28*, 147–9, *153*, 163, *165*, 186–7, *190*, 365–7
Odum, H.T., 19, *28, 212, 360,* 365–7, *368*
Odum, W.E., 148, *153*, 355, 357, *359, 360, 368*
Ohba, T., 12, *28*
Ohno, T., 287, *291*
Oklowski, W., 90, *108*
Olson, T.A., *368*
Omura, M., *291*
O'Neil, T., 194, *212*
Oney, J., 96–7, 105–6, *108*
Ono, Y., *291*
Orebamjo, T.O., 117, *154*
O'Reilly, H., 139, *153*
Orians, G.H., 269, *270*
Orr, A.P., *343*
Osborn, J.H.S., 297, *343*
Osborn, T.B.G., 295, 305–6, 314, *342*
Ostenfeld, C.H., 120–1, *153, 343*
Oviatt, C.A., 161, 163, *165*
Oviedo, G.F., 347, *361*
Owen, M., 100, *108*
Ownbey, M., *188*

Packham, J.R., 11, *28*, 33, *59*, 140, *153*
Pagan-Font, F.A., 353, *361*
Paijmans, K., 267, *270*
Pakrasi, B.B., 353, *361*
Palmisano, A.W., 103, *108, 211,* 365, *368*
Pancer, E., *152*
Pantin, G., 139, *153*
Paradiso, J.L., 102, *108*
Parham, J.W., 288, *291*
Park, R.B., *191*
Parke, M., 109, *153*
Parker, F.L., 82, *108*
Parkinson, C.E., 257, *259*
Parodi, L.R., 195, *212*
Parsons, R.F., 295, *343*
Patrick, R., 365, *368*
Patrick, W.H., 147, *153*, 365, *368*
Patton, R.T., 297, 305–6, 314, *343*
Paulsen, O., *343*
Paviour-Smith, K., 80, *108*, 147, *153*, 181, 186, *190*
Payne, K.T., 90, *108*, 146, *153*
Pearman, R.W., 355, *361*
Pearse, A.S., 87, 105, *108*
Peck, M.E., 169, 172, *190*
Pedley, L., *342*
Peña, G.M., 202, *212*
Penfound, W.T., *165*, 194, *213*, 224, 228, *231*
Perales, P., 205, *231*

Percival, M., 305, *343*
Perring, F.H., 137, *153*
Perry, R.A., *343, 344*
Pestrong, R., 33, 40, *59*, 71, 77, 168, 173, *190*
Peterson, R.T., 182, 184, *190*
Pethick, J., 33, 42–3, 45, *59*, *153*
Pfeiffer, E.W., 269, *270*
Philip, G., 121, *153*
Phipps, C.V.G., *344*
Phleger, F.B., 69–76, 77, 81–3, 105, *108*, 146, *153*, 168, 176–7, *188, 190*
Pidgeon, I.M., *59*, *343*
Pignatti, S., 127, *153*
Pigott, C.D., 117, *153*
Pilgrim, S.A.L., 161, *165*
Pillay, T.V.R., 353, *360, 361*
Piotrowska, H., 111, 127, 135–6, *153*
Pires, J.M., 199, *213*
Pitman, M.G., *339*
Pizzey, J.M., 126, *154*
Pohle, R., *67*
Polderman, P.J.G., 142–3, *153*
Polunin, N., 170, *190*
Pomeroy, L.R., 148, *153*, 159, *165*
Ponder, H., 367, *368*
Pons, L.J., 207–9, *213*, 356, *361*
Poore, M.E.D., 4, *28*
Pope, E.C., *340*
Pope, R.M., 26, *28*, 365–7, *368*
Porsild, A.E., 170, *190*
Post, E., 306–7, 319, *343*
Postma, H., 70, *77*
Poulson, T.L., 102, *108*
Prain, D., 243, *259*
Preuss, J., 135, *153*
Prud'homme van Reine, W.F., 142–3, *153*
Pugh, G.J.F., 118, *153*
Purer, E.A., 173, *190*
Puri, H.S., 74, *76*
Pustelnik, W., 355, *361*
Pynaert, L., 206, *213*, 354, *361*

Queen, W.H., *28, 29,* 107, *108*, 118, 149, *154,* 159, *164, 165, 189,* 271, 363, *368*
Qureshi, I.M., 243, *259*

Raabe, E.W., 24, *28*
Ragotzkie, R.A., 148, *154*
Rains, D.W., *343*
Rajagopalan, V.R., 253, *259*
Rand, A.L., *270*
Ranwell, D.S., 11, *28*, 34, *59*, 79, *108*, 126, 129, 140–1, 147, *154*, 181, *190*
Rao, R.S., 253, *259*

AUTHOR INDEX

Rao, T.A., 243, 252, *259*
Rao, V.S., 243, *259*
Rasmussen, E., 126–7, *154*
Raunkiaer, C., 23, 226, *231*, 273, 276, *291*
Recher, H.F., 18–9, *28*, 185, *190*, 304, 308, 324, 326–7, 330, *342*
Rechinger, K., 271, *291*
Redfield, A.C., 38, *59*, 161, *165*, 168, *190*, 365, *368*
Reeder, W.A., 185, *190*
Rees, T.K., 142, *154*
Reimnitz, E., 168, *190*
Reimold, R.J., 12, *28, 29, 107, 108*, 148–9, *153, 154*, 157, 159, 163, *164, 165, 189, 271, 368*
Reinecke, F., *291*
Reish, D.J., 182, *190*
Remane, A., 115, 126, *154*
Reye, E.J., 308, *343*
Reynolds, S., *344*
Rhode, F.A., *358*
Richards, F.J., 37, *60*, 140, *154*
Richards, P.W., 263, *270*
Ricketts, E.F., 182, *190*
Riggert, T., *343*
Riley, C.M., 72, *76*
Riley, G.A., 185, *190*
Ringuelet, E.J., 195–6, *213*
Rippingale, O.H., 324, *343*
Robas, A.K., 357, *351*
Robbins, C.S., 97, *108*
Robbins, R.G., 210, *211*, 267, *270*
Rochford, D.J., 297, *343*
Rodriguez, M., 355, *361*
Roe, E.M., *344*
Roedel, P.M., 182, *190*
Roffman, B., 148, *153*
Rogers, J.S., 355, *361*
Rogers, R.W., 307, 320
Romney, D.N., 198, *213*
Ronaldson, J.W., 18, 22, *28*, 46, 54, *59, 67*
Rorison, I.H., *151, 153*, 356, *361*
Rosevear, D.R., 206, *213*, 235, *240*, 351, *361*
Rounsefell, G.A., 364, 366, *368*
Roxburgh, W., 243, *259*
Royal Netherlands Meteorological Institute, 297, *343*
Rubel, E., 281, *291*
Rubin, M., *59*
Ruhland, W., *149*
Ruiz, S.O., 347, 356, *358*
Russell, A., 355, *361*
Russell, F.S., *360*
Russell, G.B., *342*
Russell, R.J., *212*
Russell-Hunter, W.D., 86, 106, *108*
Ryther, J.H., *154*

Ryukyu Seifu Bunkazai Hogo Iinkai, *291*

Saavedra, B.L., *358*
Sabnis, T.S., *258*
Sack, 354, *361*
Saddler, H.D., *339*
Saenger, P., 293, 297, 300–2, 305, 312, 314, 319–20, *344*
Safir, S.R., 88, *108*
Sahni, K.C., *259*, 350
Saint-Yves, A., 196–7, *213*
Salt, R.B., *344*
Salvoza, F.M., 199, *213*, 288, *291*
Sandee, A.J.J., 144, *155*
Sarma, A.V.N., 347, *361*
Sass, W., 148, *154*
Sastrakusumah, S., 357, *361*
Sastry, A.R.K., 243, 252, *259*
Sauer, J.D., 13, *28*, 54, 202, *213*, 295, 304, *344*
Savage, T., 349, *361*
Savory, H.J., 229, *231*, 234, *240*
Saxena, S.K., 256, *259*
Schmeisky, H., 111, *154*
Schmelz, G.W., 92, *108*
Schneider, E.E., 350, *361*
Schnell, R., 229, *231*, 235, *240*
Scholander, P.E., 225, *231*
Scholl, D.W., 70–1, *77*
Schou, A., 40, 41, *60*
Schroeder, D., 112, *150*
Schultz, G.A., 183, *190*
Schuster, W.H., 353, 355, *361*
Schwartz, B., 88, *108*
Scott, D.B., 182, *190*
Segura, L.R., 69, *76*, 194
Sekizuka, R., 282–3, *291*
Sellmann, P.V., 188
Seneca, E.D., 159, *165*
Sernander, R., 124
Serventy, D.L., 146, *154*
Seth, S.K., 243, 247, 250–1, *258*
Shaler, N.S., 72, *77*
Shanholtzer, G.F., 79, *108*, 159, *165*, 364, *368*
Shanware, P.C., 252, *259*
Shapiro, M.A., *344*
Shaw, S.P., 157–8, *165*, 364, *368*
Shenton, L.R., 148, *153*
Shimada, Y., 279, *290*
Shimizu, K., 282–3, *291*
Shimwell, D.W., 109, 120, *154*
Shinde, S.D., 255, *259*
Shine, R., 308–9, 320, 324, 326, 330, *344*
Short, A.D., 168, *190*
Shreve, F., 167, 176, *190*
Shure, D.J., 102, 104, *108*

Shuster, C.N., *190*
Sibley, C.G., 101, 104, *108*
Sidhu, S.S., 243, *259*
Simmoneau, P., 233, *240*
Simmons, E.G., 74, 77
Simons, J., 143, *153, 154*
Skottsberg, C., 196, *213*
Slager, S., 356, *358*
Slater, P., 330, *344*
Sloane, H., 347, *361*
Smalley, A.E., 149, *153*, 163, *165*, 186, *190*
Smallwood, M.E., 86, *108*
Smith, C.F., 175, *190*
Smith, C.T., 112, *152*
Smith, F., *340*
Smith, G.G., 297, 314, *344*
Smith, J.B., 88, *108*
Smith, L.S., *344*
Smith, R.I., 162, *165*, 182–3, *190*
Smith, W.G., *341*
Smythies, B.E., 269, *270*
Sneath, R.H.A., 179, *190*
Snedaker, S.C., 25–7, *28*, 301, *342, 362*
Snelling, B., *344*
Sokal, R.R., 179, *190*
Somerville, J., 297, 314, *341*
Southward, A.J., 229, *231*
Souza Sobrinho, R.J., 197–8, 202, *213*
Specht, M.M., 293, *341*
Specht, R.L., 293–5, 297, 305–7, 312, 314, 319–20, *340, 341, 342, 344*
Speck, N.H., *344*
Spencer, R.S., 297, *344*
Spender, M.A., *344*
Spetzman, L.A., 170, *190*
Spinner, G.P., 157–8, *165*
Squires, E.R., 161, 163, *165*
Stamp, L.D., 50, *60*
Stearns, L.A., 102–3, *108*, 364, *368*
Stebbings, R.E., 146, *154*
Stebbins, R.C., 182, 184, *190*
Steele, J.H., *153*
Steenson, J., *344*
Steers, J.A., 31, 33, 37, 53, *59, 60, 150, 151, 154*, 295, *344*
Steever, E.Z., 163, *165*
Steiner, M., 66, *67*, 224, *231*, 238–9, *240*
Steinführer, A., 111, *154*
Stellfeld, C., 197, *213*
Stephens, F.R., 170, *190*
Stephenson, A., 52, 53, *60*, 228–9, *231, 344*
Stephenson, T.A., 52, 53, *60*, 228–9, *231*, 295, 324, *344*

Stephenson, W., 295, 324, 327, 330, *344*
Sterner, R., 135, *154*
Steup, F.K.M., 264, *270*
Stevenson, R.A., 173, *190*
Stewart, 183
Stewart, G.A., *340*
Stewart, G.R., 117, *154*
Stewart, R.E., 96–100, 105, *108*
Stick, M.M., *344*
Stoddart, D.R., 42, 48, 52, 54, 55, *60*, 209, *213*
Stonehouse, B., 185, *190*
Story, R., *344*
Stowe, W.C., 365, *368*
Strenzke, K., 146, *154*
Stroud, L.M., 163, *165*
Sukhai, A.P., 356, *359*
Summers, C.C., 353, *361*
Swain, F.M., 76, 77
Sweet, D.C., 365, *368*
Symon, D.E., 295, *344*

Täckholm, V., 219, 221, 228, *231*
Tadros, T.M., 219, *231*
Tagawa, H., 271, 281
Tagunova, L.N., 12, *28*
Tampi, P.R., 353, *361*
Tandatemiya, M., *361*
Tandy, C., *344*
Tan Lee Wah, 353, *361*
Tansley, A.C., 120, *154*, 228–9, *231*
Taylor, B.W., 267, *270, 344*
Taylor, M.C., 11, *28*
Taylor, N., 161, *165*
Taylor, R.L., 170–1, *187*
Taylor, W.R., 330
Teal, J.M., 80, 83, 87, 88, *108*, 147–9, *154*, 181–2, 186–7, *190*, 363–5, *368*
Teal, M., 181, *190*, 363–4, *368*
Teichert, C., 53, *59*, 295, *341*
Terry, E.R., 355, *361*
Thanikaimoni, G., 243, 253–4, *258*
Thelin, P.K., *187*
Theophrastus, 347, *361*
Thom, B.G., 25, *28*, 48, 49, *60*, 71, 77, 197, 202–3, 206, 209, *213*, 295, 300, 303, 305, 312, *344, 345*
Thomas, D.P., *340*
Thomas, J.H., 170, 180, *191*
Thomas, W.H., 74, 77
Thompson, D.E., *165*
Thompson, J.W., *188*
Thompson, R.W., 194, *213*
Thomson, J.M., *339, 344*, 353, *361*
Thomson, R.C.M., 235, *240*
Thorne, R.F., 193–4, *213*
Thornton, I., 356, *361*

Thorsted, T.H., 176, 180, *190*
Thothathri, K., 257, *260*
Timmins, R.D., 324, 326, 330, *342*
Tomkins, I.R., 100, *108*
Tomlinson, T.E., 355, *361*
Tooley, H.J., 39, *60*
Trimble, H., 354, *361*
Troll, W., *344*
Troup, R.S., 255, *260*
Truman, R., *344*
Tshirley, F.H., 269
Tung-Pai Chen, 353, *361*
Turner, J.S., 295, *344*
Turner, R.M., *212*
Tutin, T.G., 101, 113, *150, 154*
Tüxen, R., 27, 137–9, *150*, 154
Tweedie, M.W.F., 265, *270*
Twenhofel, W.S., 168, *190*
Tyler, G., 117, 149, *154*

Udell, H.J., 161, 163, *165*, 181, *190*
Uhler, F.M., *165*
Umezu, Y., 12, *28*
Unwin, A.H., 350, *361*
Upson, J.E., 168, *190*
Urner, C.A., 95, *108*
Ursin, M.J., 160, *165*
Usinger, R.L., 160, *165*

Valentin, H., *345*
Valiela, I., 147–8, *154*, 365, *368*
Van Andel, Tj.H., 77
Van Bodegom, A.H., 263, *270*
Van Breeman, N., 356, *358, 361*
Van der Maarel, E., 109, 120, 122, *155*
Van Goor, A.C.J., 121, 126, *154*
Van Leeuzen, Chr.G., 138, *155*
Vann, J.H., 205, 207–8, *213*
Van Raalte, C.D., 147–8, *154*
Van Regteren Altena, C.O., 146, *151, 153*
Van Royen, P., 294, 306, *345*
Van Steenis, C.G.G.J., 21, 24, 26, *29*, 261, *270*, 300, *342, 345*
Van Straaten, L.M.J.U., 70, 77
Van Wijngaarden, A., 144, *154*
Vazquez, C., 197–8, *213*
Vazquez-Yanes, C., 352, *361*
Venkatesan, K.R., 253, *260*, 354, *361*
Venkateshwarlu, V., 253, *260*
Verdcourt, B., 217, 223, 228, *231*
Verhoeven, B., 112–3, *154*
Vermeer, D.E., 209, *213*
Viano, *28*
Vieillefon, J., 356, *361*
Vierack, L.A., *188*
Vogl, R.J., 173, 175–6, *191*
Vogt, W., 97, *108*

Voltzkow, A., 238, *240*
Vroman, M., 143, *154*
Vu Van Cuong, 241, *260*

Wadsworth, F.H., 350, 356, *361*
Wagner, D., 194, *212*
Wagner, D.T., 119, *154*
Wagner, K., 159, *164*
Waisel, Y., 114, 118, *154*
Waits, E.D., 162–3, *164, 165*
Walker, D., 300, *345*
Walker, F.S., 348–9, *361, 362*
Walker, R., 366, *368*
Wallace, C., *340*
Walls, W.J., 90
Walsh, G.E., 1, 2, *29*, 162, *165*, 270, *345, 356, 362*
Walter, H., 64–5, 67, 225, *231*, 238–9, *240*
Walters, S.M., 137, *153*
Walton, W.R., 81, *108*
Warburg, E.F., 109, *150*
Warlen, S.M., 92, *108*
Warme, J.E., 71, 77, 173, 181–3, *191*
Warming, E., 34, *60*
Warren, A.K., *212*
Wass, M.L., 162–3, *165*, 365–6, *368*
Waters, R.J., 126, *154*
Watson, J.G., 24, 50, *60*, 262–3, *270*, 348–9, *362*
Watts, J.C.D., 353, *362*
Webb, L.J., *345*
Webber, E.E., 159, *165*
Webster, L., *345*
Wegener, A.L., 56
Wendelberger, G., *155*
West, K.R., *339*
West, O., 236, *240*
West, R.C., 50, *60*, 205, *213*, 355, *362*
Westhoff, V., 109, 120, 122, 135–8, *154, 155*
Westing, A.H., 269, *270*
Wharton, C.H., 357, *362*
White, C.T., *345*
White, J., 115, *152*
Whitlatch, R.B., 183, *191*
Whitley, G.P., *345*
Whitlock, F.L., *345*
Whittaker, R.H., *155*
Whittard, W.F., 77
Wiehe, P.O., 140, *155*
Wieser, W., 83, *108*, 182, *190, 191*
Wiggins, I.L., 167, 170, 176–7, 180, *190, 191*
Wilimovsky, N.J., *189*
Williams, R.B., 163, *165, 166*
Willis, J.H., 295, 297, 312, 314, *345*
Wilson, J.M., 119, *152*

AUTHOR INDEX

Wilson, P.G., *345*
Wilson, R.F., 19, *28*
Wobber, F.J., 161, *164*
Wolf, P.L., 97, 105, *107*, 160, *165*, 182, *189*
Wolfe, J.N., *189*
Wolff, W.J., 127, 144, *155*
Wollaston, E.M., 307, 320
Womersley, H.B.S., 306–7, 312, 319, 324, *345*, 350, 352, *362*
Womersley, J.S., 267, *270*, 305, *343*
Wood, E.J.F., *339, 345*

Wood, J.F., 305–6, 312, 314, *343, 345*
Woodell, S.J.R., 90, *108*, 146, *155*
Wright, J.O., 157, 162–3, *166*
Wright, L.D., 168, *190, 344, 345*
Wright, T.D., 162–3, *165*, 365–6, *368*
Wright, W.W., *191*
Wyer, D.W., 126, *154*

Yamashiro, M., 287, *291*
Yapp, R.H., 33, *60*

York, H.H., 147, *152*, 159, *164*, 224, *230*
Yoshino, T., 182

Zahran, M.A., 215, 217–20, 223–6, 228–30, *230, 231*
Zarudsky, J., *165*
Zedler, J.B., *191*
Zilberberg, M.H., 92, *108*
Zim, H.S., *152, 189*
Zobell, C.E., 185, 187, *191*
Zonneveld, I.S., 112, 116, *155*
Zucca, J.J., 184, *191*

SYSTEMATIC INDEX

Abudefduf sexfasciatus (Lacépède), 329
Acacia, 246
    A. cunninghamii Hook., 303
    A. mearnsii de Wild, 246
    A. nilotica (L.) Del. subsp. *indica* (Benth.) Brenan, 256
    A. tortilis (Forsk.) Hayne, 224
    A. translucens A. Cunn., 304
Acanthaceae, 243, 312
Acanthizidae, 336
Acanthocactus, 65
Acanthopagrus australis (Günther), 330
Acanthorhynchus tenuirostris (Latham) (eastern spinebill), 337
Acanthozostera gemmata (Blainville), 321
Acanthus, 266, 267, 288
    A. ebracteatus Vahl., 243, 273, 288, 348
    A. ilicifolius L., 17, 243, 252, 254, 255, 266, 270, 288, 312, 351
    A. volubilis Wall., 243
Acarina (mites), 90, 91, 146
Accipiter cirrocephalus (Vieillot) (collared sparrowhawk), 331
A. fasciatus (Vig. and Horsf.) (brown goshawk), 331
A. novaehollandiae (Gmelin) (grey goshawk), 332
Accipitridae, 331, 332
Achillea borealis Bong., 175
Acridotheres tristis L. (Indian mynah), 337
Acrocephalus australis (Gould) (read warbler), 336
Acrostichum, 4, 22, 203, 235, 238, 266, 267, 288
    A. aureum L., 4, 20, 65, 198, 199, 211, 235–237, 244, 251, 262, 264, 268, 273, 275, 281, 288, 289, 348
    A. speciosum Willd., 17, 264, 268, 273, 275, 312
Actinocythereis subquadrata Puri, 76
Aedes, 18, 160
    A. amesii (Ludlow), 270
    A. butleri Theo., 270
    A. cantator (Coquillett), 88, 89
    A. detritus (Haliday), 79
    A. fumidus Edwards, 270
    A. littoreus Colless, 270
    A. niveus Ludlow, 270
    A. sollicitans (Walker), 88, 89
Aegialitis, 20, 52, 53
    A. annulata B.Br., 293, 310–303, 311, 339, 351
    A. rotundifolia Roxb., 244, 250

Aegiceras, 20, 22, 266, 268, 288
    A. corniculatum Blanco, 17, 20, 244, 250, 254, 258, 266–268, 287, 293, 303, 304, 311, 350, 351
    A. majus Gaertn., 348
Aegintha temporalis (Latham) (red-browed finch), 337
Aeluropus, 218, 220–222, 224–228, 230
    A. brevifolius Nees ex Steud., 228
    A. lagopoides (L.) Trin. ex. Thw., 220–223, 228, 244, 252, 254–256
    A. littoralis (Govan) Parl., 233
    A. massauensis (Fres.) Mattei, 221
    A. villosus Trin., 255
Aerva persica (Burm. f.) Merrill, 304
Agabus bipustulatus L., 79
Agelaius phoeniceus (L.) (red-winged blackbird), 160, 185
Aglossorhyncha micronesiaca Schltr., 276
Agnewia pseudamygdala (Reeve), 322
Agrostis alba L., 173, 180
A. billardieri R.Br., 316
A. exarata Trin., 171
A. stolonifera L., 11, 67, 100, 114, 115, 124, 125, 130, 135, 138, 140, 145
– – L. forma *subarenaria* Westhoff., 131
– – L. var. *compacta* Hartm. subvar. *salina* J. and W., 122, 134
– – L. subvar. *salina* J. and W., 131, 135, 137
Ailuroedus melanotus (Gray) (spotted cat bird), 339
Aix sponsa (L.) (wood duck), 98
Aizoaceae, 243, 313, 315
Alauda arvensis L. (skylark), 144
Alcedinidae, 335
Alderia modesta Lov., 146
Alhagi maurorum Medic., 218, 219, 222, 224, 225–228, 230
Allenrolfea, 13, 33
    A. occidentalis (S. Wats.) Ktze., 177, 179, 194
Alligator mississippiensis (Dandin), 160
Alnus, 110
    A. glutinosa (L.) Gaertn., 110
Alopecurus alpinus J.E. Smith, 67
A. geniculatus L., 124
Alpheidae, 325
Alpheus crassimanus Heller, 236, 237
A. edwardsi (Andouin), 325
Amaryllidaceae, 312

Ambassidae, 328
Ambassis macleayi (Castelnau), 328
A. marianus (Günther), 328
Ambonyx cinerea, 269
Amblystegium serpens var. *salinum* Carr., 137
Ammoastuta inepta (Cushman and McCulloch), 74, 75, 81, 83
Ammobaculites, 74
    A. dilatatus (Cushman and McCulloch), 82
    A. exiguus Cushman and Brönnimann, 182
Ammonia beccarii L., 74, 75, 81, 182
Ammospiza caudacuta (Gmelin) (sharp-tailed sparrow), 159
A. maritima macgillivraii (Wilson) (seaside sparrow), 100, 159
Ammotium salsum (Cushman and Bronnimann), 73, 75, 81, 82
Amoora cuculata Roxb., 247
Amphibia, 184
Amphibolidae, 322
Amphibola crenata (Martyn), 19, 310
Amphinomidae, 327
Amphipoda, 80, 86, 144, 146, 237, 324
Amyema mackayense (Blakely) Dans. ssp. *cycnei-sinus* (Blakely) Barlow, 312
A. mackayense ssp. *mackayense* (Blakely) Dans., 309, 312
Anabaena fertilissima C.B. Rao, 318
A. torulosa (Carm.) Lagerh. ex Born. and Flah., 307, 318
Anabasis, 27
    A. aphylla L., 2
Anacystis marina (Hansg.) Dr. and Daily, 307, 318
Anadara trapezia Deshayes, 323
    A. tuberculosa (Sowerby), 347
Anagallis arvensis L., 123
Anaptychia tremulans (Mull. Arg.) Kurok, 320
Anas acuta L. (pintail), 98, 99, 185
A. carolinensis Gmelin (American green-winged teal), 98, 99, 185
A. discors (Linnaeus) (blue-winged teal), 98, 99
A gibberifrons Muller (grey teal), 331
A. platyrhynchos L. (mallard), 98, 99, 185
A. rubripes Brewster (black duck), 97–99, 159
A. strepera (L.) (gadwall), 98, 99
A. superciliosa (Gmelin) (Burdekin duck), 331
Anatidae, 331
Angelica archangelica L. var. *litoralis* (Fries.) Thellung, 124, 125
A. sylvestris L., 125
Angianthus preissianus (Steelz) Benth., 313
Anhinga rufex (Daudin) (darter), 331
Anhingidae, 331
Anicetus beneficus, 287
Anisophyllea, 350
Annelida, 182, 326, 327
Anomiidae, 323
Anomocytheridea, 76
    A. cf. *floridana* (Howe and Hough), 76
Anomura, 325
Anopheles albimanus Wiedemann, 160
A. amictus Edwards subsp. *hilli* Woodhill and Lee, 270
A. faranti Lareran, 270
A. sundaicus (Rodenwaldt), 270
Anophothrips zeae, 184

Anopyxis, 350
Anous tenuirostris (Temminck) (lesser noddy), 333
Anser albifrons (Scopoli) (white-fronted goose), 100
A. anser (L.) (grey lag), 144
A. caerulescens (snow goose), 98
A. fabalis (Latham) (bean goose), 144
Anseranus semipalmata (Latham) (magpie goose), 331
Anthriscus sylvestris (L.) Bernh., 125
Anthus novaeseelandiae (Gmelin) (Australian pipit), 335
Apargidium boreale T. and G., 171
Aphalara, 183
Aphidae, 146
Apiaceae, 313, 315
Apium prostratum Labill. ex Vent., 313
A. sellowianum Wolff., 195
Aplonis metallica (Temm. and Lang.) (shining starling), 337
Apocynaceae, 243
Apopidae, 335
Aprosmictus erythropterus (Gmelin) (red-winged parrot), 334
Apus pacificus (Latham) (fork-tailed swift), 335
Arachnida, 90, 91, 144, 146, 183
Arcidae, 323
Arctophila fulva (Trin.) Anderss., 67, 170
Ardea herodias L. (great blue heron), 160, 185
A. novaehollandia Latham (white-faced heron), 331
A. pacifica Latham (white-necked heron), 331
A. sumatrana Raffles (great-billed heron), 269, 331
Ardeidae, 331
Ardeola ibis (L.) (cattle egret), 331
Arecaceae, 311
Arenaria, 236
    A. interpres (L.) (ruddy turnstone), 333
Arenariidae, 333
Arenicola, 45
    A. loveni, 236
Arenigobius frenatus (Günther), 328
Arenoparrella mexicana (Kornfeld), 73, 75, 81–83, 182
Armadillidium vulgare (Latreille), 183
Armeria, 22
    A. maritima (Mill.) Willd., 67, 115, 117–119, 122, 124, 131, 134, 135, 138, 145
Armorella spherica Heron-Allen and Earland, 81
Arothron hispidus (L.), 330
Arses telescopthalmus lorealis Deviz (frilled-neck flycatcher), 336
Artamidae, 339
Artamus leucorhynchus (L.) (white-breasted wood swallow), 339
Artemisia, 147
    A. maritima L., 35, 117, 119, 122, 124, 131, 132, 134–136, 145
    A. vulgaris L., 125
Arthonia cf. *fissurina* Nyl., 319
Arthopyrenia cf. *cinereopruinosa* (Schaer.) Korb., 319
Arthrocnemum, 4, 8, 12, 13, 22, 238, 302–305
    A. africanum Moss, 236
    A. arbusculum (R.Br.) Moq., 13, 302–306, 313
    A. australasicum (Moq.) Moss, 236
    A. bidens Nees, 313
    A. decumbens Tolken, 237

SYSTEMATIC INDEX

A. *glaucum* (Del.) Ung. Stern., 65, 218–221, 224–228, 230, 233
A. *halocnemoides* Nees, 13
– – – var. *pergranulatum* J.M. Black, 294, 302–306, 313
A. *indicum* (Moq.) Tand., 244
A. *leiostachyum* (Benth.) Paulsen, 294, 302–306, 313
A. *macrostachyum* (Moric.) Moris. et Delp., 233
A. *natalensis* Moss, 236
A. *perenne* (Mill.) Moss, 236, 237
A. *pillansii* Moss, 236
Arthropoda, 22, 86–91, 144, 148, 149
Asclepiadaceae, 243, 253, 312
Ascochytula obiones (Jaap.) Died., 119
Asio flammeus (Pontoppidan) (short-eared owl), 101, 160
Asplenium nidus L., 88
Assiminea, 183, 186, 237
   A. *bifasciata* L., 236
   A. *grayana* Flem., 146
   A. *translucens* Carpenter, 182, 183
Aster, 11, 32, 128
   A. *novi-belgii* L., 102
   A. *subspicatus* Nees, 171
   A. *tripolium* L. (sea aster), 12, 32, 66, 67, 118, 119, 122, 124, 128, 130, 134, 141, 145, 149
Asteraceae, 313, 315
Asteroma juncaginacearum Rabh., 119
Atherinidae, 328
Atherinops affinis (topsmelt) (Ayres), 184, 186
Atrina gouldii (Reeve), 323
Atriplex, 22, 139, 145, 177, 195
   A. *barclayana* D. Dietr., 177
   A. *billardieri* (Moq.) Hook f., 313
   A. *cinerea* Poir., 315
   A. *farinosa* Forsk., 219
   A. *gmelini* C.A. Mey., 12
   A. *halimus* L., 120
   A. *hastata* L., 115, 122, 130, 135–139, 195
   A. *hortensis* L., 174, 180
   A. *julacea* S. Wats., 176
   A. *lampa* Gill ex Moq., 196
   A. *latifolia* Whlnbg., 67, 124
   A. *littoralis* L., 117, 123–125, 136, 138, 139
   A. *macrostyla* Speg., 196
   A. *montevidensis* Spreng., 196
   A. *paludosa* R.Br., 13, 305, 313
   A. *patula* L., 173, 174, 313
   – – L. ssp. *hastata* (L.) Gray, 175, 178
   A. *rosea* L., 175
   A. *sagittifolia* Speg., 196
   A. *semibaccata* R.Br., 174, 180, 305, 315
   A. *stipitata* Benth., 315
   A. *stocksii* Boiss, 244, 256
   A. *watsonii* A. Nels., 175
Atydidae, 322
Atys naucum L., 322
Australonereis ehlersi (Augener), 326
Austrocochlea, 307
   A. *adelaidae* (Philippi), 321
   A. *concamerata* (Wood), 321
   A. *constricta* (Lam.), 321
   A. *obtusa* Dillwyn, 321

Austrosarepta, 323
Aves (birds), 18, 95–102, 144, 148, 149, 160, 184–187, 309, 311, 330–339, 364, 366
Avicennia, 2, 4, 15, 20–22, 47–49, 51, 53, 54, 65, 197, 199, 201–203, 205–210, 234–239, 245–247, 250–256, 258, 263–266, 268, 270, 349, 350, 356
   A. *africana* P. Beauv., 16, 20, 234, 235, 269
   A. *alba* Blume, 20, 50, 243, 252, 255, 256, 262, 263, 266, 267, 348
   A. *bicolor* Standl., 16, 199, 201
   A. *eucalyptifolia* Zipp. ex Miq., 293, 303, 305, 312, 351
   A. *germinans* (L.) L., 2, 3, 16, 20, 177, 179, 198, 199, 201, 203, 204, 210, 349, 350, 352, 354
   A. *intermedia* Griff., 50, 262, 347
   A. *lanata* Ridl., 348
   A. *marina* (Forsk.) Vierh., 2, 16, 20, 52, 53, 66, 217, 218, 225, 227, 228, 236–238, 243, 246, 252–254, 263, 265–268, 273, 282, 288, 302–305, 351
   – – (Forsk.) Vierh. var. *acutissima* Stapf. and Mold., 246, 255, 256
   – – (Forsk.) Vierh. var. *australasica* Walp., 293, 294, 297, 300–305, 310, 312
   – – (Forsk.) Vierh., var. *resinifera* (Forsk.) Bakh., 17, 18, 20, 65, 66, 310
   A. *nitida* Jacq. (red mangrove), 158, 162, 349, 351, 352
   A. *officinalis* L., 20, 243, 246, 247, 252, 265, 267, 270, 279, 280, 287–289, 348
   A. *resinifera* Forst., 267, 356
   A. *schaueriana* Stapf. and Leechman, 16, 20, 199, 201, 202
   A. *tomentosa* Willd., 210
   A. *tonduzii* Moldenke, 199, 201
Avicenniaceae, 243
Axix axix (Erxleben), 251
Aythya collaris (Donovan) (ring neck duck), 98
Azima tetracantha Lamk., 254

Baccharis, 22
   B. *halimifolia* (L.), 162, 164, 193, 315
Bactris, 203, 211
   B. *subglobosa* H. Wendl., 206
Bactronophorus subaustralis (Iredale), 324
Bairdella icistia (Jordan and Gilbert), 95
Balanidae, 324
Balanus amphitrite Darwin, 236, 324
   B. *amphitrite* Darwin var. *denticulata* Brock, 237
   B. *elizabethae* Barnard, 236
Barringtonia, 238
   B. *racemosa* Roxb., 273, 282
Bassia astrocarpa F. Muell., 304, 313
   B. *hirsuta* (L.) Aschrs., 138
   B. *muricata* (L.) Murr., 233
Bathygobius kreffti (Steindachner), 328
Batidaceae, 313
Batillaria, 18, 186
Batillaria zonalis (Bruguière), 183
Batis, 2, 57, 73, 177, 194
   B. *argillicola* van Royen, 306, 313
   B. *maritima* L. (saltwort), 13, 22, 158, 162, 175–177, 179, 193, 194, 198, 199

Batissa triquetra Deshayes, 323
Batrachoididae, 328
Baumea juncea (R.Br.) Palla, 303, 314
Bedeva hanleyi (Angas), 322
Belonidae, 328
Bembicium, 307, 308
    B. auratum Quoy and Gaimard, 321
    B. melanostomum (Gmelin), 321
Berula erecta (Huds.) Coville, 218
Beta maritima L., 139
Bidens trichosperma Britton (tick seed sunflower), 102
Bivalvia, 323
Bledius arenarius Paykull, 144
B. spectabilis Kraatz, 144
Blennidae, 328
Blidingia minima (Näg. ex. Kütz.) Kylin, 143
Blysmetum rufi, 6
Boleophthalmus boddaerti (Pallas), 263, 265
Bombacaceae, 311
Boraginaceae, 243
Borrichia, 163
    B. frutescens DC., 162, 193
Bostrychia, 17, 18, 142, 235–237, 306, 310, 317
    B. binderi Harv., 317
    B. flagellifera Post, 317
    B. kelanensis Grun. ex Post, 317
    B. mixta J.D. Hook and Harv., 317
    B. moritziana (Sond.) J. Ag., 307, 317
    B. radicans Mont., 317
    B. scorpioides (Huds.) Mont., 33, 122, 127, 128, 130, 143, 317
    B. simpliciuscula Harv. ex J.Ag., 307, 317
    B. tenella (Vahl.) J.Ag., 317
    B. tenuis Post, 307, 317
    B. vaga J.D. Hook. and Harv., 317
Botaurus lentiginosus (Rackett) (American bittern), 160
B. poiciloptilus (Wagler) (brown bittern), 331
Bothriochloa parviflora (R.Br.) Ohwi, 286
Branta bernicla (L.) (brant), 97, 144
B. canadensis (L.) (Canada goose), 98, 99
B. nigricans (Lawrence) (black brant), 185
Breynia officinalis Hemsl., 279
Bromelia humilis L., 65
Bromus mollis L., 23, 122
Brownlowia argentata Kurz., 267
B. lanceolata Benth., 245, 258, 268
Bruguiera, 4, 20, 22, 236–238, 245, 255, 258, 264, 266–268, 270, 273, 275, 283, 288, 290, 350, 352, 354, 355
    B. caryophylloides Burm., 262, 264, 348, 349, 351, 354
    B. conjugata Merr., 273, 311
    B. cylindrica (L.) Bl., 20, 244, 253, 255, 264–268, 293, 311, 349
    B. eriopetala W. and A., 262, 289, 350, 351, 354
    B. exaristata Ding Hou, 293, 311
    B. gymnorrhiza (L.) Lamk., 16, 20, 53, 217, 225, 227, 228, 236, 237, 244, 250, 251, 255, 258, 262, 264, 265, 267, 268, 271, 273–275, 279, 280, 282–284, 287, 288, 290, 293, 301, 302, 311, 347–351, 354
    B. hainesii C.G. Rogers, 264
    B. parviflora W. and A., 2, 20, 244, 258, 262, 264–268, 288, 293, 311, 351, 352, 354

    B. rheedii Bl., 311
    B. sexangula Lour., 17, 245, 266, 267, 288, 293, 311
Bryum, 67
Buellia callispora (Kn.) Stirton, 319
B. punctata (Hoffm.) Mass., 319
Bufo boreas Baird and Girard (western toad), 184
Bulbophyllum gibbonianum Schltr., 276
B. profusum Ames, 276
B. volkensii Schltr., 275, 276
Bullina lineata Gray, 322
Bupleurum, 138
    B. gracile DC., 12
    B. tenuissimum L., 137
Burhinidae, 333
Burhinus magnirostris (Latham) (southern stone-curlew), 333
Butorides striatus (L.) (mangrove heron), 331
B. virescens (L.) (green heron), 185

Cacatua galerita (Latham) (sulphur-crested cockatoo), 334
Cacomantis castaneiventris (Gould) (chestnut-breasted cuckoo), 334
C. pyrrophanus (Vieillot) (fan-tailed cuckoo), 334
C. variolosus (Vig. and Horsf.) (brush cuckoo), 334
Caesalpinia, 288
    C. crista L., 244, 252, 284
    C. nuga Ait., 252, 266
Caesalpiniaceae, 244, 311
Calamagrostis canadensis (Michx.) Beauv., 171
C. deschampsioides Trin., 67, 170, 171
C. neglecta (Ehrh.) (Gaertn.), 170
C. nutkaensis (Presl) Stend., 171
Calamus, 236
Calicium cf. *pallidellum* Willey, 319
Calidris acuminata (Horsfield) (sharp-tailed sandpiper), 333
C. canutus (L.) (knot), 333
C. ferruginea (Brunnich) (curlew sandpiper), 333
C. mauri (Cabanis) (western sandpiper), 184
C. minutilla (Vieillot) (least sandpiper), 184
C. ruficollis (Pallas) (red-necked stint), 333
C. tenuirostris (Horsfield) (grey knot), 333
Callianassa australiensis Dana, 325
C. californiensis Dana, 183
C. kraussi Stebb., 236
Callianassinae, 325
Caloglossa, 17, 235, 237, 310
    C. adnata (Zan.) de Toni, 306, 318
    C. bombayensis Boorg., 318
    C. leprieurii (Mont.) J.Ag., 17, 307, 318
    C. ogasawaraensis Okam., 318
Calophyllum cholobtaches Lant., 273
C. inophyllum L., 238
Caloplaca byrsonimae Melme, 319
C. pyracea (Ach.) Th.Fr., 319
Calothrix crustacea Thuret., 307, 318
C. pilosa Harv., 318
Caltha arctica R.Br., 67
Campnosperma, 275, 276
    C. brevipetiolata Volkens, 277
    C. panamensis Standl., 211
Camptostemon philippinensis (Vidal) Becc., 17, 267, 268, 288

C. schultzii Mast., 20, 267, 268, 293, 311
Canavalia maritima (Aubl.) Thouars, 304, 315
Candona, 76
Canis latrans Say. (coyote), 185
Capitella, 186
    C. capitata (Fabricius), 182, 186, 187, 327
Capitellidae, 327
Caprimulgidae, 334
Caprimulgus macrurus Horsfield (white-tailed nightjar), 334
Carallia, 350
Carangidae, 328
Caranx sexfasciatus Quoy and Gaimard, 328
Carapa, 245
    C. guianensis Aubl., 211
    C. moluccensis Lamk., 250, 262
    C. obovata Bl., 250, 262
Carex, 11, 158, 172, 178
    C. distans L., 122, 124, 134, 135
    C. disticha Huds., 125
    C. extensa Good., 122, 135, 136
    C. glareosa Wg., 67, 170, 171, 178
    C. lyngbyei Hornem, 171, 173, 174, 178, 179
    C. mackenziei V. Krecz., 124
    C. maritima Gunn, 11
    C. nigra Reich, 125
    C. oederi auct., non Retz, 125
    C. paleacea Wahlenb., 11, 124
    C. pluriflora Hultén, 171
    C. punctata Gaud., 135
    C. ramenskii Komarov, 12, 170, 171, 181
    C. rariflora (Wg.) Sm., 170, 171, 178
    C. recta Boot., 124
    C. stricta (Lam.), 163
    C. subspathacea Wormskj., 11, 170, 171, 178
    C. ursina Dewey, 67, 170
Carpobrotus rosii (Haw.) Schwantes, 315
Caryophyllaceae, 313, 315
Casmerodius albus (L.) (common egret), 160, 185
Cassidula augulifera Petit, 322
    C. aurisfelis (Bruguière), 263, 265
    C. labrella Desh., 237
    C. mustellina (Derh.), 263, 266
    C. rugata Menke, 322
Casuarina equisetifolia L., 349
    C. glauca Sieber ex Spreng., 303, 315
Casuarinaceae, 315
Catapodium, 138
    C. marinum (L.) Hubbard, 137
Catenella, 17, 18, 142, 237, 310
    C. nipae Zanard., 306, 318
    C. repens (Lightf.) Batt., 143
Catoptrophorus, 101
    C. semipalmatus (Gmelin) (willet), 97, 101, 160, 184
Caulerpa fastigiata Mont., 317
Cellana tramoserica (Sowerby), 321
Cellathus discoidale, 182
Celtis spinosa Spreng., 195
Cenchrus, 256
    C. ciliaris L., 256
    C. setigerus Vahl., 256

Centaurium, 125
    C. littorale (Turner) Gilmoure, 122
    C. pulchellum (Sw.) E.H.L. Krause, 122–124, 315
    C. spicatum (L.) Fritsch, 233, 315
    C. vulgare auct. (?Rafn.), 124
Centrolepidaceae, 315
Centrolepis polygyna (R.Br.) Hieron, 315
Centropogon australis (Shaw), 329
Centropus phasianus (Latham) (pheasant coucal), 334
Cerastium holosteoides Fr., 124, 125
Ceratonereis erythraensis Fauvel, 326
Cerbera manghas L., 243, 258, 350
    C. odollam Gaertn., 266
Cerberus rhynchops (Schmid.) (water snake), 263
Cercopidae (froghoppers), 146
Cercopithicus mitis Wolf. (Sykes monkey), 239
Ceriops, 22, 237–239, 245, 250, 255, 266–268, 270, 350
    C. candolleana Arn., 262, 348, 349, 354, 355
    C. decandra (Griff.) Ding Hou, 245–247, 251, 253, 265, 266, 293, 311, 349, 355
    C. erectus, 355
    C. roxburghiana Arn., 273, 349, 355
    C. tagal (Perr.) C.B. Rob., 16, 20, 237, 238, 245, 255, 258, 264–269, 279, 280, 348, 349, 351, 352
    – – (Perr.) C.B.Rob. var. *australis* MacNae, 293, 311
    – – (Perr.) C.B. Rob. var. *tagal* MacNae, 293, 301–303, 305, 312
Cerithidea, 18, 186, 265, 307
    C. californica Haldeman, 183
    C. cingulata (Gmelin), 265
    C. decollata (L.), 236, 237
    C. kieneri Hombron and Jacquinot, 322
    C. largillierti (Philippi), 322
    C. obtusa (Lam.), 265, 266, 322
Cerithiidae, 322
Cerithium dorsuosum Mencke, 322
    C. jannelli Hombron and Jacquinot, 322
Ceroplastes rubens Maskell, 287
Ceryx azureus (Latham) (azure kingfisher), 335
    C. pusillus (Temminck) (little kingfisher), 135
Chaetodon flavirostris Günther, 328
Chaetodontidae, 328
Chaetomorpha capillaris (Kutz.) Boerg., 307, 317
    C. linum (O.F.Müll.) Kütz., 121
Chaetopteridae, 327
Chaetopterus, 327
Chaetura caudacutus (Latham) (spine-tailed swift), 335
Chalcophaps indica (L.), 334
Chanos chanos Forskal (Milk fish), 353
Characeae, 120
Charadriidae, 332
Charadrius asiaticus veredus Gould (oriental dotterel), 332
C. bicinctus Jard. and Selb. (double-banded dotterel), 332
C. cinctus (Gould) (red-kneed dotterel), 332
C. leschenaultii Lesson (large dotterel), 332
C. melanops Vieillot (black-fronted dotterel), 332
C. mongolus Pallas (Mongolian dotterel), 332
C. ruficapillus Temminck (red-capped dotterel), 332
C. vociferus L. (killdeer), 184
Chelmon rostratus (L.), 328

Chen hyperborea (Pallas) (greater snow goose), 89, 97
Chenolea diffusa Thunb., 236
Chenopodiaceae, 144, 147, 244, 247, 253, 254, 313, 315
Chenopodium ambrosioides L., 174
C. glaucum L., 124
– – L. spp. *ambiguum* (R.Br.) Murr. and Thell. ex Thell., 315
C. macrospermum Hook, 195
C. rubrum L., 196
Chitonidae, 321
Chlamydera cerviniventris Gould (fawn-breasted bower bird), 339
C. nuchalis (Jard. and Selb.) (great bower bird), 339
Chlidonias hybrida Pallas (whiskered tern), 333
Chloris, 268
Chlorococcum humicola (Naeg.) Rabenhorst, 317
Chlorophyceae (Chlorophyta), 17, 18, 119, 142–144, 317
Chordata, 91–105
Chorinemus tolooparah (Rüppel), 328
Chorthippus albamarzinatus (De G.), 90
Chroococcaceae, 318
Chroococcus turgidus (Kütz.) Naeg., 306, 318
Chrysanthemum arcticum, 170, 171
Chrysococcyx lucidus (Gmelin) *plagosus* (golden bronze cuckoo), 334
C. malayanus (Raffles) *russatus* (rufous breasted bronse cuckoo), 334
Chrysomeris ramosa N. Carter, 143
Chrysophyta, 319
Chthamalidae, 325
Chthamalus caudatus Pilsbry, 325
C. malayensis Pilsbry, 325
C. rhizophorae Oliveira, 325
Cicadellidae, 146
Ciconiidae, 331
Ciliata, 81
Cinnamomum japonicum Siebold, 286
Circe tumidum (Bolten), 324
Circus approximans Peale (swamp harrier), 332
C. cyaneus (L.) (marsh hawk), 101
C. hudsonius (L.) (marsh hawk), 160
Cirripedia (barnacles), 235–237, 310, 324
Cirsium arvense (L.) Scop., 115, 123, 125
Cistanche lutea Hoffgg. et Link., 119, 120
Cisticola exilis (Vig. and Horsf.) (golden backed cisticola), 336
Cladium filum (Labill.) R.Br., 315
C. junceum R.Br., 314
Cladophora, 172, 317
C. gracilis, 172
Cladophorella calcicola Fritsch, 317
Claviceps purpurea (Fr.) Tul., 119
Cleistostoma edwardsi McLeay, 236
C. wardi Rathbun, 325
Clerodendron inerme Gaertn., 245, 252, 254, 268, 273, 284, 287, 290, 316
C. trichotomum Wall, 286
Clevelandia ios (Jordan and Gilbert), 184
Clibanarius virescens Krauss, 326
Clistocoeloma merguiense de Man, 263–265, 325
Clupea harengus pallasi Val. (Pacific herring), 184

Clupeidae, 328
Clypeomorus carbonarius Sowerby, 322
Cnidoglanis macrocephalus (Cuv. and Val.), 329
Coccocarpia, 320
Cochlearia danica L., 124, 137, 138
C. groenlandica L., 67
C. officinalis L., 67, 123, 124, 144, 170
Coenobia canpes Stimps, 237
C. rugosa M. Edw., 237
Coleoptera (beetles), 80, 144, 146, 160
Collema glaucophthalmum Nyl., 320
Collembola, 90, 146
Collocalia spodiopygia (Peale) (grey-swiftlet), 335
Colluricincla brunnea Gould (brown shrike thrush), 337
C. harmonica Latham (grey shrike thrush), 337
C. megarhyncha Quoy and Gaim. (rufous shrike thrush), 337
C. parvula Gould (little shrike thrush), 337
Colpodium, 67
Colpomenia sinuosa (Roth.) Derbes and Sol., 317
Columbidae, 334
Combretaceae, 244, 254, 311
Combretocarpus, 350
Commelina communis L., 286
Conocarpus, 48, 199, 206, 209, 210
C. erectus L. (button mangrove), 16, 20, 158, 164, 199, 203, 234, 352
Conocephalus dorsalis Latreille, 90, 146
Conopophila albogularis (Gould) (rufus-banded honey-eater), 338
Conuber, 322
C. melanostoma Swainson, 322
C. sordida Swainson, 322
C. strangei (= *C. sordida*), 322
Convolvulaceae, 244, 252, 313
Convolvulus hystrix Vahl., 223
Coracina lineata (Swainson) (barred cuckoo shrike), 335
C. novaehollandiae (Gmelin) (black-faced cuckoo shrike), 335
C. papuensis (Gmelin) (little cuckoo shrike), 336
C. tenuirostris (Jardine) (cicada bird, Jardine triller), 336
Cordiosome carniflex (Herbst.), 237
Cordylanthus maritimus Nutt. ex Benth in DC., 174
C. mollis Gray, 175
Cornulaca ehrenbergii Asch. ap. Schweinf., 221
Coronopus squamatus (Forsk.) Aschers., 122
Corophium volutator (Pallas), 80
Corticaria, 184
Corvidae, 339
Corvus cecilae Mathews (Australian crow), 339
Cotula, 22, 172
C. coronopifolia L., 134, 136, 174, 175, 180, 185, 195, 311, 313
C. dioica Hook. f., 86–88, 311
Cracticidae, 339
Cracticus mentalis Salv. and d'Alb. (black-backed butcher bird), 339
C. nigrogularis (Gould) (pied butcher bird), 339
C. quoyi (Lesson) (black butcher bird), 339
C. torquatus (Latham) (grey butcher bird), 339
Crassostrea, 18
C. rhizophorae (= *Ostrex rhizophorae*), 353

SYSTEMATIC INDEX

Crassula maritima Schon., 236
Cressa cretica L., 218, 219, 222–228, 230, 252
C. truxillensis H.B.K., 175
Crinum, 199
    C. pedunculatum R.Br., 312
Crocodylus porosus Schneider, 19, 235, 269, 310
Crotalus viridis Rafinesque (western rattlesnake), 184
Crudia cynometroides Hosok., 273
Crustacea, 89, 144
Crypsis aculeata (L.) Ait., 233
Cuculidae, 334
Cudrania cochinchinensis Masamune var. *gerontogea*
    Masamune, 279
Culex salinarius Coquillett, 89
Culicoides, 18, 160
Cuscuta salina Engelm., 174–176, 178
Cutandia memphitica (Spreng.) Benth., 233
Cyanophyceae (Cyanophyta) (blue-green algae), 74, 116,
    118, 119, 138, 143, 148, 159, 318
Cyclograpsus punctata M. Edw., 236
Cydista equinoctialis Miers, 22
Cymbidium, 287
    C. canaliculatum R.Br., 312
Cymodocea, 227
    C. ciliata (Forsk.) Ehr. ex Asch., 217
    C. rotundata (Ehr. and Hempr.) Asch. and Schweinf., 218
    C. serrulata (R.Br.) Asch. and Magn., 218
Cynanchum carnosum (R.Br.) Domin., 312
Cynodon dactylon (L.) Pers., 316
Cynometra bijuga (Span.) Prain, 273
    C. ramiflora L., 350
    – – L. subsp. *bijuga* (Span.) Prain, 293, 311
Cyperaceae, 244, 256, 288, 314, 315
Cyperus, 3, 252, 267
    C. articulatus L., 218
    C. conglomeratus Rottb., 221–223
    – – Rottb. var. *effusus* (Rottb.) Kükenth, 220
    C. corymbosus Hook., 195
    C. dives Del., 218
    C. laevigatus L., 221
    C. littoralis R.Br., 315
    C. mundtii (Nees) Kunth, 218
    C. polystachyos Rottb., 315
    C. rotundus L., 256
    C. stoloniferus Retz., 315
    C. vaginatus R.Br., 315
Cyprideis, 76, 182
    C. floridana Puri, 76
    C. miguelensis Benson, 76
    C. stewarti Benson, 76
Cypridopsis, 76
    C. vidua (O.F. Miller), 76
Cyprinodon variegatus (Lacépède) (sheepshead minnow), 92,
    160
Cystopus lepigoni de Bary, 119
Cytherella, 76
    C. cf. *lata* G.S. Brady, 76
Cytherura, 76
    C. forulata Edwards, 76
    C. johnsoni Mincher, 76

Dacelo gigas (Boddaert) (laughing kookaburra), 335
D. leachii Vig. and Horsf. (blue-winged kookaburra), 335
Dactyloctenium aristatum Link, 222
Dalbergia, 199
    D. candenatensis (Dennst.) Prain, 273, 288
    D. spinosa Roxb., 244, 252, 154
Dasson variabilis (Cantor), 328
Davallia solida Sw., 275, 276
Decapoda, 325, 326
Dendrobium, 287
    D. carolinense Sohltr., 276
    D. discolor Lindl., 312
    D. elongaticolle Schltr., 275, 276
    D. implicatum Fukuyama, 276
Dendrocygna arcuata (Horsfield) (water whistle duck), 331
Dendronereides zululandica Day, 237
Dendrophthoe falcata (L.f.) Etting, 254
Dendrostomum signifer Selenko and de Man, 327
Dendryphiella salina (Sutherland) Pugh and Nicot, 118
Derris, 22
    D. heterophylla (Willd.) Back., 90
    D. trifoliata Lour., 244, 254, 273, 288, 312
Deschampsia, 171, 172
    D. brevifolia R.Br., 67
    D. caespitosa (L.) Beauv., 170, 171, 173, 178, 179
Desmoschoenus bottnica, 11
Detracia floridana Pfeiffer, 160
Diala rufilabris (Adams), 322
Dicaeum hirundinaceum (Shaw and Nodder) (mistletoe bird),
    309, 337
Dichirotrichus pubescens, 79
Dicruridae, 338
Dicrurus bracteatus Gould (spangled drongo), 338
Dictylichthys myersi Ogilby, 328
Dicyathifer caroli Iredale, 324
Didelphis marsupialis L., 185
    – – *virginia* Kerr (opossum), 160
Diodon holocanthus L., 328
D. punctulatus (Kaup), 328
Diodontidae, 328
Diopatra dentata Kinberg, 327
Diplanthera uninervis (Forsk.) William, 218
Diptera, 160
Dirinaria applanta (Fee) Awasthi, 320
D. aspera (Magn.) Awasthi, 320
Dischidia hahliana Volkens, 275, 276
D. nummularia R.Br., 312
D. saccata Warb., 288
Discorinopsis aguayoi (Bermudez), 73, 75, 83, 182
Disphyma blackii R.J. Chinnock, 313
Distichlis, 13, 22, 172, 176, 193, 195
    D. distichophylla (Labill.) Fassett, 306, 314
    D. spicata (L.) Greene (salt grass), 84, 86, 88–90, 97, 99,
    102, 158–163, 173, 179, 181, 193, 194, 196, 198
Distichum, 67
Distylium racemosum Sieb. and Zucc., 286
Dolichandrone longissima K. Schum, 350
D. spathacea Seem., 266–268
Dolichonabis lineatus Dahlbom, 146
Donax canniformis K. Schum, 277

Dotilla fenestrata, 236
Drepanocarpus lunatus G.F.W. Meyer, 235
Drynaria quercifolia (L.) J.Sm., 288
Drypetes formosana Kanehira, 279
D. karapinensis (Hay.) Pax and Hoffn., 284
Ducula spilorrhoa (G.R. Gray) (Torres Strait pigeon), 334
Dupontia fischeri R.Br., 67, 170

Echiura, 327
Eggerella advena Cushman, 81
Egretta alba (L.) (white heron), 269, 331
E. eulophotes (Swinh.) (little egret), 269
E. garzetta (L.) (little egret), 269, 331
E. intermedia (Wagler) (plumer egret), 331
E. sacra (Gmelin) (reef heron), 331
E. thula (Molina) (snowy egret), 160, 185
Elaeagnus oldhami Hance, 279
Elaeocarpus sylvestris Blanco var. *ellipticus*, 286
Elanus notatus Gould (black-shouldered kite), 332
E. scriptus Gould (letter-winged kite), 332
Eleocharis, 172, 194
    E. geniculata (L.) R. and S., 315
    E. kamtschatica Komarov, 171
    E. palustris (L.) R. and S., 195
    E. parvula (R. and S.), 124, 128, 173
    E. pauciflora (Lightf.) Link, 124
    E. uniglumis (Link) Schult., 124, 125
Eleotridae, 328
Eleusine compressa (Forsk.) Asch. and Schweinf. ex Christens., 222, 223
Ellobiidae, 322
Ellobium aurisjudae L., 265, 266, 322
E. aurismidae L., 265
Elminius modestus Darwin, 310
Elphidium, 75
    E. articulatum (d'Orbigny), 74, 82
    E. incertum (Williamson), 82
Elymus arenarius L., 171
Elysia, 264
Elytrigia juncea (L.) Nevski, 131
E. pungens (Pers.) Tutin, 119, 122, 131, 136, 138–141, 145
E. repens (L.) Desv., 122, 124, 145
Encelia californica Nutt., 176
Enchylaena tomentosa R.Br., 313
Enhalus acoroides Rich., 275
Enigmonia rosea (Gray), 265
Entada scandens Benth., 290
Enteromorpha, 17, 73, 121, 129, 142, 143, 186, 306, 307, 317
    E. clathrata (Roth.) Grev., 317
    – – (Roth.) Grev. var. *prostrata* LeJol., 17
    E. compressa L., 172
    E. intestinalis L., 172
    E. nana (Sommer.) Bliding, 18, 142
    E. prolifera (Muell.) J.Ag., var. *pilifera* (Kütz.) Chapm., 17
Entomyzon cyanotis (Latham) (blue-faced honey eater), 338
Entophysalis deusta (Meneg.) Dr. and Daily, 318
Ephydra cinerea, 183
E. riparia Fallén, 183

Epinephelus gilberti (Richardson), 329
Equisetum arvense L., 125
Eremopogon foveolatus (Del.) Stapf., 220
Ereunetes mauri Cabanis (western sandpiper), 101
Eriachne pallescens R.Br., 305
Eriophorum augustifolium Horck., 125
Erolia minutella (Vieillot) (least sandpiper), 101
Erysinum hieraciifolium L., 138, 139
Erysiphe, 118
    E. lamprocarpa (Wallr.) Duby, 119
    E. polygoni DC., 119
Esacus magnirostris (Vieillot) (beach stone-curlew), 333
Eucalyptus, 301, 305
    E. viminalis Labill., 305
Euchelus atratus (Gmelin), 321
Eucyclogobius newberryi (Girard), 184
Eudynamys scolopacea L. (koel), 334
Eulabeornis castaneoventris Gould (chestnut rail), 332
Eulimine, 265
Eunice antennata Savigny, 327
Eunicidae, 327
Euphorbia monacantha Pax, 222
Euphorbiaceae, 52, 244, 254, 258, 311
Euplax tridentata Milne Edwards, 325
Eurhynchium stokesii B. and S., 172
Eurostopodus guttatus (Vig. and Horsf.) (spotted nightjar), 334
Eurya japonica Thunb., 286
Eurycarcinus integrifrons de Man, 326
E. maculatus (Milne Edwards), 326
Eurydice, 236
Eurythoe complanata (Pallas), 327
Eurytium limosum (Say), 87
Euscelis obsoletus Kirsch., 90, 146
Euterpe, 211
Excoecaria, 247, 252, 254, 266–268
    E. agallocha L., 20, 53, 244–247, 250, 252, 254, 255, 258, 266–269
Exogone verugera, 182
Exosphaeroma, 324

Fabaceae, 312, 315
Fabricia limnicola, 182
Falco cenchroides (Vig. and Horsf.) (nankeen kestrel), 332
F. peregrinus Tunstall (peregrine falcon), 332
Falconidae, 332
Farsetia aegyptiaca Turra, 223
Festuca arundinacea Schreb., 123
F. pratensis Huds., 124, 125
F. rubra L., 11, 90, 91, 100, 117, 119, 124, 135, 136, 138, 140, 141, 145, 171, 178
– – L., f. *toralis* Hackel, 122, 133–135
Ficus, 268
    F. ampelas Burm., 282
    F. benguetensis Herr., 282
    F. microcarpa L., 282
    F. septica Burm., 282
        Fimbristylis, 163
    F. densa S.T. Blake, 315

SYSTEMATIC INDEX

F. ferruginea (L.) Vahl., 315
F. obtusifolia Kunth, 235
F. punctata R.Br., 315
F. rara R.Br., 315
F. spathacea Roth, 252
Finlaysonia maritima Backer, 273
Fischerella muscicola (Thuret) Gom., 318
Fissurellidae, 321
Florida caerulea (L.) (little blue heron), 160
Foraminifera, 73–75, 81–83, 146, 182
Frankenia, 177, 194, 195
    F. grandifolia Cham. and Schelcht., 174–177, 179
    F. laevis L., 90, 138
    F. microphylla Cav., 196
    F. palmeri Wats., 176, 177, 179
    F. patagonica Speg., 196
    F. pauciflora DC., 13, 294, 304, 305, 314
Frankeniaceae, 314
Fucaceae, 18, 142
Fucus, 18, 172
    F. distichus L., 172
    F. spiralis L., 129
    F. vesiculosus L. forma (ecad) *volubilis* Powell, 122, 127, 143
Fulica americana Gmelin (American coot), 98, 185
F. atra L. (coot), 332
Fundulus, 160, 161, 186
    F. heteroclitus (L.) (mummichog), 92
    F. majalis (Walbaum) (striped killifish), 92
    F. parvipinnus Girard (California killifish), 184
Furcellaria fastigiata (L.) Lamour, 120

Gafrarium timidum (= *Circe tumidum*), 324
Gahnia filum (Labill.) F. Muell., 315
Galeopsis bifida Boenn., 125
Galium aparine L., 115, 124
G. palustre L., 125
G. trifidum L., 171
Gallinago hardwickii (Gray) (Japanese snipe), 333
Gallinula porphyrio (L.) (swamphen), 332
G. tenebrosa Gould (dusky moorhen), 332
Gambusia affinis (Baird and Girard) (mosquito fish), 184
Gammarus duebeni Liljeborg, 79
Gardenia jasminoides Ellis, 284, 286
Gasoul crystallinum (L.) Rothmaler, 315
Gasterosteus aculeatus L. (threespine stickleback), 184
Gastropoda, 85, 86, 321
Gelidium, 318
    G. pusillum (Stackh.) LeJol., 307, 318
Geloina coaxans (Gmelin), 323
Geloinidae, 323
Gentiana detonsa Rottböll, 136
Gentianaceae, 315
Geopelia humeralis (Temminck) (bar-shouldered dove), 334
G. striata (L.) (peaceful dove), 334
Gerres oyena (Forskål), 328
Gerrhonotus multicarinatus (Blainville) (southern alligator lizard), 184
Gerridae, 328

Gerygone chloronata Gould (green-backed warbler), 336
G. flavida Ramsay (fairy warbler), 336
G. laevigaster Gould (mangrove warbler), 336
G. magnirostris Gould (large-billed warbler), 336
G. olivacea (Gould) (white-throated warbler), 336
G. palpebrosa Wallace (black-throated warbler), 336
G. tenebrosa (Hall) (dusky warbler), 336
Gillichthys, 186
    G. mirabilis Cooper (long-jawed mudsucker), 184, 186
Girella tricuspidata (Quoy and Gaimard), 328
Girellidae, 328
Glabratella, 73, 182
Glaucomya rugosa Hanley, 323
G. virens L., 323
Glaucomyidae, 323
Glaux, 12, 136, 172, 178
    G. latifolia, 67
    G. maritima L., 114, 115, 118, 119, 122, 124, 130, 131, 133–135, 145, 171, 173, 175
Gliocladium roseum Bain, 118
Glochidion littorale Blume, 287
Gloeocapsa alpicola (Lyngb.) Born., 307, 318
G. atrata (Turp.) Kütz., 318
Glomospira, 73
Gobiidae, 328
Gobius durbanensis Barnard (goby), 237
G. microps Kroyer, 80
Goerlichia, 182
Goodeniaceae, 314
Gracilaria confervoides (L.) Grev. forma *gracilis*, 318
  – – (L.) Grev. forma *tumida*, 318
G. lichenoides (L.) Harv. forma *lemanea*, 318
G. secundata Harv., 18
  – – Harv. var. *pseudoflagellifera*, 310
Grallina cyanoleuca (Latham) (magpie lark), 336
Grallinidae, 335, 336
Gramineae, 244, 251–253, 256
Graphis anguiliformis Tayl., 319
G. cf. *caesiella* Wain., 319
G. cf. *longula* Kremplh., 319
Grapsidae, 325
Grapsus strigosus (Herbst), 325
Grindelia cuneifolia Nutt., 101, 104
G. humilis H. and A., 174
G. integrifolia DC., 171, 173, 174
G. robusta Nutt., 175
Gruidae, 332
Grus antigone (L.) (Sarus crane), 332
G. rubicunda (Perry) (brolga), 332
Gymnanthera nitida R.Br., 312
Gymnosporia diversifolia Maxim., 279
Gynotroches, 350

Haematomma, 319
Haematopodidae, 332
Haematopus fuliginosus Gould, 332
H. ostralegus L., 144, 332
Halacaridae, 146
Halcyon chloris (Bodd.) (forest kingfisher), 269, 335

H. macleayi Jard. and Selb. (mangrove kingfisher), 268, 335
H. sancta Vig. and Horsf. (sacred kingfisher), 335
H. torotoro (Lesson) (yellow-billed kingfisher), 335
Haliaetus leucogaster (Lath.), 332
Haliastur indus (Bodd.) (Brahminy kite), 269, 332
H. sphenurus (Vieillot) (whistling kite), 332
Halimione, 17, 33, 131, 136
    H. pedunculata (L.) Aell., 124, 131, 133
Halocnemon, 22
    H. strobilaceum (Pallas) M. Bied., 12, 65, 218–220, 223, 225–228, 230, 233
Halopeplis amplexicaulis (Vahl) Ung.-Sternb., 233
H. patagonica Ung.-Sternb., 196
H. perfoliata (Forsk.) Bge. et Schweinf., 218–221, 223–228, 230, 233
Halophila, 1, 3
    H. ovalis (R.Br.) Hook. f., 218
    H. stipulacea (Forsk.) Asch., 218
Halophryne diemensis (Le Sueur), 328
Haloxylon, 27
    H. salicornicum, 223, 256
Haminea, 263, 265
Haminoea, 307, 322
Haplopappus venetus ssp. *furfurascens* (Greene) Hall, 177
Haplophragmoides, 74
    H. hancocki Cushman and McCulloch, 82
    H. subinvolutum Cushman and McCulloch, 182
Haploscoloplos, 327
Haplothrips, 184
Harengula castelnau (Ogilby), 328
Hedysarum alpinum L., 171
Helice, 19, 264
    H. crassa Dana, 19, 340
    H. leachii Hess, 325
Heliotropium, 50
    H. curassavicum L., 243, 252, 254
Heloecius cordiformis Milne Edwards, 325
Helograpsus haswellianus (Whitelegge), 325
Helophorus viridicollis Steph., 79
Hemichroa diandra R.Br., 304, 313
H. pentandra R.Br., 13, 313
Hemigrapsus nudus (Dana), 183
H. oregonensis (Dana), 183, 187
Hemiptera, 80, 146, 160
Hemirhamphidae, 328
Hemirhamphus regularis Günther, 328
Hemiscyllium ocellatum (Bonaterre), 329
Heniochus acuminatus (L.), 328
Heritiera, 249–251, 255, 265, 268
    H. fomes Buch.-Ham., 20, 50, 245, 247, 251, 258
    H. littoralis (Dryand.) Ait., 17, 20, 238, 258, 265–267, 273, 282, 284, 287, 288, 290, 293, 312, 350, 351
    H. minor Roxb., 16
Heterostachys, 13
Hibiscus, 252, 268
    H. hamabo Sieb. et Zuc., 286, 287
    H. moscheutos L., 164
    H. oculiroseus Britten ex L.H. Bailey (marsh mallow), 102
    H. tiliaceus L., 203, 206, 210, 236, 237, 244, 252, 268, 287, 290, 316

Himantopus himantopus (L.) (black-winged stilt), 333
H. leucocephalus Gould (pied stilt), 311
Hippocratea macrantha Korth., 273
Hippophae rhamnoides L., 122, 133
Hippuris tetraphylla L., 170
Hirundinidae, 335
Hirundo neoxena Gould (welcome swallow), 335
Homoptera, 160
Honckenya peploides (L.) Ehr., 133
Hordeum brachyantherum Nevskii, 171, 180
H. jubatum L., 180
H. leporinum Link, 180
H. marinum Huds., 123, 134, 137, 138, 314
H. secalinum Schreb., 100, 123
Hormidium subtile (Kütz.) Heering, 306, 317
Hormosira banksii Berquist, 18, 310
Horsfieldia amklaal, 277
Humata gaimardiana J.Sm., 276
H. ophioglossa Cav., 276
H. trukensis H. Ito, 276
Hutchinsia procumbens (L.) Desvaux, 138
Hydatinadae, 322
Hydnophytum, 22, 288, 290
H. formicarum Jack, 309, 312
Hydranassa picata (Gould) (pied heron), 331
Hydrobia ulvae (Pennant), 19, 126, 128, 146
Hydrocotyle capillaris F. Muell., 315
Hyla regilla Baird and Girard (Pacific tree-frog), 184
Hymenoptera, 160
Hyperlophus vittatus (Castelnau), 328
Hypnea, 318
Hypochrysops apollo Miskin, 309

Ibla cumingi Darwin, 325
Ichthyophaga ichthyaetus (Horsfield) (sea eagle), 269
Ilex integra Thunb., 286
Ilyograpsus, 264, 265
    I. paludicola Rathbun, 237, 325
Ilyoplax, 19, 263, 265
Imperata cylindrica (L.) Beauv., 218, 224–228, 230
Indigofera argentea Burm. f., 220
I. semitrijuga Forsk., 220
I. spinosa Forsk., 223
Insecta, 18, 19, 88–90, 183, 308, 309
Intsia bijuga (Colebr.) O.Ktze., 273
I. retusa Kuntze, 262
Inula crithmoides L., 120, 134, 233
Iresina portulacoides Moq., 198
Ischnostemma cranosum (R.Br.) Merr. and Rolfe, 312
Isognomon isognomon (L.), 323
Isognomontidae, 323
Isopoda, 86, 144, 146, 183, 324
Iva, 2, 22
    I. frutescens (L.), 164, 193
Ixobrychus flavicollis (Latham) (American black bittern), 331
I. minutus (L.) (little bittern), 331

Jadammina macrescens (Brady), 73, 82
J. polystoma (Bartenstein and Brand), 73, 75, 82, 83, 182
Jaumea, 172
    J. carnosa (Less) Gray, 173–175, 177, 178

SYSTEMATIC INDEX

Juncaceae, 315
Juncaginaceae, 314, 316
Juncus, 4, 21, 22, 47, 159, 162, 164, 172, 193, 195
J. acutus L., 12, 176, 179, 195, 196, 233
— — var. *sphaerocarpus* Engelm., 175
J. articulatus L., 125
J. balticus Willd. (baltic rush), 158, 159, 161, 174
— — Willd. var. *littoralis*, 12
J. biglumis L., 67
J. bufonius L., 11, 123, 124, 137
J. gerardi Loisel. (black rush), 11, 35, 89, 100, 114, 115, 122, 124, 131, 133–136, 140, 143, 145, 158, 159, 161, 162
J. krausii Hochst., 237
J. lesueurii Bol., 172–174
J. maritimus Lam., 35, 91, 114, 119, 121, 122, 136, 139–141
— — Lam. var. *australiensis* Buch., 310, 315
J. rigidus C.A. Mey., 218, 223, 225, 227, 228, 230
J. roemerianus Scheele (needle rush), 90, 91, 99, 102, 103, 158, 159, 162, 163, 193

Kandelia, 22, 287, 350
    K. candel (L.) Druce, 17, 20, 66, 245, 247, 255, 266, 269, 280–284, 286, 287, 355
    K. rheedii W. et A., 55
Kosteletzkya virginica Presl. ex A. Gray (salt-marsh mallow), 102

Labyrinthula, 119, 127
Lachesilla pacifica, 184
Lactuca lanciniata Makino, 286
L. raddeana Maxim., 286
Laguncularia, 2, 48, 49, 235, 236, 199, 202, 206, 208, 210
    L. racemosa Gaertn., 16, 164, 177, 179, 199, 202–205, 234, 235, 350, 352, 354, 355
Lalage leucomela (Vig. and Horsf.) (varied triller), 336
L. sueurii (Vieillot) (white-winged triller), 336
Lambis lambis (L.), 322
Lampranthus tegens (F. Muell.) N.E.Br., 313
Lampropeltis getulus (L.) (California king snake), 184
Laomedia healyi Yaldwyn and Wear, 325
Laonome, 327
Lapididae, 325
Laridae, 333, 334
Larus altricilla L. (laughing gull), 100
L. californicus Lawrence (ring-billed gull), 101
L. delawarensis Ord, 101
L. novaehollandiae Stephens (silver gull), 333
Lasius flavus F., 146
Lasthenia glabrata Lindl., 175
Laterallus jamaicensis (Gmelin) (black rail), 184
Laternula creccina Reeve, 324
L. tasmanica (Reeve), 324
L. vagina Reeve, 324
Laternulidae, 324
Launaea spinosa (Forsk.) Sch.Bip, 220
Laurencia, 194
    L. nidifica J.Ag., 318
Lawrencia spicata Hook, 314
Loxoconcha, 76
    L. cf. *levis* Brady, 76

Lecanora cf. *hageni* (Ach.) Ach., 319
L. pallida (Schreb.) Rabenh., 319
L. subfusca (L.) Ach., 319
L. varia (Hoffm.) Ach., 319
Leersia, 267
L. oryzoides (L.) Sw., 163
Leiopecten sorbidulum Latreille, 263–265, 325
Lemna gibba L., 218
Leontodon autumnalis L., 122, 124, 125, 133, 135
L. nudicaulis (L.) Banks ex Lowe, 123
Lepidium latifolium L., 138, 139
Lepidophyllum, 195, 196
    L. cupressiforme Cass., 196
Leptadenia pyrotechnica (Forsk.) Deene., 224
Leptocarpus, 47, 310
Leptoconops, 160
Leptocottus armatus Girard (Pacific staghorn sculpin), 184
Leptogium, 320
Leptospermum scoparium J.R. and G. Forst., 310
Leptosphaeria discors (Saccardo and Ellis) Saccardo and Ellis, 119
L. juncaginacearum (Schröt.) Munk, 119
Leptotrema, 319
Lepturus repens (Forst.) R.Br., 316
Lepus, 186
    L. californicus Gray, 185
    L. capensis (L.) (brown hare), 144
Leucophoyx thula (Molina), 185
Lichmera indistincta (Vig. and Horsf.) (brown honey-eater), 338
Ligia, 183, 186, 324
    L. australiensis (Dana), 324
Ligidae, 324
Ligusticum scoticum L., 67
Ligustrum japonicum Maxim., 286
Lilaeopsis occidentalis Coult. and Rose, 172
Limapontia depressa A. and H., 146
Limicola falcinellus (Pontoppidan) (broad-billed sandpiper), 333
Limnodromus, 184
Limoniastrum guyonianum Dur. ex Boiss., 233
Limonium, 11, 12, 21, 22, 133, 179, 226, 233
    L. australe (R.Br.) O.Kuntze, 314
    L. axillare (Forsk.) Ktze., 218–222, 224–228, 230
    L. bellidifolium (Gouan) Dum., 138
    L. binervosum (G.E. Smith) C.E. Salmon, 314
    L. californicum (Boiss.) Heller, 174–176, 179
    — — (Boiss.) Heller var. *mexicanum* (Blake) Munz, 177
    L. ferulaceum (L.) Kuntze, 134
    L. humile Mill., 124, 129
    L. japonicum (Zucc.) O. Kuntze, 12
    L. linifolium (L.) O. Kuntze, 236
    L. pruinosum (L.) O. Kuntze, 218, 220, 223, 225, 227, 228, 230
    L. psilocladon (Boiss.) O. Kuntze, 314
    L. salicorneacea (F. Muell.) O. Kuntze, 314
    L. sebkarum (Pomel), 233
    L. sinuatum (L.) Miller, 233
    L. spathulatum (Desf.) O. Kuntze, 233
    L. tetragonum (Thunb.) Bullock, 12

L. vulgare Mill., 119, 124, 130–135, 143, 145, 146
— — Mill., ssp. *vulgare*, 122
Limosa lapponica (L.) (bartailed godwit), 333
Linaria vulgaris Mill., 125
Linum catharctium L., 125
Liocranium praepositum Ogilby, 329
Littorina, 186
    L. carinifera Menke, 265
    L. irrerata (Say), 96, 160
    L. littorea L., 126, 128, 146, 149, 160
    L. melanostoma (Gray), 265
    L. newcombiana (Hemphill), 182
    L. saxatilis Ol., 128, 146
    L. scabra (L.), 18, 237, 265
    L. sitkana Philippi, 183
Littorinidae, 321
Liza vaigiensis (Quoy and Gaimard), 329
Lola, 17
    L. capillaris (Kütz.) Hamel, 17
Lolium loliaceum (Bory and Chaub.) Hand., 316
L. perenne L., 100, 122, 135, 137
Lonchura castaneothorax (Gould) (chestnut-breasted finch), 338
L. flaviprymna (Gould) (yellow rumped finch), 338
Lophodytes cucullatus (L.) (hooded merganser), 98
Loranthacoae, 312
Loranthus quinquinervis Hochst, 237
Lotus corniculatus L., 124, 125
L. tenuis Walst. and Kit., 122, 124, 135
Lucinidae, 323
Lulworthia halima (Diehl and Mounce) Cribb and Cribb, 119
L. medusa (Ell. and Ev.) Cribb and Cribb, 119
Lumnitzera, 237, 245, 268, 288, 350
    L. coccinea W. and A. Burk., 348
    L. littorea (Jacq.) Voigt, 244, 264–266, 268, 269, 271, 273, 287, 293, 301, 311, 348, 351
    L. racemosa Presl., 20, 237, 238, 244, 252, 254, 255, 265, 268, 279, 282, 288–290, 293, 301, 311, 348
Lutjanidae, 329
Lutjanus argentimaculatus (Forskål), 329
L. carponotatus (Richardson), 329
L. fulviflamma (Forskål), 329
Lutra canadensis (Schreber), 160
L. maculicollis Licht. (otter), 239
L. perspicillata Geoff., 269
Lycastis indica, 237
Lycium arabicum Schweinf. ex Boiss., 222
L. brevipes Benth., 176
Lyngbya, 17
    L. aesturii (Mart.) Liebmann, 318
    L. epiphytica Hieron, 319
    L. lutea (Ag.) Gom., 319
    L. majuscula (Dillw.) Harv., 319
Lynx rufus (Schreber), 185
Lythraceae, 316

Macaca irus Aw. (Macaque monkey), 269
Machaerirhynchus flaviventer Gould (boat-billed flycatcher), 336
Machaerium lunatum Ducke, 199

Machilus, 278
    M. thunbergii Sieb. and Zucc., 286
Macrophthalmus, 264, 265
    M. crassipes Milne Edwards, 325
    M. depressus (Rüppel), 237, 265, 325
    M. latreillei (Desmarest), 263, 325
    M. pacificus Dana, 265, 325
    M. punctulatus Miers, 326
    M. setosus Milne Edwards, 326
Macropygia phasianella (Temminck) (brown pigeon), 334
Mactridae, 323
Mactua eximia Reeve, 267
Maireana brevifolia (R.Br.) P.G. Wilson, 294, 313
M. oppositifolia (F. Muell.) P.G. Wilson, 313
Malaclemys terrapin centrata (Schoepff) (diamondback terrapin), 160
Maluridae, 336
Malurus amabilis Gould (lovely wren), 336
M. cyaneus (Latham) (superb blue wren), 336
M. melanocephalus (Latham) (red-backed wren), 336
Malvaceae, 244, 252, 314, 316
Mammalia, 18, 102–105, 148, 149, 185, 310
Mareca americana (Gmelin) (American widgeon), 98, 99, 185
Marphysa sanguinea (Montagu), 327
Matricaria ambigua Maxim ex Kom., 67
M. inodora L., 67, 122, 123
M. maritima L., 138, 139
Maytenus, 194
Mecinus collaris Germ., 146
Mecodium polyanthos Copel., 276
Megalurus timoriensis Wallace (tawny grassbird), 336
Megapodiidae, 332
Megapodius freycinet Gaimard (scrub fowl), 332
Melaleuca, 268, 301, 304
    M. acacioides F. Muell., 316
    M. ericifolia Sin., 305, 316
    M. halmaturorum F. Muell. ex Miq., 316
    M. leucodendron L., 269
    M. quinquenervia (Cav.) S.T. Blake, 316
    M. squarrosa Donn. ex Sm., 316
Melampus, 18, 186
    M. bidentatus Say (coffee bean snail), 85, 86, 160
    M. olivaceus Carpenter, 183
Melanerita melanotragus Smith, 321
Melanotaenia nigrans (Richardson), 328
Melanotaenium ruppiae G.Feldm., 119
Melarapha infans Smith, 321
M. scabra L., 307, 308, 321
M. undulata Gray, 321
Meliaceae, 244, 255, 311
Meliosma rigida Sieb. et Zucc., 286
Melilotus indica (L.) All., 315
Meliphaga fasciogularis (Gould) (mangrove honey-eater), 338
M. flava (Gould) (yellow honey-eater), 338
M. flaviventer (Lesson) (tawny-breasted honey-eater), 338
M. gracilis (Gould) (graceful honey-eater), 338
M. lewinii (Swainson) (Lewin honey-eater), 338
M. notata (Gould) (lesser Lewin honey-eater), 338
M. versicolor (Gould) (varied honey-eater), 338
Meliphagidae, 337, 338

SYSTEMATIC INDEX

Melithreptus albogularis Gould (white-throated honey-eater), 338
M. laetior (Gould) (golden-backed honey-eater), 338
Melosidula nucleus Gmelin, 322
M. zonata (H. and A. Adams), 322
Melospiza melodia (Wilson) (song sparrow), 184
– – *maxillaris* Grinnell (salt-marsh song sparrow), 101, 102
Menidia, 160
    M. menidia (L.), 95
Mephitis mephitis (Schreber), 185
Mergus merganser L. (common merganser), 98
Meropidae, 335
Meropus ornatus Latham (rainbow bird), 335
Mesembryanthemum, 180
    M. crystallinum L., 176
    M. nodiflorum L., 180
Mesochaetopterus minuta Potts, 327
Mesostigmata, 91
Metaplax crenulatus (Gerstaecher), 263, 265
M. elegans de Man, 263–265
Metapograpsus frontalis Miers, 264, 265, 325
M. gracilipes (de Man), 325
M. latifrons (White), 264, 325
M. messor (Forskål), 325
Microcanthus strigatus (Cuvier and Valenciennes), 329
Microcoleus, 17
    M. chthonoplastes Thuret ex. Gom., 194, 319
    M. lyngbyaceus (Kütz.) Crouan, 319
    M. tenerrimus Gom., 319
Microeca flavigaster Gould (lemon-breasted flycatcher), 336
M. griseoceps Devis (yellow flycatcher), 336
Microtus californicus (Peale) (Californian meadow mouse), 104, 185
M. oregoni (Bachman) (Oregon meadow mouse), 185
M. pennsylvanicus (Ord) (meadow mouse), 102, 104, 160
M. ratticeps (Keys and Blasius) (root vole), 144
M. townsendii (Bachman) (Townsend meadow mouse), 185
Mictyridae, 325
Mictyris longicarpus (Latreille), 325
Miliammina fusca (Brady), 73, 75, 81–83, 182
Milvus migrans (Boddaert) (black kite), 332
Mimulus repens R.Br., 314
Miscanthus sinensis Anders., 284, 285
Mitridae, 322
Modiolus demissus Dillwyn, 84, 183, 187
M. inconstans (Dunker), 323
M. neozelanicus Iredale, 310
M. pulex (Lam.), 323
Mollusca, 18, 19, 22, 84–86, 146, 148, 149, 168, 182, 187, 237, 265, 307, 308, 310, 321–324
Monanthochloe, 176
M. littoralis Engelm. (salt-flat grass), 158, 175–177, 179, 194
Monarcha frater Sclater (black-winged flycatcher), 336
M. leucotis Gould (pied flycatcher), 336
M. melanopsis (Vieillot) (black-faced flycatcher), 336
M. trivirgata (Temminck) (spectacled flycatcher), 336
Monerma cylindrica (Willd.) Coss and Durieu., 174, 180, 233
Monodactylidae, 329
Monodactylus argenteus (L.), 329
Monodonta labio L., 321

Monostroma, 306, 317
    M. crepidinum Farlow, 317
Montfortula rugosa Quoy and Gaimard, 321
Montia fontana L. spp. *lamprosperma*, 125
Montrichardia arborescens Schott, 211
Mora megistosperma Britten and Rose, 211
Morula marginalba (Blainville), 322
Motacilla cinerea (Tunshall) (grey wagtail), 335
M. flava L. (yellow wagtail), 335
Motacillidae, 335
Mucuna gigantea DC., 290
Mugil, 353
    M. cephalus L., 329
    M. georgii Ogilby, 329
Mugilidae, 329
Mullidae, 329
Muntiacus muntjak (Zimmermann), 251
Muricidae, 322
Murrayella, 237
Mus musculus L., 102
Muscicapa rufigastra Raffles, 269
Muscicapidae, 336, 337
Mustela frenata Lichtenstein (long-tailed weasel), 185
M. vison (Schreber) (mink), 160
Myiagra alecto (Temm. and Lang.) (shining flycatcher), 337
M. rubecula (Latham) (leaden flycatcher), 337
M. ruficollis (Vieillot), 337
Mylio australis (= *Acanthopagrus australis*), 330
Myoporaceae, 316
Myoporum acuminatum R.Br., 316
M. bontioides A. Gray, 287
M. inerme, 287
M. insulare R.Br., 316
Myosotis laxa Lehm. ssp. *caespitosa* (K.F. Schultz) Hyl., 124
Myosurus minimus L., 174
Myriostachia wightiana Hk.f., 144, 247, 251–253
Myristica, 267, 268
Myrmecodia, 22, 267, 288, 290
    M. antoinii Becc., 309, 312
    M. tuberosa Jack, 288
Myrsinaceae, 244, 254, 311
Myrsine umbellata Mart., 266, 267
Myrtaceae, 311, 316
Mytilidae, 323
Myxophyceae, 17, 18, 142
Myzomela erythrocephala Gould (red-headed honey-eater), 338
M. obscura Gould (dusky honey-eater), 338

Nabidae, 146
Namalycastis abiuma Mueller, 326
Nannosesarma minuta (de Man), 263
Nasalio larvatus (Wurmb.) (proboscis monkey), 269
Nassarius burchardi (Philippi), 322
Nassidae, 322
Natica violacea Bruguière, 322
Naticidae, 322
Neanthes vaallii Kinberg, 326
Nectarinia jugularis (L.), 337
Nectariniidae, 337

Nematoda, 83, 144, 146, 149, 182
Neocaudites, 76
    N. cf. *nevianii* Puri, 76
Neolitsea sericea Koidz., 286
Neomysis vulgaris Leach, 80
Neottopteris nidus J.Sm., 276
Nephrolepis, 236
    N. hirsutula Presl., 275, 276
Nephtyidae, 327
Nephtys australiensis Fauchald, 327
Nereidae, 326
Nereis diversicolor Müller, 80, 149
N. uncinula (Russel), 326
Nerita, 19, 307
    N. albicilla L., 321
    N. atramentosa Reeve, 321
    N. birmanica Phil., 266
    N. chamaeleon L., 321
    N. lineata Gmelin, 321
    N. planospira Anton, 321
    N. plicata L., 321
    N. polita L., 321
Neritidae, 321
Nesogobius pulchellus (Castelnau), 328
Ninox connivens (Latham) (barking owl), 335
Nitraria, 222
    N. retusa (Forsk.) Asch., 218, 220, 222–228, 233
    N. schoberi L., 294, 316
Nostoc, 18, 67
    N. commune Vaucher, 318
    N. entophytum Born. and Flah., 318
    N. linckia (Roth.) Bornet ex Born. and Flah., 318
    N. muscorum Ag. ex Born. and Flah., 318
Nostocaceae, 318
Notesthes robusta (Günther), 329
Notamastus hemipodus Hartman, 327
Notospisula parva (Angas), 323
Numenius madagascariensis (L.) (eastern curlew), 333
N. minutus Gould (little whimbrel), 333
N. phaeopus (L.) (whimbrel), 269, 333
Nuphar advena Ait., 163
Nycticorax caledonicus (Gm.) (nankeen night heron), 269, 331
N. nycticorax (L.) (black-crowned night heron), 160
Nypa, 22, 25, 65, 249, 251, 258, 265, 266, 268, 269, 270, 281, 350
    N. fruticans Wurmb., 4, 20, 244, 247, 251, 258, 265–267, 269, 273, 281, 282, 288, 293, 300, 311, 348, 351, 352

Ochetostoma australiense Edmonds, 327
O. capense Jones and Stephen, 236
Ochrolechia subpallescens Vers., 319
Ocypode ceratophthalmus (Pallas), 325
Ocypodidae, 325, 326
Odocoileus hemionus (Rafinesque) (mule deer), 185
O. virginianus (Zimmermann) (white-tailed deer), 160
Odontites, 125
    O. littoralis (Fr.) Lange, 124, 125, 135
Oecophylla, 270
    O. smaragdina (Fabricius), 308

Oligochaeta, 146, 149, 327
Olor colombianus (Ord) (whistling swan), 98
Onchidiidae, 323
Onchidina, 307
    O. australis Semper, 323
    O. buetschli Stantshinsky, 323
    O. damellii Semper, 323
    O. verruculatum Cuvier, 323
Oncosperma filamentosa Blume, 266
O. tigillaria Ridl., 266, 267
Ondatra zibethicus (L.) (muskrat), 102, 103, 144, 160, 364
Ophiobolus maritimus (Saccardo) Saccardo, 119
Ophiocardelus, 307
    O. ornatus (Férussac), 322
    O. quoyi H. and A. Adams, 322
    O. sulcatus H. and A. Adams, 322
Opuntia, 203
    O. wentiana Britten and Rose, 65
Orbiniidae, 327
Orchestia, 186
    O. chiliensis Milne Edwards, 80
    O. gammarella (Pallus), 79, 86
    O. palustris Smith, 86
    O. traskiana Stimpson, 183, 187
Orchestoidea, 186
    O. californica (Brandt), 183
Orchidaceae, 267, 287, 312
Orectolobidae, 329
Oriolidae, 338
Oriolus flavocinctus (Vigors) (yellow oriole), 338
O. sagittatus (Latham) (olive-backed oriole), 338
Orobanchaceae, 119, 120
Orthoptera, 146, 160
Oryctolagus cuniculus (L.), 144
Oryzomus palustris (Harlan) (rice rat), 160
Osbornia, 16, 53
    O. octodonda F.V.M., 52, 293, 311, 351
Oscillatoria, 17
    O. corallinae (Kütz) Gom., 319
    O. earlei Gardner, 319
    O. laete-virens (Cronan) Gom., 319
    O. nigro-viridis Thwaites ex Gom., 319
Oscillatoriaceae, 318
Oscinella, 183
Ostracoda, 73, 74, 76, 182
Ostrea nomades Iredale, 323
    O. rhizophorae Guilding, 353
    O. tulipa Lamarck, 235
Ostreidae, 323
Otospermophilus beecheyi (Richardson) (Beechey ground squirrel), 185
Ovatella myosotis Drap, 146
Owenia fusiformis Delle Chiaje, 327
Oweniidae, 327

Pachycephala cinerea (Blyth), 269
    P. griseiceps Gray (gray whistler), 337
    P. lanioides Gould (white-breasted whistler), 337
    P. melanura Gould (mangrove golden whistler), 337
    P. rufiventris (Latham) (rufous whistler), 337
    P. simplex Gould (brown whistler), 337

SYSTEMATIC INDEX

Pachycephalidae, 337
Pachycornia tenuis (Benth.) J.M. Black, 294
Pachygrapsus, 186
    P. crassipes Randall, 183
Paguridae, 326
Paguristes, 326
Palaemonetes, 161
Paliurus ramossissimus Poir, 286, 287
Palmae, 244
Palmerinella palmerae Bermudez, 74, 75
Pandanaceae, 244
Pandanus, 65, 235, 236, 282, 283
    P. affinis Kurz., 266, 267
    P. aimiriikensis Martelli, 277
    P. japensis Martelli, 275
    P. kanehirae Martelli, 273
    P. odoratissia L.f., 281
    P. tectorius Soland, 244, 282, 283
Pandion haliaetus (L.) (osprey), 332
Pandionidae, 332
Panicum, 194
    P. turgidum Forsk., 220, 222, 223
    P. virgatum (L.), 89, 161, 164
Panopeus herbsti Milne Edwards, 87
Paphia hiantina (Lam.), 324
Papilionaceae, 244
Paracleistostoma, 264
    P. depressum de Man, 265
    P. fossula Brnrd., 237
    P. longimanum Tweedie, 265
    P. microcheirum Tweedie, 265
Paracypris, 76
Paradisaeidae, 339
Paragnathia maxillaris (Montagu), 79, 86
Paragrapsus laevis (Dana), 325
Paralichthys californicus Ayres (California halibut), 184
Parapholis, 138
    P. incurva (L.) C.E. Hubbard, 174, 180, 311, 314
    P. strigosa (Dum.) C.E. Hubbard, 114, 122, 131, 137
Parioglossus rainfordi McCulloch, 328
Parmelia dilatata Wain., 320
    P. maura Pers., 320
    P. sydneyensis Gyelnik, 320
    P. tinctorum Nyl., 320
Parupeneus jeffi (Ogilby), 329
Parus major L. (grey tit), 269
Paspalum distichum L. var. *littorale* R.Br., 316
P. vaginatum Sw., 195, 198, 199, 235, 273
Passerculus sandwichensis (Gmelin) (savannah sparrow), 101, 102
  – – *beldingi* (Gmelin), 184
Patellidae, 321
Patro australis (Gray), 323
Pavonia, 199
Peasiella tantilla (Gould), 321
Pelargopsis capensis (Shaw), 269
Pelates quadrilineatus (Bloch), 330
Pelecanidae, 330
Pelecanus conspicillatus Temminck (pelican), 330
Pelecypoda, 84–86

Pelliciera, 16
    P. rhizophora Planch and Triana, 16, 20, 199, 204, 205, 210
Pelvetia, 18, 142
    P. canaliculata (L.) Dec. and Thur. var. *libera*, 33
Pemphis acidula Forst. and Forst.f., 52, 273, 316
Penaeidae, 326
Penaeus, 326
    P. plebejus Hess, 326
Peneonanthe pulverulenta (Bonaparte) (mangrove robin), 337
Periballia minuta (L.) Ashers and Graebn., 316
Periophthalmidae, 329
Periophthalmodon schlosseri (Pallas), 263, 265
Periophthalmus australis Castelnau, 329
P. chrysospilos Blkr., 263, 265
P. kalolo Lesson, 237, 238, 265, 329
P. koelreuteri (Pallas), 235, 329
P. schlosseri (Pallas), 329
P. sobrinus Eggert., 237
Perissocytheridea brachyforma Swain, 76
Perna nucleus Lam, 323
Peromyscus leucopus (Rafinesque) (white-footed mouse), 102
P. maniculatus (Wagner) (deer mouse), 185
Peronia peroni (Cav.), 237
Pertusaria, 320
    P. bispora (Farlow) Linder, 319
    P. multipuncta (Turn.) Nyl., 319
    P. cf. *oahuensis* Magn., 320
    P. thiospoda Knight, 320
Petrochelidon ariel (Gould) (fairy martin), 335
P. nigricans (Vieillot) (tree martin), 335
Petroscirtes anolius (Cuv. and Val.), 328
Petrosimonia crassiflora Bunge, 12
Pezoporus wallicus (Kerr) (ground parrot), 334
Phaeococcus, 18, 142
Phaeophyceae (Phaeophyta), 113, 128, 141, 317
Phalacrocoracidae, 331
Phalacrocorax carbo (L.) (black cormorant), 331
P. melanogaster, 269
P. melanoleucos (Vieillot) (little pied cormorant), 331
P. niger King, 269
P. sulcirostris (Brandt) (little blcak cormorant), 331
P. varius (Gmelin) (pied cormorant), 331
Phascolosoma dunwichi Edmonds, 327
P. lurco Selenke and de Man., 308, 327
Phedale myrmecodiae, 309
Philemon argenticeps (Gould) (silver-crowned friar bird), 338
P. citreogularis (Gould) (little friar bird), 338
P. corniculatus (Latham) (noisy friar bird), 338
P. gordoni Mathews (Melville Island friar bird), 338
P. novaeguineae (Muller) (helmetted friar bird), 338
Philoscia richardsonae Holmes and Gay, 183
Philoxerus vermicularis P. Beauv., 235
Phippsia algida (Phipps) R.Br., 67
Phoenix canariensis Hort. ex Chabaud, 180
P. dactylifera L., 180
P. paludosa Roxb., 244
P. reclinata Jacq., 286
Phoma exigua Desm., 119
P. neglecta Desm., 119

P. statices Tassi, 119
Phomopsis achilleae var. *asteris* Grove, 119
Phonygammus keraudrenii (Less. and Garn.) (manucode), 339
Phormidium, 17
    P. angustissimum W. and G.S. West, 306, 319
    P. autumnale (Ag.) Gom., 142, 319
    P. corium (Ag.) Gom., 319
    P. foveolarum (Mont.) Gom., 319
    P. luridum (Kütz.) Gom., 319
    P. molle (Kütz.) Gom., 319
    P. tenue (Menegh.) Gom., 319
Phormium tenax J.R. and G. Forst., 310
Phragmites, 3, 12, 57, 111, 130, 237
    P. australis (Cav.) Trin. ex Steud., 218, 223, 225, 227, 228, 230, 237, 316
    P. communis (Trin.), 25, 115, 122, 124, 141, 145, 159, 161, 163
    P. japonica Steddal., 286, 287
    P. karka (Retz.) Trin. ex Steud., 316
Phreatia palowensis Tuyama, 276
Phyllodoce malmgreni Gravier, 327
P. novaehollandiae Kinberg, 327
Phyllodocidae, 327
Phymatodes scolopendria Ching, 275, 276
Physma byrsinum (Ach.) Müll. Arg., 320
Physocypria pustulosa (Sharpe), 76
Phytia, 186
    P. myosotis Draparnaud, 182, 186
Pilumnus, 326
Pimephales promelas Rafinesque (fathead minnow), 184
Pinctada margaritifera (L.), 323
Pinnidae, 323
Piper betle L., 47
Pisces (fish), 19, 22, 26, 74, 91–95, 160, 184, 186, 237, 249, 309, 328–330, 353, 357, 364–367
Pistacia chinensis Bunge, 278
Pisum maritimum L., 67
Pithecolobium umbellatum Benth., 350
Pitta iris Gould (rainbow pitta), 335
Plagianthus divaricatus J.R. and G. Forst., 310
P. spicatus (Hook.) Benth., 314
Planaxidae, 321
Planaxis sulcatus (Born), 321
Planchonella obovata H.J. Lam,, 267, 277
Plantaginaceae, 314
Plantago, 4, 22, 133, 158, 172
    P. coronopus L., 114, 122, 124, 137, 144, 311, 314
    P. juncoides (Lam.) Hutt., 170
    P. macrocarpa Cham. and Schlect., 171
    P. major L., 122–124, 137
    P. maritima L., 18, 26, 35, 67, 117, 119, 122, 124, 130, 131, 134, 135, 138, 145, 146, 171, 173, 174, 178
Plasmodiophora bicaudata J. Feldm., 119
P. maritima J. Feldm., 119
Platalea flavipes (Gould) (yellow-billed spoonbill), 331
P. regia Gould (royal spoonbill), 331
Platichthys stellatus (Pallas) (starry flounder), 184
Platycerium, 236
Platycercus adscitus (Latham) (pale-headed rosella), 334
P. venustus (Kulh.) (northern rosella), 334

Platynereis cf. *dumerilii antipoda* Hartman, 326
Plectorhinchus multivittatus (Macleay), 329
P. pictus (Thunberg), 329
P. sordidum (Klunzinger), 329
P. unicolor (Macleay), 329
Plegadis falcinellus (L.) (glossy ibis), 331
Pleospora herbarum (Pers.) Rabenh., 119
Pleuropogon sabinii R.Br., 67
Plotosidae, 329
Plotosus lineatus (Thunberg), 329
Pluchea, 22, 158
P. camphorata (L.) DC., 158
P. purpurascens DC., 162
Plumbaginaceae, 244, 250, 311, 314
Pluvialis dominica (Muller) (golden plover), 332
P. squatorola (L.) (grey plover), 333
Poa annua L., 122
P. caespitosa Spreng., 80
P. eminens Presl., 171
P. pratensis L., 123, 124
P. trivialis L., 123
Poaceae, 314, 316
Podargidae, 335
Podargus papuensis Quoy and Gaimard (Papuan frogmouth), 335
Poecilodryas cerviniventris (Gould) (buff-sided robin), 337
P. superciliosa (Gould) (white-browed robin), 337
Poga, 350
Poliera marina, 79
Polinices conicus (Lam.), 322
Polychaeta, 326, 327
Polydora, 327
P. nuchalis, 182
Polygonum aviculare L., 122, 124, 137
P. blumei Meisn., 286
P. maritimum L., 196
Polyplacophora, 321
Polypodium, 236
    P. sinuatum Wall., 288
Polypogon monspeliensis (L.) Desf., 180
Polysiphonia macrocarpe Harv., 318
    P. pacifica Hollenberg, 172
Pomacentridae, 329
Pomadasyidae, 329
Pomatomidae, 329
Pomatomus saltator (L.), 329
Pontodrilus bermudensis Beddard, 327
Pontoscolex corethrurus (Bahl), 327
Porcellio dilatatus Brandt and Ratzeburg, 183
P. laevis Latreille, 183
P. scaber Latreille, 183
Porphyra umbilicalis (L.) J.Ag., 121
Porphyrosiphon murzii (Zeller) Dr., 319
Porteresia coarctata (Roxb.) Tateoka, 244, 251, 253
Portunidae, 326
Portunion conformis, 183
Portunus pelagicus L., 326
Porzana carolina (L.) (sora rail), 101
P. cinerea (Vieillot) (white-browed crake), 332
Posidonia, 1

SYSTEMATIC INDEX

Potamididae, 321
Potamogeton pectinatus L., 120
P. pusillus L., 124
Potentilla, 172
    P. anserina L., 67, 122, 124, 137
    P. egedii Wormsk., 170
    P. pacifica Howell, 170, 171, 173–175, 178
Pottia heimii Fuernr., 114, 137
Presbytis cristatus (Raffles) (leaf monkey), 269
Primula borealis Duby, 170
Primulaceae, 314
Prioria copaifera Griseb., 211
Probosciger aterrimus (Gmelin) (palm cockatoo), 314
Procyon lotor (L.) (racoon), 160, 185
Prokelisia, 183
Prosopis juliflora DC., 256
P. spicigera Linn. Mant., 252
Prostigmata, 91
Protelphidium anglicum, 74
P. tisburyense (Butcher), 74, 82, 84
Proteonina atlantica Cushman, 81
Protococcus viride Ag., 317
Protoschista findens (Parker), 74, 83, 182
Prunus anasakura Nakai, 286
Psephotus chrysopterygius Gould (golden-shouldered parrot), 334
Pseudendoclonium submarinum Wille, 317
Pseudoclavulina, 74
Pseudococcus, 184
Pseudoeponides andersoni Warren, 73, 75, 83
Pseudosmittia, 183
Pseudupeneus indicus (Shaw), 329
P. signatus (Günther), 329
Psittacidae, 334
Psophodes olivaceus Latham (eastern whipbird), 336
Pteridaceae, 312
Pteriidae, 323
Pterocarpus officinalis Jacq., 211
Pteroptyx malaccae Gorham, 270
Pteropus alecto Peters (fruit bat/flying fox), 310
Ptilinopus regina Swainson (red-crowned pigeon), 334
P. superbus (Temminck) (purple-crowned pigeon), 334
Ptilonorhynchidae, 339
Ptiloris magnificus (Vieillot) (magnificent rifle bird), 339
P. victoriae Gould (Victoria rifle bird), 339
Ptittidae, 335
Puccinellia, 2, 12, 21, 22, 32, 67, 118, 136, 137, 140, 144, 147, 170–172, 178
    P. americana Th. Sørensen, 10, 12
    P. borealis Swallen, 170
    P. capillaris (Liljebl.) Jansen, 136
    P. convoluta (Hornem.) Fourn., 10
    P. distans (L.) Parl., 115, 122, 124, 136, 145, 233, 310
    P. fasciculata (Torr.) Bicknell, 122, 136
    P. festucaeformis (Host) Parl., 10
    P. gigantea Grossheim, 12
    P. kurilensis (Takeda) Honda, 2, 10
    P. limosa (Schur.) Holmb., 10, 27
    P. lucida Fern. and Weath., 174
    P. maritima (Huds.) Parl., 10, 11, 90, 91, 100, 111, 117, 118, 120, 122, 124, 128–130, 132, 133, 136, 140, 141, 145

    P. nutkaensis (Presl.) Fern. and Weath., 171
    P. palustris (Seenus) Podpera, 233
    P. peisonis (Beck.) Jav., 10
    P. phryganodes (Trin.) Scribn. and Merr., 10, 11, 67, 170, 171, 178, 181
    P. pseudodistans (Crép.) Jans. and Wacht., 136
    P. pumila (Vasey) Hitchc., 171–173
    P. retroflexa Holmb., 122, 124, 136
    P. stricta (Hook.f.) C. Blom., 311, 314
    P. triflora Swallen, 171
Puccinia absinthii DC., 119
P. asteris Duby, 119
Punctoribates quadrivertex, 91
Puriana, 76
Pycnonotus goiaver (Scop.), 269
P. plumosus Blyth, 269
Pyrazus ebeninus Bruguière, 307, 321
Pythia, 19, 265, 320
    P. scarabeus L., 266, 322
Pyxine berterana (Fee) Imshaug, 320
P. coceos (Sw.) Nyl., 320
P. physciiformis (Malme) Imshaug, 320
P. subcinerea Stirt., 320

Quararibaea, 211
Quercus glauca Thunb. var. *amamiana*, 284, 286
Querquedula discors (L.) (blue-winged teal), 159
Quoyia decollata (Quoy and Gaimard), 321

Rallidae, 332
Rallus elegans Aubudon (king rail), 96, 97
R. limicola Vieillot (Virginia rail), 101
R. longirostris Boddaert (clapper rail), 159, 184
  – – *crepitans* Gmeln., 96
  – – *waynei* Brewster, 96
R. phillipensis L. (banded landrail), 332
Ramalina calicaris (L.) Röhl., 320
R. leiodea Nyl., 320
R. pusilla Le Prev., 320
R. reagens (B. de Lesd.) Culb., 320
R. cf. *sideriza* Zahlbr., 320
Ramsayornis fasciatus (Gould) (bar-breasted honey-eater), 338
R. modestus (Gray) (brown-backed honey-eater), 338
Ramularia asteris (Phill. and Plowr.) Lind., 119
Rana cancrivora Gravenhorst, 19, 264, 265, 269
Ranunculus cymbalaria Pursh, 171
R. hyperboreus Rottb., 67
R. sceleratus L., 122–124
Rapanaea nereifolia Mez., 286
Raphia, 211, 235, 236
Raphiolepis umbellata Makino, 286
Rattus norvegicus (Berkenh.) (brown rat), 104, 144
Recurvirostra novaehollandiae Vieillot (red-necked avocet), 333
Recurvirostridae, 333
Reithrodontomys megalotis (Baird) (western harvest mouse), 104, 185
R. ravientris (Dixon) (salt-marsh harvest mouse), 104, 185
Retama raetam (Forsk.) Webb, 224
Rhabdadenia biflora Muell., 199

Rhabdosargus sarba (Forskål), 330
Rhagodia baccata (Labill.) Moq., 313
Rhinanthus serotinus (Schönh.) Oborny, 124, 125
Rhipidura fuliginosa (Sparrman) (grey fantail), 337
R. leucophrys (Latham) (willy wagtail), 337
R. javanica (Sparrman), 269
R. rufifrons (Latham) (rufous fantail), 337
R. rufiventris (Vieillot) (northern fantail), 337
Rhizoclonium, 17, 311
R. capillare Kütz., 306, 317
R. implexum (Dillw.) Kütz., 317
Rhizophora, 2, 4, 15, 21, 22, 26, 31, 48, 49, 51, 54, 162, 197, 199, 200, 202, 203, 205–209, 234, 235, 237–239, 245, 243–255, 258, 263–268, 283, 288, 290, 347, 349, 350, 352, 354–356
   R. apiculata Blume, 17, 20, 245, 253, 258, 263, 265–268, 270, 273–275, 287, 293, 305, 312, 348–352, 354
   R. brevistyla Salvosa, 199, 200
   R. candelaria DC., 349
   R. conjugata (non L.) Arn., 262, 346–351, 354
   R. harrisonii Leechman, 16, 20, 199, 200, 205, 209, 234–236
   R. mangle L., (red mangrove), 16, 20, 25, 26, 47, 71, 158, 162, 177, 179, 197, 199, 200, 202–204, 209, 210, 234–236, 288, 290, 349, 350, 352, 354–356
   R. mucronata Lam., 16, 20, 66, 217, 218, 225, 227, 228, 236–238, 245, 250, 253, 255, 258, 262, 263, 265–267, 271, 273–275, 279, 280, 282, 283, 287–290, 302, 347–352, 354, 355
   R. racemosa G.F.W. Meyer, 16, 20, 199, 200, 209, 234–236, 269
   R. samoensis Salvoza, 199
   R. stylosa Griff., 20, 52, 53, 264, 268, 288, 290, 293, 301–305, 312, 351
Rhizophoraceae, 244, 251–256, 258, 311
Rhodophyta, 317
Rhynchospora pterochaeta F. Muell., 315
Rhynchostegiella compacta var. *salina* (Bryhn) Podp., 137
Rivularia, 18, 142
Roccella montagnei Bel., 320
Rosa californica Cham. and Schlecht., 175
Rubiaceae, 245, 312, 313
Rumex crispus L., 124, 125, 139
R. obtusifolius L., 115
R. occidentalis Wats., 174
Ruppia, 120
R. maritima L. (widgeon grass), 72, 99, 110, 119, 124, 158
R. spiralis Dum., 121, 124

Sabellidae, 327
Saccostrea, 307
S. amasa (Iredale), 323
S. commercialis (Iredale and Roughley), 323
S. cucullata (Quoy and Gaimard), 18, 310, 323
Sacoglossa, 146
Sagina maritima G. Don., 114, 122–124, 137, 315
S. nodosa (L.) Fenzl., 124
S. procumbens L., 124
S. subulata (Sw.) C. Presl., 124
Sagittaria, 194

Salicornia, 2, 4, 8, 9, 11, 12, 18, 19, 21, 22, 26, 32, 40, 57, 66, 71, 73, 75, 109, 127, 128, 131, 140, 141, 159, 161, 162, 172, 175, 178, 183, 184, 187, 193–196, 198, 209
   S. ambigua Michx., 194
   S. arabica L., 233
   S. australis Soland. and Forst., 80, 213, 288, 289, 310, 311
   S. bigelovii L. (dwarf saltwort), 158, 175, 177, 179
   S. blackiana Ulbrich, 313
   S. brachiata Roxb., 244, 254, 256
   S. brachystachya (G.F.W. Meyer) Konig., 12
   S. corticosa Speg., 196
   S. europaea L., 111, 114, 117, 119, 122, 124, 127, 129–131, 145, 149, 158, 171, 174
   S. fruticosa L., 12, 65, 119, 120, 134, 196, 218, 219, 225–228, 230, 233
   S. gaudichaudiana Moq., 195, 196
   S. herbacea L., 12, 161, 233
   S. mucronata Bigel., 66
   S. pachystachya Bunge, 237
   S. pacifica Standl., 194
   S. perennis Mill., 120, 128, 129, 134, 194
   S. quinqueflora Bunge ex Unq. Sternb., 294, 304–307, 313
   S. radicans Sm., 128
   S. ramosissima Woods, 2, 137
   S. stricta Willd. ex Steud., 2, 119, 128, 140
   S. strictissima Gram., 124
   S. subterminalis Parish, 175–177, 179
   S. utahensis Tidestrom, 27
   S. virginica L. (perennial saltwort), 158, 173–179, 181–183, 186, 187
Salinator, 307
   S. fragilis (Lam.), 322
   S. solida von Martens, 322
Salix, 110
   S. arctica Pall., 171
Salmacina, 327
Salsola, 27
   S. baryosma (Schult.) Dandy, 220, 256
   S. foetida Delile, 245
   S. forskalei Schweinf., 220
   S. kali L., 245, 294, 313, 314
   S. soda L., 12
   S. vermiculata L., 220
   S. villosa Del., 223
Salsolaceae, 245
Salvadora, 255
   S. oleoides Dcne., 245, 255
   S. persica L., 245, 255, 256
Salvadoraceae, 245
Samadera indica Gaertn., 273
Samolus junceus R.Br., 314
S. repens (Forst. and Forst. f.) Pers., 13, 80, 310, 311, 314
S. valerandi L., 123, 136, 218
Sapindus mukorossi Gaertn., 279
Sarcobatus, 27
   S. carinatus Wall., 243, 253
   S. globulus Wall., 243
   S. sulphureus Schltr., 273

SYSTEMATIC INDEX

Sarmatium, 264
Saurida gracilis (Quoy and Gaimard), 330
Scaevola lobelia Murr., 236
Scartelaos viridis (H. Buchanan), 263, 265
Scatophagidae, 329
Sceloporus occidentalis Baird and Girard (western fence lizard), 184
Schoenus nitens (R.Br.) Poir., 315
Scirpodendron costatum Kurz., 290
Scirpus, 11, 25, 130
    S. acutus Muhl., 100, 175
    S. ambigua Griseb., 8
    S. americanus Pers., 102, 163, 172, 173, 311
    S. antarcticus L., 315
    S. australis (Murr. ex L.) Koch, 8
    S. caldwellii V.J. Cook, 311
    S. californicus Britton, 100, 175, 176, 195
    S. campestris Britton, 101
    S. cernuus Vahl., 80, 172, 310
    S. foliosa, 100, 101, 104
    S. lacustris L., 130, 311
    S. littoralis Schrad., 244, 247, 255, 315
    S. maritimus L., 66, 115, 119, 124, 129, 130, 137, 144, 145, 171, 172, 174, 178, 198, 314
    – – L. var. *compactus* (Hoffm.) Junge, 122, 129
    S. mucronata L., 218
    S. nodosus Rottb., 80, 310, 315
    S. olneyi Gray (Olney threesquare), 97, 99, 102, 103, 158, 193, 195
    S. riparius Presl., 195
    S. robustus Pursh., 79, 86, 88, 97, 99, 102, 162
    S. rubra, 8
    S. rufus (Huds.) Link, 124, 135, 136
    S. stricta Poir, 8
    S. tabernaemontani C.C. Gmel., 124
    S. triqueter L., 130
    S. tuberosus Desf., 218
    S. validus Vahl., 172, 173
Scolecolepis indica Fauvel, 327
Scolopacidae, 335
Scolopia oldhami Hance, 279
Scorpaenidae, 329
Scorzonera parviflora Jacq., 121
Scrophulariaceae, 314
Scylla serrata (Forskål), 237, 310, 326, 353
Scyphiphora hydrophyllacea Gaertn., 245, 258, 273, 293, 312
Scytonema rhizophorae Zeller, 318
Sebaea albidiflora F. Muell., 315
Sedum acre L., 124
Selenotoca multifasciata (Richardson), 329
Selliera, 22
    S. radicans Cav., 80, 311, 314
Semecarpus venenosa Volk., 275
Senecio lautus Forst.f. ex Willd., 315
S. vulgaris L., 123
Septoria junci Desm., 119
Serranidae, 329
Sesarma, 19, 237, 264, 265, 283
    S. bidens (de Haan), 325
    S. borneensis Tweedie, 325
    S. catenata Ort., 236
    S. cinereum Say, 87
    S. crassimana de Man, 265
    S. dehaani, 286
    S. dussumieri H.M.-Edw., 264
    S. erythrodactyla Hess, 325
    S. eulimine de Man, 237
    S. eumolpe de Man, 264
    S. guttata Milne Edwards, 237, 325
    S. indiarum Tweedie, 264
    S. indica A.M. Edw., 264
    S. mederi H.M.-Edw., 264
    S. meinertii de Man, 237, 325
    S. messa Campbell, 325
    S. moeschii de Man, 265
    S. moluccensis de Man, 325
    S. onchophora de Man, 264
    S. ortmanii Crosnier, 237
    S. palawanensis Rathbun, 264
    S. reticulatum Say, 87, 88
    S. sediliensis Tweedie, 265
    S. semperi longicristatum Campbell, 325
    S. singaporensis Tweedie, 264
    S. smithii Milne Edwards, 325
    S. tetragona Fabr., 264
    S. versicolor Tweedie, 264
    S. villosa (Milne Edwards), 325
Sesuvium, 2, 57, 59, 238, 268
    S. portulacastrum L., 13, 22, 195, 198, 235, 237, 243, 254, 268, 302, 303, 305, 313
    S. verrucosum Raf., 177, 179
Sevada schimperi Moq., 220, 221
Silene maritima With., 124
Sillaginidae, 330
Sillago ciliata (Cuv. and Val.), 330
S. maculata Quoy and Gaimard, 330
Siphonaria bifurcata Reeve, 323
S. denticulata Quoy and Gaimard, 323
S. marza (Iredale), 323
Siphonaridae, 323
Sipuncula, 327
Solanum dulcamara L., 15, 122
S. nigrum L., 123
Solen alfredensis Bartsch, 236
Solidago sempervirens A. Gray (seaside golden rod), 102
Sonchus arvensis L., 122–125
Sonneratia, 16, 22, 238, 239, 245, 247, 252, 254, 263, 264, 266, 275, 288, 349
    S. acida L., 348, 350
    S. alba Sm. in Rees, 16, 20, 238, 245, 252, 255, 263, 265–268, 282, 287, 288, 293, 300, 305, 312, 348–351, 354
    S. apetala Buch.-Ham., 16, 20, 50, 245, 246, 250, 251, 253, 255, 258
    S. caseolaris (L.) Engler (kambala), 20, 265–268, 271, 273–275, 281, 293, 300, 305, 312
    S. griffithii Kurz., 262, 265, 348
    S. ovata Backer, 17, 266, 267, 348
Sonneratiaceae, 245, 312
Sorex bendirii (Merriam) (marsh shrew), 185
S. cinereus Kerr (masked shrew), 102

S. ornatus Merriam (ornate shrew), 184, 185
S. sinuosus Grinnell (Suisun shrew), 185
S. vagrans Baird, 104
Sparidae, 330
Spartina, 4, 9–11, 18, 21–23, 32, 37, 40, 41, 59, 71, 73, 75, 109, 111, 118, 119, 128–130, 139, 141, 146–149, 161, 164, 175, 176, 179, 183, 184, 236
    S. alterniflora Loisel. (salt-marsh cord grass), 10, 26, 79, 86, 88–91, 96, 97, 129, 147, 148, 158–164, 180, 181, 187, 193, 194, 196, 197, 202, 364, 365
    S. anglica C.E. Hubbard, 10, 11, 129, 314
    S. brasiliensis Raddi, 10, 13, 195–198
    S. cynosuroides (L.) Roth. (big cord grass), 86–89, 97, 99, 102, 161–163, 194
    S. densiflora Brongn., 196
    – – Brongn. var. *patagonica* St.-Y., 196, 197
    – – Brongn. var. *typica* St.-Y., 196
    S. foliosa Trin., 174–176, 179, 181, 183, 186, 187
    S. glabra Muhl. ex Ell., 81, 86
    S. maritima (Curtis) Fern., 10, 17, 22, 117, 120, 122, 129, 131, 132, 236
    – – (Curtis) Fern. var. *brasiliensis* St.-Y., 195
    S. montevidensis Arech., 13, 27, 195, 196
    S. patagonica Speg., 196
    S. patens Muhl. (salt-meadow cord grass), 79, 81, 86, 88–90, 97, 99, 102–104, 158–163, 181, 193, 364
    S. pectinata Link., 161
    S. polystachya (Michx.) Beauv., 86
    S. spartineae (Trin.) Merr., 194, 198
    S. stricta (Ait.) Roth., 122
    S. townsendii H. and J. Groves, 2, 9–11, 40, 115, 119, 122, 126–130, 140, 145, 147, 180
Spatula clypeata (L.) (shoveler), 98, 185
Speocarnicus californiensis Stimpson, 183
Spergularia, 131, 141, 145, 178
    S. canadensis (Pers.) G. Don, 172, 173, 180
    S. macrotheca (Hornem.) Heynh., 172, 173, 180
    S. marginata (DC.) Kittel, 66, 313
    S. marina (L.) Griseb., 144, 145, 124, 136, 137, 145, 172–175, 178, 180
    S. media (L.) C. Presl., 114, 119, 122, 124, 128, 130, 135, 313
    S. rubra (L.) J. and C. Presl., 313
    S. salina J. and C. Presl., 122, 136
Sphyraenidae, 330
Sphaeroides hamiltoni (Richardson), 330
    S. pleurostictus (Günther), 330
Sphaeroma, 324
    S. rugicauda Leach, 86, 146
Sphaeromidae, 324
Sphecotheres flaviventris Gould (yellow fig bird), 338
S. vieilloti Vig. and Horsf. (southern fig bird), 338
Spinifex longifolius R.Br., 304
Spionidae, 327
Spirodela polyrrhiza (L.) Schleid., 218
Spirostachys olivascens Speg., 196
S. ritteriana Ung.-Sternb., 196
Spirulina subtilissima Kütz. ex Gom., 319
Sporobolus, 222, 256, 268
    S. contractus Hitchc., 177
    S. helveticus (Trin.) Dur. and Schintzt., 256

S. marginatus Hochst. ex A. Rich., 256
S. spicatus (Vahl.) Kunth, 218, 220–223, 225–228, 230
S. virginicus Kunth, 194, 198, 235, 236, 268, 302, 303, 306, 307, 314
Sphyraena barracuda (Walbaum), 330
S. obtusata Cuv. and Val., 330
Spyridia filamentosa (Wulf.) Harv. in Hook., 318
Stagonosporopsis salicorniae (Magn.) Died., 119
Statice, 195
    S. brasiliensis Boiss., 196
Staticobium limonii Conterini, 146
Stellaria humifusa Rottb., 67, 170, 171, 178
S. media (L.) Vill., 124, 125
Stemonurus ellipticus Sleumer, 273
Sterculiaceae, 245, 251, 255, 312
Sterna albifrons (Pallas) (little tern), 333
S. bergii Lichtenstein (crested tern), 333
S. bengalensis Lesson (lesser-crested tern), 333
S. caspia Pallas (Caspian tern), 333
S. hirundo L. (common tern), 333
S. hybrida (= *Chlidonias*) (whiskered tern), 333
S. leucoptera Temminck (white-winged black tern), 334
S. nilotica Gmelin (gull-billed tern), 334
Stichococcus bacillaris Naeg., 317
Stichocardia tiliaefolia Hallier f., 244, 252
Stipa teretifolia Stend., 310, 316
Stipiturus malachurus (Shaw) (southern emu wren), 336
Streblospio, 186
    S. benedicti Webster, 182, 186, 187
Streptera graculina (White) (pied currawong), 339
Strigatella retusa Lam., 322
Strigidae, 335
Strombidae, 322
Strombus luhuanus L., 322
Sturnella magna (L.), 160
S. neglecta Aubudon, 185
Suaeda, 9, 12, 21, 22, 27, 117, 131, 178, 194, 195, 223, 237, 238, 250, 302
    S. altissima Pall., 12
    S. arbusculoides L.S. Smith, 313
    S. australis (R.Br.) Moq., 13, 302, 306, 307, 313
    S. calcarata, 220
    S. californica S. Wats., 9, 175–177, 179
    S. depressa (Pursh) Wats., 171, 175
    S. fruticosa, 12, 120, 134, 138, 196, 218, 219, 221–223, 227, 228, 230, 233, 236, 244, 256
    S. japonica Makino, 12
    S. linearis Ell., 161
    S. maritima (L.) Dum., 9, 32, 114, 117, 119, 122, 123, 127, 128, 130, 134, 138, 139, 141, 144, 145, 158, 196, 236, 244, 252, 254
    – – (L.) Dum. var. *australis* (R.Br.) Domin., 313
    S. monoica Forsk., 218, 220, 222–228, 230, 244, 254
    S. novae-zelandiae Allan, 9, 13
    S. patagonica Speg., 196
    S. pruinosa Lange, 225
    S. splendens (Pourr.) G. and G., 120
    S. cf. *taxifolia* (Standl.) Munz., 177
    S. vera J.F. Gmel., 138
    S. vermiculata Forsk., 218–220, 223, 225–228, 230, 233
    S. volkensii C.B. Cl., 220

SYSTEMATIC INDEX

Sylviidae, 336
Sylvilagus, 186
    S. audubonii (Baird), 185
    S. bachmani (waterhouse), 185
Symploca laete-viridis Gom., 319
Syncera brevicula (Pfr.), 263, 265
Synodontidae, 330
Syringodium filiforme Kutz, 47
Syzygium buxifolium Hook. and Arn., 286

Tabanidae, 89
Tadorna radjah (Garnet) (Burdekin duck), 331
Talitridae, 324
Talorchestia, 324
    T. ancheidos Barnard, 236
    T. australis Barnard, 236
    T. malayensis Tatt., 237
Tamarix, 226
    T. africana Poir., 233
    T. aphylla (L.) Karst., 224
    T. dioica Roxb., 256
    T. mannifera (Ehrenb.) Decne., 218, 221–228, 230
    T. passerinoides Del. ex Desv., 218, 224–228
Tanysiptera sylvia Gould (white-tailed kingfisher), 335
Tapes watlingi Iredale, 324
Taraxacum, 123–125
Taxidea taxus (Schreber) (badger), 84
Tecticornia australasica (Moq.) P.G. Wilson, 294, 305, 306, 313
T. cinerea (F. Muell.) Hook.f. and Jackson, 313
Telescopium, 265
    T. mauritsi Butot, 263, 265
    T. telescopium (L.), 263, 265, 321
Telmatodytes palustris (Wilson) (marsh wren), 101, 102, 160
– – *griseus* (Brewster) (long-billed marsh wren), 102
Teloschistes flavicans (Sw.) Norm., 320
Terapon jarbua (Forskål), 330
Teraponiidae, 330
Terebralia palustris Brug., 237, 265, 321
T. sulcata Born, 265, 321
Teredinidae, 308, 324
Teredo tristi Iredale, 324
Ternstroemia japonica Thunb., 286
Tetraclita caerulescens (Spengler), 324
    T. squamosa (Bruguière), 324
    T. vitiata Darwin, 324
Tetraodontidae, 330
Tettigonidae, 146
Textularia earlandi Parker, 73, 75
Thaisidae, 322
Thalamita danae Stimpson, 326
Thalassia, 1
    T. testudinum Kon., 47
Thalassina, 19, 264, 265
    T. anomala (Herbst) (mud lobster), 264, 284, 326
Thalassinidae, 326
Thamnophis, 184
Theleophyton billardieri (Moq.) Moq., 313
Themista lageniformis Baird, 327
Thespesia acutiloba (E.G. Bak.) Exell. and Mendonca, 237
T. lampas (Cav.) Dalz. ex Dalz. and Gibs. var. *thespesioides* (R.Br. ex Benth.) P.A. Fryxell, 316

T. populnea (L.) Soland. ex Correa, 52, 268, 287
T. populneoides (Roxb.) Kostel, 316
Thomomys bottae (Eydoux and Gervais) (Botta pocket gopher), 185
Threlkeldia diffusa R.Br., 313
Threskiornis molucca (Cuvier) (white ibis), 331
T. spinicollis (Jameson) (straw-necked ibis), 331
Threskiornithidae, 331
Tiliaceae, 245
Tilapia mossambica (Peters), 353
Timaliidae, 336
Tiphotrocha comprimata (Cushman and Brönnimann), 74, 75, 81–83
Tolypothrix tenuis (Kütz.) Johs. Schmidt, 318
Tortella flavovirens (Bruch) Broth., 137, 138
Traganum nudatum Del., 233
Trainthema turgidifolia F. Muell., 313
Tricerma phyllanthoides, 177
Trichoglossus moluccanus (Gmelin) (rainbow lorikeet), 334
Trifolium, 172
    T. fragiferum L., 122, 124, 135
    T. repens L., 122, 124
    T. wormskjoldii Lehm., 171
Triglochin, 11, 143, 168, 171, 177, 179
    T. bulbosum L., 236
    T. concinnum Davy, 172, 174, 175
    T. maritima L., 12, 66, 117, 119, 121, 122, 124, 130, 137, 143, 145, 171–178
    T. mucronata R.Br., 316
    T. palustris L., 124, 137, 171
    T. striata Ruiz and Pav., 3, 310, 314
Trimeresurus purpureomaculatus (Gray), 269
T. wagleri Boie, 269
Tringa brevipes (Vieillot) (grey-tailed tattler), 333
T. cinerea (Guldenstaedt) (Terek sandpiper), 333
T. glareola L. (wood sandpiper), 333
T. hypoleucos L. (common sandpiper), 333
T. nebularia (Gunnerus) (greenshank), 333
T. terek Latham (Terek sandpiper), 269
T. toanus (L.) (redshank), 144, 269
Triodia pungens, 304
Tripleurospermum maritimum (L.) Koch., 124
Tristania conferta (R.Br.), 303
Trochammina inflata (Montagu), 73, 75, 81–83, 182
T. macrescens Brady, 75, 81–83
Trochidae, 321
Turritella cerea Reeve, 321
T. terebra L., 321
Turritellidae, 321
Tylodiplax tetratylophorus de Man, 265
Tylophora polyantha Volk., 273
Tylosaurus macleayanus (Ogilby), 328
Typha, 3, 25, 89
    T. angustifolia L., 97, 163, 174, 310
    T. domingensis Pers., 218, 223, 225, 227, 228, 230
    T. latifolia L., 100, 163, 175
Tyto longimembris (Jerdon) (grass owl), 335
Tytonidae, 335

Uca, 96, 159, 160, 186, 237
    U. annulipes (H.M.-Edw.), 237
    U. arcuata (de Haan), 326

U. bellator (White), 326
U. coarctata Milne Edwards, 263, 265, 326
U. crenulata (Lockington), 183, 187
U. dussumieri Milne Edwards, 263, 265, 326
U. gaimardi (Milne Edwards), 237, 238
U. inversa (Hoffmann), 237
U. lactea (de Haan), 326
– – (de Haan) forma *annulipes* M. Edw., 238
U. longidigitum Kingsley, 326
U. manii (Rathbun), 265
U. marionis (Desmarest), 326
U. minax (Le Conte), 87, 88, 97
U. pugilator (Bosc.), 87, 88
U. pugnax (S.I. Smith), 87, 88
U. rhizophorae Tweedie, 263
U. rosea (Tweedie), 263, 265
U. signata Hess, 326
U. triangularis (A.M.-Edw.), 263, 265
U. urvillei (M. Edw.), 237, 238
Ulothrix, 17, 142, 143
Ulva, 186
    U. lactuca L., 121, 317
    U. linza L., 172
Upeneus tragula Richardson, 329
Upogebia, 265
    U. africana (Ort.) (mud prawn), 236, 237
Urochondra setulosa (Trin.) Hubbard, 222, 244, 256
Urocyon cinereoargenteus (Schreber) (grey fox), 185
Urocystis agropyri (Reuss.) Schroet., 119
Uromyces, 118
    U. armeriae Lév., 119
    U. caryophyllacearum (Wallr.) Cif. and Biga, 119
    U. chenopodii (Duby) Schroet., 119
    U. giganteus Speg., 119
    U. limonii (DC.) Lév., 119
    U. lineolatus (Desm.) Schroet., 119
    U. salicorniae Lév. ex Cooke, 119
    U. sparsus (Schum. and Kunze) Lév., 119
Urothoe, 236
Usnea cf. *scabrida* Tayl., 320
Ustilago hypodytes (Schlecht.) Fr., 119
Uta stansburiana Baird and Girard (side-blotched lizard), 184
Utica borneensis de Man, 264, 265

Vanellus miles (Boddaert) (masked plover), 333
V. novaehollandiae Stephens (spur-winged plover), 333
Vanheurckia lewisiana Breb., 319
Varanus, 310
    V. salvator (Laur.), 269
Vaucheria, 127, 131, 143, 146
    V. sphaerospora Nordst., 18, 143
    V. thuretii Wor., 18, 143
Velacumantus australis (Quoy and Gaimard), 307, 321
Veneridae, 324

Verbenaceae, 245, 312, 316
Vetiveria elongata (R.Br.) Stapf ex C.E. Hubbard, 316
Vicia cracca L., 124, 125
Vittaria elongata Sw., 276

Wallabia agilis (Gould) (agile wallaby), 310
Wilsonia backhousei Hook.f., 313
W. humilis R.Br., 313
W. rotundifolia Hook., 313
Wolffia hyalina (Del.) Hegelm., 218

Xanthidae, 326
Xanthoria ectanea (Ach.) Raes. ex R. Fils, 320
Xenorhynchus asiaticus (Latham) (jabiru), 331
Xenylla baconae, 183, 184
Xerochloa barbata R.Br., 305, 314
Xestoleberis, 186
    X. aurantia, 182
Xylocarpus, 237, 238, 245, 255, 266, 275, 288, 350
    X. australasicus Ridl., 17, 293, 301, 311
    X. benadirensis Mattei, 6
    X. granatum Koenig, 237–239, 244, 250, 258, 264, 266, 267, 269, 273, 275, 287, 290, 293, 311, 351, 354
    X. moluccensis (Lamk.) Roem., 238, 244, 250, 266, 267, 287, 354
    X. obovatus Bl., 20

Zannichellia palustris L., 120, 121
– – L., ssp. *pedicellata* Wahlenb. and Rosen, 124
Zanthoxyllum setosum Hemsl., 279
Zapus hudsonius Zimmerman (jumping mouse), 102
Zeacumantus lutulentus, 310
Zenarchopterus buffonis (Cuv. and Val.), 328
Z. gilli Smith, 328
Zilla spinosa (Turra) Prantl, 223
Zizania aquatica L., 97, 163
Zizaniopsis miliacea (Michx.) Doell and Aschers (giant cut grass), 97
Zonotrichia leucophrys (Forster) (white-crowned sparrow), 101
Zostera, 1, 3, 34, 109, 120, 126, 127, 144, 148, 236
    Z. capensis Setchell, 236
    Z. marina L., 97, 113, 119–121, 126, 140
    Z. nana Roth., 121, 129
    Z. noltii Hornem., 119, 121, 122, 126
Zosteropidae, 337
Zosterops chloris Buonaparte, 269
Z. lateralis (Latham) (grey-breasted silvereye), 337
Z. lutea Gould (yellow silvereye), 337
Zoysia macrostachya Franch. and Stav., 12, 287
Zygophyllaceae, 316
Zygophyllum album L., 218–228, 230, 233
Z. coccineum L., 220, 221, 223

GENERAL INDEX

Aberdeen (Scotland), 135
Abrolhos Islands (Australia), 295, 297
Abu Minqar Island (Egypt), 217
acarine populations, 91
– zonation, 146
Acanthetum ilicifoliae, 15
Acanthion, 15
accretion (*see also* levees; silt deposition)
– in mangals, 31, 50, 59, 208, 209, 264
– – salt marshes, 31, 40, 42, 111, 113, 114, 126, 129, 140, 141, 229, 233
–, process of, 31, 37, 69–76, 111
–, rate of, 37, 47
–, – –, related to *Avicennia* establishment, 208, 209
–, sources of, 69
–: stratification of sediments, 40, 48, 72
–, succession determined by, 140
acidity, *see* mangal soils, pH in; salt marshes, soils of, pH of
Acrostichetea, 15
Acrostichetum aureae, 15
– *speciosae*, 15, 16
Adelaide (Australia), 299
Aden, Gulf of, 215
Adour River (France), 129
Adriatic Sea, 111, 126, 139
Aegean Sea, 126
Aegiceretalia, 15
Aegiceretum corniculatae, 15
Aegicerion, 15
aeolian transport, *see* wind-borne deposits
aeration of substrate, 114, 131, 135, 206, 297
aerial photography, 11, 44
aerosynusiae, 275
Afars and Issas, French Territory of the, *see* French Somaliland
affinity between salt marshes in different regions, 24
Africa (*see also different territories and localities*), 21, 50, 199, 206, 209, 215–240, 354–355
–, coasts of, 215–240
–, East, 13–15, 20, 65, 215, 233, 236, 307
–, mangals in, 215–218, 227–229, 233–239
–, North, 233
–, Red Sea coast of, 20, 215–231
–, salt marshes in, 215–230, 236

–, South, *see* South Africa
–, West, 13–15, 20, 50, 199, 206, 209, 235, 254, 255
aggregation, *see* spatial pattern
agricultural use (*see also* grazing)
– – of mangal soil, 250, 355, 356
– – – salt-marsh soil, 363, 364
Agropyrion pungentis, 110, 138, 139
Agropyro-Rumicion crispi, 121, 132, 133, 135, 136
Ain Sokhna (Egypt), 218, 223, 227
Aira, volcano (Japan), 287
Alabama (U.S.A.), 193, 194
Alaska (U.S.A.) (*see also particular localities*), 11, 12, 167–170, 177
–, Gulf of, 168, 169, 177
– Peninsula, 171
Albany (Western Australia), 299
alcohol from mangroves, 348, 350
Alcony River (Ga., U.S.A.), 357
algae (*see also* blue-green algae; diatoms; green algae; red algae), 3
–, communities of, 12–18, 141–144
– in food chain, 148
– – lagoons, 194
– – mangals, 18, 235, 236, 306, 307, 310, 317–319
– – salt marshes, 32–34, 109, 113, 120, 127–129, 139, 141, 144, 159, 194, 306, 307, 310, 311, 317–319
– on mud flats, 18, 109
–, tidal effects on, 141, 143, 173
Algeria, 233
alligators, 19, 160
aluminium toxicity, 356
alpheid prawns, 264
Amami-Oshima (Ryukyu Islands), 283
Amapa (Brazil), 207, 208
Amazon River, 207, 208
Ambergris Cay (Belize), 198, 209, 210
Ambon Island (Indonesia), 267
America, *see also* Caribbean; North America; Central America; South America; United States of America; West Indies; *and particular territories and localities*
–, uses of mangroves in, 352, 354, 355
amphibians (*see also* frogs, toads), 184
amphipods, 80, 86, 144, 146, 237
anaerobic conditions in marsh soils (*see also* aeration; oxidation–reduction; oxygen), 112, 127, 208, 209

Anatolia (Turkey), 63
Anchorage (Alaska, U.S.A.), 169, 177, 178
Andaman Islands (India), 241, 247, 257, 258, 349, 350
Andes, 50
Andros Island (Bahama), 210
Angelicion litoralis, 121, 133
Angola, 234–236
animal communities in mangals, 18, 19, 105, 237, 263, 265, 269, 270, 307–310
– – in salt marshes, 19, 79–108, 144–147, 181–185
– – – – –, tides affecting, 105, 106, 146
– –, zonation of, 79, 106, 147, 307
Anlandungsgebiete, 113
annual plants in salt marshes, 117, 127, 138, 139, 294
Antarctic, 65
ant hills as substrate for plant growth, 90
ant plants, 22, 267, 309
ants, 22, 146
anthropogenic effects, *see* human influence
Antwerp (Belgium), 40, 116
aphrodisiac effect of mangroves, 347
Appalachian Mountains (U.S.A.), 162
Appoquinimink Creek (Del., U.S.A.), 92–95
Apra Harbour (Guam), 273
Aqaba, Gulf of, 63, 66
aquaculture, *see* mariculture
arachnids (*see also* acarine populations; spiders), 90, 91, 144, 146, 183
Arakan Coast (Burma), 251
Aranangua River (Brazil), 202
Arcachon, Bay of (France), 144
Arctic climate, 63–65, 67
–, salt marshes in, 4–8, 10–12, 121, 131, 135–137, 170, 177, 178, 181
Argentina (*see also particular localities*), 27, 194–196
Armerieto-Festucetum, 8
Armerieto-Limonietum, 8
Armerietum, 37–39
Armerion, 8, 114, 115, 131, 136–138
Armerion maritimae, 109–110, 121–125, 131, 134
Arnhem Land (Australia), 296, 307, 330
Arno (Marshall Islands), 278
Artemisietea vulgaris, 5, 121, 138
Artemisieto-Limonietum virgati, 5
Artemisietum, 135
– *maritimae,* 5, 10, 115, 123, 125, 132, 135, 143
– – *agrostidetosum,* 135
– – *armerietosum,* 135
Arthocnemetum, 233
– *africani,* 4
– *glauci,* 4
– *halocnemoidis,* 4
– *perenne,* 4
Arthrocnemion, 4
arthropods (*see also particular taxa*), 22, 86–91, 144, 148, 149
Asia (*see also particular territories and localities*), 19, 51, 215, 300
aster, sea (*Aster tripolium*), 12, 32, 66, 67, 118, 119, 122, 124, 128, 130, 134, 141, 145, 149

Asteretea tripolii, 121, 122, 124, 125
Astereto-Salicornietum, 4
Astereto-Spartinetum maritimae, 7
Astereto-Triglochinetum, 7, 121
Asteretum, 38, 39, 128, 142
– *subulati,* 7
Asuncion, Bahia (Mexico), 177
Atake River (Japan), 287
Atlantic [biogeographical region], 243, 245
– coast of Africa, 233–236
– – – America, 4–8, 10, 12, 25, 34, 69, 147–149, 157–166, 181–182, 193–211
– – – Europe, 4–8, 10–12, 109–144
– –, blue-green algae on, 18
– Ocean as barrier to distribution, 21, 22
– –, temperature conditions in, 2
– phase of post-glacial period, 56
Atriplexitalia, 6
Atriplici-Cirsietum arvensis, 133
Atriplicetum littoralis, 6, 133, 138, 139
Atriplici-Elytrigietum pungentis, 115, 132, 136, 138, 139
Atriplicion, 6
– littoralis, 110, 121, 132, 138
Auckland (New Zealand), 46, 62, 65
Australasia (*see also* Australia; New Zealand; Papua-New Guinea), 10, 17, 21, 22, 293–345
Australia (*see also particular states and localities*)
–, algae in, 306, 307, 317–319
–, biogeography of, 294, 297, 300–310
–, birds in, 309, 330–339
–, Brahminy kite as scavenger in, 269
–, climate of, 62, 66, 296, 298, 299
–, coastal hydrology of, 297
–, fish in, 309, 328–330
–, Foraminifera in, 83
–, geological history of, 21
–, geomorphology of coasts of, 295
–, insects in, 308, 309
–, lichens in, 307, 319, 320
–, mammals in, 310
–, mangals in, 13–17, 54, 264, 267, 293, 294, 311, 312, 315–339, 357, 358
–, mangrove species in, 20
–, marginal areas between mangal and salt marsh in, 1, 46
–, marine invertebrates in, 307, 308, 321–327
–, reptiles in, 310
–, salt flats in, 294
–, salt marshes in, 4–8, 294, 295, 313–319, 321–324, 330–339
–, soils of mangals and salt marshes in, 295, 296
–, uses of mangroves in, 350, 351
–, vascular plants in, 297, 300–306, 311–316
Australian Conservation Foundation, 355, 357
Avicennieto-Excoercarietum, 14
Avicennietum africanae, 14, 15
– *albae,* 14
– *albae-marinae,* 14
– *germinansae,* 14
– *marinae,* 14
– *officinalae,* 14
– *schauerianae,* 14

GENERAL INDEX

Avicennion, 15
— occidentalis, 14
— orientalis, 14
avocet, red-necked (*Recurvirostra novaehollandiae*), 333
Ayr (Scotland), 131

Bab el-Mandab Strait, 215, 228
Baccarion, 6, 10
Bacchareto-Ivetum orariae, 6
Baccharetum halimifoliae, 6
back-mangals, 250, 252–254, 256
bacteria, 116, 118, 147, 148
badger (*Taxidea taxus*), 185
Bahama Islands, 209, 210
Bahia Blanca (Argentina), 195, 196
Baja California (Mexico), 13–15, 71, 83, 167, 168, 170, 176, 177, 179, 194, 202
Ballenas Bay (Mexico), 202
Baltic Sea, salt marshes of, accretion in, 113, 141
— —, — — —, animals in, 19, 115, 146
— —, — — —, types of, 4–8, 10, 11, 110, 111, 120, 126, 127, 130, 131, 135, 137, 139
— —, shore-meadows of, nutrients in, 118, 188
— —, tides in, 39, 40
— —, vegetation zonation in, 110
Banaba (Gilbert Islands), 272
Bandra (India), 255
Bangka Island (Indonesia), 264
Bangla Desh, 241, 246–249, 251
Bank End Marsh (England), 229
banyans (*Ficus* spp.), 282
Baram River (Indonesia), 261
Barentsburg (Spitsbergen), 64
barnacles (Cirripedia), 235–237, 310
Barnstable Estuary or Harbor (Mass., U.S.A.), 38, 39, 56, 81
barrier beach as geomorphological feature, 31, 57, 168, 295
— —, mammal distribution on, 102
Barrow, Point (Alaska, U.S.A.), 167, 169, 170
Barwon Heads (Australia), 297
basalt as source of mangal soils, 255
Bashee River (South Africa), 236
basins in salt marshes, 132
Bathurst (Gambia), 234, 235
Bathurst, Cape (N.W. Terr., Canada), 170
Bato-Salicornietum, 4
bats, fruit, 310
Baumetum juncei, 7
beach banks as salt-marsh substrate, 131, 135, 138
—, barrier, *see* barrier beach
— plains as salt-marsh substrate, 111, 114, 133, 136, 141
Beaufort (Calif., U.S.A.), 90, 96, 159
Beckmannion eruciformis, 121
beetles (Coleoptera), 80, 144, 146, 160
Bega (N.S.W., Australia), 299
Belize, 54, 198, 209, 210
Bellingham Bay (Wash., U.S.A.), 169
Bengal (India), 241, 250
—, Bay of, 20, 244, 249, 257, 258
Bering Sea, 168
Bermuda Islands, 199

Bidassoa Estuary (Spain), 129
Big Calabash Cay (Belize), 54
Bimini Island (Bahama), 210
biogeography (*see also* distribution)
—: fossil evidence, 300
—: New World and Old World origins of species, 22
— of Australia, 294, 297, 300–310
— of *Halopeplis perfoliata*, 219, 220
— of Indian halophytes, 243–245
— of lichens, 307
— of mangals, 227, 228, 297, 300–310
— of mangrove species, 19–22, 200, 201
— of marine invertebrates, 307, 308, 321–327
— of North America, 21
— of North Sea, 141
— of reed-swamp vegetation, 228
— of salt marshes, 23–25, 228, 297, 300, 305–310
biological oxygen demand (B.O.D.), 365
biomass measurements, 80
birds (*see also particular species*)
— as a resource, 366
—, drinking requirements of, 101, 102
— droppings, nutrients from, 127, 144
—, feeding habits of, 96–102, 160, 185, 186, 309
—, habital requirements of, 96–102
— in Australia, 309, 330–339
— in European salt marshes, 144
— in mangals, 18, 309, 330–339
— in New Zealand, 311
— in North American salt marshes, 95–102, 184, 185, 364, 366
—, migration of, 184, 185, 187
—, nesting behaviour of, 95–101, 184
—, role in food chain of, 148, 149, 186
—, territorial behaviour of, 100, 101
—, tides affecting, 95, 96, 100, 101, 144
—, zonation in salt marshes of, 95
bittern, American (*Botaurus lentiginosus*), 160
—, black (*Ixobrychus flavicollis*), 331
—, brown (*Botaurus poicilopticus*), 331
—, little (*Ixobrychus minutus*), 331
blackbird, red-winged (*Agelaius phoeniceus*), 160, 185
Blackbird Cay (Belize), 54
Blackbird Creek (Del., U.S.A.), 87
Black Sea, 126, 127
Blakeney (England), 31, 43, 111
— Point, 111
Block Island (R.I., U.S.A.), 63
blue-green algae (Cyanophyceae), 74, 116, 118, 119, 138, 143, 148, 159, 318
Blue Mud Bay (N.T., Australia), 296
bobcat (*Lynx rufus*), 185
Bodega Bay (Calif., U.S.A.), 167, 174, 184
bog, transition to, 12
— type of salt marsh, 111
Bolboschoenetea maritimi, 6
Bolboschoenion maritimi, 6, 10
Bolinas Bay (Calif., U.S.A.), 167, 174
Bombay (India), 243, 246, 255
Bonin Islands, 271, 272

boreal climate, 63
– phase of post-glacial period, 56
Borneo (Indo-Malesia), 19, 261, 263, 266, 267, 269, 300, 354
Bornholm (Denmark), 126
Boschplaat (Netherlands), 111
Bostrychia–Catenella community, 17
Bothnia, Gulf of (Finland, Sweden), 126, 127, 135
bower bird, fawn-breasted (*Chlamydera cerviniventris*), 339
– –, great (*C. nuchalis*), 339
brackish-water marshes, 98, 99, 162, 218, 311
– –, mangroves in, 65, 265, 266, 278
– –, muskrat in, 102
– –, snails in, 86
– –, water birds in, 97–99
– ponds in cleared mangal areas for fish culture, 353
– submergence, 115, 126
Brahmaputra River (India), 241, 247
brant (*Branta bernicla*), 97, 144
–, black (*B. nigricans*), 185
Brazil, fresh-water swamps in, 211
–, *Laguncularia* leaves as source of tannin in, 355
–, mangal in, 2, 13–15, 20, 198, 199, 202, 207, 208
–, salt marshes in, 4–8, 194, 197–199
Bremerhaven (Germany), 64
Brest, Rade de (France), 129
Bridelieto-Albizzietum procerae, 278
Bridgewater Bay (England), 140
Brisbane (Qld., Australia), 299, 307
Brisbane River (Qld., Australia), 317, 318
Bristol Bay (Alaska, U.S.A.), 167, 171
Bristol Channel (England), 32, 37
British Columbia (Canada) (*see also particular localities*), 169, 170, 172
British Guiana, *see* Guyana
British Honduras, *see* Belize
British Isles (*see also* England; Great Britain; Ireland; Scotland; Wales; *and particular localities*), 120, 126
Brittany (France), 112, 113, 130, 131, 135, 137
Broadkill River (Del., U.S.A.), 95
brolga (*Grus rubicunda*), 332
Broome (W.A., Australia), 298
brown algae (Phaeophyceae), 113, 128, 141, 317
Bruguiereto-Xylocarpetum, 13
Bruguieretum cylindricae, 13
– *gymnorrhizae*, 13, 15
– *sexangulae*, 13, 16
Bruguieron, 13, 15
Buenos Aires (Argentina), 195
bulbuls (*Pycnonotus* spp.), 269
bulrush (*Scirpus robustus*), 79, 86, 88, 97, 99, 102, 162
Bundaberg (Qld., Australia), 295, 297, 299
Burdekin River (Qld., Australia), 295
Burma, 13–16, 251, 253
burning of salt marshes, 25, 97
Burton Point (England), 58
butcher bird, black (*Cracticus quoyi*), 339
–, black-backed (*C. mentalis*), 339
–, grey (*C. torquatus*), 339
–, pied (*C. nigrogularis*), 339
buttresses on mangroves, 284, 285

Cairns (Qld., Australia), 295, 298
Cakiletea maritimae, 121
Cakiletalia maritimae, 121
Calamagrostetum neglectae maritimum, 121
Calapan (Philippines), 287
calcareous substrates (*see also* carbonates), 210
– – formed by Foraminifera, 73, 74
calcium carbide manufacture using mangrove charcoal, 352
– in marsh soils, 147
Calcutta (India), 26, 249, 251
Caldwalder Tract (N.J., U.S.A.), 89
California (U.S.A.) (*see also particular localities*), 12, 16, 23, 70–76, 168, 172, 173, 175, 179, 185
– Current, 169
–, Gulf of, 194
–, Lower, *see* Baja California
Calystegietalia sepii, 121
Cambay, Gulf of (India), 256
Cambodia, 356
Cambridge (England), 57
Cambridge Gulf (W.A., Australia), 303
Cameroun, 235
Campeche (Mexico), 203
Camp Lejeune (N.C., U.S.A.), 159
Canada (*see also particular provinces and localities*), 4–8, 10, 159
Canary Creek (Del., U.S.A.), 85
Cape Ann (Mass., U.S.A.), 86
Cape Cod Bay (Mass., U.S.A.), 81
Cape May County (N.J., U.S.A.), 97
Cape York Peninsula (Qld., Australia), 293, 296, 330
carbonates (*see also* calcareous substrates)
– in mangal soils, 218
– in salt-marsh soils, 112, 113

Cardigan Bay (Wales, U.K.), 32
Cardwell (Qld., Australia), 298
Careel Bay (N.S.W., Australia), 18, 19, 304, 308, 326, 327, 330
Caribbean (*see also particular territories and localities*), 16, 20, 48, 209, 210
Cariceto-Puccinellietalia, 121
Caricetum glareosae, 136
– *mackenziei*, 6, 121, 125, 136
– *paleaceae*, 6, 124, 134
– *ramenskii*, 6
– *rectae*, 125, 134
– *salinae*, 134
– *subspathaceae*, 6, 131
Caricion glareosae, 121, 136
Carmen, Laguna de (Mexico), 49
Carnarvon (W.A., Australia), 297, 299
Caroline Islands (*see also particular islands*), 271–273
Carpentaria, Gulf of (Australia), 296, 302, 317, 319, 330
Carpinteria (Calif., U.S.A.), 167, 175
carrying capacity of salt marshes for sheep, 19
Caspian Sea, 4–8, 10, 12
Castro Creek (Calif., U.S.A.), 174
catbird, spotted (*Ailuroedus melanotus*), 339
cat-tails (*Typha* spp.), 102

cattle grazing salt marshes, 34, 118, 136, 137, 144, 145, 149, 352
Cauvery Delta (India), 241, 246, 247, 252, 253, 258
cay, *see* sand cay
Cayenne (French Guiana), 207
Cedar Keys (Fla., U.S.A.), 202
Ceduna (S.A., Australia), 297
Celebes (Indonesia), 16, 264, 267
cellulose from mangrove pulp, 352
Central America (*see also particular territories and localities*), 13–16, 198, 199, 205, 211
Ceriopeto-Aegiceretum corniculatae, 14
Ceriopetum tagalae, 14, 16
Ceriopion, 14
chalcid flies, 287
chamaephytes (*see also* life forms), 276
channels in mangals, 4, 51, 198, 206, 208, 258
– – –, algae in, 18
– – –, animals in, 18, 19, 237, 263, 265, 269
– – –, banks of, 48
– – –, development of, 48
– – –, fish in, 22, 237, 249
– – –, geomorphology of, 48
– – –, salinity of, 50, 237
– – –, sediments in, 71
– – –, vegetation of, 238, 250, 263, 287, 288
– – salt marshes, 4, 113, 171, 173, 177, 230
– – – –, algae in, 18
– – – –, animals in, 19, 79, 84–95, 103, 105, 182, 354
– – – –, banks of, 22, 32–35, 42–44, 128, 161, 171
– – – –, bird distribution related to, 96, 98–100
– – – –, erosion in, 44
– – – –, fish in, 22, 91–95, 184
– – – –, Foraminifera in, 81
– – – –, formation of, 32
– – – –, geomorphology of, 32–36, 42–45, 110, 112
– – – –, head cutting, 42, 43
– – – –, muskrats in, 103
– – – –, Ostracoda in, 182
– – – –, oxygen content in, 76
– – – –, patterns, 43, 72, 99, 136, 173, 215
– – – –, salinity in, 76, 92–95
– – – –, sediments in, 36, 71, 76
– – – –, tidal flow in, 44, 76
– – – –, vegetation characteristics in (*see also* – – – –, banks of), 40, 161, 162, 170, 174, 193, 194
–, sinuosity of, 43
Channon Creek (Qld., Australia), 302
charcoal from mangroves, 348, 352
Chatram (India), 254
Chesapeake Bay (Md., U.S.A.), 97–100, 159, 161, 162
Cheshskaya Guba (U.S.S.R.), 126
Chesil Beach (England), 111
Chester (England), 58
Chester Dee Estuary (England), 57
Chicagof Island (Alaska, U.S.A.), 167
Chichi (Bonin Islands), 272
Chichirivichi (Venezuela), 65
Chile, 195–197, 202
Chi-lung (Taiwan), 281

Chincoteague (Va., U.S.A.), 96
chloride (*see also* salinity), 115, 118
Chuckhi Sea (Alaska, U.S.A.), 169
cicada bird (*Coracina tenuirostris*), 336
cigarette wrappers, mangrove leaves as, 348
cisticola, golden-backed (*Cisticola exilis*), 336
Cladieto-Plagianthetum, 7
Cladieto-Cyperetum ustulati, 7
Clare Island (Ireland), 141
Clarence Heads (N.S.W., Australia), 299
clay content in mangal soils, 246
– – – salt-marsh soils, 71, 117
–, deposition of, in salt marshes, 40, 70
–: relation to vegetation types in mangals, 203, 205, 208, 209, 253, 264, 265
–: – – – – – salt marshes, 128, 131, 173
–, role in nutrient fixation of, 115, 118
–, winning of, from salt marshes, 149
climate (*see also* cyclones; evaporation; precipitation; temperature; wind), 61–67, 113, 245, 296, 298, 299
–, Arctic, 63, 64, 67
–, arid, 61, 65, 66
–, –, salt marshes in, 179
–, boreal, 63, 64
– classification: Köppen system, 296
– –: – –: **Af**, 296
– –: – –: **Afi**, 271, 272, 281
– –: – –: **Afw**, 278
– –: – –: **Am**, 296
– –: – –: **Amwi**, 271, 272
– –: – –, applied to Taiwan, 279
– –: – –: **Aw**, 170, 281, 296
– –: – –: **Awa**, 271, 272, 278, 279
– –: – –: **Awi**, 296
– –: – –: **BS**, 170, 296
– –: – –: **BSh**, 296
– –: – –: **BShs**, 296
– –: – –: **BShw**, 296
– –: – –: **BSk**, 296
– –: – –: **BSwi**, 271, 272
– –: – –: **BW**, 170, 296
– –: – –: **BWh**, 296
– –: – –: **BWhw**, 296
– –: – –: **BWk**, 170
– –: – –: **Cf**, 296
– –: – –: **Cfa**, 296
– –: – –: **Cfah**, 272
– –: – –: **Cfaw**, 278
– –: – –: **Cfb**, 169
– –: – –: **Cs**, 170, 296
– –: – –: **Csa**, 296
– –: – –: **Csb**, 296
– –: – –: **Cw**, 281, 296
– –: – –: **Cwa**, 296
– –: – –: **Cwah**, 278, 279
– –: – –: **Eh**, 72, 73
– –: – –: **ET**, 169
– –: Walter system, 61–67
–, cold temperate, 63
–, desert, 63

–, equatorial, 61, 62, 271, 272
–, – moderate, 272
–, humid, 65–67
–, –, equatorial, 62
–, –, moderate, 63
– in Australia, 62, 66, 296, 298, 299
–, Mediterranean, 61
–, oceanic boreal, 64
– of equatorial arid zone, 271, 272
– – – humid zone, 271, 272
– – mangals, 64–66, 294, 305
– – Northeast Trade-wind Zone, 271, 272
– – Southeast Trade-wind Zone, 271
– on the U.S. Atlantic coast, 63
–, subtropical dry, 61
–, temperate, 61
–, tropical summer-rainfall, 61
–, warm temperate, 61, 62, 66
clumping, see spatial pattern
coastal hydrology of Australia, 298, 299
coastline stabilization, 349, 365
coasts, changes in (see also eustasy; isostasy; sea-level changes), 3, 56–59
–, progradation of, 49–51
Cocal, Laguna de (Mexico), 49
cockatoo, palm (Probosciger aterrimus), 334
–, sulphur-crested (Cacatua galerita), 334
Cockle Bight (England), 43
coconut palm, 65
Cocos Keeling Island, 307
Cod, Cape (Mass., U.S.A.), 160
Coeno-Puccinellietalia, 5, 9
Collembola, 90, 146
Colombia (see also particular localities), 20, 50, 51, 199, 204, 205, 211, 355
colonization (see also mud flats, colonization of pioneer species)
– of salt marshes by Spartina spp., 2, 10, 11, 40, 129, 130, 140, 180
Colorado River Delta (Argentina), 196
– – – (Mexico), 194
Columbia River (U.S.A.), 169
Combretalia, 14, 15
Combretetea, 14
commerce, see trade
commercial use of mangals, 347–358
– – – salt marshes, 363–366
Comodoro Rivadavia (Argentina), 196
Comoro Islands, 238
competition (see also halophytes, competition with glycophytes; light interrelationships)
– in mangals, 144, 197, 284, 287, 288
– in salt marshes, 129
Conception, Point (Calif., U.S.A.), 167
Congo River (Africa), 235
Connah's Quay (England), 58
Connecticut (U.S.A.) (see also particular localities), 157, 159, 163
Conocarpetum erectae, 14, 15, 203
Conocarpion, 14

consumers, see trophic relations
continental drift, 21, 56
Cook Inlet (Alaska, U.S.A.), 167, 171
Cooktown (Qld., Australia), 298, 318
Coorong (South Australia), 296
Coos Bay (Ore., U.S.A.), 167, 168
coot(s), 332
–, American (*F. americana*), 98, 185
Copper River Delta (Alaska, U.S.A.), 167, 169, 171
coral as mangal substrate, 2, 51–54, 209–220, 222, 239, 267, 274, 295
– reefs, 215, 217, 239, 267, 276, 295
cord grass, big (*Spartina cynosuroides*), 97
– –, salt-marsh (*S. alterniflora*), 99, 158
– –, salt-meadow (*S. patens*), 99
Coringa (India), 246, 251, 252
cormorant(s) (*Phalacrocorax* spp.), 269
–, black (*P. carbo*), 331
–, little black (*P. sulcirostris*), 331
–, little pied (*P. melanoleucos*), 331
–, pied (*P. varius*), 331
Corner Inlet (Vic., Australia), 295, 297, 307
Corral (Chile), 197
Corrientes, Cabo (Colombia), 50
Costa Rica, 199, 206, 207
Cotabato (Philippines), 287
coucal, pheasant (*Centropus phasianus*), 334
coyote (*Canis latrans*), 185
crab(s), 18, 19, 72, 86–88, 96, 236–238, 263–265, 267, 269, 283
–, blue, 357
–, fiddler (*Uca* spp.), 85–88, 96, 97, 159, 160, 183, 186, 187, 237, 238, 263, 265
– mounds, see mounds of Crustacea
–, mud (*Hemigrapsus oregonensis*), 183
crake, white-browed (*Porzana cinerea*), 332
crane, Sarus (*Grus antigone*), 332
creeks, see channels
Cretaceous, geography of the, 21, 294
Crithmo-Staticetum, 2
crocodile bird, 235
crocodiles (*Crocodylus* spp.), 19, 235, 269, 310
crow, Australian (*Corvus cecilae*), 339
Crustacea (see also amphipods; copepods; crabs; isopods; lobsters; Ostracoda; prawns; shrimps), 89, 144, 269
Ctenideto-Ficetum cuspidatocaudatae, 278
ctenophores, 95
Cuanza (Angola), 236
Cuba, 353
cuckoo, brush (*Cacomantis variolosus*), 334
–, chestnut-breated (*C. castaneiventris*), 334
–, fan-tailed (*C. pyrrophanus*), 334
–, golden bronze (*Chrysococcyx lucidus plagiosus*), 334
–, rufous-breasted bronze (*C. malayanus russatus*), 334
– shrike, barred (*Coracina lineata*), 335
– –, black-faced (*C. novaehollandiae*), 335
– –, little (*C. papuensis*), 336
curlew, eastern (*Numinus madagascariensis*), 333
–, stone, see stone-curlew
currawong, pied (*Streptera graculina*), 339

cut grass, giant (*Zizanopsis miliacea*), 97
Cuxcuchapa, Rio (Mexico), 49
cyanin formation in *Salicornia*, 66
cyclones (hurricanes), 7, 27, 54, 57, 248, 249, 296
Cypero-Spergularion salinae, 121

Daintree River (Qld., Australia), 53, 301, 312
Dakar (Sénégal), 246
Dalapon [herbicide], 269
darter (*Anhinga rufex*), 331
Darwin (N.T., Australia), 298
decalcification of marsh soils, 112
Deception Bay (Qld., Australia), 318
Dee River (England), 11
Dee Estuary, 58, 59
deer, 185, 251, 347
–, mouse, 269
–, mule (*Odocoileus hemionus*), 185
–, white-tailed (*Odocoileus virginianus*), 100
deer mouse (*Peromyscus maniculatus*), 185
Delaware (U.S.A.) (*see also particular localities*), 79, 84, 87, 88, 92–95, 97, 157–159, 161, 163
– Bay (U.S.A.), 161, 365
deltas (*see also particular deltas*), 69, 110, 241, 245, 295
–, mangals in, 48–50, 54, 203, 206
–, salt marshes in, 138
–, soils in, 246
dendritic pattern of marsh channels, 43
Denmark (*see also particular localities*), 3, 23, 37, 40, 111, 120, 127, 131, 135, 138, 139, 143, 146
Derby (W.A., Australia), 298
desalination (*see also* precipitation, leaching effect of), 114, 140, 249
desiccation hazard for marine animals in marshes, 85, 105
detritus (*see also* organic matter)
–, accumulation of, 187
–, amphipod role in, 86
– as substrate for microorganisms, 187
– causing bare areas of salt marshes, 34
–, communities developing on, 138, 139
–, export of, 149
– food chains, 148, 149, 187, 357
– imported by storm floods, 139, 147
– – – tidal drift, 34, 118, 139, 219
–, nutrient supply by, 26, 113, 117, 118, 139
–, production of, by *Spartina*, 147, 187
– protecting soil surface, 113
– transfer to and from aquatic system, 113, 147
dew as source of water for marsh birds, 102
diatoms, 17, 149
die-back in *Spartina*, 129
diking, *see* drainage
dispersal of plants by animals, 136
– – – by water, 2
– – –, intercontinental, 22
Distichlidetum spicatae, 7, 10
Distichlideto-Frankenietum, 7
Distichlideto-Limonietum, 7
Distichlideto-Salicornietum, 7
Distichlion, 7

distribution (*see also* biogeography; spatial pattern)
–: geological history, 300
–: latitudinal limits of salt-marsh species, 178
– of algae in Australia, 306, 317–319
– of algal communities, 17, 18
– of birds in Australia, 309, 330–339
– – – within salt marshes, 96, 98–100
– of *Camptostemon* spp., 268
– of community types on Red Sea coast, 227, 228
– of fish in Australia, 309, 328–330
– of halophytes in India, 243–245
– of insects in Australia, 308, 309
– of lichens in Australia, 307, 319, 320
– of *Lumnitzera* spp. in Indonesia, 265
– of mammals in Australia, 310
– of mangal communities, 2, 13–17, 293, 300
– of mangrove species, 201, 202, 300, 301, 305
– of marine invertebrates in Australia, 307, 308, 321–327
– of molluscs, 19, 84, 86, 308
– of reptiles in Australia, 310
– of salt-marsh types, 2–13, 109, 120–139, 143, 305, 306
– of spiders in Australia, 308
– of vascular plants in Australia, 297, 311–316
ditching, *see* drainage
diversity, specific, of Foraminifera, 82, 83
–, –, of mangals, 16, 66
–, –, of plants, 170, 178, 300, 356
Dominican Republic, 353
Don, Cape (N.T., Australia), 298
Dorset (England), 11, 37, 58, 90, 111, 146
dotterel, black-fronted (*Charadrius melanops*), 332
–, double-banded (*C. bicinctus*), 332
–, large (*C. leschenaultii*), 332
–, Mongolian (*C. mongolus*), 332
–, oriental (*C. asiaticus veredus*), 332
–, red-kneed (*C. cinctus*), 332
dove, bar-shouldered (*Geopelia humeralis*), 334
–, peaceful (*G. striata*), 334
Dovey Estuary (Wales), 37, 141
dowicher (*Limnodromus* spp.), 184
drainage (*see also* channels; run-off; water-level regulation)
–, animals affected by, 80, 89
– as means of human manipulation, 27, 149, 159, 161
– capacity of channels, 113
–, ditches as, 19, 180
– for agriculture, 79, 355, 357, 363, 364
– for insect control, 79, 89
–, natural patterns of, 43, 72, 99, 136, 173, 215
–, succession affected by, 25
–, vegetation affected by, 89, 90, 263
Drake's Estero (Calif., U.S.A.), 167, 174
drinking water requirements of birds, 102, 106
– – – of mammals, 103–105
drongo, spangled (*Dicrurus bracteatus*), 338
drum, red, 357
duck, 144
–, black (*Anas rubripes*), 97–99, 159
–, –, (*A. superciliosa*), 331
–, Burdekin (*Tadorna radjah*), 331
–, ring neck (*Aythya collaris*), 98

—, water whistle (*Dendrocygna arcuata*), 331
—, wood (*Aix sponsa*), 98
Dumbarton (Scotland), 135
Dumbea River (New Caledonia), 288
dune(s) (*see also* sand), 129, 138, 147
— as source of fresh-water seepage, 136
— flora of Cretaceous as source of Eremaean floristic element in Australia, 294
— protecting marshes, 37, 57, 111
— slacks as sites for marsh development, 112, 113, 131, 135, 137
—, vegetation of, 133, 138, 139
dwarf forest [physiognomic type of mangal], 25
dye from mangroves, 347, 348, 352
Dyfi Estuary, *see* Dovey Estuary
dynamics, *see* accretion; succession

eagle, fish, 235
—, sea (*Icthyophaga ichthyaetus*), 269
—, white-breasted sea (*Haliaetus leucogaster*), 332
East Anglia (England) (*see also* particular localities), 10, 38
ecotones, *see* transitions, zonation
Ecuador, 20, 198, 205, 211, 347
Eden (N.S.W., Australia), 299
eelgrass (*Zostera* spp.), 1, 3, 34, 109, 120, 126, 127, 144, 148, 236
Egg Island (N.J., U.S.A.), 88
egret, cattle (*Ardeola ibis*), 331
—, common (*Casmerodius albus*), 160, 185
—, little (*Egretta garzetta*), 331
—, plumed (*E. intermedia*), 331
—, snowy (*E. thula*), 160, 185
—, white (*E. alba*), 331
Egypt, 215–230
Eh, *see* oxidation–reduction conditions
Eighty-Mile Beach (W.A., Australia), 294
Elat (Israel), 63
Elatostemeto-Machiletum kusanoi, 278
Eleochareto-Juncetum, 6
Eleocharetum, 7
— *parvulae*, 124
Elephanta Island (India), 246, 255
Elkhorn Slough (Calif., U.S.A.), 167, 197
El-Mallaha (Egypt), 224, 228
El Salvador, 205, 206
Elytrigietum repentis maritimum, 110, 125
energy flow in salt marshes, 86, 147, 148
England (*see also* particular localities)
—, coastal movement in, 3, 56
—, salt marshes in, algae in, 141
—, — — —, insects in, 146
—, — — —, molluscs in, 146
—, — — —, soils of, 72
—, — — —, types of, 11, 35, 111, 127–129, 137, 138
—, — — —, uses of, 364
—, — — —, zonation in, 229
—, tides in, 31
English Channel, 11
Eocene in Australia, mangroves in, 300
epilia, 275–278

epiphyte quotient (Ep-Q), 272, 277
epiphytes (*see also* life forms), 22, 275–279, 287, 290, 293
—, algae as, 119, 121
—, communities of (epilia), 275–278
—, life forms of, 276–278
Eremaean flora in Australia, origin of, 294, 300
erosional processes (*see also* channels; coastal stability; soil erosion), 42, 44, 129, 365
Esbjerg (Denmark), 40
espartillal, 195
Esperance (W.A., Australia), 299
Essex (England), 34
estuaries (*see also* particular estuaries), 2, 3, 65, 69, 71, 110, 148, 184, 295, 309, 365
—, animals in, 80, 90, 146
—, eutrophication in, 128
—, fisheries in, 1, 357, 364
—, Foraminifera in, 82
—, mangals in, 197, 235, 237, 261, 263, 273, 282–284
—, nutrients in, 116
—, Ostracoda in, 76
—, productivity in, 348
—, reed swamps in, in Egypt, 218
—, salinity in, 102, 112, 115, 126
—, sediment deposition in, 70
—, salt marshes in, 99, 106, 120, 126–131
—, water fowl in, 98, 99
Ethiopia, 215, 216, 219–221, 223, 227
Eucla (W.A., Australia), 299
Eugenia, Punta (Mexico), 167
euhaline [defined], 112
Euphorbietalia peplis, 121
Euphorbion peplis, 121
Europe (*see also* Baltic Sea; Mediterranean; North Sea; *and particular territories and localities*), 10, 17, 18, 22, 23, 74, 109–156, 223, 364
eustasy (*see also* sea-level changes), 39, 54–56, 295
Eu-Staticion, 121
eutrophication in estuaries, 128
evaporation, 65, 114, 115, 169, 296, 298, 299
Everglades (Fla., U.S.A.), 209, 355
Exe Estuary (England), 115
Excoecarietalia, 15
Excoecarieto-Xylocarpetum australasicae, 15
Excoecarietum agallochae, 15
extinction of mangrove species, 245, 247, 253, 255
Eyre (S.A., Australia), 299
Eyre Peninsula (S.A., Australia), 295

faecal pellets modifying soil structure, 72
faeces of livestock, effects in salt marshes, 118, 144
Faeroe Islands, 22, 131
falcon, peregrine (*Falco peregrinus*), 332
Fanning Island, 272
fantail, grey (*Rhipidura fuliginosa*), 337
—, northern (*R. rufiventris*), 337
—, rufous (*R. rufifrons*), 337
Farakka Barrage (India), 249
fauna (*see also* animals; *and particular groups and taxa*)
— of mangals, 18, 19, 237, 269, 270, 307–309, 321–339

GENERAL INDEX

– of salt marshes, 19, 74, 79–105, 159, 160, 181–187, 307–309, 321–324, 330–339
Fehmarn (West Germany), 135
Fenning Island (England), 140
Fens (England), 57, 112
ferns, 4, 22, 203, 211, 235, 251, 266, 267, 288
Fernando Poó, 235
ferro-alloys, mangrove charcoal in manufacture of, 352
Festuceto-Agrostidetum stoloniferae, 8
Festuceto-Caricetum glareosae, 8
Festuceto-Glaucetum, 8
Festuceto-Poetum, 8
Festucetum, 37
– *arundinaceae*, 8
– *rubrae*, 8
Festucion, 8
– maritimae, 10
Ficetum cuspidatocaudatae, 278
fidelity of mangal species, 273
fig bird, southern (*Sphecotheres vieillotti*), 338
– –, yellow (*S. flaviventris*), 338
Fiji Islands, 288
finch, chestnut-breasted (*Lonchura castaneothorax*), 338
–, red-browed (*Aegintha temporalis*), 337
–, yellow rumped (*Lonchura flaviprymna*), 338
Finland, 128, 142
fireflies, 270
firewood (*see also* charcoal), mangroves for, 246, 253, 255, 256, 348–352
firmes on Colombian coast, 51
fish (*see also* mariculture; *and particular taxa*), 22, 74, 91–95
–, commercial production of, 26, 249, 353, 357, 364–367
– for mosquito control, 184
– in food chain, 186
– in mangals, 19, 22, 237, 249, 309, 328–330, 353, 357
– in salt marshes, 19, 22, 26, 91–95, 160, 184, 186, 364–367
–, poison for, from mangrove bark, 348
–, reproduction of, 92–95, 184
–, tides affecting, 92
– tolerance of salinity, 92–95
fisheries, 1, 249, 357, 364, 367
fishing floats, mangrove material for, 348
– nets, use of bark extracts for preservation of, 247, 351, 355
– traps, mangrove poles for, 349
Fitzroy Estuary (W.A., Australia), 54
Flandrian phase of post-glacial period, 47
flatworms, 144, 146, 149
fleabane, salt-marsh (*Pluchea camphorata*), 158
Fleet (England), 111
Flinders Range (S.A., Australia), 294
flood water, nutrients in, 115, 116, 148
– –, phosphorus in, 148
– –, salinity of, 26, 114, 116, 128, 135, 137
flooding, adaptation of animals to, 86, 104
– causing animal death, 146
–, debris imported by, 115, 139
–, frequency of, affecting mangal succession, 268
–, – –, – soil salinity, 114
Florianopolis (Brazil), 202

Florida (U.S.A.) (*see also particular localities*)
–, birds in, 102
–, mangals and salt marshes intermingled in, 1, 12, 162, 193
–, – in, 13–16, 46–48, 51, 157, 158, 199, 202, 206, 209
–, – –, substrates of, 48, 71
–, – –, succession in, 48
–, mangrove species in, 20, 21
–, – – –, salinity affecting productivity of, 26
–, salt marshes in, 10, 12, 157–159, 163, 193, 194, 197
–, – – –, fish in, 91, 92
–, temperature relations of *Avicennia* in, 2
Florida Keys (U.S.A.), 209
floristic composition, of salt marshes, 120, 127–130, 135–137, 139
flounder, starry (*Platichthys stallatus*), 184
flycatcher (*Muscicapa rufigastra*), 269
–, black-faced (*Monarcha melanopsis*), 336
–, black-winged (*M. frater*), 336
–, boat-billed (*Machaerirhynchus flaviventer*), 336
–, broad-billed (*Myiagra ruficollis*), 337
–, frilled-neck (*Arses telescopthalmus lorealis*), 336
–, leaden (*Myiagra rubecula*), 337
–, lemon-breasted (*Microeca flavigaster*), 336
–, pied (*Monarcha leucotis*), 336
–, shining (*Myiagra alecto*), 337
–, spectacled (*Monarcha trivirgata*), 336
–, yellow (*Microeca griseoceps*), 336
flying fox (*Pteropus alecto*), 310
– squirrels (Phalangeridae), 310
Fly River (Papua-New Guinea), 261, 267
Fonseca, Gulf of (Central America), 198
food, *see* human food
– chain, *see* trophic relations
Foraminifera, 73–75, 81–83, 146, 182
forest management in mangals, 253, 254, 348–352
Formosa, *see* Taiwan
Foryd Bay (Wales), 140
Fowlers Bay (S.A., Australia), 299
fox, gray (*Urocyon cinereoargenteus*), 185
France, 12, 126, 127, 129, 135, 137–139, 144, 146
Frankenio-Staticetum lychnidifoliae, 138
Fremantle (W.A., Australia), 299
French Guiana, 198, 207–209
French Somaliland, 215, 217, 223, 227
Fresco (Ivory Coast), 235
fresh-water influence in marshes, 162, 172, 174, 177, 195, 209, 310
– swamps (*see also* transitions), 3, 25, 102, 111, 194, 211, 266
fresh water [defined], 112
friar bird, helmetted (*Philemon novaeguineae*), 338
– –, little (*P. citreogularis*), 338
– –, Melville Island (*P. gordoni*), 338
– –, noisy (*P. corniculatus*), 338
– –, silver-crowned (*P. argenticeps*), 338
Friendly Islands (Tonga), 290
fringe forest [physiognomic type of mangal], 25
frog(s), 19, 264, 265, 269, 270
–, Pacific tree, 184

froghoppers (Cercopidae), 146
frogmouth, Papuan (*Podargus papuensis*), 335
frost, 46, 126, 129, 131, 134, 169, 298, 299
Fruti-Suaedion, 5
fucoids, 18, 128, 141, 142
fuel, *see* firewood
Fundy, Bay of (N.B., Canada), 12, 32, 37, 160, 161
Fungi, 118, 119, 127, 146, 148

Gabon, 235
gadwall (*Anas strepera*), 98
Galgeschoor (Belgium), 116
Gallegos, Rio (Argentina), 196
Galveston Bay (Tex., U.S.A.), 81
Ganges (India), 241, 246–249, 251, 253, 258
Gangetic Delta (India), 245–247, 249–251, 258
Gastropoda (*see also* snails), 85, 86
gelatinous Myxophyceae community, 18
geographical distribution of plants and animals, *see* biogeography; distribution
geological history (*see also* sea-level changes), 21, 45, 168, 193, 255, 294, 295, 300
geomorphological effects of tides, 32, 110, 111
geomorphology, *see also* barrier beaches; channels; dunes; lagoons; land forms; physiography; salt flats; sand cays; sand flats; sand dunes
– of mangals and salt marshes, 30–59, 113, 203, 295
geophytes (*see also* life forms), 23, 276
Georges River (N.S.W., Australia), 317
Georgia, Strait of (B.C., Canada), 167, 169
Georgia (U.S.A.) (*see also* particular localities)
– salt marshes, 157, 162
– – –, birds in, 96
– – –, crabs in, 87, 88
– – –, fisheries in, 357, 367
– – –, productivity of, 25, 163, 365
– – –, types of, 80
Geraldton (W.A., Australia), 299
Germany, 11, 113, 128, 135, 138, 141, 146
germination and seedling establishment (*see also* mangroves, seedlings of)
– – – –: germination activated by desalination, 114
– – – –: *Salicornia* seedlings requiring three days for establishment, 140
– – – –: seasonal factors affecting germination, 173
– – – –: wave action preventing establishment, 2, 140
Ghana, 234
Ghodbunder (India), 255
Gibraltar, 126
Gibraltar Point (England), 118
glacial action in N.E. Pacific, 168
– periods, 56
Glacier Bay (Alaska, U.S.A.), 167, 171
glades in mangals, *see* channels
Gladstone (Qld., Australia), 298, 306, 307, 317–320
Glauco-Puccinellietalia, 121, 122, 124, 125
Goa (India), 255
goanna (*Varanus* spp.), 310
goby (*Gobius durbanensis*), 184, 237
–, arrow (*Clevelandia ios*), 184

–, tidewater (*Eucyclogobius newberryi*), 184
Godavari River (India), 241, 245–247, 251, 252, 254
godwit, bar-tailed (*Limosa lapponica*), 333
golden rod, seaside (*Solidago sempervirens*), 102
Goleta (Calif., U.S.A.), 167, 175
Goleta Slough (Calif., U.S.A.), 182–184
Gondwanaland, 21
goose, 144, 149
–, bean (*Anser fabalis*), 144
–, brent, *see* brant
–, Canada (*Branta canadensis*), 98, 99
–, greater snow (*Chen hyperborea*), 97
–, magpie (*Anseranus semipalmata*), 331
–, snow (*Anser caerulescens*), 89, 98
–, white-fronted (*A. albifrons*), 100
gopher, Botta pocket (*Thomomys bottae*), 185
Gosaba Island (India), 249
goshawk, brown (*Accipiter fasciatus*), 331
–, grey (*A. novaehollandiae*), 331
Göteborg (Sweden), 66
Gotland (Sweden), 131, 135
Gove (N.T., Australia), 298
Gower (Wales), 146
Grande, Rio, Delta (Mexico, U.S.A.), 194, 202
grasses (Gramineae) (*see also particular species*), 244, 251–253, 256
grassbird, tawny (*Megalurus timoriensis*), 336
grasshoppers, 90, 146
grazing of mangals by livestock, 246, 256
– – salt marshes affected by tides, 144
– – – – – by camels, 223, 245
– – – – – by cattle and sheep, 11, 112, 115, 130, 131, 135, 136, 140, 144, 145, 256, 364, 367
– – – – – by native animals, 149, 186
– – – – –, food chains involving, 186
– – – – – limited by climate, 144
– – – – –, vegetation modified by, 11, 112, 115, 130, 131, 135, 136, 140, 145
Great Australian Bight, 300
Great Barrier Reef (Australia), 51, 52, 295, 300
Great Britain (*see also* England; Scotland; Wales; *and particular localities*)
– –, algal communities in, 141
– –, floods of 1953 in, 56
– –, grazing in salt marshes of, 144
– –, reclamation of marshes in, 57, 58
– –, salt-marsh types in, 4–8, 11, 137, 139
Great Salt Lake (Utah, U.S.A.), 27
Great Yarmouth (England), 38
green algae, 17–18, 119, 142–144, 317
greenhead flies (Tabanidae), 89
Greenland, 11
greenshank (*Tringa nebularia*), 333
Grevelingen (Netherlands), 113, 143
grey-lag (*Anser anser*), 144
Grijalva River Delta (Mexico), 204
Groene Strand (Netherlands), 130
Groote Eylandt (N.T., Australia), 298, 320
ground squirrel, Beechey (*Otospermophilus beecheyi*), 185
growth initiation in marsh plants, 173

Guadalupe River (Tex., U.S.A.), 71
Guam, 271, 272, 274, 275, 278
Guanabara Bay (Brazil), 198, 203
Guayaquil, Gulf of (Ecuador), 198
Guerrero Negro Lagoon (Mexico), 71, 72, 83, 167, 168, 176
Guiana (*see also* French Guiana, Guyana), 50, 197, 199, 206, 208, 209
gull, laughing (*Larus atricilla*), 100
–, ring-billed (*L. californicus*), 101
–, silver (*L. novaehollandiae*), 333
Gusuku (Ryukyu Islands), 284
gypsum in mangal soils, 295
– – salt-marsh soils, 71, 256
Guyana, 197, 207

halacarids, 146
halibut, California (*Paralichthys californicus*), 184
Halimioneto-Artemisietum monogynae, 8
Halimioneto-Crypsidetum schoenodis, 8
Halimioneto-Limonietum, 142
Halimioneto-Puccinellietum, 142
Halimionetum, 113, 131, 142
– *portulacoidis,* 8, 12, 115, 122, 127, 128, 130–132, 143
Halimionion, 8, 10
Hallstatt peat in Norfolk salt marshes, 56
Halo-Acrostichion, 15
Halo-Artemision, 5, 10
Halo-Caricetalia, 6, 10
Halocladion, 7, 10
Halocnemetum, 233
– *caspiae,* 5
– *strobilacei,* 5, 10
Halo-Phragmitetalia, 6
Halo-Phragmitetum, 6
halophytes (*see also* salinity)
–, competition of, with glycophytes, 2, 25, 26, 117, 136, 137
–, cumulative, 225
–, distribution of, in India, 243–245
–: euhalophytes, 225
–, excretion of salt by, 225
–, facultative, 26, 225
–, fungi associated with, 118, 119
–, germination of, 114
–, grazing effects on, 145
–, in mangals, 249, 250
– in salt marshes, 65, 109, 113, 130, 135, 136, 167, 176, 195, 221, 223, 224, 228
–, mycorrhizae in, 118
–, obligate, 2, 26
–, physiology of, 118
–, succulent, 225
–, types of, 225
Halo-Scirpetum maritimi, 123, 129
Halo-Scirpion, 123, 124, 129
Halostachyetalia, 8, 10, 121
Halostachyetea, 5
Halostachyon, 10
Halo-Typhetum, 6, 10
Hamford Water (England), 34
Harbor Island (Tex., U.S.A.), 202

hare, brown (*Lepus capensis*), 144
harpacticoid copepods, 146
harrier, swamp (*Circus approximans*), 332
harvest mouse, salt-marsh (*Reithrodontomys raviventris*), 104, 185
– –, western (*R. megalotis*), 104, 185
Hatteras, Cape (N.C., U.S.A.), 162
Hawaii, 356
hawks, 101, 160
–, marsh (*Circus* spp.), 101, 160
hay production in *Spartina* marshes, 364
Hayes Inlet (Qld., Australia), 303
Heiligenhafen (Germany), 111
hemicryptophytes (*see also* life forms), 23, 276
– dominant in salt marshes, 23
Henry, Cape (Va., U.S.A.), 162
herbicides, effects on mangals, 1, 2, 27, 269
herbivores, *see* trophic relations
Heritieretum littorali, 15, 16
– *minori,* 15, 16
Heritietalia, 15
Herition, 15
hermit crabs, 19
heron, black-crowned night (*Nycticorax nycticorax*), 160
–, great-billed (*Ardea sumatrana*), 269, 331
–, great blue (*A. herodias*), 160, 185
–, green (*Butidores virescens*), 185
–, little blue (*Florida caerulea*), 160
–, mangrove (*Butidores striatus*), 331
–, night (*Nycticorax caledonicus*), 269, 331
–, pied (*Hydranassa picata*), 331
–, reef (*Egretta sacra*), 331
–, white (*E. alba*), 269, 331
–, white-faced (*Ardea novaehollandiae*), 331
–, white-necked (*A. pacifica*), 331
herring, Pacific (*Clupea harengus pallasi*), 184
Hervey Bay (Qld., Australia), 295
Heswell (England), 58
Hiddensee (Germany), 135
Hobart (Tas., Australia), 299
Ho Bugt (Denmark), 40
Hokianga (New Zealand), 310, 311
Holland, *see* Netherlands
Honduras, British, *see* Belize
honey from mangal vegetation, 247
honey-eater, bar-breasted (*Ramsayornis fasciatus*), 338
–, blue-faced (*Entomyzon cyanotis*), 338
–, brown (*Lichmera indistincta*), 338
–, brown-backed (*Ramsayornis modestus*), 338
–, dusky (*Myzomela obscura*), 338
–, golden-backed (*Melithreptus laetior*), 338
–, graceful (*Meliphaga gracilis*), 338
–, lesser Lewin (*M. notata*), 338
–, Lewin (*M. lewinii*), 338
–, mangrove (*M. fasciogularis*), 338
–, red-headed (*Myzomela erythocephala*), 338
–, rufous-banded (*Conopophila albogularis*), 338
–, tawny-breasted (*Meliphaga flaviventer*), 338
–, varied (*M. versicolor*), 338
–, white-throated (*Melithreptus albogularis*), 338
–, yellow (*Meliphaga flava*), 338

Hong-mao (Taiwan), 280
Hook of Holland, 40
Hope Isles (Qld., Australia), 51
Houtman Abrolhos, see Abrolhos Islands
human food from mangals, 247, 347, 350, 352
– – – salt marshes, 148, 149, 180
– influence (see also firewood; grazing; herbicides; mariculture), 1, 27, 144, 157, 159, 180, 243, 246, 247, 249–256, 273, 347–368
– – by bark harvesting, see mangrove bark, uses of
– – – burning, 25, 97
– – – clay-winning, 149
– – – clearing, 246, 355, 356, 363, 364
– – – felling of mangroves (see also firewood), 25, 246, 252–254, 283, 288, 348–352
– – – fertilizers, 147
– – – modifying succession, 25
– – – mowing, 97, 149
– – – oil spillage, 1
– – – open-cast mining, 149
– – – turf removal, 137, 149
– – – waste disposal, 159, 356, 363, 365–367
– – – water-level regulation, see drainage
– –: forest management of mangals, 253, 254, 348–352
Humboldt Bay (Calif., U.S.A.), 167, 174
Humboldt Current, 59, 202
humification, 115, 117, 137
humus, see peat; soil, organic matter in
Hunstanton (England), 31
Hurghada (Egypt), 216, 217, 219, 227, 228
hurricanes, see cyclones
Hurricane Hattie, 54
Hut Marsh (England), 36, 43
hydrophytes (see also life forms), 277

Iberian Peninsula (see also Portugal; Spain; and particular localities), 11, 109, 127, 129, 135
ibis, glossy (*Plegadis falcinellus*), 331
–, straw-necked (*Threskiornis spinicollis*), 331
–, white (*T. molucca*), 331
Iceland, 11, 17, 22, 131
ichthyoplankton, 92–95
Ifars and Assas, French Territory of the, see French Somaliland
Imperateto-Bombacetum ceibae, 278
incense wood from mangroves, 348
India (see also particular localities), 13–17, 241–258, 353, 355
–, algal communities in, 17
–, cyclones in, 249
–, fisheries in, 249
–, mangals in, area of, 241–243
–, – –, associations of, 13–16, 247–258
–, – –, ecological conditions of, 246
–, – –, flora of, 243–245, 252, 253, 255
–, – –, human influence on, 247, 249–256
–, – –, local distribution of, 247–258
–, – –, management of, 253, 254
–, – –, soils of, 249, 254
–, – –, succession in, 247

–, – –, use of, 246, 247, 249–251, 253, 255, 256
–, pollen spectra in, 254
–, river flow in, 246, 248, 249
–, salt marshes in interior of, 256
Indian Ocean, 215
Indio, Punta del (Argentina), 195
Indo-China (see also Vietnam), 13–15, 350
Indo-Malesia (see also Indonesia; Malaysia; and particular islands and localities), 15, 19, 20, 25, 261–267, 269, 271, 288, 300
Indonesia (see also particular islands and localities), 16, 66, 26: 267, 350, 353, 354
Inhaca Island (Moçambique), 237
Inhambene (Moçambique), 238
Innisfail (Qld., Australia), 51
insects (see also particular groups and species), 18, 19, 88–90, 183, 270, 308, 309
– affected by tides, 88–90
introduced plant species in salt-marsh flora (see also colonization of salt marshes by *Spartina* spp.), 140, 180
– – –: *Kandelia candel* in Taiwan, 280
– – –: mangroves in Pacific Islands, 347
– – –: *Spartina alterniflora* in Europe, 10, 129
– – –: transplantation of *Zostera* spp., 126
Ireland, 11, 12, 111, 112, 115, 129, 131, 137, 139, 141, 144
Iriomote-shima (Ryukyu Islands), 282
iron in salt-marsh soils, 71, 115, 147, 149
Irrawaddy River and Delta (Burma), 20, 48, 50
Ishigaki-shima (Ryukyu Islands), 283
isopods, 86, 144, 146, 183
isostasy (see also sea-level changes), 39, 40, 48, 54–56
Inverness (Scotland), 11
Izmir, 63
Izmir, Bay of, 66

Jabiru (*Xenorhynchus asiaticus*), 331
Jaltepeque, Estero de (El Salvador), 205, 206
Jaluit (Marshall Islands), 272
Jamaica, 3, 27, 54, 202, 353
Jamnagar (India), 256
Japan, mangals and salt marshes intermingled in, 1
–, mangals in, 16, 17, 269, 281–287
–, mangrove pulpwood exported from Philippines to, 352
–, mariculture of shrimps in, 365
–, *Melarapha scabra* in, 307
–, salt marshes in, 4–8
Java (Indonesia), 353
Judy Hard (England), 56
juncal, 195
Juncetalia maritimi, 121
Juncetea, 17
– maritimi, 121
Juncetum, 37, 38, 142
– *acuti,* 6, 10
– *baltici,* 5, 10
– *bufoni,* 6, 10
– *gerardii,* 5, 10, 23, 24, 67, 110, 123, 125, 133, 135, 136, 143, 147
– – *leontodontetosum autumnalis,* 135, 136
– – *pannonicum,* 121

GENERAL INDEX

— *maritimi*, 5, 10
— *roemeriani*, 6, 10
Junceto-Caricetum extensae, 6, 123, 133, 135, 136
Junceto-Festucetum rubrae, 6
Junceto-Fimbristyletum, 6
Junceto-Galietum, 121
Junceto-Puccinellietum, 6
Junceto-Sporoboletum virginici, 6
Junceto-Stipetum textifoliae, 6
Junceto-Trifolietum, 121
Juncion, 5, 10
Juncion gerardi, 121
Jutland (Denmark), 41

Kagoshima (Japan), 62
Kairuku (Papua-New Guinea), 261
Kalidietum caspici, 5
kambala (*Sonneratia apetala*), 16, 20, 50, 245, 246, 250, 251, 253, 255, 258
kanazo (*Heritiera fomes*), 20, 50, 245, 247, 251, 258
Kandelietum candeli, 14, 16
Kandelion, 14
Kandla (India), 256
Kangaroo Island (S.A., Australia), 295, 317, 339
Kao-hsiung (Taiwan), 279
Karumba (Qld., Australia), 298
Kathiawar (India), 245, 247, 255
Kedah (Malaysia), 50
Kei River (South Africa), 236
Kent (England), 112
Kentani (South Africa), 236
Kenya, 348, 349
Kerala (India), 245
Kesaji River (Japan), 283
kestrel, nankeen (*Falco cenchroides*), 332
Khashm El-Galala (Egypt), 215
Kiel (Germany), 141
Kiel Bay (Germany), 126
Kiire Beach (Japan), 287
Kikori River (Papua-New Guinea), 261
killdeer (*Charadrius vociferus*), 184
killifish (*Fundulus* spp.), 92, 160, 184
—, California (*Fundulus parvipinnus*), 184
—, striped (*Fundulus majalis*), 92
Kimberley Region (W.A., Australia), 296, 330
Kincardine (Scotland), 11
Kinderzee (Indonesia), 266
kingfisher(s), 269
—, azure (*Caryx azureus*), 335
—, forest (*Halcyon macleayi*), 335
—, little (*Ceryx pusillus*), 335
—, mangrove (*Halcyon chloris*), 268, 335
—, sacred (*H. sancta*), 335
—, white-tailed (*Tanysiptera sylvia*), 335
—, yellow-billed (*Halcyon tototoro*), 335
Kinto Bay (Japan), 287
Kissing Point (Qld., Australia), 318
kite, black (*Milvus migrans*), 332
—, black-shouldered (*Elanus notatus*), 332
—, Brahminy (*Haliastur indus*), 269, 332

—, letter-winged (*Elanus scriptus*), 332
—, whistling (*Haliastur sphenurus*), 332
knot (*Calidris canutus*), 333
—, grey (*Calidris tenuirostris*), 333
Kochio-Suaedetum, 12
Kodiak Island (Alaska, U.S.A.), 171
koel (*Eudynamys scolopacea*), 334
Koelerion, 138
Kolguyev Island (U.S.S.R.), 67
kookaburra, blue-winged (*Dacelo leachii*), 335
—, laughing (*D. gigas*), 335
Kotel'nyy Island (U.S.S.R.), 65, 67
kotoh (*Sonneratia caseolaris*), 20, 265–269, 271, 273–275, 281, 293, 300, 305, 312
Kotzebue (Alaska, U.S.A.), 167, 170
Krabbendijke (Netherlands), 117
Krishna River (India), 241, 245, 251, 252
Kuala Lumpur (Malaya), 62
Kurio River (Japan), 284
Kusaie (Caroline Islands), 271, 272, 274–276, 278
Kuskokwim Delta (Alaska, U.S.A.), 167, 171
Kutch (India), 245, 256
Kyushu (Japan), 20, 281, 287

Labrador (Que., Canada), 158
lagoons (*see also particular lagoons*)
—, algae in, 194
—, conditions in, 76, 169
—, fauna of, 19
—, Foraminifera in, 83
—, geomorphology of, 2, 51, 72, 295
—, human modification of, in southern California, 180
—, mangal in, 203, 206, 217, 218, 234, 235, 267, 273, 274, 286, 287, 295
— of coral islands, 51–54
—, salt marshes in, 69, 110, 111, 128, 129, 136, 168, 177, 195
— type of salt marsh, 111
Lagos (Nigeria), 235
Lagos (Portugal), 120
Laguncularietum racemosae, 14, 15, 203, 209
Lamu (Kenya), 237, 239
Lancashire (England), 19
land forms, *see* geomorphology
landrail, banded (*Rallus phillipensis*), 332
Lane Cove (N.S.W., Australia), 318
La Plata Estuary (South America), 195
larks, *see* meadowlark, skylark
Laurasia, 21
Lawas (Sarawak), 266
leaching of soil, *see* desalination; rainfall, leaching effect of
Learmonth (W.A., Australia), 299
Lepidietum latifoli, 139
Leptocarpeto-Baumetum juncei, 6
Leptocarpeto-Juncetum, 6
Leptocarpeto-Plagianthetum, 6
Leptocarpion, 6, 10
Lesser Antilles, 199
levees, animals on, 84, 101, 104, 146
—, conditions in, 70, 71
—, formation of, 48

– in mangals (*see also* channels in mangals, banks of), 49, 203, 204
– – salt marshes (*see also* channels in salt marshes, banks of), 45, 110, 112
– – – –, vegetation of, 80, 104, 135, 162, 171, 174, 177, 193
–, soil characters of, 70, 71
lianes (*see also* life forms), 22, 199, 288, 290
Liberia, 235
lichens, 307, 319, 320
life forms, 23, 139, 273, 276–278
life-form spectra of vascular epiphytes, 276, 278
light interrelationships in mangals and salt marshes, 26, 197, 199, 247, 275–278, 284, 287, 288
Lillo (Belgium), 116
Lima, Rio, Estuary (Portugal), 135
Limeburners Creek (Vic., Australia), 297
limnicolous Fucaceae community, 18, 142
Limoniastrion monopetali, 121
Limonietalia, 7, 10, 121
Limonieto-Artemisietum, 12, 17
Limonieto-Limoniastretum, 12
Limonietum, 38, 39, 233, 236
– *bellidifoliae*, 7, 10
– *brasiliensi*, 7
– *carolinae*, 7, 10
– *ferulacei*, 7
– *japonici*, 8
– *latifolii*, 7
– *linifolii*, 10
– *nashii*, 7, 10
– *pacifici*, 7, 10
– *vulgari*, 7, 10
Limonion ferulacei, 121
– galloprovincialis, 7
– gmelinii, 7, 10
– occidentale, 7
– septemtrionale, 7
Limonio-Spartinetum maritimae, 7, 128
Limonio-Spartinetum townsendii, 7
Lincolnshire (England), 57
Lisbon (Portugal), 63
litter, *see* detritus
Little Sippewisset (Mass., U.S.A.), 86
Littorellion, 128
Liverpool (England), 58
Liverpool Bay (England), 57
lizards (*see also* monitors)
–, side-blotched (*Uta stansburiana*), 184
–, southern alligator (*Gerrhonotus multicarinatus*), 184
–, western fence (*Sceloporus occidentalis*), 184
lobster, mud (*Thalassina anomala*), 19, 264, 266, 267
Lofoten Islands (Norway), 127, 131, 137
London (England), 21
Long Island (N.Y., U.S.A.), 161, 163, 181
lorikeet, rainbow (*Trichoglossus moluccanus*), 334
Los Angeles (Calif., U.S.A.), 183
Loto-Trifolion, 121, 133, 135
Louisiana (U.S.A.), 22, 72, 97, 103, 193, 194, 199, 202, 365
Lourenço Marques (Moçambique), 238

Low Countries (*see also* Belgium, Netherlands), 144
Lower Hut Marsh (England), 36
Low Isles (Qld., Australia), 52, 53
Lucinda (Qld., Australia), 298
Lumnitzeretum littorale, 14
– *racemosae*, 14
Lumnitzereto-Xylocarpetum obovatum, 14

Machilion kusanoi, 278
Mackay (Qld., Australia), 295, 298
Mackenzie River Delta (N.W.T., Canada), 170
Madre, Laguna (Tex., U.S.A.),
Madagascar, 18, 237, 238
Maera River (Ryukyu Islands), 282
Magdalena, Bahia de la (Mexico), 167
Magellan, Strait of (Magallanes, Estrecho de), 196
magnesium in mangal soils, 246, 254
Magnocaricion paleaceae, 6
magpie lark (*Grallina cyanoleuca*), 336
Mahanadi River (India), 245, 250
Maine (U.S.A.), 157, 158
Malaya (*see also* Indo-Malesia)
–, climate of, 62
–, mangals in, 16, 50
–, – –, animals of, 263–266, 269, 308
–, – –, area of, 349
–, – –, regeneration of, 349
–, – –, species diversity of, 16, 66
–, – –, vegetation of, 264, 265
–, – –, zonation of, 262, 265
–, mangroves in, bark from, 354
–, – –, growth rate of, 349
–, – –, silviculture of, 348
–, – –, uses of, 348
Malden Island, 272
mallard (*Anas platyrhynchos*), 98
mallow, marsh (*Hibiscus ociliroseus*), 102
–, salt-marsh (*Kosteletzkya virginica*), 102
mammals (*see also particular species*)
– adaptation to flooding, 104
– – – salinity, 104, 105
– in mangals, 18, 310
– in salt marshes, 102–105, 148, 149, 185
man, *see* human food; human influence
management, *see* human influence
mangals (*for regional discussions, see under* Africa; Australia; Brazil, Florida; India; Japan; Malaya; Mexico; New Zealand; Philippines; Queensland; South America; *see also* mangroves)
–, agricultural use of, 250, 355, 356
–, algae in, *see* algae
–, animal communities in, *see* animal communities
–, associations of, 13–15
– as shoreline stabilizers, 349
– as source of human food, 247, 347, 350, 352
–: back-mangals, 250, 252–254, 256
–, biogeography of, *see* biogeography
–, channels in, *see* channels
–, climatic régime of, 64–66, 294, 305
–, clearing of, 246, 355, 356

GENERAL INDEX

–, commercial uses of, 347–358
–, competition in, 144, 197, 284, 287, 288
–, depressions in, caused by taro pits, 273
–, distribution of, 2, 13–17, 293, 300
–, environmental conditions affecting, 3
–, epilia in, 275, 276
–, exploitation of, 347–358
–, fauna of, *see* fauna
–, fidelity of species in, 273
–, fish in, *see* fish
–, floristics of, 293, 311, 312
–, forest management of, 253, 254, 348–352
–, geographic variation in, 305
–, grazing of, by livestock, 246, 256
–, halophytes in, 249, 250
–, herbicide effects on, 1, 2, 27, 269
–, human food from, 247, 347, 350, 352
– in deltas, 48–50, 54, 203, 206
– – estuaries, *see* estuaries
– – Indo-Malesia, 261–267, 269, 270
– – lagoons, *see* lagoons
– – North America, 16
– – the Pacific, 271–290
– – Papua-New Guinea, 267, 268
– – Taiwan, 278–281
– – Vietnam, 268, 269
–, inland, 218, 294
– intermingled with salt marshes, 1, 12, 162, 193, 202, 256
–, levees in, 49, 203, 204
–, life-form spectra compared with other forest types, 277
–, light effects in, 197, 199, 247, 284
–, mammals in, 18, 310
–, management of, 253, 254, 348–352
–, modification of, by man, *see* human influence
–, molluscs in, *see* molluscs
–, mud flats in, 209
–, peat in, 46, 48, 203, 209, 266
–, physiography of, 46–54
–, pioneer species of, *see* pioneer
–, preservation of, 1, 357, 358
–, productivity of, 25, 26
–, protective effects of, 54, 349
–, reclamation of, 250, 256
–, salinity as factor in, 50, 202, 209, 229, 237, 249, 263, 267, 268, 273, 297
–, sand in, *see* sand
–, siltation in, 50, 208, 272, 284
–, soil building by, 162
– soils, 2, 71, 202, 295, 296
– –, chemistry of, 218, 295, 356
– –, pH in, 199, 237, 246, 249, 254
– structure (synusiae), 275
– succession, *see* succession
–: survey needs, 357
–, temperature effects in, *see* temperature
–, tidal effects in, 202, 255, 297
–: transitions to other vegetation types, *see* transitions
–, understorey of, 312
–, value of, 1, 356, 357
– vegetation, 217, 271, 275, 297, 300–306, 311–316
–, zonation of, *see* zonation

mangroves (*see also* Australia, mangrove species in; biogeography of mangrove species; distribution of mangrove species; Florida, mangrove species in; Philippines, mangrove species in)
–, age structure of populations, 282
–, aphrodisiac effects of, 347
–, bark, uses of, 217, 234, 247, 347, 349, 354, 355
–, black (*Avicennia nitida*), 158, 162, 164, 349, 351, 352
–, button (*Conocarpus erecta*), 16, 20, 158, 164, 199, 203, 234, 352
–, buttresses on, 284, 285
–, coppicing of, 252, 254, 255
–, density of, 284, 286
–, felling of (*see also* firewood), 25, 246, 252–254, 283, 288, 348–352
–, fidelity of species of, 273
–, forest reserves in Philippines, 352
–, growth rate of, 349
– in brackish-water marshes, 65, 265, 266, 278
–, inland location of, 26
–, introduction of, to Pacific Islands, 347
–, origins of, 19–22
–, productivity of, 254
–, red (*Rhizophora mangle*), 1–2, 16, 19, 20, 25, 26, 47, 71, 158, 162, 164, 177, 179, 197, 199, 200, 202–204, 209, 210, 234–236, 288, 290, 349, 350, 352, 354–356
–, regeneration of, 252, 255, 267, 288, 349
–, roots of (*see also* pneumatophores), 26, 204, 284, 285, 287
– seedlings, abundance of, 22, 209
– –, adequacy of, for regeneration, 252, 267
– – collected for planting, 254
– – eaten by cattle, 255
– –, establishment of, depth of water affecting, 46
– –, – –, herbicides preventing, 1, 2, 269
– –, – –, seasonal effects on, 287
– –, – –, wave action preventing, 279
– – removed by crabs, 267, 270, 283
– – transported by water, 54, 264, 281
– –, types of, 22
–, silviculture of (*see also* forest management of mangals), 348
–, size distribution of, 282
– timber, strength of, 349
–, uses of, 234, 251, 348–355
–, vegetative reproduction of, 247, 254
manucode (*Phonygammus keraudrenii*), 339
Mapoon (Qld., Australia), 298
Maputo, Rio (Moçambique), 238
Maracaibo (Venezuela), 347
Mariana Islands, 271–273
mariculture, 249, 352, 353, 365
marl as marsh substrate, 48, 209
Marsa Alam (Egypt), 218, 221, 227, 228
Marsa Cuba (Ethiopia), 220
Marsa el-Madfa (Egypt), 217
Marsa Gulbub (Ethiopia), 227
Marsa Halaib (Egypt), 217, 227
Marsa Kileis (Egypt), 219, 227
Marseille (France), 354
Marshall Islands, 271, 273, 278

marshes, brackish-water, *see* brackish-water marshes
—, fresh-water, *see* fresh-water swamps
—, salt, *see* salt marshes
martin, fairy (*Petrochelidon ariel*), 335
—, tree (*P. nigricans*), 335
Maryland (U.S.A.) (*see also particular localities*), 23, 87, 88, 96, 157, 158, 161–163
Massachusetts (U.S.A.) (*see also particular localities*), 38, 39, 72, 86, 157–160
Massawa (Ethiopia), 216, 217, 227
Mauritania, 126
Mauritius, 54
MCPA used against mangroves, 269
meadowlark (*Sturnella* spp.), 160, 185
meadow mouse (*Microtus pennsylvanicus*), 102, 104, 160, 185
— —, Californian (*M. californicus*), 185
— —, Oregon (*M. oregoni*), 185
— —, Townsend (*M. townsendii*), 185
Mediterranean (*see also particular territories and localities*)
—, climate in, 66, 110
—, geological history of, 21
—, salt-marsh associations in, 4–8, 10, 12, 126–128, 134, 137, 139, 233
Mekong River (Vietnam), 241
Melanesia, 288
Melbourne (Vic., Australia), 62, 299
Melville Bay (N.T., Australia), 320
Mendocino, Cape (Calif., U.S.A.), 169
merganser, common (*Mergus merganser*), 98
—, hooded (*Lophodytes cucullatus*), 98
Merimbula Estuary (N.S.W., Australia), 297
Mersey Estuary (England), 58
mesohaline [defined], 112
2-methyl-4-chlorophenoxyacetic acid, *see* MCPA
Mexico (*see also particular localities*)
—, Foraminifera in, 75, 82
—, Gulf of, burning of marshes in, 25
—, — —, isostatic changes in, 48
—, — —, mangals and salt marshes intermingled in, 1, 202
—, — —, wind tides in, 76
—, mangals in, 48, 49, 198, 202, 203, 204
—, — —, use of, 352
—, mangrove species in, 20
—, — — —, associated with landform and substrate, 203
—, Ostracoda in, 76
—, salt marshes in, 70–72, 194, 198
Micronesia, 271, 272, 274
midges, 18, 19, 160, 270, 308
Midway Island, 272
migration of species, 19, 21, 27, 294, 300
milkfish (*Chanos chanos*), 353
Millingimbi (N.T., Australia), 298
Minato River (Japan), 287
mineralisation of nutrients, 115, 117, 137
minimal area of algal communities, 141, 142
mink (*Mustela vison*), 160, 364
minnow, fathead (*Pimephales promelas*), 184
—, sheepshead (*Cyprinodon variegatus*), 92, 160
Mispillion River (Del., U.S.A.), 87
Missel Marsh (England), 36, 43

Mission Bay (Calif., U.S.A.), 71–76, 83, 167–169, 175
Mississippi [state](U.S.A.), 193, 194
— Delta (U.S.A.), 74, 193
— River (U.S.A.), 48
— Sound (Miss., U.S.A.), 83
mistletoe, 237, 309
— bird (*Dicaeum hirundinaceum*), 309, 337
mites (Acarina), 90, 91, 146
Miyara River (Japan), 283
Moçambique, 50, 237
Molokai (Hawaii), 356
molluscs (*see also particular groups and species*), 22, 84–86
—, air-gaping of, 84, 85
—, clumping of, 84
—, dehydration tolerance, 85
—, fossils of, 168
— in mangals, 18, 237, 308, 310, 321–324
— — —, density of, 18
— — —, distribution of, 19, 308
— — —, species numbers, 308
— — —, zonation of, 265, 307
— in salt marshes, 19, 146, 182, 308, 321–324
— — — —, avoidance of submersion, 86
— — — —, density of, 84, 86
— — — —, distribution of, 85
— — — —, growth rate of, 84
— — — —, harvested by man, 149
— — — —, reproductive cycle, 86
— — — —, trophic relations of, 148, 182, 187
— — — —, vertical distribution of, 84–86
—, oxygen needs of, 85
—, salinity range, 85, 86
—, temperature relations, 85
Moluccas (Indonesia), 267
Monastirian interglacial phase, 47
monitors, 19
—, water, 269
monkey, leaf (*Presbytis cristatus*), 269
—, Macaque (*Macaca irus*), 269
—, Mona, 235
—, proboscis (*Nasalis larvatus*), 269
—, Sykes (*Cercopithicus mitis*), 239
Monrovia (Liberia), 235
Montevideo (Uruguay), 195
Montpellier (France), 4, 13, 109, 120, 142
moorhen, dusky (*Gallinula tenebrosa*), 332
Morecambe Bay (England), 39, 58
Moreton Bay (Qld., Australia), 295, 307, 308, 326, 327, 330
Morocco, 128
Morro Bay (Calif., U.S.A.), 167, 175, 182
mosquitoes, 18, 19, 160, 270
—, control of, 88, 184, 235
—, density of, 88
— eaten by fish, 91, 92, 184
—, effect of diking on, 89
—, — — tidal inundation on, 88, 89
—, relation of vegetation to, 89
mosquito fish (*Gambusia affinis*), 184
mounds of Crustacea as special habitat, 264, 267, 282–284
Mount Isa (Qld., Australia), 307

mouse (see also harvest mouse; meadow mouse), 102, 104, 160, 185
–, deer (*Peromyscus maniculatus*), 185
–, house (*Mus musculus*), 102
–, jumping (*Zapus hudsonius*), 102
–, white-footed (*Peromyscus leucopus*), 102
mowing of salt marshes, 97, 149
mud flats, algal communities on, 18, 109
– –, birds on, 269
– – bordering marshes to seaward, 3, 45, 109
– – – – – landward, 177
– –, colonization of, 171, 174, 196, 197, 205, 208, 266
– – in Australia, 295
– –, salinity in, 114, 117
– –, soil composition of, 117
– –, succession on, 132
– –, vegetation of, 117, 127, 128, 137, 170, 203
– – within mangal stands, 209
mud in mangal substrates, 46, 50, 51, 53, 54, 80, 86, 209, 218, 237, 254, 263, 265, 273, 287, 308
– – salt-marsh substrates, 32, 34, 35, 38, 40, 42, 67, 71, 80, 85, 129, 138, 195, 236
– patches in marshes, 9
mudskippers, 19, 235, 237, 238, 263, 309
mudsucker, long-jawed (*Gillichtlys mirabilis*), 184
Mugu Lagoon (Calif., U.S.A.), 71, 167, 168, 175, 179, 182
mullet (*Mugil* spp.), 353
–, striped, 357
mummichog (*Fundulus heteroclitus*), 92
Mundra (India), 256
Murik Lagoon (Papua-New Guinea), 267
Murray River (Australia), 295
musk grass (*Chara* sp.), 99
muskrat (*Ondatra zibethicus*), 102–104, 144, 160, 364
mussels, 84, 85, 160
Muttupet (India), 246, 252–254
Myaungmya (Burma), 50
mycorrhizae, 118
mynah, Indian (*Acridotheres tristis*), 337
Myos Hormos Bay (Egypt), 217

Nakara River (Japan), 282
Nansei-shoto (Japan), 281
Natal (South Africa), 236
Nauru Island, 272
Navlakhi (India), 256
Negro, Rio (Argentina), 196
Nehrungen in Baltic, 111
nematodes, 83, 144, 146, 149, 182
Nerang River (Qld., Australia), 317
nereids, 19
nesting pits of crabs, 286
Neston (England), 58
Netherlands, The (see also particular localities)
–, –, floods of 1953 in, 56
–, –, fungi associated with halophytes in, 119
–, –, salt marshes in, algae of, 141, 143
–, –, – – –, Amphipoda in, 146
–, –, – – –, associations of, 120, 122–137
–, –, – – –, changes in, 126, 128, 130, 131, 134

–, –, – – –, draining of, 363
–, –, – – –, effects of grazing on, 145
–, –, – – –, Foraminifera in, 146, 147
–, –, – – –, polderland type of, 112
–, –, – – –, salinity in, 115
–, –, – – –, sedimentation in, 40, 42, 113
–, –, – – –, soils of, 112, 113
–, –, sea-level changes in, 40
–, –, tidal range in, 113, 114, 143
New Caledonia, 288
Newcastle (England), 299
New England (U.S.A.) (see also particular states and localities), 12, 23, 111, 159, 161, 364
Newfoundland (Canada), 158
New Guinea (see also Papua-New Guinea; and particular localities), 19, 66, 267, 269, 293, 305
New Hampshire (U.S.A.), 81, 157, 161
New Jersey (U.S.A.) (see also particular localities), 23, 89, 97, 157–161, 163
Newport Bay (Calif., U.S.A.), 167, 168, 175, 176
New Siberian Islands (U.S.S.R.), 67
New South Wales (Australia) (see also particular localities), 18, 295, 304, 306, 308, 312, 314, 317, 326, 327, 330
New York [state] (U.S.A.), 157, 158, 163
New Zealand (see also particular localities)
– –, algae in, 18, 310, 311
– –, climate of, 62, 65, 66
– –, Foraminifera in, 83
– –, mangals in, 13–15, 17, 22, 310, 311
– –, – –, reclamation of, 356
– –, – –, temperature limits of, 2, 65, 66
– –, salt marshes and meadows in, 4–8, 13, 23, 310, 311
– –, – – – – –, animal communities of, 80
– –, – – – – –, energy flow in, 80, 147
New Zealand flax, 310
Nicobar Islands, 241, 247, 257, 350
Nicoya, Gulf of (Costa Rica), 206, 207
Niger River (Africa), 235
Niger Delta, 72
Nigeria, 234
Night Island (Qld., Australia), 57
nightjar, spotted (*Eurostopodus guttatus*), 334
–, white-tailed (*Caprimulgus macrurus*), 334
Nile Valley (Egypt), 220
Nishino-omote (Japan), 287
nitrate reductase in salt marshes, 117
nitrification in salt marshes, 176
night-heron, nankeen (*Nycticorax caledonicus*), 269, 331
nitrogen (see also nutrients)
– cycle in salt marshes, 1, 147, 148, 365, 366
– deficiency in salt marshes and meadows, 117, 118
– fertilizers, effects of, 117, 118, 147
– fixation of, 116, 147
– in flooding water, 116
– in soils of salt marshes, 117, 176
–, mineralization of, 115, 117
–, removal of, from waste water by salt marshes, 148, 365
–, seasonal changes in, in salt marshes, 116
–, sources of, in salt marshes, 116
– status of salt marshes, 117, 118

nitrophilous plants, 113, 136
noddy, lesser (*Anous tenuirostris*), 333
nodules in salt-marsh soils, 71
Noosa (Qld., Australia), 317, 318, 326, 330
Norfolk (England), 31–37, 40, 43, 45, 56, 57, 128, 141, 142, 146
Normandy (France), 112, 131
Normanton (Qld., Australia), 298
North America (*see also* Canada; Mexico; United States of America; *and particular localities*)
– –, as origin of "wasting disease" of *Zostera*, 126, 127
– –, biogeography of, 21
– –, cold currents affecting vegetation in, 66
– –, isostasy in, 54
– –, mangals in, 16
– –, salt marshes in, 157–187
– –, – –, algal communities of, 17, 18
– –, – –, Foraminifera of, 74
– –, – –, nitrogen and phosphorus in, 147, 148
– –, – –, productivity of, 163
– –, – –, trophic structure of, 147–149
North Cape (Norway), 310
North Carolina (U.S.A.), 90, 96, 157, 159, 160, 162, 163
Northern Territory (Australia), 296, 307, 309, 312, 314, 320, 330
North Pacific Current, 169
North Pine River (Qld., Australia), 317, 318
North Sea (*see also* Europe; *and neighbouring territories and localities*)
– –, biogeography of, 141
– –, salt marshes of, 40
– –, – –, algae in, 141
– –, – –, Collembola in, 146
– –, – –, plant communities, in, 10, 11, 40, 110, 130, 131
– –, – –, vertical distribution of organisms in, 115
– –, sedimentation in, 69, 113
North Stradbroke Island (Qld., Australia), 317
Northumberland (England), 131
Norton Sound (Alaska, U.S.A.), 167, 171
Norway, 64, 66, 126, 127, 130, 135, 137, 139
Nova Scotia (Canada), 158, 163
Nukumi (Japan), 287
numerical analysis of salt marshes of North American Pacific coast, 172, 179, 180
nutria, 364
nutrient(s) (*see also* iron; magnesium; nitrogen; phosphorus; potassium; sulphur), 115–118
– cycles, 118, 147, 148, 357
– deficiencies in Arctic salt marshes, 170
–, fixation of, in soil, 115, 116, 147
– from bird droppings, 127
– from flooding water, 115, 116, 148
– from plant debris, 26, 113, 117, 118, 139
– from run-off, 26, 69, 115
– in waste water fixed by marshes, 116, 365
–: mineralization and humification, 115, 117, 137
–, plant responses to, in marshes, 117, 135
– removal in ebb tides, 148, 175, 176
– turnover, 149

Oak Island (N.C., U.S.A.), 159
Oceania, 271, 274, 288, 350
Ogooué River (Gabon), 235
oil spills, effects on ecosystems, 1, 27
Ojo de Liebre (Mexico), 167
Ojo de Liebre, Laguna (Mexico), 71, 74
Okinawa (Japan), 283
oligocheates, 146, 149
oligohaline [defined], 112
Oman, Gulf of, 347
Onslow (W.A., Australia), 299
Oostvoorne (Netherlands), 130
open-cast mining in marshes, 149
opossum (*Didelphis virginiana*), 160, 185
Oran (Algeria), 233, 236
orchids, 267, 287
Oregon (U.S.A.), 168, 169, 172
Öresund (Baltic), 39, 40
organic matter, *see* detritus; peat; soil, organic matter in
Orinoco River Delta (Venezuela), 205
oribatid mites, 91
oriole, olive-backed (*Oriolus sagittatus*), 338
–, yellow (*O. flavocinctus*), 338
osmotic effects on animals, 80, 105
– gradient in mangal soils, 26
– potential of *Salicornia*, 26
– – – soils, 65, 116, 224
osprey (*Pandion haliaetus*), 332
Ossendrecht (Netherlands), 116
Ostracoda, 73, 74, 76, 182
otter (*Lutra maculicollis*), 160, 239
Oura River (Japan), 287
over-wash forest [physiognomic type of mangal], 25
owl, 101
–, barking (*Ninox connivens*), 335
–, grass (*Tyto longimembris*), 335
–, short-eared (*Asio flammeus*), 101, 160
oxidation–reduction conditions in marsh soils
– – – – – –: acid formation, 112
– – – – – – affecting iron compounds, 147
– – – – – – affecting nutrient cycling, 147
– – – – – –: Eh values at different depths, 73
– – – – – –, nitrification affected by, 147
– – – – – –: oxidation favoured on levees, 71, 203, 209
– – – – – –: oxidizing conditions in *Avicennia* stands, 208, 209
– – – – – –: reduced mud as substrate for *Rhizophora mangle*, 203, 209
– – – – – –: sulphide accumulation under reducing conditions, 112, 127
oxygen (*see also* aeration of substrate)
– affecting distribution of Foraminifera, 71
– demand, biological (B.O.D.), 365
– in salt-marsh sediments, 73, 76
–: supplied to mussels by air-gaping, 84, 85
–: transport to mangrove roots, 26
oysters, 18, 26, 235, 347, 353, 364, 365
oyster-catchers (Haematopodidae), 144, 311

GENERAL INDEX

Pacific Islands (see also Oceania; and particular islands and groups), 13–15
– –, Western [as zoogeographical region], 19
– Ocean, 10, 12, 18, 19, 81–83
– –, coasts of, see neighbouring territories
packing material from salt marshes, 364
Pagbilao Bay (Philippines), 287
pajon grass, 49
Pakistan, 16
Palau Islands, 18, 271, 272, 274–278
palm forest in Borneo, 267
Pampas (Argentina), 196
Panama, 20, 199, 211
pans, see clay pans, salt pans
paper pulp, mangrove material for, 351
Papua-New Guinea (see also New Guinea; and particular localities), 16, 20, 261, 263, 264, 271, 288, 350
Parana River (South America), 195
Parana [state] (Brazil), 197
parasitic angiosperms, 119, 120, 237, 254
– fungi, 118, 119
Parkgate (England), 58
Parramatta River (N.S.W., Australia), 317
parrot, ground (*Pezoporus wallicus*), 334
–, golden-shouldered (*Psephotus chrysopterygius*), 334
–, red-winged (*Aprosmictus erythropterus*), 334
Paspalo-Agrostidion, 121
Patagonia (Argentina), 196
peat in mangals, 46, 48, 203, 209, 266
– – –, depth of, 209
– – salt marshes, 113, 161, 162, 168
– – – –, age of, 38, 46, 56
– – – –, depth of, 38, 46, 71
– – – –, formation of, 40, 69, 111, 112
– – – – indicating sea-level changes, 38, 40, 46–48, 56
– – – – interlayered with mineral sediments, 40
– – – –, pollen analysis of, 56
peccaries, 347
pelican (*Pelecanus conspicillatus*), 235, 330
Pellicierietalia, 15
Pellicierion, 15
Pellicieretum rhizophorae, 15, 16
Pennsylvania (U.S.A.), 365
periwinkle (*Littorina littorea*), 96, 149
permafrost in Alaskan salt marshes, 170
Persian Gulf, 16, 347
Peru, 195, 202
phanerophytes (see also life forms), 23
Philippines, mangals in, 13–15, 17, 267, 287, 288
–, – –, zonation of, 288
–, mangrove bark from, tannins in, 354
–, – pulpwood production in, 352
–, – species in, 20
–, – – –, uses of, 350
–, – wood from, value of, 352
–, mariculture in, 353
phosphorus cycling, 147
– exchanged between flood water and sediments, 148
– in salt-marsh soil related to organic matter, 116
– in waste water can be fixed by salt marshes, 365

–, losses of, in ebb tide, 175
–, mobility of, 147
–, plant requirements for, in salt marshes, 115, 116
–, translocation of, by *Spartina*, 148
Phragmitetalia, 130
Phragmition, 6, 10
physiography (see also geomorphology) of mangals, 46–54
– – salt marshes, 31–46, 131–146
phytogeography, see biogeography; distribution
Pichavaram (India), 253, 254, 258
Picloram as poison for mangroves, 1, 269
pigeon, brown (*Macropygia phasianella*), 334
–, purple-crowned (*Ptilinopus superbus*), 334
–, red-crowned (*P. regina*), 334
–, Torres Strait (*Ducula spilorrhoa*), 334
Pigeon Island (Jamaica), 54
pintail (*Anas acuta*), 98, 99, 185
pioneer species in mangals
– – – –: *Aegiceras corniculatum*, 267
– – – –: *Avicennia* spp., 207–209, 234–237, 247, 263, 267, 268
– – – –: *Excoecaria agallocha*, 247
– – – –: herbaceous pioneers, 247
– – – –: *Laguncularia* sp., 236
– – – –: *Rhizophora* spp., 202, 203, 234, 235, 263, 275, 287, 288
– – – –: salt marsh as mangal predecessor, 197
– – – –: *Sonneratia alba*, 238, 263, 266, 267
– – – –: *Spartina* spp., 177, 194, 197
– – in salt marshes: *Arthrocnemum* spp., 233
– – – – –: *Halocnemum strobilaceum*, 8, 233
– – – – –: *Halopeplis perfoliata*, 220
– – – – –: *Puccinellia phryganodes* in Arctic, 67, 170, 171
– – – – –: *Puccinellia pumila*, 171
– – – – –: *Salicornia* spp., 8, 22, 109, 127, 233, 310
– – – – –: *Scirpus maritimus*, 130
– – – – –: *Spartina* spp., 109, 129, 174, 196, 197
– – – – –: *Suaeda* spp., 9
– – – – –: therophytes, 23
– –: random distribution of colonists, 140
Pipe de Tabac [fort] (Belgium), 116
pipit, Australian (*Anthus novaeseelandiae*), 335
pitta, rainbow (*Pitta iris*), 335
pit vipers, 269
plankton, 92–95, 116, 148, 185
Plantaginetea maioris, 121
Plantaginetum, 23, 38, 39, 142
– *maritimae*, 8
Plantagineto-Agrostidetum stoloniferae, 8
Plantagineto-Limonietum nashii, 7
Plantagini-Limonietum, 130, 131
Plantaginion crassifoliae, 121
plantain (*Plantago* spp.), 158
Pleistocene, Australian coast during, 295
–, North American Pacific coast during, 168
plover, golden (*Pluvialis dominica*), 332
–, grey (*P. squatarola*), 332
–, marked (*Vanellus miles*), 332
–, spur-winged (*V. novaehollandiae*), 333
Plover Marsh (England), 35, 43

Plucheo-Baccharetum, 6
Plum Island (Mass., U.S.A.), 72
plywood adhesive from mangroves, 352
pneumatophores, 22, 204, 208, 237, 251, 255, 263, 269, 306, 310
pocket gopher, Botta (*Thomomys bottae*), 185
Poland, 127
Polar Sea, 67
polderland type of salt marsh, 112, 128, 129, 131
pollen in sediments, 56, 254, 300
Pollen Island (New Zealand), 46
polychaetes, 149, 237
Polygonion avicularis, 121
polyhaline [defined], 112
Polynesia, 271, 290
Ponape (Caroline Islands), 271, 272, 274–276, 278
Pond Lagoon (Mexico), 167, 177
Pontevedra, Rio de (Spain), 129
Pony Slough (Calif., U.S.A.), 168
Poole Harbour (England), 11, 37, 58, 90, 146
pools in marshes, 19, 33, 72, 92, 99, 129, 162, 170
– – –, animals in, 79–81, 92, 264, 353
Poropotank River (Va., U.S.A.), 86–88
Port Adelaide (S.A., Australia), 307, 319, 320
Port Broughton (S.A., Australia), 296
Port Curtis (Qld., Australia), 309, 326
Port George IV (W.A., Australia), 298
Port Hacking (N.S.W., Australia), 304
Port Hedland (W.A., Australia), 299, 304
Port Kembla (S.A., Australia), 299
Portland (Vic., Australia), 299
Port Langdon (N.T., Australia), 298
Port Lincoln (S.A., Australia), 299
Port McArthur (N.T., Australia), 298
Port Moresby (Papua-New Guinea), 261
Port Phillip Bay (Vic., Australia), 294, 295, 297, 307, 318
Port Pirie (S.A., Australia), 296, 299
Port Sudan (Sudan), 217, 227
Portugal, climate of, 63
–, salt marshes in, 63, 66, 119, 120, 127, 128, 130, 131, 135, 136, 138, 139, 144
Port Wakefield (S.A., Australia), 305
Port Warrender (W.A., Australia), 298
Posidonion, 121
potassium in salt-marsh soils, 115, 116, 149
Potentillo-Caricetum rariflorae, 136
Potomac River (Va., Md., U.S.A.), 162
prawns, 246, 264
–, mud (*Upogebia africana*), 236, 237
precipitation as climatic factor, 61–67
–, leaching effects of (*see also* desalination), 65, 114, 174, 225, 229
– in Australia, 296, 298, 299
– in Micronesia, 272
– in Taiwan, 280
– on Pacific coast of North America, 169, 170
predators, *see* trophic relations
Prince Edward Island (Canada), 85
Prince William Sound (Alaska, U.S.A.), 169
productivity depending on nutrient import, 69

–, Foraminiferal populations reflecting, 74
– in estuaries, 348
– of mangals, 25, 26
– of salt marshes, 25, 26, 74, 161, 163, 181, 364–367
prop-roots, 204, 284, 285
protein from mangrove leaves, 352
Protozoa (*see also* Foraminifera), 81–83, 127
Puccinellietalia, 10, 121
Puccinellietum, 24, 37, 113, 147
– *americanae*, 5
– *coarctatae*, 5, 131
– *distantis*, 5, 123, 136, 137
– – *agrostidetosum*, 132
– – *atriplicetosum*, 137
– – *juncetosum*, 137
– – *pholiuretosum*, 137
– *fasciculatae*, 123, 136, 137, 143
– *gigantis*, 5
– *kurilensae*, 5
– *maritimae*, 5, 24, 66, 110, 113, 115, 122, 124, 127, 130–133, 143, 146
– – *pholiuretosum*, 133
– – *typicum*, 110
– *phryganodis*, 5, 67, 131
– *retroflexae*, 123, 136, 137
Puccinellio-Asteretum, 5, 23
Puccinellio-Festucetum, 5
Puccinellio-Halimionetum, 5
Puccinellio-Limonietum, 5
Puccinellio-Plantaginetum, 23
Puccinellio-Salicornietum, 5
Puccinellio-Spartinetum, 5
Puccinellio-Spergularion salinae, 123, 124, 136
Puccinellio-Spergularietum salinae, 5, 122
Puccinellio-Suaedetum, 5
Puccinellion, 10, 136, 137
– limosae, 121
– martimae, 5, 110, 121, 122, 124, 130, 131
– maritimae–Spergularion salinae, 121
– peisonis, 121
– phryganodis, 5, 121, 131
Puerto Chale (Mexico), 167, 177
Puerto Jesús (Costa Rica), 207
Puerto de Lobos (Mexico), 202
Puerto Madryn (Argentina), 196
Puerto Montt (Chile), 197
Puerto Pirámides (Argentina), 196
Puerto Rico, 26, 350, 353, 356
Puget Sound (Wash., U.S.A.), 167–169, 183
pulpwood, mangroves as, 352
Puna Island (Ecuador), 198
Punta Banda, Esterode (Mexico), 167, 176, 182

Quaternary glaciations, effect on coastlines, 54
Quebec (Canada), 158
Queen Charlotte Islands (B.C., Canada), 167, 171, 172, 178
Queensland (Australia) (*see also particular localities*)
–, algae in, 306, 307, 317–319
–, climate of, 296, 298
–, coasts of, 295, 300

–, mangals in, fauna of, 307–310, 313, 321–339
–, – –, flora of, 293, 307, 311, 312, 315, 316, 319, 320
–, – –, hurricane damage to, 57
–, – –, physiography of, 51–54
–, – –, zonation of, 301–303
–, salt marsh flora of, 313–319

rabbit, 19, 118, 144
racoon (*Procyon lotor*), 160, 185, 364
radiation, solar, in Australia, 298, 299
rail, black (*Laterallus jamaicensis*), 184
–, chestnut (*Eulabeornis castaneoventris*), 332
–, clapper (*Rallus longirostris*), 96, 97, 101, 159, 184
–, king (*R. elegans*), 96, 97
–, sora (*Porzana carolina*), 101
–, Virginia (*Rallus limicola*), 101
rainbow bird (*Meropus ornatus*), 335
rainfall, *see* precipitation
rain-forest epilia compared with those of mangals, 275, 276
raised beaches as habitat for *Limonium axilare*, 220
Ras Gharib (Egypt), 227
rat, brown (*Rattus norvegicus*), 144
–, Norway (*R. norvegicus*), 104
–, rice (*Oryzomus palustris*), 160
Ratnagiri (India), 255
rattlesnake, western (*Crotalus viridis*), 184
Raunkiaerian school of phytosociology, 120
reclamation of marshes (*see also* human influence), 1
– of mangals, 250, 356
–, use of *Spartina* for, 129
rectangular pattern of marsh channels, 43
red algae (Rhodophyta), 128, 130, 143, 317
Redcliffe (Qld., Australia), 317
redox conditions, *see* oxidation–reduction
Red Sea, 2, 16, 20, 215–231
redshank (*Tringa toanus*), 144, 269
reed, common, 159
– swamp, 217, 218, 228, 230
reproduction of arthropods, temperature effects on, 80
– – birds, 95–101
– – fish, 92–95, 184
– – molluscs, 86
– – plants, *see* germination; mangrove seedlings; salt marshes, seedling establishment in; vegetative reproduction
reptiles (*see also* alligators; crocodiles; lizards; snakes), 160, 310
Rewa River (Fiji Islands), 288, 290
Rey, Rio del (Cameroun), 235
Rhizophoretalia, 13
Rhizophoretum apiculatae, 13
– *harrisonii*, 13, 15
– *manglae*, 13, 15, 202
– *mucronatae*, 13
– *racemosae*, 13, 15
– *stylosae*, 13
Rhizophorion, 15
– occidentale, 13, 15
– orientale, 13
Rhode Island (U.S.A.), 157, 163
Rhône Delta (France), 12

Riau Islands (Indonesia), 261
rice cultivation on mangal soils, 355, 356
rifle bird, magnificent (*Ptiloris magnificus*), 339
–, Victoria (*P. victoriae*), 339
Rio de Janeiro (Brazil), 198, 203
riverine forest [physiognomic type of mangal], 25, 27
robin, buff-sided (*Poecilodryas cerviniventris*), 337
–, mangrove (*Peneoenanthe pulverulenta*), 337
Rockhampton (Qld., Australia), 318
Romara, Cape (Florida, U.S.A.), 74
Romney Marshes (England), 112
root(s), *see also* mangrove species, roots of; pneumatophores
– elongation in *Kandelia*, 287
– extent in *Alhagi*, 225
– of *Aeluropus* binding substrate, 221
– types of *Kandelia*, 285
Rosebank Peninsula (New Zealand), 46
rosella, northern (*Platycercus venustus*), 334
–, pale-headed (*P. adscitus*), 334
rot holes in mangroves as animal habitat, 18
Rota (Mariana Islands), 271, 272
rotten spots in salt marshes, 34, 162
Rufiji River (Tanzania), 238
Rügen (Germany), 135
Rupelmonde (Belgium), 112, 116
Ruppietalia, 124
Ruppietea, 124
Ruppietum maritimae, 66
Ruppion maritimae, 124
rush, Baltic (*Juncus balticus*), 158
–, black (*J. gerardi*), 158
–, needle (*J. roemerianus*), 158
Russia, 11, 126
Ryukyu Retto [archipelago] (Japan), 20, 282

Sabah (*see also* Borneo), 350
sacoglossans, 146
Sacramento–San Joaquin Delta (Calif., U.S.A.), 174
Safaga Island (Egypt), 217
Saginetalia maritimae, 121, 125
Saginetea maritimae, 121, 125
Sagineto maritimae-Cochlearietum danicae, 133, 137
Saginion, 138
– maritimae, 110, 121, 125, 137
Sagino maritimae–Cochlearietum danicae, 125, 137
Sagino-Phippsietum algidae, 137
Saint Augustine (Florida), 199
Saint Lawrence River (Canada, U.S.A.), 10, 12
Saint Louis (Mauritania), 233
Saint Vincent Gulf (S.A., Australia), 297
Saipan (Mariana Islands), 271, 272, 274–276, 278
Salado River (Argentina), 195
Salaya (India), 256
Salicornietalia fruticosae, 121
Salicornietea fruticosae, 121
Salicornieto-Distichlidetum, 4
Salicornieto-Spartinetum strictae, 4
Salicornieto-Suaedetum californicae, 4
Salicornietum, 38, 39, 66, 127, 142, 233
– *ambiguae*, 4

– *australi,* 4
– *brachystachyae,* 4
– *europaeae,* 124, 132
– *fruticosae,* 4
– *patulae,* 4, 127
– *radicantis,* 4, 121, 128
– *rubrae,* 4
– *strictae,* 4, 23, 122, 127, 129, 132, 143
– *strictissimae,* 110, 124, 127
Salicornion fruticosae, 121
Salicornio-Sesuvietum, 4
salinas, *see* salt flats
Salinas Grandes (Argentina), 27
salinity (*see also* brackish-water marshes; estuaries, salinity in; mangal, salinity as factor in)
–, adaptation of mammals to, 104, 105
–, algae affected by, 143
–, animals affected by, 87, 115, 146, 364
– as eliminator of plant competition (*see also* halophytes), 25
–, cattle grazing affected by, 144
–, changes in, due to flooding water, 26, 114
–, – –, due to precipitation, 114
–, distribution of, with depth, 225
–, diurnal changes in, 72, 76, 92, 94
–, Foraminifera affected by, 82
–, fish tolerance of, 92–95
– in channels of mangals, 50, 237
– – – – salt marshes, 76, 92–95
– in different mangal types, 237, 249, 267, 268
– – – salt-marsh types, 115, 117, 137, 169, 175, 177, 193, 221, 223–225
– – salt flats, 177
–, molluscs affected by, 85, 86
– of marsh flood-water, 114, 116, 128, 135, 137
– – mud flats, 114, 117
– – oceans, variation in, 169, 208
–, plant resistance and response to (*see also* halophytes), 114, 115, 170, 220, 221, 224, 227, 251, 252
–, potassium in soil related to, 116
– ranges defined, 112
–, seasonal changes in, 76, 93, 127, 175, 224, 251
–, tidal effects on, 76, 95, 114–116, 173, 229, 297
–, zonation of mangals determined by, 65, 237
–, – – salt marshes determined by, 229
Salsoletum sodae, 4, 138
Salsolo-Honckenyon peploidis, 121
salt flats (pans), 4, 161
– – behind mangal in South Africa, 236
– – colonized by *Halopeplis perfoliata* in Ethiopia, 220
– – – – *Salicornia* spp., 8, 66
– –, crabs in, 237
– –, formation of, 32–34, 45
– – in Australia, 294
– – in Egypt, plant communities of, 222, 224, 226, 229, 230
– –, molluscs in, 19
– –, salinity in, 177
salt-flat grass (*Monanthochloe littoralis*), 158
salt grass (*Distichlis spicata*), 99, 158
Salt Island (Jamaica), 54
salt marshes (*for regional discussions, see under* Africa; Arctic; Australia; Baltic Sea; Brazil; England; Europe; Florida; Georgia; Great Britain; Mediterranean; Mexico; Netherlands; New Zealand; North America; North Sea; Portugal; South America)
– –, age of, 38, 43, 46, 56–58, 68, 169
– –, algae in, *see* algae
– –, animals of (*see also* fauna), 19, 79–105, 144–147, 181–187
– –, annual plants in, 117, 127, 138, 139, 294
– –, basins in, 132
– –, biogeography of, 23–25, 228, 297, 300, 305–310
– –, burning of, 25, 97
– –, carrying capacity of, for sheep, 19
– –, channels in, *see* channels
– –, climatic relations of, 66
– –, coastal changes and, 57–59
– –, competition in, 129
– –, detritus deposits causing bare areas in, 34
– –, distribution of, 2–13, 109, 120–139, 143, 305, 306
– –, fish in, *see* fish
– –, floristics of, in Australia, 294, 305, 313–319
– –, grazing of, *see* grazing
– –, growing season in, factors determining, 170, 173
– –, halophytism in, *see* halophytes
– –, human food from, 148, 149, 180
– – in deltas, 138
– – – estuaries, 99, 106, 120, 126–131
– – – lagoons, *see* lagoons
– –, inland development of, 27, 127, 198, 220, 256, 294
– – intermingled with mangals, 1, 12, 162, 193, 202, 256
– –, levees in, *see* levees in salt marshes
– –, mammals in, 102–105, 148, 149, 185
– –, molluscs in, *see* molluscs
– –, mowing of, 97, 149
– –, mud in, *see* mud
– –, nitrogen in, 1, 115–118, 147, 148, 365, 366
– –, numerical analysis of, on North American Pacific coast, 179, 180
– – on beach banks, 131, 135, 138
– – – plains, 111, 114, 133, 136, 141
– –, oxygen in, 73, 76
– –, peat in, *see* peat
– –, phosphorus in, 116, 147, 148
– –, physiography of, 131–146
– –, pioneer species in, *see* pioneer
– –, preservation of, 1, 56, 57, 363–367
– –, productivity of, *see* productivity
– –, rate of development of, 38
– –, reclamation of, by *Spartina,* 57, 58, 129
– –, regional groupings of, 11–13, 178, 179
– –, relationship of, to mangals, 300, 306
– –, rotten spots in, 34, 162
– –, salinity, *see* salinity
– –, sand in, *see* sand
– –, seedling establishment in, 140, 173
– –: seeds imported by tides, 32
– –, silt deposition in, *see* silt
– –, soil building by, 72, 129, 194, 211, 212
– –, soils of, 69–76, 112, 113, 256, 295, 296
– –, – –, agricultural use of, 363, 364

– –, – –, carbonates in, 112, 113
– –, – –, depth of, 43
– –, – –, iron in, 71, 115, 147, 149
– –, – –, pH of, 72, 73, 76, 91, 112, 218
– –, – –, texture of, 2, 70, 113
– –, succession in, *see* succession
– –, temperature effects on, 2, 170
– –, tidal effects on, *see* tides
– –: transition to other ecosystem types, *see* transitions
– –, treading effects on, 136, 144
– –, uses of, 79, 363–368
– –, value of, 26, 364–367
– –, vegetation in, 4–8, 40, 122–125, 161, 162, 170, 174, 193, 194
– –, vegetative reproduction in, 170
– –, zonation in, *see* zonation
salt spray, vegetation influenced by, 138
saltwort (*Batis maritima*), 13, 22, 158, 162, 175–177, 179, 193, 194, 198, 199
–, dwarf (*Salicornia bigelovii*), 158, 175, 177, 179
–, perennial (*S. virginica*), 158, 173–179, 181–183, 186, 187
Samborombon Bay (Argentina), 195
Samoa, 271, 290
samphire (*Salicornia* spp.), 111, 114, 117, 119, 122, 124, 127, 129–131, 145, 149, 158, 171, 174
San Antonio Bay (Tex., U.S.A.), 70, 71, 76
San Blas Bay (Argentina), 196
sand, *see also* dunes
– barriers leading to marsh development, 111, 112, 135, 177
– cays, 51, 295
– flats, 3, 22, 25, 50, 45, 109, 112, 131, 137, 180
– hillocks around plants in marshes, 218, 220, 221, 226
– in mangal substrates, 2, 3, 47, 54, 71, 246, 253, 263, 264, 275
– in salt-marsh substrates, 2, 3, 31–33, 37, 40, 43, 72, 80, 88, 97, 111, 135, 140, 221, 226, 236
– patches (*firmes*) in swamps, 51
– plains behind marshes, 177
– ridges in marshes, 54, 177, 219, 305
–, succession on, contrasted with salt substrate, 172
–, wind-blown, as source of marsh sediments, 37, 72, 219–222, 225, 230
San Diego (Calif., U.S.A.), 183
San Diego Bay (Calif., U.S.A.), 167, 175, 176
Sand Lake (Ore., U.S.A.), 172
sandpipers, 235
–, broad-billed (*Limicola falcinellus*), 333
–, common (*Tringa hypoleucos*), 333
–, curlew (*Calidus ferruginea*), 333
–, least (*C. minutilla, Erolia minutilla*), 101, 184
–, sharp-tailed (*Calidris acuminata*), 333
–, Terek (*Tringa cinerea*), 269, 333
–, western (*Calidris mauri, Ereunetes mauri*), 101, 184
–, wood (*Tringa glareola*), 333
Sandwich Moraine (Mass., U.S.A.), 38
Sandy Neck (Mass., U.S.A.), 38
San Francisco (Calif., U.S.A.), 40, 169
San Francisco Bay (Calif., U.S.A.), 40–46, 71, 74, 97–102, 104, 167, 168, 174, 183, 185
San Ignacio, Laguna (Mexico), 117, 167

San Joaquin–Sacramento River system (Calif., U.S.A.), 40, 174
San Juan River Delta (Colombia), 204, 205
San Julian (Argentina), 196
San Lucas, Cabo (Mexico), 167, 169
San Pablo Bay (Calif., U.S.A.), 167, 174
San Quintin, Bahia de (Mexico), 167, 168, 176, 182
Santa Catarina [state] (Brazil), 197, 202
Santa Cruz Mountains (Calif., U.S.A.), 40
Santa Lucia River (Uruguay), 195
Santo Domingo, Laguna (Mexico), 177
São Paulo [state] (Brazil), 197, 202
Sapelo Island (Ga., U.S.A.), 72, 83, 84, 365
Sarawak (*see also* Borneo), 354
Saurashtra (India), 243
Scandinavia, 4–8, 11, 40, 54, 112, 128, 144
Scheldt Estuary (Belgium, Netherlands), 112, 114, 115, 117, 126, 128
Schleier community, 139
Schleswig-Holstein (Germany), 11, 24
Scirpeto-Phragmitetum, 6
– *americani*, 6
– *maritimi*, 6, 124
– *rufii*, 125, 135, 136
Scolt Head Island (England), 33, 35–38, 43, 44, 56, 90, 111, 146
Scotland, 11, 39, 54, 56, 112, 129, 131
scrub fowl (*Megapodius freycinet*), 332
sculpin, Pacific staghorn (*Leptocottus armatus*), 184
seablite (*Suaeda maritima*), 9, 32, 114, 117, 119, 122, 123, 127, 128, 130, 134, 138, 139, 141, 144, 145, 158, 196, 236, 244, 252, 254
sea-level changes (*see also* eustasy; isostasy; tectonic movements), 3, 37–40, 46–48, 54–57, 209
Sealand (England), 58
seatrout, spotted, 357
sedges, 102, 158
sedimentation, *see* accretion
sediments in channels, 36, 71, 76
seed availability affecting succession, 140
Segara Anakan (Indonesia), 266
Selangor (Malaya), 349
Sénégal, 233–235
Sepik (Papua-New Guinea), 261, 267
seres, *see* succession
Serpentine Creek (Qld., Australia), 308
Severn Estuary (England), 11
sewage effluent treatment, 26, 148
Seychelles [islands], 238
shading, *see* light interrelationships
sheep grazing salt marshes, 19, 118, 136, 144, 145, 149
Shetland Islands (Scotland), 127
Shiira River (Ryukyu Islands), 282
shingle barriers leading to marsh development, 31, 51, 54, 110–112, 295
shipworms (Teredinidae), 308
shore-line stabilization, 54, 349
Shotwick (England), 58
shoveler (*Spatula clypeata*), 98, 185

shrew(s), 102
–, marsh (*Sorex bendirii*), 185
–, masked (*S. cinereus*), 102
–, ornate (*S. ornatus*), 185
–, Suisun (*S. sinuosus*), 185
shrike, cuckoo, *see* cuckoo shrike
– thrush, brown (*Colluricincla brunnea*), 337
– –, grey (*C. harmonica*), 337
– –, little (*C. parvula*), 337
– –, rufous (*C. megarhyncha*), 337
shrimp(s) as food for ducks, 160
– associated with Foraminifera, 74
–, mangrove (*Thalassina anomala*), 285
–, mariculture of, in Japan, 365
–, migration of, 364
–, production of, 246, 357
Siberia, 64
    Sierra Leone, 235, 269, 353, 355
silt deposition, 2, 43, 50, 67, 76
– in salt-marsh substrates, 3, 34, 71, 115, 131, 140, 161, 173, 182
– in mangal substrates, 208, 272, 284
–, succession on, contrasted with sand substrate, 172
silting, *see* accretion
silvereye, grey-breasted (*Zosterops laterolis*), 337
–, yellow (*Z. lutea*), 337
silversides (*Menidia* spp), 95, 160
Sinai (Egypt), 217
Singapore, 50, 261, 353
Skallingen [peninsula] (Denmark), 40, 129, 131, 135, 139
skylark (*Alauda arvensis*), 144
snail(s), 19, 86, 160, 263, 266
–, coffee bean (*Melampus bidentatus*), 85
snake(s), 19, 184, 269
–, California king (*Lampropeltis getulus*), 184
–, garter (*Thamnophis* sp.), 184
–, water (*Cerberus rhynchops*), 263
snipe, Japanese (*Gallinago hardwickii*), 333
sodium chloride (*see also* halophytes; salinity)
– –, layer of, in Indian salt-marsh soils, 256
– in exchange complex of mangal soils, 246
soil (*see also* clay; mangals, soils of; mud; nitrogen; oxidation–reduction conditions; peat; salt marshes, soils of; sand; silt)
–, base saturation of, 246, 249
–, calcareous material in, 69–71, 112, 113, 205, 209, 210, 218
–, chloride content of, 109
–, classification of, 295, 296
–, compaction of, 12, 199, 269
–, composition of (*see also* specific constituents), 69, 117, 218, 246, 249, 254, 295, 356
–, depth of, 43, 53, 168, 209
–, development of, depending on allochthonous or autochthonous sources of material, 113
– erosion favoured by frost death of *Spartina*, 129
– – – – crab burrows, 88
– –: surfaces colonized by *Carex subspathacea* in Arctic, 170
– exchange complex, 115, 116, 246, 249, 264
– fauna (*see also* Foraminifera; mites), 83, 91, 146, 147, 149, 182–184

– fixed by vegetation, 72, 240, 365, 244
–, grain size of (*see also* clay; sand; silt), 70, 71, 82, 218
–, gypsum in, 71, 256, 295
– in deltas, 246
– – levees, 70, 71
–, mechanical analysis of, 70, 71, 82, 218
–, microbial processes in, 147, 148
–, nodules in, 71
– of mangals used for agriculture, 355, 356
–, organic matter in, 71, 72, 81, 112, 113, 116, 117, 161, 126, 172, 203, 205, 209, 218, 249
–, – – –, related to phosphorus content, 116
–, osmotic potential of, 26, 224
–, permeability of, 224
–, pH in, *see* mangal soils; salt-marsh soils
–, profiles of, 112, 256, 296
–, salinity of, *see* salinity
–, saline crust of, 219, 220, 225, 256
–, sources of material, 69
–, stratification in, 40, 48, 72
–: structure modified by faecal pellets, 72
–, sulphides in, 26, 112, 127
–, surface cracking of, 221, 256
–, surface modifications, *see* ant hills; mangals, depressions in; mounds of crustacea; sand hillocks
– – – protected by detritus, 113
– – – by plants, 129
–, texture of, 70, 113, 254, 296
–, vegetation related to, 2, 126–139, 203, 206–209, 222, 223, 229, 256, 263, 295, 296
soil-building role of salt-marsh plants, 72, 129, 194, 211, 212
– – – mangal plants, 162
Solomons Island (Md., U.S.A.), 87, 88
Solontchak, 233
Solway Firth (U.K.), 39, 56, 57
Somaliland, 215, 239
Somaliland, French, 215, 217, 223, 227
Somerset (England), 140
Sonneratietalia, 14
Sonneratietum albae, 14, 16
– *apetalea*, 14, 16
– *caseolariae*, 14
Sonneratio-Camptostemonetum, 14
Sonneration, 14
Sonora (Mexico), 194
South Africa, 4–8, 21, 134, 233, 236, 237
South America, mangals in, 13–16, 18, 50–59, 199–211
– –, mangrove migration to, 21
– –, salt marshes, 9, 16, 18, 193–199
Southampton Water (England), 58, 59, 129
South Australia (*see also* particular localities), 66, 294–296, 305–307, 312, 314, 317
South Carolina (U.S.A.), 97, 102, 157, 159, 160, 162, 163, 365
Southport (N.C., U.S.A.), 159
Southport (Qld., Australia), 317, 318
Soviet Union, *see* Russia; Siberia; *and particular localities*
Spain (*see also* particular localities), 21, 125, 128, 129, 139, 146
sparrow, savannah (*Passerculus sandwichensis beldingi*), 101, 102, 184

GENERAL INDEX

–, seaside (*Ammospiza maritima*), 100, 159
–, song (*Melospiza melodia*), 101, 102, 184
–, sharp-tailed (*Ammospiza caudacuta*), 159
–, white-crowned (*Zonotrichia leucophrys*), 101
sparrowhawk, collared (*Accipiter cirrocephalus*), 331
Spartina hybrids, *see* colonization
Spartinetalia, 121, 122
– maritimae, 7, 9
Spartinetea, 121, 122
– maritimae, 9
Spartineto-Distichlidetum, 7
Spartinetum, 233, 236
– *alterniflorae,* 7, 23, 128
– *brasiliensis,* 7
– *cynosuroidis,* 7, 10
– *foliosae,* 7, 10
– *glabrae,* 10
– *gracilis,* 23
– *maritimae,* 122, 127–129, 132, 233
– *patentis,* 7, 10
– *pilosae,* 10
– *townsendii,* 122, 128, 129, 132, 143
Spartinion, 110, 121, 122, 128
– americanae, 7
– europeae, 7
spatial patterns of molluscs, 84
– – of plants, 140, 141
species, origins of, *see* biogeography
Spencer Gulf (South Australia), 297
Spergularion salinae, 121
spiders, 90, 91, 146, 149, 159, 160, 308
spinebill, eastern (*Acanthorhynchus tenuirostris*), 337
Spitsbergen, 64, 66, 67
spoonbill, royal (*Platalea regia*), 331
–, yellow-billed (*P. flavipes*), 331
Springersgors (Netherlands), 143
spray communities, 295
squirrel, Beechey ground (*Otospermophilus beecheyi*), 185
–, flying (Phalangeridae), 310
Scri Lanka, 20
stability of ecosystems, 118, 140, 149
stabilization of shore-lines, 349, 365
– – soil, *see* soil fixed by vegetation; soil-building
Stanley (Tas., Australia), 299
starling, shining (*Aplonis metallica*), 337
Staticion galloprovincialis, 121
– orientale, 121
stickleback, threespine (*Gasterosteus aculeatus*), 184
Stillaguamish Bay (Wash., U.S.A.), 180
stilt, black-winged (*Himantopus himantopus*), 333
–, pied (*Himantopus* spp.), 311, 333
stint, red-necked (*Calidris ruficollis*), 333
stone-curlew, beach (*Esacus magnirostris*), 333
–, southern (*Burhinus magnirostris*), 333
Stony Point (Vic., Australia), 299
Stradbroke Island (Qld., Australia), 303, 317, 320, 326, 327
Suaedetalia, 4
Suaedeto-Asteretum, 5
Suaedeto-Atriplicetum, 5
Suaedeto-Limonietum vulgari, 7

Suaedeto-Salicornietum, 4
Suaedeto-Spergularietum, 5
Suaedetum californicae, 5
– *depressae,* 5
– *fruticosae,* 5, 139
– *maritimae,* 4, 23, 132, 138, 139
– *splendentis,* 138
Suaedion brevifoliae, 121
Suaedo-Kochietum hirsutae, 138, 139
Suakin (Sudan), 215–217, 221, 227
submergence (*see also* tides), tolerance by plants of, 32, 171, 174, 175
succession (*see also* pioneer species; transitions; zonation)
– affected by accretion rate, 140
– – – climatic extremes, 140
– – – drainage pattern, 25
– – – human activities, 25
– – – introduction of exotics, 140
– – – seed availability, 140
– – – soil type, 25, 172
– in mangals, 48, 198, 205, 247, 264, 268, 275, 284
– – salt marshes, in Africa, 222, 229
– – – –, in Europe, 127, 129–133, 139, 140
– – – –, – North America, 171, 172, 174, 175, 178, 193
– – *Zostera* communities, 126
–, plant pattern in, 140, 141
–, rate of, 140, 141
–, role of crustacean mounds in, 284
–, – – light in, 197
–, thresholds for, 140
–, zonation not always indicating, 25, 139, 229
Sudan, 215–217, 219, 227, 228
Suez, Gulf of (Egypt), 215, 217, 227, 228
Suisun Bay (Calif., U.S.A.), 167, 174
sulphides in marsh soils, 26, 112, 127
sulphur bacteria affecting phosphorus availability, 118
– cycle, 147
Sumatra (Indonesia), 50, 206, 261, 266, 353
Sumiyo Bay (Japan), 284
Sunderbans, 247
sunflower, tick seed (*Bidens trichosperma*), 102
Surinam, 197, 207–209
Suva (Fiji), 288
swallow, welcome (*Hirundo neoxena*), 335
–, white-breasted wood (*Artamus leucorhynchus*), 339
swamp forest, transition to, 3, 25, 266, 282
swamphen (*Gallinula porphyrio*), 332
swan, whistling (*Olor colombianus*), 98
Sweden, 23, 24, 39, 40, 120, 124, 135, 138, 139
swift, fork-tailed (*Apus pacificus*), 335
–, spine-tailed (*Chaetura caudacutus*), 335
swiftlet, grey (*Collocalia spodiopygia*), 335
Sydney (N.S.W., Australia), 46, 299, 306, 308, 318
Sylt (Germany), 141

Tabasco (Mexico), 48, 49, 71, 197, 202–204
Tadjoura, Gulf of (East Africa), 216
Taiwan, 278–282, 353
Tallebudgera (Qld., Australia), 308, 317, 319, 320, 326, 330
Talsarnau (Wales), 32, 39

Tamaulipas (Mexico), 194
Tamiahua, Laguna de (Mexico), 202
Tamil Nadu (India), 253
Tampa Bay (Fla., U.S.A.), 194, 202
Tanega-shima (Japan), 286, 287
Tanga (Tanzania), 50, 65, 238, 239
tannins, 246, 255, 347, 348, 352–355
Tarrant Point (Qld., Australia), 302
Tasmania (Australia), 294, 295, 309, 314
tattler, grey-tailed (*Tringa brevipes*), 333
Tauranga (New Zealand), 310
teal, 160
–, American green-winged (*Anas carolinensis*), 98, 99, 185
–, blue-winged (*Querquedula discors, Anas discors*), 98, 99, 159
–, green-wing (*A. carolinensis*), 98, 99
–, grey (*A. gibberifrons*), 331
temperature (*see also* climate; frost)
–, animals affected by, 1, 80–83, 85, 86, 104, 105, 269
– as a climatic factor, 61–64, 76
– – – – – in Australia 296–299
– – – – – in India, 245–249
– – – – – on North American Pacific coast, 169, 170, 179
– – – – – on Red Sea coast, 216
–, mangal development affected by, 2, 64, 66, 193, 202, 234
– of oceans, 202, 297–299
–, reproduction of arthropods affected by, 80
–, salinity related to, 194
–, salt-marsh development affected by, 2, 170
Tenasserim (Burma), 251
Ten Thousand Islands (Fla., U.S.A.), 47, 51, 71, 202, 209
Teppich community, 139
tern, 144, 235
–, Caspian (*Sterna caspia*), 333
–, common (*S. hirundo*), 333
–, crested (*S. bergii*), 333
–, gull-billed (*S. nilotica*), 334
–, little (*S. albifrons*), 333
–, lesser-crested (*S. bengalensis*), 333
–, whiskered [*Chlidonias (Sterna hybrida)*], 333
–, white-winged black (*S. leucoptera*), 334
terrapin, diamondback (*Malaclemys centrata*), 160
Terschelling (Netherlands), 111
Tethys Sea, 21
Texas (U.S.A.) (*see also particular localities*), 70, 71, 76, 194, 202
Thailand, 348, 349, 356
Thames Estuary (England), 38, 39, 141
thatching material from mangals and salt marshes, 247, 348, 364
therophytes (*see also* life-forms), 23, 137, 139
Thero-Salicornietalia, 122, 124
Thero-Salicornietea, 121, 122, 124
Thero-Salicornion, 4, 110, 121, 122, 124, 127
Thero-Suaedetalia, 121
Thero-Suaedion, 4, 110, 121, 138
Thevenard (S.A., Australia), 299
Thompson, Cape (Alaska, U.S.A.), 167, 170
Thornham (England), 57
Three Isles (Qld., Australia), 52

threesquare, Olney (*Scirpus olneyi*), 102, 158
thrush, shrike, *see* shrike thrush
Thursday Island (Qld., Australia), 298
tides, *see also* channels; zonation
–, algae affected by, 141, 143, 173
–, animals affected by, 105, 106, 146
–, birds affected by, 95, 96, 100, 101, 144
–, crab behaviour and distribution related to, 87, 88
–, fish affected by, 92
–, frequency and range of, 31, 32, 40, 48, 51, 54, 113, 114, 143, 144, 196, 298, 299
–, fungi affected by, 118
–, geomorphological effects of, 32, 110, 111
– importing detritus, 34, 118, 139, 147, 219
– in Baltic Sea, 39, 40
–, insects affected by, 88–90
–, mangal inundation by, 297
–, – vegetation affected by, 202, 255
–, muskrat behaviour and distribution related to, 103, 104
–, salinity changes resulting from, 76, 95, 114–116, 173, 229, 297
–, salt-marsh conditions determined by, 26, 45, 109, 112
–, – – vegetation affected by, 113–115, 126–131, 135, 137, 173, 196, 310
–, velocity of, in channels, 43, 44, 76
–, wind effects on, 39, 69, 70, 76
Tierra del Fuego (Chile), 194
Tigre, Punta del (Uruguay), 195
Tijuana Estuary (Mexico), 184
Tijuana Slough (Mexico), 167, 175
timber from mangroves, 246, 348–351
Tinian (Mariana Islands), 271, 272
Tinicum Marsh (Pa., U.S.A.), 365
Titi (Bonin Islands), 272
toad, western (*Bufo boreas*), 184
Todos Santos Bay (Mexico), 76
Tokar (Sudan), 221, 227
Tomales Bay (Calif., U.S.A.), 167, 174
Tonga, 290
topsmelt (*Atherinops affinis*), 184
Townsville (Qld., Australia), 51, 298, 301
trace elements supplied to marshes by flood water, 115
trade in mangrove products (*see also* mangrove products as exports), 234, 350–355
Traherne Island (New Zealand), 46
transitions between mangal and salt marsh, 46, 162, 177, 193
– to beach communities, 25
– to fresh-water swamp or marsh, 3, 12, 16, 25, 119, 210, 211, 223, 235, 251
– to inland vegetation, 65, 67, 80, 223
– to swamp forest, 3, 25, 266, 282
– to xerosere, 65, 138
treading of salt marshes by livestock, effects of, 136, 144
tree-frog, Pacific (*Hyla regilla*), 184
trellis pattern of marsh channels, 43
Trieste Lagoon, 111
Trifolieto-Cynodontion, 121
Triglochineto-Potentilletum pacificae, 8
Triglochinetum maritimi, 7
Triglochino-Caricetum subspathaceae, 131

GENERAL INDEX

Triglochino-Puccinellietum coarctatae, 131
Triglochino-Puccinellietum phryganodis, 131
triller, Jardine (*Coracina tenuirostris*), 336
–, varied (*Lalage leucomela*), 336
–, white-winged (*L. sueurii*), 336
Trinidad, 347, 353
Trondheim (Norway), 64
trophic relations (*see also* birds, feeding habits of; detritus; grazing; human food; nitrogen cycle; nutrient cycles; productivity)
– –: aerial and aquatic phases, 113, 147–149
– –: carrying capacity, 19
– –: detritus food chains, 148, 149, 187, 357
– –: energy flow, 27, 147–149, 186
– –: exchanges with flood waters, 113, 147, 148
– –: export from marshes, 148
– –: food webs, 147–149, 357
– –: grazing food chains, 148, 149, 186, 187
– –: microbial activities, *see* bacteria; fungi
– –, model of, 27
– – of insects, 183
– –: sulphur cycle, 147
– –: turnover rates, 149
Truk (Caroline Islands), 272, 274, 276, 278
Tsuro River (Japan), 283
Tuléar (Madagascar), 18
Tumbes (Peru), 202
Tunisia, 233
turbellarians, 146, 149
Turneffe Island and Reefs (Belize), 54, 55
turnstone, ruddy (*Arenaria interpres*), 333
Tyne River (England), 146

Ulenge Island (Tanzania), 238
Union of Socialist Soviet Republics (U.S.S.R.), *see* Russia; Siberia; *and particular localities*)
United States of America (U.S.A.) (*see also particular states and localities*)
– – – –, climate on East Coast of, 63
– – – –, mangals in, *see* Florida
– – – –, salt marshes on Atlantic coast of, 4–8, 10, 12, 19, 22, 25, 26, 34–37, 157–164
– – – –, – – – – –, distribution of, 157
– – – –, – – – – –, ecosystem studies of, 147–149
– – – –, – – – – –, productivity of, 25
– – – –, – – – – –, sedimentation rate in, 37
– – – –, – – – – –, value of, 26, 364–367
– – – –, – – on Pacific coast of, 4–8, 12, 167–187
– – – –, utilization of mangroves in, 352, 355
Upper Hut Marsh (England), 36
Upper Newport Bay (Calif., U.S.A.), 176
Uppland (Sweden), 135
Uppsala school of phytosociology, 120
Uruguay, 13–15, 195

Valdivia (Chile), 197
value of mangals to man, 347–362
– – salt marshes to man, 363–368
Vancouver (B.C., Canada), 177
Vancouver Island (B.C., Canada), 182

Vaucherietum, 18
vegetative reproduction of plants, 170
Venezuela, 65, 199, 347, 349, 352, 353, 355–357
–: origin of name, 347
Venice, 111, 128
Veracruz (Mexico), 197
Verlandungsgebiete, 113
vernal *Ulothrix* community, 17
Victoria (Australia), 294, 295, 305–308, 312, 314, 317
Vietnam, 1, 268, 269
Virginia (U.S.A.), 86–88, 157, 158, 161–163
Virgin Islands, 199
Viti Levu (Fiji), 288
Vlissingen (Netherlands), 113, 116
vole, root (*Microtus ratticeps*), 144

Wadden, 69, 70, 110–113, 135, 141
– type of salt marsh, 110
– Sea, 42
Wadi Ambagi (Egypt), 227
Wadi El-Ghweibba (Egypt), 218, 223
Wadi Gimal (Egypt), 218
Wadi Hommath (Egypt), 224
wagtail, grey (*Motacilla cinerea*), 335
–, willy (*Rhipidura leucophrys*), 337
–, yellow (*Motacilla flava*), 335
Wales (*see also particular localities*), 3, 31, 37, 39, 128, 140
wallaby, agile (*Wallabia agilis*), 310
warbler, black-throated (*Gerygone palpebrosa*), 336
–, dusky (*G. tenebrosa*), 336
–, fairy (*G. flavida*), 336
–, green-backed (*G. chloronata*), 336
–, large-billed (*G. magnirostris*), 336
–, mangrove (*G. laevigaster*), 336
–, reed (*Acrocephalus australis*), 336
–, white-throated (*Gerygone olivacea*), 336
Warham (England), 43
Wash (England), 37, 45, 57, 72
Washington [state] (U.S.A.), 168, 169, 172
wasting disease of *Zostera,* 126, 127
Watam (Papua-New Guinea), 267
water level as an edaphic factor (*see also* tides), 237, 278
– – regulation in salt marshes (*see also* drainage), 79, 102, 161
waterlogging, effects (*see also* aeration), 2, 31, 264
weasel, long-tailed (*Mustela frenata*), 185
weaver ant (*Oecophylla* spp.), 270, 308
Weipa (Qld., Australia), 298, 307, 319, 320
Wellington Point (Qld., Australia), 317
Western Australia, 19, 294–296, 300, 303, 304, 312, 330
Western Desert of Egypt, 220
Westernport Bay (Vic., Australia), 294, 295, 297, 299, 305, 312, 314, 317, 318, 339
West Indies (*see also* Caribbean; *and particular islands and localities*), 13–16, 199, 209, 347
whimbrel (*Numenius phaeopus*), 269, 333
–, little (*N. minutus*), 333
whipbird, eastern (*Psophodes olivaceus*), 336
whistler (*Pachycephala cinerae*), 269
–, brown (*P. simplex*), 337

–, grey (*P. griseiceps*), 337
–, mangrove golden (*P. melanura*), 337
–, rufous (*P. rufiventris*), 337
–, white-breasted (*P. lanioides*), 337
White Sea (U.S.S.R.), 67, 126, 127
Whitewater Bay (Fla., U.S.A.), 71
widgeon, American (*Mareca americana*), 98, 99, 185
– grass (*Ruppia maritima*), 99, 158
wild rice (*Zizania aquatica*), 97
Willapa Bay (Wash., U.S.A.), 167, 180
willet (*Catoptrophorus semipalmatus*), 97, 101, 160, 184
Williamstown (Vic., Australia), 299
wind amplifying tidal effects, 39, 69, 70, 76
– as climatic element in Taiwan, 280
– determining mangal distribution, 274
wind-borne deposits of sediment, 37, 69, 72, 219–222, 225, 230
wine made from mangrove propagules, 347
Witham (England), 45
Wollongong (N.S.W., Australia), 299
wood swallow, white-breasted (*Artamus leucorhynchus*), 339
worms, 83, 119, 144, 146, 149, 182, 237
wren(s), 102, 160
–, long-billed marsh (*Telmatodytes palustris grisens*), 102
–, lovely (*Malurus amabilis*), 336
–, marsh (*Telmatodytes palustris*), 160
–, red-backed (*Malurus melanocephalus*), 336
–, southern emu (*Stipiturus malachurus*), 336
–, superb blue (*Malurus cyaneus*), 336
Wrightsville Beach (N.C., U.S.A.), 159
Würm glaciations, 47
Wyndham (W.A., Australia), 298
Wynnum (Qld., Australia), 318

xerosere, transitions to, 65, 138
Xylocarpetalia, 14
Xylocarpetum australasicae, 14
– *benadirensae*, 14
– *granatae*, 14
– *moluccensae*, 14
Xylocarpion, 14

Yaku-jima (Japan), 284, 287
Yakutat (Alaska, U.S.A.), 167, 169
Yap (Caroline Islands), 271, 272, 274–276, 278
Yaquina Bay (Ore., U.S.A.), 167, 173
Yaringa (Vic., Australia), 305
Yarmouth (England), 39
Yashiminato River (Japan), 282

York, Cape (Qld., Australia), 53
Yucatan (Mexico), 198
Yukon Delta (Alaska, U.S.A.), 167, 171

Zambesi River (Africa), 237, 238
Zanzibar (Tanzania), 354
zonation (*see also* pioneer species; succession; tides; transitions)
–, geographical variation in, 301
– in mangals, 66
– – – depending on salinity differences, 65, 237
– – – in Africa, 229, 235–237
– – – in America, 202, 203, 205, 206, 209–211
– – – in Australia, 301–303
– – – in India, 253, 255, 256, 258
– – – in Japan, 283
– – – in Malaya, 262, 265
– – – in New Caledonia, 289
– – – in the Philippines, 287, 288
– – – of molluscs, 265, 307
– – salt marshes determined by salinity, 229
– – – – – tides, 113
– – – – in Africa, 218, 223, 228–230, 236
– – – – in Australia, 305, 306
– – – – in Europe, 110, 139–141
– – – – in North America, 161, 174, 195
– – – – of birds, 95
– – – – of Foraminifera, 74, 81–83
– – – – of insects, 90
– – – – of mites, 91
– – – – of molluscs, 183
– – – – of spiders, 91
– of animal communities, 79, 106, 147, 307
– related to succession, 25, 139, 140, 229
–, "reversals" of, 195
–, sharpness of boundaries in, 25
zoogeography, *see* biogeography; distribution
Zostera, destruction by wasting disease, 126
Zosteretalia, 121
Zosteretea, 121
Zosteretum, 236
– *marinae*, 126, 132
– – *stenophyllae*, 126
– *nanae*, 126, 132
Zosterion, 110, 120, 121
Zürich–Montpellier school of phytosociology, 109, 120, 142
Zuyderzee, 126
Zwartcops Estuary (South Africa), 236